Organic Azides

Organic Azides
Syntheses and Applications

Editors

STEFAN BRÄSE

Institute of Organic Chemistry,
Karlsruhe Institute of Technology (KIT), Germany

KLAUS BANERT

Institute of Chemistry,
Chemnitz University of Technology, Germany

WILEY

A John Wiley & Sons, Ltd., Publication

This edition first published 2010

Registered office
John Wiley & Sons Ltd, The Atrium, Southern Gate, Chichester, West Sussex, PO19 8SQ, United Kingdom

For details of our global editorial offices, for customer services and for information about how to apply for permission to reuse the copyright material in this book please see our website at www.wiley.com.

Reprinted October 2010

Library of Congress Cataloging-in-Publication Data
Organic azides : syntheses and applications / editors, Stefan Bräse, Klaus Banert.
p. cm.
Includes bibliographical references and index.
ISBN 978-0-470-51998-1 (cloth)
1. Azides. I. Bräse, Stefan II. Banert, Klaus.
QD305.N84O74 2009
547′.04–dc22

2009021624

A catalogue record for this book is available from the British Library.

ISBN 978-0-470-51998-1 (H/B)

Set in 10 on 12 pt Times by Toppan Best-set Premedia Limited
Printed and bound in Great Britain by CPI Antony Rowe, Chippenham, Wiltshire

Contents

Foreword

Whenever the significance of a chemical discovery is ubiquitously recognized, a swarm of research groups will quickly join (and compete) in developing the 'new territory'. Take the finding of [60]fullerene as a new modification of elemental carbon: the evolution of a new research area boomed, an exponential growth in the number of publications being the consequence.

In the early history of organic chemistry, new discoveries – there were so many! – struck less of an echo; there were so few chemists. Peter Griess, a German chemist in a British brewery, prepared the first aromatic diazonium salts (1858) and, on his own, unveiled their rich reactivity, the azo coupling with phenolates included. The reaction of arenediazonium perbromides with ammonia provided aryl azides (1864).[1] After Griess' death (1888), three obituaries by August Wilhelm von Hofmann, Emil Fischer, and Heinrich Caro praised the 'single combatant'.[2–4]

The azide story vigorously unfolded when Theodor Curtius, the grandseigneur of nitrogen chemistry, entered the scene. The preparation of ethyl diazoacetate, the first aliphatic diazo compound (1883),[5] paved the way to hydrazine *via* 'bisdiazoacetate' (1887).[6] Several steps in Curtius' career – moves from Munich to Erlangen and further to Kiel (later to Bonn and Heidelberg) – hardly curbed the momentum of basic discoveries. The reaction of benzoyl hydrazine with nitrous acid provided benzoyl azide, and alkaline hydrolysis gave sodium azide. By acidification of the latter, 'azoimid' HN_3 was set free (1890),[7] a gas of 'highly peculiar, dreadfully pungent smell'. The preparation of alkyl azides from AgN_3 and alkyl iodides was likewise found in Curtius' laboratory.

In hot ethanol, benzoyl azide was converted to ethyl *N*-phenylcarbamate (1894). This 'Curtius rearrangement' was recognized by its discoverer as a general method of degrading carboxylic acids to amines,[8] whereas the intermediacy of isocyanates escaped him.

There was no overlooking the explosive character of azides. Curtius described a sample of aqueous hydrogen azide on local heating giving rise to a 'formidable detonation and disintegration of the thick-walled glass tube to dust'. And furthermore: 'The detonation which a few mg of silver azide generate on impact or heating is unparalleled.' In World War I lead azide replaced the mercury fulminate as initiator.

Thus, azides gained the bad reputation of being dangerous in handling rather early (see Chapter 13). Explosives are 'energy-rich' or 'high-energy' compounds in technical jargon, since they serve as a source of energy. More correctly, there is an increase of bond energy due to the formation of N_2 in the explosion, and the pressure of the produced gases contributes to the destructive force. A rule of thumb in the preparative use of organic azides:

the explosion danger decreases with diminishing fraction of N_3 in the molecular mass; e.g., phenyl azide is easier to handle than methyl azide.

Whereas P. Griess abstained from drafting a structure for the N_3 unit, Th. Curtius and contemporaries formulated azides as cyclic 1*H*-triazirines. A. Angeli[9] and J. Thiele[10] found an open-chain azide group more consistent with the reactivity spectrum; still, the open formula bore the blemish of a pentacovalent middle nitrogen. This flaw was overcome by the resonance description of $R{-}N_3$: N_β appears as iminium function, whereas N_α and N_γ share the anionic charge (Sutton, 1931).[11] X-ray analyses and many physical methods confirmed this open-chain formula of azides. In modern terminology, by the way, 1*H*-triazirines constitute antiaromatic 4π systems.

Curtius dreamed of HN_5 and derivatives, the more so as pentazole was the 'missing link' in the azole series. The reaction of benzenediazonium chloride and sodium azide furnished phenyl azide and N_2 instead of the desired phenylpentazole (Arthur Hantzsch, 1903).[12] In fact, part of the reaction passes through phenylpentazole, as unambiguously shown by a combination of kinetics and ^{15}N-labeling (Clusius, Huisgen, & Ugi, 1956).[13] Some arylpentazoles were obtained crystalline, but extrude N_2 in solution to give aryl azides. What about the parent HN_5? In a study of 2008, Richard Butler *et al.* oxidatively dearylated *p*-anisylpentazoles, which differed in the position of the ^{15}N-label, and the ^{15}N-distribution in N_3^-, appears to be in harmony with the fleeting occurrence of HN_5/N_5^-.[14]

There is no place for pentazole in Beilstein's *Handbook of Organic Chemistry*. What is the organic compound with the highest nitrogen content? Tetraazidomethane with 93% N merits this reputation; the recently prepared CN_{12} is a highly explosive liquid (Banert, 2007).[15]

The application of azides as carboxy-activating group in peptide synthesis goes back to Curtius, too. On treating *N*-benzoylglycyl azide with glycine in aqueous alkali, *N*-benzoylglycylglycine was obtained. Renewed conversion to the azide allowed a repetitive procedure by which the *N*-benzoyltetrapeptide was achieved (1902).[16] The actual significance of the azide method is based on avoiding the feared racemization.

In 1893, Arthur Michael observed the formation of a 1,2,3-triazole derivative in the reaction of 'diazobenzolimid' (i.e. $Ph{-}N_3$) with dimethyl acetylenedicarboxylate;[17] Michael – a future Harvard professor – worked with R. Bunsen and A.W. von Hofmann. Numerous cycloadditions of organic azides and HN_3 to alkynes and alkenes were described in the sequel. In the general definition and classification azides belong to the 1,3-dipoles of propargyl-allenyl type (R. Huisgen, 1960).[18] 1,3-Dipolar cycloadditions share the 6π-electron balance with Diels-Alder reactions – and the wide synthetic application. Albert Padwa edited monographs on 1,3-dipolar cycloaddition chemistry in 1984 and 2003 – substantial chapters on azides were included.[19]

Rate constants (k_2) for cycloadditions of phenyl azide to substituted ethylenes and acetylenes stretch over seven magnitudes; high values were observed for enamines, moderate k_2 for the acrylic ester type, and the rate minimum was found for common alkenes and alkynes (Huisgen, Szeimies, & Möbius, 1967).[20] In the PMO treatment of concerted cycloadditions, Reiner Sustmann found the key to the understanding of substituent effects; e.g., a plot of k_2 for the cycloadditions of $Ph{-}N_3$ versus the ionization potential of substituted ethylenes and acetylenes furnished a degenerate U shape (1971).[21] Such a plot is a distinguishing feature for each 1,3-dipole and reflects the specific mix of nucleophilic and electrophilic activity, modified by steric effects.

The 1,3-cycloaddition of alkyl azides to terminal alkynes is very slow, but can be catalyzed by Cu(I) (mechanism: Straub, 2007).[22] This formation of 1,2,3-triazoles, popularized as 'click reaction', was used by Sharpless and Meldal (both 2002)[23,24] for the selective and biocompatible ligation of peptides, proteins, and especially for the introduction of biomarkers. *In vivo* applications in aqueous medium are feasible. The bioresearch community applauded this new tool which aroused fresh enthusiasm in azide chemistry.

In preparative and synthetic application, organic azides unfold an astonishing versatility and witnessed a renaissance in recent decades; a renaissance to which the two editors successfully contributed. A recent review is entitled 'An Exploding Diversity of a Unique Class of Compounds' (Bräse, 2005),[25] and a yearly increase by more than a thousand publications on organic azides is mentioned. This profusion is intimidating and demonstrates the necessity of a multi-authored monograph. The editors succeeded in dividing the abundance in handy packages and in persuading competent experts to write the chapters. A certain overlap among the chapters is not harmful, is even desirable, since not every user will devour the whole book. The monograph offers access to the most recent state of research. The faster such a monograph may become obsolete, the higher has been its benefit to the chemical community.

Rolf Huisgen
LMU München

References

[1] P. Griess, *Proc. R. Soc. London* **1864**, *13*, 375–84. P. Griess, *Liebigs Ann. Chem.* **1866**, *137*, 39–91.
[2] A.W. von Hofmann, *Ber. Dtsch. Chem. Ges.* **1891**, *24*, 1007–57.
[3] E. Fischer, *Ber. Dtsch. Chem. Ges.* **1891**, *24*, 1058–78.
[4] H. Caro, *Ber. Dtsch. Chem. Ges.* **1891**, *24*, I–XXXVIII.
[5] Th. Curtius, *Ber. Dtsch. Chem. Ges.* **1883**, *16*, 2230–1.
[6] Th. Curtius, *Ber. Dtsch. Chem. Ges.* **1887**, *20*, 1632–4.
[7] Th. Curtius, *Ber. Dtsch. Chem. Ges.* **1890**, *23*, 3023–33.
[8] Th. Curtius, *Ber. Dtsch. Chem. Ges.* **1894**, *27*, 778–81; Th. Curtius, *J. Prakt. Chem.* **1894**, *50*, 275–94.
[9] A. Angeli, *Atti Reale Accad. Lincei* **1907**, *16 II*, 790.
[10] J. Thiele, *Ber. Dtsch. Chem. Ges.* **1911**, *44*, 2522–4.
[11] L.E. Sutton, Nature **1931**, *128*, 639. N.V. Sidgwick, L.E. Sutton, W. Thomas, *J. Chem. Soc.* **1933**, 406–12.
[12] A. Hantzsch, *Ber. Dtsch. Chem. Ges.* **1903**, *36*, 2056–8.
[13] R. Huisgen, I. Ugi, *Angew. Chem.* **1956**, *68*, 705–6. I. Ugi, R. Huisgen, K. Clusius, M. Vecchi, *Angew. Chem.* **1958**, *68*, 753–4.
[14] R.N. Butler, J.M. Hanniffy, J.C. Stephens, L.A. Burke, *J. Org. Chem.* **2008**, *73*, 1354–64.
[15] K. Banert, Y.H. Joo, T. Rüffer, B. Walfort, H. Lang, *Angew. Chem.* **2007**, *119*, 1187–90; *Angew. Chem. Int. Ed.* **2007**, *46*, 1168–71.
[16] Th. Curtius, *Ber. Dtsch. Chem. Ges.* **1902**, *35*, 3326–8. Th. Curtius, A. Benrath, *Ber. Dtsch. Chem. Ges.* **1904**, *37*, 1279–84.
[17] A. Michael, *J. Prakt. Chem.* **1893**, *48*, 94–5.
[18] R. Huisgen, Centenary Lecture 1960; *Proc. Chem. Soc.* **1961**, 357–69. R. Huisgen, *Angew. Chem.* **1963**, *75*, 604–37; *Angew. Chem. Int. Ed.* **1963**, *2*, 565–98.

[19] A. Padwa (ed.), *1,3-Dipolar Cycloaddition Chemistry*, John Wiley & Sons, Inc., New York, **1984**. A. Padwa, W.H. Pearson (eds.), *Synthetic Applications of 1,3-Dipolar Cycloaddition Chemistry Towards Heterocycles and Natural Products*, John Wiley & Sons, Inc., New York, **2003**.
[20] R. Huisgen, G. Szeimies, L. Möbius, *Chem. Ber.* **1967**, *100*, 2494–2507.
[21] R. Sustmann, H. Trill, *Angew. Chem.* **1972**, *84*, 887–8; *Angew. Chem. Int. Ed.* **1972**, *11*, 838–9. R. Sustmann, *Pure Appl. Chem.* **1974**, *40*, 569–93.
[22] C. Nolte, P. Mayer, B.F. Straub, *Angew. Chem.* **2007**, *119*, 2147–9; *Angew. Chem. Int. Ed.* **2007**, *46*, 2101–3. B.F. Straub, *Chem. Commun.* **2007**, 3868–70.
[23] V.V. Rostovtsev, L.G. Green, V.V. Fokin, K.B. Sharpless, *Angew. Chem.* **2002**, *114*, 2708–11; *Angew. Chem. Int. Ed.* **2002**, *41*, 2596–9.
[24] C.W. Tornøe, C. Christensen, M. Meldal, *J. Org. Chem.* **2002**, *67*, 3057–64.
[25] S. Bräse, C. Gil, K. Knepper, V. Zimmermann, *Angew. Chem.* **2005**, *117*, 5320–74; *Angew. Chem. Int. Ed.* **2005**, *44*, 5188–240.

Foreword

It is a privilege to write a few words here about our favorite functional group – organic azides. Both of us were fortunate to read (in truth memorize) the two-volume monograph 'Chemistry of Open-Chain Organic Nitrogen Compounds' by P.A.S. Smith; K.B.S. in 1966, its year of publication, and V.V.F almost four decades later. Despite this time gap, Smith's great style and deep knowledge – both physical and descriptive, made for a fantastic story for both of us, start to finish. We are convinced it is still the only guaranteed way to get injected with 'the right stuff' – assuming it's mastery of nitrogen reactivity you seek!

What Smith had liked, and thence taught best about nitrogen, was that it was the only element in the first row which engaged in fast and loose atom transfer redox events, thanks to the panoply of fast reactions open to its oxidation states III and lower. In other words, nitrogen species could be slippery and redox-able like the transition metals we know so well.

Among reactive organic groups, azides are near the top energetically, and yet paradoxically they are *kinetically* locked up. Nevertheless, the fear of the energy these small energetic groups pack, the sort of 'azidophobia', has curtailed the ideas and experiments needing organic azides. Although they have been known for over 100 years, the utility of organic azides has been often limited to the facile introduction of the amino group into organic molecules. Other facets of their razor-sharp reactivity remained largely unexplored until relatively recent years. Yet organic azides are versatile sources of nitrenes, amines, and nitrogen heterocycles containing three contiguous nitrogen atoms. The latter are a foreign territory for nature and hence, unique tools for studying it.

Bertozzi recognized nearly bioorthogonal properties of azides and the ease of their introduction into the biological molecules, and pioneered their reaction with phosphines in her studies of biological processes. Around the same time, Finn and Sharpless 'saw' that Rolf Huisgen's 1,3-dipolar cycloadditions of azides and alkynes forming triazoles was *the cream of the crop* among all known organic transformations. The alkyne and the azide groups are nearly completely *orthogonal* to all terrestrial environments – including *inter alia,* the fluids and tissues of live organisms. They either react with each other or not at all, so are in effect invisible, which endows them for stealth-like uses, such as discovery of enzyme inhibitors through target guided '*in situ* click chemistry.'

Shortly thereafter, Fokin's discovery of the reactivity of in situ-generated copper acetylides with azides in aqueous solutions made the copper-catalyzed azide-alkyne cycloaddition (CuAAC) one of the most widely utilized reactions involving organic azides.

We are often asked if there are more 'orthogonal' click reactions like the CuAAC and the thiol-ene addition lurking out there. The answer is a confident 'yes'. However, when people want to replace the azide group with another 1,3-dipole, this is hard to imagine, assuming one demands identical, or better reactivity parameters. The simple reason is that the other dipoles are not adequately 'invisible' in the acid-base world. Hence, the azide functionality should remain a rich source of new reactivity discoveries for many years.

From our own experience, and wisdom gleaned from the likes of Peter Smith, Thomas Archibald, and Alfred Hassner, we say get ready for the arrival of more azides in applied chemistry. They may be late bloomers, but they're coming on strong. In fact, the best evidence for the ascendancy of organic azides in synthesis is right here, in this outstanding collection of reviews on the topic edited by Stefan Bräse and Klaus Banert.

K. Barry Sharpless and Valery V. Fokin
La Jolla, California

Preface

This book is aimed at graduate students or researchers, who have basic knowledge in organic chemistry and want to approach the field of organic azides from a historical perspective through to the state-of-the-art. The material will be suitable for supplementing a graduate course in organic syntheses.

The contributing authors are leading scientists in their field. Each individual was asked to contribute 20–30 printed pages putting their own research in the context of the development of the chemistry of organic azides. This material has been organized into 16 chapters.

List of Contributors

Concepción Alonso, Departamento de Química Orgánica I, Facultad de Farmacia, Universidad del País Vasco. Apartado 450. 01080 Vitoria, Spain

Maria José Alves, Departamento de Química, Campus Gualtar, Universidade do Minho, P-4710057 Braga, Portugal

Domitila Aparicio, Departamento de Química Orgánica I, Facultad de Farmacia, Universidad del País Vasco. Apartado 450. 01080 Vitoria, Spain

Jeffrey Aubé, Department of Medicinal Chemistry, University of Kansas, School of Pharmacy, Malott Hall, 1251 Wescoe Hall Drive, Room 4070, Lawrence, KS 66045-7582, USA

Klaus Banert, Institute of Chemistry, Chemnitz University of Technology, Strasse der Nationen 62, 09111 Chemnitz, Germany

Henning S.G. Beckmann, Fachbereich Chemie, Universität Konstanz, Universitätsstr. 10, D-78457 Konstanz, Germany

Stefan Bräse, Institute of Organic Chemistry, Karlsruhe Institute of Technology (KIT), Fritz-Haber-Weg 6, Karlsruhe, D-76131, Germany

Erick M. Carreira, Laboratorium für Organische Chemie, HCI H 335, Wolfgang-Pauli-Str. 10, 8093 Zürich, Switzerland

Stéphanie Durot, Institut de Chimie, CNRS et Université de Strasbourg, UMR 7177 Laboratoire de Chimie-Organo-Minérale, 4 rue Blaise Pascal, BP 1032, F-67070 Strasbourg cedex, France

Julien Frey, CEA Saclay, iBiTec-S, Service de Bioénergétique, Biologie Structurale et Mécanismes. 91191 Gif-sur-Yvette, France

Thomas L. Gilchrist, Cunningham Drive, Wirral, CH63 0JX, UK

Scott Grecian, Lacamas Laboratories, 3625 North Suttle Road, Portland, Oregon 97217, USA

Nina Gritsan, Institute of Chemical Kinetics and Combustion of Siberian Branch of Russian Academy of Sciences, Novosibirsk, Russia

Jürgen Haase, Dynamit Nobel GmbH, Explosivstoff und Systemtechnik, Kalkstrasse 218, 51377 Leverkusen, Germany

Ciril Jimeno, University of Bern, Department of Chemistry and Biochemistry, Freiestrasse 3, CH-3012 Bern, Switzerland

Nicole Jung, Institute of Organic Chemistry, Karlsruhe Institute of Technology (KIT), Fritz-Haber-Weg 6, Karlsruhe, D-76131, Germany

Thomas Keicher, Fraunhofer Institut für Chemische Technologie ICT, Pfinztal, Germany

Thomas M. Klapötke, Department Chemie und Biochemie, Ludwig-Maximilians-Universität München, Butenandtstr. 5-13(D), D-81377 Munich, Germany

Burkhard Krumm, Department Chemie und Biochemie, Ludwig-Maximilians-Universität München, Butenandtstr. 5-13(D), D-81377 Munich, Germany

Stefan Löbbecke, Fraunhofer Institut für Chemische Technologie ICT, Pfinztal, Germany

Morten Meldal, Carlsberg Laboratory, Gamle Carlsberg Vej 10, DK-2500 Valby, Denmark

Francisco Palacios, Departamento de Química Orgánica I, Facultad de Farmacia, Universidad del País Vasco. Apartado 450. 01080 Vitoria, Spain

Teresa M.V.D. Pinho e Melo, Department of Chemistry, University of Coimbra, 3004-535 Coimbra, Portugal

Matthew Platz, Chemistry Department, the Ohio State University, Columbus, Ohio, USA

Philippe Renaud, University of Bern, Department of Chemistry and Biochemistry, Freiestrasse 3, CH-3012 Bern, Switzerland

Gloria Rubiales, Departamento de Química Orgánica I, Facultad de Farmacia, Universidad del País Vasco. Apartado 450. 01080 Vitoria, Spain

Jesús M. de los Santos, Departamento de Química Orgánica I, Facultad de Farmacia, Universidad del País Vasco. Apartado 450. 01080 Vitoria, Spain

Jean-Pierre Sauvage, Institut de Chimie, CNRS et Université de Strasbourg, UMR 7177, Laboratoire de Chimie-Organo-Minérale, 4 rue Blaise Pascal, BP 1032, F-67070 Strasbourg cedex, France

Christine Schilling, Institute of Organic Chemistry, Karlsruhe Institute of Technology (KIT), Fritz-Haber-Weg 6, Karlsruhe, D-76131, Germany

Werner R. Thiel, Fachbereich Chemie, Technische Universität Kaiserslautern, Erwin-Schrödinger-Str. Geb. 54, D-67663 Kaiserslautern, Germany

Christian Tock, BASF SE, 67056 Ludwigshafen, Germany

Jérôme Waser, Laboratory of Catalysis and Organic Synthesis, EPFL SB ISIC LCSO BCH 4306 (Bâtiment de chimie UNIL), CH-1015 Lausanne, Switzerland

Christian Wenzel Tornøe, H. Lundbeck A/S, Ottiliavej 9, DK-2500 Valby, Denmark

Valentin Wittmann, Fachbereich Chemie, Universität Konstanz, Universitätsstr. 10, D-78457 Konstanz, Germany

Abbreviations

Chemical abbreviations, ligands and radicals

Aa	Amino acid
AAE	Azidoacetic acid ethyl ester
ABL	Allegheny Ballistics Laboratory
Ac	Acetyl
acac	Acetylacetonate
ACCN	1,1′-Azobis(cyclohexanecarbonitrile)
AD-mix	Asymmetric dihydroxylation-mix
AIBN	2,2′-Azobisisobutyronitrile
aq	Aqueous
Ar	Aryl
ARC	Accelerated Rate Calorimetry
Asc	Ascorbate
AW-IC	Aza-Wittig reaction/intramolecular cyclization
AW-IEC	Aza-Wittig/intramolecular electrocyclic ring closure
AW-NA-IC	Aza-Wittig/intermolecular nucleophilic addition/intramolecular cyclization
AZT	Azidothymidine
BAM	German Federal Institute for Materials Research and Testing
BEMP	2-*tert*-Butylimino-2-diethylamino-1,3-dimethylperhydro-1,3,2-diazaphosphorine
bmim	1-Butyl-3-methyl-imidazolium
Bn	Benzyl
Boc	*t*-Butyloxycarbonyl
BPC	Biphenylcarboxylic acid
BPT	Biphenyl tetrazole
Bu	Butyl
Bz	Benzoyl
CAN	Ammonium cerium(IV) nitrate
CASPT2	Complete active space self-consistent field second-order perturbation theory
CASSCF	Complete active space self-consistent field
cat	Catalytic
CB	Cucurbituril

CBS-QB3	Complete Basis Set
Cbz	Carboxybenzyl
CCSD(T)	Coupled-cluster singles and doubles
CD	Cyclodextrin
CD4	Cluster of differentation 4
CuAAC	Copper(I)-Catalyzed Azide-Alkyne Cycloaddition
DABCO	1,4-Diazabicyclo[2.2.2]octane
DANP	1,3-Diazido-2-nitro-2-azapropane
DATH	1,7-Diazido-2,4,6-trinitro-2,4,6-triazaheptane
DBU	1,8-Diazabicyclo[5.4.0]undec-7-ene
DCC	Dicyclohexylcarbodiimide
DCM	CH_2Cl_2 or dichloromethane
DCR	Dipolar cycloaddition reaction
DEA	Diethylamine
DEAD	Diethyl azodicarboxylate
DFT	Density functional theory
DIAD	Diisopropyl azodicarboxylate
DIBALH	Diisobutylaluminium hydride
DIPEA	Diisopropylethylamine
DLP	Dilauroyl peroxide
DMAP	4-(Dimethylamino)pyridine
DMF	Dimethylformamide
DMPU	1,3-Dimethyl-3,4,5,6-tetrahydropyrimidin-2(1H)-one
DMSO	Dimethyl sulfoxide
DNA	Deoxyribonucleic acid
DNJ	Deoxynojirimycin
DNP	Dioxynaphthalene
DPPA	Diphenylphosphoryl azide
DSC	Differential Scanning Calorimetry
EGA	Evolved Gas Analysis
EPA	Diethyl ether-isopentane-ethanol 5 : 5 : 2
EPR	Electron paramagnetic resonance
ESD	Electrostatic Discharge
ESR	Electron spin resonance
Et	Ethyl
FAB	Fast atom bombardment
Fmoc	Fluorenylmethyloxycarbonyl
FRET	Fluorescence resonance energy transfer
FTIR	Fourier transform infra red
GAP	Glycidyl azide polymer
HEPES	2-(4-(2-Hydroxyethyl)-1-piperazinyl)ethanesulfonic acid
HIV	Human Immunodeficiency Virus
hpyr	1-Hexylpyridinium
IC	Inhibitory concentration
Im	Imidazoyl
KHMDS	Potassium hexamethyldisilazanide

LC	Liquid crystal
LDA	Lithium diisopropylamide
LOVA	Low-vulnerability ammunition
MBQ	Methoxybenzoquinone
m-CPBA	3-Chloroperoxybenzoic acid
Me	Methyl
MIP	2-Methoxy-isopropyl
MMDOc	MMDOC, *N*,*S*-dimethyldithiocarbamoyl-*N*-oxycarbonyl
MMP	Matrix metalloprotease
MOM	Methoxymethyl
MP2	Møller-Plesset
MPDOC	*S*-Methyl-*N*-phenyl-1,3-dithiocarbamoyloxycarbonyl
Ms	Mesyl
MW	Microwave/μW
NBS	*N*-Bromosuccinimide
NHE	Normal hydrogen electrode
NMP	*N*-Methylpyrrolidone
Ns	2-Nitrobenzenesulfonyl
PEG	Polyethyleneglycol
PEGA	Polyethylene glycolpoly-(*N,N*-dimethylacrylamide)
PETN	Pentaerythrittetranitrate
PFP	Pentafluorophenyl
Ph	Phenyl
PMHS	Poly(methylhydrosiloxane)
PMPA	*N,N′*-[3-Phenylenebis(methylene)]dipropargylamine
PNA	Peptide nucleic acid
PPG	α,ω-Bisazidopropylene glycol
PPTS	Pyridinium *p*-toluenesulfonate
psi	Pound per square inch (1 psi = 6 894.75729 pascals)
PT	Phenyl tetrazole
PTOC	1*H*-Pyridine-2-thione-*N*-oxycarbonyl
py	Pyridine
PyBOP	Tradename of benzotriazol-1-yl-oxytripyrrolidinophosphonium hexafluorophosphate
RDX	Research Department Explosive / Royal Demolition Explosive (Hexahydro-1,3,5-trinitro-1,3,5-triazine)
rfx	Reflux
RNA	Ribonucleic acid
ROESY	Rotating Frame Overhauser Enhancement Spectroscopy
$scCO_2$	Super critcial CO_2
SEM	[β-(Trimethylsilyl)ethoxy]methyl
SHR	Self-heating rate
STAT3	Signal transducers and activators of transcription 3
TAH	Triazidoheptazine
TAM	Thermal Activity Monitor
TANA	Thioacetamido nucleic acids

TAP	2,4,6-Triazidopyrimidine
TAP-Ac	Triazido pentaerythrite acetate
TASP	Template-assembled synthetic proteins
TAT	*trans*-Activating transcriptional activator
TBAA	Tetrabutylammonium azide
TBAF	Tetrabutylammonium fluoride
TBAI	Tetrabutylammonium iodide
TBDMS	*tert*-Butyldimethylsilyl
TBS	Tributylsilyl or *tert*-butyldimethylsilyl
TBSnA	Tributyltin azide
TEA	Triethylamine
TEMPO	2,2,6,6-Tetramethylpiperidine-1-oxyl
TES	Triethylsilyl
Tf	Trifluoromethylsulfonyl
TFAA	Trifluoroacetic anhydride
TGA	Thermogravimetric Analysis
THF	Tetrahydrofurane
TMDSO	Tetramethyldisiloxane
TMEDA	*N,N,N′,N′*-Tetramethylethylenediamine
TMGA	Tetramethylguanidinium azide
TMS	Trimethylsilyl
TMSA	Trimethylsilyl azide
TNT	Trinitrotoluene
TPP	Tetraphenylporphyrin
Ts	Tosyl
TTF	Tetrathiafulvalene
VEGF	Vascular Endothelial Growth Factor
VEGFR1	Vascular Endothelial Growth Factor receptor 1
YAG	Yttrium aluminium garnet

PART 1

Synthesis and Safety

1

Lab-scale Synthesis of Azido Compounds: Safety Measures and Analysis

Thomas Keicher and Stefan Löbbecke

Fraunhofer Institut für Chemische Technologie ICT, Pfinztal, Germany

1.1 Introduction

More than a century after their discovery organic azides have received renewed interest in synthetic chemistry and are becoming established as an important and versatile class of chemical compounds.[1–3] In particular, in the past two decades an increasing interest in organic azides could be observed due to their vast synthetic utility in conjunction with their easy accessibility via various synthetic routes. Among others, organic azides are currently considered as powerful precursors for reactive species such as nitrenes and nitrenium ions as well as nitrogen-rich compounds such as aziridines, azirines, triazoles, triazolines and triazenes. Moreover, organic azides can be easily transformed into amines, isocyanates and other functional molecules and have more recently received an increasing interest as valuable and versatile reagents within the concept of 'Click Chemistry'.[4–5]

However, alongside their huge utility in organic synthesis the potential hazardous properties of organic azides must be also carefully taken into account. Organic and inorganic azides are first and foremost energy-rich molecules which often exhibit explosive properties. The azido group is a highly energetic functional group. The N_3 π-bond can be easily polarized which consequently results in strong exothermic dissociation reactions under release of molecular nitrogen and reactive nitrene groups. In general, the introduction of an azido group into an organic compound increases its energy content by approx. 290–355 kJ/mol.[6–7] This is one of the reasons why organic azides are considered and partly

Organic Azides: Syntheses and Applications Edited by Stefan Bräse and Klaus Banert

used as energetic materials such as energetic polymers or high-energy-density-materials (HEDM) in explosives or propellant formulations.[8–9] However, many organic compounds containing azido groups have not found wide application as practical energetic materials because of their high sensitivity to heat and shock stimuli. This poor thermal and mechanical stability of many organic azides is probably the most crucial property that has prevented chemists in the past from extending their research on azido compounds.

Therefore, in times of increasing worldwide interest in organic azides, it is of utmost importance that their hazardous potential (arising from their energetic and/or toxic properties) and the corresponding safety issues are adequately emphasized. Here, we report on safety precautions and practical measures for the safe handling of azides at laboratory scale. Analytical techniques and other test methods are described that allow characterizing the hazardous potential of organic azides qualitatively and quantitatively at an early stage of small-scale preparation.

1.2 Properties that Impose Restrictions on Lab-scale Handling of Azides

1.2.1 Hydrazoic Acid and Its Metal Salts

Hydrazoic acid, HN_3, and its salts are very poisonous compounds with a toxicity comparable to hydrogen cyanide. Pure hydrazoic acid is a colorless strong-smelling liquid that tends to spontaneous explosion. Because of its high vapour pressure (b.p. 36 °C) lab-scale handling of HN_3 is conducted either in water solution or by diluting the acid in organic solvents.

Metal salts of hydrazoic acid from lead, silver, mercury, copper and other heavy metals are very sensitive to mechanical stimulus and explode easily. Heating of these metal azides causes strong explosions. Table 1.1 summarizes some deflagration temperatures of well-known metal azides that decompose explosively when small samples embedded within a metal sleeve are dipped into a hot Wood metal bath.[10] Although these temperatures are relatively high other new metal azide complexes might explode at much lower temperatures.

All heavy metal azides run very quickly into detonation. This specific property has established the use of silver azide and lead azide as primary explosives in detonators.

Remarkably more stable in terms of safe handling are lithium and sodium azide which can be more reliably used in the laboratory since they are hard to initiate explosion by impact or friction. For most laboratory conditions, alkali metal azides are not considered as explosives.[11] However, if ignited or when exposed to strong heat, alkali metal azides decompose rapidly with the evolution of large volumes of nitrogen gas.

Table 1.1 *Deflagration temperatures of selected metal azides*

Metal azide	Deflagration temperature
Silver azide	297 °C
Lead azide	327 °C
Mercury azide	281 °C
Cadmium azide	291 °C

In general, metal azides show an increase in mechanical sensitivity when going from earth alkali metal azides to heavy metal azides which are highly sensitive and explosive. The impact sensitivity of metal azides decreases in the following order:[12]

Copper > Lead, Mercury > Nickel > Cobalt > Manganese > Barium > Strontium > Calcium > Silver > Thallium > Zinc > Lithium = no explosion

However, this sequence of impact sensitivities represents only one type of mechanical stability. In case of friction sensitivity (see Section 1.4.2) this order changes. For example, silver azide is approx. 10 times more friction-sensitive than lead azide.[13]

1.2.2 Organic Azides

Organic azides are considered as explosives whenever the azido content is remarkably high. Of course, there is no sharp threshold at which the explosive hazard starts. However, as a rule of thumb violent decomposition reactions are expected for azido compounds having a (C + O)/N ratio of <3.[14]

Organic compounds with high azido content are very sensitive to friction and impact, causing strong explosions. For example, cyanuric azide is very sensitive against mechanical stimulus and thus decomposes very easily by detonation. Although the initiation power of this detonation exceeds that of classical primary explosives no technical application has been found so far for this compound due to its high vapour pressure.

Organic azides also show remarkable lower ignition temperatures in comparison to inorganic metal azides. Most of the organic azides decompose at approx. 180 °C.

Some organic azides also show light sensitivity[15] and strong incompatibility with certain chemicals. Several examples have been published where azides exploded when they were brought in contact with sulfuric acid or other compounds.[16–18]

1.3 Laboratory Safety Instructions for the Small-scale Synthesis of Azido Compounds

Scientific papers publishing the syntheses of azido compounds usually include in the experimental section certain safety instructions and hints of adequate protection. However, rarely specific instructions for enhancing the safety or detailed methods for protection measures are given.

In general, azido compounds have to be considered and handled as explosive materials. An additional hazard might be caused by their toxicity. In the following the most relevant safety instructions and measures are summarized for the synthesis and handling of azido compounds in the lab:

- Separate the experimental setup with proper shielding and an additional safety screen in the fume hood. Keep the screen of the fume hood always closed during critical operations like heating, distilling and vigorous stirring.
- Safety screens should be made from laminated glass with one ore more layers of plastic film embedded between the glass layers. In case of an explosion the broken pieces of glass should remain sticking to the plastic interlayers and not shoot through the laboratory.

- Cover the glassware with adhesive films to reduce the fragmentation in case of explosions, as it is usually done on rotary evaporators and desiccators.
- In addition to the usual protection outfit (lab coat, gloves, safety glasses) wear a face protection shield, ear protection, a leather jacket or a bullet proof vest with arm protection.
- For hand protection use leather gloves (welding type), ideally in combination with steel interwoven Kevlar® gloves! (Klapötke *et al.* have recently published a systematic investigation on the stability of protective gloves against explosion impact. They have found that none of the tested protective gloves could withstand all different kinds of explosion impacts. For example, leather gloves showed best protection against small glass fragments whereas steel interwoven Kevlar® gloves protected well against larger glass splinters.[19])
- Start the first experiments on a small scale of only a few mg to allow determination of first sensitivity data. Increase the scale only when the scale-up is in accordance with the sensitivity data obtained.
- Keep hazardous azides in solution as long as possible. Solvents desensitize explosives by reducing the sensitivity to mechanical stress.
- Keep solid material wet or soaked with solvent as long as possible. In mixture with liquids explosives are normally desensitized (phlegmatized). Favoured solvents for this purpose are nonflammable solvents like water or halogenated hydrocarbons. Only highly sensitive compounds like primary explosives can detonate without obvious reason even when they are stored under water. It is supposed that these unexpected explosions are caused by internal stress of larger crystals leading to crystal cracks.
- Try to obtain solid products of small particle sizes. Smaller particles/crystals are less sensitive to mechanical stress than larger ones. Only very small particles of $<10\,\mu m$ sometimes show enhanced mechanical sensitivity. Recrystallization experiments should thus be cooled down very quickly under stirring.
- All experiments containing azide compounds that are set up under vacuum should not be vented at elevated temperature. The vent should be remote controlled.
- Do not use metal spatulas because they transfer stronger mechanical stress to the material than spatulas made of wood or Teflon®.
- Keep sufficient distance between the azide material and your body. Do not touch the potential explosive material directly and use, whenever it is possible, gripping devices to manoeuvre the container that encloses the azide compound. Figure 1.1 shows examples of suitable gripping tools. The peak pressure for the explosion decreases at the rate of $1/R$ for large distances from the explosion centre (R is the safety distance from the charge). However, when the distance is close to the explosion center, the rate of decrease is between $1/R$ and $1/R^2$. Consequently, in this case the safety distance is a more critical parameter since every small additional distance to the charge will drop the pressure significantly.[20]
- Sensitive explosive azides can be also easily ignited by electrostatic discharge (ESD). Therefore, wear ESD protective clothing (or at least cotton clothes) and antistatic shoes. The laboratory floor should be ESD conductive or there should be at least static dissipative mats (ESD mats) installed in front of the fume hood where the azides will be synthesized and handled.
- Apart from all mentioned energetic hazards, the synthetically working chemist must also keep in mind the toxic nature of azides!

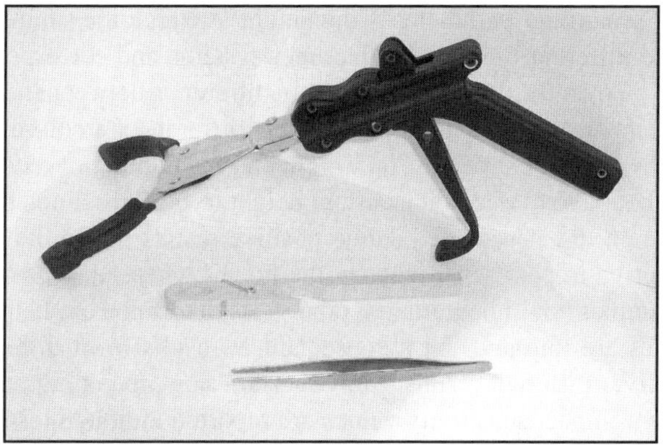

Figure 1.1 *Examples for gripping devices*

1.4 Analyzing Safety-related Properties of Azides

Because of the described hazardous potential of organic azides the analysis of their safety-related properties is of utmost importance to ensure safe synthesis and subsequent processing. Unfortunately, in literature only very few data can be found describing thermal properties of organic azides. Data on impact, friction or ESD sensitivity are also mostly not available as well as any reliable information on (long-term) stability and energy content. One of the reasons for these missing data might be that most scientific reports on organic azides are provided by research groups focusing on the preparation and subsequent conversion of organic azides. One can assume that the specific analytical techniques that are required to measure all relevant safety-related properties are not fully available in these synthesis labs.

Therefore, in the following we describe the most important and relevant analytical methods and characterization techniques that are required to evaluate the hazardous potential of organic azides qualitatively and quantitatively. From our experience, it is essential to conduct sensitivity tests, thermoanalytical measurements and stability tests of energy-rich compounds at an early stage of every small-scale preparation. Naturally, such safety analysis is mandatory for the synthesis of new organic azides to decide whether the scale of synthesis can be enlarged and subsequent processing of this new compound is possible in a safe manner. Moreover, we strongly recommend conducting permanent safety analyses also for such energy-rich azides whose syntheses have already been established in the labs. Slight differences in the experimental procedure might result in products of different heat/shock sensitivity and stability, for example due to different particle sizes and crystal morphologies. Consequently, sensitivity and stability data of an energetic azide must be strictly rechecked after each synthesis campaign.

1.4.1 Impact Sensitivity Testing

The impact sensitivity of energetic compounds is tested with a so-called fall hammer equipment. Samples are exposed to the impact of falling weights from variable heights and the measured sensitivity parameter is the height at which the samples decompose or explode. There are different types of fall hammer systems and corresponding test procedures in operation all over the world.[21–24] The main difference between them is in the design of the sample holders or sample confinements. Other differences are the amount of samples that are used for the tests, the type of drop weight and the number of recorded decompositions/explosions that have to occur at a certain height to produce a positive result. Unfortunately, the different test procedures define positive results not in the same way. Some tests describe positive impact sensitivity by the height where at least one of the six (or sometimes ten) samples could be initiated; other tests determine the height where 50% of at least 20 samples are initiated. As a consequence, results from different fall hammer systems might deliver different results. Moreover, for new upcoming energetic materials sometimes different impact sensitivity values are reported although tests were conducted by using the same type of fall hammer but were operated by different laboratories. These varying sensitivity data might be either caused by different sample qualities (purity, particle size, crystal density, liquids with/without gas bubbles that act as 'hot spots', etc.[24–26]) or by different operators of the test system. In case of strong explosions initiation can be easily recognized, but sometimes the decomposition of the sample starts with weak smouldering, which is hard to notice for the operator of the fall hammer system. Consequently, to obtain a more reproducible detection level for positive responds some laboratories use microphones to measure the explosion bang during impact sensitivity tests.[22,27–30]

In literature, there are several attempts described to predict and calculate the impact sensitivity of energetic materials.[28,31–38] Most of these papers deal only with nitro- and nitrate ester compounds, whereas the impact sensitivity of azides has not been the subject of detailed calculations so far, apart from recently published structure-sensitivity correlations on inorganic azides.[39]

Figure 1.2 shows the set-up of the fall hammer equipment as it has been defined by the German Federal Institute for Materials Research and Testing (BAM).[40] There are two versions of different sizes in operation. The small fall hammer is for testing sensitive explosives such as primary explosives and is operated with weights up to 1 kg. The large hammer is used for more insensitive explosives that can be impacted by hammer weights of 1.5 and 10 kg. The test sample has a volume of 40 mm^3 and it is placed between two steel cylinders that are fixed by a steel ring (Figure 1.2). The cylinders have a diameter and a height of 10 mm and are made from ground and hardened steel. The reported impact sensitivity value is the fall energy, given in Nm, at which at least one sample from a series of six has been initiated.

In Table 1.2 impact sensitivity values of different azido compounds according to the BAM fall hammer procedure are listed and compared with the corresponding values of the well-known explosives trinitrotoluene (TNT) and nitroglycerine. For the BAM procedure it is necessary to have at least one positive event within a series of six trials (probability of at least 16.7%). In the case of the US drop hammer tests (according to the Bruceton procedure) the required probability level is often 50% initiation within a series of at least 25 trials.

Therefore, the impact sensitivity of an energetic compound is not a strictly fixed absolute value like its melting point but is subject to certain fluctuations depending on the sample characteristics, the test equipment and testing procedure as well as the operator. However, impact sensitivity values provide clear safety information and can be used in practice in particular as a comparative method.

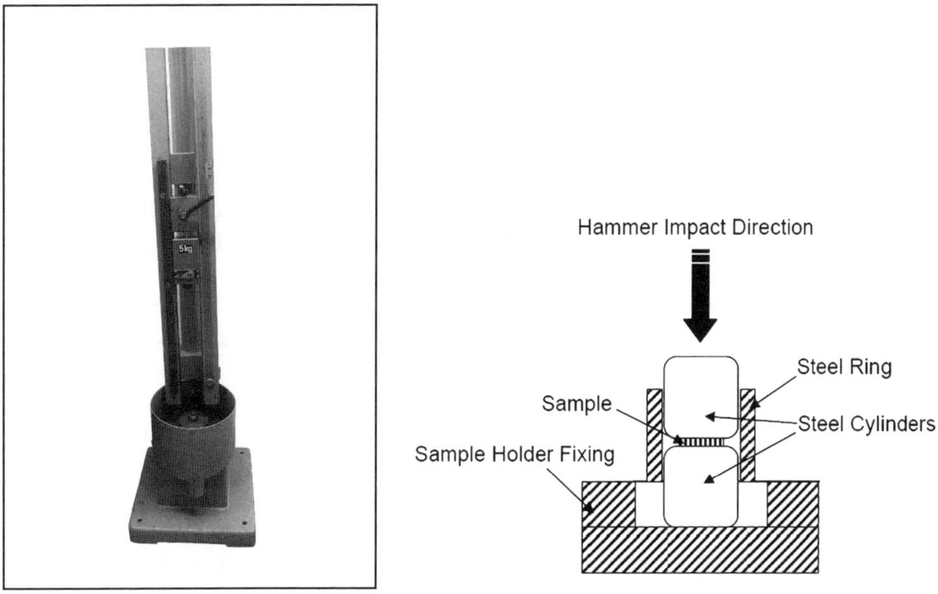

Figure 1.2 *BAM fall hammer set-up (left) and corresponding sample holder (right)*

Table 1.2 *Impact sensitivity values of selected azide compounds compared with trinitrotoluene and nitroglycerine*

Explosive	Fall hammer weight / kg	Fall height / m	Impact sensitivity (BAM procedure) / Nm
Lead azide[24]	5	0.15	7.5
Trinitrotoluene[24]	5	0.30	15
Nitroglycerine[24]	0.1	0.20	0.2
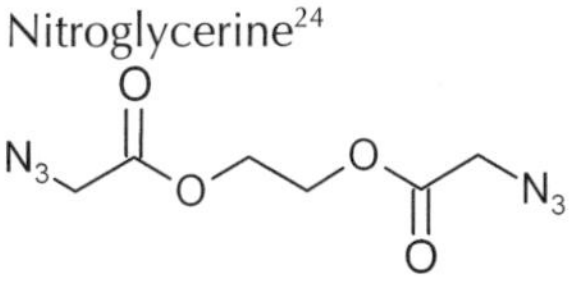	5	0.1	5
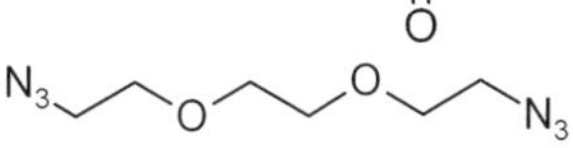	1	0.2	2
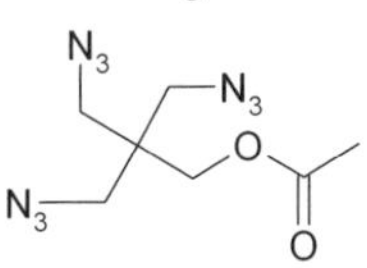	0.1	0.2	0.2

1.4.2 Friction Sensitivity Testing

For measuring and testing the friction sensitivity, samples of energetic compounds are exposed to friction forces that are generated by different setups and methods.[22–24,41] In the ABL (Allegheny Ballistics Laboratory) Sliding Friction Test the sample is pressed by a steel roll with an adjustable force on an anvil which is accelerated by the impact of a pendulum. The measurement of the compressive force is done when from 20 samples 50% are initiated. In another setup developed by the Bureau of Mines, a pendulum with different shoes (steel or fibre reinforced plastic) grazes over the sample (7 g) that is spread on an anvil with three grooves. This test is passed when 20 trials are not giving any initiation.

In the Roto-Friction Test developed at the American Naval Surface Warfare Center a friction rod is rotating on the sample that is placed into a recess bored sample holder. The normal force weights that press the friction rod on the sample can be varied and torque measurement equipment records the force transmitted through the sample to the sample holder. The friction energy value is calculated from the measured torque and from the exposure time that the sample is stimulated by rotating friction till any decomposition or explosion occurs.

The friction test setup defined by the German Federal Institute for Materials Research and Testing (BAM) measures the sensitivity of samples that are exposed to a friction stimulus generated between two roughened porcelain surfaces.[40] 50 mg of a sample resting on a porcelain plate is stimulated by a porcelain pin with adjustable down-pressing force. For stimulation, the sample table is driven by a motor horizontally forwards and backwards for one full cycle of reciprocating motion. Figure 1.3 shows the setup of the BAM friction test.

There are two versions of the friction test apparatus in operation, a standard size apparatus and a small size version. The small device is particularly designed to test sensitive materials such as primary explosives. It can be operated with different weights on the porcelain bolt holder allowing loads in the range of 0.1 to 10 N. On the standard BAM friction test apparatus higher forces on the pin varying from 5 to 360 N can be applied.

The results obtained by the BAM friction test refer to the smallest load on the pin under which deflagration, crackling or explosion of the sample is observed, at least once in six consecutive trials. Other test procedures provide friction energy values on the basis of a 50% initiation probability. Therefore, absolute friction sensitivity values that are measured might be subject to certain fluctuations depending on the specific properties of a sample (e.g. purity, particle size, etc.) and the test equipment used. Nevertheless, friction sensitivity measurements provide useful safety information and allow direct comparison with other sensitive or less sensitive materials, and are thus of the same importance as impact sensitivity data.

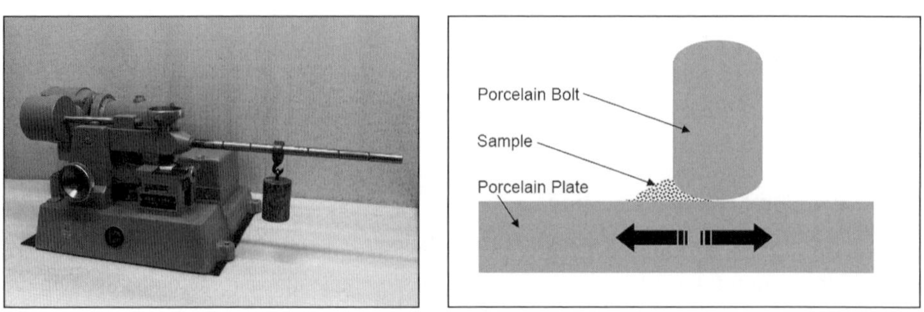

Figure 1.3 *BAM friction test apparatus (left) and scheme of the measuring principle (right)*

Table 1.3 *Friction sensitivity values of selected azide compounds compared with TNT and nitroglycerine*

Explosive	Friction sensitivity (BAM procedure) / N
Lead azide[24]	0.1
TNT[24]	up to 360 no reaction
Nitroglycerine[24]	up to 360 no reaction
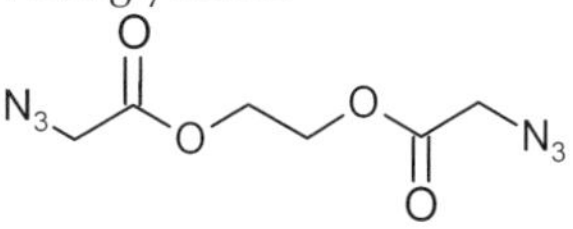	128
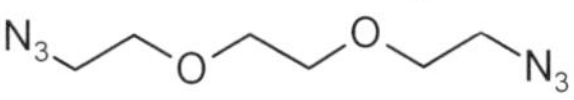	128
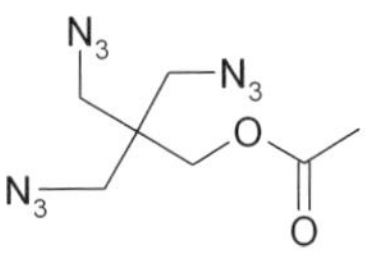	84

In Table 1.3 exemplary friction sensitivity values of azido compounds according to the BAM procedure are listed and compared with the corresponding values of TNT and nitroglycerine. A comparison with the sensitivity values listed in Table 1.3 points out that the sensitivity to friction and impact of a specific energetic compound might differ significantly.

1.4.3 ESD Testing

In electrostatic discharge (ESD) tests, the amount of energy that is required to ignite explosives by electrostatic stimuli is determined. Most explosives have low electrical conductivity. Therefore, the potential of the electrostatic pulse has to be high to generate a sparkover. Figure 1.4 illustrates the principle setup of an ESD testing device.

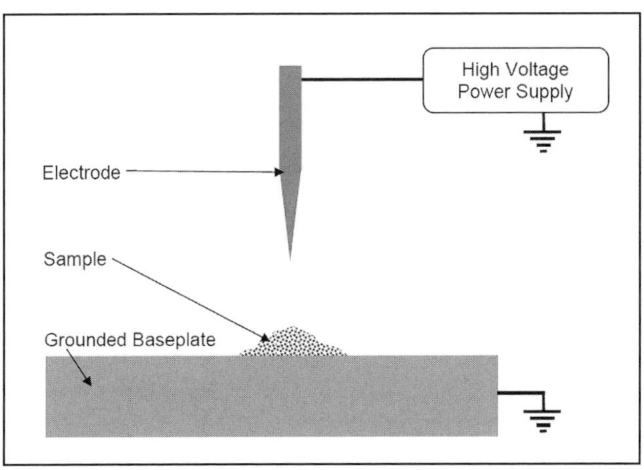

Figure 1.4 *Schematic setup of an ESD testing apparatus*

First, a capacitor is charged up to a high voltage level followed by the release of the electric energy via a discharge pole through the sample which is placed on an earthed plate or pole. Samples are tested by varying the intensity of the released electrostatic discharge. A positive result is defined whenever a flash, spark, burn, or specific noise is detected.

There are different ESD test systems and procedures in worldwide operation.[22–23,42–44] Main differences are in the design of the sample holder, the size and shape of the discharge electrode, the voltage level before discharge and the number of experiments that are required to define a positive or negative ESD test result. Consequently, different ESD sensitivity values can be found in literature for the same energetic compound. Besides the influence of different ESD test apparatus the actual constitution of a sample has mostly a more significant impact on its ESD sensitivity. For example, it is well known that differences in particle size, grain shape, temperature and moisture content provide different ESD sensitivity values.[13,23,45-49]

In Table 1.4 exemplary ESD test results are listed that are reported for different energetic materials including inorganic azides (no comparable data are available for organic azides). It can be clearly seen that different test setups give different sensitivity values for the same material. Tremendous differences arise when samples of different particle size are tested. As expected, smaller particles are more sensitive to electrostatic ignition than larger ones. Another parameter that influences the ESD test result is the confinement of the sample. Fine powders are more sensitive to electrostatic ignition in the unconfined state and coarse material gets more sensitive if it is exposed to the electrostatic discharge in the confined state.

These partly huge fluctuations in ESD sensitivity of energetic compounds arising from different sample constitutions should sensitize every person practically working with energetic compounds like organic azides in the lab. One person can store up to 100 mJ by wearing insulating shoe soles in a dry environment. The maximum electrostatic discharge energy in a spark is up to 20 mJ and is thus high enough to initiate sensitive materials.[50]

Table 1.4 *Exemplary ESD test results for some azide compounds and other energetic materials*

Sample	50% ignition probability at varied voltage / mJ (according to[45])	Zero ignition probability at 5000 volts / mJ (according to[46])	
		Unconfined sample	Confined sample
Lead azide	0.06	7	7
Lead azide/dextrin	23 / 112[a]	n.a.	n.a.
Lead azide/dextrin	23	n.a.	n.a.
Sodium azide	>79,433	n.a.	n.a.
TNT	22,387	62 / >11,000[b]	4,380 / 4,680[b]
PETN	2,630	62 / >11,000[b]	210
Black powder	2,692 / 4,074[b]	>12,500	800

[a] two different ESD test devices;
[b] different particle sizes; n.a.: not available.

1.4.4 Thermoanalytical Measurements

In addition to measuring the mechanical and electrical sensitivity of energetic compounds it is essential to also analyze their thermal and caloric properites thoroughly. In particular, the data on the thermally induced decomposition behavior are required to evaluate the hazardous potential of energy-rich compounds such as organic azides.

In thermal analysis, physical parameters like mass, heat flow, heat capacity and enthalpy are measured as a function of temperature and time, while the sample is subjected to a controlled temperature programme (which in most cases is the application of linear heating rates or isothermal conditions).[51–53] The two most common thermoanalytical techniques to investigate thermal and caloric properites of energy-rich compounds are Differential Scanning Calorimetry (DSC) and Thermogravimetric Analysis (TGA).

Differential Scanning Calorimetry (DSC) is a technique for measuring the energy necessary to establish a nearly zero temperature difference between a sample and an inert reference material, as the two specimens are subjected to identical temperature regimes in an environment heated or cooled at a controlled rate. Two different types of DSC instruments are known: power-compensation DSC and heat-flux DSC.[54] In power-compensation DSC the temperatures of the sample and reference are controlled independently using separate, identical furnaces. Both, sample and reference are kept at an identical temperature by varying the power input to the two furnaces; the energy required to do this is a measure of the enthalpy or heat capacity changes in the sample relative to the reference.

Today, heat-flux DSC is more commonly used. Here, sample and reference are connected by a low-resistance heat-flow path which is mostly a metal disc. This assembly is enclosed into a single furnace. Enthalpy or heat capacity changes in the sample cause a difference in its temperature relative to the reference. The temperature difference and thus the resulting heat flow are recorded and related to enthalpy changes in the sample. Figure 1.5 shows a schematic cross-section of a typical heat-flux DSC cell. The sample (up to

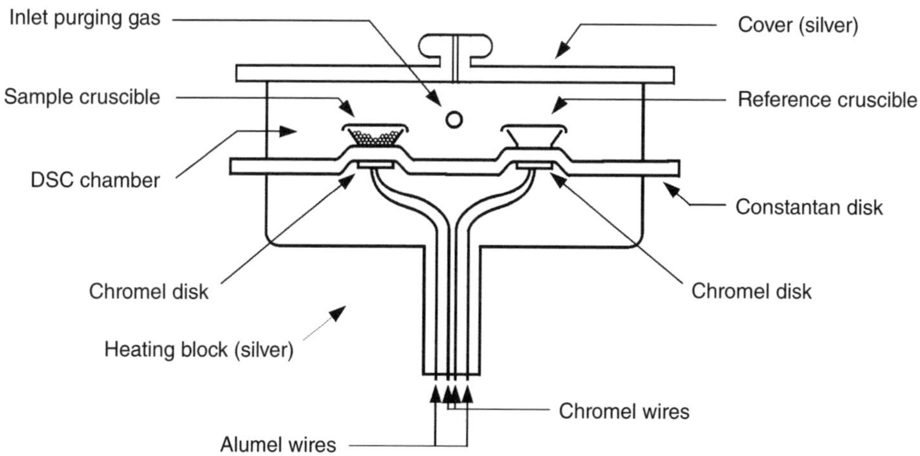

Figure 1.5 *Schematic setup of a heat-flux DSC cell (DSC 2920, TA Instruments Inc.)*

several mg) is placed in a small cruscible (for example, an alminum pan). In most cases an empty cruscible is used as reference sample. The entire DSC cell can be permanently purged by inert or reactive gases. In case of analyzing energetic materials, argon or nitrogen gas is usually used to remove volatile substances and decomposition gases during the measurement. In case of azido compounds the use of argon is recommended since nitrogen is also one of the decomposition products. Samples that are expected to exhibit strong exothermic decomposition are usually analyzed in non-hermetically sealed cruscibles to avoid uncontrolled pressure built-up and allow decomposition gases to be released (in most cases aluminum pans with pierced lids are employed). Moreover, small sample sizes of partly <1.0 mg are used and only slow heating rates up to 5.0 K/min are applied to avoid uncontrolled decomposition.[55]

In general, DSC measurements allow the recording of all types of chemical and physical transformations of a sample that involve exothermic and endothermic processes or changes in heat capacity.[54] In particular, exothermic decomposition reactions as well as endothermic phase transitions (melting, boiling, sublimation, solid-solid phase transition between different crystal morphologies, and glass transition of polymers) are the most relevant processes which are considered in energetic materials analysis. DSC measurements provide both, the characteristic temperature values of all endothermic and exothermic processes (onset temperature, peak temperature) and the corresponding enthalpies.

As an example, Figure 1.6 shows the DSC measurement of triphenylmethyl azide (trityl azide) applying a linear heating rate of only 1.0 K/min (sample size: 3.35 mg, Argon atmosphere, Al pan with pierced lid). Endothermic processes are displayed by negative

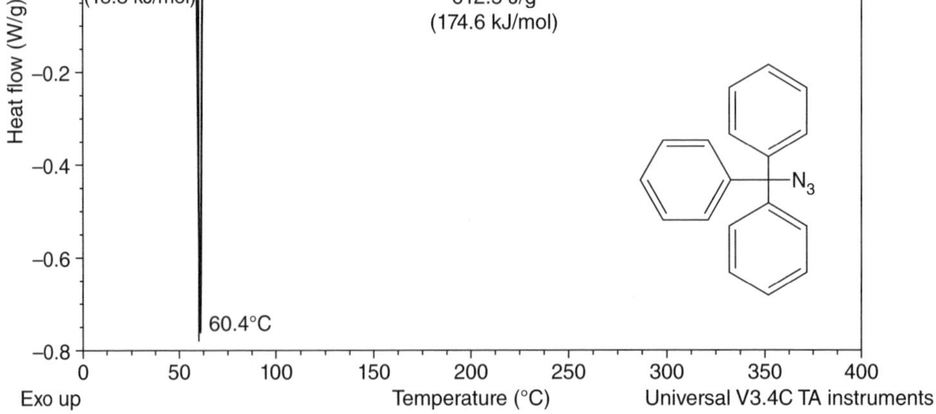

Figure 1.6 *DSC measurement of triphenylmethyl azide (1.0 K/min, 3.35 mg, open Al pan)*

heat flow values, whereas exothermic processes show a positive heat flow. Three main processes can be identified in the DSC graph of trityl azide. First, a sharp endothermic melting peak occurs at a calculated onset temperature of 59.8 °C (peak maximum at 60.4 °C). The melting enthalpy, calculated from the integral of the melting peak, is approx. 18.8 kJ/mol (66 J/g). Further heating of trityl azide leads to its exothermic decomposition starting at approx. 150 °C. However, the calculated onset temperature of the decomposition is 168 °C and the peak maximum temperature is 198 °C. The decomposition enthalpy under the chosen experimental conditions is 174.6 kJ/mol (612.5 J/g), which is already remarkable but not hazardous. Finally, after completion of the exothermic decomposition, a huge endothermic process can be observed, which can be assigned to the slow sublimation of decomposition products.

Besides temperature and enthalpy values DSC measurements provide additional safety-related information on the strength and intensity of decomposition reactions. In particular the steepness and width of the exothermic decomposition peaks are qualitative indicators for reactivity and thus vehemency of the thermally induced decomposition reaction. As an example, Figure 1.7 shows the DSC curve of 2,5,8-triazido-s-heptazine (TAH),[9,56] a nitrogen-rich energetic compound, recorded at a linear heating rate of 5.0 K/min (argon atmosphere, Al pan with pierced lid). A sample mass of only 0.71 mg was used to detect the strongly exothermic decomposition at a calculated onset temperature of 190 °C. The steep and relatively narrow exothermicity indicates a more vehement decomposition in

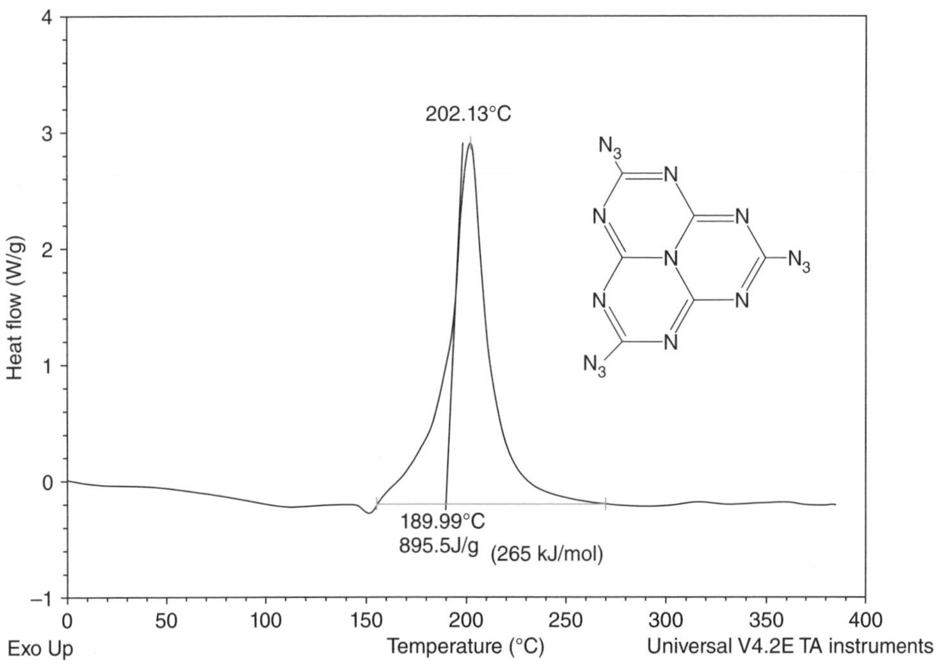

Figure 1.7 *DSC measurement of 2,5,8-triazido-s-heptazine (TAH) (5.0 K/min, 0.71 mg, open Al pan)*

comparison to the previous example by releasing 265 kJ/mol under the chosen experimental conditions.

Almost threefold the amount of heat is released by a sample of triazido pentaerythrite acetate (TAP-Ac[57–58]), confined in a hermetically sealed aluminum pan; its DSC measurement is shown in Figure 1.8. A small sample of only 0.35 mg was analyzed applying a linear heating rate of 5.0 K/min. The DSC graph shows no phase transition or any other transformation of TAP-Ac until decomposition starts at approx. 190 °C (calculated onset temperature: 222 °C). A steep increase in heat flow combined with a narrow exothermic peak is a clear indication for a violent and rapid decomposition of TAP-Ac. In fact, a high decomposition enthalpy of approx. 730 kJ/mol was measured under the chosen experimental conditions.

TAP-Ac is a good example to emphasize the importance of thoroughly analyzing the hazardous potential of energy-rich compounds. In spite of its three azido groups TAP-Ac shows a surprisingly high stability over a wide temperature range in DSC experiments. This makes one believe that TAP-Ac is a thermally stable and thus nonhazardous compound under ambient conditions. However, in huge contrast to its thermal robustness fall hammer tests of TAP-Ac have revealed its high impact sensitivity of only 0.2 Nm (see Table 1.2).

In many laboratories DSC measurements of energetic materials are complemented by *Thermogravimetric Analysis (TGA)*. In TGA experiments the sample mass is recorded

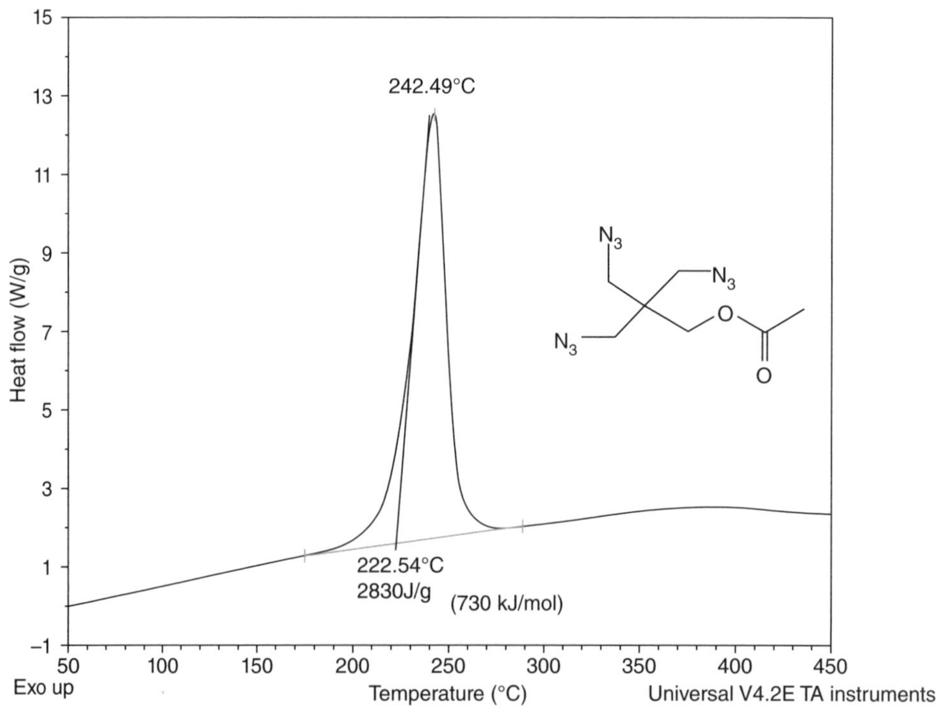

Figure 1.8 *DSC measurement of triazido pentaerythrite acetate (TAP-Ac) (5.0 K/min, 0.35 mg, sealed Al pan)*

as a function of temperature and time while the sample is subjected to a linear heating rate or an isothermal treatment.[51] The samples (up to several mg) are filled into an open cruscible made of platinum or alumina which is attached to the arm of a recording microbalance, the so-called thermobalance. The sample is heated in a temperature-controlled furnace according to a pre-programmed temperature/time profile. During the experiment both the furnace and the thermobalance are purged independently with inert gas (usually argon or nitrogen). Figure 1.9 shows schematically a typical setup of a TGA furnace.

Since the mass changes of a sample are recorded in TGA experiments, the method is predominantly used to investigate the thermal decomposition behavior of compounds. It typically provides information on the decomposition onset temperature and the mass loss that occurs during the decomposition reaction. TAP-Ac, for example, degrades completely during its strongly exothermic decomposition forming only gaseous products within one total mass loss step. However, other energetic azides degrade stepwise.[59] For example, Figure 1.10 shows the DSC and TGA data of 1,3,5,7-tetrakis(4-azidophenyl)adamantane, a recently synthesized compound.[60] The organic azide decomposes stepwise during slow heating at 5.0 K/min. However, the main exothermicity of the thermal decomposition (as measured by DSC) is related only to the first mass loss step of 18.59% which corresponds well with the release of four equivalents of molecular nitrogen.

This example also shows that the combined use of different thermoanalytical methods allows a more detailed analysis of decomposition processes. Moreover, whenever DSC and TGA are combined with *Evolved Gas Analysis (EGA)* – which allows an in-situ

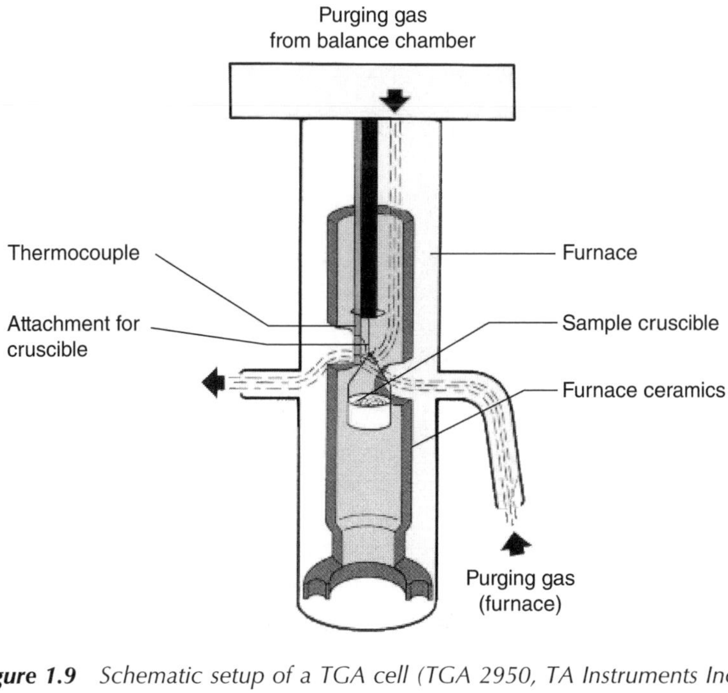

Figure 1.9 *Schematic setup of a TGA cell (TGA 2950, TA Instruments Inc.)*

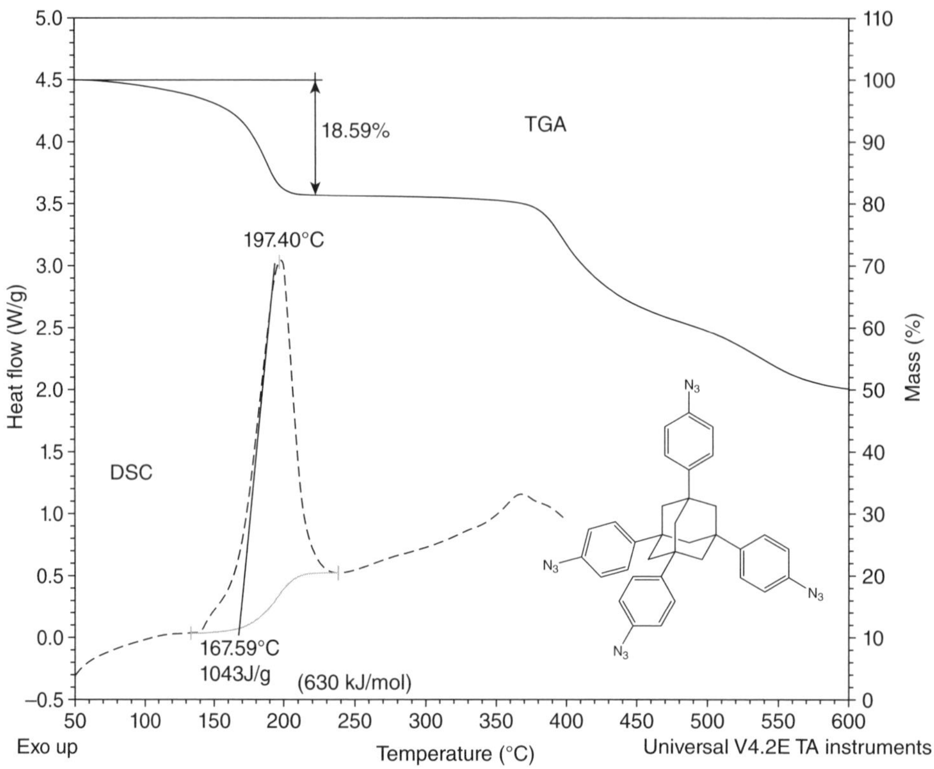

Figure 1.10 *DSC and TGA measurement of 1,3,5,7-tetrakis(4-azidophenyl)adamantane (DSC: 5.0 K/min, 1.05 mg, open Al pan; TGA: 5.0 K/min, 0.89 mg, Pt pan)*

detection of gaseous decomposition products by infrared spectroscopy or mass spectrometry – chemical pathways and mechanisms of thermal decomposition can be revealed.[51,61] As an example, Figure 1.11 shows the infrared spectroscopic detection of decomposition gases during the linear heating of 4.0 mg TAP-Ac at 5.0 K/min. The EGA waterfall-plot illustrates the temperature-resolved release of carbon dioxide (characteristic infrared absorption bands at 2360 cm^{-1}, 2322 cm^{-1}, and 700 cm^{-1}), water (broad absorption centered at 3750 cm^{-1} and 1600 cm^{-1}) and methyl acetate (2964 cm^{-1}, 1778 cm^{-1}, 1450 cm^{-1}, 1375 cm^{-1}, 1247 cm^{-1}, and 1050 cm^{-1}) as well as the formation of traces of ammonia (double band at 965 and 931 cm^{-1}) in subsequent gas phase reactions (Note: molecular nitrogen is also a main decomposition product of TAP-Ac that, however, cannot be detected by infrared spectroscopy but by mass spectrometry in coupled TGA-MS setups.) From the individual gas evolution profiles kinetic data can be derived as they can be also obtained from DSC and TGA experiments conducted under different heating rates.

In conclusion, thermoanalytical methods are powerful tools to determine safety-related thermal properties of azides. Whenever possible, we strongly recommend performing DSC measurements of potentially energetic azides as soon as a few milligrams of substance are available. The combination of DSC data on decomposition temperature and

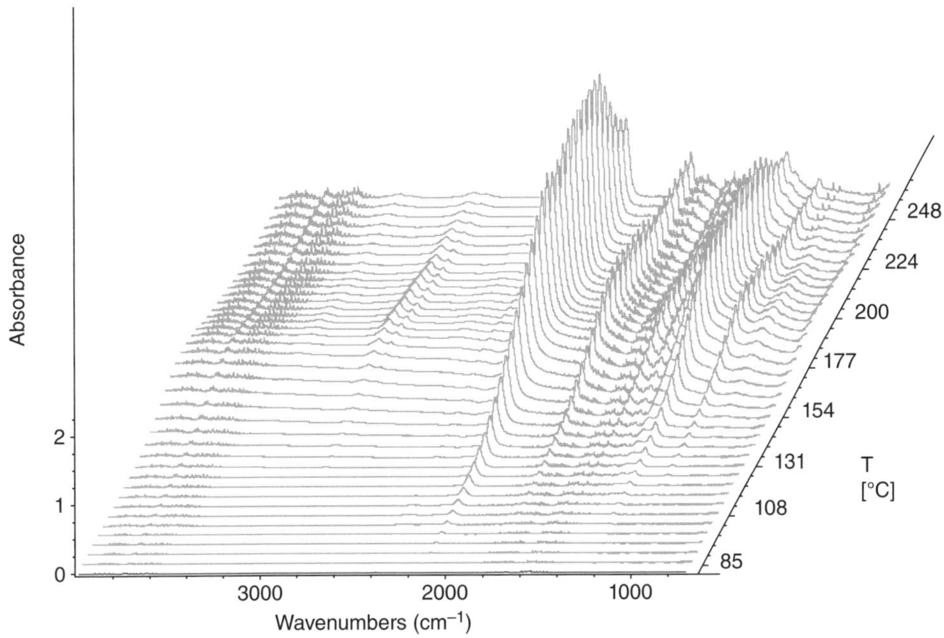

Figure 1.11 *Infrared spectroscopic Evolved Gas Analysis of triazido pentaerythrite acetate (TAP-Ac) (5.0 K/min, 4.0 mg, open Al pan)*

enthalphy with data on mechanical and electrical sensitivity will provide a first safety evaluation on the basis of small sample sizes.

1.4.5 Calorimetric and Gravimetric Stability Tests

So far, analytical methods and characterization techniques have been described which provide relatively fast information on the shock, temperature and heat sensitivity of energetic compounds. However, besides the short-term sensitivity to temperature and heat as measured by thermoanalytical techniques, also the mid- and long-term sensitivity and stability of energetic compounds must be considered. In particular, stability becomes an important safety issue whenever energetic compounds like azides are stored in larger quantities for further processing.

Therefore, gravimetric and different calorimetric methods have been established to investigate stability and aging behavior of energetic compounds.[62–64] Here, mass loss tests and the analysis by adiabatic and isothermal heat flow calorimetry are briefly described.

Mass loss tests of solid energetic materials are carried out under isothermal conditions in precise temperature controlled furnaces. Usually, samples of 1–2 g are stored in special, open sample tubes at 75 °C or 90 °C for at least 18 days. During this period the sample mass is constantly recorded. A mass loss of >3% after 18 days at 90 °C is usually an indicator for restrictions in long-term stability. However, stability standards are only specified for specific energetic materials and compositions. For example, stable nitrocellulose-based propellants have to exhibit a mass loss of <2% after 18 days storage at 90 °C.[65] Mass loss data of energetic azides have been only rarely published so far. Only

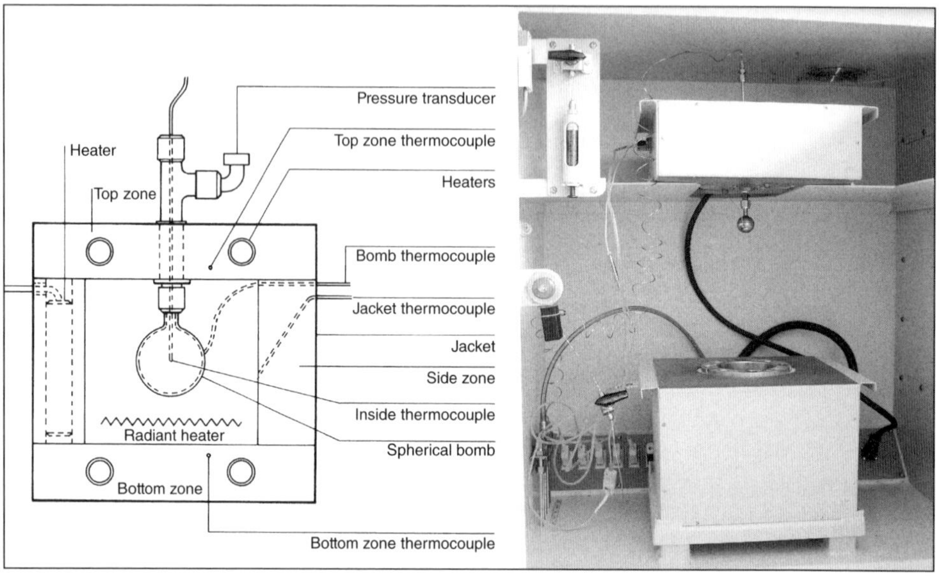

Figure 1.12 *Setup of an accelerating rate calorimeter (Thermal Hazard Technology, GB)*

few data are available for azido polymers which are used as binders or plastizisers in propellant formulations.[66]

Other techniques to investigate the thermal stability of energetic materials are employing calorimetric methods. For example, adiabatic self-heating of samples is measured by *Accelerating Rate Calorimetry (ARC)*.[64,67–69] In ARC experiments a sample is placed in a spherical metal cell of 10 cm^3 volume which may hold several grams. The sample cell is mounted in the center of a well-isolated furnace whose temperature is precisely adjusted and controlled. Figure 1.12 shows a typical ARC setup. Pressure within the cell can be monitored during the measurement via a direct connection to an external pressure sensor (pressure range: 1–200 bar). Adiabatic conditions are realized by adjusting the furnace temperature to the temperature of the sample. This allows an active control of potential heat losses.

The ARC system is often operated in a stepwise 'heat-wait-search' modus. After heating to a certain temperature, the system is stabilized for a pre-defined time until the calorimeter starts seeking for a temperature increase caused by first decomposition processes. If the temperature increase surpasses a pre-defined threshold (e.g. 0.01 K/min) the furnace temperature follows the sample temperature in the adiabatic mode and the calorimeter tracks the adiabatic temperature rise due to the self-heating of the sample. If the threshold is not surpassed after a certain period of time, the calorimeter proceeds with the next temperature step. In comparison to DSC analysis ARC measurements are significantly more sensitive, usually by a factor of 100 or more. Sensitivity is as low as 0.5 mW/g and self-heating rates of 0.01 K/min can be detected.

The most relevant safety and stability information obtained from ARC experiments are the self-heating rate, the pressure rate and the adiabatic temperature rise of energetic materials as a function of temperature. As an example, Figures 1.13 and 1.14 show such data for GAP diol, an energetic glycidyl azide polymer based on polyether diol and grafted with energetic azido groups in the polymer chain. The ARC measurement confirms the overall good stability of the polymer showing a transition to deflagration at >200 °C.[66,70]

Another highly sensitive calorimeter is the *Thermal Acitivity Monitor (TAM)*, an isothermal heat flow calorimeter which was originally developed for the investigation of biological systems.[71–72] The thermal acitivity monitor is a differential calorimeter working with reference samples. It measures heat flows induced by slow decomposition reactions of samples stored under precisely controlled isothermal conditions. The high sensitivity

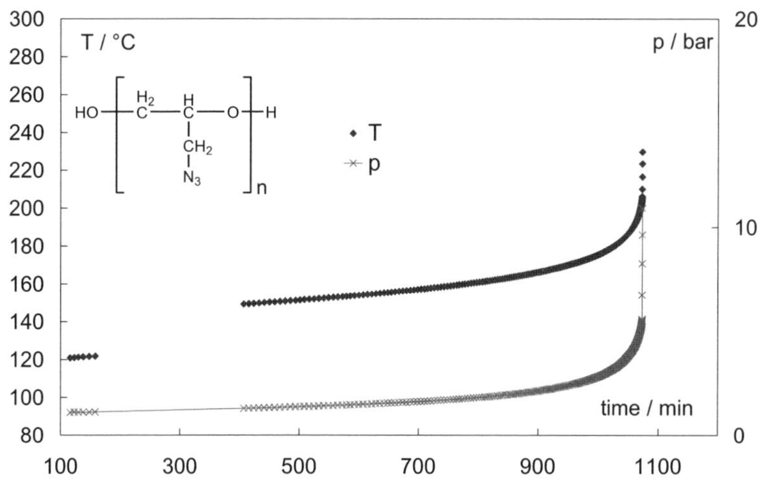

Figure 1.13 *ARC measurement of GAP diol: self-heating until deflagration*

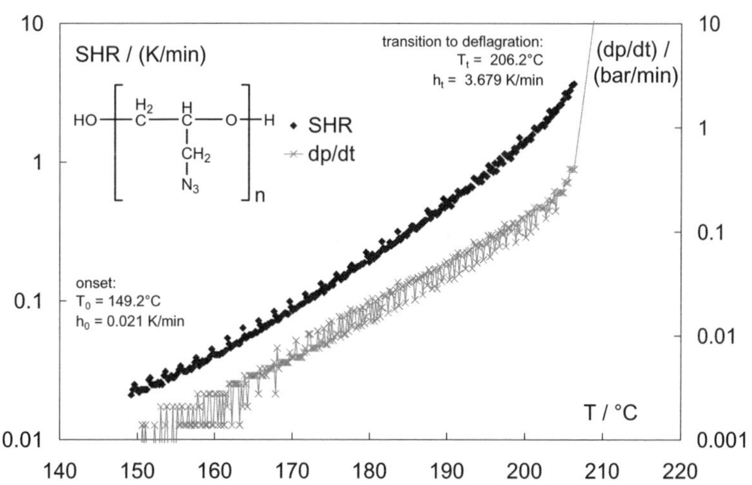

Figure 1.14 *ARC measurement of GAP diol: self-heating rate (SHR) and pressure rate until deflagration*

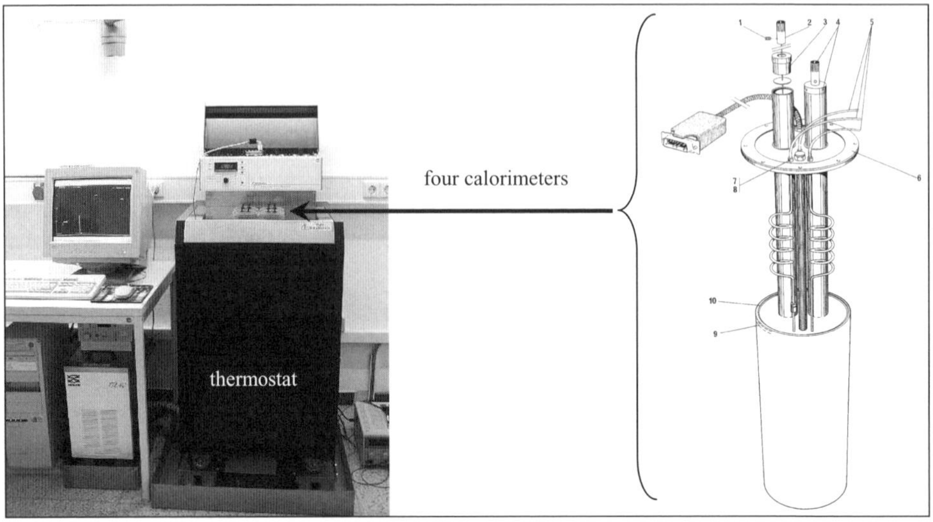

Figure 1.15 Setup of a Thermal Activity Monitor, TAM (Thermometric/TA Instruments, Sweden/USA); right: one of up to four calorimeters insertable into the thermostat

of a TAM system allows the detecting of very weak thermal effects even in the μW and nW range. This high sensitivity is achieved by a high precision temperature bath and a series of thermocouples controlling the temperature of sample and reference with an accuracy of at least 10^{-4} K. Therefore, TAM systems are ideally suited for the investigation of long-term stabilities and compatibilities of energetic materials.[64,73–74] Figure 1.15 shows the setup of a typical TAM system. Up to four independently working calorimeters can be inserted into one high-precision thermostat. Each of them contains a sample of up to 3 g filled in special glass or steel ampoules.

As an example, Figure 1.16 shows the TAM measurement of the azido polymer GAP triol (the corresponding three-functional analog to GAP diol). The absolute heat and the heat flow rate were recorded for a period of 10 days at 89 °C. The data show a typical equilibration process at the beginning of the measurement as it is often observed in TAM experiments. After inserting the calorimeter in the thermostat a certain time for thermal equilibration is required due to differences in heat capacity but also due to moisture or other impurities in the sample, and for other reasons. After equilibration only a low heat flow rate of 10–20 W/g is measured. Likewise, only a weak heat of 18 J/g was recorded after 10 days' storage at 89 °C. Therefore, the GAP triol sample exhibits a sufficiently high thermal stability for storage and further processing. In case of thermally unstable compounds heat flow rates may reach values of several hundred W/g.

Experimental data obtained from isothermal mass loss experiments, adiabatic and isothermal heat flow calorimetry can be used for kinetic modeling and the prediction of life and storage time of energetic materials under different environmental conditions. However, the models, that such predictions are based on, are often very complex and thus not a result of simple extrapolation procedures. For example, different chemical pathways and mechanisms of decomposition reactions as well as aspects of autocatalysis must be considered.[75]

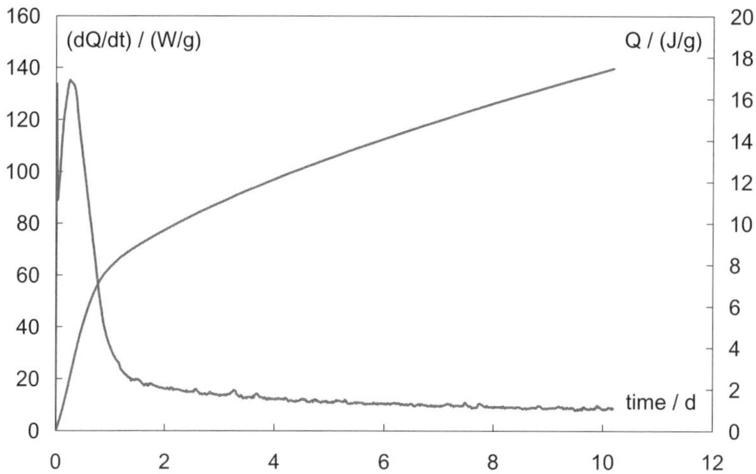

Figure 1.16 *TAM measurement of GAP triol at 89 °C*

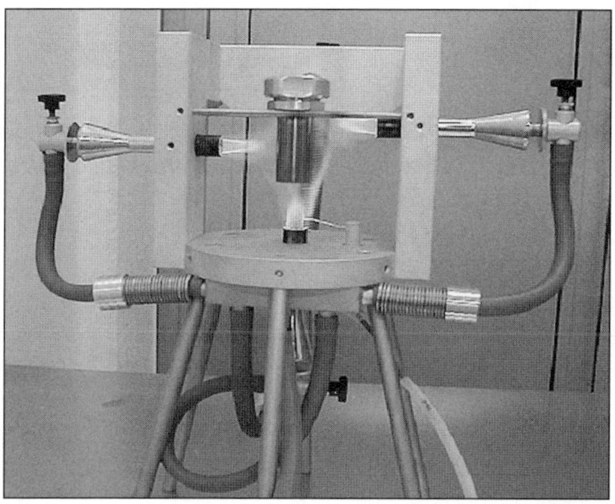

Figure 1.17 *Koenen test setup. Courtesy of Prof. Dr Thomas Klapötke, Ludwig-Maximilians University, Munich, Germany*

1.4.6 Koenen Test

Besides analyzing the thermal sensitivity of energy-rich compounds under conditions of slow heating and pyrolysis by employing thermoanalytical techniques as described in other chapters, additional test procedures are available to determine the sensitivity of larger sample quantities to intense heating while being under confinement. The so-called Koenen Test (Steel Sleeve Test) is also used to determine the shipping classification of energetic (and non-energetic) compounds and to evaluate the degree of venting required to avoid an explosion during processing operations.[24,76]

A typical Koenen test setup is shown in Figure 1.17. The sample is filled into a non-reusable cylindrical steel sleeve which is closed by a metal plate with a variable orifice

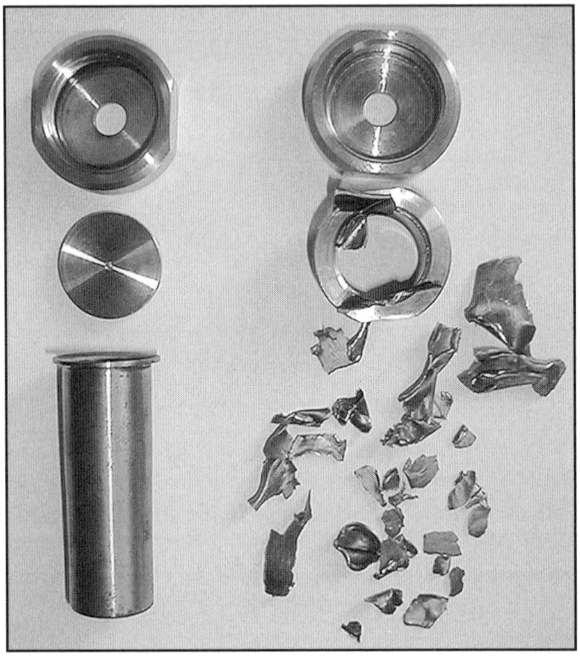

Figure 1.18 Steel sleeves used in Koenen tests: before (left) and after the test (right). Courtesy of Prof. Dr Thomas Klapötke, Ludwig-Maximilians University, Munich, Germany

Table 1.5 Exemplary Koenen test results[24]

Explosive	Limiting diameter of the orifice / mm	Time until ignition / s	Time of combustion / s
Nitroglycerine	24	13	0
Pentaerythritol tetranitrate (PETN)	6	7	0
TNT	5	52	29
Ammonium nitrate	1	43	29

through which the decomposition gases can escape. The closing plate is secured with a nut. The diameter of the orifice can be varied between 1 and 20 mm and in case of sensitive materials the sample holder is not closed. The dimension of the steel sleeve is 25 mm OD × 24 mm ID × 75 mm length. The sample is loaded up to a filling height of 60 mm (sample volume: 27 mL). For the test the charged sample holder is heated simultaneously by four atmospheric burners.

The test is completed upon rupture of the steel sleeve or after heating the tube for a minimum of 5 minutes with no reaction. The elapsed time till ignition and the duration of the combustion are measured. With the variable orifice the limiting diameter is determined at which at least one explosion within a series of three consecutive experiments occurs and the sleeve is ruptured into three or more fragments (Figure 1.18). In Table 1.5 exemplary Koenen test results of common explosives are listed.[24]

References

[1] S. Bräse, C. Gil, K. Knepper, V. Zimmermann, *Angew. Chem. Int. Ed.*, **2005**, *44*, 5188–240.
[2] E.F.V. Scriven, K. Turnbull, *Chem. Rev.*, **1988**, *88*, 297–368.
[3] E.F.V. Scriven (ed.), *Azides and Nitrenes: Reactivity and Utility*, **1984**, Academic Press, Orlando, FL, USA.
[4] H.C. Kolb, M.G. Finn, K.B. Sharpless, *Angew. Chem. Int. Ed.*, **2001**, *40*, 2004–21.
[5] W.H. Binder, C. Kluger, *Curr. Org. Chem.*, **2006**, *10*, 1791–815.
[6] I.J. Dagley, R.J. Spear, in: *Organic Energetic Compounds* (ed.: P.L. Marinkas), **1996**, Nova Science Publishers Inc., New York, USA.
[7] R. Haiges, A. Boatz, A. Vij, M. Gerken, S. Schneider, T. Schroer, K.O. Christie, *Angew. Chem. Int. Ed.*, **2003**, *42*, 5847–51.
[8] J.P. Agrawal, R.D. Hodgson, *Organic Chemistry of Explosives*, **2007**, John Wiley & Sons, Inc., New York, USA.
[9] M.H.V. Huynh, M.A. Hiskey, D.E. Chavez, D.L. Naud, R.D. Gilardi, *J. Am. Chem. Soc.*, **2005**, *127*, 12537–43.
[10] R. Escales, A. Stettbacher, *Initialexplosivstoffe*, **1917**, Verlag von Veit & Comp., Leipzig, Germany; p. 167.
[11] M.E.C. Biffin, J. Miller, D.B. Paul, in: *The Chemistry of the Azido Group* (ed.: S. Patai), **1971**, Interscience Publishers, New York, USA.
[12] F. Martin, *Über Azide und Fulminate*, **1913**, Darmstadt, Germany; cited in: T. Urbański, *Chemistry and Technology of Explosives*, **1964**, Pergamon Press, Oxford, Great Britain, Vol. *III*; p. 164.
[13] M.B. Talawar, A.P. Agrawal, M. Anniyappan, D.S. Wani, M.K. Bansode, G.M. Gore, *J. Hazard. Mater.*, **2006**, *137*, 1074–8.
[14] P.A.S. Smith, *The Chemistry of Open-Chain Organic Nitrogen Compounds, Vol. 2*, **1966**, W.A. Benjamin Inc., New York, USA, pp. 211–56.
[15] G. Abbenante, G.T. Le, D.P. Fairlie, *Chem. Commun.*, **2007**, 4501–3.
[16] G.R. Harvey, K.W. Ratts; Synthesis of azirenes from allenic esters; *J. Org. Chem.*, **1966**, *31*, 3907–10.
[17] J.H. Boyer, F.C. Canter, *Chem. Rev.*, **1954**, *54*, 1–57.
[18] E.E. Gilbert, *1,2,4,5-tetrakis (Diazidomethyl) benzene energetic polyazide*, United States Patent H000428.
[19] T.M. Klapötke, B. Krumm, N. Mayr, F.X. Steemann, G. Steinhauser, *Safety Testing of Protective Gloves*, Proceedings of 11th International Seminar on New Trends in Research of Energetic Materials, **2008**, Pardubice, Czech Republic, pp. 597–605.
[20] J.S. Rinehart, J. Pearson, *Explosive Working of Metals*, **1963**, Pergamon Press, Oxford, Great Britain; p. 38.
[21] NATO STANAG 4489 Document Information, Explosives, Impact Sensitivity Tests, **1999**.
[22] Department of Defence Test Method Standard, *Safety and Performance Tests for the Qualification of Explosives (High Explosives, Propellants and Pyrotechnics)*, MIL-STD-1751A, **2001**; Superseding MIL-STD-1751(USAF), **1982**.
[23] P.W. Cooper, S.R. Kurowski, *Introduction to the Technology of Explosives*, **1996**, John Wiley & Sons, Inc., New York, USA.
[24] R. Meyer, J. Köhler, A. Homburg, *Explosives*, **2007**, 6th revised edition, Wiley-VCH, Weinheim, Germany.
[25] C.-O. Leiber, B. Dobratz, *Assessment of Safety and Risk with a Microscopic Model of Detonation*, **2003**, Elsevier, Amsterdam, The Netherlands.
[26] F.P. Bowden, A. Yoffe, Hot spots and the initiation of explosion, *Proceed. Symposium on Combustion and Flame, and Explosion Phenomena*, Vol. *3*, **1949**, Cambrige, MA, USA.
[27] T.M. Klapötke, C.M. Rienäcker, *Propellants Explosives and Pyrotechnics*, **2001**, *26*, 43–7.
[28] M.J. Kamlet, H.G. Adolph, *Propellants Explosives and Pyrotechnics*, **1979**, *4*, 30–4.
[29] J. Mullay, *Propellants Explosives and Pyrotechnics*, **1987**, *12*, 60–3.

[30] M.H.V. Huynh, M.A. Hiskey, T.J. Meyer, M. Wetzler, *Proc. Natl. Acad. Sci. USA*, **2006**, *103*, 5409–12.
[31] M.H. Keshavarz, *J. Hazard. Mater.*, **2007**, *148*, 648–52.
[32] M.H. Keshavarz, H.R. Pouretedal, A. Semnani, *J. Hazard. Mater.*, **2007**, *141*, 803–7.
[33] M.H. Keshavarz, H.R. Pouretedal, *J. Hazard. Mater.*, **2005**, *124*, 27–33.
[34] L. Türker, *J. Mol. Struct.: THEOCHEM*, **2005**, *725*, 85–7.
[35] S. Ye, K. Tonokura, M. Koshi, *Combust. Flame*, **2003**, *132*, 240–6.
[36] G. Su-Hong, C. Xin-Lu, W. Li-Sha, Y. Xiang-Dong, *J. Mol. Struct.: THEOCHEM*, **2007**, *809*, 55–60.
[37] R. Sundararajan, S.R. Jain, *Indian J. Technol.*, **1983**, *21*, 474–7.
[38] M. Vaullerin, A. Espagnacq, L. Morin-Allory, *Propellants Explosives and Pyrotechnics*, **1998**, *23*, 237–9.
[39] M. Cartwright, J. Wilkinson, *Correlation of Structure and Sensitivity in Azides*, **2008**, Cranfield CERES Publication (https://dspace.lib.cranfield.ac.uk/handle/1826/2584).
[40] H. Koenen, K. H. Ide, *Explosivstoffe*, Bd. *9*, **1961**, Erwin Barth Verlag KG, Mannheim, Germany, pp. 4–13 and 30–42.
[41] NATO STANAG 4487 Document Information, *Explosives, Friction Sensitivity Tests*, **2002**.
[42] NATO STANAG 4490 Document Information, *Explosives, Electrostatic Discharge Sensitivity Tests*, **2001**.
[43] NATO STANAG 4239 Document Information, *Electrostatic Discharge, Munitions Test Procedures*, **1997**.
[44] NATO AOP-24 Document Information, *Electrostatic Discharge, Munition Assessment and Test Procedures*, **1998**.
[45] S. Amari, F. Hosoya, Y. Mizushima, T. Yoshida, *Electrostatic Spark Ignitability of Energetic Materials*, **1995**, 21st International Pyrotechnic Seminar, Moscow, Russia, 13–31.
[46] B.T. Fedoroff, O.E. Sheffield, *Encyclopedia of Explosives and Related Items PATR 2700*, Vol. *5*, **1972**, Picatinny Arsenal, Dover, N.J., USA.
[47] D. Skinner, D. Olson, A. Block-Bolten, *Propellants Explosives and Pyrotechnics*, **1998**, *23*, 34–42.
[48] M. Roux, M. Auzanneau, C. Brassy, *Propellants Explosives and Pyrotechnics*, **1993**, *18*, 317–24.
[49] M. Auzanneau, M. Roux, *Propellants Explosives and Pyrotechnics*, **1995**, *20*, 96–101.
[50] C.J. Dahn, B.N. Reyes, A. Kashani, J. Finkelstein, Electrostatic hazards of explosive, propellant and pyrotechnic powders, **1998**, *Proceed. 20th Electrical Overstress/ Electrostatic Discharge Symposium*, Reno, Nevada, USA; 139–50.
[51] M.E. Brown, *Introduction to Thermal Analysis: Techniques and Applications*, **2001**, Kluwer Academic Publishers, Norwell, MA, USA.
[52] B. Wunderlich, *Thermal Analysis*, **1990**, Academic Press, San Diego, CA, USA.
[53] W.W. Wendlandt, *Thermal Analysis*, **1986**, John Wiley & Sons, Inc., New York, NY, USA.
[54] G.W.H. Höhne, W.F. Hemminger, H.-J. Flammersheim, *Differential Scanning Calorimetry*, **2003**, Springer, Heidelberg, Germany.
[55] S. Löbbecke, M. Kaiser, G.A. Chiganova, in: *Energetic Materials: Particle Processing and Characterization* (ed.: U. Teipel), **2004**, Wiley, Weinheim, Germany.
[56] D.R. Miller, D.C. Swenson, E.G. Gillan, *J. Am. Chem. Soc.*, **2004**, *126*, 5372–3.
[57] T. Keicher, G. Unkelbach, H. Krause, Synthesis and characterization of new triazidoplasticizers, **2005**, *Proceed. 36th Int. Annual Conference ICT*, Karlsruhe, Germany; pp. 49/1–8.
[58] D. Röseling, G. Unkelbach, T. Keicher, H. Krause, Synthesis, characterization and first formulations of new triazidoplasticizers, **2007**, *Proceed. NTREM Conference – New Trends in Research of Energetic Materials*, Pardubice, Czech Republic, 943–50.
[59] A. Pfeil, S. Löbbecke, *Propellants Explosives and Pyrotechnics*, **1997**, *22*, 137–42.
[60] C.I. Schilling, S. Bräse, *Org. Biomol. Chem.*, **2007**, *5*, 3586–8.
[61] S. Löbbecke, H. Schuppler, W. Schweikert, *J. Therm. Anal. Calorim.*, **2003**, *72*, 453–63.
[62] B. Vogelsanger, *Chimia*, **2004**, *58*, 401–8.
[63] F. Stoessel, *Thermal Safety of Chemical Processes*, **2008**, Wiley-VCH, Germany.

[64] M.W. Whitmore, J.K. Wilberforce, *J. Loss Prev. Process Ind.*, **1993**, *6*, 95–101.
[65] Bundesamt für Wehrtechnik und Beschaffung, *Arbeitsvorschriften für die chemische und physikalische Untersuchung von Treibladungspulver (TLP) 2.31.1 Bestimmung der chemischen Beständigkeit bei 90 °C und 75 °C*, **1999**, Technische Lieferungsbedingung TL 1376-0600/430.
[66] M.A. Bohn, Decomposition behaviour of azido based and nitric acid ester based plasticizers and binders determined by adiabatic selfheating, **1998**, *Proceed. 11th Symposium on Chemical Problems Connected with the Stability of Explosives*, Båstad, Sweden; 61–88.
[67] D.I. Townsend, *Accelerating Rate Calorimetry*, **1981**, I.Chem.E. Symposium Series 68.
[68] D.I. Townsend, J.C. Tou, *Thermochim. Acta*, **1980**, *37*, 1–30.
[69] X.-R. Li, H. Koseki, *J. Loss Prev. Process Ind.*, **2005**, *18*, 455–9.
[70] M.A. Bohn, Heat generation of propellants & explosives, **1994**, *Proceed. Int. Symp. on Energetic Materials Technology*, Orlando, USA.
[71] J. Suurkuusk, I. Wadsö, *Chemica Scripta*, **1982**, *20*, 155–63.
[72] P. Bäckman, M. Bastos, L.E. Briggner, *et al.*, *Pure Appl. Chem.*, **1994**, *66*, 375–82.
[73] NATO STANAG 4582 Document Information, *Explosives, single, double and triple base propellants'*, **2004**.
[74] NATO STANAG 4147 Document Information, *Chemical compatibility of ammunition components with explosives and propellants*, **2001**.
[75] M.A. Bohn, Modelling of the stability, ageing and thermal decomposition of energetic components and formulations using mass loss and heat generation, **2000**, *Proceed. 27th Int. Pyrotechnics Seminar*, Grand Junction, Colorado, USA; 751–70.
[76] United Nations, *Recommendations on the Transport of Dangerous Goods, Manual of Tests and Criteria*, **2003**, 4th revised edition, New York and Geneva.

2

Large-scale Preparation and Usage of Azides

Jürgen Haase

Dynamit Nobel GmbH, Explosivstoff und Systemtechnik, Kalkstrasse 218, 51377 Leverkusen, Germany

2.1 Introduction

The beginnings of azide chemistry already date back almost 150 years. During this period, the application options and limits of azide chemistry have been examined in great detail by many authors and the results have been summarized in numerous overview articles.[1–5] Particularly during the last 20 years, azide chemistry seems to have experienced a renaissance on the grounds of its applications in medicine, biology and materials sciences.

As with nearly all research work, decades often go by until the results of basic research translate into commercial production processes.

The risks described in Chapters 1 and 13 (Keicher and Löbbecke as well as Klapötke and Krumm) regarding the safe handling of azides have contributed to the fact that, outside of the production of explosives, azide chemistry has only reached a commercial production volume of 1000 t/a during the last 30–40 years. The safety risk involved in the handling of azides refers to their toxicity, sometimes to their thermal instability, but also to their possible sensitivity to shock and friction. A particular risk is found where the formation of free, extremely shock-sensitive hydrazoic acid (HN_3) must be reckoned with. The explosion of a few tenths of a milliliter of free, liquid HN_3 can destroy – or more precisely pulverize – a complete laboratory-scale production unit. The detonation speed of HN_3 lies in the range of 8000 m/s. The explosion of grams, kilograms or even tons of HN_3 would be a disaster for employees and plant equipment alike. The explosion-like decomposition can be triggered by the slightest

Organic Azides: Syntheses and Applications Edited by Stefan Bräse and Klaus Banert

commotion, exposure to extremely small friction- or shock-energies and/or flowing over rough surfaces. Consequently, the greatest care must be exercised when handling azides both in the laboratory scale and in the technical production scale.

The production processes quoted in the following relate to batch or semi-batch processes. To reduce the safety risk, also in azide chemistry increasing thought is given to continuous and micro-reaction technology processes.

The disposal of toxic, azide containing solvent and effluent streams is a very important aspect. Environmental protection and safety considerations now play a significant role in the chemical industry in western countries. The same also holds true for the azide-processing companies. All waste streams must be either carefully and laboriously freed from azide residues by way of chemical processes or treated in a dedicated incineration unit.

Unlike with basic research, information about processes translated into commercial production scale is not readily available. For this reason and to keep within the limits of this chapter, reference is made to a selection of processes that have in the meantime reached the commercial production stage. Consequently, this paper does not claim to be complete.

Specifically, the well-known commercial use of inorganic azides in the field of energetic materials (priming explosives, see Chapter 13 by Klapötke and Krumm) or as propellants in airbags and seat belt pre-tensioning systems in the automotive industry will not be dealt with in more detail.

Section 2.2 of this chapter describes the main azides that are available for azidation reactions not only to the R&D chemist but also to the industrial chemist. For a description of their full application range, reference is made to the numerous review articles.[1–5]

Section 2.3 gives some examples for the most common technical applications of azide chemistry.

2.2 Precursor Azides, Technical Production and Properties

2.2.1 Sodium azide (NaN_3)

Sodium azide is the most frequently used precursor azide for both, laboratory scale and technical production. In commercial production, it is obtained in two ways:

In 1892, Wislicenus[6] developed a two-step process:

$$2\,Na + 2\,NH_3 \longrightarrow 2\,NaNH_2 + H_2$$

$$2\,NaNH_2 + N_2O \longrightarrow NaN_3 + NaOH + NH_3$$

Scheme 2.1 *Wislicenus sodium azide synthesis*

At 350 °C, molten sodium is converted with ammonia to sodium amide in a closed steel reactor. In a second step, this product reacts with nitrous oxide to give sodium azide. The gas composition and the extreme reactivity of sodium and sodium amide with traces of water represents a particular safety risk of this synthesis method.

Variations of this production process are described by Bretschneider and Abe.[7,8]

Curtius and Thiele[9–11] developed another production process where a nitrite ester is converted to sodium azide using hydrazine:

$$2\,NaNO_2 + 2\,C_2H_5OH + H_2SO_4 \longrightarrow 2\,C_2H_5ONO + Na_2SO_4 + 2\,H_2O$$

$$C_2H_5ONO + N_2H_4{\cdot}H_2O + NaOH \longrightarrow NaN_3 + C_2H_5OH + 3\,H_2O$$

Scheme 2.2 *Curtius synthesis of sodium azide*

Worldwide, around 20 suppliers in China, India, Japan and the USA are currently producing NaN_3 at volumes of 1000 t/a. But only little is known about their used route of NaN_3 production.[12] At t-scale the NaN_3 price is explicit lower than 30 €/kg.

Of this production volume, the major part is used in the automotive industry.[13] Since this application is on the decline due to the toxicity of NaN_3 and the disposal problems encountered when recycling the cars at the end of their useful life, we will probably have excess capacities in future unless substitutes such as 5-aminotetrazole are developed involving the use of NaN_3 (see Section 2.3.1.4). The growing need for NaN_3 in organic syntheses described in the following chapters cannot compensate for the decline in sales in the automotive sector.

Table 2.1 shows the main characteristics and risks associated with the handling[14] of NaN_3.

2.2.2 Trimethylsilyl Azide (TMSA)[14]

TMSA is produced and marketed by ten companies in Europe, Japan and USA with total production in the range of >10 t/a. Given its toxicity and sensitivity to hydrolysis, extensive safety measures are required for the processing of this substance.

TMSA is much more expensive than sodium azide since it is produced from the latter. In addition, for the application of this substance it must be borne in mind that only ~37% of its molar mass can be used for the azidation reaction.

The easiest synthesis[15,16] is based on the technically accessible trimethylsilyl chloride, which is converted with sodium azide in a two-phase reaction to yield TMSA:

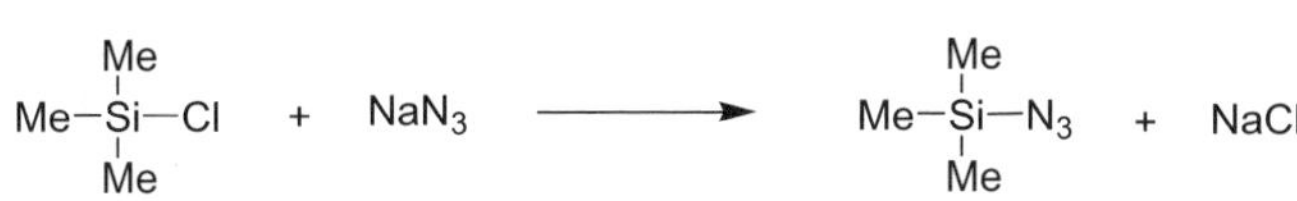

Scheme 2.3 *Synthesis steps to produce TMSA*

Unlike sodium azide, which – owing to its ionic structure – is only soluble in highly polar solvents and is converted in such solvents or in a two-phase system by phase transfer catalysis, TMSA is a covalently bonded azide that is stable and miscible with many aprotic organic solvents. This is why it can be used in azide syntheses for which water-sensitive organic substances are to be used.

Table 2.1 *Sodium azide (NaN_3): physical and chemical properties*

CAS No.	26628-22-8
Molecular weight	65.01 g/mol
Density	approx. 1.85 g/cm^3
Solubility, H_2O, 17 °C	420 g/l, completely dissociated, pH > 12
Solubility in liquid NH_3	Good
Solubility in DMF and DMSO	Good
Solubility in ethanol	Little
Solubility in ether	not soluble
Melting point	275 °C
Boiling point	>300 °C, decomposition
Appearance	Colorless crystals, odorless under anhydrous conditions; dull, fishy smell in the presence of even the smallest traces of water (HN_3 formation)
Toxicity	– T+: highly toxic, – Inhaling small quantities of gaseous HN_3 leads to immediate reddening of the eyes, vertigo and drowsiness, drop in blood pressure to the point of death – Threshold limit value: 0.2 mg/m^3; 0.007 ppm – N: hazardous to the environment
Properties	Risk of explosive decomposition caused by strong heating, blow, shock or friction with release of metallic sodium and nitrogen
Hazardous reactions	– In aqueous solutions and with acids formation of extremely shock-sensitive hydrazoic acid (HN_3), which is highly toxic and volatile at a boiling point of 36 °C – Potentially explosive metal azides form upon contact with heavy metals – Explosion risk upon contact with chromyl chloride – Explosion risk when reacting with chlorinated hydrocarbons, in particular dichloromethane – Explosion risk upon contact with nitric acid – Explosion risk upon contact with carbon disulfide – Strongly exothermal reaction with potassium nitrate, barium carbonate

Examples for TMSA additions[17] to olefins,[18] isocyanates,[19] acetylenes and nitriles[20,21] to form triazoles, tetrazolinones, tetrazoles and α,β-unsaturated carbonyl compounds[22] complement the applications of TMSA listed in Sections 2.3.1.2 and 2.3.4 of this chapter.

TMSA quickly and exothermally decomposes upon contact with water to give hydrazoic acid (HN_3) and siloxane compounds.

The other properties of TMSA have been compiled in Table 2.2.

2.2.3 Diphenylphosphoryl Azide (DPPA)[14]

On account of the higher price as compared to TMSA and the even more unfavorable ratio of azide mass to total DPPA mass, the technical-scale use of DPPA is restricted to a few special applications such as the direct conversion of carboxylic acids to the cor-

Table 2.2 *Trimethylsilyl azide (TMSA): physical and chemical properties*

CAS No.	4648-54-8
Molecular weight	115.21 g/mol
Density	0.875 g/cm^3
Solubility in toluene	Miscible
Solubility in methylene chloride	Miscible
Solubility in diethyl ether	Miscible
Melting point	6 °C
Boiling point	95 °C at 1013 hPa
Appearance	Colorless liquid, odorless under anhydrous conditions; dull, fishy odor in the presence of even the smallest traces of water (decomposition with formation of HN_3)
Toxicity	T: toxic, F: flammable, N: hazardous for the environment
Properties	Toxic, due to its low boiling point, high vapor pressure there is a high risk of inhaling the vapors that hydrolyze to form HN_3 which could in the worst case be fatal due to its hypotensive effect.
Hazardous reactions	Reaction with water results in the formation of hydrazoic acid. May form explosive compounds with heavy metal salts

responding carboxylic acid azides without having to go through the activating intermediates like carboxylic acid chlorides, anhydrides or esters.[23–26] In some cases variations of the Mitsunobu reaction involving DPPA are described.[27–30] In Section 2.3.1.5 of this chapter, a commercial-scale example for the use of DPPA is shown.

Due to the high price and the above-mentioned unfavorable mass ratio the demand for technical-scale quantities of DPPA will probably only amount to about 10 t/a. The database of the Directory of World Chemical Producers (DWCP) lists 15 companies in Asia, Europe and the USA as producers of DPPA.

Compared to TMSA, safe technical-scale handling (sensibility to hydrolysis, formation of HN_3) of DPPA is easier given the low azide content in relation to the relatively high molecular weight. However, the toxicity should always be borne in mind.

DPPA can be produced as specified in the literature.[31–33]

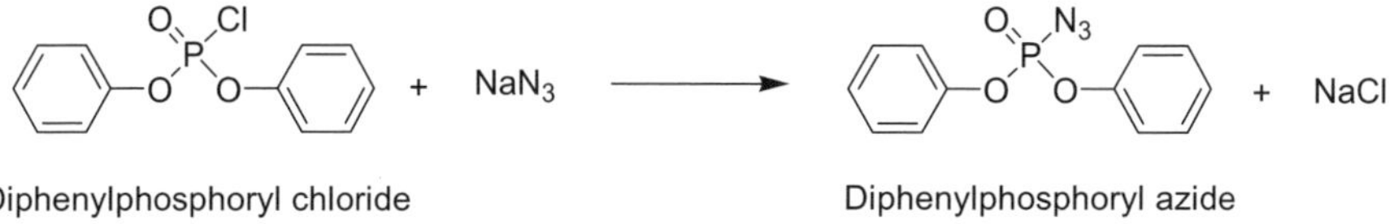

Scheme 2.4 *Synthesis for the production of DPPA*

Table 2.3 shows additional properties of DPPA.

Table 2.3 *Diphenylphosphoryl azide (DPPA): physical and chemical properties*

CAS No.	26386-88-9
Molecular weight	275.22 g/mol
Density	1.277 g/cm^3
Solubility in DMF	Miscible
Solubility in Toluene	Miscible
Solubility in THF	Miscible
Solubility in t-butyl alcohol	Miscible
Solubility in acetone	Miscible
Solubility in acetonitrile	Miscible
Solubility in hexane	Immiscible
Solubility in water	Immiscible
Boiling point	157 °C; 0.5 hPa
Appearance	Colorless liquid, odorless in the complete absence of water
Toxicity	T: toxic
Hazardous reactions	May form explosive compounds with heavy metal salts, acids and strong oxidants

2.2.4 Tributyltin Azide (TBSnA)

Like TMSA and DPPA, tributyltin azide is a covalently linked azide that allows for an azidation reaction in relatively homopolar solvents. In view of the high azide concentrations achievable in the solution, the use of TBSnA is interesting in terms of time (tetrazole-yielding reactions are often slow on the basis of the two-phase reaction system: NaN_3 and nitriles) and thus for a technical realization (see also the examples in Sections 2.3.1.1 and 2.3.4).

TBSnA is preferably produced in situ. Also with a view to the environmental risk potential involved in the recovery of tributyltin compounds the commercial-scale application should be left in the hands of specialized producers.

On account of its excellent accessibility[34–36] from the corresponding tributyltin chloride (disinfectant to prevent fungi in textiles, leather, paper, wood; algae and snails on boat paints (antifouling paint); seed pickling agent in pest management: against fungi, mites, available at a volume of 50,000 t/a), TBSnA will probably be used in organic synthesis in the scope of several 100 t/a.

A downside of TBSnA is its tin concentration because in nearly all products heavy metal concentrations are only acceptable in the ppm range.

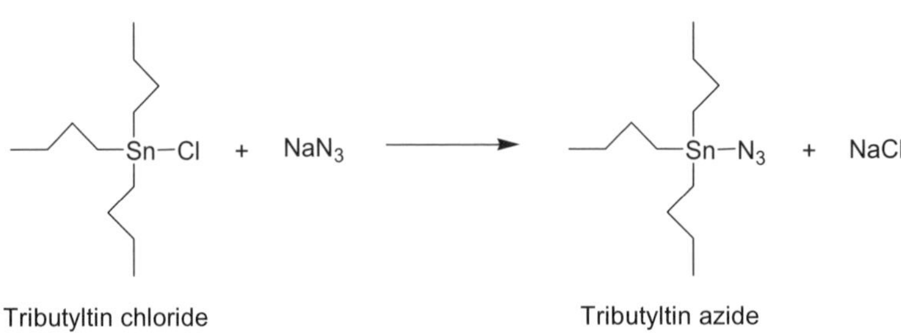

Scheme 2.5 *Synthesis to produce TBSnA*

Table 2.4 *Tributyltin azide (TBSnA): physical and chemical properties*

CAS No.	17846-68-3
Molecular weight	332.06 g/mol
Density	g/cm^3
Solubility in toluene	Miscible
Boiling point	126 °C/1.3 mbar
Appearance	Colorless liquid, odorless under anhydrous conditions
Toxicity	T: toxic
Remark: by analogy with other tributyltin compounds	Threshold limit value: 0.05 mg/m^3 N: hazardous for the environment
Properties	Sensitive to hydrolysis
Hazardous reactions	May form explosive compounds with heavy metal salts

The physical and chemical properties are described in Table 2.4.

2.2.5 Azidoacetic Acid Ethyl Ester (AAE)[14]

According to Hemetsberger-Knittel and others, AAE is widely used for the synthesis of heterocycles such as aziridines, pyrroles and indoles.[37–47] Due to its explosive properties, AAE may only be transported in special containers as a solution in various solvents. To reduce the safety risks and yield losses, an immediate conversion (possibly at the production facility) without transport is recommended. The production scale might be in the range of 100 kg/a.

The product can be obtained via substitution in a reaction of chloroacetic acid[48,49] with NaN_3:

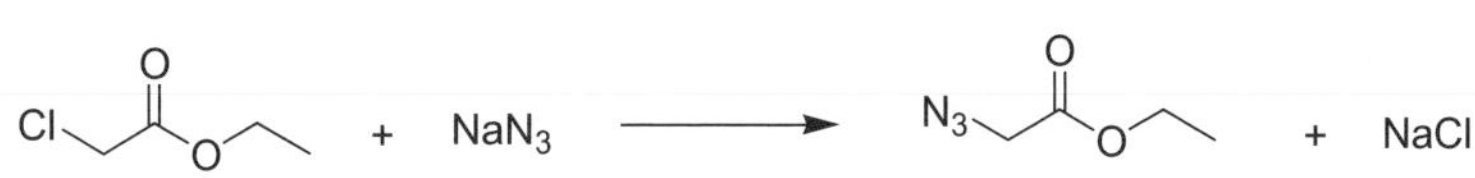

Scheme 2.6 *Synthesis for the production of AAE*

The properties are shown in Table 2.5.

2.2.6 Tetrabutylammonium Azide (TBAA)[14]

Since up to now only very little is known about the advantageous use of TBAA[50] the market volume will be in the range of < 1 t/a. The product is marketed as a 15% solution in THF.

For many conversions, such as that of nitriles with NaN_3 to give tetrazole, ammonium chloride or triethylammonium chloride (TEA*HCl) are used as Lewis acid and N_3-tranfer reagent (TEA*HN_3). The reaction yields the intermediates ammonium azide or triethylammonium azide, which complete the azide transfer.

TBAA, which is an even better phase transfer reagent, could show a similar effect – without having similar acidic properties – and thus also be used in aprotic or less polar

Table 2.5 *Azidoacetic acid ethyl ester (AAE): physical and chemical properties*

CAS No.	637-81-0
Molecular weight	129.12 g/mol
Density	1.119 g/cm^3, 25 °C
Solubility in methylene chloride	Miscible
Solubility in ethanol	Miscible
Boiling point	63 °C, 16 hPa
Appearance	Colorless liquid
Toxicity	T: toxic
Remark: by analogy to azide anion	Threshold limit value: not described N: hazardous for the environment
Properties	Explosive, thermal decomposition >127 °C Sensitive to hydrolysis
Hazardous reactions	– May form explosive compounds with heavy metal salts – With acids and strong oxidants – Highly explosive azidoacetic acid forms during hydrolysis of the ester

Table 2.6 *Tetrabutylammonium azide (TBAA): physical and chemical properties*

CAS No.	993-22-6
Molecular weight	284.49 g/mol
Density	0.9 g/cm^3
Solubility in THF	150 g/l
Solubility in water	good
Melting point	>180 °C
Appearance	Yellow-brownish powder
Toxicity	Xi: irritant, toxic
Hazardous reactions	May form explosive compounds with heavy metal salts. May release highly toxic gases (hydrazoic acid) upon contact with acids

solvents than water such as THF, or it could also be used in halogen substitution reactions if the molecule contains other particularly reactive functional groups.

TBAA is produced according to the instructions described in the following literature:[51]

Scheme 2.7 *Synthesis for the production of TBAA*

The properties are summarized in Table 2.6.

2.2.7 Others

A series of other azidation reagents and diazotransfer reagents are described in academic research papers, including sulfonyl azides[5] such as benzenesulfonyl azide, trifluoromethanesulfonyl azide, pyridine-3-sulfonyl azide or also iodine azide, nosyl azide, tributylhexadecylphosphonium azide, tetramethylguanidinium azide, t-butoxycarbonyl azide.

Should further examinations confirm their advantages over the existing reagents, it will be possible to also safely produce them on a technical scale just like p-toluenesulfonyl azide[14] and 4-acetamidobenzenesulfonyl azide.[14]

2.3 Examples for the Use of Azides on a Technical Scale

2.3.1 Addition of NaN_3 to Multiple CC- or CN-Bonds

2.3.1.1 Addition of NaN_3 to Convert Nitriles to Tetrazoles: Sartans

Even if the following is a description of the addition of sodium azide to nitriles to afford tetrazoles, there may also be variances from this in specific cases where for example TBSnA[52] or TMSA are produced in an upstream reaction with sodium azide, which are then used for addition to the nitrile.

Sartans are a class of substances where the biosteric replacement of the carboxylic acid group into the tetrazole group comes in useful. This means that both compounds exhibit a similar biological in-vivo effect. According to A. Burger, compounds and molecule groups with nearly identical molecular form and volume as well as the same distribution of electrons and similar physical properties are referred to as biosteric functional groups.[53,54]

By replacing the carboxyl group with a tetrazole system, the bio-availability is substantially increased while maintaining a similar acidity:

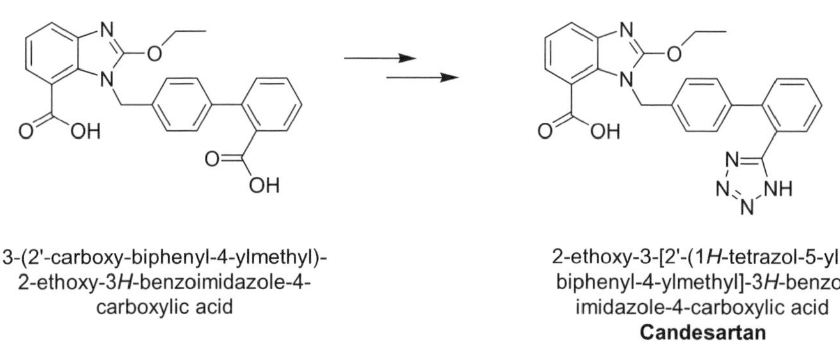

Scheme 2.8 *Biosteric replacement of the carboxylic acid group into the tetrazole group*

This bioisosteric property of the tetrazoles is utilized in relation to the active ingredient group of the sartans.

Sartans are a class of compounds used as specific inhibitors for the treatment of hypertension, chronic heart insufficiency (e.g. Candesartan, Irbesartan, Losartan, Valsartan),

Table 2.7 *Sartans, originators and commercial relevance*[55]

Sartan name	Originator	Biosteric functional groups[1)]	Patented since[55]	Dosage[mg/d][55]	Sales2006 [USD mn][7)56]	Drug / Marketed by
Candesartan	Takeda	BPT	1990	8–16	3864	Blopress® / Takeda; Atacand® / AstaZeneca
Elisartan	GE Healthcare	BPT				2)
Eprosartan	GSK	BPT	1989	300–400	119	Teveten® / Solvay; Emestar® / Trommsdorff
Fimasartan	Boryung Pharm	BPT	2001			2)
Forasartan	Pfizer	BPT	1991			3)
Irbesartan	Sanofi	BPT	1990	150–300	2336	Aprovel® / Sanofi-Aventis; Karvea® / Bristol-Myers Squibb
Losartan	DuPontMerck	BPT	1986	50–100	3163	Lorzaar® MSD
Milfasartan	Menarini	BPT	1991			3)
Olmesartan	Daiichi/Sankyo	BPT	1991	>20 mg	1237	Olmetec® / Sankyo; Votum® / Berlin-Chemie Mencord® / Menarini Pharma
Pratosartan	Kotobuki	BPT	1992			4)
Valsartan	Novartis	BPT	1990	80–160	4343	Diovan® / Novartis; Provas® / Schwarz Pharma; Cordinate® / AWDPharma
Tasosartan	Wyeth	BPT	1991			5)
Telmisartan	Boehringer Ingelheim	BPC	1991	40–80	1639	Micardis® / Boehringer Ingelheim; Kinzalmono® / Bayer
Zolasartan	SKB	PT	1992			6)

1) BPT = biphenylyltetrazole; BPC = biphenylcarboxylic acid; PT = phenyltetrazole
2) according to database55 in phase II
3) according to database55 development stopped
4) according to database55 in clinical development phase
5) 1998 development stopped and application for approval withdrawn (toxicity)
6) according to databases: phase status not given
7) USD mn = million US-Dollar

condition after a heart attack (Valsartan) and diabetic nephropathy (in the context of hypertension treatment: Losartan, Irbesartan). The ingredient group represents a further development of the ACE inhibitors.

Since hypertension is one of the most common human diseases in western industrialized nations, numerous pharmaceutical companies have developed active ingredients of this class, which are available on the market since the mid-1990s. Some of these products have reached a market volume of several 100 t/a with an upward trend and are therefore considered as blockbusters. More exact dates about production volume are not available.

Table 2.7 gives an overview of the most important sartans.

The above-mentioned active ingredients are complex elements with sometimes one or more stereo centers that have to be produced in a multi-step synthesis process. Most sartans share the common structural element biphenyl tetrazole. For this reason, the structure of this element will only be briefly describe without going into more detail regarding the other special steps of the sartan synthesis. Depending on the specific sartan in question, the biphenyl tetrazole element can also be created and incorporated at the start of the synthesis or at a later stage or towards the end of the synthesis process, depending on the other synthesis steps required and their corresponding reaction conditions. The selected synthesis strategy will also determine the adequate tetrazole formation variant.

Some of these variants are quoted in the following literature.[18,52,57–59]

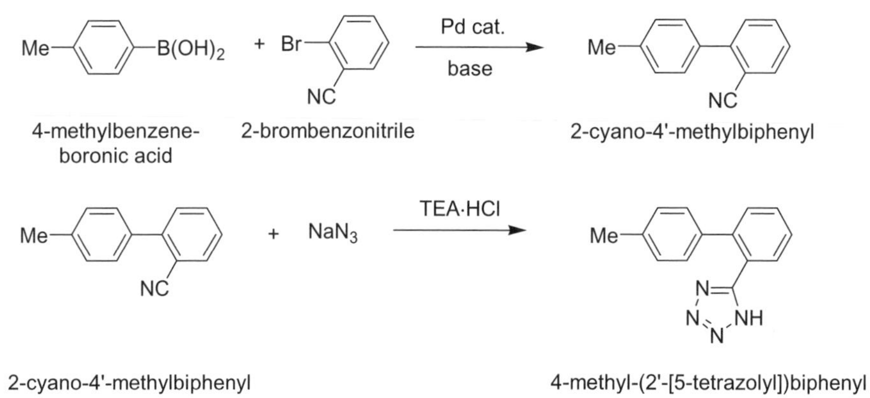

Scheme 2.9 *Example of the synthesis process for the production of the biphenyl tetrazole group*

2.3.1.2 Addition of NaN_3 or TMSA to Isocyanates to Yield Tetrazolinones

Different synthesis routes have been proposed for phenyl tetrazolinones.[19,60] In both cases, different azides (NaN_3 or TMSA) are added to an isocyanate.

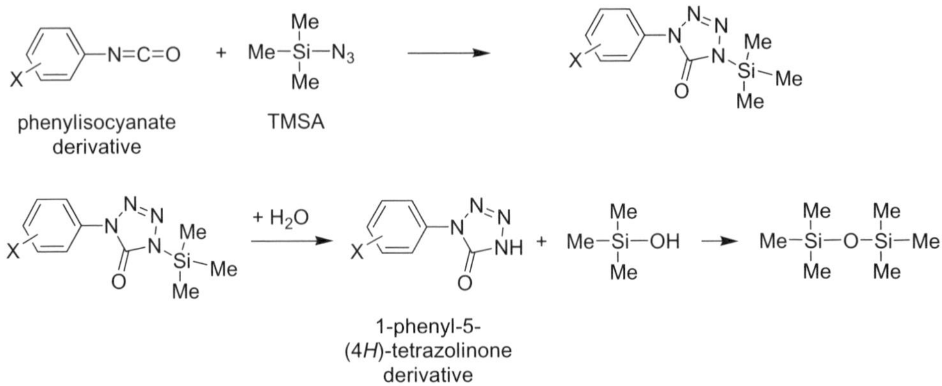

Scheme 2.10 *Synthesis for production of phenyl tetrazolinones taking the example of TMSA*

Phenyl tetrazolinone derivatives are used as herbicides against weeds in rice fields, for example.[61,62]

2.3.1.3 Addition of NaN_3 to Isothiocyanates to Yield Mercaptotetrazoles: Cephalosporins[63–64]

By analogy with the addition of azides to isocyanates, azides can also be added to isothiocyanates.

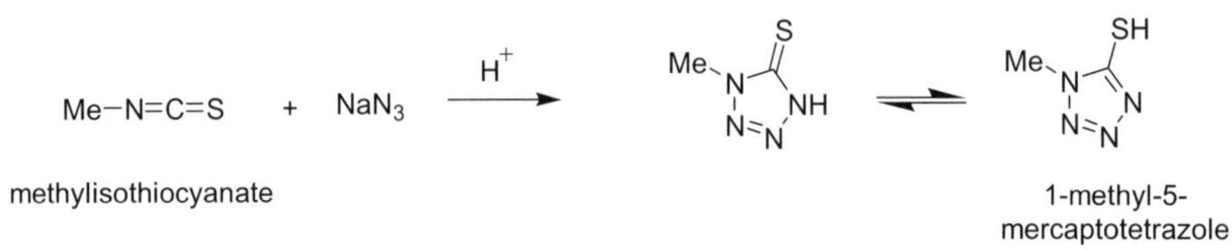

Scheme 2.11 *Synthesis for the production of mercaptotetrazoles*[65]

This type of reaction is used for the technical-scale production of some lateral chains of Cephalosporins.

It was developed in the 1950s as a broad-spectrum antibiotic (penicillin) for human medicine. It kills bacteria by destroying their cell wall synthesis. Naturally, Cephalosporins are found in the mold *Cephalosporium acremonium* in the form of Cephalosporin C. Second and third generation cephalosporins were developed by variation of the side chains so that we now have drugs offering good compatibility and effectiveness.

Cephalosporins all have ß-lactam as their common structural element.

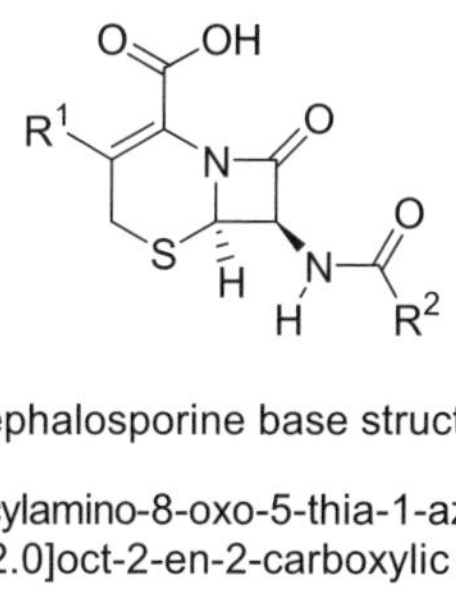

Scheme 2.12 *Cephalosporin base structure*

The fact that most of these tetrazole side chain elements for Cephalosporins since years are now produced worldwide at a volume of several 100 t/a demonstrates that azide chemistry – whose evolution to commercial scale was originally a source for concern – has come of age. It is now also offered as a standard production process by manufacturers that have specialized on the safe handling of the risk potential, also in custom synthesis. The safety risks associated with the handling of azides should not be underestimated, however. Its toxicity and the latent hazard of formation of highly explosive hydrazoic acid intermediates require expertise and plants with specific safety features for the safe handling of azides.

Cephalosporins vary in regard to their two side chains (R^1 and R^2). Table 2.8 shows Cephalosporins[66–74] whose side chains were synthesized using azide. For more examples, please refer to the literature:[75–79]

2.3.1.4 Addition of NaN_3 to Cyanoguanidine to Yield 5-Aminotetrazole (5-AT)[14]

As already mentioned in Section 2.2.1, 5-AT is a substituent for NaN_3 in the automotive industry for use in airbags.[80]

Besides the diazotization of aminoguanidine,[81] 5-AT can also be synthesized by addition of cyanoguanidine / dicyandiamide:[82]

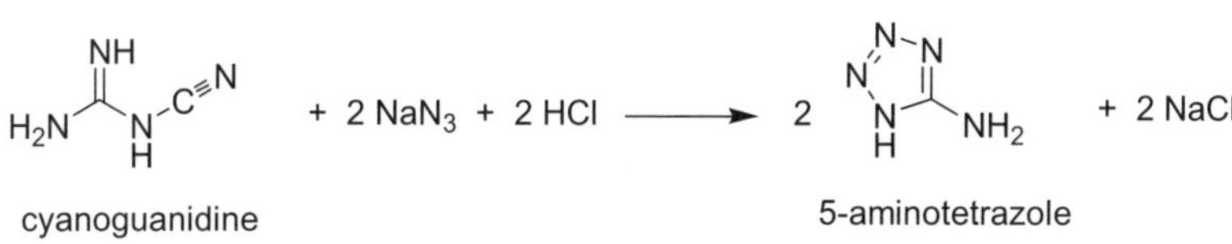

Scheme 2.13 *Synthesis of 5-aminotetrazole*

2.3.1.5 Addition of DPPA to Enamines to Yield α-aryl-carboxylic Acids

Naproxen is a non-steroidal, anti-inflammatory active ingredient used in numerous drugs. Only the (S)-(+)-enantiomer is therapeutically effective. Naproxen can for example be produced in the following four synthesis steps with a yield of approximately 60%.

Table 2.8 *Cephalosporins with tetrazole or mercaptotetrazole element in the side chains*

	R^1	Shortcut	R^2	Shortcut
Cefoperazone[74]	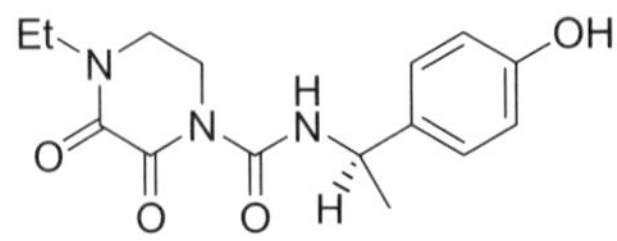		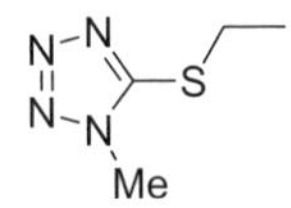	MMT
Cefamandole[66]	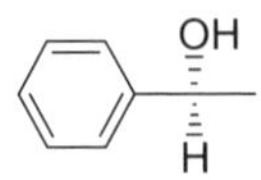		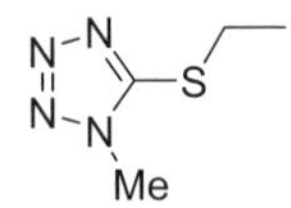	MMT
Cefmenoxime[67]	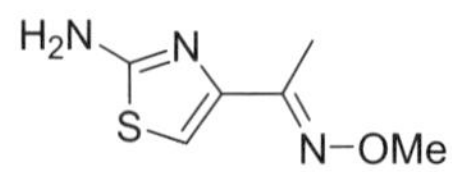		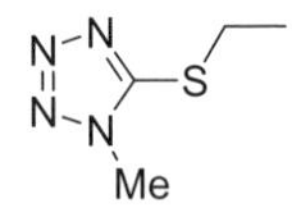	MMT
Cefotiam[68]	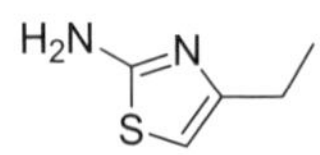		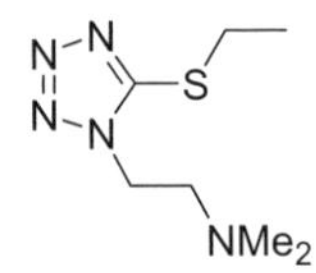	MTDMAE
Ceftezole[69]	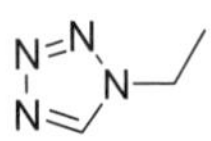	MT	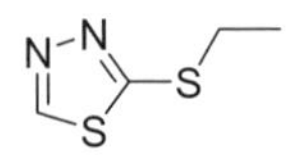	
Cefazoline[70]	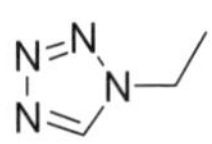	MT	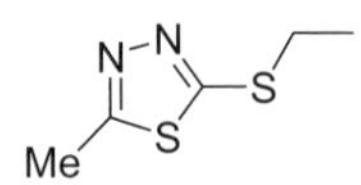	
Cefazaflur[71]	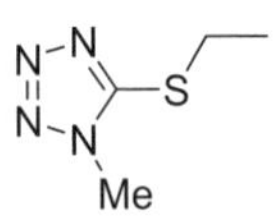	MMT	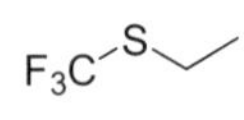	
Cefonicide[72]	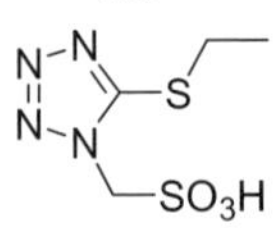	MTMS	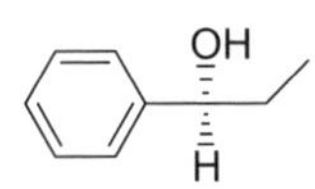	
Ceforanide[73]	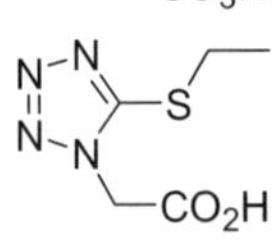	MTAA	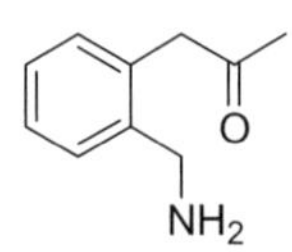	

MMT = 1-methyl-5-mercaptotetrazole
MT = 1-methyltetrazole
MTDMAE = 1-(*N*,*N*-dimethylamino)-2-(5-mercaptotetrazole-1-yl)-ethane
MTMS = (5-mercaptotetrazole-1-yl)-methanesulfonic acid
MTAA = (5-mercaptotetrazole-1-yl) acetic acid

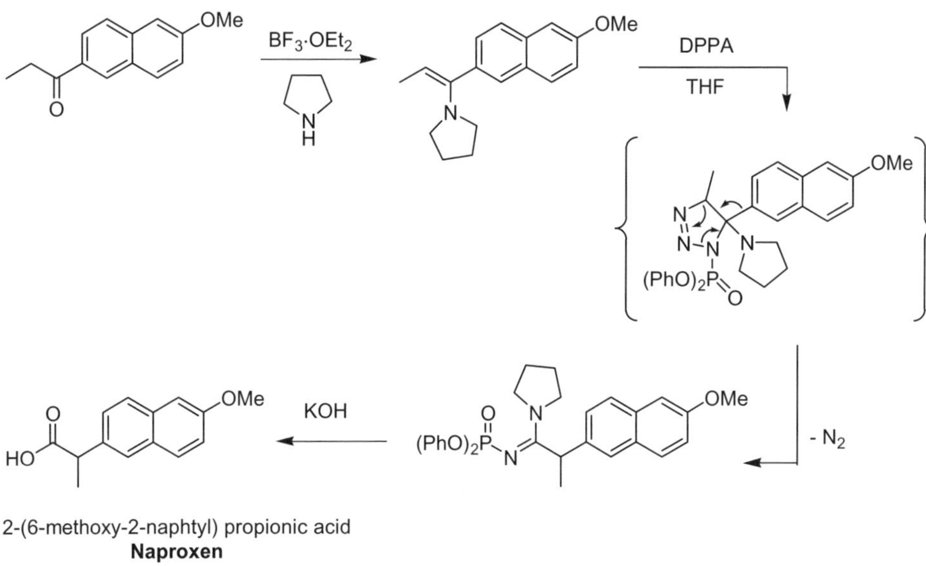

Scheme 2.14 *Synthesis to produce Naproxen*[83]

2.3.2 Addition of Alk-N_3 and Ar-N_3 to Multiple CC- and/or CN-Bonds

The preceding sections described the technical application of ionic bond azide (TBAA, NaN_3) or covalent bond silicon (TMSA), phosphor (DPPA) or tin (TBSnA) azide. In addition, numerous reactions of azides with a covalent carbon bond and the most varied multiple bonds have been examined and published over the years.[5, 84]

The use of alkyl azides (Alk-N_3) or aryl azides (Ar-N_3) can yield the corresponding substituted 1,5- or 2,5-tetrazolene,[85] 1,4-tetrazolinones[86] and/or the corresponding mercaptotetrazoles.[87] Moreover, substituted 1,2,3-triazoles[88] are feasible and also aziridines[89] provided that the azide was thermally or photochemically converted by separation of the nitrogen or by decomposition of the resulting dihydro-1,2,3-triazole.

2.3.3 Carboxylic Acid Azides: Precursors for Isocyanates

Also on the commercial scale, carboxylic acid azides are produced according to the traditional synthesis methods described below. Reacting *via* Curtius rearrangement,[90–91] they serve as precursors for isocyanates, which represent an important and frequently used element for the production of tetrazolinone mentioned under Section 2.3.1.2 and for all other addition or hydrolysis reactions described in the literature.

2.3.4 Organic Azides: Ring Opening Reaction on Oxiranes and Aziridines: Paclitaxel, Tamiflu®

The literature describes numerous active pharmaceutical ingredients where the synthesis uses the ring opening reaction of oxiranes or aziridines with organic or inorganic azides. Two well-known examples of commercial-scale application shall be mentioned here.

Table 2.9 *Commercial examples for carboxylic acid and its derivatives as an isocyanate precursor*

Original substance / reagent	Commercial examples	Originator	Literature
Carboxylic acid / DPPA	MIV 150	Medivir	92
	Tecadenoson	CV Therapeutics	93
Carboxylic acid / ClCOOEt / NaN_3	Nefazodone®	Mead Johnson	94
Carboxylic acid chloride / NaN_3	Teglicar	Sigma-Tau	95
Carboxylic acid hydrazide / HNO_2	Terguride	Spofa	96

The natural substance Paclitaxel (Taxol®)[97] can be extracted from the bark of Pacific yew and exhibits a positive effect when used in treating various types of cancer[98] such as ovary, breast, lung and prostate carcinoma. As the above-mentioned natural source is not sufficient to cover global demand, presently a combination of extraction (Baccatin III from the needles of English yew) and synthesis[99] labor- and cost-extensive chromatographic purification processes[100–102] is used.

In the course of the ten-step synthesis, the heterocyclic ring of the 2-phenyl-3-carbethoxyoxiran is opened upon reaction with NaN_3. Depending on the original compound used, this ring opening reaction is either enantioselective or leads to the racemate so that sometimes a chromatographic separation of enantiomers may be necessary:

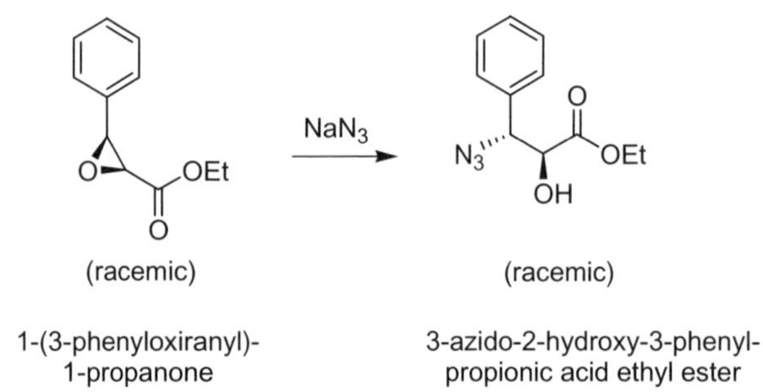

Scheme 2.15 *On the formation of 3-azido-2-hydroxy-3-phenylpropionic acid ethyl ester as intermediate in the Paclitaxel synthesis*[99]

Table 2.10 illustrates the commercial relevance of Paclitaxel by presenting its global sales figures:[103]

Table 2.10 *Global sales of Paclitaxel*

Year	2000	2001	2002	2003	2004	2005	2006
Sales[USD mn/a]	1561	1112	857	934	991	747	563

Table 2.11 Global sales of Oseltamivir[111]

Year	2003	2004	2005	2006
Sales[USD mn/a]	319	721	1246	2102

A similarly significant cure produced worldwide at a volume of several 100 t/a is Tamiflu®,[104] whose active pharmaceutical ingredient, Oseltamivir [(3*R*,4*R*,5*S*)-4-acetamido-5-amino-3-(1-ethylpropoxy)-cyclohex-1-en-1-carboxylic acid ethylester] for the treatment of influenza (influenza A or influenza B) was developed by the biotech company Gilead Sciences Inc. and Roche and was commercialized[105] for the first time in Switzerland in 1999.

A rapidly rising demand for Oseltamivir phosphate occurred in response to the risk of an avian influenza pandemic (virus H5N1) in mid- to late 2006 because the WHO recommended all countries to keep a sufficiently large stockpile of this product so that 25% of their population could be treated. According to present findings, the product has a virostatic effect on the H5N1 virus, i.e. it inhibits the replication of the virus but does not destroy the virus (viricidal).

From the viewpoint of azide chemistry, the synthesis of Oseltamivir is very interesting in many aspects because azides are used at more than one point and the azide synthesis processes used have prevailed over azide-free alternatives.[106]

In most cases, the Oseltamivir phosphate syntheses are based on (-)shikimic acid (fermentative production from glucose using a strain of *e.coli*). A novel synthesis route[107–108] has been published recently that obviates this raw material and uses TMSA on the easily accessible component 1,4-cyclohexadiene.

Below please find an illustration of the total synthesis steps that involve azide chemistry[109–110] to produce Oseltamivir.

It shows that an epoxide ring is opened stereospecifically with NaN_3 to azido hydroxy cyclohexene, which undergoes ring closure with TPP to give the corresponding aziridine, and is opened again with NaN_3 to the amino azido cyclohexene derivative with an exactly defined stereochemistry. This azide function remains in the molecule as a masked amine until the second last reaction step.

Table 2.11 illustrates the commercial relevance of Oseltamivir by presenting its global sales figures:[111]

2.3.5 Organic Azides: Protective Group, Masked Amines

As already demonstrated taking the example of the Oseltamivir (Section 2.3.4, second last step of the reaction sequence), the azide group can act as a synthetic equivalent and protective group for an amino function and thus be introduced into a synthesis at an earlier or later stage as required and depending on the chemical reaction conditions.

In the production of Loracarbef,[112] azidoacetic acid chloride is used for the formation of a variation of Cephalosporin ring system:

Other examples are Aprepitant (Merck & Co, 1995), Clopidogrel (Sanofi 1998), Linezolid (Upjohn 1995) and Mosapride (Dainippon 1987).[113–116]

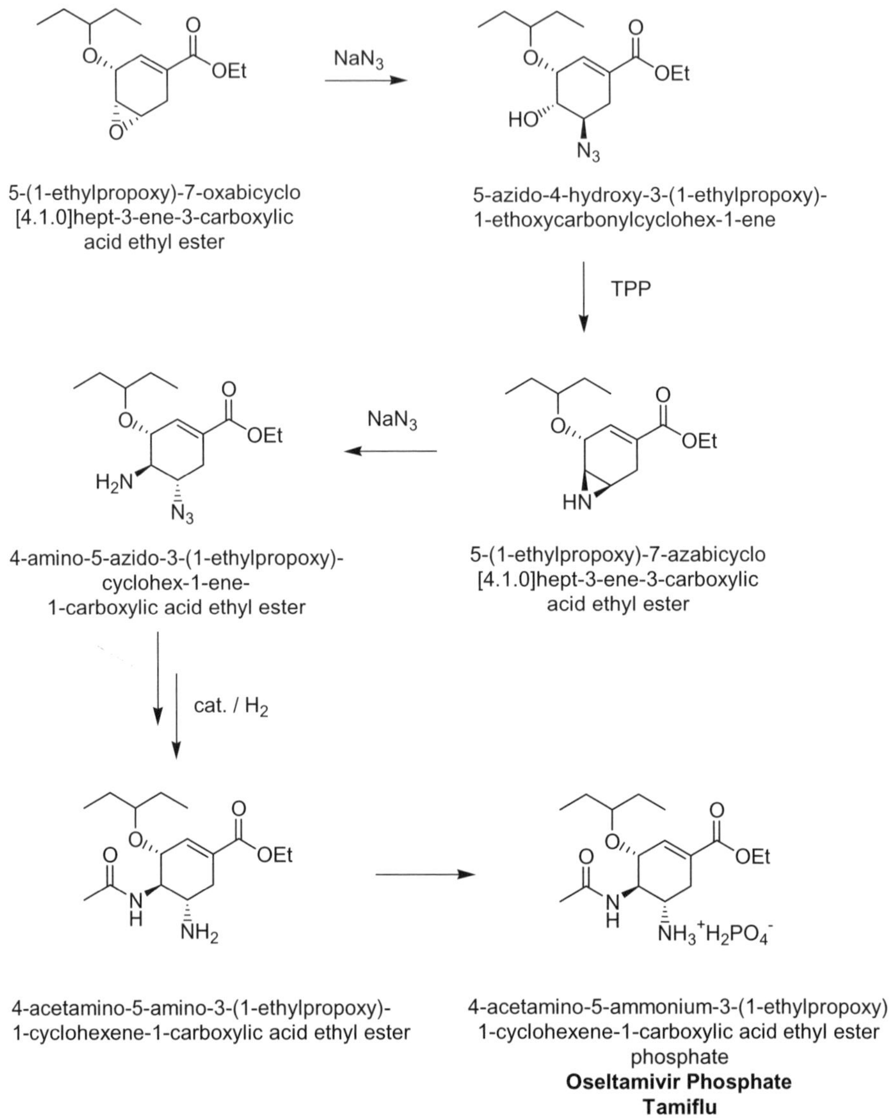

Scheme 2.16 *Selected steps of the total synthesis to produce Oseltamivir phosphate*

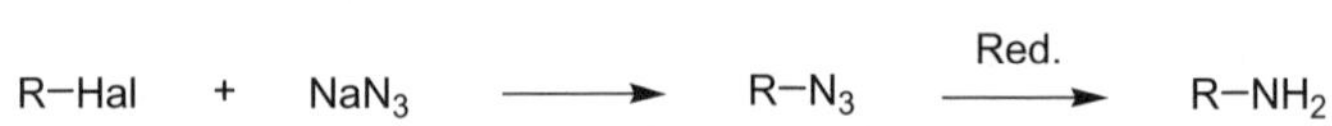

Scheme 2.17 *Azide as protective group for amino function*

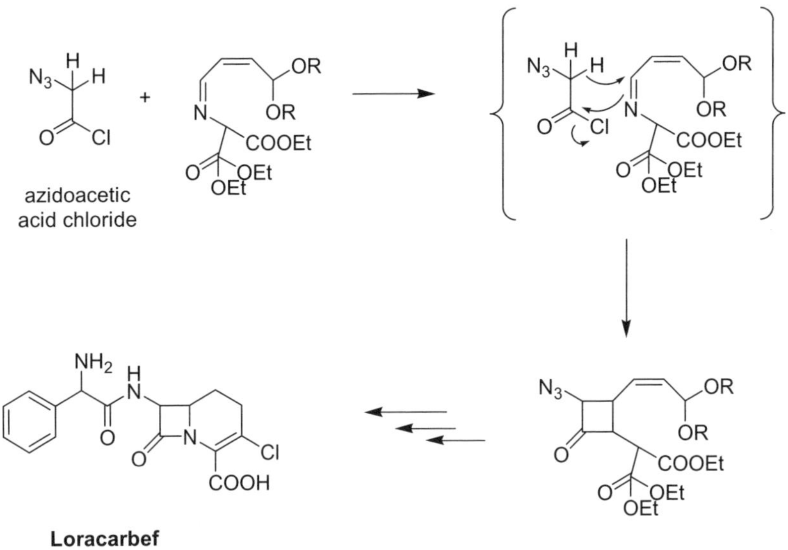

Scheme 2.18 Use of azidoacetic acid chloride for a masked amine

2.3.6 Organic Azides: Cross-linking Agents for Polymers

In the technical literature,[117–120] aromatic bisulfonyl azides are described as cross-linking agents / vulcanizing agents in polyamide butadiene block polymers and in thermoplastic polymers such as polypropylene.

2.4 The Future of Commercial-scale Azide Chemistry

The above sections have clearly demonstrated that, despite the latent safety risks involved, azide chemistry has found a broad, commercial-scale application over the past forty years. This is attributable to the fact that

- some companies have acquired the safety knowhow and developed special plants in which azide reactions can be operated safely and reproducibly; and
- compared with other synthesis routes leading to the same result, reactions involving azides allow us to realize cost savings (shorter synthesis routes, stereochemically exact conditions).

According to numerous publications[5,121–124] attempts have been made to minimize the safety risks of azide chemistry by way of immobilization. However, this route will only be successful if the concomitant drop in space-time-yield can be compensated by a continuous reaction process.

One can safely assume that new azide reactions, which are presently tested and described in academic research (see the following chapters of this book, including reference),[5] will also be realized on a commercial scale in the future as long as they prove to be more cost-efficient than alternative routes. The safety risks will be measurable and thus also manageable for those select companies that have the expertise and equipment required to this effect.

References

[1] J.H. Boyer, F.C. Canter, *Chem. Rev.* **1954**, *54*, 1–57.
[2] C. Grundmann, *Houben-Weyl* **1965**, *10/3*, 777–836.
[3] H.H. Jobelius. H.-D. Scharf, *Ullmann' Encyclopedia of Industrial Chemisty, 5th Ed.* **1989**, *A13*, 193–7.
[4] M. Regitz, G. Maas, *Diazo Compounds. Properties and Synthesis*. Academic Press, Orlando, **1986**.
[5] S. Bräse, C. Gil, K. Knepper, V. Zimmermann, *Angew. Chem. Int. Ed.* **2005**, *44*, 5188–240.
[6] W. Wislicenus, *Ber. Dtsch. Chem. Ges.* **1892**, *25*, 2084–7.
[7] G. Bretschneider, G. Deising, H. Klöpfer, F. Sperr, H. Schmidt (Degussa), DE 1206405 **1964**; *Chem. Abstr. 64*:33559.
[8] S. Abe, T. Kawakami, *Sci. Rep. Res. Inst. Tohoku Univ. Ser.* **1952**, *A4*, 105.
[9] T. Curtius, *Ber. Dtsch. Chem. Ges.* **1890**, *23*, 3023–33.
[10] J. Thiele, *Ber. Dtsch. Chem. Ges.* **1908**, *41*, 2681–3.
[11] No inventor name available, *DE 619753* **1935** (Dynamit-Nobel); *Chem. Abstr. 30*:11558.
[12] www.alkalimetals.com, DRH Prospectus; Business Overview; p. 78.
[13] E.A. Betterton, *Critical Reviews in Enviromental Science and Technology* **2003**, *33*, 423–58.
[14] Supply and Safety Data Sheet available by *Dynamit-Nobel GmbH, Explosivstoff und Systemtechnik* Leverkusen, Germany.
[15] L. Birkofer, P. Wegner, *Org. Synth.* **1970**, *50*, 107–10.
[16] T. Kofukuda, S. Nakazawa (Toyo Kasei Kogy Com. Ltd.), *WO 2006038329* **2006**; *Chem. Abstr. 144*:370241.
[17] C. Moberg, H. Adolfsson, *Organometallics: Compounds of Group 15 (As,Sb,Bi) and Silicon Compounds*, (Editor: I. Fleming), Thieme, Stuttgart **2002**, *Bd. 4*, 435–49.
[18] E. Ettenhuber, K. Rühlmann, *Chem. Ber.* **1968**, *101*, 743–50.
[19] O. Tsuge, S. Urano, K. Oe, *J. Org. Chem.* **1980**, *45*, 5130–6.
[20] L. Birkofer, P. Wegner, *Chem. Ber.* **1966**, *99*, 2512–17.
[21] S.S. Washburne, W.R. Peterson, Jr., *J. Organomet. Chem.* **1970**, *21*, 427–30.
[22] D.J. Guerin, T.E. Horstmann, S.J. Miller, *Org. Lett.* **1999**, *1*, 1107–9.
[23] B.L. Kedrowski, *J. Org. Chem.* **2003**, *68*, 5403–6.
[24] J. Lutz, H.-J. Musiol, L. Moroder, *Houben Weyl* **2001**, *Bd. E22a*, 427–42.
[25] H. Shao, M. Colucci, S.J. Tong, H.S. Zang, A.L. Castelhano, *Tetrahedron Lett.* **1998**, *39*, 7235–8.
[26] S. Sunami, T. Sagara, M. Ohkubo, H. Morishima, *Tetrahedron Lett.* **1999**, *40*, 1721–4.
[27] B. Jiang, C.-G. Yang, J. Wang, *J. Org. Chem.* **2002**, *67*, 1396–8.
[28] K.C. Nicolaou, N. Winssinger, D. Vourloumis, *et al.*, *J. Am. Chem. Soc.* **1998**, *120*, 10814–26.
[29] D.L. Hughes, *Org. React.* **1992**, *42*, 335–56.
[30] P. Magnus, K.S. Matthews, V. Lynch, *Org. Lett.* **2003**, *5*, 2181–4.
[31] O. Wolf, S.R. Waldvogel, *Synthesis* **2004**, 1303–5.
[32] T. Shioiri, K. Ninomiya, S. Yamada, *J. Am. Chem. Soc.* **1972**, *94*, 6203–5.
[33] T. Shioiri, S. Yamada, *Org. Synth.* **1984**, *62*, 187–90.
[34] J. Lorberth, H. Krapf, H. Noeth, *Chem. Ber.* **1967**, *100*, 3511–19.
[35] A. Kumar, M.M. Nimbalkar, S.G. Barve, *et al.* (Ipca Laboratories Ltd., India), *EP 1714963* **2006**; Chem. Abstr. 145:455017.
[36] J. Wiss, A. Zilian, *Org. Process Res. Dev.* **2003**, *7*, 1059–66.

[37] K. Isomura, S. Kobayashi, H. Taniguchi, *Tetrahedron Lett.* **1968**, 3499–502.
[38] H. Hemetsberger, D. Knittel, H. Weidmann, *Monatsh. Chem.* **1969**, *100*, 1599–603.
[39] H. Hemetsberger, D. Knittel, *Monatsh. Chem.* **1972**, *103*, 194–204.
[40] H. Hemetsberger, I. Spira, W. Schoenfelder, *J. Chem. Res. Synop.* **1977**, 247–9.
[41] D.L. Boger, R.S. Coleman, *J. Am. Chem. Soc.* **1987**, *109*, 2717–27.
[42] R.E. Bolton, C.J. Moody, C.W. Rees, G. Tojo, *J. Chem. Soc. Perkin Trans. 1*, **1987**, 931–6.
[43] F. Hong, J. Zaidi, B. Cusack, E. Richelson, *Bioorg. Med. Chem. Lett.* **2002**, *12*, 3849–58.
[44] I. Borza, S. Kolok, A. Gere, *et al.*, *Bioorg. Med. Chem. Lett.* **2003**, *13*, 3859–61.
[45] P. Molina, P.M. Fresneda, S. Delgado, *J. Org. Chem.* **2003**, *68*, 489–99.
[46] K.L. Milkiewicz, D.J. Parks, T. Lu, *Tetrahedron Lett.* **2003**, *44*, 4257–60.
[47] P.E. Brandish, N. Brandon, W. Zheng, *et al.* (Merck Sharp Dohme), *WO 2007039773* **2007**; Chem. Abstr. 146:421961.
[48] A.S. Katner, S.J. Bogard (Eli Lilly), *EP 48 167* **1982**; Chem. Abstr. 97:23797.
[49] A.J. Papa, *J. Org. Chem.* **1966**, *31*, 1426–30.
[50] D.R. Tortolani, S.A. Biller, *Tetrahedron Lett.* **1996**, *37*, 5687–90.
[51] R.A. Moss, J. Terpinski, D.P. Cox, D.Z. Denny, K. Krogh-Jespersen, *J. Am. Chem. Soc.* **1985**, *107*, 2743–8.
[52] A. Enhsen, H. Glombik, W. Kramer, G. Wess (Hoechst AG), *EP 624596* **1994**; Chem. Abstr. 122:187872.
[53] R.J. Herr, *Bioorg. Med. Chem.* **2002**, *10*, 3379–93.
[54] C.D. Siebert, *Chem. Unserer Zeit* **2004**, *38*, 320–4.
[55] Becker Associates, *B.I.C 3000 database Version* 09/2007.
[56] Prous Science Integrity, http://integrity.prous.com **2008**.
[57] P.K. Kadaba, *Synthesis* **1973**, 71–84.
[58] K. Sisido, K. Nabika, T. Isida, S. Kozima, *J. Organomet. Chem.* **1971**, *33*, 337–46.
[59] J.L. Kraus, *Synth. Commun.* **1986**, *16*, 827–32.
[60] J.P. Horwitz, B.E. Fisher, A.J. Tomasewski, *J. Am. Chem. Soc.* **1959**, *81*, 3077–8.
[61] T. Goto, S. Ito, Y. Watanabe, S. Narabu, A. Yanagi (Nihon Bayer Agrochem), *EP 612735* **1994**; Chem. Abstr. 122:81375.
[62] T. Goto, H. Hayakawa, Y. Watanabe, S. Narabu, A. Yanagi (Nihon Bayer Agrochem), *EP 578090* **1994**; Chem. Abstr. 121:57514.
[63] A. Sammes, *Chem. Rev.* **1976**, *76*, 113–55.
[64] O. Masaji, O. Masami, Y. Morimasa, *et al.*, *Ullmann' Encyclopedia of Industrial Chemisty, 5th Ed.* **1985**, *A2*, 467–557.
[65] R.R. Raap, *Can. J. Chem.* **1971**, *49*, 2139–42.
[66] C.W. Ryan, (Eli Lilly) *US 3 641 021* **1972**; Chem. Abstr. 74:13171.
[67] H. Natsugari, I. Mikami, M. Ochiai (Takeda), *US 4298607* **1981**; Chem. Abstr. 93:239438.
[68] M. Numata, I. Minamida, M. Yamaoka, M. Shiraishi, T. Miyawaki (Takeda), *FR 2255 077* **1975**; Chem. Abstr. 84:74284.
[69] T. Takano, M. Kurita, H. Nikaido, M. Mera, N. Konishi, R. Nakagawa (Fujisawa), *GB 1206 305* **1970**; Chem. Abstr. 72:100724.
[70] T. Takano, M. Kurita, H. Nikaido, M. Mera, N. Konishi, R. Nakagawa (Fujisawa), *ZA 6804513* **1969**; Chem. Abstr. 72:100724.
[71] R.M. DeMarinis, J.R.E. Hoover (Smithkline), *US 3943131* **1976**; Chem. Abstr. 85:46714.
[72] D.A. Berges (Smithkline), *GB 1547 473* **1979**; Chem. Abstr. 86:29854.
[73] W.J. Gottstein, M.A. Kaplan, A.P. Granatek, P. Alphonse (Bristol-Myers), *US 4100 346* **1978**; Chem. Abstr. 90:54955.
[74] I. Saikawa, S. Takano, C. Yoshida, *et al.* (Toyama), *BE 837682* **1976**; Chem. Abstr. 87:6002.
[75] I. Saikawa, S. Takano, K. Momonoi, *et al.* (Toyama), *DE 2841706* **1979**; Chem. Abstr. 91:57036.
[76] H. Nakao, H. Yanagisawa, M. Nagano, *et al.* (Sankyo), *DE 2455 884* **1975**; Chem. Abstr. 83:97330.
[77] K. Iwamatsu, S. Inoue, K. Miyauchi, *et al.* (Meiji), *DE 2950 990* **1980**; Chem. Abstr. 93:220759.

[78] M. Iwanami, T. Maeda, Y. Nagano, *et al.* (Yamanouchi), *DE 2824 559* **1978**; Chem. Abstr. 90:137844
[79] H. Yamada, K. Okamura, H. Tobiki, *et al.* (Sumitomo), *BE 833063* **1975**; Chem. Abstr. 85:94384.
[80] R.D. Tayor, I.V. Mendenhall (Autoliv Asp, Inc.), *WO 2006047085* **2006**; Chem. Abstr. 144:435497.
[81] J. Arient, I. Voboril (no company mentioned), *CS 190055* **1955**; Chem. Abstr. 96:85565.
[82] R.M. Herbst, J.A. Garrison, *J. Org. Chem.* **1953**, *18*, 941–5.
[83] J. Riegl, M.L Maddox, I.T. Harrison, *J. Med. Chem.* **1974**, *17*, 377–8.
[84] C.-K. Sha, A.K. Mohanakrishnan in *Synthetic Applications of 1,3-Dipolar Cycloaddition Chemistry towards Heterocycles and Natural Products* (A. Padwa, W.H. Pearson, eds) John Wiley & Sons, Inc., New York **2002**, p. 623.
[85] G.I. Koldobskii, R.B. Kharbash, *Russ. J. Org. Chem.* (Translation of Zhurnal Organicheskoi Khimii), **2003**, *39*, 453–70.
[86] J.M. Vandensavel, G. Smets, G. L'abbé, *J. Org. Chem.* **1973**, *38*, 675–8.
[87] G. L'abbé, G. Verhelst, S. Toppet, *J. Org. Chem.* **1977**, *42*, 1159–63.
[88] K. Sugita, M. Otsuka, H. Oki, *et al.* (Daiichi Pharm.), *WO 2007055093* **2007**; Chem. Abstr. 147:9956.
[89] A. Mordini, L. Sbaragli, M. Valacchi, F. Russo, G. Reginato, *Chem. Commun.* **2002**, *7*, 778–9.
[90] P.A.S. Smith, *Org. React.* **1947**, *3*, 267–49.
[91] E.F.V. Scriven, *Azides and Nitrenes, New York; Academic Press* **1984**.
[92] S. Cai, M. Dimitroff, T. McKennon, *et al.*, *Org. Process Dev.* **2004**, *8*, 353–9.
[93] T.D. Ashton, K.M. Aumann, S.P. Baker, C.H. Schiesser, P.J. Scammells, *Bioorg. Med. Chem. Lett.* **2007**, *17*, 6779–84.
[94] D.L. Temple, G.W. Lobeck (Mead Johnson), *US 4487773* **1984**; Chem. Abstr. 102:166780.
[95] O. Piccolo, R. Castagnani, P. De Witt Scalfaro (Sigma-Tau Industrie Farmaceutiche Riunnite S.p.A.), *WO 2003010129* **2003**; Chem. Abstr. 138:137589.
[96] J. Benes, L. Markovic, A. Cerven, M. Schreiberova, (GALENA A.S.) *CZ 279929* **1995**; Chem. Abstr. 124:343777.
[97] D.M. Vyas, *Taxol: Science and Applications* (Editor: Matthew Suffness, National Cancer Institute), CRC Press **1995**.
[98] H.-P. Lipp, C. Brokmeyer, *Pharm. Unserer Zeit*, **2005**, *34*, 128–37.
[99] Prous Science Integrity (Paclitaxel) http://integrity.prous.com **2008**.
[100] A. Ghassempour, M. Noruzi, M. Zandehzaban, *et al.*, *J. Liq. Chromatogr. Related Technol.* **2008**, *31*, 382–94.
[101] P. Sun, X. Wang, L. Alquire, C.A. Maryanoff, *J. Chromatogr. A* **2008**, *1177*, 87–91.
[102] S.K. Rijhwani, Y.Y. Chan (no company mentioned), *US 2007190623* **2007**; Chem. Abstr. 147:275832.
[103] Becker & Associates, *B.I.C. 3000 Database: Paclitacel*, **2007**.
[104] V. Farina, J.D. Brown, *Angew. Chem. Int. Ed.* **2006**, *45*, 7330–43.
[105] T. Käding, *Kölner Stadt-Anzeiger* **2006**, *65*, 33.
[106] U. Jahn, *Nachr. Chem.* **2005**, *54*, 524–6.
[107] A. Thayer, *Chem. Eng.* **2006**, *84*, 29–30.
[108] M. Shibasaki, M. Kanai, T. Mita, N. Fukuda, Y. Fukuta (Uni. Tokyo, Japan), *WO 2007099843* **2007**; Chem. Abstr. 147:322624.
[109] M. Ferderspiel, R. Fischer, M. Henning, *et al.*, *Org. Process Res. Dev.* **1999**, *3*, 266–74.
[110] S. Abrecht, P. Harrington, H. Iding, *et al.*, *Chimia* **2004**, *58*, 621–9.
[111] Becker & Associates, *B.I.C. 3000 Database: Oseltamivir*, **2007**.
[112] T. Ogasa, H. Saito, Y. Hashimoto, K. Sato, T. Hirata, *Chem. Pharm. Bull.* **1989**, *37*, 315–21.
[113] C.P. Dorn, J.J. Jeffrey, M. MacCoss, S.G. Mills (Merck & Co), *WO 9523798* **1995**: Chem. Abstr. 124:146177.
[114] B. Castro, J.-R. Dormoy, A. Previdero (Sanofi), *WO 9839322* **1998**; Chem. Abstr. 129: 245036.

[115] M.R. Barbachyn, S.J. Brickner, D.K. Hutchinson (Upjohn), *WO 9507271* **1995**; Chem. Abstr. 123:256742.
[116] T. Kon, S. Kato, T. Morie, *et al.* (Dainippon), *EP 243959* **1987**; Chem. Abstr. 108:94575.
[117] T. Tagami (*Tomoegawa Paper*), *JP 06345962* **1994**; Chem. Abstr. 122:316107.
[118] M. Walters, M.W. Sorenson, F.M. Finlayson, R.J. Lee, C.J. Clark (Dow Chem.), *WO 2001083605* **2001**; Chem. Abstr. 135:344929.
[119] M.F. Finlayson, M.E. Walters, M.W. Sorenson, *et al.*, (Dow Chem.), *WO 2003040229* **2003**; Chem. Abstr. 138:386308.
[120] K. Sehanobish, T.H. Ho (Dow Chem.), *WO 2003082971* **2003**; Chem. Abstr. 39:308337.
[121] A. Kirschning, H. Monenschein, C. Schmeck, *Angew. Chem. Int. Ed.* **1999**, *38*, 2594–6.
[122] H. Shao, M. Colucci, S. Tong, H. Zhang and A.L. Castelhano, *Tetrahedron Lett.* **1998**, *39*, 7235–8.
[123] S. Loeber, P. Rodriguez-Loaiza, P. Gmeiner, *Org. Lett.* **2003**, *5*, 1753–5.
[124] L. Leeb, P. Gmeiner, S. Löber, *QSAR & Combinatorial Science*, **2007**, *26*, 1145–50.

3

Synthesis of Azides

Teresa M.V.D. Pinho e Melo

Department of Chemistry, University of Coimbra, 3004-535 Coimbra, Portugal

3.1 Introduction

The chemistry of azides starts with the preparation of the first organic azide, phenyl azide, by Peter Griess in 1864[1] and with the discovery of hydrogen azide and the rearrangement of acyl azides to the corresponding isocyanate reported by Curtius in 1890 (Curtius rearrangement).[2] However, only in the 1950s and 1960s did organic azides receive considerable attention pushed by the reviews of Smith[3] and Boyer *et al.*[4] on the chemistry of the acyl, aryl, and alkyl azides.

Since then numerous syntheses and applications of organic azides have been developed.[5] These energy-rich molecules became valuable intermediates in organic synthesis, in particular in the synthesis of various nitrogen-containing heterocycles, in peptide chemistry and in combinatorial chemistry. They found application as blowing agents and as pharmaceuticals. It is worthwhile to mention the international interest on azidonucleosides in the treatment of AIDS[6] and their application for the preparation of bioconjugates via Staudinger ligation.[7]

Here an overview is provided of the more relevant synthetic methods for the preparation of organic azides.

3.2 Synthesis of Alkyl Azides

3.2.1 Classic Nucleophilic Substitutions: Azides from Halides, Sulfonates, Sulfites, Carbonates, Thiocarbonates and Sulfonium Salts

Among the various possibilities for preparing alkyl azides, classic nucleophilic substitutions are of special importance. Alkali metal azides, mainly sodium azide, are commonly

Organic Azides: Syntheses and Applications Edited by Stefan Bräse and Klaus Banert

used as the azide source and substrates bearing leaving groups such as halides or sulfonates offer a simple and high yield access to alkyl azides.[5]

The preparation of the highly unpredictable ethyl azide (**2**) with minimum handling can be achieved by a phase-transfer reaction with sodium azide, catalyzed by tetra-*n*-butylammonium tetrafluoroborate.[8] Besides primary alkyl halides, secondary and even some tertiary azides can be prepared by the reaction with azide ion.[9] One example of the former is the reaction of bromohydrin **3** with NaN_3 in DMSO at 40 °C giving the expected product with complete inversion of configuration.[10a] α-Chloro methyl ester **5** and α-chloro alcohol **7** are also transformed into the corresponding azides by treatment with NaN_3 in DMF at 60 °C without loss of *ee*.[10b] In the case of the dibromo ester **9** the reaction with a slight excess of sodium azide in DMF gives the monoazide **10** in high yield and about 15% of the diazide species.[10c] Trimethylsilyl analogues have been similarly prepared.[10d] A combination of NaN_3 and $ZnCl_2$ often gives rise to more efficient syntheses particularly for the generation of tertiary azides (e.g. **12**).[10e,10f] 3-Bromo-3-phenyl-3*H*-diazirine (**13**) reacts with tetrabutylammonium azide to afford benzonitrile in 90% yield via 3-azido-3-phenyl-3*H*-diazirine (**14**) as intermediate[10g,10h]. Kool *et al.* reported functionalizations of DNA oligonucleotides including the conversion in high yield of 5′-iodinated oligonucleotides to 5′-azido derivatives via halide displacement with NaN_3 in DMF.[10i] Solid-phase synthesis of aliphatic azides via substitution of alkyl halides on solid supports can be carried out with sodium azide or tetrabutylammonium azide.[10j–10l] The nucleophilic substitution of the resin-bound 1-chloro-2-alkanol **16** with sodium azide gives the azido alcohol **17** in high yield[10k] (Scheme 3.1).

The displacement of sulfonates by azide ion can be used as an indirect conversion of an alcohol to an azide. It is a strategy used for the preparation of azido carbohydrates, which are particularly interesting considering the possibility for reductive generation of amino sugars.[5b] A carbohydrate-based enantiospecific synthesis of (*R*)-proline has been described.[11a] Azido-substitution reaction of the tosylate derivative of **18** gave the corresponding *D-threo*-azide **20** (91%), which was further transformed into (*R*)-proline. Baran *et al.* reported the total synthesis of (±)-sceptrin, which involved the synthesis of diazide **23** via mesylation of the diol **22** followed by displacement with NaN_3.[11b] In the reported total synthesis of (–)-ephedradine A the displacement of a mesylate group with NaN_3 is also explored.[11c] The triflate **25**, obtained selectively from the reaction of diol **24** with triflic anhydride, affords γ-azido ester **26** on reacting with NaN_3 in an overall yield of 46%.[11d] A similar approach was applied to the manipulation of the C2 hydroxy group of the glycoside of hydroxyproline **27**.[11e] Boger *et al.* described the synthesis of methyl (2*S*,3*S*)-2-azido-3-hydroxy-3-(4-iodophenyl)propionate (**31**).[11f] The Sharpless asymmetric dihydroxylation (AD-mix-α) reaction of methyl (*E*)-4-iodocinnamate was followed by the selective formation of α-hydroxy sulfonate **30** resulting from the reaction of the more acidic alcohol with 4-nitrobenzenesulfonyl chloride. Subsequent NaN_3 displacement of the nosylate group gave the desired product in high yield. A similar synthetic strategy was applied in the oxidative cyclization of tryptophan derivatives.[11g] Solid-phase synthesis of alkyl azides from alcohols has also been reported.[11h] The support-bound secondary alcohols **32** activated to the corresponding nosyl derivatives followed by displacement with sodium azide at 50 °C afford the azides **33** (Scheme 3.2).

Synthetic approaches to the synthesis of azides directly from alcohols are known and include the use of $NaN_3/BF_3{\cdot}Et_2O$, $HN_3/TiCl_4$ and NaN_3/triphosgene.[12a–12c] Conversion of

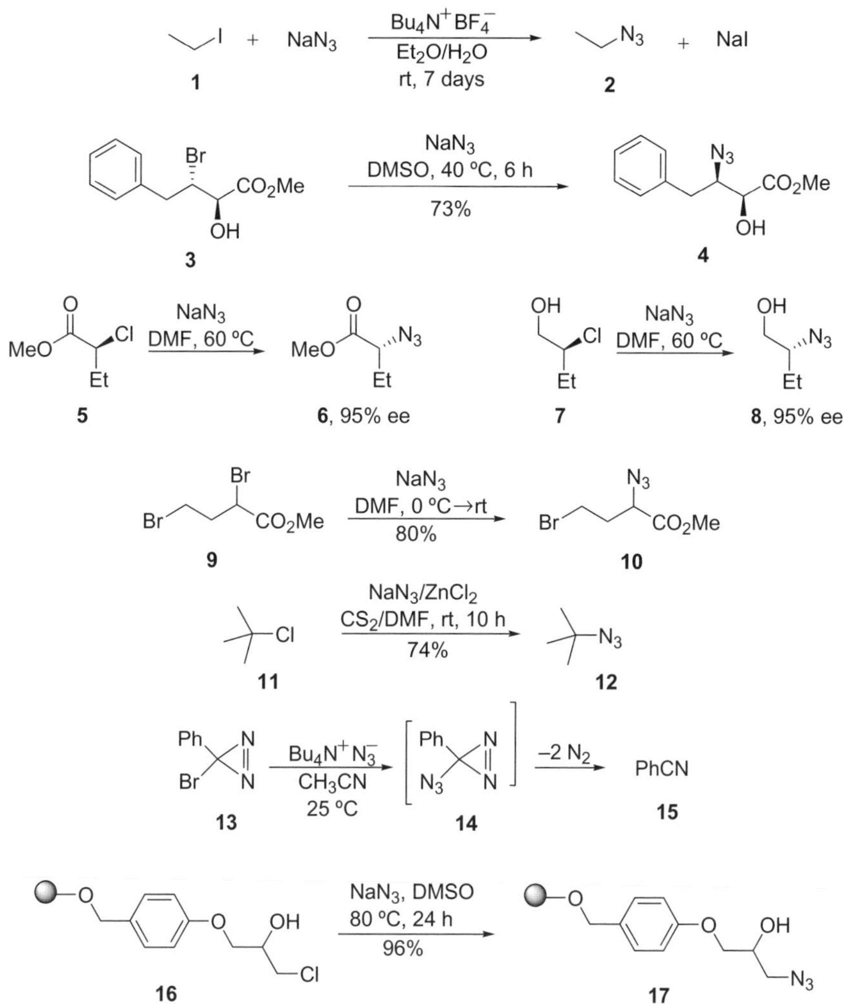

Scheme 3.1 *Synthesis of alkyl azides from halides*[8,10a-10h,10k]

alcohols to azides can be carried out in the presence of NaN_3, triethylamine, *N*-(*p*-toluenesulfonyl)imidazole (TsIm) and catalytic amounts of tetra-*n*-butylammonium iodide (TBAI) in DMF.[12d] The authors suggest that the process occurs through reaction of the base-activated alcohol with TsIm giving alkyl tosylate, which undergoes nucleophilic substitution on reacting with azide ion (Scheme 3.3).

Organic azides can be obtained from cyclic sulfates and from cyclic sulfites[13] (Scheme 3.4). Avenoza *et al.* studied the nucleophilic ring-opening reactions of *gem*-disubstituted cyclic sulfates with sodium azide, which lead to the synthesis of azido alcohols **37** and **38**.[13a] They observed that the regioselectivity depends on the substituent present on the cyclic sulfate. Amide substituents lead preferentially to products arising from nucleophilic attack at the least substituted C_β position, whereas reverse regioselectivity is obtained

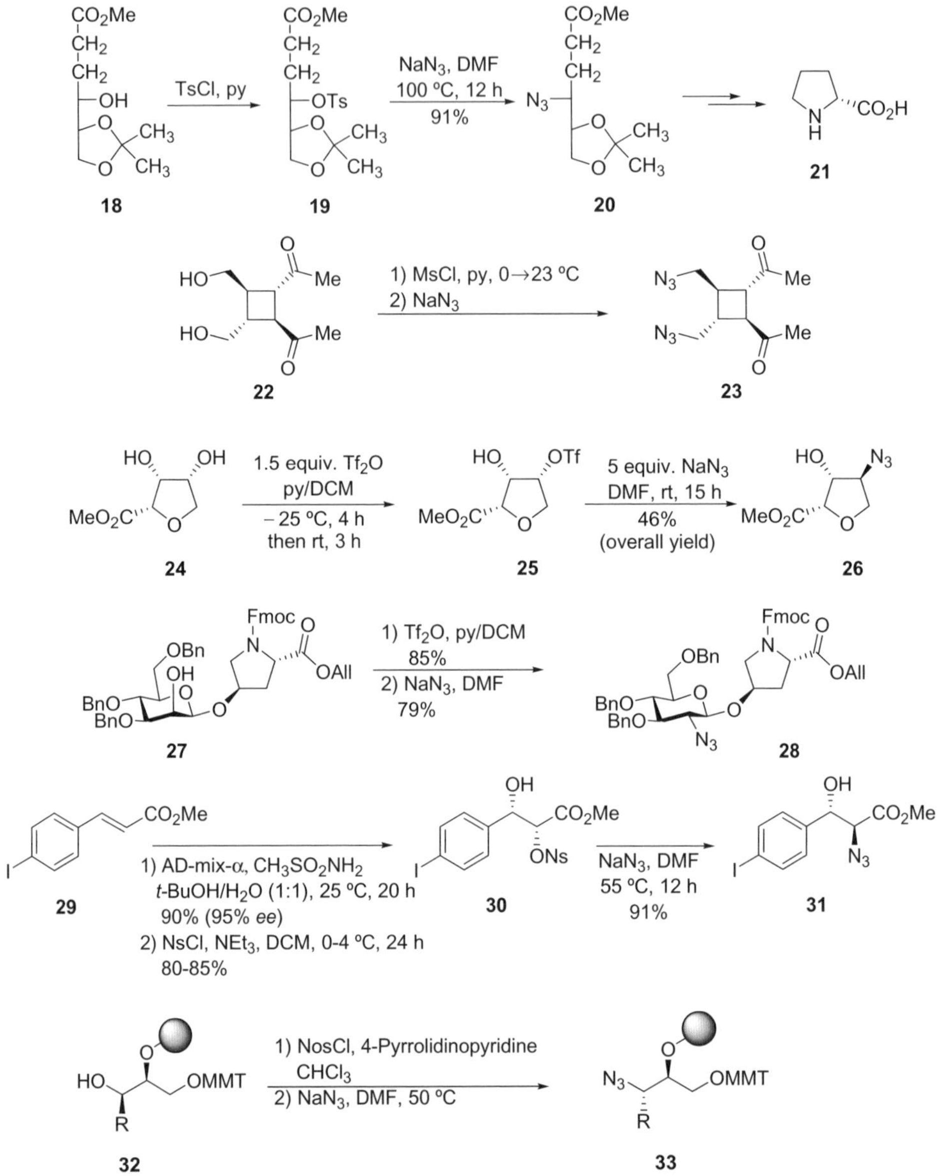

Scheme 3.2 *Synthesis of alkyl azides from sulfonates*[11]

with ester substituents. The azido alcohols were also obtained by the ring-opening reaction of cyclic sulfite **39**, under the same reaction conditions giving a product mixture in 85% yield (**37a/38a** ratio, 5.5 : 1). The reaction of vicinal diols **40** with $SOCl_2$ in presence of triethyl amine gives cyclic sulfites **41** as diastereomeric mixtures in a 1 : 1 ratio and these compounds are converted in high yield to the corresponding azido alcohols **42** by reacting with sodium azide (Scheme 3.4).[13b]

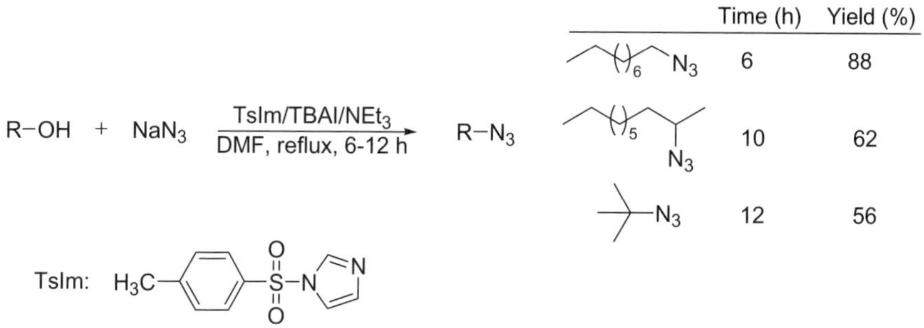

Scheme 3.3 *Alkyl azides from alcohols*[12d]

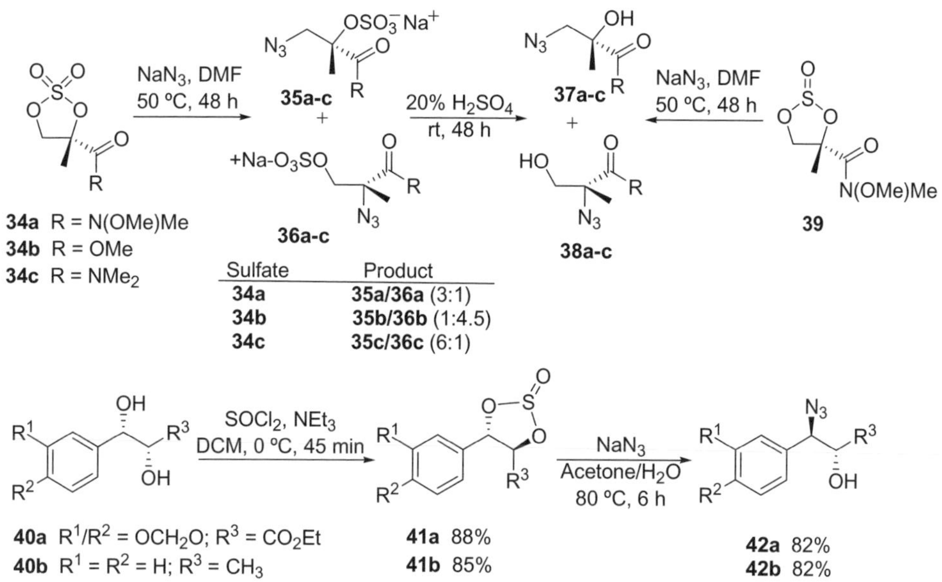

Scheme 3.4 *Synthesis of alkyl azides from cyclic sulfates and cyclic sulfites*[13]

Cyclic carbonates and thiocarbonates are convenient starting material for the synthesis of α-azido alcohols (Scheme 3.5). The nucleophilic ring opening of optically active cyclic carbonate **43** affords the α-azido alcohol **44** stereoselectively.[14a] Diols **45** are converted into the corresponding cyclic thienocarbonates on reacting with thiophosgene in the presence of a catalytic amount of DMAP. These compounds undergo ring-opening reaction with NaN_3. In the case of **46b** and **46c** the nucleophilic attack occurs exclusively at the α-position in high yields.[14b] The synthesis of glycofuranosyl azides **50** can be achieved by the regioselective ring-opening of cyclic 1,2-thiocarbonate sugars **49** with NaN_3.[14c]

Bicyclic sulfonium salt **52** obtained by a 5-*exo*-tet cyclization of the 1,3-dithiane tosyl derivative can be used to generate alkyl azides[15] (Scheme 3.6). The authors observed that the nucleophilic attack on **52** by azide ion kinetically favours a ring-opening reaction

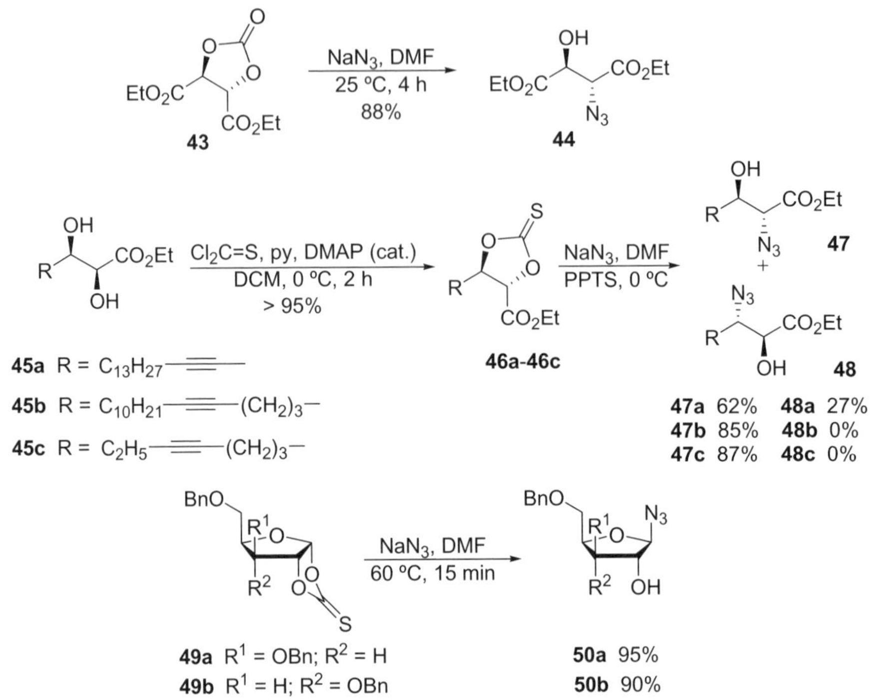

Scheme 3.5 *Synthesis of alkyl azides from cyclic carbonates and thiocarbonates*[14]

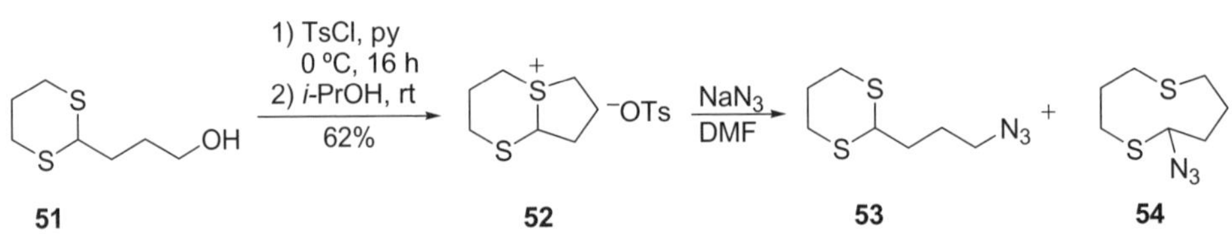

Scheme 3.6 *Synthesis of alkyl azides from a sulfonium salt*[15]

giving a nine-membered α-azidosulfide **54**, being 2-(3-azidopropyl)-1,3-dithiane (**53**) the thermodynamic product.

The classic reaction conditions to carry out the synthesis of alkyl azides with alkali azide, mainly sodium azide, involve the use of polar solvents (typically DMF or DMSO) to provide homogeneity. However, this type of solvents in some cases leads to difficulties regarding azide isolation. In some cases the higher solubility of lithium azide in organic solvents may be an advantage compared to other alkali metal azides. In the reported total synthesis of ustiloxin D to introduce the C3 nitrogen functionality, the authors converted the alcohol **55** into the azide compound by activation of the hydroxyl group with trifluoromethanesulfonate followed by the reaction with anhydrous lithium azide in 1,3-dimethyl-3,4,5,6-tetrahydro-2(*H*)-pyrimidinone (DMPU)[16] (Scheme 3.7).

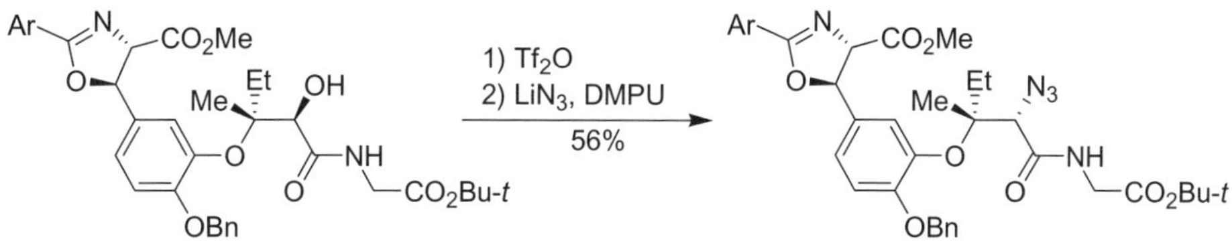

55 **56**

Scheme 3.7 *Synthesis of alkyl azides using LiN_3*[16]

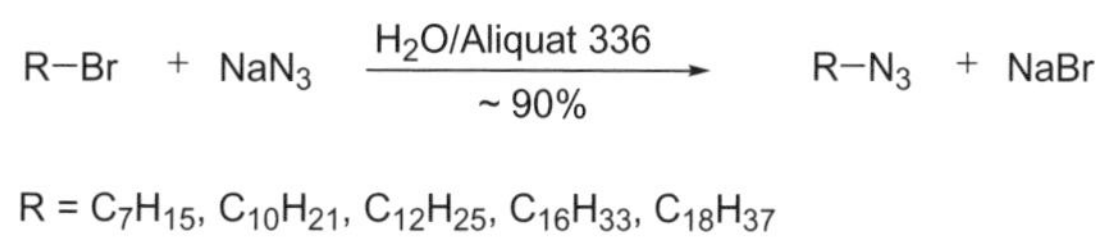

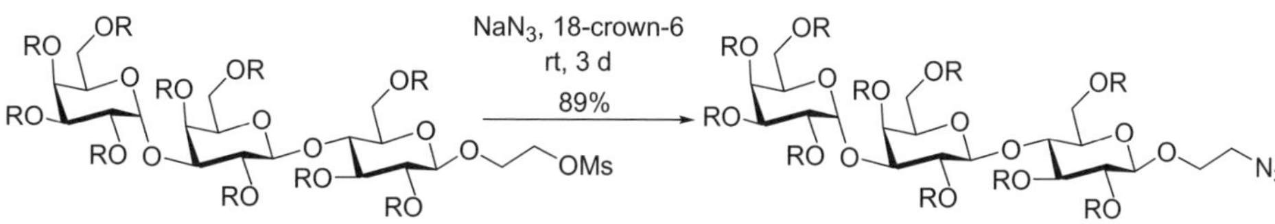

57 **58**

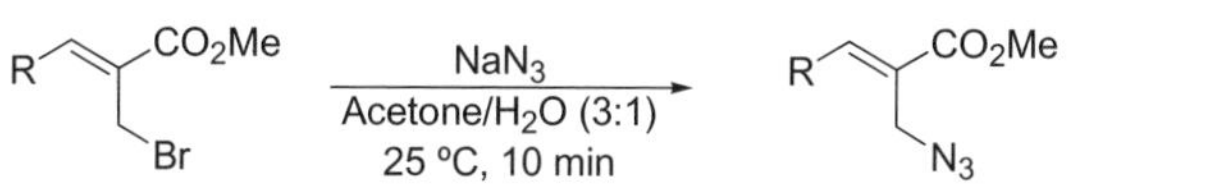

59 **60** 92-97%

Scheme 3.8 *Synthesis of alkyl azides via nucleophilic substitution carried out in various reaction media*[17]

Another approach to the synthesis of alkyl azides is the use of a phase-transfer catalyst (see Scheme 3.8). For instance, high yields are obtained when alkyl bromides are treated with NaN_3 in the presence of 'aliquat 336'.[17a] The use of crown ethers has also been described.[5b] An illustrative example is the reported synthesis of glycoside derivatives, which explores the mesylate displacement with sodium azide in 18-crown-6 to give azide **58**.[17b] More recently the synthesis of alkyl- and aryl-substituted (*E*)-2-(azidomethyl)alkenoates from the corresponding allylic bromides **59** in aqueous acetone has been reported[17c] (Scheme 3.8).

The use of organic azides (e.g. acetyl azide or trimethylsilyl azide) as the azide source allows to carry out azide synthesis under nonbasic conditions since they are soluble in organic solvents.[18a,18b] Trimethylsilyl azide ($TMSN_3$) reacts with activated primary halides (e.g. benzyl chloride, benzyl bromide, allyl bromide, chloroacetonitrile and ethyl chloroacetate) in HMPA at 60 °C to give alkyl azides in good yield.[5b] Tertiary and secondary cyclic azides (e.g. **62** and **64**) can be obtained from the chlorides or bromides by using $TMSN_3$ in the presence of stannic chloride.[18a] Acetyl azide (a DCM solution generated from the treatment of an aqueous solution of NaN_3 with acetyl chloride in DCM followed by the separation of the organic layer) has been used in the synthesis of the trinitro azide **66** from the bromo derivative **65**[18b] (Scheme 3.9).

The selection of the azide source and reaction conditions may influence the stereochemical outcome of reactions when chiral substrates are used. In fact, when α-bromo carboximide **67** was treated with sodium azide in DMSO at 0 °C, 9% epimerization was observed during the course of azide displacement. However, in the reaction of **67** with tetramethylguanidinium azide (TMGA) in DCM at 0 °C to room temperature, less then 1% epimerization was observed[18c] (Scheme 3.9).

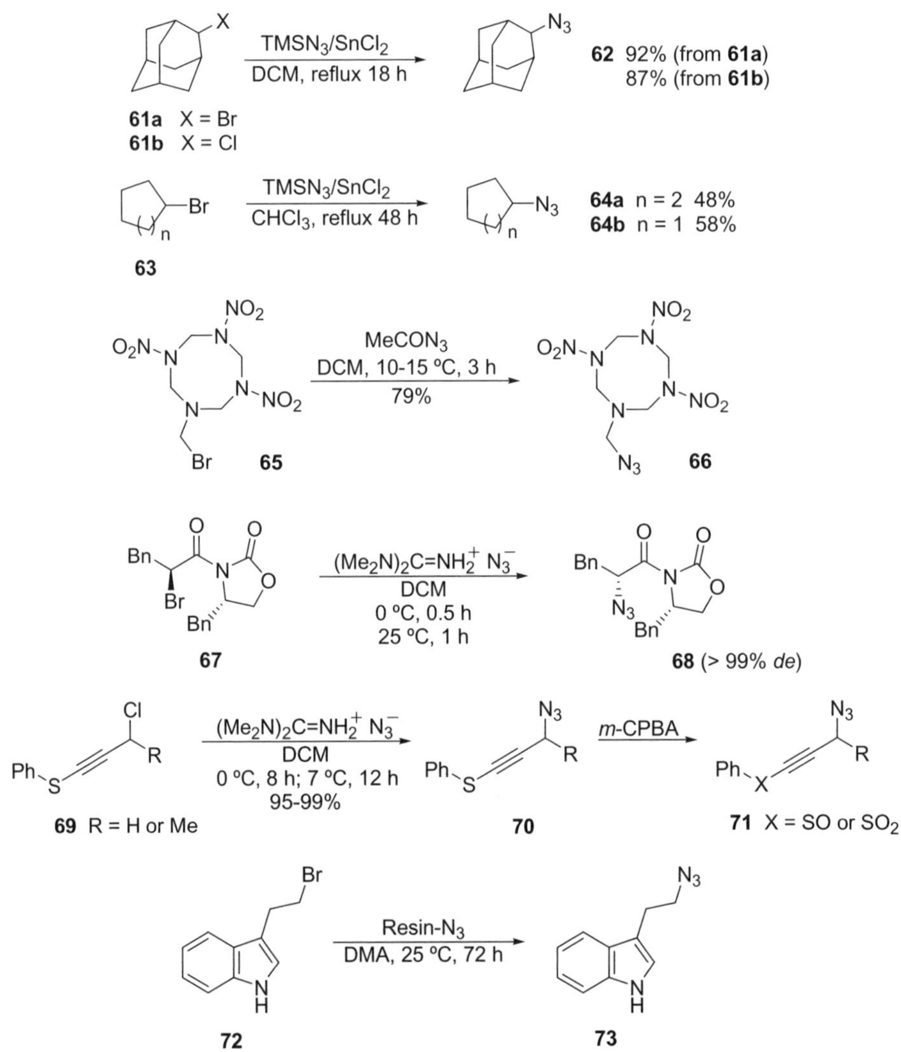

Scheme 3.9 *Synthesis of alkyl azides via nucleophilic substitution using various azide sources*[18]

Propargyl azides can be prepared by direct substitution of propargyl precursors although they isomerize easily to allenyl azides by [3,3]-sigmatropic migration of the azido group and by prototropic rearrangements which in turn can undergo further reactions.[18d,18e] However, in the case of propargyl azides bearing an acceptor substituent, an alternative synthetic approach must be used. Attempts to carry out the direct substitution reaction of propargyl precursors led to the synthesis of vinyl azides since the presence of the acceptor group increases the acidity of the propargyl hydrogen favouring prototropic isomerization of the starting material giving an allenic derivative, which then reacts with the azide source.[18f] To overcome this problem, the sulfur(II)-containing propargyl azides **69** were prepared by the reaction of the corresponding chloride with TMGA. Oxidation of the thioethers with *m*-CPBA allows the synthesis of sulfoxides or sulfones **71**.[18f] The use of polymer-bound azide sources is also known.[18g] Either Amberlite azide ion exchange resin or Merrifield resin supported tetra-alkyl ammonium azide can be used to convert alkyl bromides into the corresponding azides as illustrated by reaction of the indole derivative **72** which affords azide **73** in high yield in both cases. Removal of the resin by filtration gives a solution of the desired azide, which can be used for subsequent reactions (Scheme 3.9).

Trimethylsilyl azide undergoes reaction with tetrabutylammonium fluoride to generate the hypervalent trimethylfluorosilicate *in situ*, which is an efficient source of nucleophilic azide. The reaction of the silicate anion with acetoxy or benzyl ether protected sugars **74** bearing leaving groups such as bromide, chloride, triflate, tosylate, trichloroacetimidate and oxazoline affords glycosyl azides in good yields and predominantly with inversion of configuration.[19a] This methodology was applied to the synthesis of the *N*-glycan trisaccharide building block **77** with a terminal azide group.[19b] The stereoselective synthesis of this anomeric organic azide was achieved through oxazoline ring opening. Glycosyl azide **79** can also be obtained from thioglycoside **78** using 1-fluoro-2,6-dichloropyridinium triflate (**80**) and $TMSN_3$. This S_N1 reaction involves the generation of a glycosyl cation[19c] (Scheme 3.10).

Ionic liquids (IL) can be used as solvents for nucleophilic substitution reactions of alkyl halides or tosylates with NaN_3.[20a] The authors studied three ionic liquids (**84** and **85**), [bmim][PF_6], [bmim][$N(Tf)_2$], [hpyr][$N(Tf)_2$] (where bmim = 1-butyl-3-methyl-imidazolium, hpyr = 1-hexylpyridinium, PF_6 = hexafluorophosphate, $N(Tf)_2$ = bis(trifluoromethy lsulfonyl)imide). It was observed that nucleofugacity scales for this reaction are similar to those reported for the same process in cyclohexane. It was also observed that elimination reaction does not compete with substitution even in cases with sterically hindered substrates such as the triflate ester of diacetone-*D*-glucose **81**. The nucleophilic displacement on *n*-octyl mesylate (**86**) with potassium azide in a biphase system of supercritical carbon dioxide ($scCO_2$) and water, in the presence of catalyst Bu_4PBr is also an adequate medium for the synthesis of the corresponding azide **87**[20b] (Scheme 3.11).

Microwave radiation can be used in the synthesis of alkyl azides.[21] The microwave-assisted synthesis of β- and γ-azidoarylketones **89** from haloarylketones **88** and NaN_3 leads to acceleration in reaction rates and yields comparable to the ones using conventional heating.[21a] The microwave-enhanced nucleophilic substitution approach to alkyl azides (**91**, **93** and **95**) in aqueous medium from halides or tosylates and NaN_3 is also known.[21b] The authors observed that a variety of reactive functional groups are tolerated, namely ester, carboxylic acid and imide (Scheme 3.12).

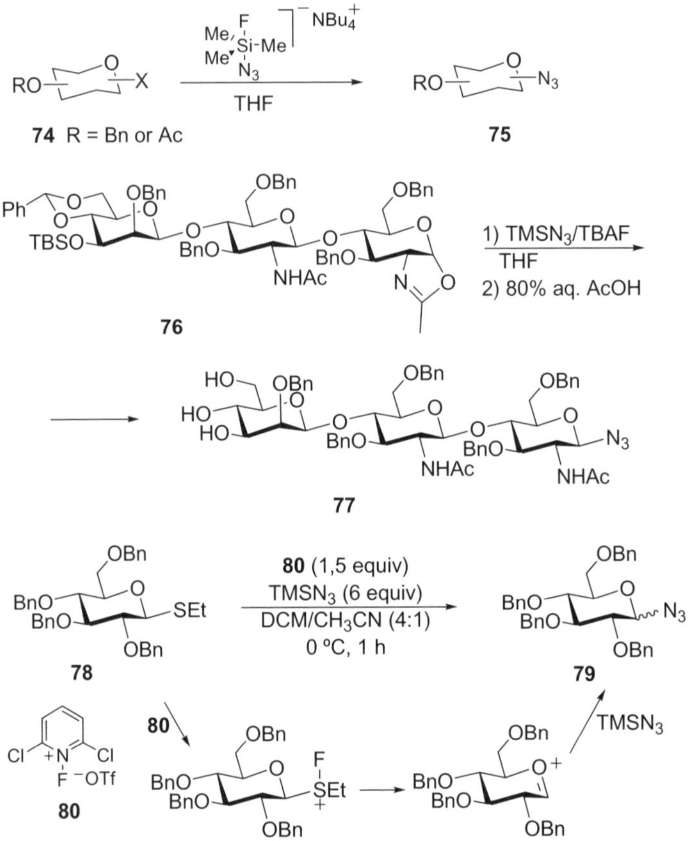

Scheme 3.10 *Synthesis of glycosyl azides*[19]

Multi-component reactions have recently received much attention as an elegant and rapid approach to functionalized molecules from simple building blocks. This synthetic strategy is particularly useful in azide chemistry since the *in situ* generation of organic azides circumvents the problems encountered with the handling of these compounds. Three-component reactions were used to prepare 1,4-disubstituted-1,2,3-triazoles from the corresponding alkyl halides, sodium azide and alkynes.[22] In the selected example shown in Scheme 3.13 the one-pot procedure for the direct conversion of α-bromo esters to 1,4-disubstituted-1,2,3-triazoles **97** was carried out in neutral aqueous solutions (pH = 7–8) at room temperature.[22a] The process involves the copper(I) catalyzed 1,3-dipolar cycloaddition between the *in situ* generated azide and terminal acetylenes. A microwave-assisted copper(I)-catalyzed three-component reaction was also applied to the synthesis of 1,4-disubstituted-1,2,3-triazoles[22b] (Scheme 3.13).

An extensive review on the use of hexadecyltributylphosphonium azide [n-$C_{16}H_{33}$(n-C_4H_9)$_3PN_3$] for the synthesis of unusual azides has been published recently.[23] This highly potent reagent allows extremely rapid nucleophilic substitution reactions and is appropriate for the synthesis of azides from unstable starting materials and compounds with steric hindrance or with internal strain.

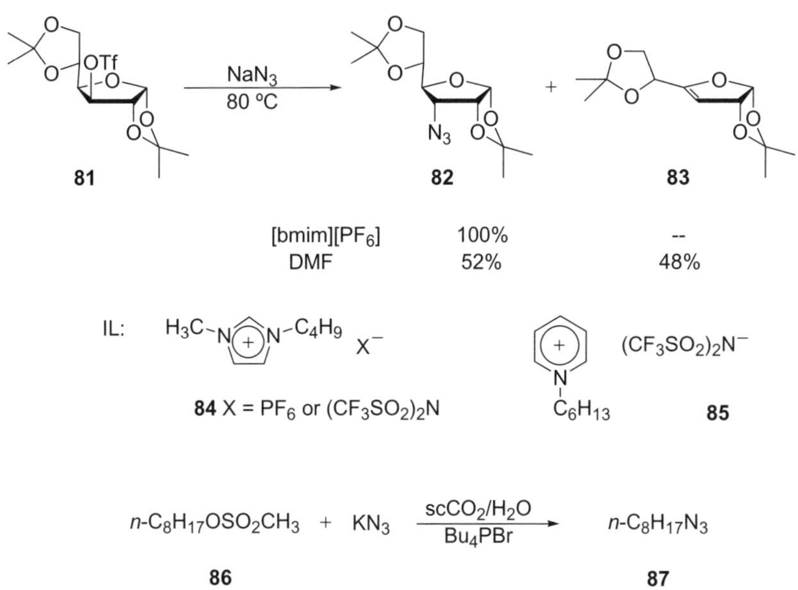

Scheme 3.11 *Synthesis of alkyl azides carried out in ionic liquids and supercritical carbon dioxide*[20]

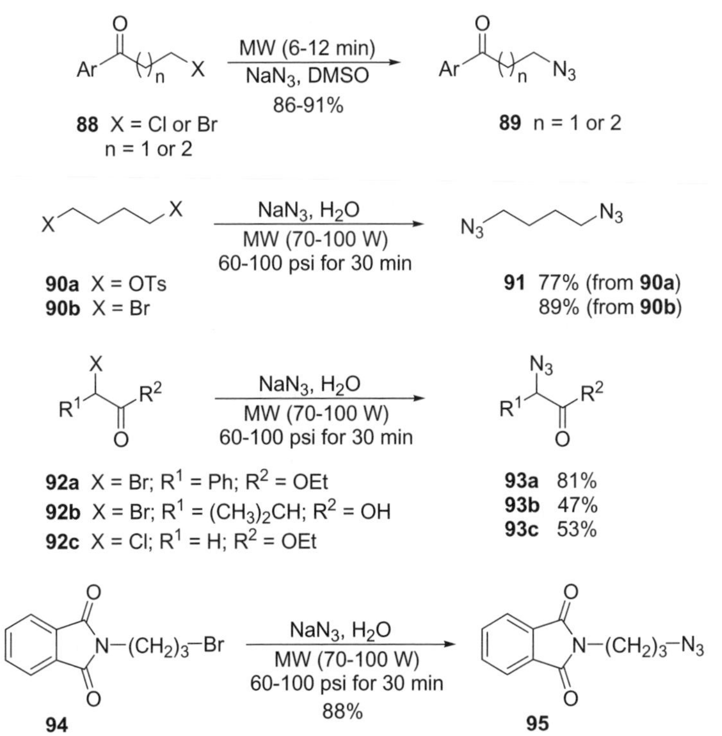

Scheme 3.12 *Microwave-assisted synthesis of alkyl azides*[21]

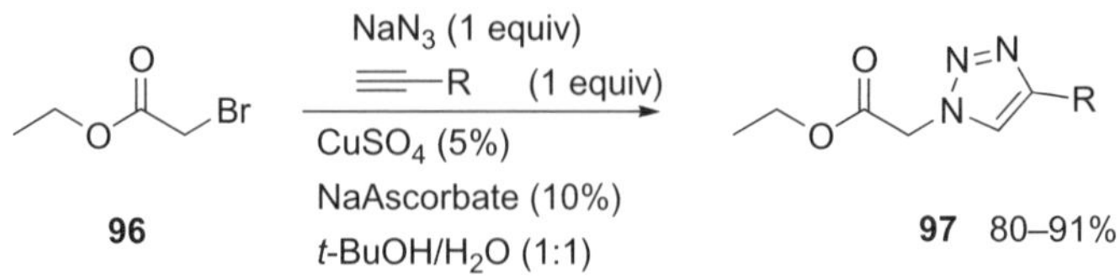

Scheme 3.13 In situ generation of alkyl azides[22]

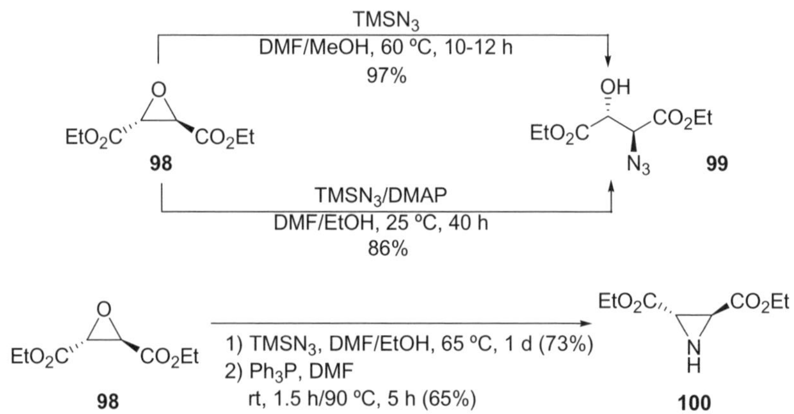

Scheme 3.14 Azides by ring opening of epoxides[24]

3.2.2 Azides by Ring Opening of Epoxides and Aziridines

The ring opening of epoxides represents an interesting route to β-azido alcohols, which in turn can lead to β-amino alcohols and aziridines[24] (Scheme 3.14). Thus, epoxysuccinate **98** is converted into β-azido alcohol **99** in 97% yield by cleavage with hydrazoic acid (HN_3) generated *in situ* from trimethylsilyl azide and methanol in DMF.[24a] The (2*S*,3*S*)-(+)-aziridine-2,3-dicarboxylate **100** was prepared in two steps via ring opening of epoxide **98** with $TMSN_3$ and EtOH in DMF, followed by treatment with triphenylphosphine in DMF.[24b] The observation by Moriwake *et al.* that the ring opening of the epoxide could be accelerated in the presence of amines led to the modification of this synthetic procedure.[24b,24c] In the presence of DMAP the reaction can be carried out at 25 °C giving the desired product in 86% yield. The ring opening of polymer-bound epoxides with sodium azide has also been reported.[24e]

The ring opening reaction of *meso*-epoxides can be carried out enantioselectively.[25] The first examples of enantioselective ring opening of achiral epoxides by azide nucleophiles were reported by Yamashita using metal(II) (2*R*,3*R*)-tartrate as heterogeneous chiral Lewis acid catalysts.[25c] An illustrative example is shown in Scheme 3.15. Epoxide **101** reacted with trimethylsilyl azide in the presence of Zn(II) (2*R*,3*R*)-tartrate to give *trans*-*O*-trimethylsilyl-2-azido alcohol **102** in 96% yield; the optical purity, however, was only 42% *ee*. Jacobsen *et al.* described a particularly interesting enantioselective ring opening of achiral epoxides involving (salen)Cr(III) complexes (e.g.

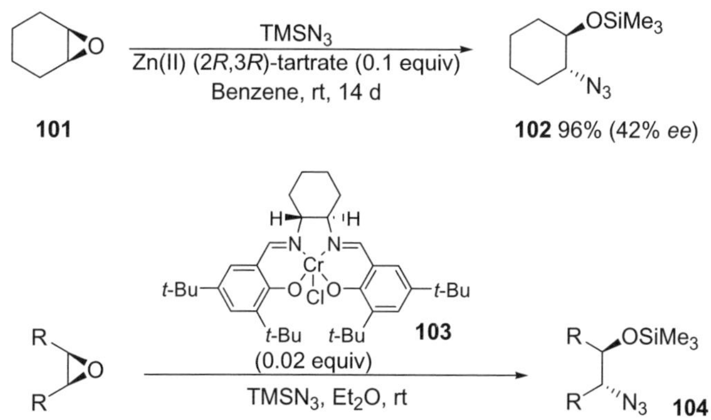

Epoxide	Time (h)	Yield (%)	*ee* (%)
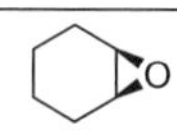 **101**	18	80	88
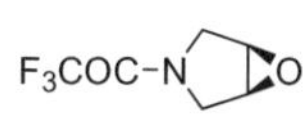 **105**	16	90	95
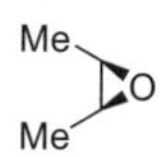 **106**	30	65	82

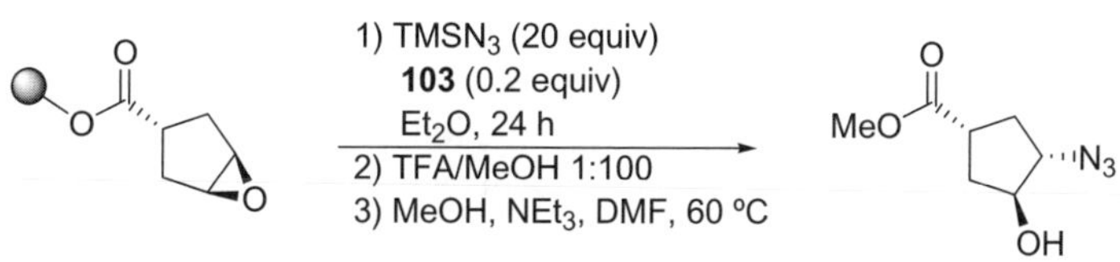

107 **108** 95% (92% *ee*)

Scheme 3.15 *Azides by ring opening of meso-epoxides*[25]

103).[25d] The alkyl azides **104** are obtained in good yield (55–90%) and high *ee* (81–98%). The authors observed that epoxides fused to five-membered rings undergo ring opening with higher levels of enantioselectivity then six-membered rings and monocyclic substrates were slightly less efficient. The asymmetric ring opening of *meso*-epoxides in the solid-phase synthesis of cyclic azido alcohols **108** is also known.[25h] The metal-catalysed asymmetric ring opening of epoxides is now a well-established method for the preparation of azides, the catalysts of choice being salen complexes with chromium as the central metal.

Kinetic studies on the mechanism of asymmetric (salen)Cr(III) catalysed ring opening of epoxides by $TMSN_3$ provide strong support for a mechanism involving catalyst activation of both nucleophile and electrophile by two different catalyst molecules.[25e] This observation led Jacobsen *et al.* to construct covalently linked dimeric salen complexes which would allow cooperative asymmetric catalysis.[25f] Accordingly, complexes **109** were found to catalyze the asymmetric ring opening of cyclopentene oxide by $TMSN_3$ producing the corresponding azide in high enantiomeric excess (90–94% *ee*) and were 1–2 orders of magnitude more reactive than the monomeric analogues. An allosteric catalyst with a Cr(III)-salen base active site, made possible through supramolecular coordination chemistry, was also used in the asymmetric ring opening of cyclohexene oxide with $TMSN_3$. This catalyst showed an increase in the rate and selectivity of the reaction, compared to the monomeric analogue.[25g]

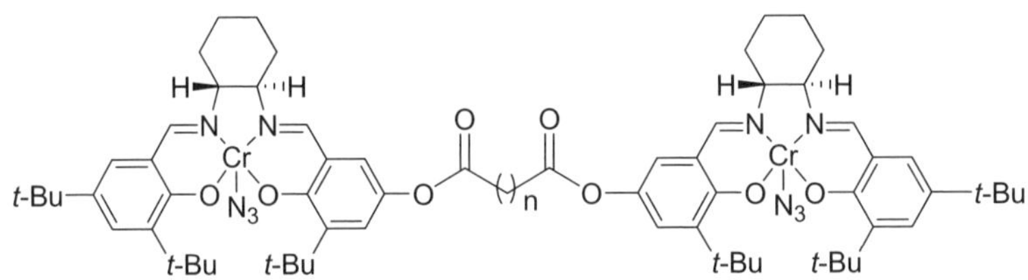

109 n = 2, 4, 5, 6, 7, 8 or 10

Scheme 3.15a

The use of immobilized (salen)Cr(III) complexes has been reported. The impregnation of the complex on silica resulted in a heterogeneous catalyst for the asymmetric ring opening of epoxides with $TMSN_3$.[26]

The halohydrin dehalogenase from Agrobacterium radiobacter AD1 acts as catalyst for the enantioselective and β-regioselective azidolysis of styrene oxides **110**.[27a] The kinetic separation of the racemate is achieved, leading to azido alcohol **111** in high *ee* together with the remaining epoxide also in high *ee*. The reaction of diastereoisomeric mixtures of bicyclic 1,2-epoxy-terpenes bearing C4-substituents with $TMSN_3$ and racemic (salen) Cr(III) complexes was also studied.[27b] One selected example is shown in Scheme 3.16.

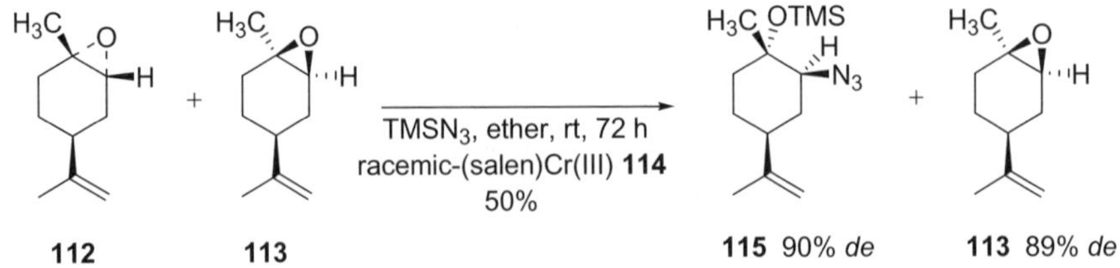

Scheme 3.16 *Ring opening of cyclic (−)-limonene 1,2-epoxide*[27b]

The reaction of the mixture **112** and **113** leads to the corresponding azido-product **115** together with 1,2-epoxy-terpene **113**. Therefore, **112** was selectively transformed and epoxide **113** was recovered. On the other hand, the authors could conclude that the diastereoselectivity observed was due to the presence of the C4-substituent at the starting epoxides, which forces the substrate into the more stable conformation and consequently blocks the approach of the Cr–N_3 species from one side.

Zeolite-bound sodium azide was also used to carry out the regioselective epoxide ring opening as illustrated by the reaction of **116**.[28] The reaction of epoxides with NaN_3 using 4 Å molecular sieves acting as catalyst has also been reported. Alkyl terminal epoxides undergo azide nucleophilic attack at the less hindered position whereas with α-phenylepoxides the attack occurs preferentially at the benzylic position (e.g. reaction of **119**). This methodology was applied to the synthesis of (−)-cytoxazone (**121**), a natural product possessing cytokine modulating activity.[29] Epoxide ring opening in a regioselective manner can also be achieved with $Ti(OPr\text{-}i)_2(N_3)_2$.[30] Using aluminum reagents such as diethylaluminum azide Markovnikow regioselectivity is observed. In fact, trisubstituted epoxides react with Et_2AlN_3 to form tertiary azides as the major product as illustrated by the synthesis of **123**[31] (Scheme 3.17).

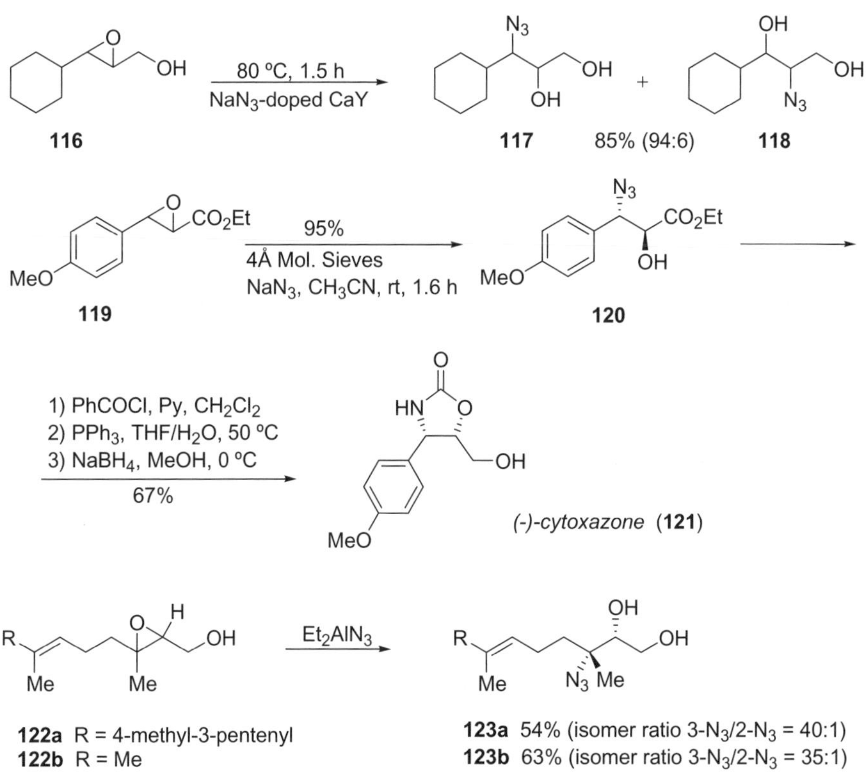

Scheme 3.17 *Ring opening of 1,2-epoxides*[28,29,31]

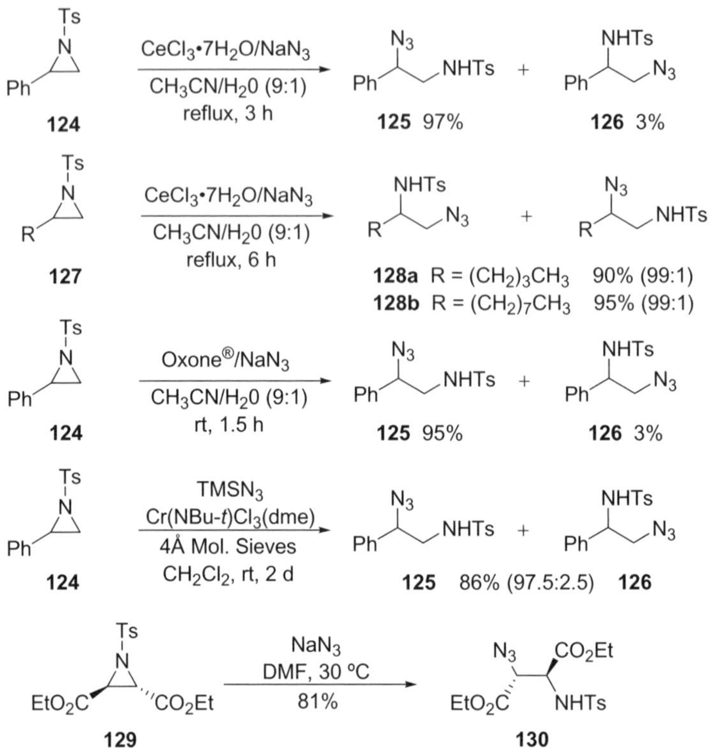

Scheme 3.18 *Ring opening of aziridines*[33]

The ring opening of aziridines can lead to valuable 1,2-diaminoalkanes.[32] The synthesis of 1,2-azidoalcohols and 1,2-azidoamines in high yields and regioselectivity has been achieved by using cerium(III) chloride and sodium azide in an acetonitrile and water mixture (9 : 1) for ring opening of epoxides and aziridines (e.g. aziridine **124**).[33a] 1,2-Azidoalcohols and 1,2-azidoamines are also obtained carrying out the reaction in the presence of copper ions[33b] or Oxone® in aqueous acetonitrile under mild reaction conditions.[33c] It has also been shown that 4 Å molecular sieves increase both the yield and the regioselectivity of imidochromium complex catalyzed addition of $TMSN_3$ to *N*-tosylaziridines.[33d] Tanner *et al.* carried out the opening of the non-racemic C_2-symmetric aziridines derived from tartaric acid (e.g. **129**) with sodium azide, which gave single adducts in high yield[33e] (Scheme 3.18).

Jacobsen *et al.* described the enantioselective ring opening of meso aziridines by $TMSN_3$ catalysed by tridentate Schiff base chromium complexes. Using **133** as the catalyst, high yields of conversion to azidoamines and high levels of enantioselectivity are obtained[34] (Scheme 3.19).

Bicyclic aziridines can also undergo ring opening on reacting with an azide source[35] (Scheme 3.20). In the case of bicyclic aziridines **133** the reaction with $TMSN_3$ gives the corresponding pyrrolidines in high yield and high diastereoselectivity. The initially

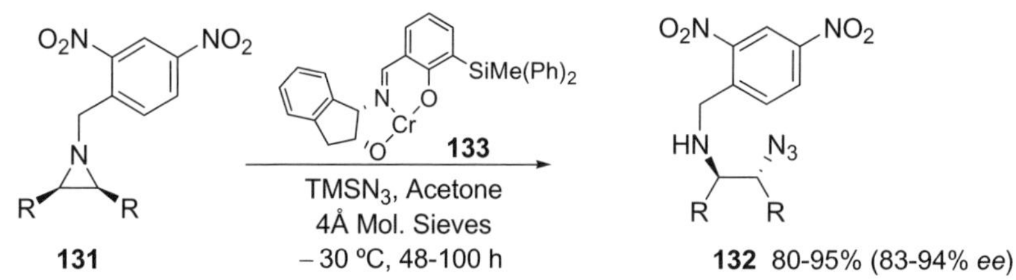

Scheme 3.19 *Asymmetric ring opening of aziridines*[26]

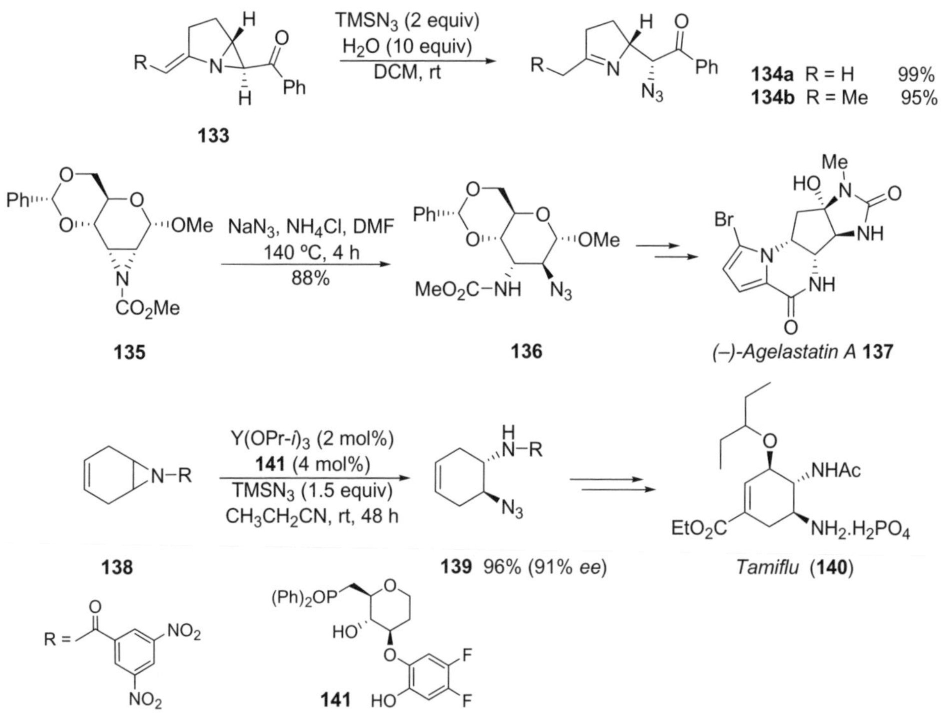

Scheme 3.20 *Ring opening of bicyclic aziridines*[35]

formed enamines are tautomerized *in situ* into the cyclic imines.[35b] Hale *et al.* described the total synthesis of alkaloid (–)-Agelastatin A (**137**), which included the regioselective *trans*-diaxial ring opening of aziridine **135** with sodium azide in hot DMF.[35c] Recently an asymmetric synthesis of Tamiflu using the catalytic enantioselective ring opening of a *meso*-aziridine with $TMSN_3$ and a yttrium complex of ligand **141** has been reported.[35d] The reaction is thought to proceed through the generation of a reactive yttrium azide formed from $TMSN_3$ by transmetalation and intramolecular transfer of the azide to the acylaziridine activated by a Lewis acidic yttrium in the same poly yttrium catalyst.

3.2.3 Azides by the Mitsunobu Reaction

Alkyl azides can be obtained from alcohols via the Mitsunobu reaction in the presence of hydrogen azide.[36] This transformation involves the reaction of primary and secondary alcohols treated with hydrogen azide, triphenylphosphane and diethyl azodicarboxylate (DEAD). It is worth while noticing that secondary alcohols undergo inversion of stereochemistry. Lee *et al.* reported the asymmetric synthesis of 2,3-diamino-3-phenylpropanoic acid derivative **144**, an important building block for a variety of biologically active compounds. Starting with alcohol **142** the second amino group was introduced using the Mitsunobu reaction.[36c] Spino *et al.* described the synthesis of chiral allylic azides, which were used to prepare homochiral α-amino acids, heterocycles and carboxycles.[36d] The Mitsunobu reaction of allylic alcohols **145** with hydrogen azide gave exclusively the allylic azides **146** with stereoselectivity ranging from 92 : 8 to 98 : 2.[36d] The authors rationalized this result considering that the products are formed from a normal S_N2-selective Mitsunobu reaction followed by a [3,3]-sigmatropic rearrangement rather than a S_N2'-displacement. However, Carell *et al.* demonstrated that in the case of allylic alcohol **147** the [3,3]-sigmatropic rearrangement of the initially formed allylic azide intermediate could be suppressed.[36e] In fact, they observed that although the allyl azide equilibration was very fast at room temperature, it was suppressed at 0 °C. Thus, carrying out the reaction at 0 °C the chiral allylic azide **148**, a building block for the transfer-RNA nucleoside quenosine, was efficiently obtained. The Mitsunobu reaction can be used to modify solid supports in order to widen their application for the construction of combinatorial peptide/chemical libraries. Therefore, although solid-phase synthesis is possible with the hydroxyl group of poly(ethylene glycol)-grafted polystyrene (PS-g-PEG) resin, its replacement by an amino group leads to a more useful solid support. In fact, the PS-g-PEG-OH **149** was converted into PS-g-PEG-NH_2 **150** via Mitsunobu/Staudinger reaction followed by hydrolysis of the corresponding iminophosporane (Scheme 3.21).[36f]

The explosive hydrogen azide can be replaced by less dangerous diphenylphosphoryl azide (DPPA).[37] In the presence of this reagent, triphenylphosphane and diisopropyl azodicarboxylate (DIAD) the *anti* homoallylic alcohol **151** is converted into the *syn* azide **152** in 95% yield.[37c] Also using DPPA, a one step synthesis of *N*-Boc-*cis*-4-azido-*L*-proline methyl ester (**154**) has been reported using protected *trans*-4-hydroxyl-*L*-proline **153** under Mitsunobu conditions.[37f] The reaction of *N*-Boc-*cis*-4-hydroxyl-*L*-proline methyl ester affords *N*-Boc-*trans*-4-azido-*L*-proline methyl ester in high yield. Jiang *et al.* reported the enantioselective total synthesis of marine indole-alkaloid hamacanthin B (**163**) which was based on the asymmetric synthesis of (*S*)-2-azido-2-(indol-3-yl)ethylamine **160** *via* Mitsunobu reaction.[37g] This alkyl azide was coupled with 2-(6-bromo-1*H*-indol-3-yl)-2-oxoacetyl chloride followed by an intramolecular Staudinger-aza Wittig cyclization to form the central dihydropyrazinone ring. The substitution of hydroxy groups under Mitsunobu conditions can also be carried out in the solid phase with DPPA.[37i,37j] The synthesis of polymer-supported diphenylphosphoryl azide is known and has been successfully applied to the conversion of carboxylic acids to urethanes and ureas through Curtius rearrangements (Scheme 3.22).[37k]

Thompson *et al.* reported that the conversion of alcohols to the corresponding azides with inversion of configuration could be achieved using DPPA and DBU, where DBU acts as a base in the conversion of the alcohol into the corresponding phosphate interme-

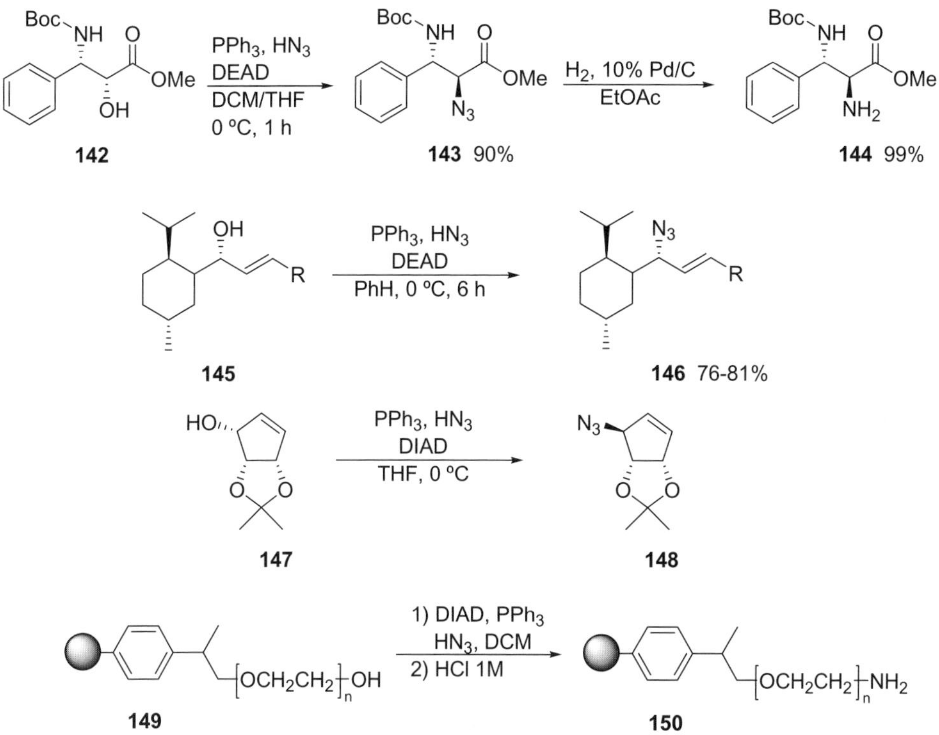

Scheme 3.21 *Alkyl azides via Mitsunobu reaction in the presence of hydrogen azide*[36c,36d,36e,36f]

diate[38a] (Scheme 3.23). Modifications using bis(2,4-dichlorophenyl)chlorophosphate/NaN_3/4-(dimethylamino)pyridine[38b] and bis(*p*-nitrophenyl)phosphoryl azide/DBU have been reported.[38c] One example is the synthesis of (*S*)-(−)-1-phenyl-ethyl azide (**165**) in good yield with inversion of configuration.[38c] Alternatively, azidation of alcohols via Mitsunobu-type substitution can be carried out with zinc azide/bis-pyridine complex as the azide source.[38d–38g] This approach was applied in the synthesis of alkyl azide **167**, which is a precursor of alkaloid (−)-Lasubine II (**168**) and was obtained from alcohol **166** in 81% yield.[38g] Another method for the preparation of alkyl azides (e.g. **170**) is the reaction of alcohols with a reagent combination of sodium azide, tetrabromomethane and triphenylphosphane.[38h,38i] Alcohols, as well as thiols and silyl ethers, are converted into alkyl azides by treatment with PPh_3/DDQ/n-Bu_4NN_3 in DCM at room temperature.[38j] Chiral *tert*-alkyl azides (e.g. **173**) are formed with inversion of configuration by treating *tert*-alkylphosphinites, prepared from the corresponding alcohols, with $TMSN_3$ in the presence of methoxybenzoquinone (MBQ).[38k]

3.2.4 Alkyl Azides from Amines

In the previously described methods for the synthesis of alkyl azides the azide group was introduced by formation of the C–N bond. However, primary aliphatic amines may be

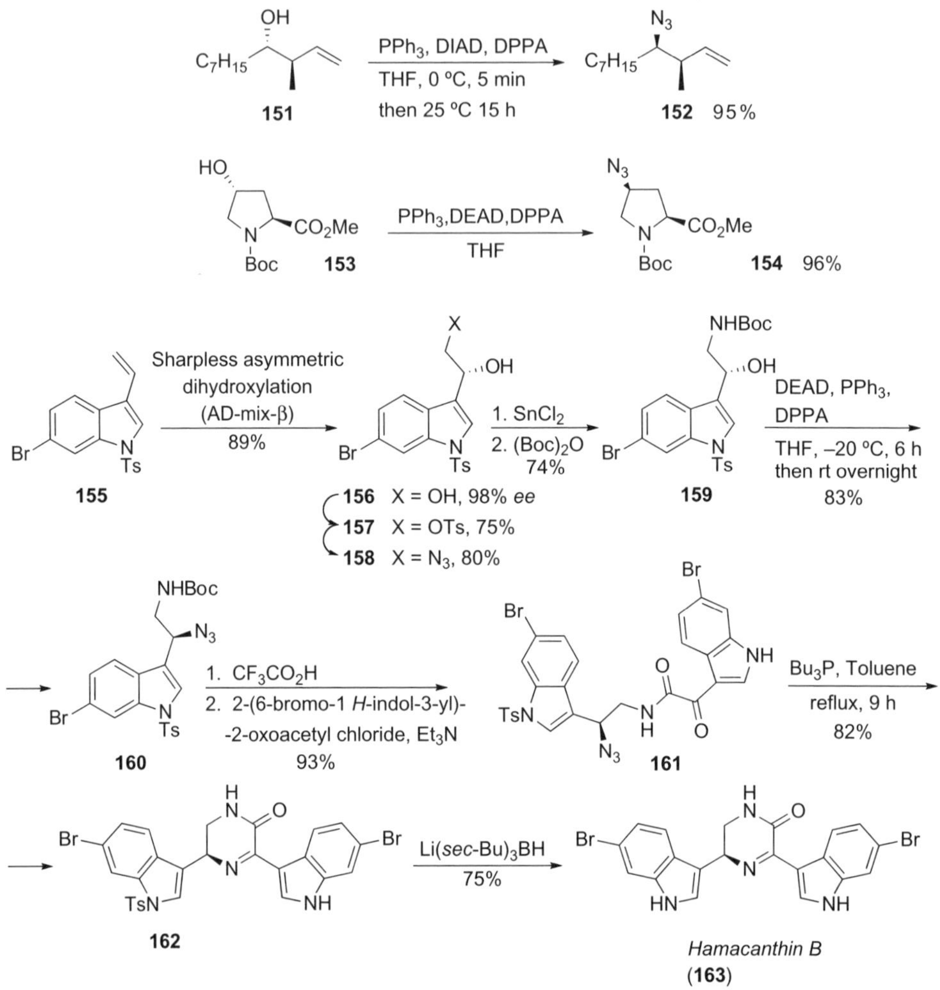

Scheme 3.22 *Alkyl azides via Mitsunobu reaction in the presence of diphenylphosphoryl azide*[37c,37f,37g]

converted into the corresponding azides by the diazo transfer reaction (Scheme 3.24). The most commonly used 'diazo-transfer reagent' is triflyl azide (TfN_3), which can be prepared from trifluoromethanesulfonic anhydride and sodium azide.[39] Triflyl azide is prepared in solution due to the explosive nature of neat TfN_3 and its short shelf life. Azido cubanes **175** have been obtained by this approach, through the treatment of the cubyl amines **174** with triflyl azide in the presence of triethylamine affording the desired alkyl azides.[39c] Diazo-transfer reaction on β-lactams have also been reported.[39d] The reaction of 6-amino penicillanate **176** with TfN_3 carried out in toluene and NEt_3 gives 6-azidopenicillanate **177** in 70% yield. Wong *et al.* reported a significant improvement of this methodology by carrying out the diazo transfer reaction using divalent Cu ions as

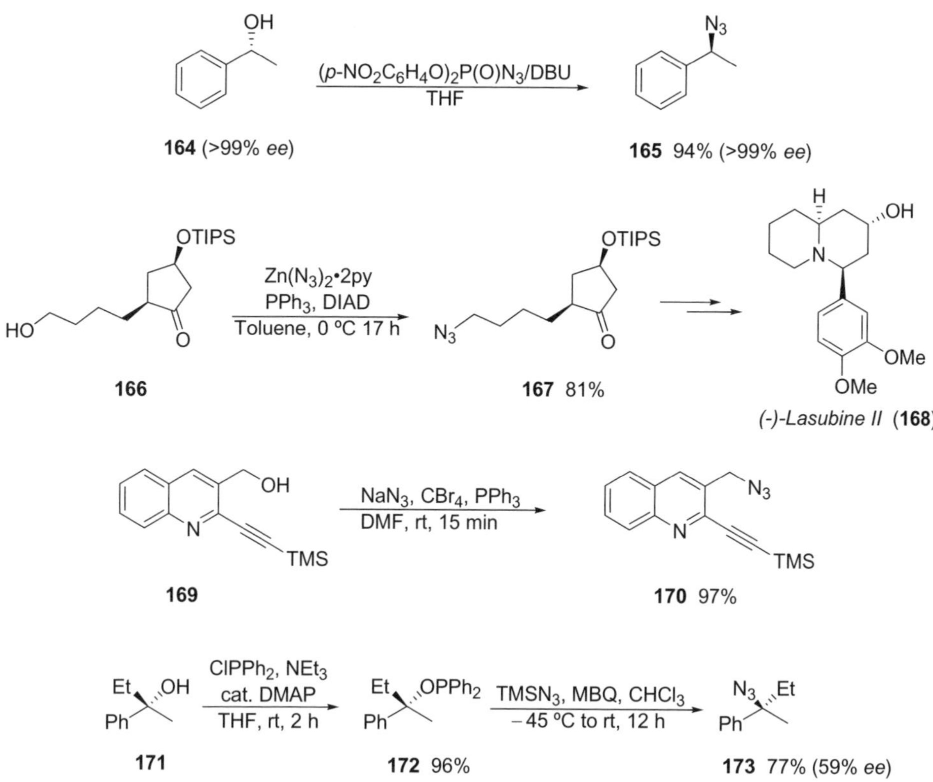

Scheme 3.23 *Alkyl azides via derivations of the Mitsunobu reaction*[38e,38f,38h,38k]

catalysts.[39e,f] The copper(II)-catalyzed diazo transfer method was applied to the synthesis of α-azido acids (**179**) from the corresponding α-amino acids with complete retention of configuration.[39g,39h] These building blocks can be used to prepare small peptides on Wang resin[39g] or can be converted into 1,2,3-triazoles on reacting with alkynes.[39h] Alkyl azides generated by this procedure have also been used in the synthesis of aminoglycosides, an important group of antibiotics.[39i,39j] In the Wong procedure the triflyl azide is prepared from trifluoromethanesulfonic anhydride and sodium azide using a biphasic DCM/H_2O mixture leading to a solution of the reagent in DCM. A three solvent system H_2O/MeOH/DCM may be used in the diazo transfer reaction in order to have a homogeneous phase. Ye *et al.* reported that triflyl azide can also be prepared in CH_3CN or pyridine and the resulting solution can be added directly to the amine solution for subsequent diazo transfer.[39k] In this way, hydrolysis of the reagent is minimized allowing the reduction of the required amount of NaN_3 and Tf_2O. The synthesis of **181** is an illustrative example of the application of this protocol. Ernt *et al.* described that, replacing DCM by toluene in the Wong procedure, the formation of hazardous side products such as azido-chloromethane and diazidomethane could be avoided.[39l] Solid supported azides have been prepared from amino-functionalized solid supports via the diazo transfer reaction.[39m,39n]

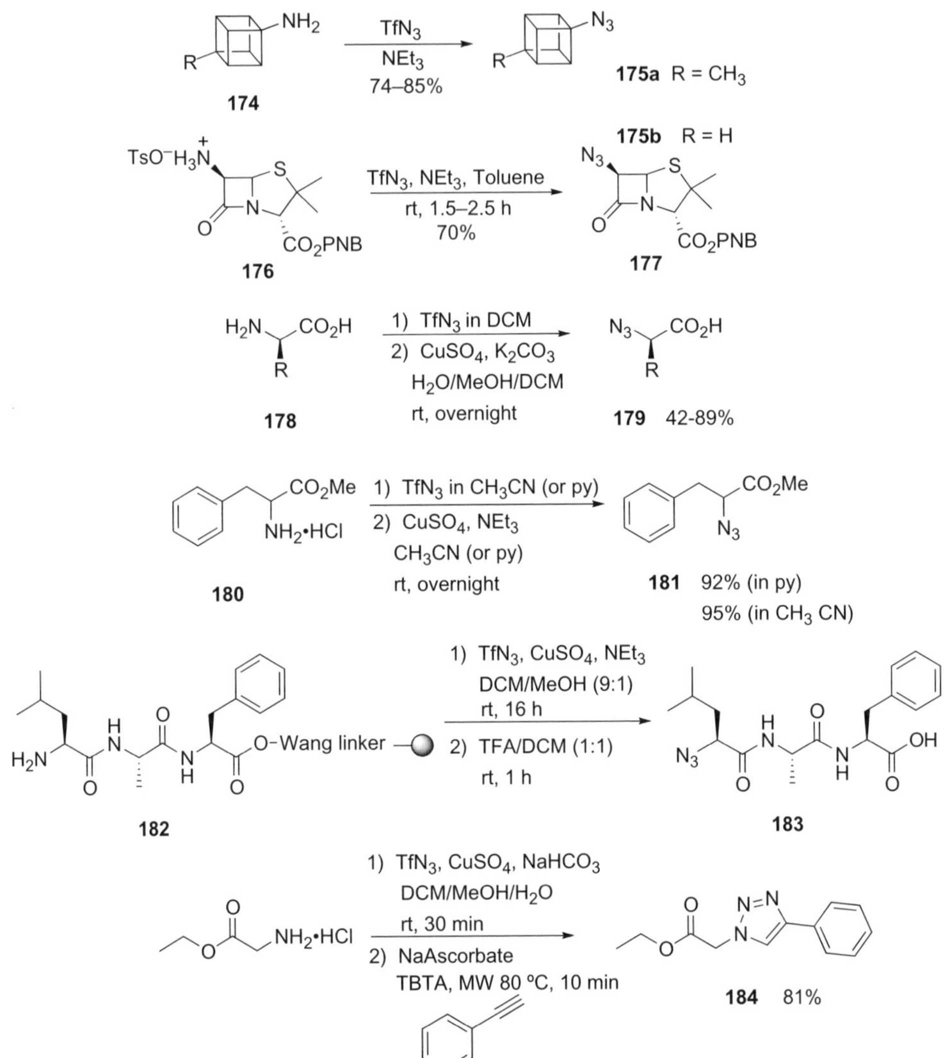

Scheme 3.24 Alkyl azides via the diazo transfer reaction using triflyl azide[39c,39f,39j,39l]

Using a derivation of the Wong methodology Liskamp *et al.* described the conversion of solid phase bound peptide amines into azides, which were cleaved from the resin to give azido peptides **183** in good yields.[39m] The generation of alkyl azides, via Cu(II)-catalyzed diazo transfer reaction to amines using TfN_3, can be carried out *in situ* followed by azide-alkyne cycloaddition. This one-pot procedure avoids the isolation of intermediate azides, giving access to a variety of triazoles (e.g. **184**).[39o]

Recently the use of imidazole-1-sulfonyl azide hydrochloride (**185**) as a diazo transfer reagent has been reported[40] (Scheme 3.25). This compound is a crystalline, shelf-stable

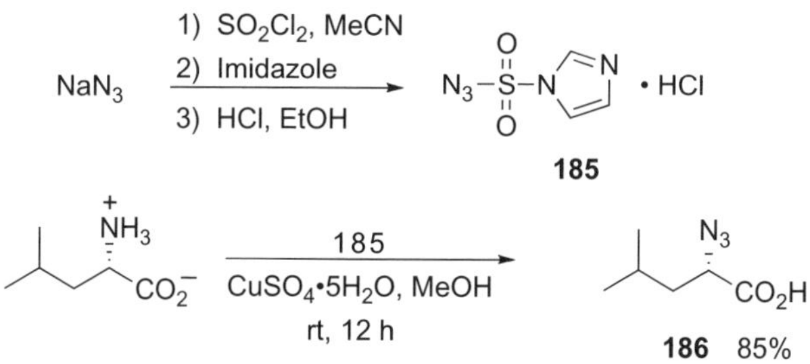

Scheme 3.25 *Alkyl azides via the diazo transfer reaction using imidazole-1-sulfonyl azide hydrochloride*[40]

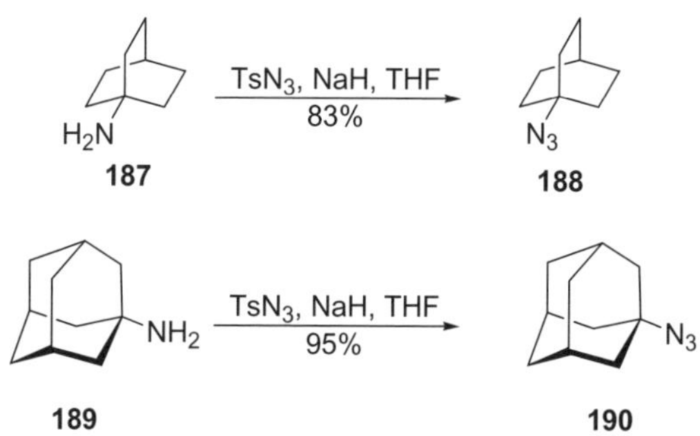

Scheme 3.26 *Alkyl azides from amines by treatment with tosyl azide*[41]

solid and was found to equal triflyl azide in its ability to act as a diazo donor in the conversion of primary alkyl and aryl amines into azides (e.g. synthesis of **186**).

Alkyl azides can alternatively be obtained from amines by treatment with tosyl azide and sodium hydride.[41] This procedure allows for the synthesis of bridgehead azides such as 1-azidobicyclo[2.2.2]octane (**188**) and 1-azidoadamantane (**190**) in 83% and 92% yield, respectively (Scheme 3.26).

3.2.5 Alkyl Azides from Carbon Nucleophiles and Electron-poor Sulfonyl Azides

Azide group transfer can be achieved starting with carbon nucleophiles and electron-poor sulfonyl azides[42] (Scheme 3.27). Thus, azido indole **192** was prepared in 62% yield by deprotonation of the malonate substituent with sodium hydride and subsequent reaction with tosyl azide.[42b] Enolates can also undergo electrophilic azidation upon reacting with arylsulfonyl azides.[42c–42e] In fact, the direct azidation of the potassium enolates bearing a chiral auxiliary, derived from **193**, with 2,4,6-triisopropylbenzenesulfonyl azide (trisyl azide) gives the corresponding alkyl azides with high stereoselectivity.[42c] The azido transfer to carbanions of β-lactams is a synthetic strategy explored to functionalize the α position, namely the stereospecific introduction of an amino group.[42f–42j] The β-lactam malonate **195** was converted into azido derivative **196** in 80% yield by treatment with LDA, followed by the reaction with TsN_3 and then TMSCl.[42g]

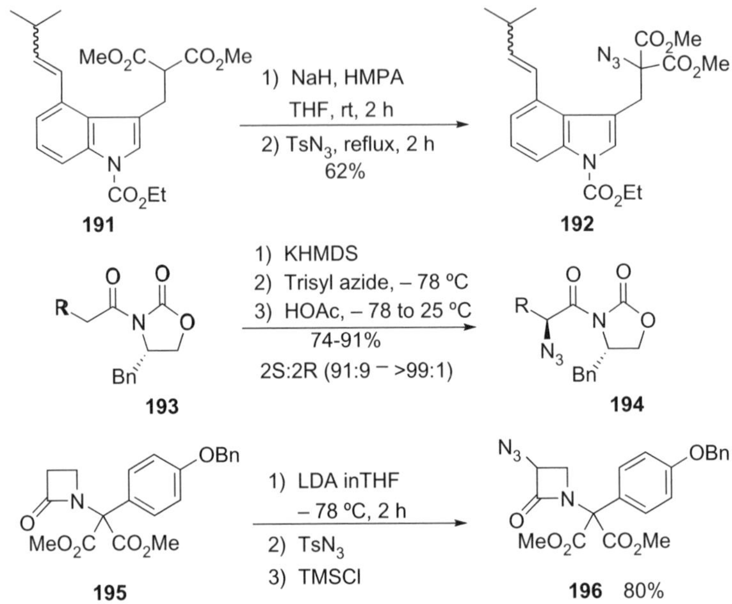

Scheme 3.27 *Alkyl azides from carbon nucleophiles*[42b,42c,42g]

3.3 Synthesis of Aryl Azides

3.3.1 Nucleophilic Aromatic Substitution: S_NAr Reactions

The synthesis of aromatic azides can be achieved via displacement with azide ions of aromatic systems bearing an electron-withdrawing group, either in the *ortho* or the *para* position relative to the potential leaving group.[43] Fluorinated aryl azides are among the most widely used photoaffinity probes to study protein structure and function and a variety of these compounds with functionalities suitable for attaching them to the desired structure have been prepared. Thus, the perfluorophenyl azides **198** were obtained in good yields from the reaction of the corresponding pentafluorophenyl compounds with sodium azide[43a] (Scheme 3.28). The perfluorophenyl azides **198d** and **198f** have been used as precursors for multi-step synthesis of *p*-azidotetrafluorophenylaniline (**199**), perfluorophenyl azides containing a chemically reactive electron-donating amino group *para* to the azido substituent.[43b]

Displacement of a leaving group (e.g. halides, sulfonates or nitro groups) in activated heteroaryl compounds utilizing azide ion is also possible as in the synthesis of azido-1,3,5-triazines **201**,[44a] azidopyridines **203**,[44b,44c] azido-quinolones **205**,[44d] azido-1,3,4-oxadiazoles **207**,[44e] 5-azido-2-acylfurans **209**[44f] and 3-azido-5-phenyl-1,2,4-oxadiazole (**211**).[44g] Still another example is the synthesis of 2,5,8-triazido-*s*-heptazine (**213**), an energetically unstable molecule due to its high azide content. This heteroaryl azide is completely conjugated, is comprised of only carbon and nitrogen, exhibits visible light photoluminescence and rapidly decomposes at 185 °C to nitrogen-rich carbon nitrides (Scheme 3.29).[44h]

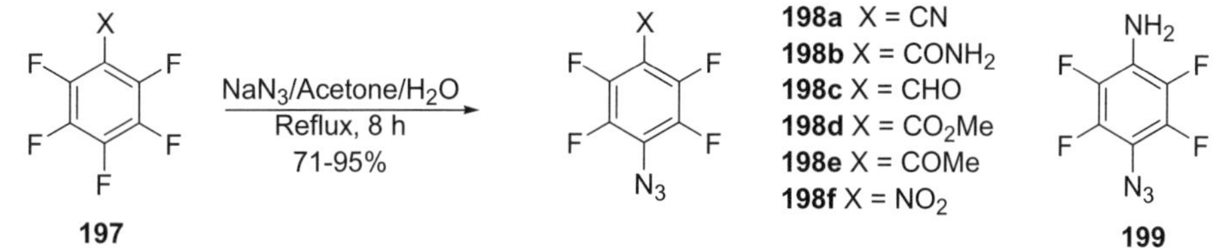

Scheme 3.28 *Aryl azides via nucleophilic aromatic substitution of electron-deficient arenes*[43]

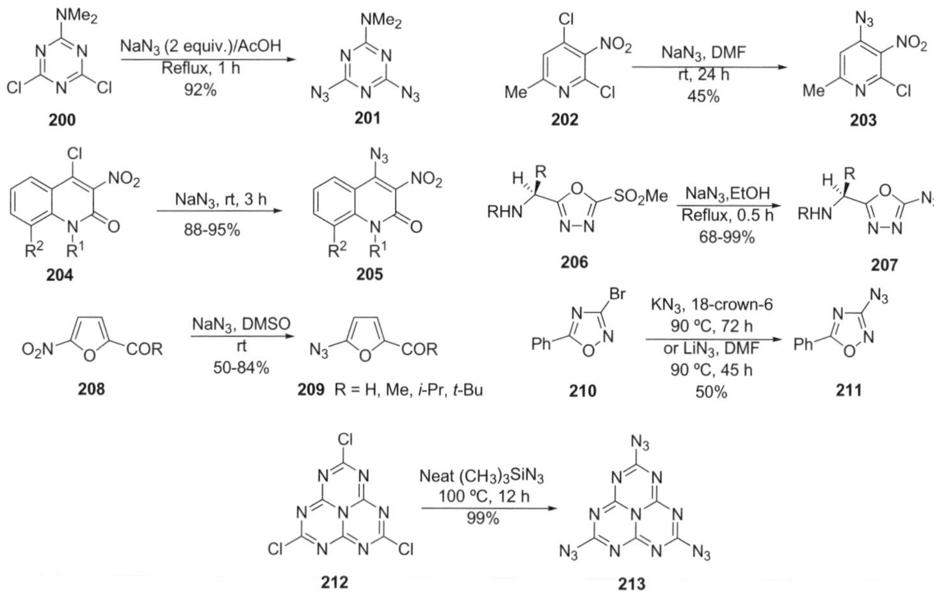

Scheme 3.29 *Heteroaryl azides via nucleophilic aromatic substitution*[44]

The cleavage of heteroaryl sulfones from polymeric supports can be achieved by reaction with azide ions. Suckling *et al.* applied this strategy to the synthesis of pteridines.[45] In fact, the starting pyrimidine was linked to polystyrene via a thioether **214**. After construction of the pteridine ring system, the activation of the sulfur linker by oxidation to the sulfone with dimethyldioxirane followed by nucleophilic substitution with sodium azide affords the target molecule **218** in 41% overall yield (Scheme 3.30).

Nucleophilic substitution can be carried out with electron-rich arenes, although it requires the presence of appropriate leaving groups in the aromatic system such as thallium substituents. This strategy was applied in the total synthesis of (–)-indolactam V (**221**), an indole alkaloid isolated from *Streptomyces blasmyceticum*[46a] (Scheme 3.31). The key steps involve regiospecific thallation of the starting acylindole, followed by copper(II)triflate-mediated displacement of thallium with sodium azide and reduction to introduce the 13-amino group. A variety of aryl azides have been obtained from aryl iodides, including deactivated aryl iodides, with sodium azide via *L*-proline-promoted

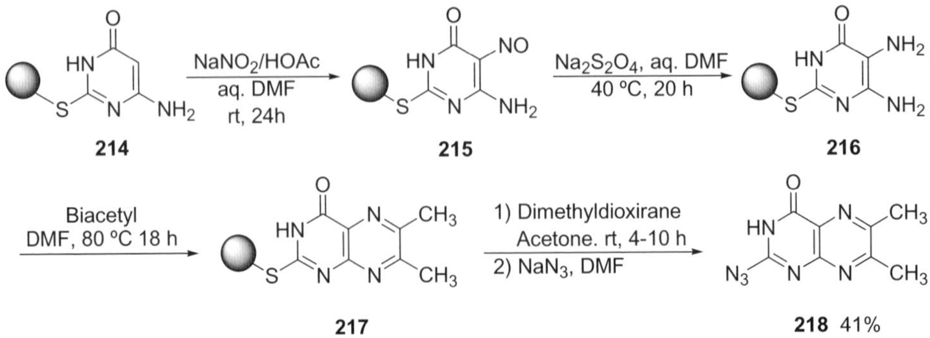

Scheme 3.30 Solid-phase synthesis of heteroaryl azides via nucleophilic aromatic substitution[45]

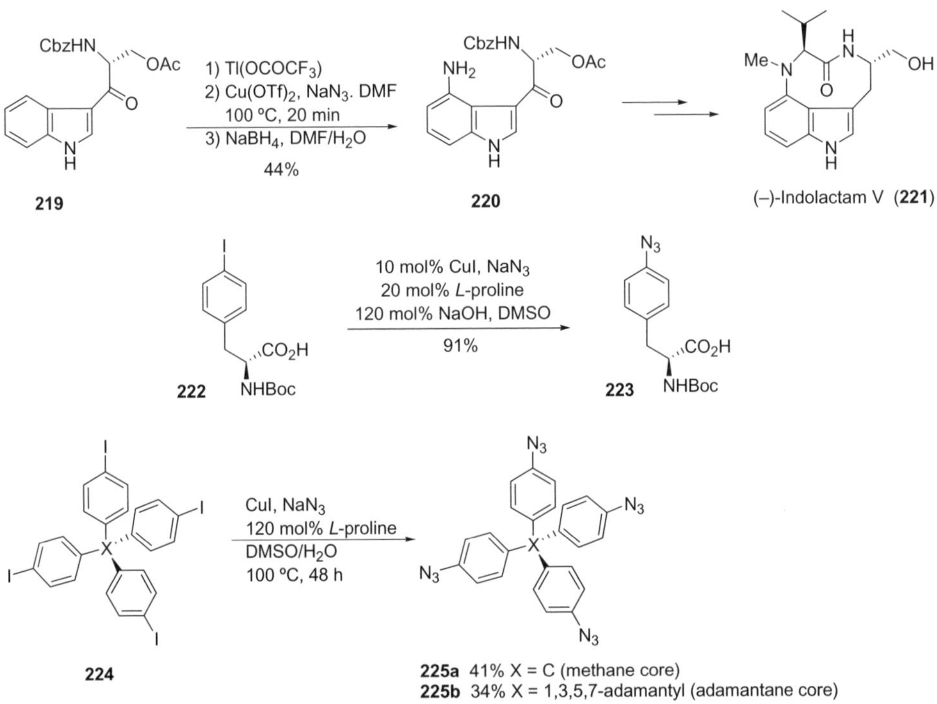

Scheme 3.31 Aryl azides via nucleophilic aromatic substitution of electron-rich arenes[46]

CuI-catalyzed coupling reactions.[46b] An example is shown in Scheme 3.31, the synthesis of aryl azide **223** from *L*-phenylalanine-derived iodide **222** in 91% yield. A similar approach allows the synthesis of stable organic polyazides, based on methane and adamantane cores **225**. Despite the high energetic properties, these rigid structures can be handled without special precautions.[46c]

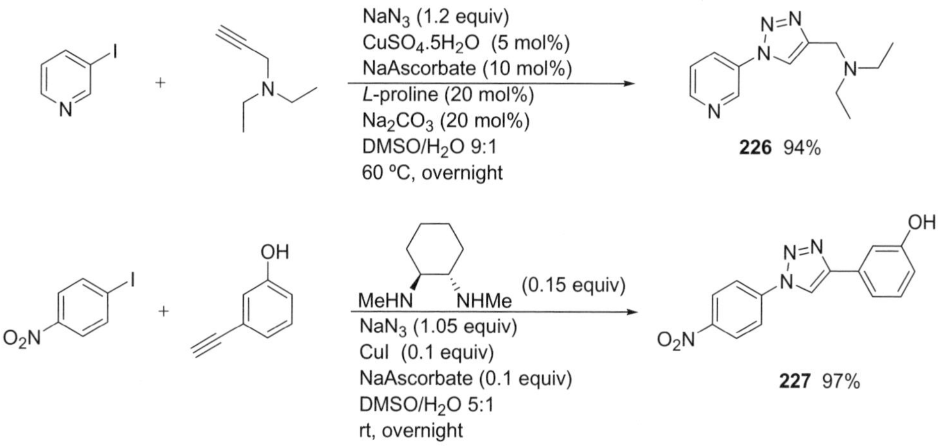

Scheme 3.32 *One-pot synthesis of 1,4-disubstituted 1,2,3-triazoles via cycloaddition of aryl and heteroaryl azides prepared in situ from the corresponding aryl halides*[47a,b]

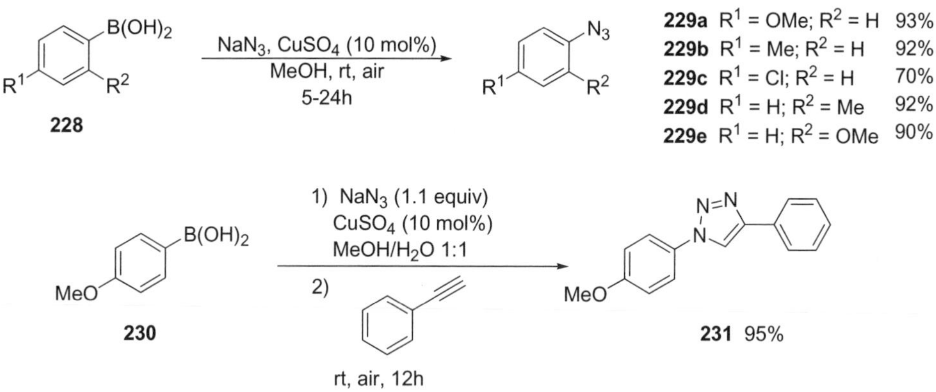

Scheme 3.33 *Aryl azides via nucleophilic aromatic substitution of aryl boronic acids*[48]

The *in situ* preparation of aryl and heteroaryl azides from the corresponding aryl halides via *L*-proline-promoted CuI-catalyzed coupling reactions in the presence of alkynes allows the one-pot synthesis of 1,4-disubstituted 1,2,3-triazoles (e.g. **226**).[47a] Liang *et al.* also reported the one-pot synthesis of 1,4-disubstituted 1,2,3-triazoles (e.g. **227**) from aryl bromides or iodides and terminal alkynes in the presence of sodium azide using *diamine*-promoted CuI-catalyzed reactions.[47b] It has also been shown that this type of synthesis can be carried out in a mixture of the ionic liquid [bmim][BF_4] and water (Scheme 3.32).[47c]

Starting with boronic acids, the catalytic approach to aryl azides and 1-aryl-1,2,3-triazoles can be carried[48] out under milder reaction conditions and improved substrate tolerance (Scheme 3.33).[48] In fact, it was demonstrated that both electron-rich and electron-poor aryl boronic acids **228** could be efficiently converted into the corresponding aryl azides (**229**) in the presence of sodium azide and $CuSO_4$. A one-pot protocol to synthesize 1-aryl-1,2,3-triazoles directly from boronic acids and alkynes has also been reported (e.g. **231**).

3.3.2 Aryl Azides from Diazonium Compounds

Diazotization of aromatic and heteroaromatic amines and subsequent treatment with sodium azide has been used for the synthesis of aryl azide (Scheme 3.34). This reaction does not involve the cleavage of the carbon-heteroatom bond. Instead, after the addition of the azide onto the diazonium ion, an open pentazene or a cyclic pentazole is formed, followed by the release of nitrogen to yield the aryl azide.[49a,49b] This approach can be applied to the high yield conversion of 3-amino-2-phenylindazole **232** into the corresponding 3-azido derivative **233**.[49c] It was also observed that sodium nitrite in 80% aqueous acetic acid was a good diazotizing agent for aminoindoles **234**, and that the resulting diazonium salts reacted with sodium azide to give azidoindoles **235a–235c** and azidotryptophane **235d**.[49d] The synthesis of 4-azido-2,3,5,6-tetrafluorobenzoic acid (**237**) can also be carried out starting with amine **236** by diazotization in TFA followed by nucleophilic displacement by azide ion.[49e] Decomposition of diazonium salts into the corresponding aryl azides was also applied in the synthesis of azido-labeled thalidomide analogue **239**, which shows activity comparable to that of thalidomide.[49f] In a similar way aryl azide **241**, the precursor of new triazole calix[6]arene ligands, was obtained in quantitative yield.[49g] Diazotization/azide treatment reaction was applied to convert amine **242** into 8-(4-azidophenyl)-4,4-difluoro-4-bora-3a,4a-diaza-*s*-indacene **243** in 77% yield.[49h]

A one-pot protocol to obtain triazoles from aromatic amines has been reported.[50] The aryl azides, generated *in situ* from the corresponding amine with *t*-butyl nitrite and $TMSN_3$, participate in Cu(I)-catalyzed azide-alkyne 1,3-dipolar cycloaddition giving 1,4-disubstituted-1,2,3-triazoles **244** (Scheme 3.35).

Diazonium ions are readily available by cleavage of triazenes. Thus, various substituted triazene resins undergo cleavage in the presence of trimethylsilyl azide to give aryl azides.[51a,51b] In the case of polymer-bound triazene **245**, the resulting aryl azides can be converted to 1,4-benzodiazepin-5-ones **247** via intramolecular aza-Wittig reactions[51b] (Scheme 3.36). The triazene resin cleavage is achieved by treatment with 5% TFA in dichloromethane at room temperature affording the corresponding diazonium salts which react immediately with trimethylsilyl azide giving the aryl azides in moderate yields (21–62%). The preparation of polyfunctional aryl azides by the reaction of aryl triazenes with NaN_3 in the presence of $KHSO_4$ or $BF_3 \cdot OEt_2$/TFA has also been reported.[51c] This methodology was applied to the synthesis of biologically active compounds namely the antitumor agent 9-methoxyellipticine (**250**).

3.3.3 Aryl Azides from Organometallic Reagents

Tosyl azide reacts with Grignard or lithium reagents to form aryl azides[52] (Scheme 3.37). In fact, mesityl azide (**252**) can be obtained from the reaction of mesitylmagnesium bromide with tosyl azide via the formation of the corresponding tosyltriazene salt, followed by its fragmentation through treatment with aqueous sodium pyrophosphate.[52a] When 2,6-dimesitylphenyl iodide (**253**) is treated with *n*-butyllithium at 0 °C, and the resulting lithium salt further reacts with *p*-toluenesulfonyl azide, the aryl azide **254** is isolated in 96% yield.[52c] A similar strategy can be applied to the synthesis of heteroaryl

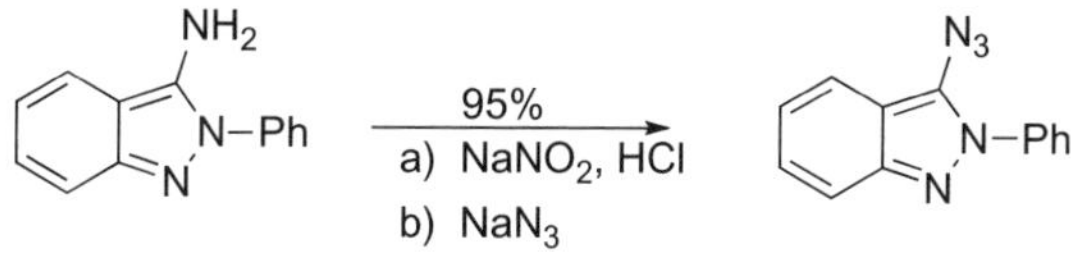

232 **233**

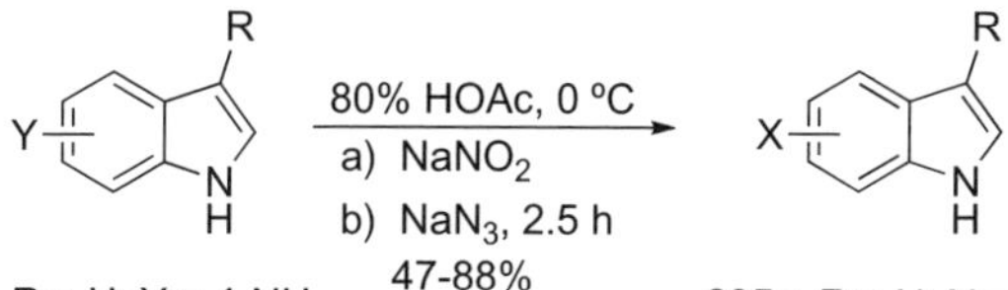

234a R = H; Y = 4-NH_2
234b R = H; Y = 5-NH_2
234c R = H; Y = 6-NH_2
234d R = $CH_2CH(NH_2)CO_2H$; Y = 6-NH_2

235a R = H; X = 4-N_3
235b R = H; X = 5-N_3
235c R = H; X = 6-N_3
235d R = $CH_2CH(NH_2)CO_2H$; X = 6-N_3

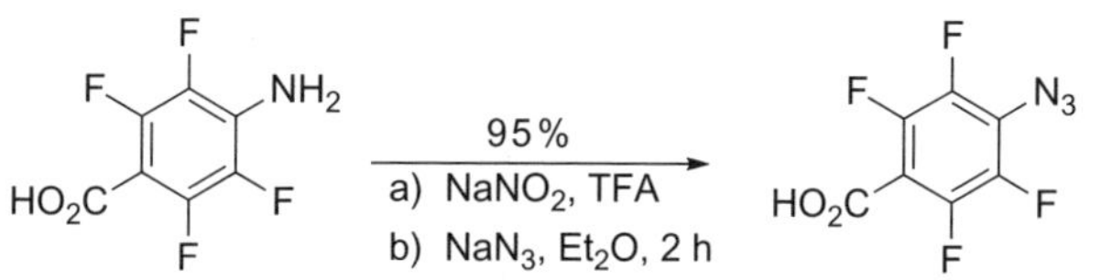

236 **237**

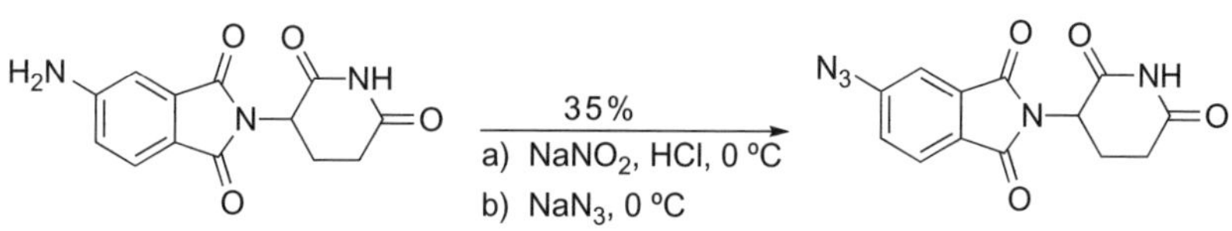

238 **239**

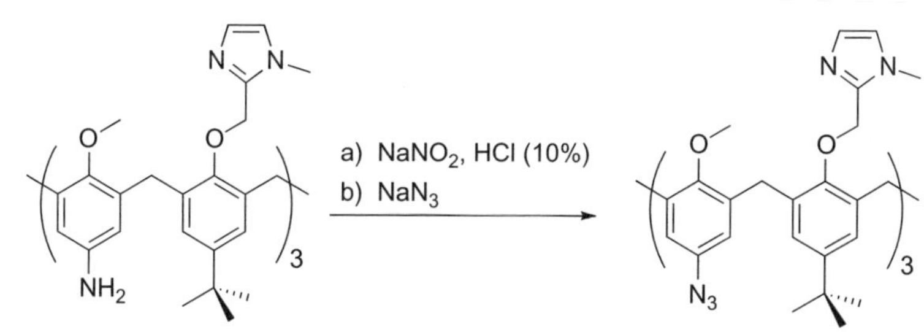

240 **241**

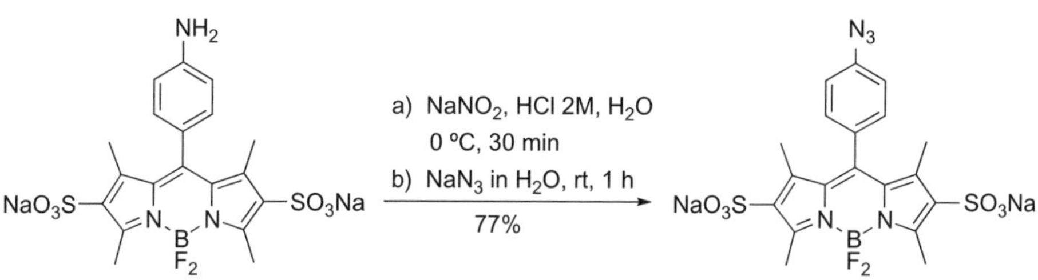

242 **243**

Scheme 3.34 *Aryl azides from diazonium compounds*[49]

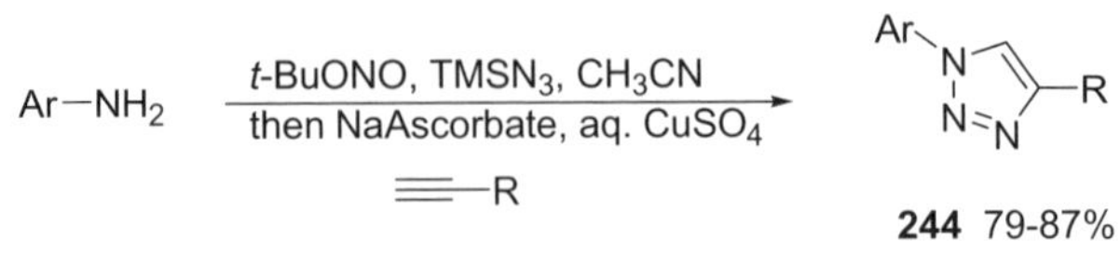

Scheme 3.35 *One-pot protocol to obtain triazoles from aromatic amines*[50]

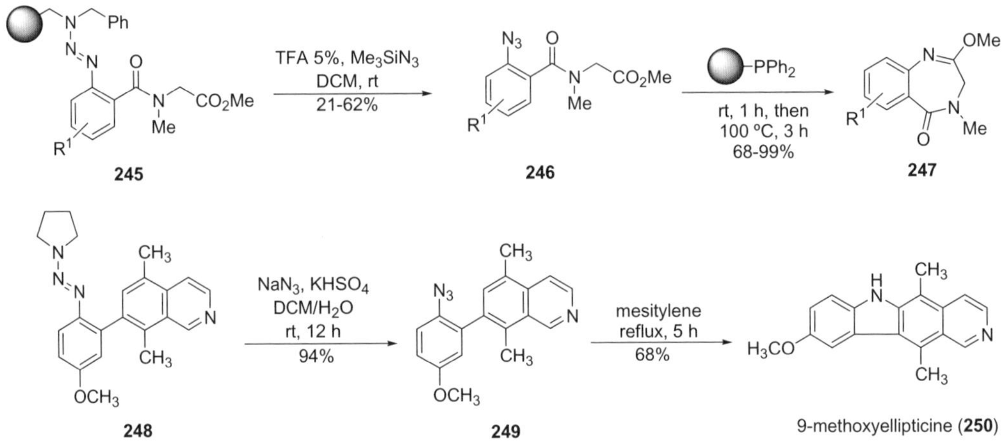

Scheme 3.36 *Aryl azides from triazenes*[51b,51c]

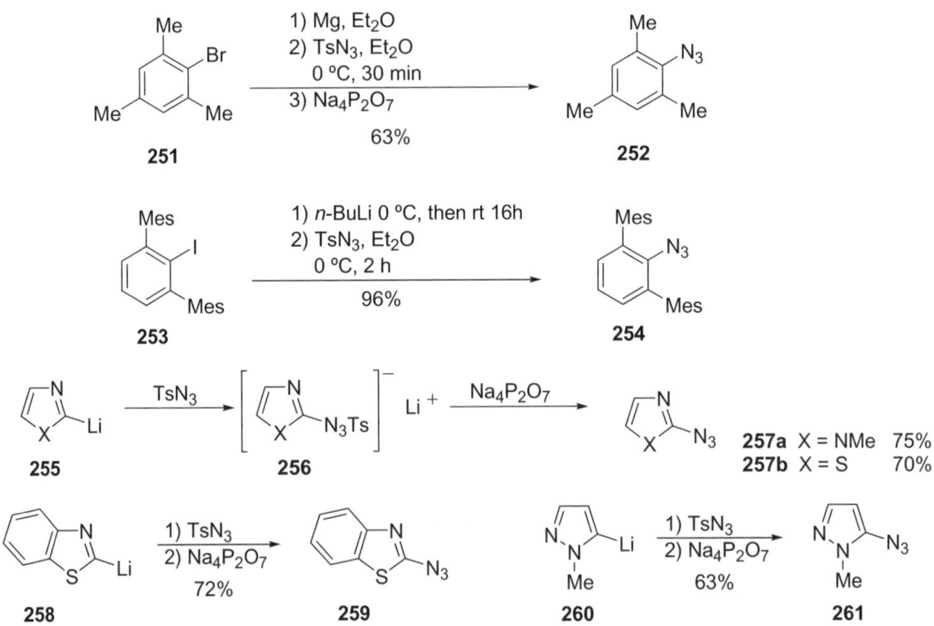

Scheme 3.37 *Aryl azides from organometallic reagents*[52]

azides. Zanirato *et al.* reported that 2-lithiated azoles (e.g. lithiated *N*-methylimidazole **255a**, 1,3-thiazole **255b** and benzo-1,3-thiazole **258**) and 5-lithiated azoles (e.g. *N*-methylpyrazole **260**) react with tosyl azide to afford the corresponding lithium triazene salt. The subsequent fragmentation of these salts produces the heteroaryl azides.[52d] The azido transfer reaction of the appropriate heteroaryllithium derivative and tosyl azide, followed by fragmentation of the corresponding tosyltriazene salt, was also used to prepare 2-azido and 3-azido-1-methylindole, azidothiophenes, 2-azido- and 3-azido selenophene.[52e–52g]

3.3.4 Aryl Azides by Diazo Transfer

Aryl azides and heteroaryl azides can be prepared by the reaction of aromatic amines with triflyl azide. This straightforward approach to aromatic azides can be carried out under mild reaction conditions. In fact, the reaction of 8-aminoquinoline (**262**) in dichloromethane/methanol occurs at room temperature in the presence of triethylamine and copper sulfate to afford 8-azidoquinoline (**263**) in 95% isolated yield.[53a] The reaction of aryl amide salts **265**, generated from the corresponding anilines **264** and strong bases, with tosyl azide also affords aryl azides **266** (Scheme 3.38).[53b]

Lithium derivatives of aromatic and heteroaromatic amines react with azidotris(diethylamino)phosphonium bromide (**267**) to give the corresponding azides (e.g. **268** and **269**) in high yield (Scheme 3.39).[54]

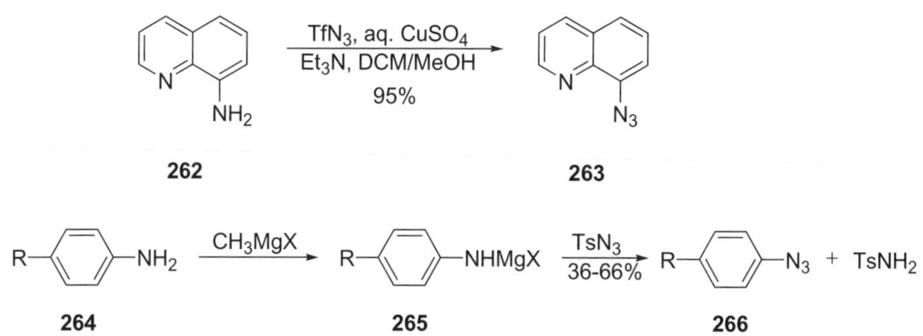

Scheme 3.38 *Aryl azides by diazo transfer*[53]

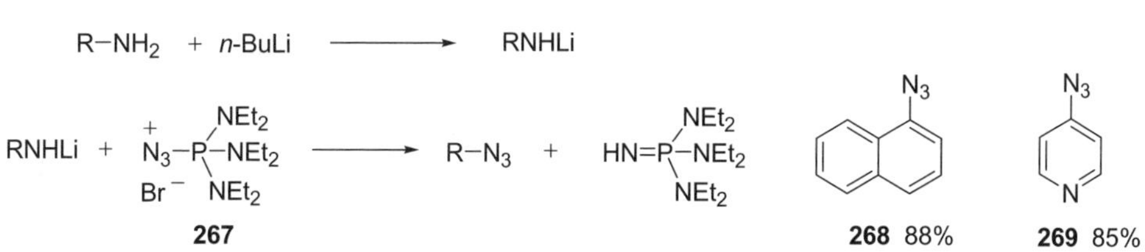

Scheme 3.39 *Aryl azides from amines by treatment with n-butyllithium and azidotris(diethylamino)phosphonium bromide*[54]

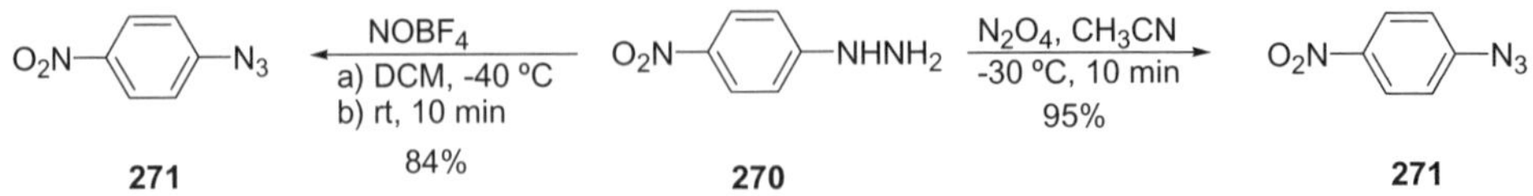

Scheme 3.40 Aryl azides by diazotization of hydrazines[55b,55c]

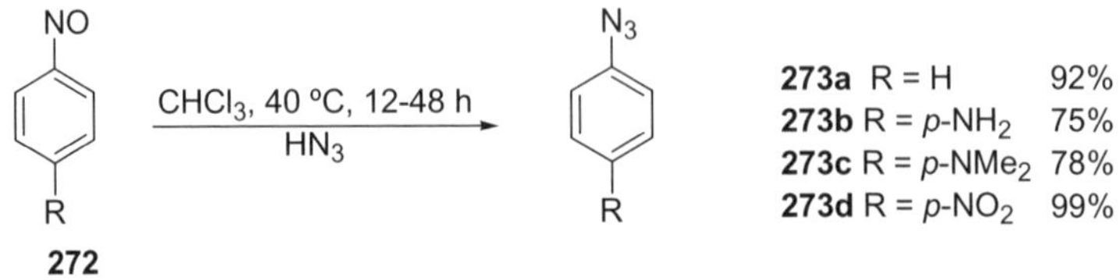

Scheme 3.41 Aryl azides from nitrosoarenes[57]

3.3.5 Aryl Azides from Hydrazines and from Nitrosoarenes

Aryl azides can be obtained from *N*-nitrosation reaction of aromatic hydrazines, using nitrosation reagents such as nitrous acid, dinitrogen tetroxide, nitrosonium tetrafluoroborate and nitric oxide in the presence of oxygen[55] (Scheme 3.40). The use of hydrazones is also possible.[56]

An efficient synthesis of aryl azides can be achieved by the reaction of nitrosoarenes with hydrogen azide[57] (Scheme 3.41). The drawback of this approach is that two equivalents of the explosive reagent is required since the process involves the initial formation of diazonium ions followed by the reaction with the azide ion.

3.4 Synthesis of Acyl Azides

3.4.1 Acyl Azides from Mixed Acid Chlorides

The synthesis of acyl azides by the acid chloride method has been used for the preparation of substituted furans.[58a,58b] The initial conversion of 2-furoic acids **274** into the corresponding acid chlorides with thionyl chloride followed by the reaction with sodium azide affords the acyl azides **275** in good overall yield. Under Curtius reactions conditions, isocyanates **277** are obtained, which can be converted into amidofurans **278**. The reaction of acyl azides **275** with alcohols affords furanamino carboxylates **276**. The synthesis of Fmoc amino acid azides **280** from the corresponding protected amino acids and sodium azide by the acid chloride method has also been reported.[58c] These are particularly interesting compounds as coupling agents in peptide synthesis. 3-(2-Thienyl)acryloyl chloride (**281**) reacts with sodium azide in a biphasic mixture of water and dioxane at 5 °C to give the acyl azide **282**.[58d] Thermally induced Curtius rearrangement of **282** into the isocyanate and subsequent electrophilic cyclization affords the bicyclic system **283**. The reaction of aroyl chlorides with $TMSN_3$ to form aroyl azides **285** can be carried out at 0 °C in DCM if a catalytic amount of ZnI_2 is used (Scheme 3.42).[58e]

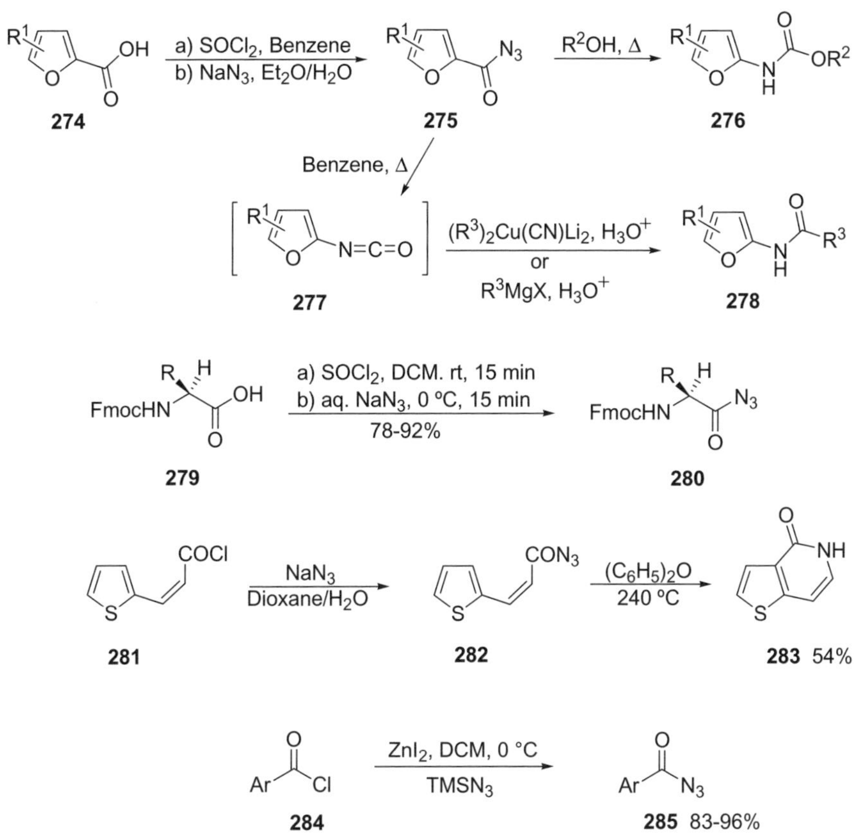

Scheme 3.42 Acyl azides from acid chlorides[58]

γ-Lactone **286** was transformed into aminomonocarba-disaccharide **288** using a five-step procedure involving the Curtius rearrangement of acyl azide **287** formed from the appropriate acid chloride (Scheme 3.43).[59]

3.4.2 Acyl Azides from Mixed Anhydrides

Kitahara *et al.* described the synthesis of enamides via Curtius rearrangement of α,β-unsaturated acyl azides and organometallic addition of the corresponding isocyanates.[60a] These enamines were used for the construction of the side chain moiety of naturally occurring enamides such as oximidines, lansiumamides A-C and lansamide-I.[60] The synthesis of lansamide-I (**292**) started with the treatment of cinnamic acid with $ClCO_2Et$ and NEt_3, followed by addition of sodium azide to give the acyl azide.[60b] The subsequent thermolysis gives the isocyanate, which reacts with styryl Grignard reagent affording **291** in 42% overall yield. Methylation of this compound leads to lansamide-I (Scheme 3.44). A similar methodology was also applied to the efficient total synthesis of the naturally occurring coscinamides, chondriamides, igzamide, salicylihalamide and apicularen.[61]

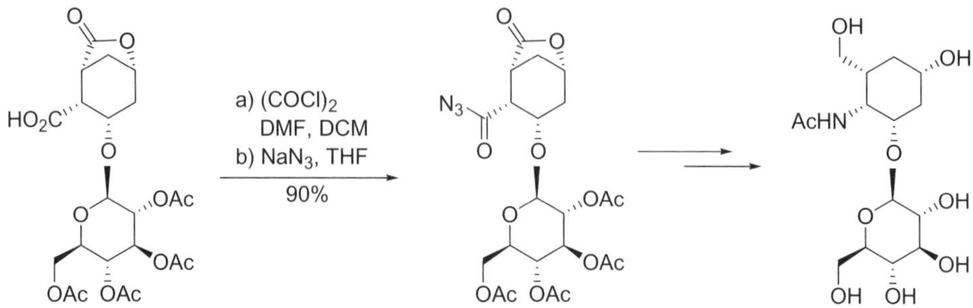

Scheme 3.43 *Synthesis of a aminomonocarba-disaccharide via Curtius rearrangement of the corresponding acyl azide*[59]

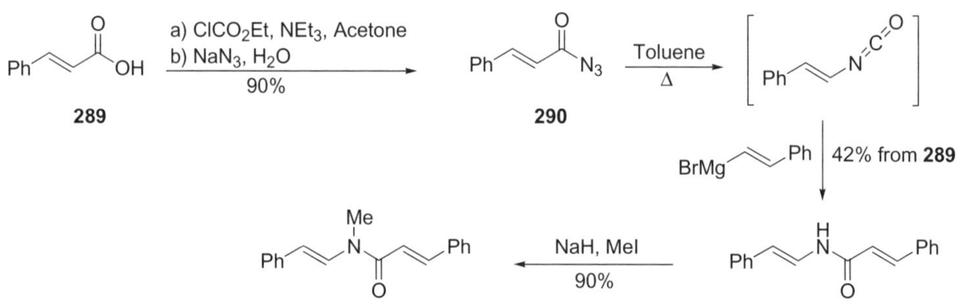

Scheme 3.44 *Synthesis of lansamide-I*[60b]

Asymmetric synthesis of unnatural β-amino acids derivatives based on azide chemistry is known.[62] Enantioselective desymmetrization of *meso*-anhydride **293** mediated by cinchona alkaloids gives optically active monomethylester **294**. This compound was converted into the acyl azide, which underwent Curtius degradation followed by alcoholysis of the intermediate isocyanate affording β-amino acid derivative **295** in high enantiomeric excess. The authors observed that Grubbs' catalyst was able to polymerize norbornene-type monomer **295** affording the corresponding polymer **296** in quantitative yield (Scheme 3.45).

3.4.3 Acyl Azides by Direct Conversion of Carboxylic Acids

Bandgar *et al.* reported a general route for the synthesis of acyl azides from aryl, heteroaryl, alkylaryl and alkyl acids with cyanuric chloride (**297**) in the presence of sodium azide and *N*-methylmorpholine.[63a] Acid activation can also be achieved using triphosgene.[63b] In fact, aromatic and aliphatic carboxylic acids react with triphosgene in the presence of sodium azide and triethylamine giving acyl azides in good yield (e.g. synthesis of **301**).

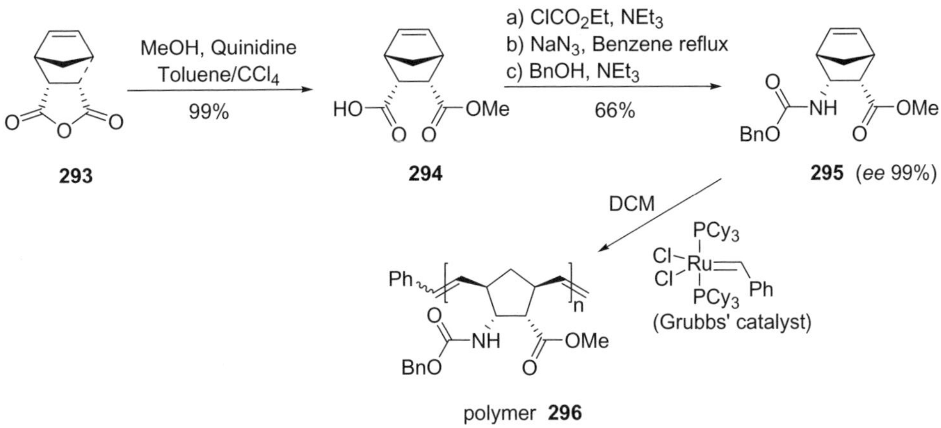

Scheme 3.45 *Acyl azides from mixed anhydrides*[62]

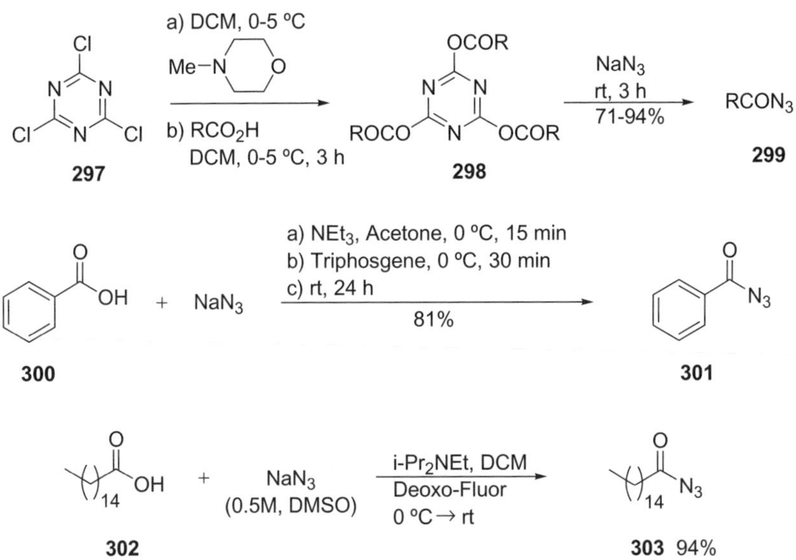

Scheme 3.46 *Acyl azides by direct conversion of carboxylic acids*[63a,63b,63d]

Deshmukh *et al.* demonstrated that triphosgene can also be used to prepare dialkylcarbamoyl azides from tertiary amines and sodium azide.[63c] Alternatively, acyl azides (e.g. **303**) can be converted *via* a one-step procedure from carboxylic acids using bis(2-methoxyethyl)aminosulfur trifluoride (Deoxo-Fluor) (Scheme 3.46).[63d]

The reaction of carboxylic acids with diphenylphosphoryl azide (DPPA) and subsequent Curtius rearrangement can be used for the conversion of malonic esters **304** into protected (*R*)-2-methylcysteine **308**.[64a] Monomethylation of dimethyl malonate followed by alkylation with *t*-butylchloromethyl sulfide gives the achiral diester **305**. Enantioselec-

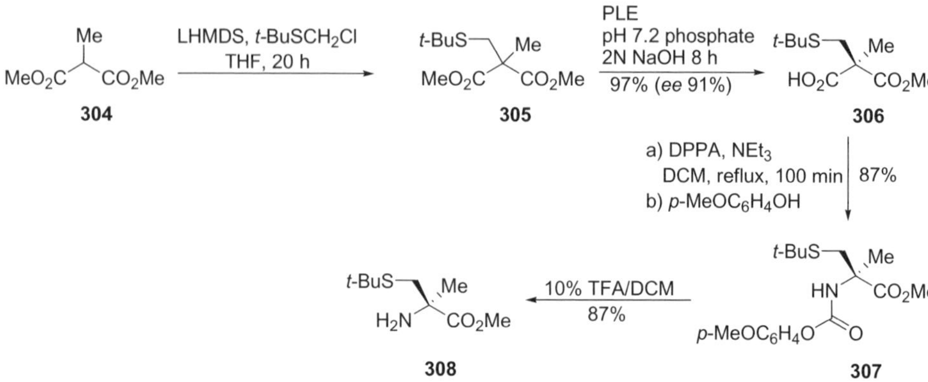

304 305 306 307 308

Scheme 3.47 *Synthesis of a derivative of (R)-2-methylcysteine*[64a]

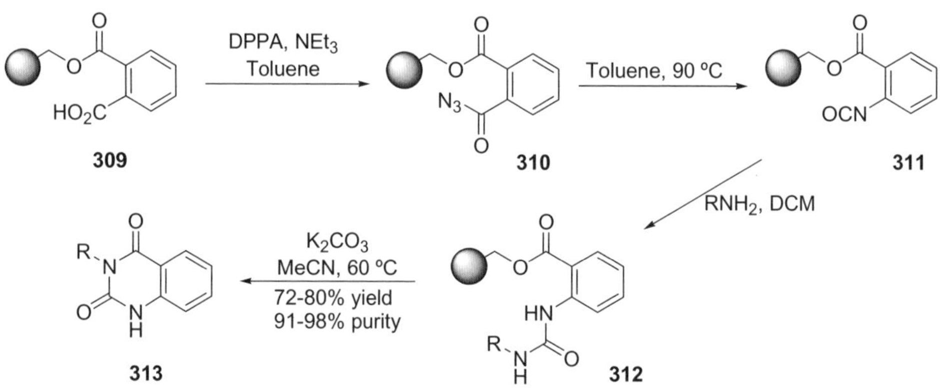

309 310 311 312 313

Scheme 3.48 *Solid-phase synthesis of quinazoline-2,4-diones*[65]

tive enzymatic desymmetrization by selective hydrolysis of one ester with pig-liver esterase (PLE) affords the chiral acid **306**. Heating this acid with DPPA, followed by reaction with 4-methoxybenzyl alcohol gives the chiral *α*-amino ester **308** (Scheme 3.47). The synthesis of (poly)peptides can also be achieved with acyl azides, directly formed from carboxylic acids and DPPA, acting as peptide coupling reagents.[64b]

Castelhano *et al.* described the solid-phase synthesis of quinazoline-2,4-diones **313** based on the chemistry of acyl azides.[65] Phthalic acid was immobilized on a PEG_4-PS resin and converted into acyl azide **310** with DPPA and triethylamine in toluene. Curtius rearrangement, followed by reaction with a primary amine and subsequent cyclocleavage with K_2CO_3 gave quinazoline-2,4-diones **313** in good yield (Scheme 3.48).

3.4.4 Acyl azides by Direct Conversion of Aldehydes

A one-step preparation of acyl azides from aldehydes using Dess-Martin periodinane and sodium azide is known.[66a] The acyl azides **315** can be isolated without Curtius rearrangement due to the mild reactions conditions. Aliphatic and aromatic aldehydes can be converted to acyl azides (e.g. **317**) by treatment with iodine azide at room temperature, presumably via a radical mechanism[66b] (Scheme 3.49). Transformation of aldehydes to acyl azides has also been carried out with $CrO_3/TMSN_3$[66c] and with $MnO_2/SiCl_4/NaN_3$.[66d]

3.4.5 Acyl Azides by Direct Conversion of Acylhydrazines

Diazotization of acylhydrazines is one route to acyl azides.[67] In fact, the diazotization of bis-hydrazide **318** can be accomplished by treatment with $NaNO_2$ under acidic conditions giving the corresponding bis-acyl azide **319** in 90% yield.[67a] A similar procedure can be applied to the synthesis of nicotinoyl azide **321** (Scheme 3.50).[67b]

3.4.6 Acyl Azides from N-acylbenzotriazoles

Recently, a synthetic approach to acyl azides from the corresponding *N*-acylbenzotriazoles has been reported.[68] The *N*-acylbenzotriazoles were accessible in good yields from

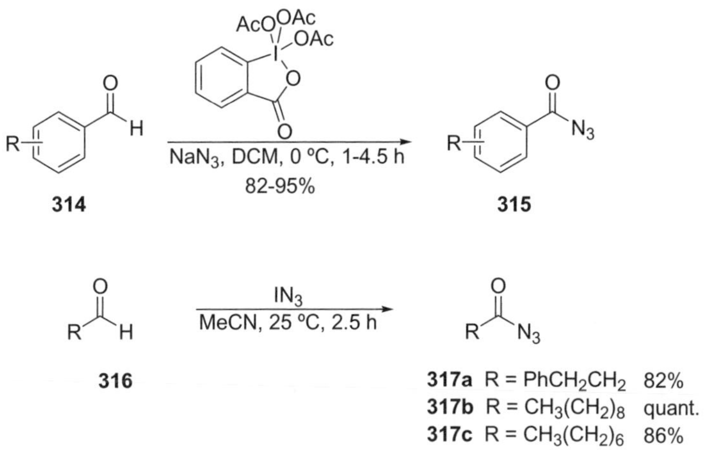

Scheme 3.49 *Acyl azides from aldehydes*[66]

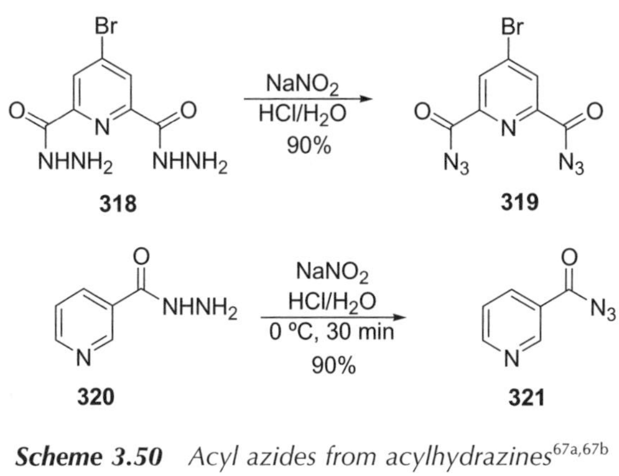

Scheme 3.50 *Acyl azides from acylhydrazines*[67a,67b]

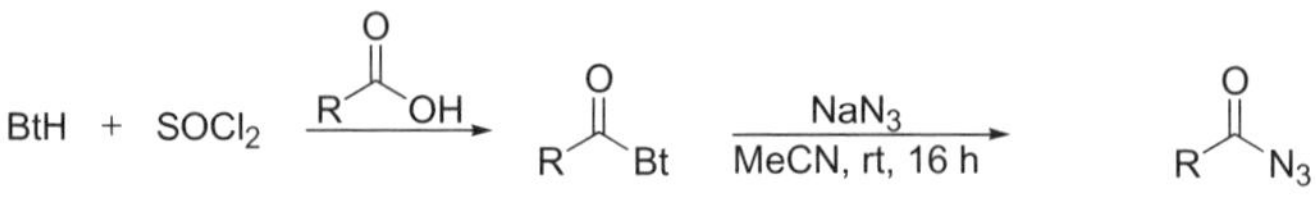

322

323a	R = C_6H_5	82%	**323f**	R = 2-pyridyl	80%
323b	R = 4-MeC_6H_4	82%	**323g**	R = 3-pyridyl	78%
323c	R = 2-thienyl	80%	**323h**	R = 4-pyridyl	82%
323d	R = 2-furanyl	75%	**323i**	R = C_6H_5CH=CH	72%
323e	R = 2-indolyl	78%	**323j**	R = *L*-Tryp	83%

Scheme 3.51 *Acyl azides from N-acylbenzotriazoles*[68]

benzotriazole (BtH), thionyl chloride and the appropriate carboxylic acid. The reaction of **322** with sodium azide in acetonitrile at room temperature for 16h affords the acyl azides **323** in 72–83% yield (Scheme 3.51).

References

[1] P. Griess, *Proc. R. Soc. London* **1864**, *13*, 375–84.

[2] T. Curtius, *Ber. Dtsch. Chem. Ges.* **1890**, *23*, 3023–33.

[3] P.A.S. Smith, *Org. React.* **1946**, *3*, 337–49.

[4] J.H. Boyer, F.C. Canter, *Chem. Rev.* **1954**, *54*, 1–57.

[5] For previous reviews on this topic, see: (a) G. L'abbé, *Chem. Rev.* **1969**, *69*, 345–63. (b) E.F.V. Scriven, K. Turnbull, *Chem. Rev.* **1988**, *88*, 297–368. (c) R. Grashey in *Comprehensive Organic Synthesis*, B.M. Trost and I. Fleming, eds; Pergamon Press, **1991**, Vol. *6*, p. 225. (d) S. Bräse, C. Gil, K. Knepper, V. Zimmermann, *Angew. Chem., Int. Ed.* **2005**, *44*, 5188–240. (e) W.H. Binder, C. Kluger, *Curr. Org. Chem.* **2006**, *10*, 1791–815. (f) F. Palacios, D. Aparicio, G. Rubiales, C. Alonso, J.M. de los Santos, *Curr. Org. Chem.* **2006**, *10*, 2371–92.

[6] T.S. Lin, W.H. Prusoff, *J. Med. Chem.* **1978**, *21*, 109–12.

[7] M. Köhn, R. Breinbauer, *Angew. Chem. Int. Ed. Engl.* **2004**, *43*, 3106–16.

[8] R.H. Smith, Jr., B.D. Wladkowski, A.F. Mehl, *et al.*, *J. Org. Chem.* **1989**, *54*, 1036–42.

[9] H. Bock, R. Dammel, *Angew. Chem. Int. Ed. Engl.* **1987**, *26*, 505–26.

[10] (a) G. Righi, C. D'Achille, G. Pescatore, C. Bonini, *Tetrahedron Lett.* **2003**, *44*, 6999–7002. (b) N. Halland, A. Braunton, S. Bachmann, M. Marigo, K.A. Jørgensen, *J. Am. Chem. Soc.* **2004**, *126*, 4790–1. (c) G. DuBois, G.A. Crosby, G.V. McGarraugh, *et al.*, *J. Org. Chem.* **1982**, *47*, 1319–23. (d) M.C. Pirrung, G.M. McGeehan, *J. Org. Chem.* **1983**, *48*, 5143–4. (e) H. Bock, R. Dammel, *J. Am. Chem. Soc.* **1988**, *110*, 5261–9. (f) J.A. Miller, *Tetrahedron Lett.* **1975**, 2959–60. (g) D.P. Cox, R.A. Moss, J. Terpinski, *J. Am. Chem. Soc.* **1983**, *105*, 6513–14. (h) R.A. Moss, J. Terpinski, D.P. Cox, D.Z. Denney, K. Krogh-Jespersen, *J. Am. Chem. Soc.* **1985**, *107*, 2743–8. (i) G.P. Miller, E.T. Kool, *J. Org. Chem.* **2004**, *69*, 2404–10. (j) D.H. Drewry, S.W. Gerritz, J.A. Linn, *Tetrahedron Lett.* **1997**, *38*, 3377–80. (k) H.S. Oh, H.G. Hahn, S.H. Cheon, D.C. Ha, *Tetrahedron Lett.* **2000**, *41*, 5069–72. (l) D.R. Tortolani, S.A. Biller, *Tetrahedron Lett.* **1996**, *37*, 5687–90.

[11] (a) C. Mazzini, L. Sambri, H. Regeling, B. Zwanenburg, G.J.F. Chittenden, *J. Chem. Soc. Perkin Trans. 1* **1997**, 3351–6. (b) P.S. Bara, A.L. Zografos, D. O'Malley, *J. Am. Chem. Soc.* **2004**, *126*, 3726–7. (c) W. Kurosawa, T. Kan, T. Fukuyama, *J. Am. Chem. Soc.* **2003**, *125*, 8112–3. (d) G.J. Sanjayan, A. Stewart, S. Hachisu, R. Gonzalez, M.P. Watterson, G.W.J. Fleet, *Tetrahedron Lett.* **2003**, *44*, 5847–51. (e) K.J. Shaffer, C.M. Taylor, *Org. Lett.* **2006**, *8*, 3959–62. (f) D.L. Boger, M.A. Patane, J. Zhou, *J. Am. Chem. Soc.* **1994**, *116*, 8544–56. (g) K.S. Feldman, A.G. Karatjas, *Org. Lett.* **2004**, *6*, 2849–52. (h) C.E. Lee, E.K. Kick, J.A. Ellman, *J. Am. Chem. Soc.* **1998**, *120*, 9735–47.

[12] (a) H.M. Sampath Kumar, B.V. Subba Reddy, S. Anjaneyulu, J.S. Yadav, *Tetrahedron Lett.* **1998**, *39*, 7385–8. (b) A. Hassner, R. Fibiger, D. Andisik, *J. Org. Chem.* **1984**, *49*, 4237–44. (c) A. Jayanthi, V.K. Gumaste, A.R.A.S. Deshmukh, *Synlett* **2004**, 979–82. (d) M.N.S. Rad, S. Behrouz, A. Khalafi-Nezhad, *Tetrahedron Lett.* **2007**, *48*, 3445–9.

[13] (a) A. Avenoza, J.H. Busto, F. Corzana, J.I. Garcia, J.M. Peregrina, *J. Org. Chem.* **2003**, *68*, 4506–13. (b) I.A. Sayyed, A. Sudali, *Tetrahedron: Asymmetry* **2004**, *15*, 3111–16.

[14] (a) S.-K. Kang, D.-C. Park, H.-S. Rho, S.-H. Yoon, J.-S. Shin, *J. Chem. Soc. Perkin Trans. 1* **1994**, 3513–14. (b) L. He, H.S. Byun, R. Bittman, *J. Org. Chem.* **2000**, *65*, 7627–33. (c) L. Alvarez de Cienfuegos, C. Rodriguez, A.J. Mota, R. Robles, *Org. Lett.* **2003**, *5*, 2743–5.

[15] M. Gibson, J.M. Goodman, L.J. Farrugia, R.C. Hartley, *Tetrahedron Lett.* **2003**, *44*, 2841–4.

[16] H. Tanaka, A.M. Sawayama, T.J. Wandless, *J. Am. Chem. Soc.* **2003**, *125*, 6864–5.

[17] (a) M.J. Marti, I. Rico, J.C. Ader, A. de Savignac, A. Lattes, *Tetrahedron Lett.* **1989**, *30*, 1245–8. (b) O.N. Yuadina, A.A. Sherman, N.E. Nifantiev, *Carbohydr. Res.* **2001**, *332*, 363–71. (c) M.M. Sá, M.D. Ramos, L. Fernandes, *Tetrahedron* **2006**, *62*, 11652–6.

[18] (a) G.K. Surya Prakash, M.A. Stephenson, J.G. Shih, G.A. Olah, *J. Org. Chem.* **1986**, *51*, 3215–17. (b) M.B. Frankel, D.O. Woolery, *J. Org. Chem.* **1983**, *48*, 611–12. (c) D.A. Evans, T.C. Britton, J.A. Ellman, R.L. Dorow, *J. Am. Chem. Soc.* **1990**, *112*, 4011–30. (d) K. Banert, *Chem. Ber.* **1989**, *122*, 911–18. (e) K. Banert, M. Hagedorn, *Angew. Chem., Int. Ed.* **1989**, *28*, 1675–6. (f) J.R. Fotsing, K. Banert, *Eur. J. Org. Chem.* **2005**, 3704–14. (g) B.E. Blass, K.R. Coburn, A.L. Faulkner, W.L. Seibela, A. Srivastava, *Tetrahedron Lett.* **2003**, *44*, 2153–5.

[19] (a) E.D. Soli, A.S. Manoso, M.C. Patterson, P. DeShong, *J. Org. Chem.* **1999**, *64*, 3171–7. (b) V.Y. Dudkin, D. Crich, *Tetrahedron Lett.* **2003**, *44*, 1787–9; (c) H. Tsukamoto, Y. Kondo, *Tetrahedron Lett.* **2003**, *44*, 5247–9.

[20] (a) C. Chiappe, D. Pieraccini, P. Saullo, *J. Org. Chem.* **2003**, *68*, 6710–15. (b) A. Loris, A. Perosa, M. Selva, P. Tundo, *J. Org. Chem.* **2003**, *68*, 4046–51.

[21] (a) P.N.D. Singh, S. Muthukrishnan, R.S. Murthy, *et al.*, *Tetrahedron Lett.* **2003**, *44*, 9169–71. (b) Y. Ju, D. Kumar, R.S. Varma, *J. Org. Chem.* **2006**, *71*, 6697–700.

[22] (a) K. Odlo, E.A. Høydahl, T.V. Hansen, *Tetrahedron Lett.* **2007**, *48*, 2097–9. (b) P. Appukkuttan, W. Dehaen, V.V. Fokin, E. Van der Eycken, *Org. Lett.* **2004**, *6*, 4223–5. (c) P. Li, L. Wang, *Lett. Org. Chem.* **2007**, *4*, 23–6.

[23] K. Banert, *Synthesis* **2007**, 3431–46.

[24] (a) S. Saito, N. Bunya, M. Inaba, T. Moriwake, S. Torii, *Tetrahedron Lett.* **1985**, *26*, 5309–12. (b) J. Legters, L. Thijs, B. Zwanenburg, *Tetrahedron* **1991**, *28*, 5287–94. (c) S. Saito, K. Komoda, T. Moriwake, *Org. Synth.* **1995**, *73*, 183–200. (d) A. Breuning, R. Vicik, T. Schirmeister, *Tetrahedron: Asymmetry* **2003**, *14*, 3301–12. (e) C. Le Hetet, M. David, F. Carreaux, B. Carboni, A. Sauleau, *Tetrahedron Lett.* **1997**, *38*, 5153–6.

[25] (a) D.M. Hodgson, A.R. Gibbs, G.P. Lee, *Tetrahedron* **1996**, *52*,14361–84. (b) E.N. Jacobsen, *Acc. Chem. Res.* **2000**, *33*, 421–31. (c) H. Yamashita, *Bull. Chem. Soc. Jpn.* **1988**, *61*, 1213–20. (d) L.E. Martinez, J.L. Leighton, D.H. Carsten, E.N. Jacobsen, *J. Am. Chem. Soc.* **1995**, *117*, 5897–8. (e) K.B. Hansen, J.L. Leighton, E.N. Jacobsen, *J. Am. Chem. Soc.* **1996**, *118*, 10924–5. (f) R.G. Konsler, J. Karl, E.N. Jacobsen, *J. Am. Chem. Soc.* **1998**, *120*, 10780–1. (g) N.C. Gianneschi, P.A. Bertin, S.T. Nguyen, C.A. Mirkin, L.N. Zakharov, A.L. Rheingold, *J. Am. Chem. Soc.* **2003**, *125*, 10508–9. (h) D.A. Annis, O. Helluin, E.N. Jacobsen, *Angew. Chem. Int. Ed.* **1998**, *37*, 1907–9.

[26] B.M.L. Dioos, P.A. Jacobs, *Tetrahedron Lett.* **2003**, *44*, 8815–17.

[27] (a) J.H. Spelberg, J.E.T. van Hylckama Vlieg, L. Tang, D.B. Janssen, R.M. Kellog, *Org. Lett.* **2001**, *3*, 41–3. (b) B.M.L. Dioos, P.A. Jacobs, *Tetrahedron Lett.* **2003**, *44*, 4715–17.

[28] (a) M. Onaka, K. Sugita, Y. Izumi, *J. Org. Chem.* **1989**, *54*, 1116–23. (b) S.E. Sen, S.M. Smith, K.A. Sullivan, *Tetrahedron* **1999**, *55*, 12657–98.

[29] J. Boruwa, J.C. Borah, B. Kalita, N.C. Barua, *Tetrahedron Lett.* **2004**, *45*, 7355–8.

[30] (a) M. Caron, P.R. Carlier, K.B. Sharpless, *J. Org. Chem.* **1988**, *53*, 5185–7. (b) N. Aguilar, A. Moyano, M.A. Pericàs, A. Riera, *J. Org. Chem.* **1998**, *63*, 3560–7. (c) K.S. Reddy, L. Solà, A. Moyano, M.A. Pericàs, A. Riera, *J. Org. Chem.* **1999**, *64*, 3969–74. (d) X. Ginesta, M. Pastó, M.A. Pericàs, A. Riera, *Org. Lett.* **2003**, *5*, 3001–4.

[31] C.E. Davis, J.L. Bailey, J. Lockner, R.M. Coates, *J. Org. Chem.* **2003**, *68*, 75–82.
[32] Reviews: (a) X.E. Hu, *Tetrahedron* **2004**, 60, 2701–43. (b) D. Lucet, T. Le Gall, C. Mioskowski, *Angew. Chem. Int. Ed.* **1998**, *37*, 2580–627.
[33] (a) G. Sabitha, R.S. Babu, M. Rajkumar, J.S. Vadav, *Org. Lett.* **2002**, *4*, 343–5. (b) A. Bisai, G. Pandey, M.K. Pandey, V.K. Singh, *Tetrahedron Lett.* **2003**, *44*, 5839–41. (c) G. Sabitha, S.R. Babu, M.S.K. Reddy, J.S. Yadav, *Synthesis* **2002**, 2254–8. (d) W.-H. Leung, M.-T. Yu, M.-C. Wu, L.-L. Yeung, *Tetrahedron Lett.* **1996**, *37*, 891–2. (e) D. Tanner, C. Birgersson, H. Dhaliwal, *Tetrahedron Lett.* **1990**, *31*, 1903–6.
[34] Z. Li, M. Fernández, E.N. Jacobsen, *Org. Lett.* **1999**, *1*, 1611–13.
[35] (a) C.U. Kim, W. Lew, M.A. Williams, *et al.*, *J. Am. Chem. Soc.* **1997**, *119*, 681–90. (b) M. Sasaki, A.K. Yudin, *J. Am. Chem. Soc.* **2003**, *125*, 14242–3. (c) K.J. Hale, M.M. Domostoj, D.A. Tocher, E. Irving, F. Scheinmann, *Org. Lett.* **2003**, *5*, 2927–30. (d) Y. Fukuta, T. Mita, N. Fukuda, M. Kanai, M. Shibasali, *J. Am. Chem. Soc.* **2006**, *128*, 6312–13.
[36] (a) O. Mitsunobu, Y. Yamada, *Bull. Chem. Soc. Jpn.* **1967**, *40*, 2380–2. (b) H. Loibner, E. Zibral, *Helv. Chim. Acta* **1976**, *59*, 2100–13. (c) S.-H. Lee, J. Yoon, S.-H. Chung, Y.-S. Lee, *Tetrahedron* **2001**, *57*, 2139–45. (d) D. Gagnon, S. Lauzon, C. Godbout, C. Spino, *Org. Lett.* **2005**, *7*, 4769–71. (e) F. Klepper, E.-M. Jahn, V. Hickmann, T. Carell, *Angew. Chem., Int. Ed.* **2007**, *46*, 2325–7. (f) S.-J. Ryoo, J. Kim, J.-S. Kim, Y.-S. Lee, *J. Comb. Chem.* **2002**, *4*, 187–90.
[37] (a) C. Simon, S. Hosztafi, S. Makleit, *Tetrahedron Lett.* **1993**, *34*, 6475–8. (b) Y. Yoshimura, K. Kitano, K. Yamada, *et al.*, *J. Org. Chem.* **1997**, *62*, 3140–52. (c) T.Q. Dinh, X. Du, C.D. Smith, R.W. Armstrong, *J. Org. Chem.* **1997**, *62*, 6773–83. (d) M.E. Maier, C. Hermann, *Tetrahedron* **2000**, *56*, 557–61. (e) T. Watanabe, Y. Tanaka, R. Shoda, R. Sakamoto, K. Kamikawa, M. Uemura, *J. Org. Chem.* **2004**, *69*, 4152–8. (f) J.A. Gómez-Vidal, R.B. Silverman, *Org. Lett.* **2001**, *3*, 2481–4. (g) B. Jiang, C.-G. Yang, J. Wang, *J. Org. Chem.* **2002**, *67*, 1396–8. (h) F. Diaba, E. Ricou, J. Bonjoch, *Tetrahedron: Asymmetry* **2006**, *17*, 1437–43. (i) S. Hanessian, F. Xie, *Tetrahedron Lett.* **1998**, *39*, 737–40. (j) K.C. Nicolaou, N. Winssinger, D. Vourloumis, *et al.*, *J. Am. Chem. Soc.* **1998**, *120*, 10814–26. (k) Y. Lu, R.T. Taylor, *Tetrahedron Lett.* **2003**, *44*, 9267–9.
[38] (a) A.S. Thompson, G.R. Humphrey, A.M. DeMarco, D.J. Mathre, E.J.J. Grabowski, *J. Org. Chem.* **1993**, *58*, 5886–8. (b) C. Yu, B. Liu, L. Hu, *Org. Lett.* **2000**, *2*, 1959–61. (c) M. Mizuno, T. Shioiri, *Chem. Commun.* **1997**, 2165–6. (d) M.C. Viaud, P. Rollin, *Synthesis* **1990**, 130–3. (e) S. Czernecki, S. Horms, J.M. Valery, *J. Org. Chem.* **1995**, *60*, 650–5. (f) S. Ma, B. Ni, *Chem. Eur. J.* **2004**, *10*, 3286–300. (g) V. Gracias, Y. Zeng, P. Desai, J. Aubé, *Org. Lett.* **2003**, *5*, 4999–5001. (h) M. Toyota, C. Komori, M. Ihara, *J. Org. Chem.* **2000**, *65*, 7110–13. (i) Z.-S. Li, R.-P. Qiao, Z.-J. Yang, L.-R. Zhang, L.-H. Zhang, *Tetrahedron: Asymmetry* **2006**, *17*, 1056–61. (j) N. Iranpoor, H. Fireouzabadi, B. Akhlaghinia, N. Nowrouzi, *Tetrahedron Lett.* **2004**, *45*, 3291–4. (k) K. Kuroda, Y. Hayashi, T. Mukaiyama, *Tetrahedron* **2007**, *63*, 6358–64.
[39] (a) C.J. Caveander, V.J. Shiner, *J. Org. Chem.* **1972**, *37*, 3567–9. (b) J. Zaloom, D.C. Roberts, *J. Org. Chem.* **1981**, *46*, 5173–6. (c) P.E. Eaton, A.M. Fisher, R.E. Hormann, *Synlett* **1990**, 737–8. (d) P.C. Chen, R.E. Wharton, P.A. Patel, A.K. Oyelere, *Bioorg. Med. Chem.* **2007**, *15*, 7288–300. (e) P.B. Alper, S.-C. Hung, C.-H. Wong, *Tetrahedron Lett.* **1996**, *37*, 6029–32. (f) P.T. Nyffeler, C.-H. Liang, K.M. Koeller, C.-H. Wong, *J. Am. Chem. Soc.* **2002**, *124*, 10773–8. (g) J.T. Lundquist, J.C. Pelletier, *Org. Lett.* **2001**, *3*, 781–3. (h) W.S. Horne, C.S. Stout, M.R. Ghadiri, *J. Am. Chem. Soc.* **2003**, *125*, 9372–6. (i) Y. Ding, E.E. Swayze, S.A. Hofstadler, R.H. Griffey, *Tetrahedron Lett.* **2000**, *41*, 4049–52. (j) W.A. Greenberg, E.S. Priestley, P.S. Sears, *et al.*, *J. Am. Chem. Soc.* **1999**, *121*, 6527–41. (k) R.-B. Yan, F. Yang, Y. Wu, L.-H. Zhang, X.-S. Ye, *Tetrahedron Lett.* **2005**, *46*, 8993–5. (l) A. Titz, Z. Radic, O. Schwardt, B. Ernst, *Tetrahedron Lett.* **2006**, *47*, 2383–5. (m) D.T.S. Rijkers, H.H. Ricardo van Vugt, H.J.F. Jacobs, R.M.J. Liskamp, *Tetrahedron Lett.* **2002**, *43*, 3657–60. (n) A.K. Oyelere, P.C. Chen, L.P. Yao, N. Boguslavsky *J. Org. Chem.* **2006**, *71*, 9791–6. (o) H.S.G. Beckmann, V. Wittmann, *Org. Lett.* **2007**, *9*, 1–4.
[40] E.D. Goddard-Borger, R.V. Stick, *Org. Lett.* **2007**, *9*, 3797–800.
[41] (a) T. Sasaki, S. Eguchi, T. Okano, Y. Wakata, *J. Org. Chem.* **1983**, *48*, 4067–72. (b) H. Quast, P. Eckert, *Liebigs Ann. Chem.* **1974**, 1727–41.

[42] (a) J.O. Reed, W. Lwowski, *J. Org. Chem.* **1971**, *36*, 2864–9. (b) A.P. Kozikowski, M.N. Greco, *J. Org. Chem.* **1984**, *49*, 2310–14. (c) D.A. Evans, T.C. Britton, J.A. Ellman, R.L. Dorow, *J. Am. Chem. Soc.* **1990**, *112*, 4011–30. (d) S. Derrer, J.E. Davies, A.B. Holmes, *J. Chem. Soc., Perkin Trans. 1* **2000**, 2957–67. (e) S. Derrer, J.E. Davies, A.B. Holmes, *J. Chem. Soc., Perkin Trans. 1* **2000**, 2943–56. (f) K. Kühlein, H. Jensen, *Liebigs Ann. Chem.* **1974**, 369–402. (g) H.H. Wasserman, D.J. Hlasta, *J. Am. Chem. Soc.* **1978**, *100*, 6780–1. (h) B.T. Golding, A.J. Smith, *J. Chem. Soc., Chem. Commun.* **1980**, 702–3. (i) P.F. Bevilacqua, D.D. Keith, J.L. Roberts, *J. Org. Chem.* **1984**, *49*, 1430–4. (j) A. Nishida, M. Shibasaki, S. Ikegami, *Tetrahedron Lett.* **1984**, *25*, 765–8.

[43] (a) J.F.W. Keana, S.X. Cai, *J. Org. Chem.* **1990**, *55*, 3640–7. (b) K.A.H. Chehade, H.P. Spielmann, *J. Org. Chem.* **2000**, *65*, 4949–53.

[44] (a) R.J. Simmonds, M.F.G. Stevens, *J. Chem. Soc., Perkin Trans. 1* **1982**, 1821–5. (b) C.K. Lowe-Ma, R.A. Nissan, W.S. Wilson, *J. Org. Chem.* **1990**, *55*, 3755–61. (c) W. Stadlbauer, W. Fiala, M. Fischer, G. Hojas, *J. Heterocycl. Chem.* **2000**, *37*, 1253–6. (d) P. Roschger, W. Fiala, W. Stadlbauer, *J. Heterocycl. Chem.* **1992**, *29*, 225–31. (e) P.N. Confalone, R.B. Woodward, *J. Am. Chem. Soc.* **1983**, *105*, 902–6. (f) G.B. Barlin, *Aust. J. Chem.* **1983**, *36*, 983–92. (g) P. Choi, C.W. Rees, E.H. Smith, *Tetrahedron Lett.* **1982**, *23*, 121–4. (h) D.R. Miller, D.C. Svenson, E.G. Gillan, *J. Am. Chem. Soc.* **2004**, *126*, 5372–3.

[45] C.L. Gibson, S. La Rosa, C.J. Suckling, *Tetrahedron Lett.* **2003**, *44*, 1267–70.

[46] (a) T.P. Kogan, T.C. Somers, M.C. Venuti, *Tetrahedron* **1990**, *46*, 6623–32. (b) W. Zhu, D. Ma, *Chem. Commun.* **2004**, 888–9. (c) C.I. Schilling, S. Bräse, *Org. Biomol. Chem.* **2007**, *5*, 3586–8.

[47] (a) A.K. Feldman, B. Colasson, V.V. Fokin, *Org. Lett.* **2004**, *6*, 3897–9. (b) J. Andersen, S. Bolvig, X. Liang, *Synlett.* **2005**, 2941–7. (c) Y.-B. Zhao, Z.-Y. Yan, Y.-M. Liang, *Tetrahedron Lett.* **2006**, *47*, 1545–9.

[48] C.-Z. Tao, X. Cui, J. Li, A.-X. Liu, L. Liu, Q.-X. Guo, *Tetrahedron Lett.* **2007**, *48*, 3525–9.

[49] (a) J.D. Wallis, J.D. Dunitz, *Chem. Commun.* **1983**, 910–1. (b) R.N. Butler, A. Fox, S. Collier, L.A. Burke, *J. Chem. Soc., Perkin Trans. 2* **1998**, 2243–7. (c) M.F. Joucla, C.W. Rees, *Chem. Commun.* **1984**, 374–5. (d) L.L. Melhado, N.J. Leonard, *J. Org. Chem.* **1983**, *48*, 5130–3. (e) C. Cismas, T. Gimisis, *Tetrahedron Lett.* **2008**, *49*, 1336–9. (f) S.M. Capitosti, T.P. Hansen, M.L. Brown, *Org. Lett.* **2003**, *5*, 2865–7. (g) B. Colasson, M. Save, P. Milko, J. Roithová, D. Schröder, O. Reinaud, *Org. Lett.* **2007**, *9*, 4987–90. (h) L. Li, J. Han, B. Nguyen, K. Burgess, *J. Org. Chem.* **2008**, *73*, 1963–70.

[50] K. Barral, A.D. Moorhouse, J.E. Moses, *Org. Lett.* **2007**, *9*, 1809–11.

[51] (a) F. Avemaria, V. Zimmermann, S. Bräse, *Synlett* **2004**, 1163–6. (b) C. Gil, S. Bräse, *Chem. Eur. J.* **2005**, *11*, 2680–8. (c) C.-Y. Liu, P. Knochel, *J. Org. Chem.* **2007**, *72*, 7106–15.

[52] (a) P.A.S. Smith, C.D. Rowe, L.B. Bruner, *J. Org. Chem.* **1969**, *34*, 3430–3. (b) H. Suschitzky, W. Kramer, R. Neidlein, P. Rosyk, T. Bohn, *J. Chem. Soc., Perkin Trans. 1* **1991**, 923–7. (c) J. Gavenonis, T.D. Tilley, *Organometallics* **2002**, *21*, 5549–63. (d) P. Zanirato, S. Cerini, *Org. Biomol. Chem.* **2005**, *3*, 1508–13. (e) E. Foresti, *Gazz. Chim. Ital.* **1995**, *125*, 151–61. (f) P. Spagnolo, P. Zanirato, *J. Chem. Soc., Perkin Trans. 1* **1996**, 963–4. (g) S. Gronowitz, P. Zanirato, *J. Chem. Soc., Perkin Trans. 2* **1994**, 1815–9.

[53] (a) Q. Liu, Y. Tor, *Org. Lett.* **2003**, *5*, 2571–2. (b) W. Fisher, J.-P. Anselme, *J. Am. Chem. Soc.* **1967**, *89*, 5284–5.

[54] S.P. Klump, H. Shechter, *Tetrahedron Lett.* **2002**, *43*, 8421–3.

[55] (a) M. de Rosa, P. Haberfield, *J. Org. Chem.* **1981**, *46*, 2639–43. (b) Y.H. Kim, K. Kim, S.B. Shim, *Tetrahedron Lett.* **1986**, *27*, 4749–52. (c) V. Pozsgay, H.J. Jennings, *Tetrahedron Lett.* **1987**, *28*, 5091–2. (d) Y. Matsuya, T. Itoh, K. Nagata, A. Ohsawa, *Tetrahedron* **1997**, *53*, 15701–10.

[56] L. Caglioti, F. Gasparrini, *Synthesis* **1979**, 207–8.

[57] S. Maffei, A.M. Rivolta, *Gazz. Chim. Ital.* **1954**, *84*, 750–2.

[58] (a) A. Padwa, M.A. Brodney, B. Liu, K. Satake, T. Wu, *J. Org. Chem.* **1999**, *64*, 3595–607. (b) A. Padwa, K.R. Crawford, P. Rashatasakhon, M. Rose, *J. Org. Chem.* **2003**, *68*, 2609–17. (c) V.V. Suresh Babu, K. Ananda, G.-R. Vasanthakumar, *J. Chem. Soc., Perkin Trans. 1* **2000**, 4328–31. (d) J.S. New, W.L. Christopher, J.P. Yevich, *et al.*, *J. Med. Chem.* **1989**, *32*,

1147–56. (e) G.K. Surya Prakash, P.S. Iyer, M. Arvanaghi, G.A. Olah, *J. Org. Chem.* **1983**, *48*, 3358–9.

[59] D.S. Larsen, R.J. Lins, R.J. Stoodley, N.S. Trotter, *J. Chem. Soc., Perkin Trans. 1* **2001**, 2204–12.

[60] (a) K. Kuramochi, H. Watanabe, T. Kitahara, *Synlett* **2000**, 397–9. (b) I. Stefanuti, S.A. Smith, R.J.K. Taylor, *Tetrahedron Lett.* **2000**, *41*, 3735–8.

[61] (a) K. Kuramochi, Y, Osada, T. Kitahara, *Tetrahedron* **2003**, *59*, 9447–54. (b) Y. Wu, L. Esser, J.K. De Brabander, *Angew. Chem. Int. Ed.* **2000**, *39*, 4308–10. (c) Y. Wu, X. Liao, R. Wang, X.-S. Xie, J.K. De Brabander, *J. Am. Chem. Soc.* **2002**, *124*, 3245–53. (d) A. Bhattacharjee, O.R. Seguil, J.K. De Brabander, *Tetrahedron Lett.* **2001**, *42*, 1217–20.

[62] (a) C. Bolm, C.L. Dinter, I. Schiffers, L. Defrère, *Synlett* **2001**, 1875–7. (b) C. Bolm, I. Schiffers, C.L. Dinter, L. Defrère, A. Gerlach, G. Raabe, *Synthesis* **2001**, 1719–30.

[63] (a) B.P. Bandgar, S.S. Pandit, *Tetrahedron Lett.* **2002**, *43*, 3413–4. (b) V.K. Gumaste, B.M. Bhawal, A.R.A.S. Desmukh, *Tetrahedron Lett.* **2002**, *43*, 1345–6. (c) V.K. Gumaste, A.R.A.S. Desmukh, *Tetrahedron Lett.* **2004**, *45*, 6571–3. (d) C.O. Kangani, B.W. Day, D.E. Kelley, *Tetrahedron Lett.* **2007**, *48*, 5933–7.

[64] (a) B.L. Kedrowski, *J. Org. Chem.* **2003**, *68*, 5403–6. (b) S.-Y. Han, Y.-A. Kim, *Tetrahedron* **2004**, *60*, 2447–67.

[65] H. Shao, M. Colucci, S. Tong, H. Zhang, A.L. Castelhano, *Tetrahedron Lett.* **1998**, *39*, 7235–8.

[66] (a) D. Subhas, A.V. Narsimha Reddy, *Tetrahedron Lett.* **2003**, *44*, 3543–5. (b) L. Marinnescu, J. Thinggaard, I.B. Thomsen, M. Bols, *J. Org. Chem.* **2003**, *68*, 9453–5. (c) J.G. Lee, *Tetrahedron Lett.* **1992**, *33*, 3165–6. (d) S.S. Elmorsy, *Tetrahedron Lett.* **1995**, *36*, 1341–2.

[67] (a) M. Nettekoven, *Synlett* **2001**, 1917–20. (b) G. Papeo, H. Posteri, P. Vianello, M. Varasi, *Synthesis* **2004**, 2886–92. (c) J.E. Macor, G. Mullen, P. Verhoest, A. Sampognaro, B. Shepardson, R.A. Mack, *J. Org. Chem.* **2004**, *69*, 6493–5.

[68] A.R. Katritzky, K. Widyan, K. Kirichenko, *J. Org. Chem.* **2007**, *72*, 5802–4.

4

Azides by Olefin Hydroazidation Reactions

Jérôme Waser[1] and Erick M. Carreira[2]

[1]*Laboratory of Catalysis and Organic Synthesis, EPFL SB ISIC LCSO BCH 4306 (Bâtiment de chimie UNIL), CH-1015 Lausanne, Switzerland;* [2]*Laboratorium für Organische Chemie, HCI H 335, Wolfgang-Pauli-Str. 10, 8093 Zürich, Switzerland*

4.1 Introduction

Azides have been recognized for a long time as versatile intermediates in organic synthesis.[1] Recent progress in the use of azides, especially in cycloaddition reactions, has further increased their utility.[2] They have also generated much interest in biochemistry due to their stability under physiological conditions combined with their unique reactivity.[3] In order to fully exploit the potential of these new methods and techniques, efficient and selective syntheses of azides are highly desired. The most common method for the preparation of alkyl azides involves the substitution reaction of primary or secondary alkyl halides with inorganic azides.[4] This approach requires usually the multi-step installation of an adequate leaving group and generates a stoichiometric amount of waste salts.

Direct introduction of HN_3, or its equivalents, onto olefins constitutes an efficient and straightforward approach. Unfortunately, the most efficient reactions are multi-step sequences, such as epoxidation followed by opening with azide ions[5] or hydroboration followed by iodination and substitution.[6] Several one-step methods, such as halo-azidation,[7] diazidation,[8] seleno-azidation,[9] nitrato-azidation,[10] formation of α-azido ketones,[11] and carboazidation,[12] have also been reported, but the conceptually simplest reaction, the hydroazidation reaction, has been much less developed (Figure 4.1).

Organic Azides: Syntheses and Applications Edited by Stefan Bräse and Klaus Banert

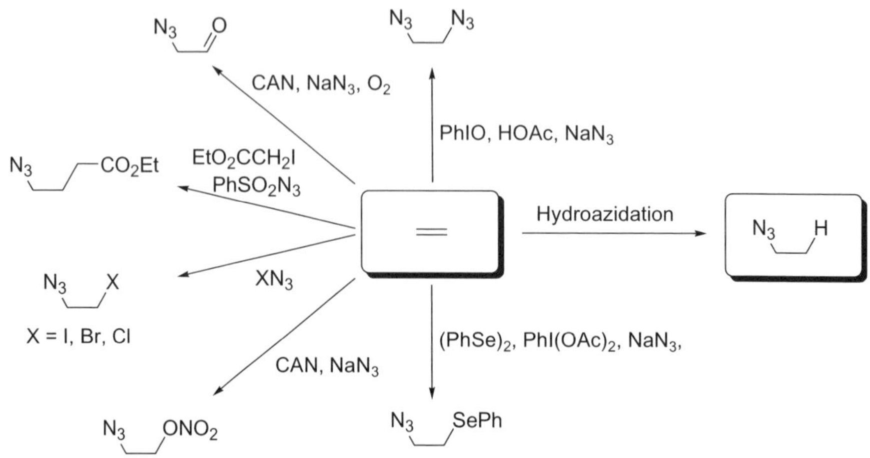

Figure 4.1 *Direct introduction of azide groups onto olefins*

The addition of hydrazoic acid itself onto unactivated double bonds is insufficiently rapid to be preparatively useful. Thus, the examples that can be found in this respect make use of the inherent nucleophilicity of the azide ion in additions to activated, electrophilic C–C bonds. This approach was successful for α,β-unsaturated carbonyl compounds for example, in which case efficient catalytic asymmetric hydroazidation reactions could be developed. In the case of non-activated olefins, however, stronger Lewis or Brønsted acids are needed, and only olefins which give rise to stabilized carbocations react efficiently (see below). Recently, a new metal-catalyzed hydroazidation reaction was developed in our laboratory, in which the inherent reactivity or polarity of the azide ion is formally inverted: olefin hydrometallation reaction leads to an organometallic intermediate that adds to an electrophilic sulfonyl azide.[13] This has led to a general and convenient hydroazidation reaction of unactivated olefins.[14]

In this chapter, we first describe the successful methods for the hydroazidation of α,β-unsaturated carbonyl compounds. The classic approach for the addition of hydrazoic acid and its derivatives onto non-activated double bonds is then examined. In the last section, we present our own work in this area and our progress towards a general method for the hydroazidation of olefins.

4.2 Conjugate Addition of Hydrazoic Acid and Its Derivatives

The conjugate addition of hydrazoic acid is of considerable interest, as the azides obtained are easily reduced, for example, to the corresponding β-amino acid derivatives. The importance of β-amino acids resides in their special conformational preferences in peptides, and their presence in many natural products and bioactive compounds.[15]

In 1915, Oliveri-Mandala first reported on the conjugate addition of hydrazoic acid onto a quinone derivative, but the product obtained was an aromatic azide.[16] The first real hydroazidation reaction of electron-deficient double bond was then reported by Boyer in 1951.[17] In these studies, conjugated aldehydes, esters, ketones, nitro alkenes, nitriles and vinyl pyridines gave *β*-azides in 19–71% yield using 1.5 equivalents of sodium azide in acetic acid. In 1966, the hydroazidation of acceptor-substituted allenes was reported by Harvey and Ratts.[18a]

After the seminal work of Boyer, progress towards more efficient methods for conjugate addition of hydrazoic acid was very slow. Only two studies reported on a significant improvement of the original procedure: Chung documented the use of diethylaluminium azide as a more effective reagent in 1988,[19] and Rao used hydrazoic acid together with triethyl amine in 1997, which led to much better yields (>90%) than Boyer's protocol.[20]

In 1999, Jacobsen reported on a catalytic asymmetric conjugated addition of hydrazoic acid to unsaturated imide derivatives (Equation 4.1).[21] This breakthrough was possible through the use of aluminium salen azide complex **1** as catalyst. The reaction proceeded in excellent yields and enantioselectivities for alkyl substituted acceptors. Two mechanisms were proposed for this reaction: activation of the azide as an aluminium azide as shown by Chung and co-workers[19] or Lewis acid activation of the imide. The first-order dependence of the rate law on catalyst **1** indicated that dual activation was improbable. In 2005, Jacobsen reported on the extension of this methodology to *α,β*-unsaturated ketones.

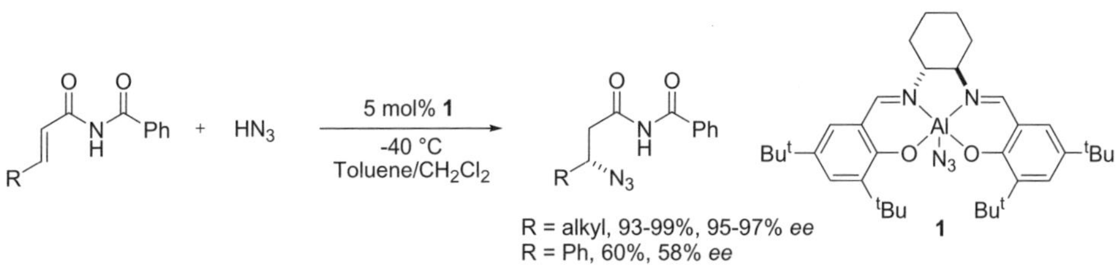

(4.1)

A different approach was followed by Miller in 1999. Based on the work of Rao,[20] trimethylsilyl azide together with acetic acid was employed as a mild and efficient method for the conjugate addition of hydrazoic acid mediated by an organic catalyst.[22a] In 2000, Miller reported an asymmetric variation of this method using histidine derived small peptides as catalysts (Equation 4.2).[22b,22c] A salient feature of the method is the use of trimethylsilyl azide which is easier to handle than hydrazoic acid.

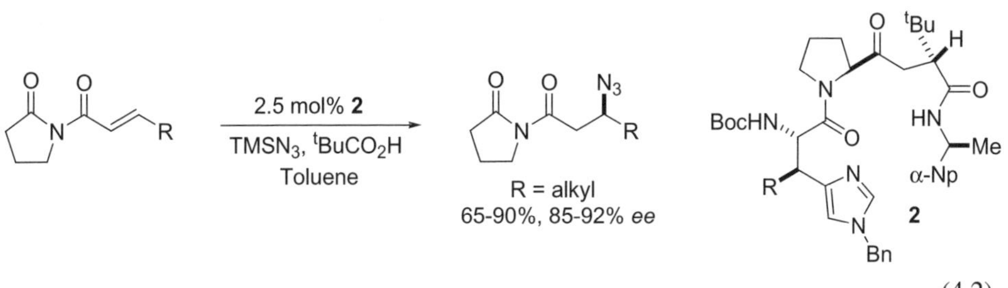

(4.2)

Since the work of Jacobsen and Miller, conjugate addition of hydrazoic acid has been reported to occur in water[23a] or in ionic liquids.[23b] Amberlite has been introduced as a catalyst.[23c] Base catalysis was also successful in the case of α,β-unsaturated aldehydes[23d] and in more complex settings.[23e,23f]

4.3 Addition of Hydrazoic Acid and Its Derivatives to Non-Activated Olefins

The addition of hydrazoic acid onto non-activated olefins is a difficult reaction that has been observed only in rare cases.[24] The main problem was the prevention of subsequent decomposition of the azide formed via nitrogen release or Schmidt rearrangement due to harsh reaction conditions that are typically employed.

The first successful approach was reported by Heathcock in 1969 using mercury(II) salts.[25a] The reaction proceeds via an olefin azidomercuration followed by a reductive demercuration in the work-up. The azides derived from terminal olefins were obtained in 50–88% yield with good Markovnikov selectivity, while non-terminal olefins gave lower yields (Figure 4.2). An obvious drawback of this procedure is the use of a stoichiometric amount of mercury salts. This procedure also leads to the formation of potentially explosive $Hg(N_3)_2$.[25b] Nevertheless, this method is unique and reliable for the hydroazidation of monosubstituted non-activated olefins.

Based on a single report from Khuong-Huu in 1974,[26] Hassner systematically examined the Lewis acid promoted hydroazidation of olefins in 1984.[27] Using $TiCl_4$ or $AlCl_3$ together with hydrazoic acid, hydroazidation of olefins with at least two geminal alkyl substituents or a phenyl substituent was possible (Figure 4.2). The limitation of this method is the requirement for stabilized carbocation intermediates.

The progress towards a mild general metal-based hydroazidation reaction has proven slow. Minor improvements have been reported by Kropp,[28] who used trimethylsilyl azide together with triflic acid on silica gel and Sreekumar,[29] who used zeolite bound sodium

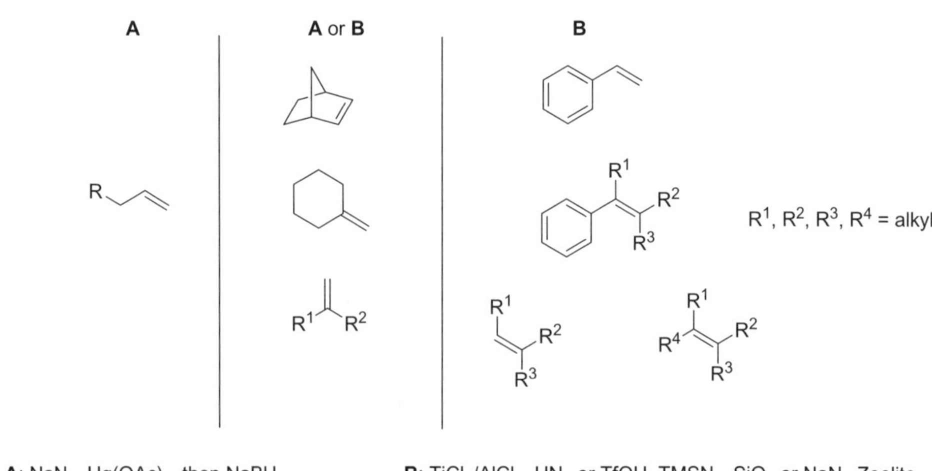

Figure 4.2 Methods for the hydroazidation of non-activated olefins

azide. Although these methods resulted in safer and more practicable procedures without the need of preparing dangerous hydrazoic acid solutions, they did not result in an expansion of the scope or efficiency of the hydroazidation reaction.

4.4 Cobalt-Catalyzed Hydroazidation

In 2004, we reported the Cobalt-catalyzed hydrohydrazination of olefins with di-*tert*-butyl azodicarboxylate (**5**) and phenylsilane (Scheme 4.1).[30] Our approach was based on a stepwise introduction of a hydride and an electrophilic nitrogen source, instead of the more classical approach based on electrophilic activation of the olefin followed by addition of a hydrazine nucleophile. This solution to override the inherently low reactivity of alkenes was first introduced by Mukaiyama for the related Cobalt-catalyzed hydroperoxidation reaction.[31] The introduction of new Cobalt-catalyst **4** was the key for an efficient hydrohydrazination reaction, as the Cobalt-complexes with acetylacetonate-derived ligands used by Mukaiyama promoted direct reduction of the azodicarboxylate.

The hydrohydrazination represented a general solution for the amination of alkenes, but the protected hydrazines obtained are sometimes difficult to transform to the free amines. At this point, we turned to sulfonyl azides as nitrogen sources, based on their capacity to react both with enolates[13] and carbon-centered radicals.[12] Mechanistic investigations of the hydrohydrazination reaction had suggested a radical character for the formed organocobalt intermediate.[14b] We were pleased to see that the Cobalt-catalyst **4** was able to promote the hydroazidation of 4-phenylbut-1-ene (**3**) with ethanesulfonyl azide (**7**), giving the product derived from the formal *Markovnikov* addition of hydrazoic acid onto the C-C double bond exclusively, albeit in moderate yields (50%).

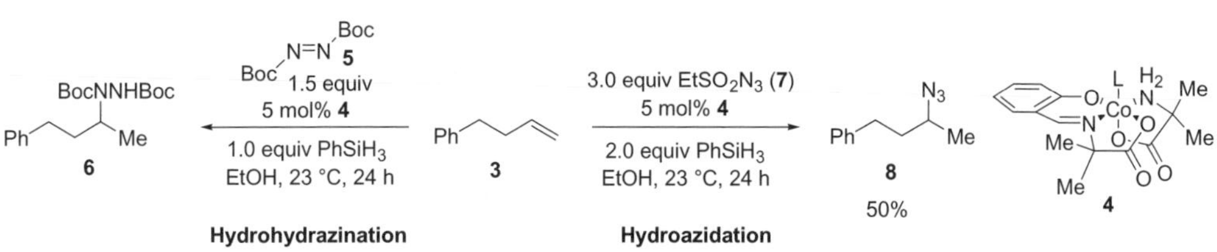

Scheme 4.1 *Hydrohydrazination and hydroazidation of 4-phenylbut-1-ene (**3**)*

4.4.1 Optimization of the Cobalt-Catalyzed Hydroazidation Reaction

At this stage of development, the hydroazidation reaction was still low yielding and very slow. Several Co-derived catalysts (Figure 4.3) for the hydroazidation of 4-phenylbut-1-ene (**3**) were examined. Interestingly, several Co-salen complexes were also able to catalyze the hydroazidation reaction, although the yields were lower. The most active of these catalysts was **9**, which afforded 45% yield of the desired azide. We also examined structural variations of ligands L for catalyst **4** and synthesized a library of related catalysts **10**. Introduction of substituents on the aromatic ring did not lead to any improvement. Interestingly, the second ligand on Co (X in **10**, Figure 4.3) did not have any influence on the reaction and 2-amino-isobutyric acid could be replaced by other amino acids, pyridine or water.

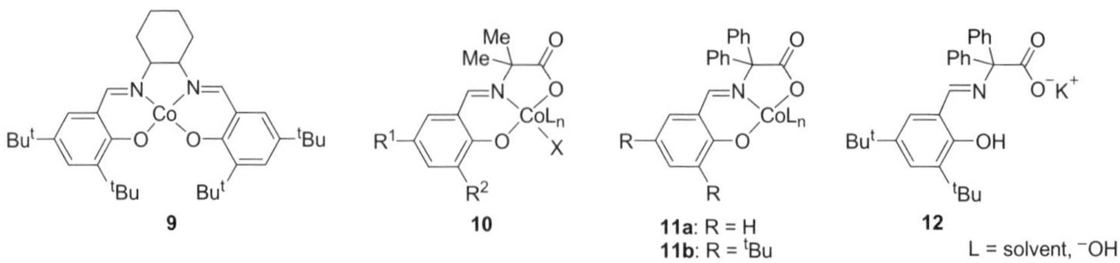

***Figure 4.3** Catalysts and ligands for the hydroazidation reaction*

We then examined complexes **11a** and **11b** derived from α,α-diphenylglycine. Complex **11b** displayed increased reactivity when compared to standard catalyst **4**, and azide **8** was obtained in 8 h in 70% yield with ethanesulfonyl azide (**7**) from 4-phenylbut-1-ene (**3**). Catalyst **11b** could be synthesized from the corresponding ligand **12** and Co(II) salts under oxidative conditions (Scheme 4.2). Ligand **12** was easily available from 3,5-di-*tert*-butylsalicylaldehyde (**13**) and α,α-diphenylglycine (**14**). Recently, ligand **12** has also become commercially available from Aldrich under the name SALDIPAC.[32]

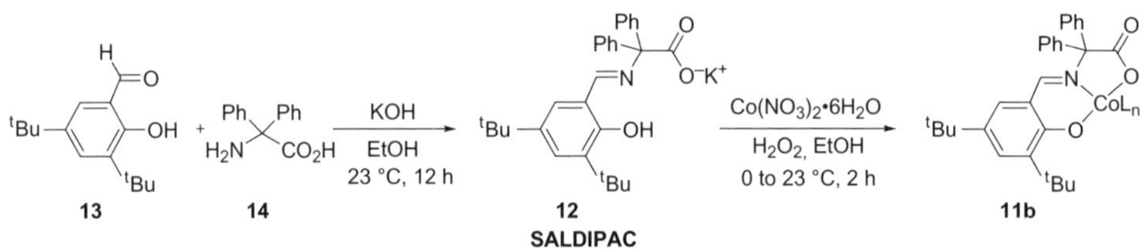

Scheme 4.2** Synthesis of catalyst **11b

The structural features of **11b** are noteworthy: even in the presence of excess amino acid, only the 1 : 1 complex of **12** and Co was detected by NMR spectroscopy in sharp contrast to complex **4** (Figure 4.4). This is probably due to the steric bulk of ligand **12** and explains its increased reactivity and lower stability. Unfortunately, we were unable to obtain reproducible results using complex **11b**, as yields (40–70%) and reaction time (8–48 h) were batch-dependent. In many cases, an initiation time was observed before the reaction started. Mukaiyama and co-workers used *tert*-butyl hydroperoxide as a cobalt-catalyst for the hydration of certain olefins when initiation of the reaction was difficult. A similar effect was observed in the hydroazidation reaction: when using catalyst **11** with ethanesulfonyl azide (**7**) for the hydroazidation of 4-phenylbut-1-ene (**3**), complete conversion was observed after 2–8 h using 30% of *tert*-butyl hydroperoxide. *In situ* formation of complex **11b** in the reaction mixture leads to reproducible reaction times (2 h) and yields (70%). $Co(BF_4)_2{\cdot}6H_2O$ was the best Co salt for this procedure, as complex formation was faster than with other salts and quick oxidation to the Co(III) complex occurred in the presence of *tert*-butyl hydroperoxide.

Careful examination of the reaction products showed that the yield was mostly limited by the formation of 4-phenylbutane (**15**) as the main side product (Equation 4.3). Consequently, improving the azide/alkane ratio was paramount to improve the efficiency of the process.

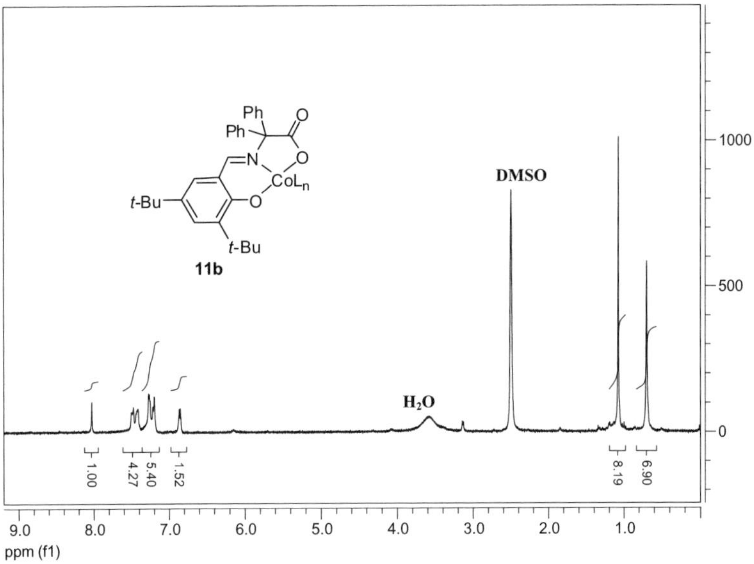

Figure 4.4 *^{1}H NMR spectra of catalyst **11b** measured at 300 MHz*

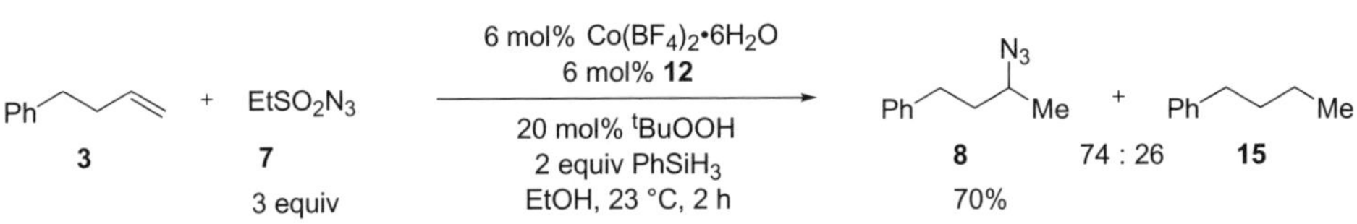

(4.3)

In order to solve this problem, we examined structural variation of the sulfonyl azide and silane (Table 4.1). Changing the nitrogen source from ethane- (**7**) to methanesulfonyl azide (**16**) showed no improvements (entry 2). The use of tosyl azide (**17**) (entry 3) gave the first increase in selectivity affording a 89 : 11 ratio in favor of the desired azide. No conversion was observed with the electron-poor nosyl azide (**18**) (entry 4).

Finally, we examined the effect of varying the silane structure on reaction rate and selectivity. The use of tetramethyldisiloxane (TMDSO) (entry 5) showed a small but significant increase in the azide/alkane ratio (84 : 16 vs. 77 : 23 with ethanesulfonyl azide (**7**)). The reaction with poly(methylhydrosiloxane) (PMHS) was too slow (entry 6). However, addition of a sub-stoichiometric amount of phenylsilane was enough to give useful conversion (entry 7). Triethylsilane and triethoxysilane (entry 8 and 9) could not be used. Finally, combining tosyl azide (**17**) and TMDSO gave full conversion of 4-phenylbut-1-ene (**3**) in 3 h with an improved azide/alkane ratio of 96 : 4 (entry 10) and 86% isolated yield.

4.4.2 Scope of the Hydroazidation of Olefins

The scope of the hydroazidation reaction was examined next, both with phenylsilane and TMDSO (General procedure **A** and **B** (Table 4.2)). All tested terminal olefins showed

Table 4.1 *Influence of silane and azide on the hydroazidation of 4-phenylbut-1-ene (**3**)*

Entry	Sulfonyl azide	Silane	Time	Conversion[a]	**8:15**[b]
1	$EtSO_2N_3$ (**7**)	$PhSiH_3$, 1.6 eq	2 h	>98%	77:23
2	$MeSO_2N_3$ (**16**)	$PhSiH_3$, 1.6 eq	2 h	>98%	77:23
3	TsN_3 (**17**)	$PhSiH_3$, 1.6 eq	4 h	>98%	89:11
4	NsN_3 (**18**)	$PhSiH_3$, 1.6 eq	24 h	<10%	nd
5	$EtSO_2N_3$ (**7**)	TMDSO, 2 eq	2 h	>98%	84:16
6	$EtSO_2N_3$ (**7**)	PMHS, 4H eq	24 h	20%	nd[c]
7	$EtSO_2N_3$ (**7**)	PMHS, 4H eq, $PhSiH_3$, 0.2 eq	18 h	81%	90:10
8	$EtSO_2N_3$ (**7**)	Et_3SiH, 4 eq	24 h	<10%	nd[c]
9	$EtSO_2N_3$ (**7**)	$(EtO)_3SiH$, 4 eq	24 h	<10%	nd[c]
10	TsN_3 (**17**)	TMDSO, 2 eq	3 h	>98%	96:4

[a] Standard conditions: 0.10 mmol 4-phenylbut-1-ene (**3**), 0.30 mmol sulfonyl azide, 30 mol% ^{t}BuOOH, 6 mol% $Co(BF_4)_2 \cdot 6H_2O$, 6 mol% ligand **12** in 0.50 mL ethanol at 23 °C under argon.
[b] Determined by gas chromatography.
[c] Not determined.

excellent *Markovnikov* selectivity. An aromatic ring in allylic or homoallylic position was well tolerated (entries 1–3). Surprisingly, styrene derivatives (not shown) were not reactive, although they had proven to be excellent substrates for the related hydrohydrazination reaction. The functionalization of safrole (**20**) (entry 3) led to azide **27** in 65% yield. Amines derived from azide **27** are a class of biologically active compounds well-known for their psychopharmacological activity.[33] In homoallylic and allylic alcohols, protection of the OH group was necessary to obtain useful yields (entries 4–6). Esters and ketones were tolerated (entries 7,8) with excellent chemoselectivity, as the carbonyl groups were not reduced.

Di- and trisubstituted olefins were also good substrates (Table 4.3). α,α-disubstituted olefins (entries 1–3) gave good yields of the tertiary azides. For these substrates, the competitive reduction of the alkene was less pronounced and use of the more reactive phenylsilane gave higher yields. Cyclooctene (**36**) also reacted, but the yield was moderate (entry 4). For trisubstituted olefins **37** and **38** (entries 5,6) full conversion of the starting material could not be achieved. Nevertheless, useful yields were obtained when using phenylsilane as reductant.

4.4.3 Further Process Optimization

Although reaction conditions are similar to the earlier developed hydrohydrazination reaction, the scope of the hydroazidation reaction appeared more limited. We hypothesized that this resulted from the lower reactivity of the tosyl azide towards a possible Co-alkyl (or radical) intermediate. Consequently, other potential azide sources were examined for the hydroazidation of 4-phenylbut-1-ene (**3**) (Figure 4.5).[34]

First, commercially available reagents **45**–**47** were examined. Phosphorous-based azide reagent **45** could not be used as azide source. Sulfonyl azides **46** and **47** gave more promising results with 4-phenylbut-1-ene (**3**), but with more sterically hindered substrates, yields were much lower. With these two reagents, precipitation and deactivation

Table 4.2 *The hydroazidation of monosubstituted olefins*

Entry	Alkene	Product	Isolated yield[a]
1	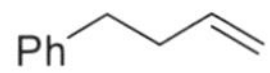 **3**	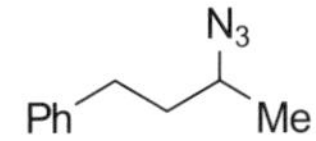 **8**	90% (86%)[b]
2	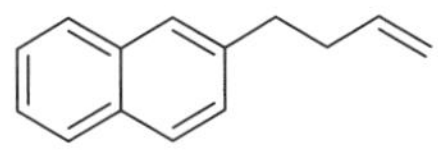 **19**	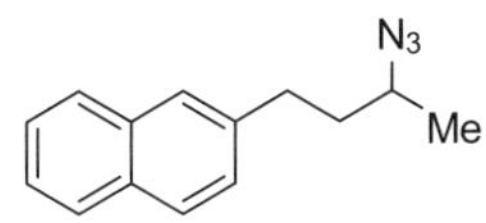 **26**	72%
3	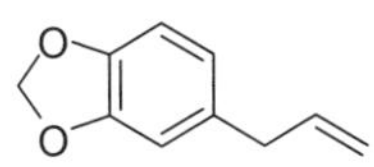 **20**	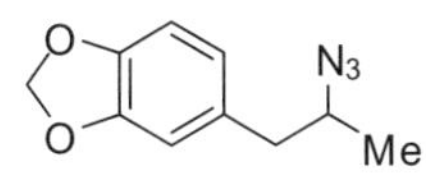 **27**	65%
4	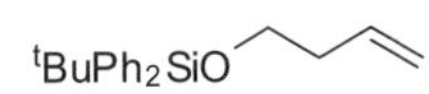 **21**	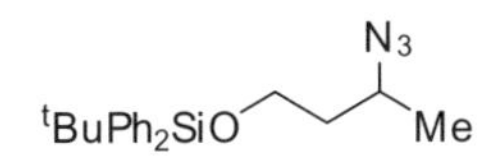 **28**	73% (85%)[b]
5	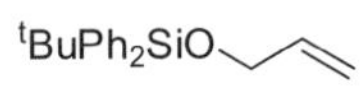 **22**	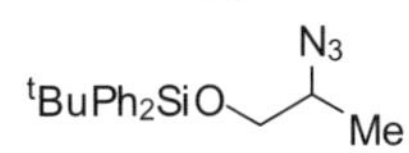 **29**	55% (67%)[b]
6	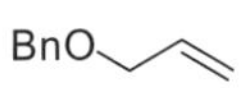 **23**	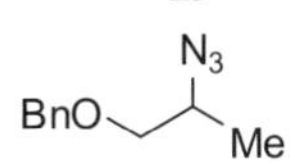 **30**	35% (39%)[b,c]
7	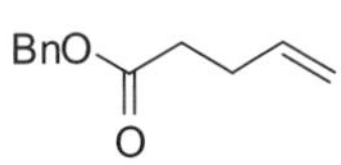 **24**	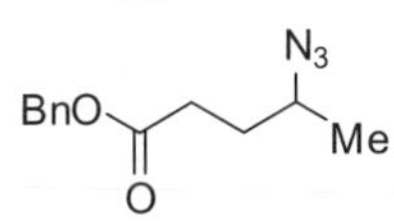 **31**	75% (77%)[b]
8	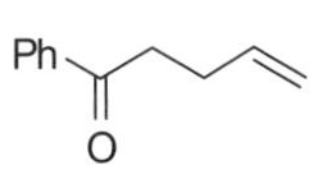 **25**	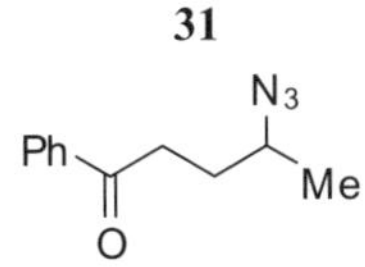 **32**	49%

[a] General procedure **A**: 0.50 mmol alkene, 0.80 mmol $PhSiH_3$, 1.5 mmol TsN_3 (**17**), 30 mol% *t*-BuOOH, 6 mol% ligand **12**, 6 mol% $Co(BF_4)_2{\cdot}6H_2O$, 2.5 mL ethanol at 23 °C under argon.
[b] General procedure **B**: 1.0 mmol TMDSO was used instead of $PhSiH_3$.
[c] 2.0 mmol TMDSO were used.

of the catalyst already occurred after 1 h, which made them inefficient for less reactive substrates. The chiral sulfonyl azide **48** showed no conversion. We then examined arylsulfonyl azides functionalized at the ortho position (compounds **49** and **50**), as we hypothesized that secondary interactions between the sulfonyl azide and the Cobalt-catalyst could prevent catalyst decomposition. Azide **49** bearing a methoxy group at the ortho position is easily synthesized from 4-methoxytoluene in two steps via chlorosulfonylation

Table 4.3 *Hydroazidation of di- and trisubstituted olefins*

Entry	Alkene	Product	Isolated yield[a]
1	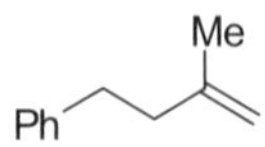 **33**	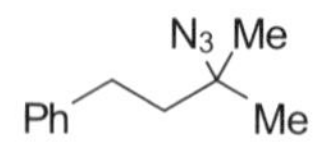 **39**	86% (90%)[b]
2	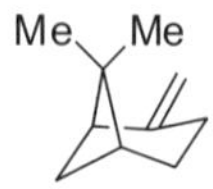 **34**	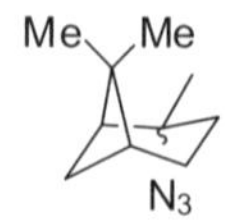 **40**	89% (dr 4:1)
3	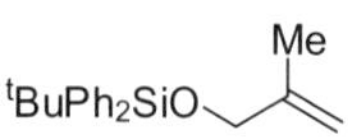 **35**	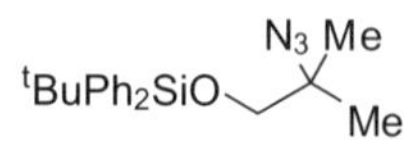 **41**	73%
4	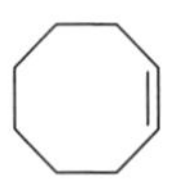 **36**	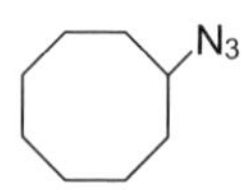 **42**	56%
5	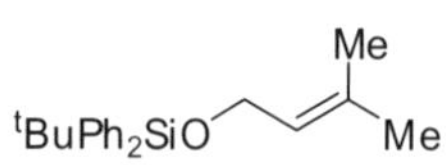 **37**	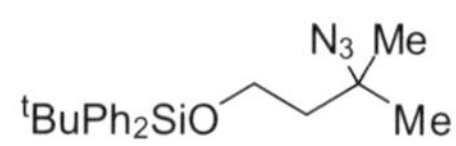 **43**	63%
6	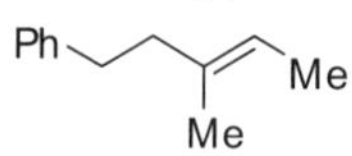 **38**	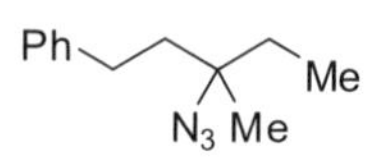 **44**	66%

[a] General procedure **A**: 0.50 mmol alkene, 0.80 mmol $PhSiH_3$, 1.5 mmol TsN_3 (**17**), 30 mol% ^tBuOOH, 6 mol% ligand **12**, 6 mol% $Co(BF_4)_2{\cdot}6H_2O$, 2.5 mL ethanol at 23 °C under argon.
[b] General procedure **B**: 1.0 mmol TMDSO was used instead of $PhSiH_3$.

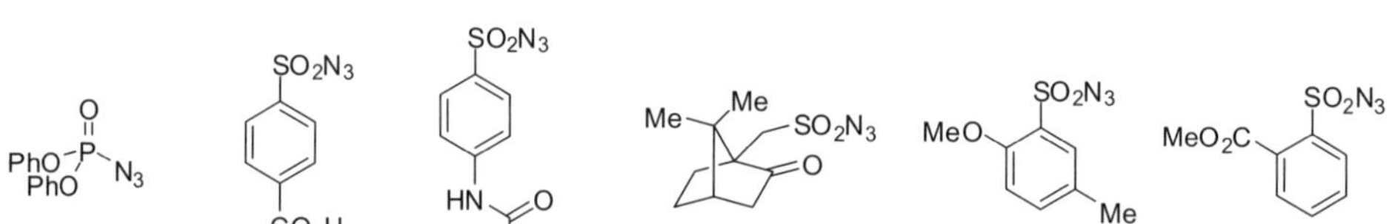

45	46	47	48	49	50
<10%	48%	82%	<10%	94%	91%

Figure 4.5 *Azides tested for the hydroazidation of 4-phenylbut-1-ene (**3**)*[35]

and reaction with sodium azide. We were pleased to see that this azide is a good nitrogen source, giving full conversion in 4 h without precipitation of the catalyst. In order to rule out a simple electron-donating effect of the methoxy group as the source of the improved stability, azide **50**, available in 3 steps from 2-sulfobenzoic anhydride, was examined and gave similar results as **49**.

In further studies, the potential of these new azide source was then analyzed (Table 4.4). For 4-phenylbut-1-ene (**3**), the use of only 1.5 equivalents of azide **49** or **50** gave

Table 4.4 *Comparison of* TsN_3 *(**17**), azide **49** and azide **50** in the hydroazidation reaction*

Entry	Alkene	Product	Yield[a] with 3 equiv TsN_3 (**17**)	Yield[b] with 1.5 equiv **49**	Yield[c] with 1.5 equiv **50**
1	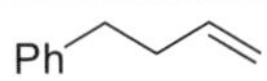 **3**	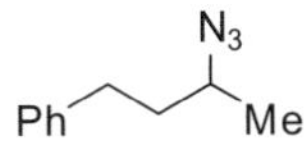 **8**	86%	94%	91%
2	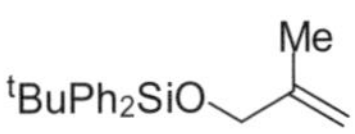 **35**	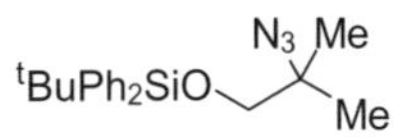 **41**	58%	89%	91%
3	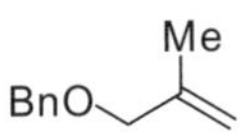 **51**	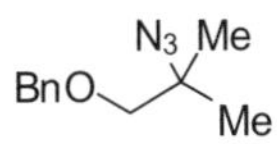 **52**	40%	64%	76%
4	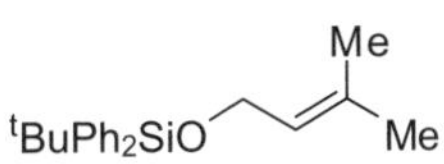 **37**	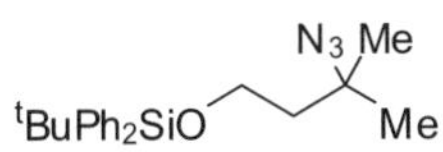 **43**	48%	83%	79%

[a]Standard conditions: 0.50 mmol alkene, 1.5 mmol TsN_3 (**17**), 1.0 mmol TMDSO, 30 mol% *t*-BuOOH, 6 mol% $Co(BF_4)_2{\cdot}6H_2O$, 6 mol% ligand **12** in 2.5 mL ethanol at 23 °C under argon.
[b]0.75 mmol sulfonyl azide **49** and 0.75 mmol TMDSO were used.
[c]0.75 mmol sulfonyl azide **50** and 0.75 mmol TMDSO were used and 1.0 mL methylene chloride was added as co-solvent.

more than 90% yield of the desired azide **8** (entry 1). In contrast, **8** was obtained in 70% yield only when using 1.5 equivalents tosyl azide (**17**). For the hydroazidation of mono-substituted alkenes, the use of **49** or **50** gave no improvement, but the yields were increased in the case of allylic ethers **35** and **51** bearing an α-methyl disubstituted double bond, leading to good conversion with half as much sulfonyl azide (entries 2,3). Finally, trisubstituted olefin **37** could also be functionalized in good yield (entry 4).

Although reagents **49** and **50** offered a more efficient access to tertiary azides, they required a multi-step synthesis. Furthermore, their safety profile has not been examined in detail, which would be important for larger scale applications. The use of commercially available azides would be more practical and safer, as these reagents are carefully being tested. We were never able to find general conditions for the hydroazidation using 4-acetamido benzenesulfonyl azide (**47**). For 4-carboxybenzenesulfonyl azide (**46**), we hypothesized that the low yield could be associated with the acidity of this reagent. Indeed, the sodium salt of **46**, obtained via deprotonation with sodium hydride, was a much better reagent for geminally disubstituted double bonds (Equation 4.4). To our surprise, it was even largely superior to tosyl azide (**17**).

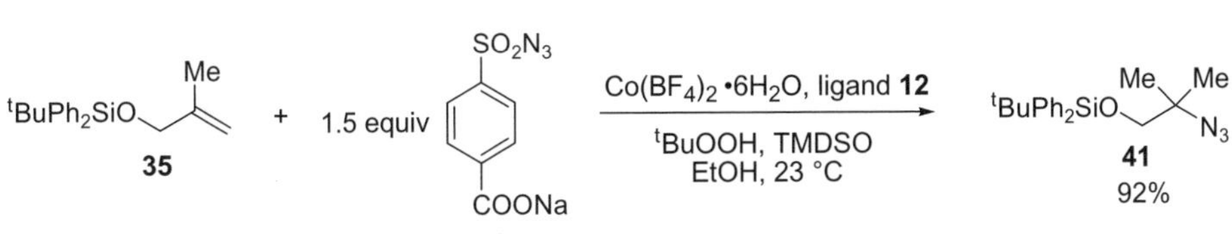

(4.4)

At this point, we decided to reexamine the functional group tolerance of the hydroazidation reaction with this class of olefins, as the tertiary azides are not easily accessed via substitution reactions and no chiral center is formed during the reaction, alleviating issues of diastereoselectivity with chiral substrates (Table 4.5).[36]

Non-functionalized alkenes and protected alcohols were good substrates for the reaction (entries 1,2). Free alcohols, which were not tolerated when using TsN_3 (**17**), could be employed, providing the corresponding azides in good yields (entries 3,11). The reaction was also successful when esters (entries 4–6) and amides (entry 7) were present. Amino acid derivatives with the alkene functionality connected to either the *O*-end through an ester linkage (entries 8,9,11) or to the *N*-end through an amide bond (entry 10) gave interesting products in useful yields. Alkenes conjugated to an ester or a phenyl showed no conversion at all and they represent the limitation of the process.

In order to avoid pre-formation and isolation of sodium 4-azidosulfonylbenzoate (**53**), we envisaged forming the reagent *in situ*. Attempts to access **53** via deprotonation of the acid **46** directly in the reaction mixture were not practicable, as it resulted in a very dense heterogeneous mixture and the stirring was not efficient. Using NEt_3 as a base in EtOH was more convenient, as it led to a clear homogeneous mixture. The efficiency of the new reagent was then compared with that of the isolated sodium salt **53** (Table 4.5).

Alkenes **33** and **35** and esters gave similar yields as with azide **53** (entries 1,2,4,5). However, for substrates where the alkene functionality is connected to an amino acid through an ester bond, slightly lower yields were observed (entries 8,9).

4.4.4 One-pot Functionalization of the Azide Products

The reduction of azides to free amines[37] or the cycloaddition reaction of azides with terminal alkynes[2] are well-established methods. The mild conditions of the olefin hydroazidation reaction permitted us to examine these processes without the need of isolation and purification of the azides themselves (Scheme 4.3). In the reduction reaction a simple extraction procedure sufficed to allow isolation of the free amine **73** in 76% yield and 95% purity, as determined by NMR (Equation 4.5).

The initially developed procedure using TsN_3 (**17**) is not convenient for reactions conducted on larger scale, because a large excess of reagents (3 equivalents TsN_3 (**17**), 2 equivalents TMDSO) is needed to achieve useful yields. The use of azide **49** was preferable, as it did not lead to decomposition of the catalyst and large excess of reagents are not needed to drive the reaction to completion. For the reaction of 4-phenylbut-1-ene (**3**) on a 5 mmol scale (Equation 4.5), the use of only 1.2 equivalents of **49** and 1.0 equivalents

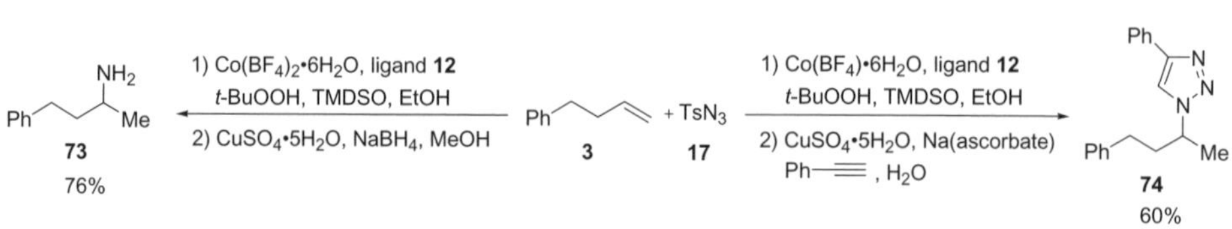

Scheme 4.3 In situ *functionalization of the hydroazidation product*

Table 4.5 *Hydroazidation reaction with deprotonated sulfonide azide* ***46***

Entry	Alkene	Product	Yield with **53**[a]	Yield with **46** and NEt_3[b]
1	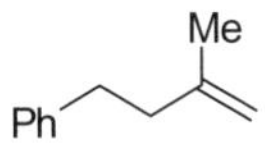 **33**	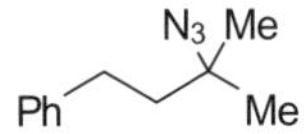 **39**	94	99
2	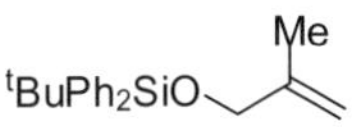 **35**	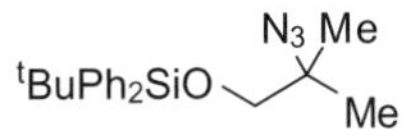 **41**	92	93
3	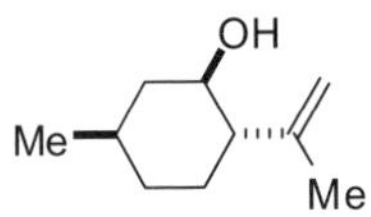 **54**	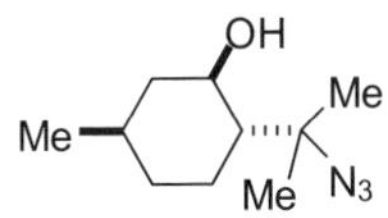 **63**	71	–
4	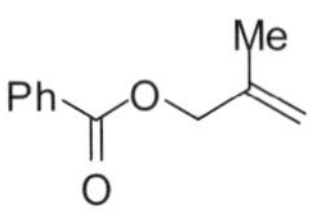 **55**	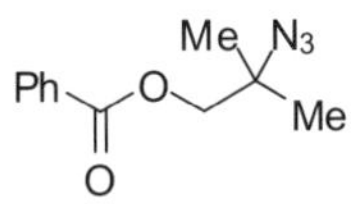 **64**	48	76
5	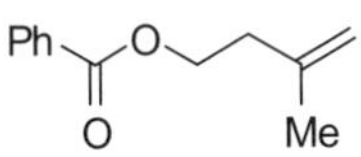 **56**	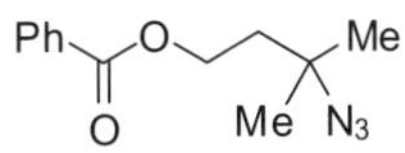 **65**	70	62
6	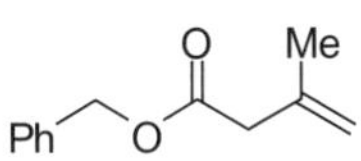 **57**	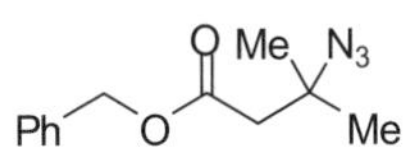 **66**	84	–
7	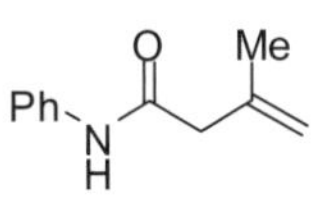 **58**	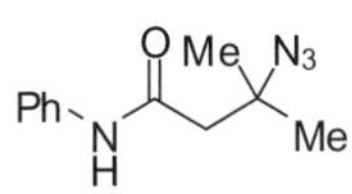 **67**	73	–
8	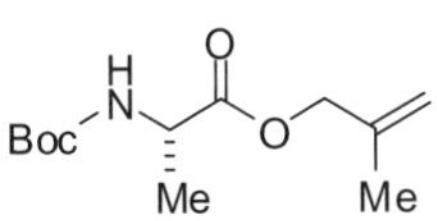 **59**	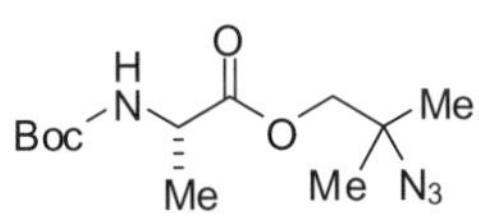 **68**	74	57
9	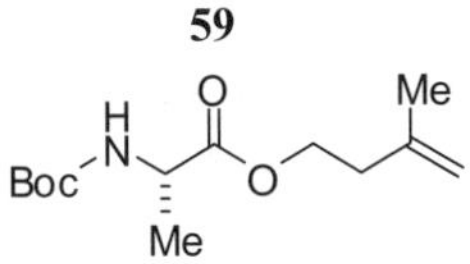 **60**	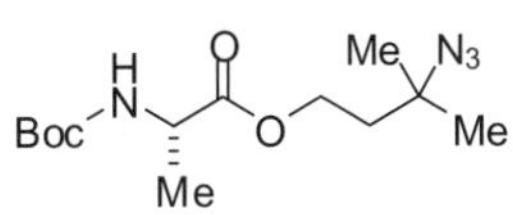 **69**	75	57
10	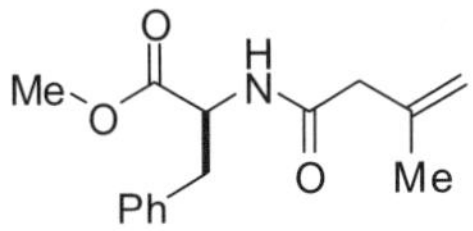 **61**	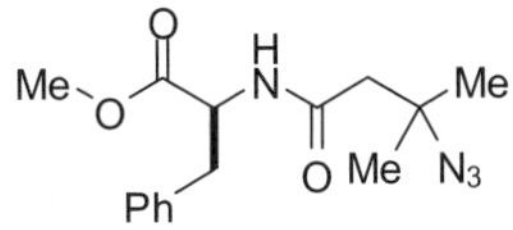 **70**	78	–
11	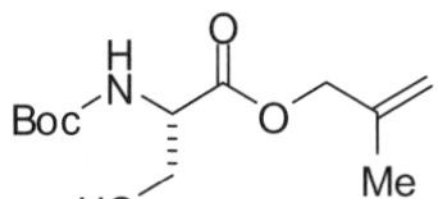 **62**	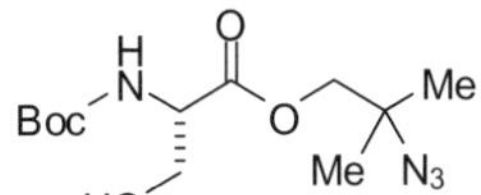 **71**	52	–

[a] General conditions: $Co(BF_4)_2 \cdot 6H_2O$ (6 mol%), ligand **12** (6 mol%), alkene (0.5 mmol), azide **53** (0.75 mmol), *t*-BuOOH (28 mol%), TMDSO (1 mmol), EtOH (2.5 ml), argon, 23 °C.
[b] As with a, but with azide **46** (0.75 mmol), and NEt_3 (0.715 mmol).

of TMDSO were sufficient to afford the desired amine **73** in 68% overall yield. Furthermore, we were able to lower the catalyst loading to 3%.

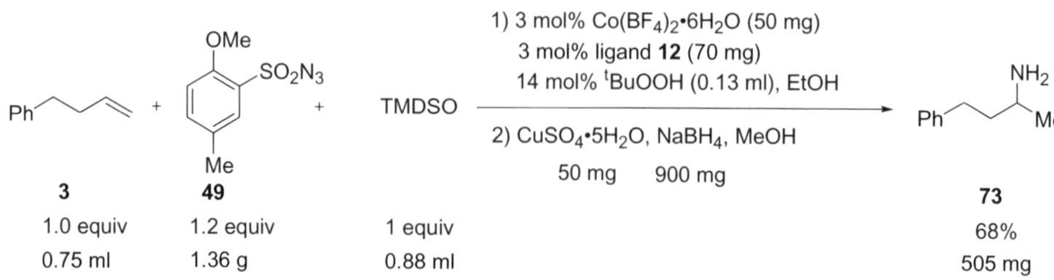

(4.5)

The importance of a one-pot protocol is particularly relevant in the case of small volatile substrates, where the isolation of the azide intermediates would be both difficult and hazardous. In contrast, the corresponding amines can be isolated as their solid hydrochloride salts, which can be purified via recrystallization. For example, hydroazidation of methylenecyclobutane (**75**) followed by *in situ* reduction and acidification yielded pure cyclobutane amine hydrochloride (**77**) in 60% overall yield after recrystallization (Scheme 4.4).

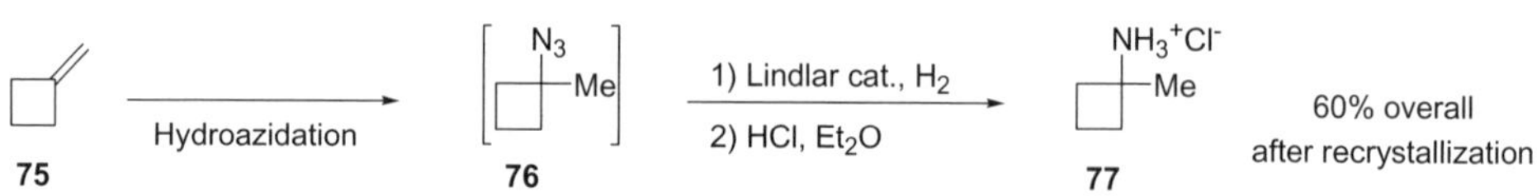

Scheme 4.4 *One-pot synthesis of cyclobutane amine hydrochloride (**77**)*

4.4.5 Mechanistic Investigations

Although we have conducted some mechanistic studies, the data gathered for the hydroazidation reaction does not allow formulation of a precise mechanism. However, a working model, which is in accordance with our observations, is presented in Scheme 4.5. We hypothesize an entry in the catalytic cycle via the Co-hydride complex **I**. Hydrocobaltation would then give the Co-alkyl complex **II**. Two pathways can be envisaged next: a free radical pathway (**A**) or direct reaction of Co-alkyl complex **II** with sulfonyl azide (**B**). The reaction of free radicals with sulfonyl azide has already been reported by Renaud and co-workers.[12] The radical adduct formed in the addition to the terminal or internal N–N bond of the sulfonyl azide could be re-captured by a Co(II) complex to give Cobalt-complex **III**, which can collapse to the alkyl azide and a sulfonyl radical. Direct reaction of the Co-alkyl complex **II** with the sulfonyl azide to form **III** is another possibility. In contrast to the hydrohydrazination reaction, the amination step is successful only if no R group with a stabilizing effect (for example phenyl, ester, alkyne) is present on the alkene.

To close the catalytic cycle, the intermediacy of a Co-sulfonato complex **IV** can be proposed, which could be formed directly from **III** via elimination. The formation of **IV**

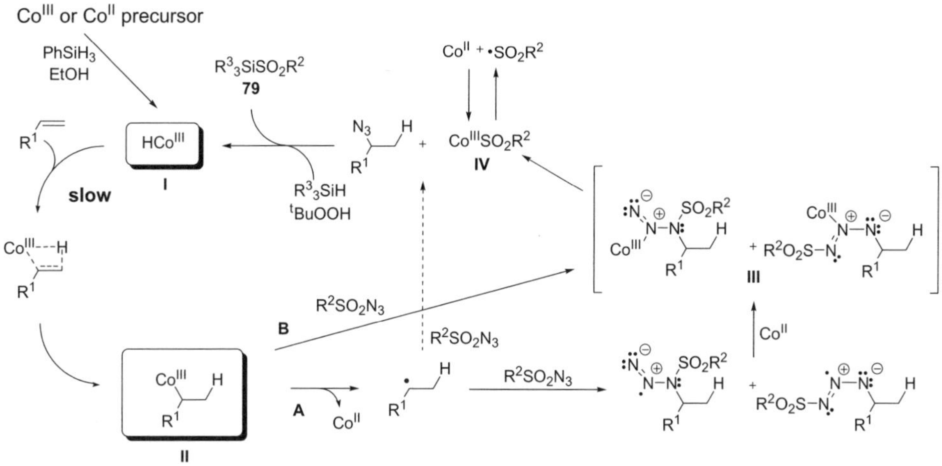

Scheme 4.5 *Working model for the mechanism of the hydroazidation reaction*

could be the origin of the inactive precipitate observed in the reaction (see above). As the order of the reaction in the sulfonyl azide was zero, the azidation step is not rate-determining, and peroxide additives could be important for the acceleration of the conversion of **III** or **IV** to Co-hydride complex **I**, allowing useful turnovers. Furthermore, silane derivatives of polymeric nature were often detected in the crude mixture. A mixed silane/sulfonate species such as **79** would be in accordance with our working model. Obviously, further efforts are needed to identify the exact nature of the observed precipitate as well as the end products resulting from the silane and sulfonyl azides in order to elucidate the exact mechanism of the hydroazidation reaction.

4.5 Conclusion

Considering the increasing number of applications for organic azides in biological and materials sciences, efficient methods for their synthesis are highly desirable. The methods we have described provide access to this important class of compounds from readily available olefins as starting materials. Future developments in this field will certainly be focused on the identification of new catalyst systems having wide substrate scope and providing experimental procedures.

References

[1] (a) E.F.V. Scriven, K. Turnbull, *Chem. Rev.* **1988**, *88*, 297–368. (b) S. Bräse, C. Gil, K. Knepper, V. Zimmermann, *Angew. Chem. Int. Ed.* **2005**, *44*, 5188–240.
[2] (a) R. Huisgen, R. Knorr, L. Möbius, G. Szeimies, *Chem. Ber.* **1965**, *98*, 4014–21. (b) Z.P. Demko, K.B. Sharpless, *Angew. Chem. Int. Ed.* **2002**, *41*, 2110–3. (c) Z.P. Demko, K.B.

Sharpless, *Angew. Chem. Int. Ed.* **2002**, *41*, 2113–6. See also the contribution of C. Schilling, N. Jung, S. Bräse (Chapter 9).

[3] (a) H.C. Kolb, K.B. Sharpless, *Drug Discov. Today* **2003**, *8*, 1128–37. (b) M. Kohn, R. Breinbauer, *Angew. Chem. Int. Ed.* **2004**, *43*, 3106–16. See also the contribution of C.W. Tornøe, M. Meldal (Chapter 10).

[4] See the contribution of T. M. V. D. Pinho e Melo (Chapter 3).

[5] See for example: L.E. Martinez, J.L. Leighton, D.H. Carsten, E.N. Jacobsen, *J. Am. Chem. Soc.* **1995**, *117*, 5897–8.

[6] See for example: D. Lertpibulpanya, S.P. Marsden, I. Rodriguez-Garcia, C.A. Kilner, *Angew. Chem. Int. Ed.* **2006**, *45*, 5000–2.

[7] (a) A. Hassner, L.A. Levy, *J. Am. Chem. Soc.* **1965**, *87*, 4203–4. (b) A. Kirschning, H. Monenschein, C. Schmeck, *Angew. Chem. Int. Ed.* **1999**, *38*, 2594–6.

[8] (a) H. Schäfer, *Angew. Chem. Int. Ed.* **1970**, *9*, 158–9. (b) W.E. Fristad, T.A. Brandvold, J.R. Peterson, S.R. Thompson, *J. Org. Chem.* **1985**, *50*, 3647–9. (c) R.M. Moriarty, J.S. Khosrowshahi, *Tetrahedron Lett.* **1986**, *27*, 2809–12. (d) P. Magnus, M.B. Roe, C. Hulme, *J. Chem. Soc., Chem. Comm.* **1995**, 263–5. (e) B.B. Snider, H. Lin, *Synth. Commun.* **1998**, *28*, 1913–22. (f) R. Chung, E.S. Yu, C.D. Incarvito, D.J. Austin, *Org. Lett.* **2004**, *6*, 3881–4.

[9] (a) M. Tingoli, M. Tiecco, D. Chianelli, R. Balducci, A. Temperini, *J. Org. Chem.* **1991**, *56*, 6809–13. (b) R.M. Giuliano, F. Duarte, *Synlett* **1992**, 419–21. (c) S. Czernecki, D. Randriamandimby, *Tetrahedron Lett.* **1993**, *34*, 7915–16. (d) M. Tingoli, M. Tiecco, L. Testaferri, R. Andrenacci, R. Balducci, *J. Org. Chem.* **1993**, *58*, 6097–102. (e) M. Tiecco, L. Testaferri, A. Temperini, L. Bagnoli, F. Marini, C. Santi, *Synth. Commun.* **1998**, *28*, 2167–79. (f) Y.C. Mang, L.L. Wu, M. Huang, *Chin. Chem. Lett.* **2003**, *14*, 451–2. (g) M. Tiecco, L. Testaferri, C. Sand, C. Tomassini, F. Marini, L. Bagnoli, A. Temperini, *Angew. Chem. Int. Ed.* **2003**, *42*, 3131–3. (h) Y.V. Mironov, A.A. Sherman, N.E. Nifantiev, *Tetrahedron Lett.* **2004**, *45*, 9107–10.

[10] W.S. Trahanov, M.D. Robbins, *J. Am. Chem. Soc.* **1971**, *93*, 5256–8.

[11] V. Nair, L.G. Nair, T.G. George, A. Augustine, *Tetrahedron* **2000**, *56*, 7607–11.

[12] (a) C. Ollivier, P. Renaud, *J. Am. Chem. Soc.* **2000**, *122*, 6496–7. (b) P. Renaud, C. Ollivier, P. Panchaud, *Angew. Chem. Int. Ed.* **2002**, *41*, 3460–2. (c) P. Panchaud, L. Chabaud, Y. Landais, C. Ollivier, C. Jimeno, P. Renaud, S. Zigmantas, *Chem. Eur. J.* **2004**, *10*, 3606–14. See also the contribution of C. Jimeno, P. Renaud (Chapter 8).

[13] Prior to our work, electrophilic azidation was limited to highly nucleophilic enol (silyl)ether derivatives. See for example: (a) D.A. Evans, T.C. Britton, J.A. Ellman, R.L. Dorow, *J. Am. Chem. Soc.* **1990**, *112*, 4011–30. (b) P. Magnus, J. Lacour, P.A. Evans, M.B. Roe, C. Hulme, *J. Am. Chem. Soc.* **1996**, *118*, 3406–18.

[14] (a) J. Waser, H. Nambu, E.M. Carreira, *J. Am. Chem. Soc.* **2005**, *127*, 8294–5. (b) J. Waser, B. Gaspar, H. Nambu, E.M. Carreira, *J. Am. Chem. Soc.* **2006**, *128*, 11693–712.

[15] (a) *Enantioselective Synthesis of β-Amino Acids*, (ed.: E. Juaristi), Wiley-VCH, New York, **1997**. (b) K. Gademann, T. Hintermann, J.V. Schreiber, *Curr. Med. Chem.* **1999**, *6*, 905–25. (c) M. Liu, M.P. Sibi, *Tetrahedron* **2002**, *58*, 7991–8035. (d) D.L. Steer, R.A. Lew, P. Perlmutter, A.I. Smith, M.I. Aguilar, *Curr. Med. Chem.* **2002**, *9*, 811–22. (e) D. Seebach, A.K. Beck, D.J. Bierbaum, *Chem. Biodivers.* **2004**, *1*, 1111–239.

[16] E. Oliveri-Mandala, E. Calderao, *Gazz. Chim. Ital.* **1915**, *45*, 307.

[17] J.H. Boyer, *J. Am. Chem. Soc.* **1951**, *73*, 5248–52.

[18] (a) G.R. Harvey, K.W. Ratts, *J. Org. Chem.* **1966**, *31*, 3907–10. (b) X. Huang, R.W. Shen, T.X. Zhang, *J. Org. Chem.* **2007**, *72*, 1534–7. For a review on acceptor-substituted allenes, see: (c) K. Banert, J. Lehmann, in *Modern Allene Chemistry, Vol. 1* (eds.: N. Krause, A.S. Hashmi), Wiley, **2005**, pp. 359–424.

[19] B.Y. Chung, Y.S. Park, I.S. Cho, B.C. Hyun, *Bull. Korean Chem. Soc.* **1988**, *9*, 269–70.

[20] P. Lakshmipathi, A.V.R. Rao, *Tetrahedron Lett.* **1997**, *38*, 2551–2.

[21] (a) J.K. Myers, E.N. Jacobsen, *J. Am. Chem. Soc.* **1999**, *121*, 8959–60. (b) M.S. Taylor, D.N. Zalatan, A.M. Lerchner, E.N. Jacobsen, *J. Am. Chem. Soc.* **2005**, *127*, 1313–7.

[22] (a) D.J. Guerin, T.E. Horstmann, S.J. Miller, *Org. Lett.* **1999**, *1*, 1107–9. (b) T.E. Horstmann, D.J. Guerin, S.J. Miller, *Angew. Chem. Int. Ed.* **2000**, *39*, 3635–8. (c) D.J. Guerin, S.J. Miller, *J. Am. Chem. Soc.* **2002**, *124*, 2134–6.

[23] (a) L.W. Xu, C.G. Xia, J.W. Li, S.L. Zhou, *Synlett* **2003**, 2246–8. (b) L.W. Xu, L. Li, C.G. Xia, S.L. Zhou, J.W. Li, *Tetrahedron Lett.* **2004**, *45*, 1219–21. (c) L. Castrica, F. Fringuelli, L. Gregoli, F. Pizzo, L. Vaccaro, *J. Org. Chem.* **2006**, *71*, 9536–9. (d) S.G. Kim, T.H. Park, *Synth. Commun.* **2007**, *37*, 1027–35. (e) K. Tsuboike, D.J. Guerin, S.M. Mennen, S.J. Miller, *Tetrahedron* **2004**, *60*, 7367–74. (f) I. Adamo, F. Benedetti, F. Berti, P. Campaner, *Org. Lett.* **2006**, *8*, 51–4.
[24] For the first report, see: S.N. Ege, K.W. Sherk, *J. Am. Chem. Soc.* **1953**, *75*, 354–7.
[25] (a) C.H. Heathcock, *Angew. Chem. Int. Ed.* **1969**, *8*, 134–5. (b) J.E. Galle, A. Hassner, *J. Am. Chem. Soc.* **1972**, *94*, 3930–3. (c) V.R. Kartashov, T.N. Sokolova, A.Y. Pavinskii, I.V. Timofeev, A.B. Radbil, *Russ. Chem. Bull.* **1995**, *44*, 2375–81.
[26] A. Pancrazi, Q. Khuong-Huu, *Tetrahedron* **1974**, *30*, 2337–43.
[27] A. Hassner, R. Fibiger, D. Andisik, *J. Org. Chem.* **1984**, *49*, 4237–44.
[28] G.W. Breton, K.A. Daus, P.J. Kropp, *J. Org. Chem.* **1992**, *57*, 6646–9.
[29] R. Sreekumar, R. Padmakumar, P. Rugmini, *Chem. Commun.* **1997**, 1133–4.
[30] J. Waser, E.M. Carreira, *J. Am. Chem. Soc.* **2004**, *126*, 5676–7.
[31] (a) S. Isayama, T. Mukaiyama, *Chem. Lett.* **1989**, 1071–4. (b) S. Isayama, *Bull. Chem. Soc. Jpn.* **1990**, *63*, 1305–10.
[32] SALDIPAC, Aldrich catalog number 676551.
[33] D.E. Nichols, A.J. Hoffman, R.A. Oberlender, P. Jacob, A.T. Shulgin, *J. Med. Chem.* **1986**, *29*, 2009–15.
[34] The use of benzenesulfonyl azide led to similar results as tosyl azide. Mesitylsulfonyl and 2,4,6-triisopropylbenzenesulfonyl azide led to a significant drop in the reaction rate.
[35] The conversions using 2 equiv of azide are given.
[36] B. Gaspar, J. Waser, E.M. Carreira, *Synthesis* **2007**, 3839.
[37] (a) E.J. Corey, K.C. Nicolaou, R.D. Balanson, Y. Machida, *Synthesis* **1975**, 590–1. (b) H.S.P. Rao, P. Siva, *Synth. Commun.* **1994**, *24*, 549–55.

PART 2

Reactions

5

The Chemistry of Vinyl, Allenyl, and Ethynyl Azides

Klaus Banert

Institute of Chemistry, Chemnitz University of Technology, Strasse der Nationen 62, 09111 Chemnitz, Germany

Dedicated to Professor Helmut Quast on the occasion of his 75th birthday

5.1 Introduction and Early Synthetic Methods for Vinyl Azides

The parent compound of vinyl azide **1** has been known for about hundred years when Forster and Newman prepared and characterized this compound (Scheme 5.1).[1] However, after publication of this first example of vinyl azides, these compounds were ignored for approximately fifty years. Perhaps, the multi-step synthesis of **1** and the assumption that dehydrohalogenation of homologous vicinal azidohaloalkanes may lead to mixtures of vinyl and allyl azides detained the investigation of the title compounds of type **1**. Furthermore, the report[2] on the explosive properties of **1** could be responsible for the reservedness against vinyl azides. These compounds are indeed more sensitive than alkyl azides but more stable than acyl azides in most cases. Thus, great amounts of vinyl azides should be handled not as a pure substance but in solution if the number of carbon atoms does not exceed that of the nitrogen atoms by a factor of three.

The only alternative method for the synthesis of vinyl azides already known in 1923 was the nucleophilic substitution of vinyl halides bearing an electron-withdrawing group in the β-position. However, this method was used at first mainly for the generation of azidoquinones as depicted in Scheme 5.2.[3–5]

In the case of other vinyl halides with an electron acceptor in the β-position, the generation of vinyl azides is often followed by rapid liberation of dinitrogen and formation of a five-membered aromatic heterocycle. In 1958, for example, treatment of nitro

Organic Azides: Syntheses and Applications Edited by Stefan Bräse and Klaus Banert

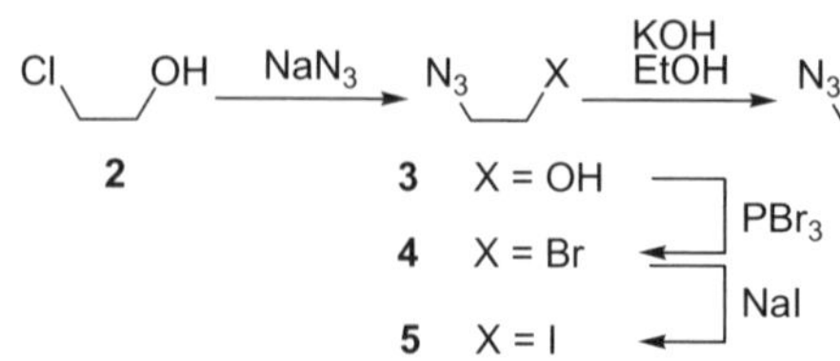

Scheme 5.1 *Synthesis of vinyl azide (**1**) as the first representative of this class of azides*[1]

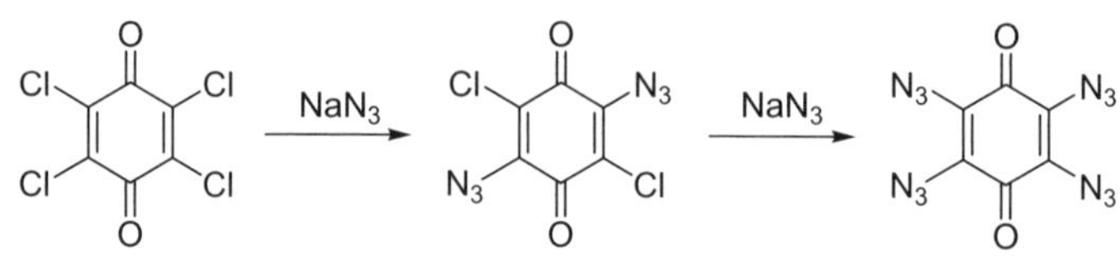

Scheme 5.2 *Synthesis of azidoquinones by nucleophilic substitution*[3]

compound **9** with sodium azide was found to give the furoxan **11**, which was explained by ring closure of the short-lived intermediate **10** (Scheme 5.3).[6] One decade later, stereochemistry of vinyl azides was recognized to have a significant influence on the stability and the sequential reactions of these compounds. Thus, (*E*)-configured vinyl azide (*E*)-**13** can be isolated after synthesis from precursor (*E*)-**12** and leads to the main product **14** only on heating. On the other hand, the nucleophilic substitution of the starting material (*Z*)-**12** afforded directly the isoxazole **15** because the primary product (*Z*)-**13** is highly unstable and tends to rapid separation of dinitrogen.[7] Stereospecifity in the transformation of vinyl halides into vinyl azides is not a rare case. This phenomenon of retention is observed, for example, also recently in the synthesis of diazides (*Z,Z*)-**17** and (*E,Z*)-**17**.[8,9] However, an interesting (*Z*) → (*E*) isomerization of a vinyl azide was studied recently in the reaction of 3-azidomethylenedihydro-(3*H*)-furan-2-one with sodium azide in aqueous acetone.[10]

In 1957, it was shown that the addition of hydrazoic acid, generated in situ from potassium azide and acetic acid, at the electron-poor alkyne **18** led to the vinyl azide **19** (Scheme 5.4).[11] Several analogous attempts to add the same reagent to acetylenes, bearing an electron-withdrawing group, were also successful in the following years.[12] Instead of hydrazoic acid, less dangerous tetramethylguanidinium azide (TMGA) was also utilized.[13] However, treatment of alkynes with trimethylsilyl azide usually gave the corresponding 1,2,3-triazoles.[14] Some exceptions, in which electron-poor alkynes were transformed into vinyl azides with the help of trimethylsilyl azide, have been reported recently.[15] If solutions of sodium azide in protic solvents are used, the reaction with acetylenes, bearing an electron-withdrawing group, can also lead to vinyl azides.[16] But 1,2,3-triazoles are formed when the pH is too high or aprotic solvents are applied.[17] Even in the case of the reagent hydrazoic acid, older articles described the exclusive formation of 1,2,3-triazoles from alkynes.[18] In two reports, for example, the heterocycle **22** was the only product mentioned and characterized after treatment of propynal (**20**) with hydrazoic acid.[19] However, nearly equal amounts of **22** and the vinyl azide **21** were obtained when the reaction was repeated

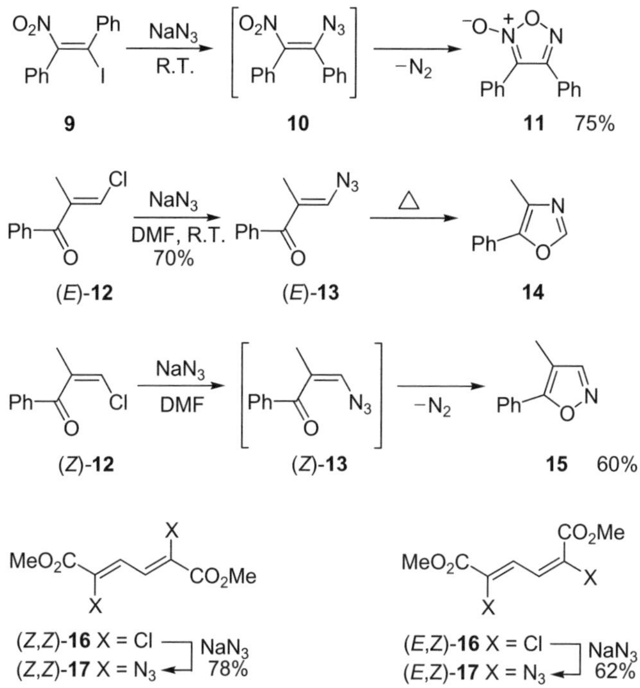

Scheme 5.3 *Synthesis of vinyl azides by nucleophilic substitution of vinyl halides bearing an electron-withdrawing group*[6–8]

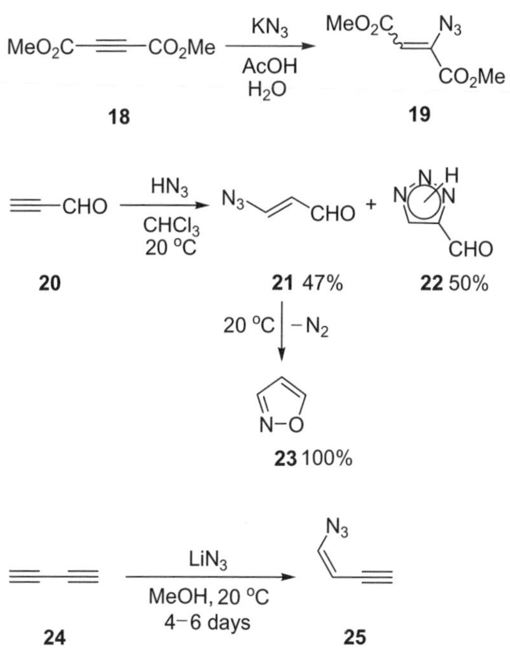

Scheme 5.4 *Synthesis of vinyl azides by nucleophilic addition of hydrazoic acid at alkynes*[11,19,20,22]

recently (Scheme 5.4).[20] The addition product **21** is a highly explosive, yellow oil, which quantitatively afforded isoxazole (**23**) when stored in solution at room temperature. The concerted as well as stepwise reaction pathways of the cyclization of 3-azidoacrolein to the heterocycle **23** have been comprehensively studied by density functional and *ab initio* methods presupposing (*Z*) configuration of the starting material.[21]

Vinyl azides were seldom prepared by nucleophilic addition at alkynes lacking in strongly electron-withdrawing groups. Thus, the reaction of butadiyne (**24**) with lithium azide in methanol is such a rare case, which allows the isolation of enyne **25** as a yellow liquid.[22]

Already in 1935, vinyl azides of type **27** were prepared by treating dibromides **26** with sodium azide (Scheme 5.5).[23] The regioselectivity of this transformation is obviously caused by the carbonyl group, which turned the bromo atom in the α-position to a highly reactive leaving group for nucleophilic substitution, and the acidity of the α-hydrogen atom favors the elimination of hydrogen bromide including separation of the β-bromo atom.[24] Thus, the α-azido-β-bromoketone or ester can be discussed as a plausible intermediate. However, in some cases, the vicinal diazide was detected as the primary product,[25] that needed a base to give the desired azide **27**.[26] Even the corresponding α-bromo-α,β-unsaturated ketone was found to be an intermediate in the transformation **26** → **27**.[27] Open-chain products are usually formed with (*Z*)-configuration as shown in Scheme 5.5.[28] But if **27** is part of a cyclic structure, (*E*)-configuration is unavoidable.[29]

Not only the dibromides originating from α,β-unsaturated ketones or esters but also those prepared from styrenes or similar compounds can be transformed directly into vinyl azides when treated with sodium azide.[30] One-pot procedures with sodium azide inducing the substitution step and sodium hydroxide to perform the elimination were also successful in the synthesis of vinyl azides **29** (Scheme 5.6).[31] In 1961, Smolinsky found the first

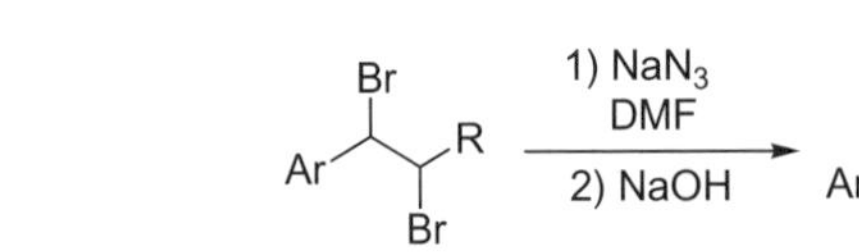

26 **27**

a	R^1 = H	R^2 = OBn	95%
b	R^1 = Ph	R^2 = Me	86%
c	R^1 = Me	R^2 = Ph	91%
d	R^1 = Ph	R^2 = Ph	87–91%
e	R^1 = alkyl	R^2 = OMe	86–87%

Scheme 5.5 *Synthesis of vinyl azides from α,β-dibromoketones or esters*[23,24,27,28b]

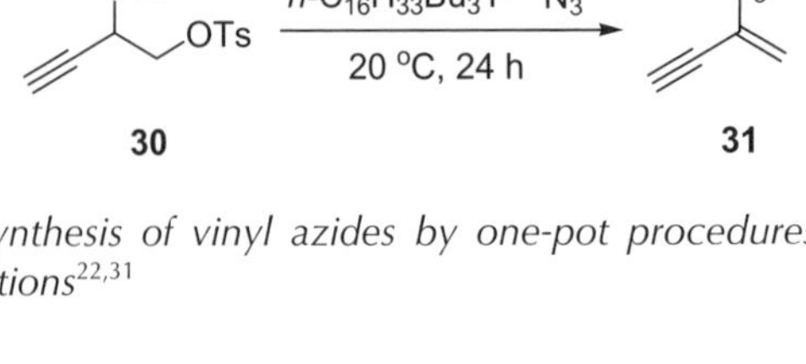

30 **31** 85–93%

Scheme 5.6 *Synthesis of vinyl azides by one-pot procedures including substitution and elimination reactions*[22,31]

access to the parent compound **29** (Ar=Ph, R=H) by treating the corresponding dibromide **28** with sodium azide in dimethylformamide followed by the reaction of the crude 1-azido-2-bromo-1-phenylethane with potassium *tert*-butoxide (76% yield for both steps).[32] Recently, it has been shown that vicinal disulfonates such as **30** are also convenient precursors for the synthesis of vinyl azides.[22,33] Thus, usage of the reagent hexadecyltributylphosphonium azide[34] led to 2-azidobut-1-en-3-yne (**31**) in a simple one-pot procedure.[22]

5.2 Routes to Vinyl Azides Developed in the Period 1965–70

In the second half of the 1960s, several new and highly effective methods to prepare vinyl azides were published.[35] This very fruitful period began in 1965 when Hassner developed the addition of iodine azide to alkenes followed by base-induced elimination of hydrogen iodide from the corresponding vicinal azidoiodoalkane (Scheme 5.7).[36] Using this method, explosive iodine azide was generated in situ from iodine chloride and sodium azide. The Hassner reaction was the breakthrough for the chemistry of vinyl azides because this access to the target compounds did not demand special structural requirements for the starting materials.[37] Thus, simple olefins like **32** and **35** were regioselectively and stereospecifically transformed into the desired products **34** and **37**, respectively, by two straightforward steps.[38] The addition step is interpreted as a polar process via an iodonium ion to explain the electronically controlled regiochemistry leading to **33** and subsequently to the inner vinyl azide **34** but not to the terminal isomer 1-azidohex-1-ene. Obviously, the formation of **40** from **38** is an exception due to the bulky *tert*-butyl group which sterically favors the attack of azide at the terminal carbon atom to yield **39**. The stereochemistry of the addition products **36** shows that an *anti* addition has taken place, and the configuration of the final product **37** is a result of an *anti* elimination. Consequently, cyclohexene (**41**) must afford the *trans*-configured addition product **42**, which cannot lead to the vinyl azide **46** by an *anti* elimination of hydrogen iodide. Thus, the allyl azide **43** was observed instead.[38] The desired compound **46** was available by the corresponding reaction of the *cis*-configured precursor **45** that can be prepared, for example, by the Mitsunobu-like inverting substitution reaction **44** → **45**.[39] Some other routes to 1-azidocyclohex-1-enes were published quite recently.[40] When cycloalkenes with larger rings were used for the Hassner reaction including an ionic addition of iodine azide, the synthesis of vinyl azides was successful as shown by the sequence **47** → **48** → **49**.[38]

Nowadays, Hassner's iodine-azide method is accepted as an efficient tool and utilized also in recent publications.[41] Bromine azide, generated in situ from *N*-bromosuccinimide and sodium azide, was likewise added at alkenes to furnish vinyl azides after treating the intermediate vicinal bromoazido compound with a base.[42] However, Hassner has demonstrated already at the beginning of his pioneering work that the regioselectivity of the addition of halogen azides, especially that of bromine azide, can be controlled by the reaction conditions.[35b,37] Thus, an ionic pathway in the addition reaction, for example the transformation **50** → **51**, is favored in a polar medium and in the presence of oxygen, which acts as a free radical inhibitor (Scheme 5.8). On the other hand, a free radical route, such as the conversion **50** → **53**, is enhanced in a solvent of low polarity, in the presence of light, and in the absence of oxygen.[43] The regiochemistry of the addition reactions is

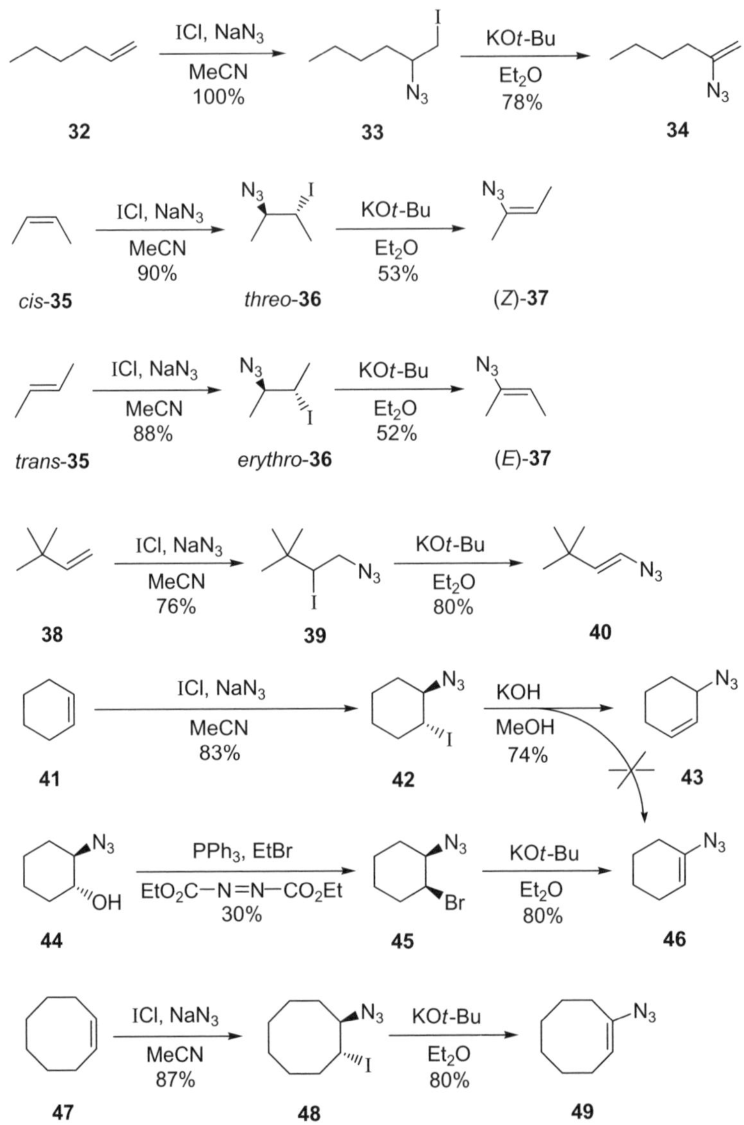

Scheme 5.7 Synthesis of vinyl azides by ionic addition of iodine azide and supplementary reactions[36–39]

reflected in the structure of the final products **52** and **54**. Moreover, in the case of geminal disubstituted or trisubstituted alkenes, ionic addition of halogen azides leads to tertiary azides, which are unsuited for the synthesis of vinyl azides. But radical addition yielded in such cases non-tertiary azides that could be transformed into the desired products. Recently, several new reagent combinations have been presented to perform the radical[44] or the ionic[45] addition of iodine azide at alkenes.

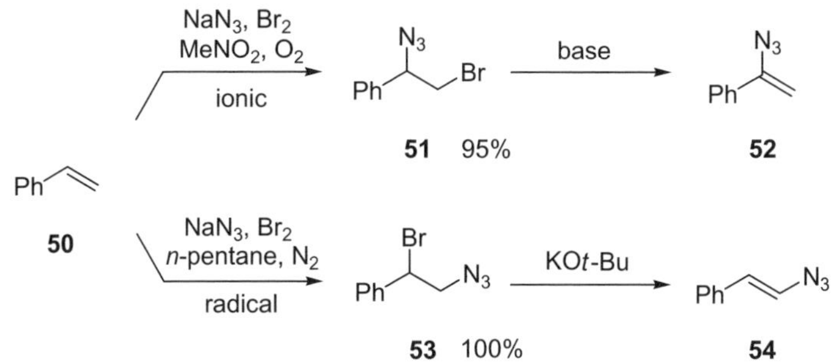

Scheme 5.8 *Synthesis of α- and β-azidostyrene via ionic or radical addition of bromine azide*[43]

In 1966, Harvey and Ratts reported on the nucleophilic addition of azide ion to conjugated allenic esters such as **55a** or the corresponding amides (Scheme 5.9).[46] These formal addition reactions of hydrazoic acid were performed with sodium azide in aqueous solvents and led, for example, to the conjugated product **57a** in 70% yield. However, it was recognized in early studies of nucleophilic addition to acceptor-substituted allenes that formation of the non-conjugated product of type **61** is a kinetically controlled reaction.[47] On the other hand, a conjugated product like **62** is the result of a thermodynamically controlled reaction. Apparently, after the attack of the azide on the central carbon atom of the allene **58**, the intermediate **59** is formed first. This has to execute a torsion of 90° to merge into the allylic carbanion **60**. Whereas **59** can only yield the product **61** by proton transfer, the protonation of **60** leads to both **61** and **62**. Recent investigations showed that the formal addition of hydrazoic acid to the simple acceptor-substituted allenes **55a–e** yielded first the isolable vinyl azides **56a–e**, which isomerized under basic reaction conditions to the conjugated products **57a–e**.[48] Some of the final products of type **57** were prepared earlier from the allenes **55** without observing the intermediates **56**.[13a,46,49] In the case of the carboxylic acid **55b**, the almost neutral reaction conditions facilitated the isolation of **56b** during the reaction with sodium azide in water or aqueous acetic acid.[48] In the presence of sodium hydroxide, **56b** rearranged to **57b** irreversibly. In general, the succeeding reaction **56** → **57** can be suppressed if the allenes **55a,c–e** were treated with sodium azide not in pure DMF but in a mixture of DMF and acetic acid or by using less basic TMGA.

Electron-deficient allenes bearing additional substituents do not behave in all cases completely analogously to the parent compounds **55a–e**.[50] Whereas **55f** reacted with TMGA first to give the isolable intermediate **56f**[48] and then to the final product **57f**, the analogous reaction of **55g** yielded only the product **57g** without a detectable intermediate (Scheme 5.10).[51] On the other hand, especially high yields of the vinyl azides **56h–k** were obtained when the sulfonylallenes **55h–k** were treated with TMGA.[52] The addition of hydrazoic acid to ethenylidenecyclopentadiene (**63**) demonstrates that the hydrocarbon **63** behaves like an acceptor-substituted allene.[53] Treatment of diallenes like **65** with TMGA allows access to the long-sought[54] 1,4-diazidobuta-1,3-dienes.[48] The product **66** is quite stable and can be prepared on a large scale, which rendered the separation of its diastereomers by simple recrystallization. In the case of the transformation **67** → **68**, the

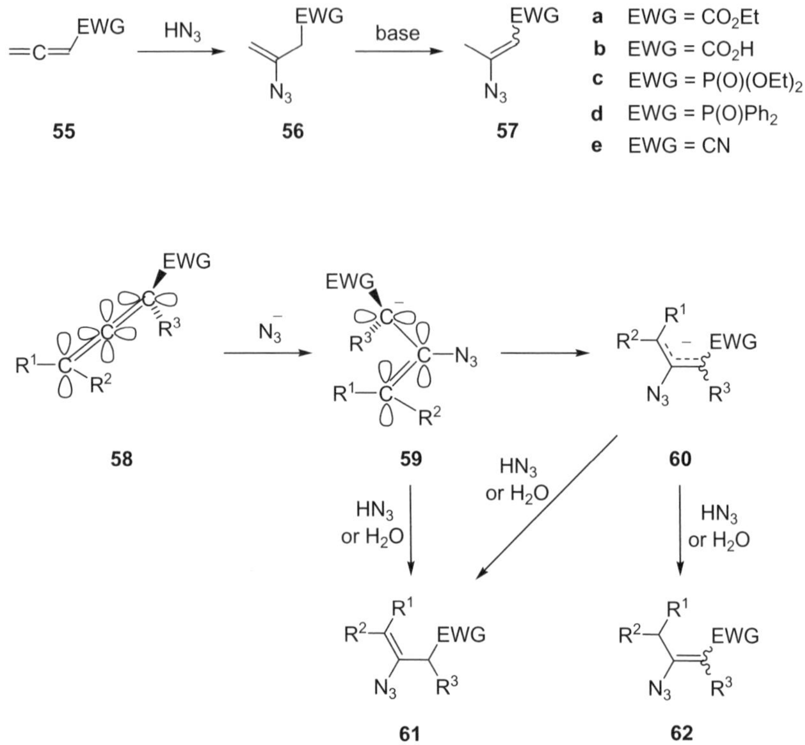

Scheme 5.9 Synthesis of vinyl azides by nucleophilic addition of hydrazoic acid to acceptor-substituted allenes (EWG = electron-withdrawing group)[13a,46–49]

acceptor-substituted allenes were most probably generated in situ by prototropic isomerization of the propargyl units. This was completed by addition of hydrazoic acid to afford three diastereomers of the diazides **68** which can be separated with 3, 34, and 15% yield.[55] The attack of the nucleophile on the electron-deficient allene usually happens at the central sp-hybridized carbon atom. This holds true also if no nucleophilic addition but a nucleophilic substitution in terms of an S_N2' reaction such as **69** → **70** occurs.[56]

In 1968, Smolinsky and Pryde prepared the terminal vinyl azides **72** by treating the tertiary alcohols **71**, which were easily available from the corresponding epoxides, with mesyl chloride and sulfur dioxide in dimethylformamide and pyridine (Scheme 5.11).[57] Similar procedures[58] but also reagents such as acetic acid/hydrogen chloride[59] or thionyl chloride/pyridine[60] were utilized to synthesize the desired dehydration products. However, these transformations often led to (*E*)/(*Z*) mixtures of the vinyl azides, and allyl azides were also formed if possible. For example, treatment of *unlike*-**73** or *like*-**73** with thionyl chloride and pyridine gave mixtures of azido-1,2,3-triphenylpropenes (*E*)-**74**, (*Z*)-**74**, and **75** as depicted in Scheme 5.11.[61]

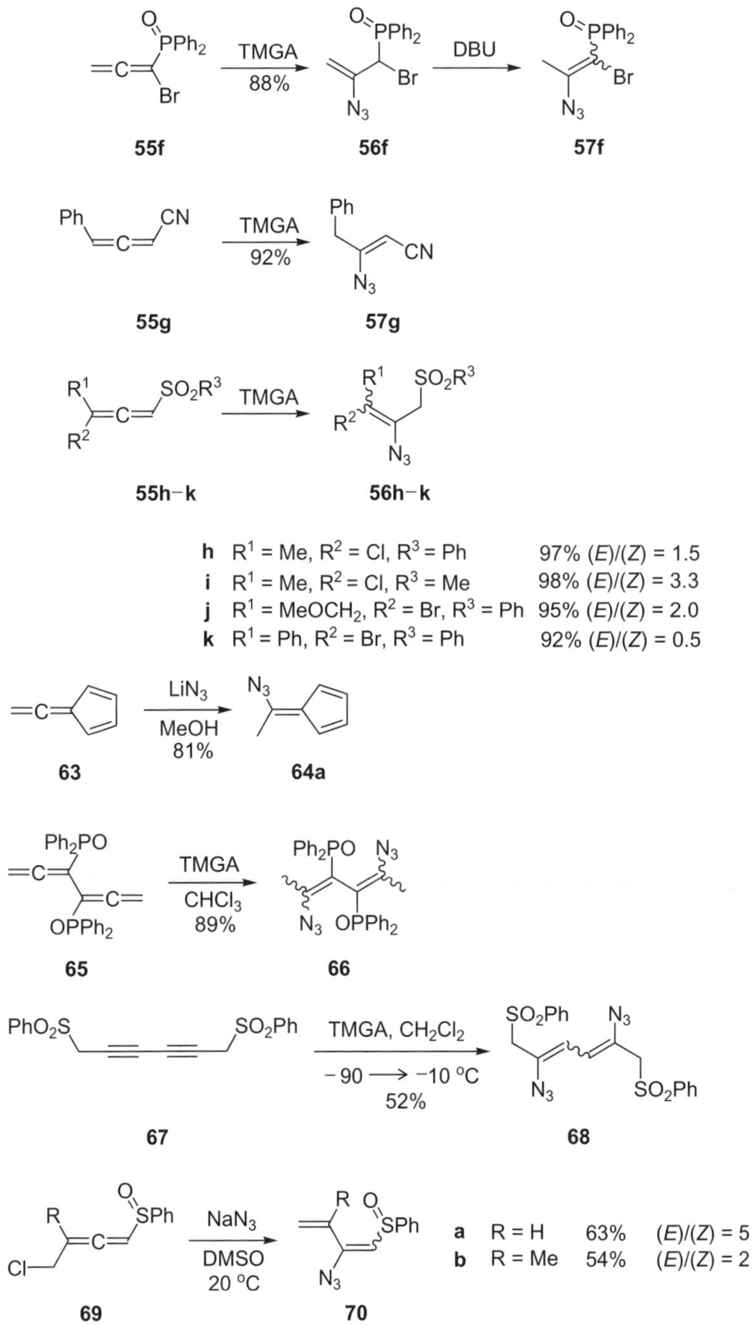

Scheme 5.10 *Vinyl azides from electron-deficient allenes bearing additional substituents*[48,51–53,55,56]

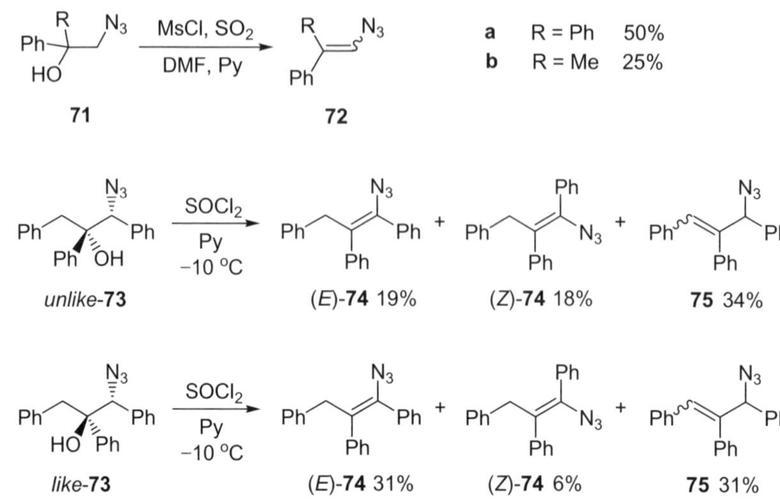

Scheme 5.11 *Synthesis of vinyl azides by dehydration of β-azidoalcohols*[57,61]

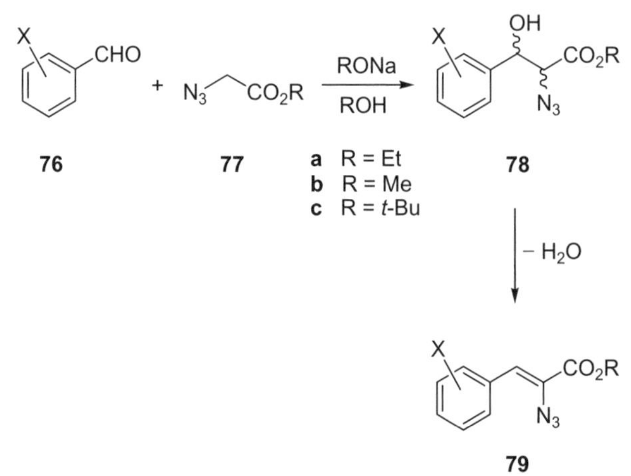

Scheme 5.12 *Synthesis of vinyl azides from aldehydes and azidoacetates by Hemetsberger–Knittel reaction*[62,63,66]

Hemetsberger and coworkers published in 1969 the special condensation reaction of benzaldehydes **76** with ethyl azidoacetate (**77a**) in the presence of sodium ethoxide to get moderate to good yields of the α-azidocinnamic esters **79a** (Scheme 5.12).[62] The temperature and the reaction time had to be controlled carefully, because the strongly alkaline reaction conditions induced the danger of decomposition of the base-sensitive azides **77** and other unwanted side reactions. Thus, Knittel reported some years later that usage of methyl azidoacetate (**77b**) and sodium methoxide offers several advantages in the synthesis of vinyl azides **79b**.[63] Nevertheless, the Hemetsberger–Knittel reaction can only be

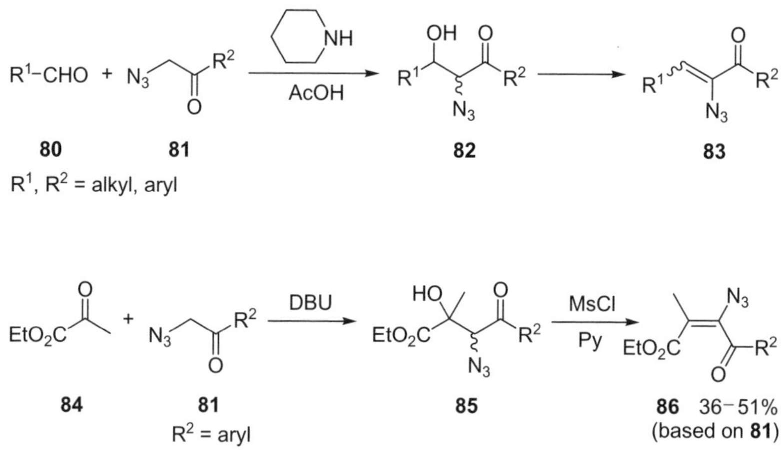

Scheme 5.13 *Synthesis of vinyl azides by condensation of azidomethyl ketones with aldehydes or ethyl pyruvate*[67–71]

performed with relatively stable aromatic precursors such as **76** or α,β-unsaturated aldehydes[64] like cinnamaldehydes whereas simple aliphatic aldehydes or ketones led only to products of aldol self-condensation.[63] In spite of these restrictions, the method of Hemetsberger and Knittel is currently the most frequently utilized access to vinyl azides because it can be applied not only to benzaldehydes such as **76** but also to all kinds of aromatic aldehydes, for example, heterocyclic derivatives and formylferrocenes.[65] Moreover, the resulting 2-azidoacrylic acid esters are highly useful starting materials especially for the synthesis of nitrogen heterocycles (see Section 5.4). Recently, it has been shown that 2-azido-3-hydroxypropionates like **78** or analogous intermediates can be isolated in some special cases when the reaction is performed at low temperature or *tert*-butyl azidoacetate (**77c**) is utilized.[66]

In 1970, Hemetsberger and coworkers reported that condensation of aromatic aldehydes of type **80** with α-azidoacetophenones **81** is also a successful method to prepare vinyl azides (Scheme 5.13).[67] Since compounds **81** are significantly more base-sensitive than azidoacetates **77**, the mild catalyst piperidinium acetate has to be used to get moderate to very good yields of the products **83**. Some years later, it was demonstrated that aliphatic aldehydes[68] **80** and even non-aromatic azidomethyl ketones[69] of type **81** are also appropriate starting materials. Intermediates **82** were isolated in several cases,[69] and the first hints, that the aldol reaction of formaldehyde (**80**, R^1=H) and α-azidoacetophenone (**81**, R^2=Ph) led to the corresponding azidohydrin **82**, date back to 1953.[70] As shown quite recently by Patonay and coworkers, even the activated ketone **84** can be treated with α-azidoacetophenones of type **81** to give the (*E*)-configured vinyl azides **86** in moderate yield.[71] However, the dehydration step **85** $\rightarrow$ **86** has to be performed with the help of mesyl chloride in pyridine.

Finally, Rens and Ghosez reported in 1970 on treatment of α-chloroenamines **87** with sodium azide, that led to 3-amino-2*H*-azirines **89** in excellent yields (Scheme 5.14, see also Chapter 6, Section 6.2 (Gilchrist & Alves)).[72] These products were explained by the

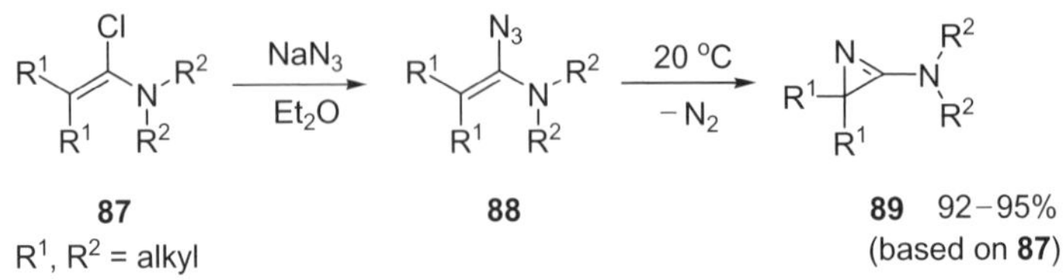

Scheme 5.14 *Generation of α-azidoenamines*[72,73]

intermediates **88**, which resulted from **87** by an elimination-addition mechanism. Obviously, the electron-rich vinyl azide **88** loses dinitrogen at relatively low temperature already (compare to Section 5.4), and thus **88** can only be detected by IR data but not be isolated. Small-ring amidines of type **89** are highly valuable synthetic equivalents of α,α-disubstituted α-amino acids and building blocks to prepare heterocycles and peptides as demonstrated by Heimgartner and coworkers.[73,74] In some special cases, short-lived α-azidoenamines similar to **88** did not lose dinitrogen but cyclized to give 5-amino-4*H*-1,2,3-triazoles which can isomerize to the corresponding 2*H*-1,2,3-triazoles.[75]

5.3 New Methods to Prepare Vinyl Azides

In the last three decades, several novel routes were developed to make accessible new kinds and substitution patterns of vinyl azides. Such routes can be based on Winstein's [3,3]-sigmatropic rearrangement of allylic azides.[76] If this simple reaction is modified, the azido group is shifted from an allylic or propargylic into a vinylic position.[77] Thus, 2-azidobuta-1,3-dienes **92**[78,79] or similar products[80] were prepared by treating buta-2,3-dienyl precursors **90** with sodium azide (Scheme 5.15). Although this one-pot procedure included nucleophilic substitution and rapid [3,3]-sigmatropic migration of the azido group, intermediates of type **91** could be isolated.[79] With the help of **91**, which was selectively ^{15}N-labeled at N-α, it was possible to prove the [3,3]-sigmatropic nature of the rearrangement **91** → **92** since **92** bore the label exclusively at N-γ.[81] After introduction of two azido groups into the starting material **93**, a sequence of two [3,3]-sigmatropic rearrangements led to 2,3-diazidobuta-1,3-dienes **96**.[78,82–84] Because the isomerization **95** → **96** is much more rapid than the first shift **94** → **95**, the proportion of the quasistationary intermediates **95** is very low. Nevertheless, the parent compound **95** (R^1=R^2=H) can be detected by NMR spectroscopy.[85,86] A cascade of two [3,3]-sigmatropic migrations was also assumed when racemic diazide *rac*-**97** was heated in solution.[87] The products (1*Z*,3*E*)-**99** and (1*E*,3*Z*)-**99** were expected due to the stereochemical results of other rearrangement reactions of but-3-yne-1,2-diyl precursors.[88] However, the intermediates **99** include the substructure of 1,2-diazidoethenes, which are well-known to split off easily two molecules of dinitrogen with cleavage of the C,C double bond and formation of two nitrile fragments.[89] Thus, hydrogen cyanide and the nitrile **100** were obtained from *rac*-**97**. But the heterocycle **101** was found to be the main product generated via an intramolecular 1,3-dipolar cycloaddition of (1*E*,3*Z*)-**99**, whereas *meso*-**97** gave only the nitriles **100** and hydrocyanic acid.[87] Treatment of α,β-unsaturated aldehydes **102** with trimethylsilyl azide

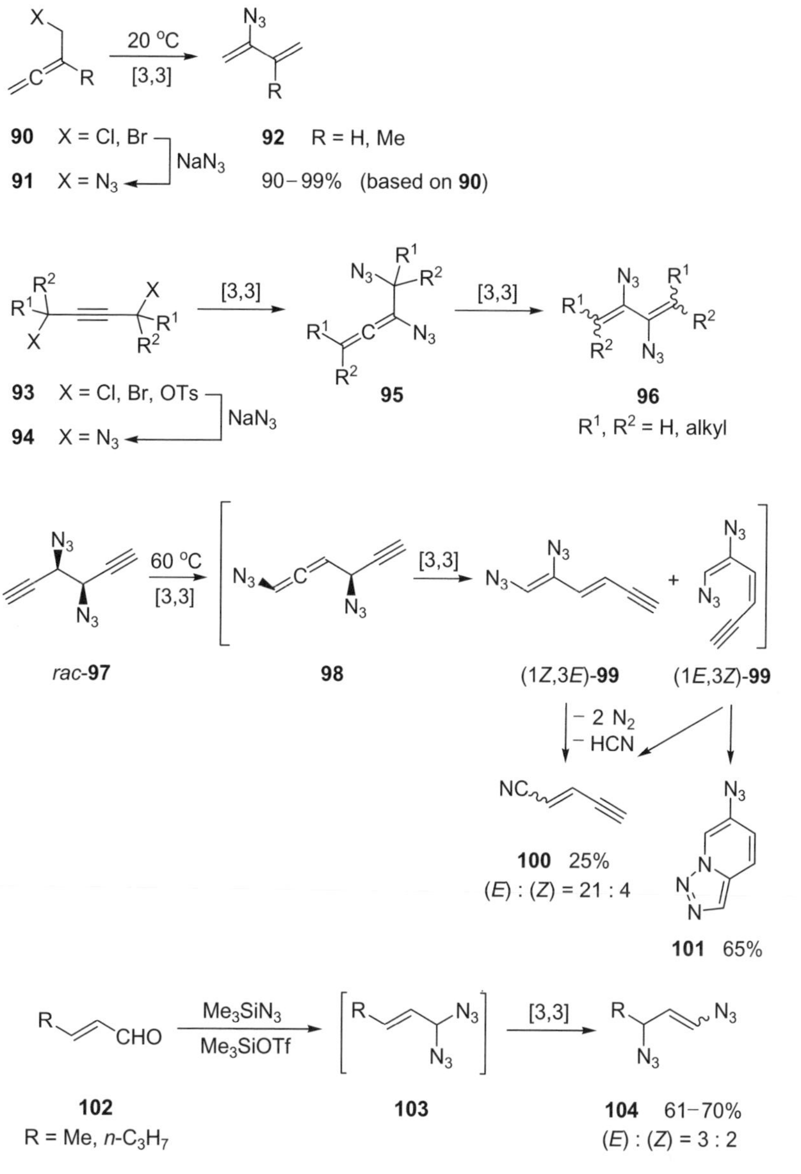

Scheme 5.15 *Synthesis of vinyl azides by [3,3]-sigmatropic migration of the azido group*[77–79,81–87,90]

in the presence of trimethylsilyl triflate afforded the products **104**, which were postulated to arise from geminal diazides **103** by a [3,3]-sigmatropic shift of an azido group.[90]

Since vinyl azides bearing acidic allyl protons can be isomerized prototropically (see, for example, **56** → **57**, Scheme 5.9), it is not surprising that base-induced rearrangement was also utilized to prepare vinylic azido compounds **106** from allyl azides **105** (Table 5.1).[48] The base DABCO proves to be appropriate for substrates with strongly

Table 5.1 *Synthesis of vinyl azides by DABCO-catalyzed isomerization of allyl azides*[48]

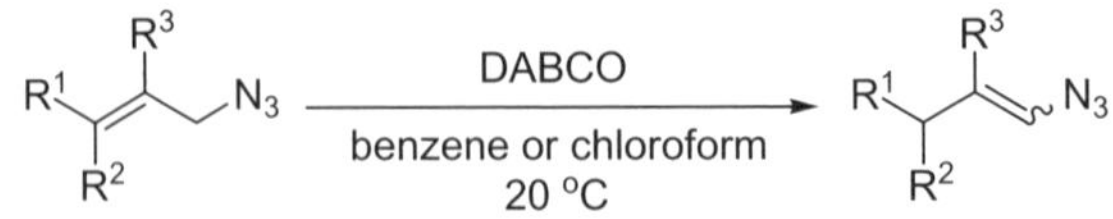

105 **106**

Entry	R^1	R^2	R^3	reaction time	isolated yield	ratio (*E*):(*Z*)
a	CO_2Me	CO_2Me	Me	2 h	95%	1:0
b	CO_2Me	H	H	33 d	80%	1:2.5
c[a)]	CN	H	H	2 d	61%	1:2.5
d	CN	H	SePh	4 d	98%	10:1
e	CN	H	Se(O)Ph	<5 min	94%[b)]	1:0

[a)] Starting material **105c** with (*E*):(*Z*) = 1.25:1.
[b)] Yield calculated by ^1H NMR data.

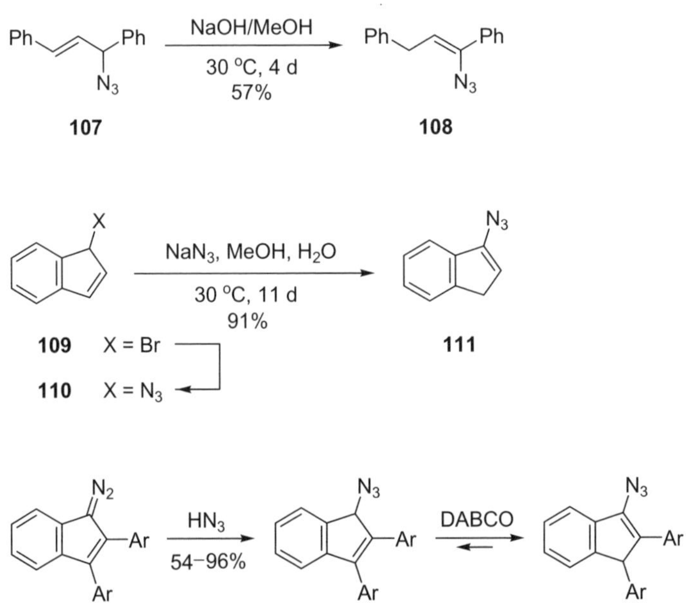

107 **108**

109 X = Br **110** X = N_3 **111**

112 **113** **114**

Scheme 5.16 *Synthesis of vinyl azides by base-induced prototropic rearrangement*[48,61]

electron-withdrawing substituents, but other reagents such as DBU,[91] DBN,[92] diisopropylamine[93a] or sodium azide[93b] were used likewise to catalyze the formation of vinyl azides.

If potent acceptor groups are lacking, stronger bases like sodium hydroxide are necessary as shown by the transformation **107** → **108** (Scheme 5.16).[61] In other similar cases, the base sensibility of the azido group limited prototropic isomerization as a synthetic method to prepare vinyl azides. However, the desired product **111** was already obtained during nucleophilic substitution **109** → **110**. When the reaction conditions were reduced to 1 h at 20 °C in order to repress the influence of sodium azide as a base, the allylic intermediate **110** was isolated in 95% yield.[61] Similar indenyl azides **113**, which were easily accessible by treatment of diazo compounds **112** with hydrazoic acid, led to the vinyl azides **114**. But in this case, DABCO-catalyzed isomerization established an equilibrium that included small amounts (3–15%) of the allylic azides **113**.[48]

Whereas 3-azido-1*H*-indenes are not only available by base-induced rearrangement but also by Hassner reaction of indenes,[38b] azidocyclopentadienes have been unknown until quite recently. The only earlier report on such a compound described the hexaazide **116**, which was prepared by treating the precursor **115** with an excess of sodium azide and depicted as a stable solid that decomposed slowly at 200 °C (Scheme 5.17).[94] This unusual stability is obviously incompatible with the claimed structure of **116** because of the high proportion of nitrogen and especially the 1,2-diazidoethene substructures.[89] When the experiment to synthesize **116** was repeated, no organic azide could be detected.[95] However, milder reaction conditions and only two equivalents of sodium azide led to the diazide **117**, which was also characterized with the help of the cycloaddition product **118**. Only decomposition occurred on treating **117** with sodium azide. Azidocyclopentadienes can be prepared not only by nucleophilic substitution but also by electrophilic transfer of an azido group as shown by the synthesis of **121** from **119** or **120**. Fully substituted cyclopentadienes such as **117** or **121** cannot be isomerized to vinyl azides. Thus, heating of **121** in DMSO gave only the pyridine derivative **122**. On the other hand, the reaction of the bromide **123** with sodium azide in dimethylformamide/acetic acid furnished directly the desired product **126** in 91% yield via prototropic rearrangement of the intermediate **125**. When this transformation was performed without acetic acid, the more basic reaction conditions induced cleavage of the azido group and exclusive formation of the imine **127** in 77% yield. However, treatment of diazo compound **124** with hydrazoic acid afforded in 87% yield the allyl azide **125**, which could be isolated as pale yellow needles but isomerized slowly in solution to produce **126** already at room temperature. The parent azidocyclopentadienes **129a** and **129b** are accessible from the iodo substrate **128** and tetramethylguanidinium azide (TMGA). They can only be handled in solution and separation of the isomers proved to be difficult. Nevertheless, heating the 1 : 1 mixture at 70 °C induced the fragmentation **129a** → **130** and thus led to an enrichment of **129b** with **129a** : **129b** = ca. 1 : 6. But prolonged thermolysis resulted in quantitative transformation of **129a** and **129b** into **130** because of the slow equilibration of **129a** and **129b**. When a 1 : 6 mixture of **129a** and **129b** was treated at room temperature with a catalytic trace of DABCO, the 1 : 1 equilibration of the azidocyclopentadienes was established instantaneously. The reaction of **129a,b** with acetone and pyrrolidine led to the azidofulvene **131**.[95]

Recently, several methods to prepare vicinal halovinyl azides were presented (see also Scheme 5.10).[52,96,97] The phosphonium ylides **132** reacted with *N*-halosuccinimides in the presence of azidotrimethylsilane to give the products **135** in moderate to excellent yields and as mixtures of (*E*)/(*Z*) isomers in most cases (Scheme 5.18).[96] The intermediates **133** and **134** were postulated to explain the obtained results. Another approach to vicinal halovinyl azides started with pyranose derivatives **136**, which were available from the corresponding glycals.[97] Oxidation of **136** in the presence of (diacetoxyiodo)benzene and iodine to generate the anomeric alkoxyl radicals **137** followed by β-fragmentation and

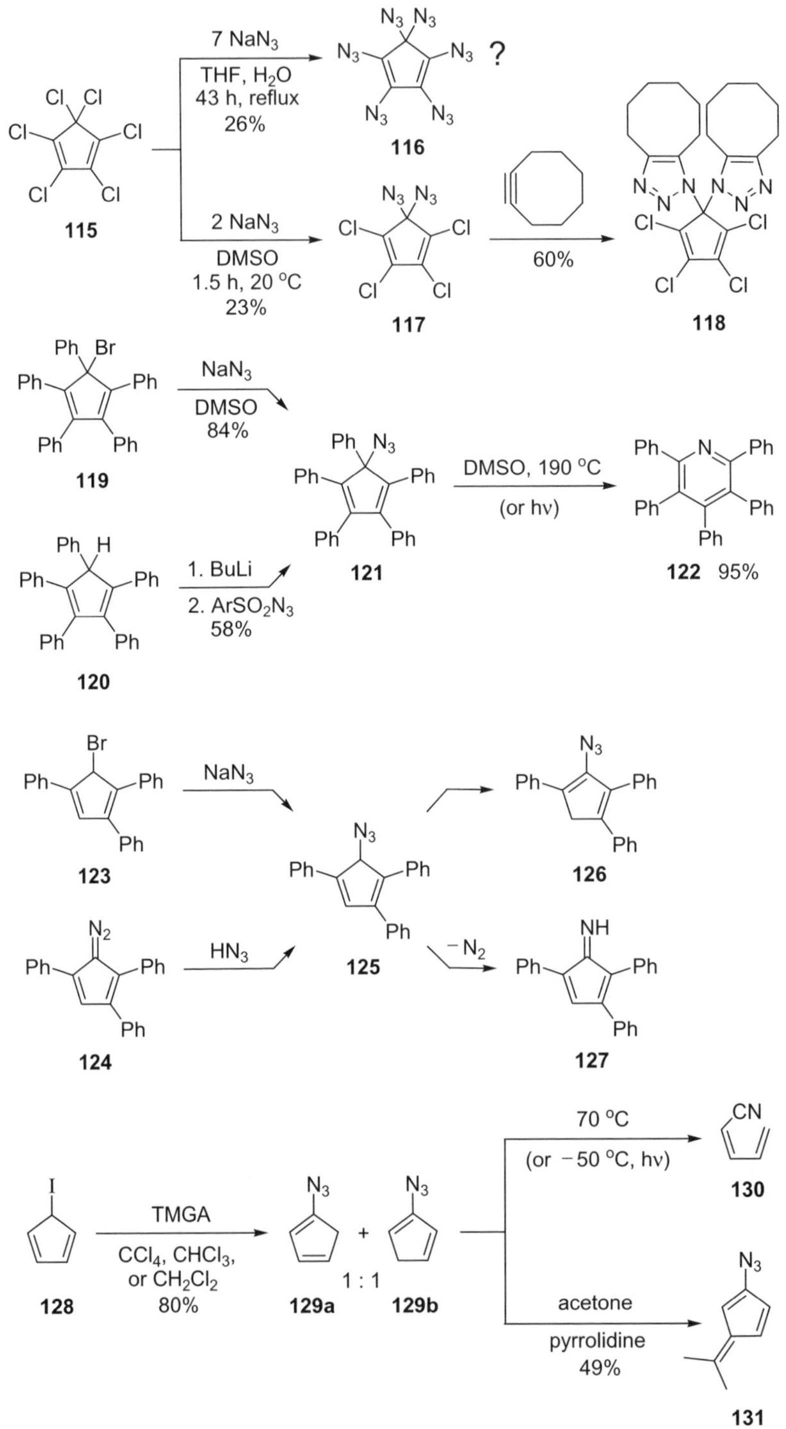

Scheme 5.17 *Synthesis of azidocyclopentadienes [Ar = 2,4,6-(i-Pr)$_3$C$_6$H$_2$]*[94,95]

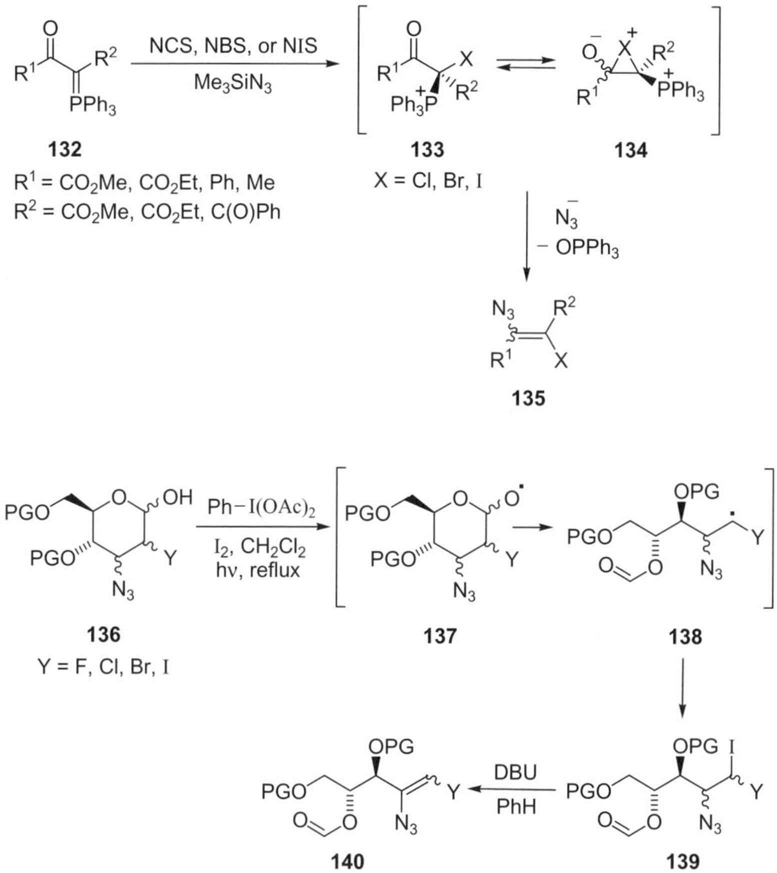

Scheme 5.18 Synthesis of vicinal halovinyl azides (PG = protecting group = acetate or acetal with benzaldehyde)[96,97]

trapping of the carbon-centered radical **138** led to the dihalo compounds **139**, which were chemoselectively dehydroiodinated to furnish the products **140**. The complete sequence could be performed with good to excellent yields, and a transfer to **136** with Y=H to get less substituted vinyl azides **140** (Y=H) is also possible.[97,98]

Starting materials, like **141**, which are accessible from alkenes by azido-selenenylation, afforded only in a few cases vinyl azides exclusively as shown by the example **141** → **142** (Scheme 5.19).[99] In most cases, after oxidation of the vicinal phenylseleno azides and elimination reaction on the intermediate selenoxide, mixtures of allyl and vinyl azides were obtained.[100] Ring opening of trialkylsilyl-substituted epoxides was utilized several times for stereoselective synthesis of vinyl azides. Thus, treatment of the *trans*-configured oxiranes **143** with azidotrimethylsilane and boron trifluoride etherate yielded *cis*-configured products **145**. This result was explained by the intermediate **144** which should undergo *anti*-elimination.[101a] On the other hand, when subjected to sodium azide in dimethylformamide, epoxides **146** were transformed into vinyl azides **148** via

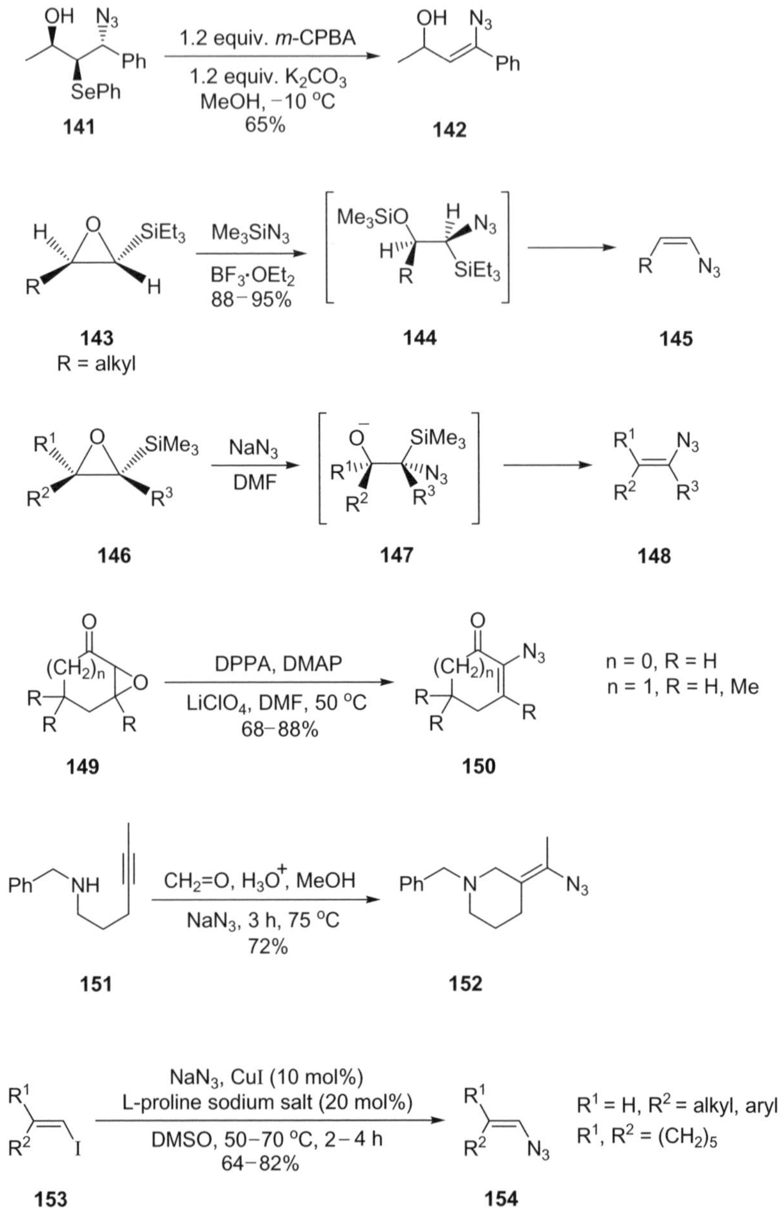

Scheme 5.19 *Miscellaneous synthetic methods for vinyl azides*[99,101,102a,103a,104]

syn-elimination reaction of intermediate **147**.[101b] Especially, the α-azido-α,β-unsaturated ketones of type **150** were produced in good yields if bicyclic oxiranes **149** were heated in the presence of diphosphoryl azide (DPPA) and 4-(dimethylamino)pyridine (DMAP).[102a] When α,β-unsaturated ketones or esters were treated with cerium(IV) ammonium nitrate in the presence of sodium azide and then with sodium acetate, the corresponding α-azido-

α,β-unsaturated carbonyl compounds were formed in moderate to good yields.[102b,c] Intramolecular electrophilic addition of formaldiminium ions to alkynes could be used, in single examples, for the synthesis of heterocyclic vinyl azides such as **152**.[103a] Vilsmeier formylation of 2-azidoacetophenones proved to be an advantageous route to (*Z*)-α-azido-β-chlorocinnamaldehydes.[103b] Quite recently, it has been found that coupling reaction of vinyl iodides **153** with sodium azide under catalysis of copper(I) iodide and proline works at relatively low temperatures to provide the desired products **154** in good to excellent yields.[104] Alkenyl azides with (*E*)-configuration were synthesized by hydroboration of alkynes with bis(1,2-dimethylpropyl)borane followed by the reaction with sodium azide, copper(II) acetate, and copper(II) nitrate.[105] Finally, single vinyl azides are available by special methods. For example, (*E*)-3-azidohex-3-ene-2,5-dione was prepared by treatment of 3-azido-2,5-dimethylfuran with bromine in the presence of pyridine and aqueous acetone.[106a,b]

The copper-catalyzed reaction of β-styrylboronic acid with sodium azide afforded (*E*)-β-styryl azide in good yield.[106c] In rare cases, the rearrangement of butatrienylidene and alkynyl complexes of iridium led to special enazides.[106d] A single example of the formation of an alkenyl azide via palladium-catalyzed cyclization of an allenic substrate has also been reported.[106e]

5.4 Reactions of Vinyl Azides

Alkenyl azides allow so numerous and manifold transformations that they cannot be described in this chapter comprehensively.[35,107] Thus, only particularly interesting or important reactions of vinyl azides are included here, which leads always to a subjective selection. Alkenyl azides show not only the rich chemistry of azides[108] and olefins. The conjugated structure, which was investigated experimentally[82,109] as well as theoretically,[110] causes a polarization of the C,C double bond by the electron-donating azido group.[111] Consequently, vinyl azides were classified as *N*-diazoenamines,[112] and they possess indeed an electron-rich vinylic unit like enol ethers, but the latter are more sensitive to hydrolysis. Therefore, enazides can attack electrophiles by the β-carbon atom but also by N-α. On the other hand, nucleophiles are able to react with nitrogen (N-γ) or carbon of vinyl azides. The conjugation in alkenyl azides increases the reactivity of the functional groups and the variety of possible transformations, for example, in photolyses, thermolyses, cycloaddition and reduction reactions.

In 1961, Smolinsky reported on vapor phase pyrolysis of α-azidostyrene (**52**, Scheme 5.8), which furnished 3-phenyl-2*H*-azirine as the main product and provided the first example of the synthesis of such strained heterocycles from vinyl azides.[32a] By analyzing the IR data of the pyrolysates produced from **52**, it was shown one year later that *N*-phenylketenimine is formed as a side product.[32b] By utilizing IR and NMR spectroscopy at low temperature, Wentrup and coworkers have studied recently the detailed structures of another azirine–ketenimine pair, generated by thermal or photochemical decomposition of an enazide.[9e] The transformation of vinyl azides into 2*H*-azirines is currently the most frequently used access to these heterocycles. The manifold chemistry of azirines was reviewed several times,[113,114] and most of the aspects of their synthesis from alkenyl azides are summarized in Chapter 6 (Gilchrist & Alwes). Therefore, only some additional certain points are included here.

Lately, procedures for the preparation of 3-substituted 2*H*-azirines from enazides were presented utilizing low boiling solvents in closed vessels at elevated temperatures, for example, dichloromethane at 150 °C.[41c,115] The same reaction was also performed by heating the starting materials with the help of microwave irradiation in solvent-free conditions.[116] Unfortunately, no safety hints were given for these experiments. In special cases,[9a] the thermal reaction of vinyl azides to yield azirines occurred already at room temperature as depicted in Scheme 5.14 with **88** → **89**.[72,73] However, Hassner demonstrated previously in 1967 that enazides bearing two bulky substituents in *cis* position like **155a** tend to split off dinitrogen at low temperatures to give **156a** (Scheme 5.20).[38a] As shown later, the compounds **155b,c** behave similar.[61] Whereas transfer of vinyl azide **157** to the corresponding 2*H*-azirine required heating at 80 °C in toluene,[31b] treatment of **157** with manganese dioxide afforded the formyl derivative **158** surprisingly at room temperature.[30b] Thermolysis or irradiation of the solutions of enazides can both lead to high yields of 2*H*-azirines, for example in the synthesis of **156b,c**,[61] **159a**,[79] **160b**,[83] and **161a**.[53] With other substitution patterns or lower numbers of substituents and especially in cases without a substituent in the 3-position of the heterocycle,[117] however, photolysis is highly advantageous as illustrated by the generation of **160a**,[83,84] **161b,c**,[53] (±)-azirinomycin[118] (**164**),[48] **165**,[119] and **167**/**168**.[120] Thus, heating the corresponding vinyl azides resulted in low yields of the 2*H*-azirines because of thermal sequential reactions or degradation of the strained heterocycles. But completely different reaction types of thermolysis and photolysis are also possible as indicated by the formation of **23** and **165** from **21** (see Schemes 5.4 and 5.20). Photoelectron spectroscopy of the spirocyclic compound **161a** showed that the lone-pair orbital n(N) of the 2*H*-azirine nitrogen atom interacts strongly with the π_1-orbital of the cyclopentadiene ring. This type of homoconjugation termed azaspiroconjugation was found to be compatible with quantum-chemical calculations.[53] By heating diluted solutions of **64b** or **161b** or by irradiation of **64b,c** in methanol, the first isocyanocyclopentadienes **162** and **163** were prepared.[121] The heterocycles **167** and **168** generated from diazides **166**[120a] could only be detected by NMR spectroscopy at low temperature or by IR data in argon matrix because fragmentation to produce the nitriles **169** and **170** occurred on warming to room temperature or prolonged photolysis.[120b] Thus, it was demonstrated for the first time that 2-azido-2*H*-azirines are intermediates in the well-known[89] photochemical or thermal transformation of vicinal vinyl diazides to yield cyano compounds.

Vinyl azides bearing a carbonyl group of a ketone or an ester in the α-position like **171** should lead to destabilized 2*H*-azirines since the acceptor substituent increases the electron deficiency at C-3 of the heterocycle (Scheme 5.21). Nevertheless, these azirines **172** were synthesized by thermolysis of **171** in solution.[63,68a,113e,115] The products **172** could be handled successfully and utilized to add stereoselectively nucleophiles NuH such as thiophenol[26] or nitrogen heterocycles[24a,122] yielding aziridines **173** (see also Chapter 6 (Gilchrist & Alwes)). Electron-poor azirines **172** are not only activated for nucleophilic addition but also for Diels–Alder reaction as shown by the stereoselective formation of **174** or similar cycloadducts.[68a,123] This cycloaddition could be performed enantioselectively using an ester of type **172** equipped with a chiral auxiliary[24b,26] or a modified diene bearing an enantiopure substituent.[25c] Small amounts of dihydropyrazine **175**, its tautomer, or the corresponding aromatic heterocycle were obtained on thermolysis of **172** in protic solvents or chromatography of this starting material.[26b,63,123a] However, heating of

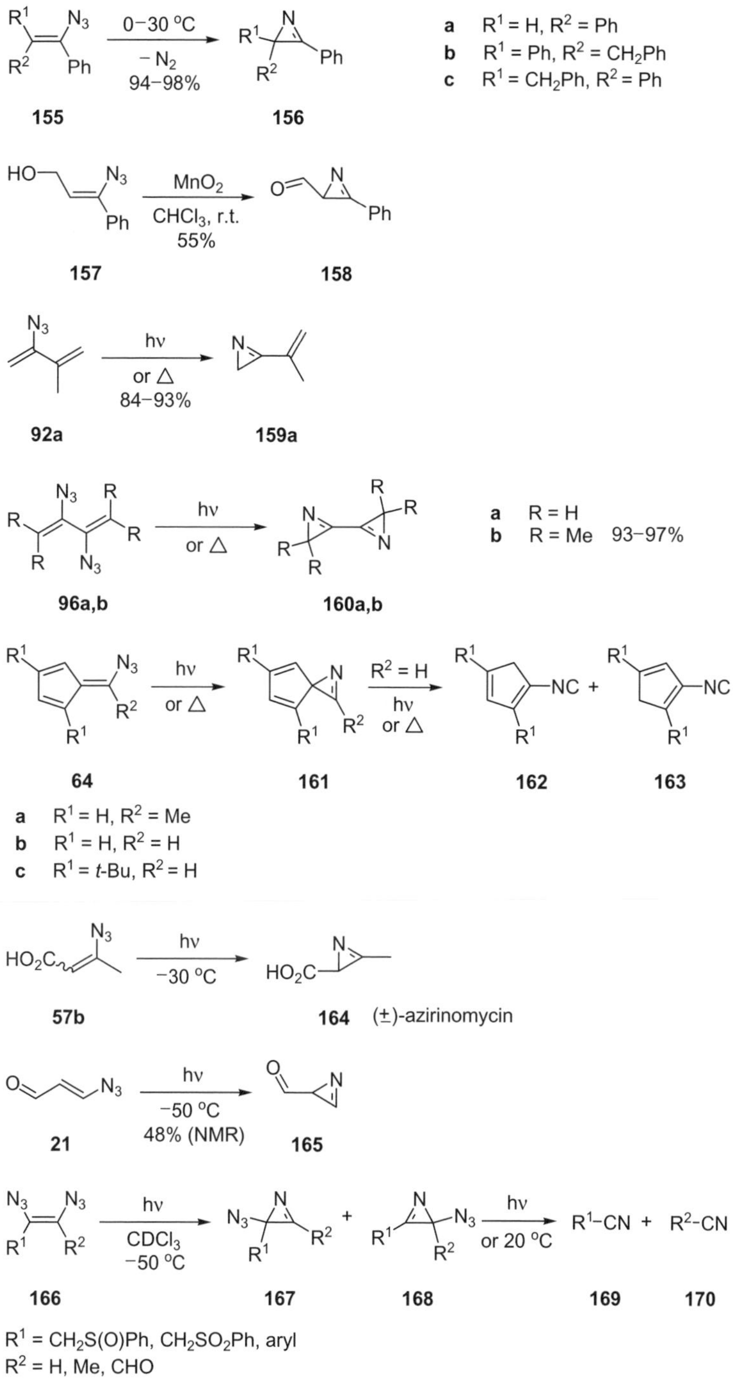

Scheme 5.20 *Generation of 2H-azirines from vinyl azides by thermolysis or photolysis*[30b,38a,48,53,61,79,83,119–121]

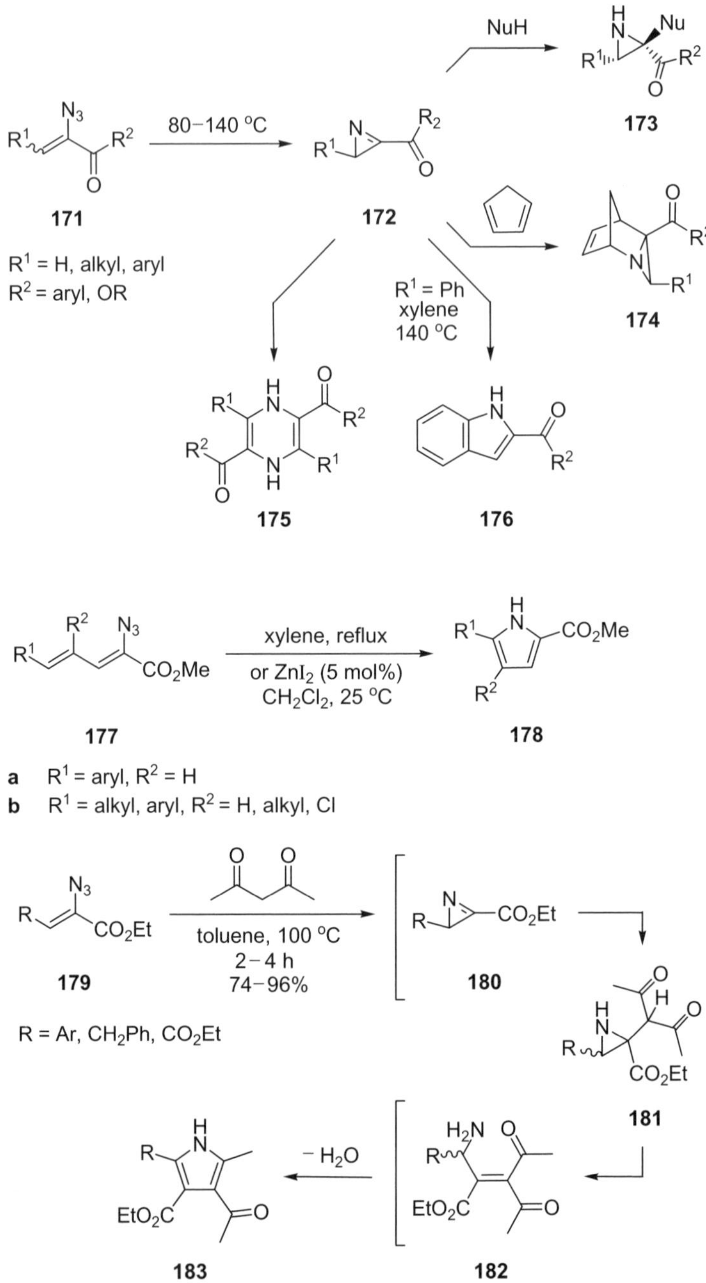

Scheme 5.21 *Reactions of α-azido-α,β-unsaturated ketones or esters via 2H-azirines*[24a,26,63,64f,67b,68a,113e,115,122,123,124,130,131]

171 (R^1=aryl) in aprotic solvents, for example, xylene at 140 °C, is more important and has been applied very frequently to prepare indoles of type **176** via intermediate **172**.[124] This method is connected with the names of Hemetsberger and Knittel[63,67b] and can be transferred to the synthesis of azaindoles,[125] fused indoles,[126] and fused pyrrol derivatives.[127] Quite recently, reactions like **171** (R^1=Ph) → **176** have been conducted in a rhodium(II)-catalyzed way, which allows to decrease the necessary temperature into a range of 25–60 °C.[128] The transformation of **171** into isoquinolines is also possible if one or both *ortho* positions of the aryl group R^1 are substituted appropriately.[129]

When the pyrroles **178a** were prepared by thermal cyclization of the precursors **177a** in boiling xylene, the corresponding 2*H*-azirines should be discussed as intermediates because the reaction conditions proved to be typical for the conversion of vinyl azides into these three-membered heterocycles.[64f] As opposed to this, the synthesis of similar pyrroles **178b** from azides **177b** was performed under mild conditions with the help of small amounts of zinc iodide. Several mechanisms were presented to explain the catalyzed formation of the five-membered products.[130] Thermolysis of vinyl azides **179** in the presence of 1,3-dicarbonyl compounds like acetylacetone yielded tetrasubstituted *N*-H pyrroles **183**, which were interpreted by nucleophilic attack of the 1,3-diketone at the 2*H*-azirine **180** followed by ring opening of the resulting aziridine **181** and condensation reaction of the intermediate **182**.[131]

Until quite recently, the highly strained heterocycle **185b** and many similar 2,3-bridged 2*H*-azirines bearing a six-membered ring were generated only in situ and proved by trapping reactions (Scheme 5.22).[132] On the other hand, the azirine resulting from cyclic vinyl azide **49** (Scheme 5.7) was distilled in vacuo.[132k,133] However, the more strained bicyclic compounds **185a–c** can be produced in solution by photolysis of **184a–c** at low temperature and detected in 83–91% yield by IR and NMR spectroscopy even at room temperature as shown quite lately.[40] But, a rapid subsequent reaction led to the formation of the dimers **186a–c**, which were quickly oxidized to the corresponding aromatic pyrazines in the case of R=H. This dimerization can be catalyzed by traces of moisture. Most probably, water induces ring cleavage of the three-membered ring and generation of an α-aminoketone, which is able to attack a second molecule of the azirine to give **186** after a final condensation reaction. Even by irradiation of **184d,e** at −80 to −120 °C and NMR spectroscopy at the same temperature, it was not possible to detect 2,3-bridged 2*H*-azirines bearing a five-membered ring such as **185d,e**. Nevertheless, the photolyzed solutions yielded dimers of type **186** or the corresponding aromatic pyrazine derivatives after thawing. As a result of their increased ring strain, 2,3-bridged 2*H*-azirines like **185a–c** can undergo addition and cycloaddition reactions that are not possible for simple azirines, which do not react, for example, with hydrogen cyanide. However, the addition of this reagent to **185a–c** takes place already at −50 °C leading to aziridines **187a–c**. Moreover, the selective *exo* cycloaddition of diazomethane to **185a,b** can be performed at −40 or −50 °C generating the tricyclic products **188a,b**, which were cleaved at −20 to +20 °C to give high yields of the azides **190a,b** and **191a,b**.[40] The slow reaction of simple 2*H*-azirines with diazo compounds to afford allylic azides has been known for more than forty years.[61,134] But short-lived 1,2,3-triazabicyclo[3.1.0]hex-2-enes, which were also discussed as intermediates of the rearrangement of allyl azides,[76] were proved with the help of **188a,b** for the first time.[40] Whereas alkyl- or aryl-substituted azirines without an electron-withdrawing group do not react with cyclopentadiene,[113b] the corresponding Diels–Alder reactions of

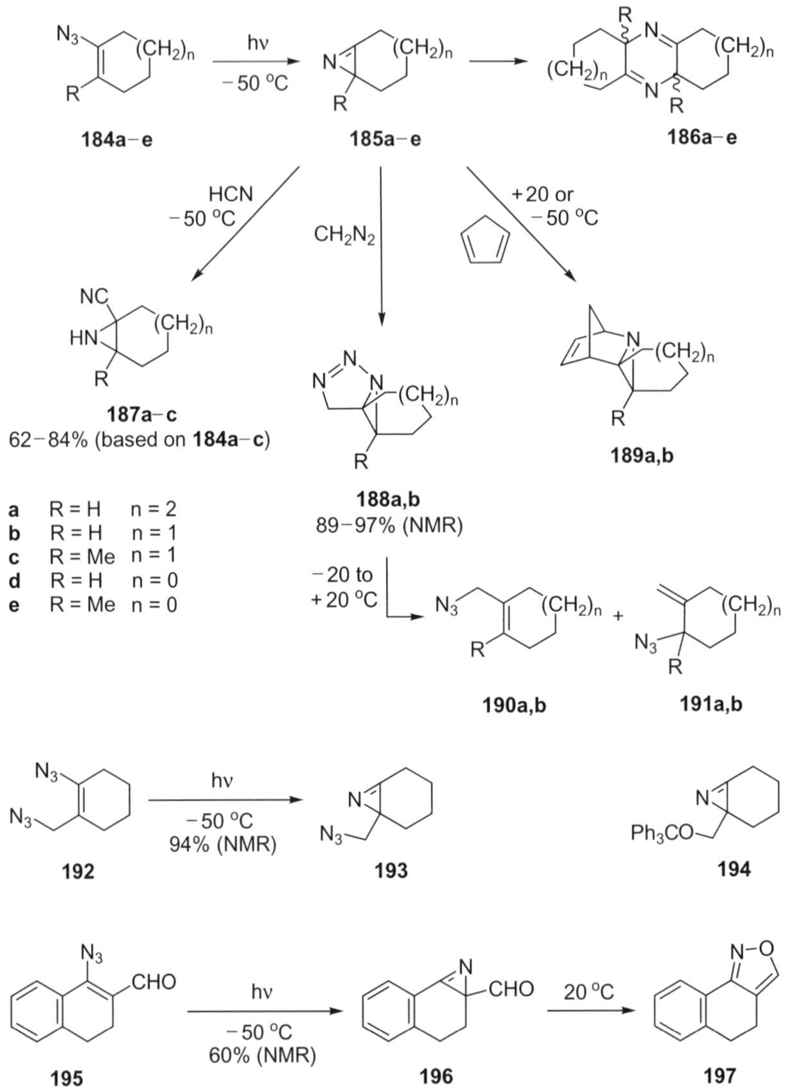

Scheme 5.22 *Generation of 2,3-bridged 2H-azirines by photolysis of cyclic vinyl azides*[40,135]

185a and **185b** were successful under mild conditions.[40] Of the four imaginable diastereomeric [4 + 2] cycloadducts (*exo-exo*; *endo-exo*; *exo-endo; endo-endo*), the *exo-endo* stereoisomers **189** were isolated exclusively. Thus, the cyclopentadiene was added to the *exo* side of the bridgehead azirines to form 5-azanorbornenes with an attached three-membered ring in the *endo* position.[40]

The priority of vinyl azides over alkyl azides in light absorption and photochemical transformation was demonstrated by the irradiation of diazide **192**, which led to the strained heterocycle **193** without degradation of the second azido group.[135] The

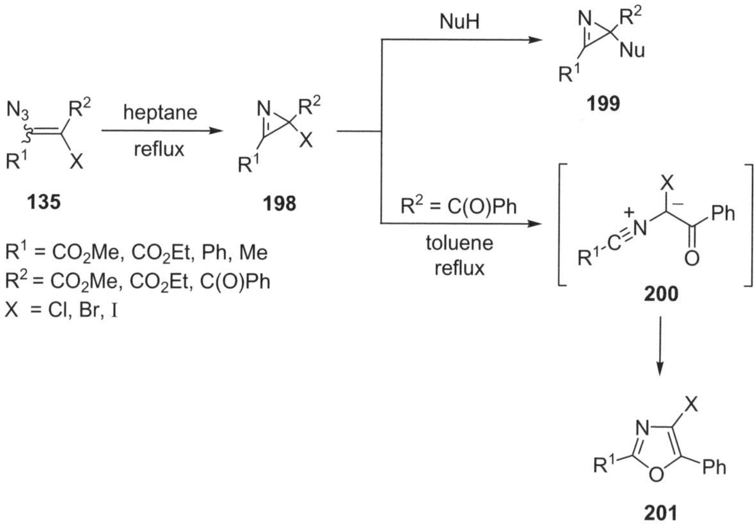

Scheme 5.23 *Reactions of vicinal halovinyl azides via 2H-azirines*[96,136–139]

2,3-bridged 2*H*-azirines are stabilized by bulky substituents in the 2-position. This was shown, for example, by liberating compound **194** from its solution to get a slight yellow solid that was stable for a short time. On the other hand, the tricyclic azirine derivative **196** could be handled only in solution at low temperature because it is transformed quantitatively into the ring enlargement product **197** at room temperature.

Heating solutions of the halovinyl azides **135** resulted in moderate (X=I) or excellent (X=Cl, Br) yields of the heterocyclic products **198** (Scheme 5.23).[96] Substitution reactions in the presence of nucleophiles NuH such as potassium phthalimide or aniline allowed the synthesis of the azirines **199**,[136] and the dehalogenation of **198** with the help of sodium borohydride or tributyltin hydride was also possible.[137] Since thermolysis of β-azido-α,β-unsaturated aldehydes, ketones, or esters led usually to isoxazoles via the corresponding azirines and vinyl nitrenes (see, for example, Schemes 5.4 and 5.22), such products were also assumed in the thermal ring expansion of **198** at first.[138] However, a few exceptions from this rule are known (see Scheme 5.3), and thus it has been shown quite recently that heating of **135** or **198** in toluene surprisingly afforded oxazoles **201** in excellent yields.[139] The formation of these five-membered heterocycles from 2*H*-azirines can be explained by cleavage of the C2-C3 bond to generate nitrile ylides like **200**. Such short-lived intermediates resulted usually from irradiation of azirines.[113]

Not only azirines but also nitriles were obtained in thermal reactions of vinyl azides (see, for example, Schemes 5.15 and 5.17). Hassner and coworkers summarized some rules,[140] which allow to expect the former or latter products after loss of dinitrogen from enazides.[35d] The 1-azidocyclopentadiene **202** turned out to be highly unstable because fragmentation to furnish the nitrile **203** was observed already at room temperature (Scheme 5.24). On the other hand, the isomer **204** was isolated as a relative stable lemon-yellow solid, which was converted to the indole **205** by refluxing in toluene or irradiation in chloroform.[95] In some cases, the formation of cyano compounds from vinyl azides was

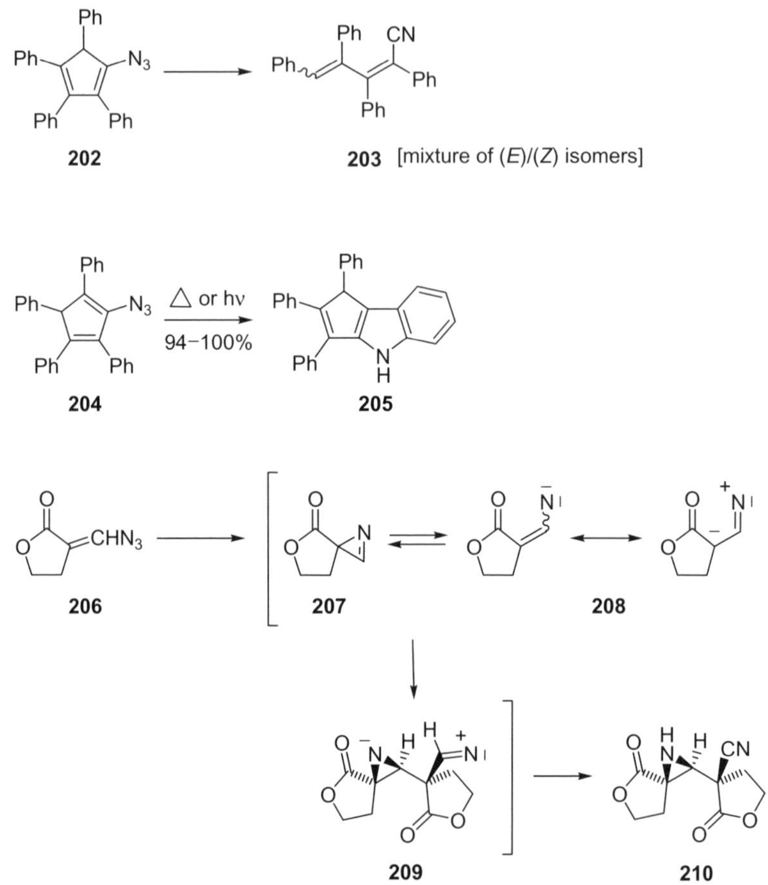

Scheme 5.24 *Formation of cyano compounds from vinyl azides*[95,141]

elucidated by short-lived azirine intermediates (see also Scheme 5.20). An interesting example was found in the thermal decomposition of the terminal azide **206** that gave the spirocyclic nitrile **210** with an excellent diastereoselectivity when heated in solution. A reaction mechanism via the intermediate 2*H*-azirine **207** and its dimerization to generate **209**, followed by a final intramolecular proton shift, was suggested.[141]

Since vinyl azides like **34** are electron-rich olefins, [2 + 2] cycloaddition with electron-deficient alkenes such as diphenylketene could lead to azidocyclobutanes.[142] The stability of the cycloadducts **211**, prepared from **34** or **52** and tetracyanoethene (TCNE), allowed characterization in solution but not isolation of these products because rapid ring-expansion regioselectively afforded the dihydropyrroles **212** already at room temperature (Scheme 5.25).[143] A similar mechanism via [2 + 2] cycloaddition and quick ring-enlargement may perhaps explain the formation of **213** from **52** and 4-phenyl-1,2,4-triazole-3,5-dione (PTAD). In this[144] and other[142,145] cases, however, different interpretations were offered. The 2-azidobuta-1,3-dienes **92a,b** underwent [4 + 2] cycloaddition in the

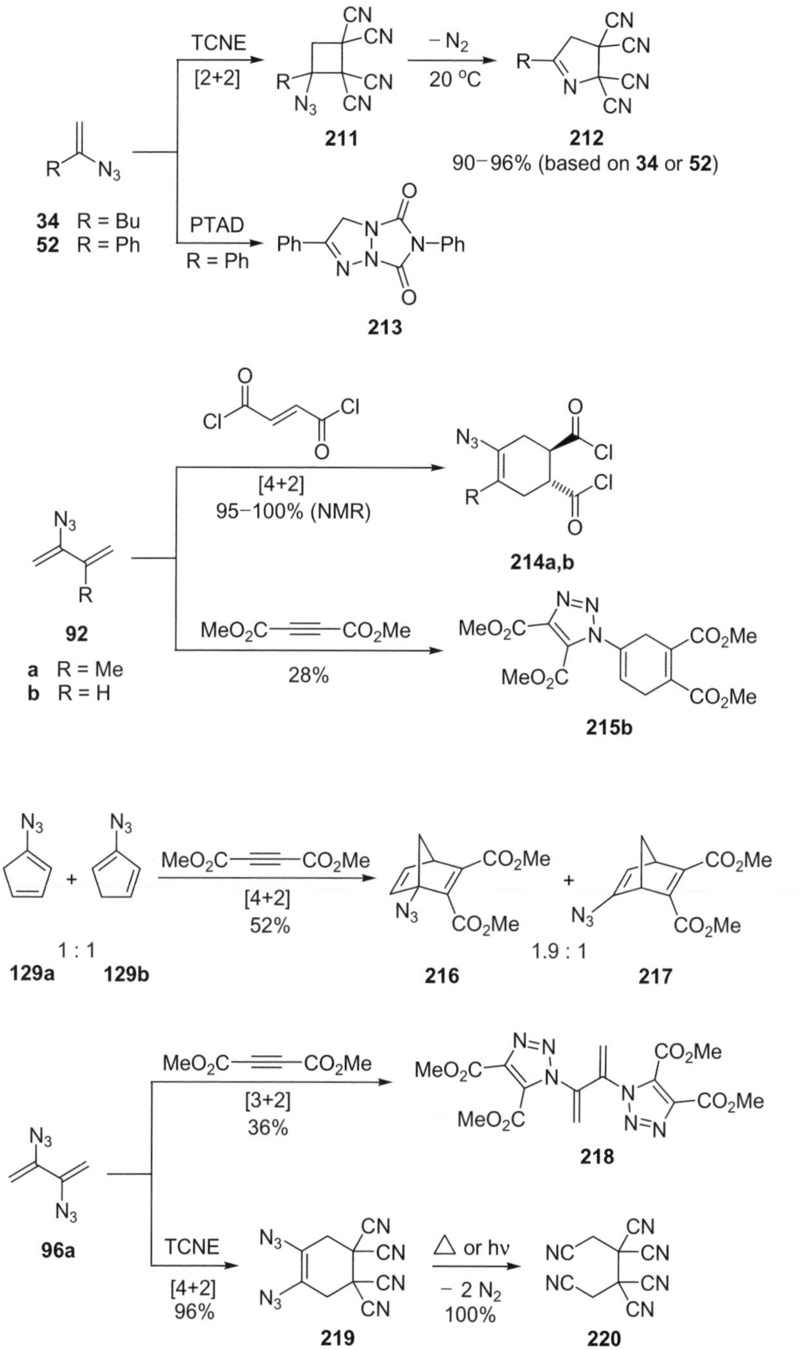

Scheme 5.25 *Cycloaddition reactions of vinyl azides*[78,95,143,144,146,149]

presence of highly electron-poor dienophiles, for example, TCNE, PTAD, or furmaryl chloride to furnish cycloadducts of type **214a,b**,[146] whereas treatment of **92b** with dimethyl acetylenedicarboxylate resulted in Diels–Alder reaction accompanied by 1,3-dipolar cycloaddition to give **215b**.[78] A mixture of the unstable products **216** and **217** was obtained by the reaction of the azidocyclopentadienes **129a,b** with dimethyl acetylenedicarboxylate.[95] An excess of the same reagent converted the diazide **96a** into triazole derivative **218**.[78] Similar transformations of **96a** via two 1,3-dipolar cycloaddition reactions were observed in the presence of hexafluorobut-2-yne or angular strained compounds, for example, norbornene, norbornadiene, or cyclooctyne.[143] The latter reagent and **96a** led finally to a 3 : 1 cycloadduct when a slower Diels–Alder reaction followed the very rapid [3 + 2] cycloaddition reactions. The quick and straightforward triazole-producing 1,3-dipolar cycloaddition of cyclooctyne can be used generally to characterize unstable vinyl or other azides (see also Schemes 5.17, 5.31, and 5.39).[61,147,148] With the help of electron-rich 2,3-diazidobuta-1,3-dienes like **96a** and highly electron-deficient dienophiles such as PTAD, TCNE, or others, [4 + 2] cycloadducts of type **219** were formed exclusively.[143,149] These products include the substructure of vicinal vinyl diazides and thus can be cleaved by thermolysis or photolysis in solution to synthesize the 1,4-dicyano compounds of type **220**. Sequences like **96a** + TCNE → **219** → **220** represent formally a *cis* addition of two cyanomethyl groups to the electron-poor dienophile.

The 1,3-dipolar cycloaddition reactions of enazides are not restricted to the reaction of their azido group with another olefin or alkyne. The addition of dipolar reagents, for example, a nitrile imine,[150] a nitrile oxide,[151] or a second azido group,[152] at the C,C double bond of the vinyl azide was shown to afford regioselectively five-membered heterocycles. In advantageous intramolecular cases, the [3 + 2] cycloaddition of the azido unit of an alkenyl azide to another C,C double bond occurred already at 0 °C.[152,153] Such an intramolecular reaction was also assumed when the vinyl azide **222** was heated to give the pyrrole derivative **225** (Scheme 5.26).[50b] The formation of the final product was explained via cleavage of the intermediate dihydro-1,2,3-triazole **223** followed by hydride shift, loss

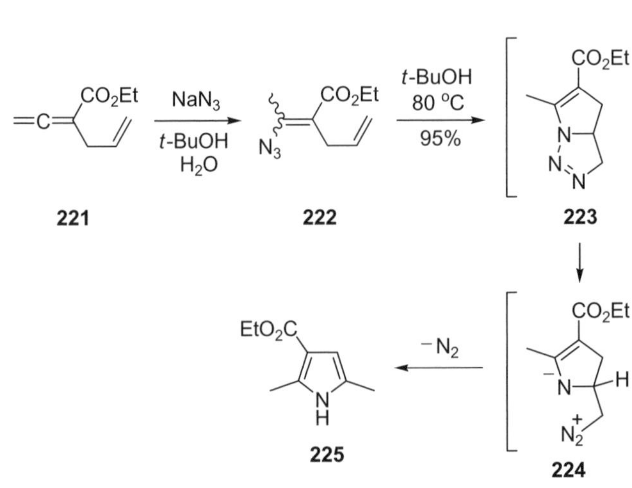

Scheme 5.26 *Intramolecular cycloaddition of a vinyl azide*[50b]

of dinitrogen, and tautomerism to get the aromatic heterocycle. Since enazides of type **222** were easily prepared by nucleophilic addition at acceptor-substituted allenes like **221** (see also Scheme 5.9), the synthesis of **225** was likewise performed in a one-pot procedure starting with **221** and sodium azide in aqueous *tert*-butanol. The vinyl azide **222** and similar compounds were also used for copper(I)-catalyzed intermolecular cycloaddition to terminal alkynes (see also Chapter 9, CuAAC reaction, Schilling, Jung & Bräse).[50b]

Cheletropic additions at the C,C double bond of alkenyl azides, for example, reactions with carbenes to generate azidocyclopropanes[154] or epoxidation[39,155] are well-known for several decades. Recently, the 1,4-addition of sulfur dioxide to 2-azidobuta-1,3-dienes **92** has been investigated (Scheme 5.27).[135] Moderate yields of the corresponding 3-sulfolenes **226** were obtained.

Because vinyl azides are electron-rich alkenes, treatment of these substrates with electrophiles was studied in detail.[35,107] In most cases, the attack of the α-nitrogen atom or the β-carbon of the olefinic part at the electrophile led to final products without an intact azido group such as amides or ketones. Only a few exceptions were reported, for instance, the reaction of the terminal enazides **40** or **54** with bromine in methanol yielding mainly the α-azido ethers **227** (Scheme 5.28).[156] Already in 1910, Forster and Newman investigated the conversion of the parent compound **1** with bromine in aqueous solution.[1a] Due to the explosion-like course of the reaction and the high sensitivity of the addition product to hydrolysis generating bromoacetaldehyde, however, the desired substance **228** could not be characterized. Quite recently, is has been shown that the transformation **1** $\rightarrow$ **228** can be performed conveniently with bromine in organic solvents at low temperature.[157]

Treatment of vinyl azides with nucleophiles was also studied comprehensively.[35,107] In most cases, the reactions were induced by a nucleophilic attack at the terminal nitrogen

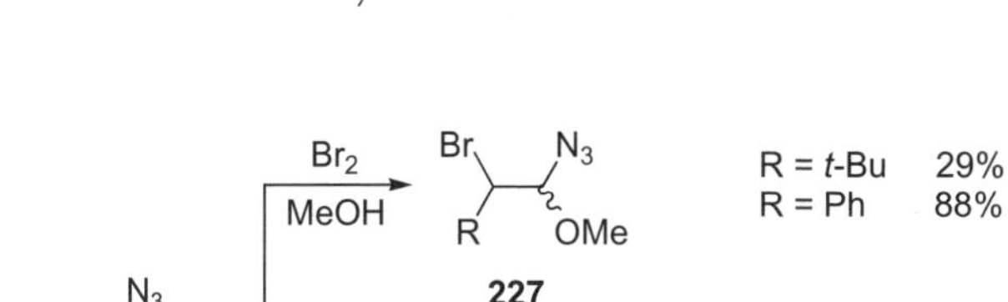

Scheme 5.27 *Synthesis of 3-azido-3-sulfolenes*[135]

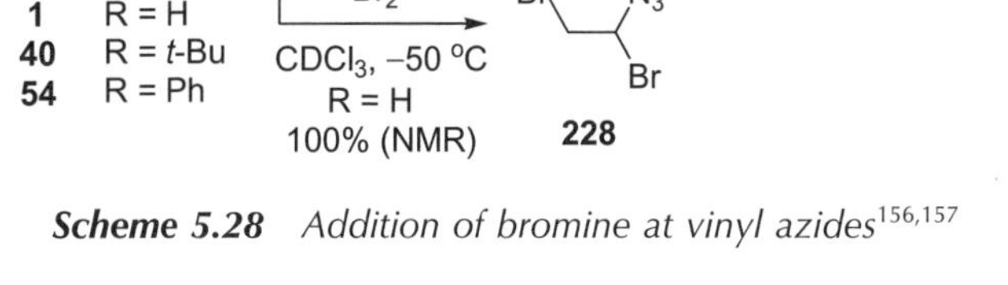

Scheme 5.28 *Addition of bromine at vinyl azides*[156,157]

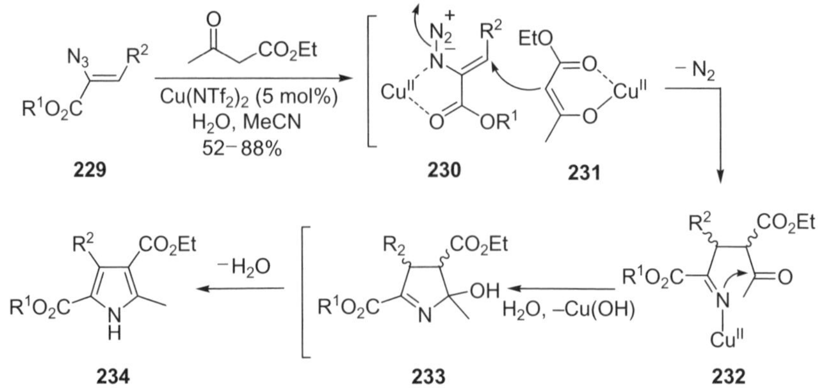

Scheme 5.29 *Synthesis of N-H pyrroles from vinyl azides and ethyl acetoacetate*[131a]

atom, for example, in the synthesis of triazenes with the help of organolithium compounds[158] or when 1,2,3-triazoles were prepared by the Dimroth method from vinyl azides and active methylene compounds under basic conditions.[159] Similar conversions of enazides using sulfonium ylides[160] or sulfoxonium ylides[161] were likewise published. However, the formation of the pyrroles **234** by copper(II)-catalyzed reaction of alkenyl azides **229** with ethyl acetoacetate was explained by an attack of the enolate **231** at the β-carbon atom of the polarized C,C double bond of intermediate **230** (Scheme 5.29).[131a] The proposed mechanism included ring closure by a second nucleophilic attack followed by hydrolysis, dehydration, and tautomerism to get the aromatic final product **234**.

The Staudinger reaction of vinyl azides and phosphorus(III) compounds, such as trialkyl or triarylphosphanes or phosphites, was mainly applied to substrates originating from Hemetsberger–Knittel synthesis (see also Scheme 5.12).[162] The resulting iminophosphoranes or iminophosphates **235** or alike ylides were treated with isocyanates or isothiocyanates to generate carbodiimides **236**[13b,c,163] or subjected to carbonyl compounds, like aldehydes,[164a–f] ketones,[164g] acyl chlorides,[165] or ketenes,[166] leading to imino compounds of type **237** or similar intermediates (Scheme 5.30). The products of these aza-Wittig reactions were not isolated in most cases but utilized in situ to prepare nitrogen heterocycles and natural products (see also Chapter 15, Palacios *et al.*).[167] Vinyl isothiocyanates were synthesized from **235** and carbon disulfide.[168] The same reagent was used to obtain the cyclopentadienyl derivative **238**.[95]

The addition of radicals to enazides was only rarely investigated. Recently, it has been shown that sulfanyl radicals generated from thiols add to the β carbon of alkenyl azides to give a mixture of the corresponding imines and the tautomeric β-sulfanylated enamines.[169]

The transformation of vinyl azides into saturated primary amines was performed with the help of sodium borohydride[170a] or by hydrogenation.[83] Thus, the esters of α-amino acids were easily available through hydrogenation of the Hemetsberger–Knittel products **229**,[170b–d] and the esters of β-amino acids were accessible in an analogous way.[9b,c] In advantageous cases, the reduction of alkenyl azides led to primary enamines.[171] An example is included in Scheme 5.31 depicting the synthesis of **242** from (*E,E*)-**66**.[172]

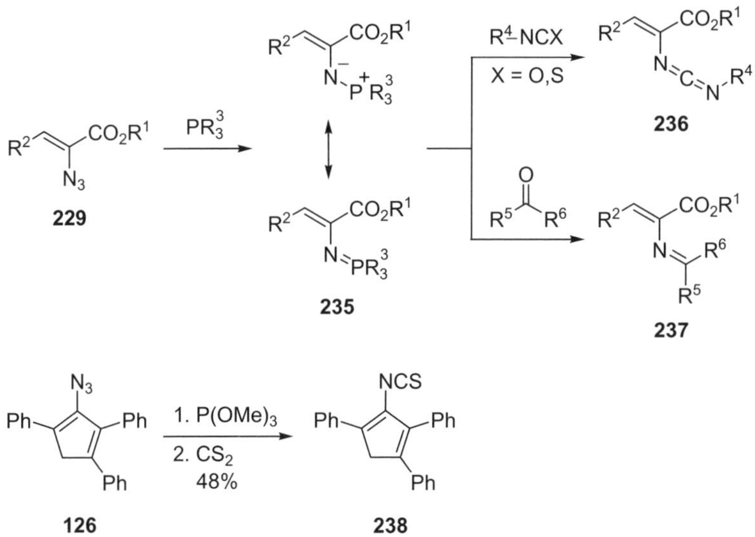

Scheme 5.30 *Staudinger reaction of vinyl azides*[13b,c,95,162–165]

Heating or irradiation of a solution of (*E*,*E*)-**66** or the diastereomeric 1,4-diazidobuta-1,3-dienes gave a 1 : 2 mixture of *meso*- and *rac*-**239** in excellent yields.[48] These bi-2*H*-azirin-2-yls are stable solids, which were separated easily by simple recrystallization. At higher temperatures, however, a fragmentation affording two molecules of acetonitrile and the alkyne **240** was obtained. In a side reaction, the pyridazine **241** was also observed. When the single isomers of **239** were thermolyzed, it was shown that an equilibrium of *meso*- and *rac*-**239** occurred and the generation of **240** and **241** started from *rac*-**239**. Photolysis of *meso*- or *rac*-**239** in dry chloroform produced 4,5-bis(diphenylphosphinoyl)-2,6-dimethylpyrimidine beside **240** and acetonitrile. The relative high thermal stability of **66** and **239** is based upon steric shielding due to the bulky substituents. Nevertheless, not only unimolecular but also bimolecular conversions, for example, the synthesis of the cycloadduct **243** from (*E*,*E*)-**66** and cyclooctyne, were successful. The transformation of the 1,4-diazidobuta-1,3-dienes of type **244**, which were prepared by electrocyclic conrotatory ring opening of the corresponding *trans*-3,4-diazidocyclobutenes, into the short-lived biazirinyls **245** was performed by photolysis in dichloromethane or chloroform at low temperature (R^1,R^2 = H or alkyl) or by thermal reaction (R^1=R^2=Ph, +5 °C).[147,173] The extremely unstable parent compounds *meso*- and *rac*-**245** (R^1=R^2=H) were generated only in low yield (8%) and degraded rapidly to undefined products. On the other hand, fully substituted biazirinyls, for instance, **245** with R^1=R^2=Me, were obtained quantitatively as 1 : 1 mixture of *meso* and *rac* diastereomers. These products underwent quantitative valence isomerization to furnish **246** with first-order kinetics. However, the tendency to this aromatization was quite different: The *meso* compound **245** (R^1=R^2=Me) was formed with $k = 1.73 \times 10^{-4}\ s^{-1}$ at +10 °C whereas its *rac* isomer gave $k = 3.65 \times 10^{-4}\ s^{-1}$ at −25 °C. These results were explained by quantum-chemical calculations, which interpreted the low-temperature valence isomerization of *rac*-**245** by simultaneous homolytic cleavage

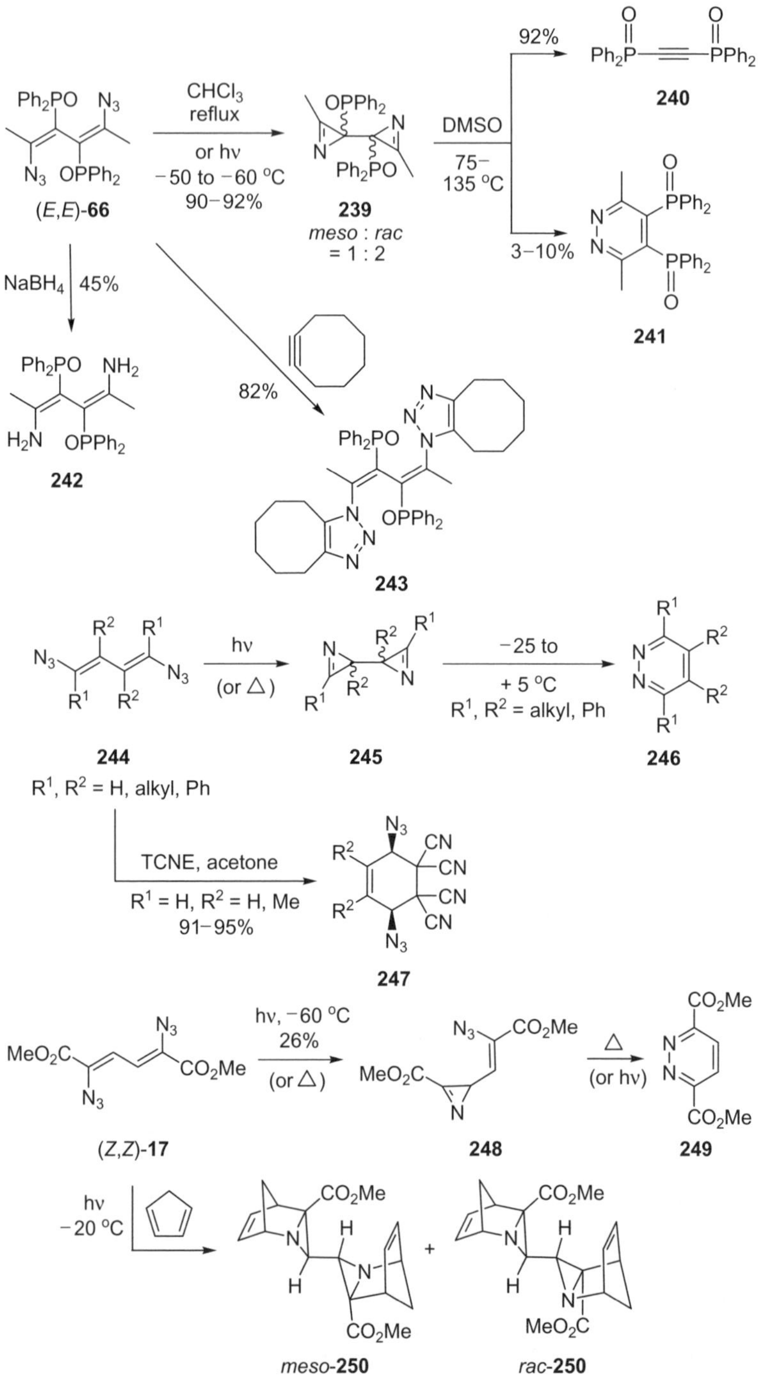

Scheme 5.31 *Reactions of 1,4-diazidobuta-1,3-dienes*[8,48,147,172,173]

of both C–N single bonds to generate energetically favorable nitrogen-centered 1,6-diradicals leading to pyridazines **246**.

The tetraphenyl-substituted diazide **244** afforded, by thermal reaction via **245**, tetraphenylpyridazine **246** (R^1=R^2=Ph) as the only product.[147,173] In preceding literature, the thermal isomerization of bi-2*H*-azirin-2-yls was postulated to yield pyrimidines and pyrazines.[54,174] But aromatic compounds different from **246** were only found if **245** (R^1=R^2=Me, Ph) was photolyzed or treated with silver(I) tetrafluoroborate.[147,173] In these cases, the corresponding pyrimidines were produced. 1,4-Diazidobuta-1,3-dienes of type **244** could be utilized in several cycloaddition reactions to prepare, for instance, the Diels–Alder product **247**. When the diazide (*Z,Z*)-**17** was irradiated at low temperature, only the mono-azirine **248** but not the corresponding biazirinyl could be detected.[8] Prolonged photolysis gave only traces of pyridazine **249** whereas heating of (*Z,Z*)-**17** in dichloromethane or chloroform resulted in the formation of **249** (27% yield) beside traces of **248**. Perhaps, the aromatic compound is generated via the corresponding biazirinyl, which is extremely unstable due to the electron-withdrawing ester groups. Thus, the photochemical reaction of (*Z,Z*)-**17** in the presence of an excess of cyclopentadiene furnished quantitatively a mixture of nearly equal parts of the trapping products *meso*- and *rac*-**250**. Thermolysis of 1,4-diazidobuta-1,3-dienes **68** (see Scheme 5.10) led also to the corresponding pyridazine derivative.[55]

5.5 The Chemistry of Allenyl Azides

Whereas vinyl azides are well-known for their manifold reactions and can be prepared by various methods,[35,107] all attempts to isolate allenyl azides were unsuccessful[175] until recently. In 1967, Shiner and Humphrey claimed to have obtained the allenyl azide **254a** together with the propargyl azide **255a** after treatment of bromides **251** or **252** with sodium azide in deuterated aqueous ethanol (Scheme 5.32).[176] However, **254a** and **255a** were only characterized in the reaction mixture by the ^{1}H NMR data of their methyl groups because they decomposed already at room temperature to unknown products. While the observation of **254a** represented the only direct reference to an allenyl azide for several decades, a detailed reinvestigation[177] of the reactions of **251** and **253** with sodium azide in aqueous alcohols showed that Shiner and Humphrey correctly recognized **255a** but the triazole **257** was taken for **254a**. The unknown succeeding products were now identified as a polymeric triazole **256** and the monomeric heterocycle **258**. In order to explain the surprising formation of **256**, **257**, and **258**, the very unstable azide **255a** was prepared by treating **253** with hexadecyltributylphosphonium azide[34] and isolated by preparative gas chromatography. When undiluted **255a** was allowed to warm to room temperature, the clear and colorless liquid became turbid and formed a white solid of **256** within a few seconds. It is very unlikely that this rapid polymerization process runs via an intermolecular 1,3-dipolar cycloaddition because such a reaction is slow even in the case of the parent propargyl azide[178] (**255b**), which is not hindered by steric effects due to methyl groups. With sodium azide in aqueous alcohols, **255a** produced **256** and **257** as well as **258**.[177] Using ^{15}N-labeled sodium azide, the label was found only in the azido group of **257** but not in the triazole part of **257** or **258**. This result excluded simple cycloaddition to give triazoles. In aqueous alcohols without sodium azide, **255a** afforded exclusively **258**.

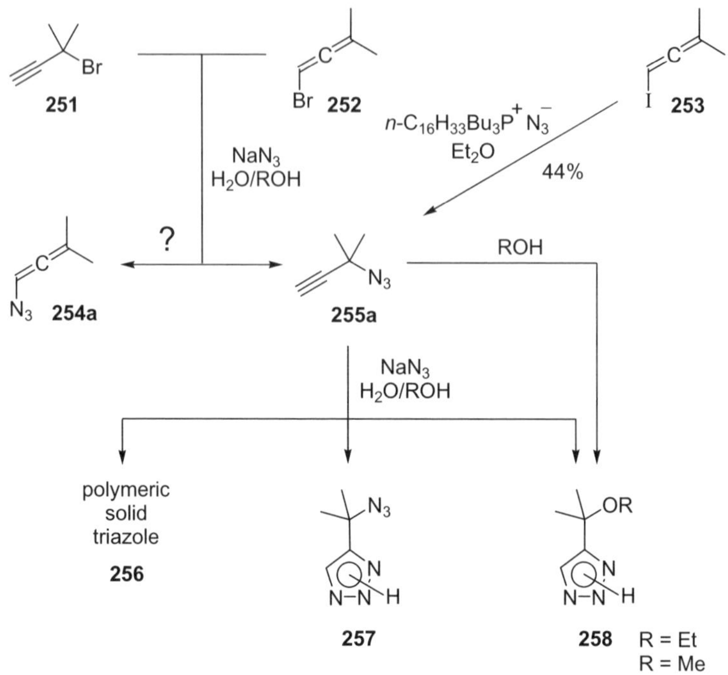

Scheme 5.32 *Synthesis and succeeding reactions of 3-azido-3-methylbut-1-yne*[176,177]

All these facts could only be explained by the [3,3]-sigmatropic rearrangement of **255** to yield **254** as a short-lived intermediate, which tends to rapid electrocyclization generating triazafulvene **259** (Scheme 5.33). Without any nucleophile, **259** led only to polymeric triazole derivatives, but trapping of **259** is possible in the presence of nucleophilic reagents NuH like methanol to furnish triazole **260** by an addition reaction. If optically active **255h** was treated with methanol, the resulting triazole was formed with complete racemization, which is compatible with an achiral intermediate such as **259h**. When propargyl azides **255i** or **255j** were heated with an excess of tetracyanoethene (TCNE) in tetrahydrofuran, the cycloadducts **261i** and **261j**, respectively, were isolated.[87] These products were easily explained by Diels–Alder reaction of short-lived allenyl azides **254i** and **254j**. Since cyclization of **254** is a rapid process (k_2) in comparison to the sigmatropic rearrangement of **255** (k_1), the maximum amount of the quasi-stationary intermediate **254** was always relative small (0.13–8%). Nevertheless, the allenyl azides **254a–h** could be characterized in the mixture by their NMR data.[85] Furthermore, the ^{1}H NMR shifts of **254a**[86] were clearly distinguished from those assumed for this compound by Shiner and Humphrey.[176] In the case of the cumulenes **254b–e**, even isolation by preparative gas chromatography was possible. This allowed the ^{15}N NMR analysis of **254d** generated from **255d**, which was selectively ^{15}N-labeled at N-α. The [3,3]-sigmatropic nature of the isomerization **255** → **254** was proved by this experiment because **254d** bore the ^{15}N label exclusively at N-γ.[81] When **254e** was treated with hexadecyltributylphosphonium azide, the maximum amount of the resulting diazide **95** (R^1=R^2=H, see Scheme 5.15) attained

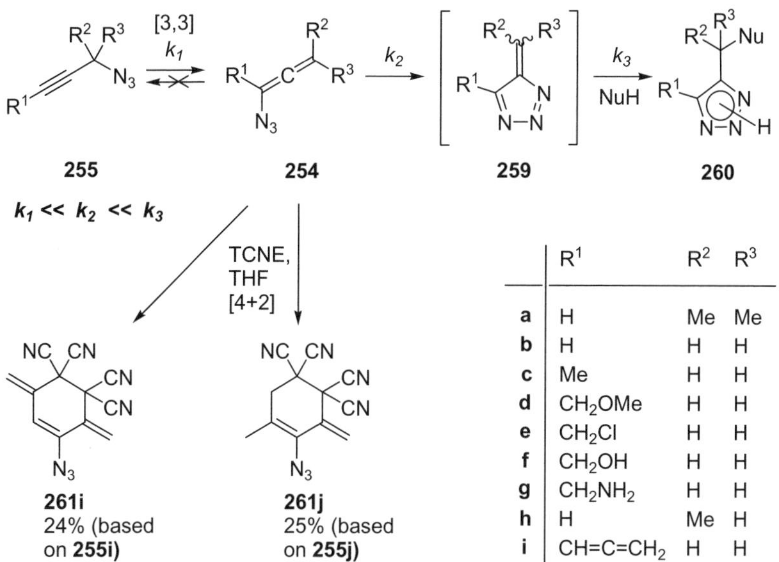

	R^1	R^2	R^3
a	H	Me	Me
b	H	H	H
c	Me	H	H
d	CH_2OMe	H	H
e	CH_2Cl	H	H
f	CH_2OH	H	H
g	CH_2NH_2	H	H
h	H	Me	H
i	$CH{=}C{=}CH_2$	H	H
j	$C(Me){=}CH_2$	H	H

Scheme 5.33 *Generation of allenyl azides by [3,3]-sigmatropic rearrangement of propargyl azides and transformation of these short-lived intermediates into 1,2,3-triazoles and Diels–Alder products (NuH = nucleophile)*[85,87,177]

30%. Thus, it was easy to detect **95** (R^1=R^2=H) as an intermediate in the rearrangement sequence **94** → **95** → **96** since **95** prepared from **254e** did not cyclize to the corresponding triazafulvene **259** but isomerized solely by a sigmatropic migration of the azido group to give **96** (R^1=R^2=H).[85]

Ring closure of vinyl azides **262** to 4*H*-1,2,3-triazoles **264** was studied by ab initio calculations[179] and discussed as the first step for the transformation of **262** to 2*H*-azirines (Scheme 5.34).[32,107,132k,140,180] However, the cyclization **262** → **264** was found only in the case of α-azidoenamines,[75,181] whilst simple vinyl azides do not give 1,2,3-triazoles by intramolecular reaction.[182] Poor nucleophilicity at C-5 of **262** or unfavorable geometry (distance between the orbitals of N-1 and C-5) may be the reason why simple vinyl azides do not cyclize to **264**. The possibility of ring closure is more favorable in the case of the carbanion **263**, and the conversion to **265** is known to be a rapid isomerization.[183] Advantageous geometry (two perpendicular p orbitals of the central carbon atom) possibly causes the fast cyclization of allenyl azide **254** to triazafulvene **259**.[86] Allenes with additional donor substituents exhibit an increased nucleophilicity at the central carbon atom and therefore should undergo an accelerated ring closure to **259**.

It was shown experimentally that donor substituents such as methyl groups speed up the [3,3]-sigmatropic rearrangement of **255a,c,h** (k_1) to produce the corresponding allenyl azides **254** and thus the rate of the entire transformation **255** → **260**, (Scheme 5.33).[85] However, the rate of the ring closure to **259** (k_2) was even more greatly increased by the donor substituents, thereby allowing only small proportions of **254a** (≤0.13%), **254c** (≤1%), or **254h** (≤0.32%) to build up as quasi-stationary intermediates.[86] By contrast, the

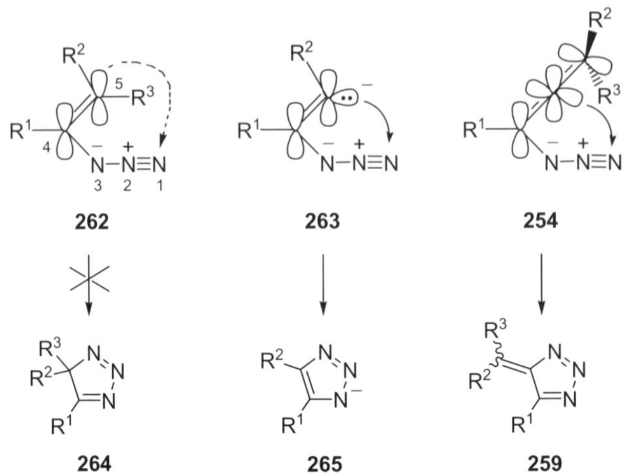

Scheme 5.34 *1,2,3-Triazole derivatives from ring closure of vinyl azides, their anions, or allenyl azides*[86]

Table 5.2 *Values of the rate constants k_2 [$10^{-5}\,s^{-1}$] for the cyclization reaction of **254a–e,h** in methanol-d4 at 29 °C*[86]

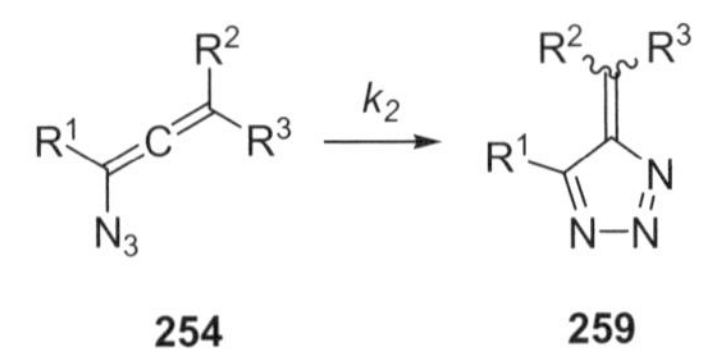

254	**a**	**b**	**c**	**d**	**e**	**h**
	4000	2.81	9.34	2.53	1.13	160

proportion of **254e**, which bears a weak acceptor substituent (R^1=CH_2Cl), reached a higher value (≤7%). A plot of the rate constants k_2 (Table 5.2) measured by varying the substituent R^1 versus the Taft σ^* values of R^1 led to $\rho^* = -0.87$ and exhibited that acceptor substituents retard the ring closure of **254** to **259** and therefore the formation of **260**.[85] The results in Table 5.2 confirm that a methyl group in the terminal position of the allene (R^2, R^3) accelerates the cyclization even more than the same group in the other position of **254** (R^1).[86] These findings demonstrate that allenyl azides are better stabilized by (weak) acceptor substituents than by the sterically demanding[175a,b] alkyl substituents which act as donors.

The knowledge of the surprising unimolecular reactions of propargyl azides shown in Schemes 5.32 and 5.33 offered the possibility to develop a one-pot procedure for the synthesis of 1,2,3-triazoles[184] of type **270** (Scheme 5.35). When a small excess of the starting material **266** was treated with sodium azide, the resulting solution of azide **267** yielded the final product **270** after heating with an excess of the nucleophile

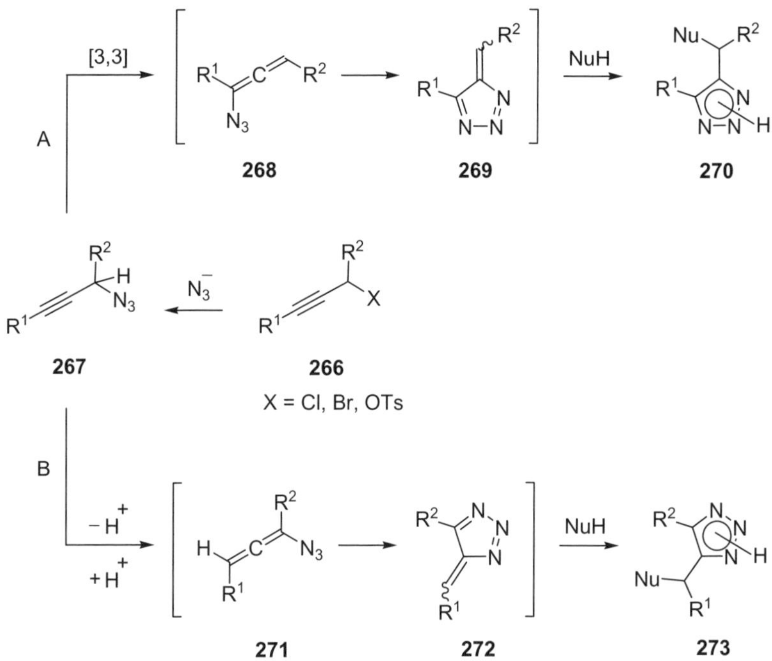

Scheme 5.35 *One-pot synthesis of 1,2,3-triazoles via [3,3]-sigmatropic (A) or prototropic (B) rearrangement of propargyl azides*[22,86,148b,177,185,186]

NuH.[22,86,148b,177,185,186] The course of the first step and the complete consumption of sodium azide could be conveniently observed by measurement of the decreasing pH. Alcohols, phenols, thiols, ammonia, primary and secondary amines as well as hydrazoic acid (sodium azide in protic solvents) and carboxylic acids served as nucleophiles NuH to get *N*-unsubstituted 1,2,3-triazoles bearing the functionality Nu in the side chain at C-4. This method utilizes easily available propargyl compounds and cheap sodium azide, avoiding expensive or violently explosive reagents and isolation of dangerous azides. Thus, the cascade approach A depicted in Scheme 5.35 has been adopted to synthesize various biologically active compounds containing a triazole ring with secondary aminomethyl substituents[187] and has been developed further by Sharpless and coworkers.[188] Furthermore, base-induced (prototropic) rearrangement of propargyl azides **267** led to isomeric allenyl azides **271**, which gave triazoles **273** in the presence of nucleophiles NuH.[86,148b,185,186] This modification could also be used in a one-pot procedure B **266** → **267** → **271** → **272** → **273**, in which the base and the nucleophile NuH may be the same or two different reagents.

Starting with the substrate **274**, for example, treatment with sodium azide and then with nucleophiles NuH, such as methanol, water, or ammonia, in the presence of the base sodium hydroxide furnished the triazoles **277** via the cascade B (Scheme 5.36).[185] On the other hand, the products **276** were formed from **274** through the sequence A without the strong base sodium hydroxide. Thus, the nucleophile ammonia was dominating over sodium hydroxide. But without the latter, base-induced generation of allenyl azides of

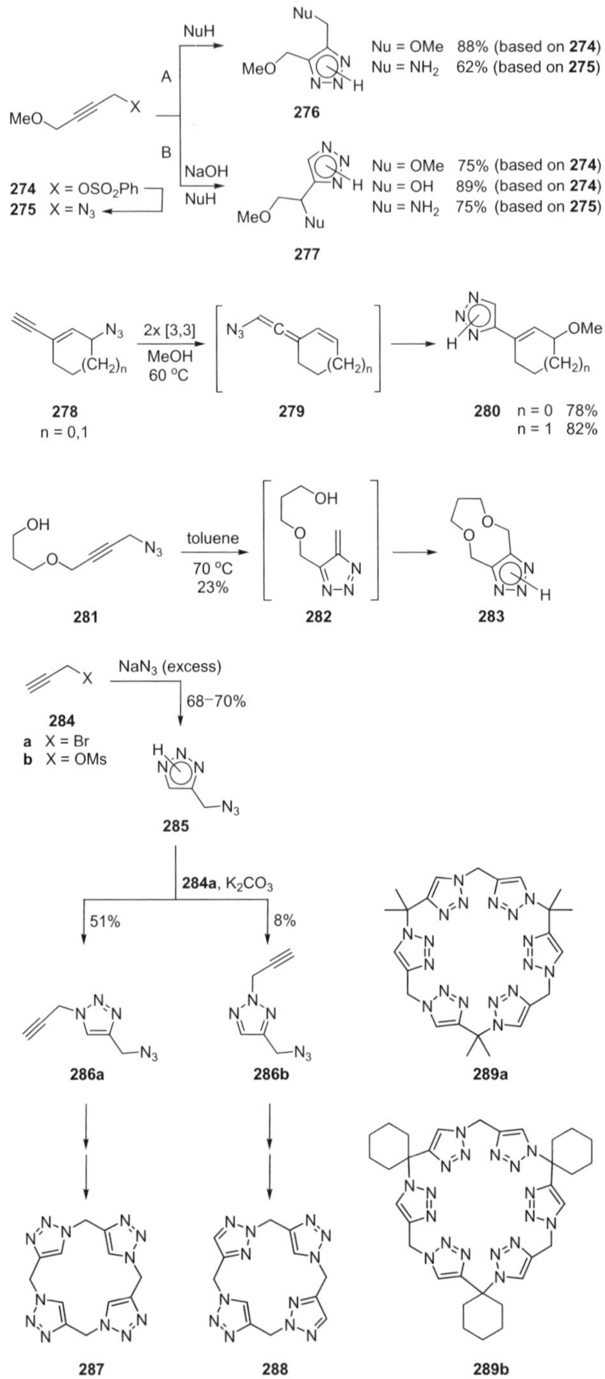

Scheme 5.36 Synthesis of triazole derivatives via short-lived allenyl azides[22,87,185,188,189]

type **271** was not possible with ammonia alone. In the case of the starting materials **278**, two [3,3]-sigmatropic successive migrations were necessary to produce the intermediate allenyl azides **279**, which led to the triazole derivatives **280**.[87] Intramolecular trapping of short-lived triazafulvenes like **282** was also possible as shown by heating **281** in non-nucleophilic toluene. However, the yield of the bicyclic product **283** was only moderate.[87] The most simple cascade reaction was performed by treating the precursors **284** with an excess of sodium azide in protic solvents.[188,189] The resulting building block **285** was *N*-alkylated to give the propargyl derivatives **286a** and **286b**, which were easily separated, dimerized by Cu(I)-catalyzed click chemistry, and finally macrocyclized to afford the triazolocyclophanes **287** and **288**.[189] Cyclophanes connecting even six triazole units in an analogous way such as **289a,b** could be prepared similarly if solubility promoting alkyl groups were incorporated into the 4-azidomethyl-1,2,3-triazoles of type **285**.

Azidocumulenes, which tend to rapid cyclization leading to triazole derivatives, are available not only by sigmatropic or prototropic rearrangement (see Scheme 5.35), but also by 1,4-elimination of hydrogen chloride from the substrate **291** (Scheme 5.37).[22] Thus, the reaction of **291** with sodium hydroxide in methanol gave 4-ethynyl-1,2,3-triazole (**295**) after aqueous workup. This result was explained by formation of short-lived azidobutatriene (**292**) followed by ring closure and deprotonation of intermediate **293**. When dichloride **290** was treated with a substoichiometric amount of sodium azide and then with sodium hydroxide in methanol, the product **295** was synthesized in a one-pot procedure with 45% yield.

Allenyl azides bearing a strong acceptor substituent like a phenylsulfonyl group were conveniently generated by prototropic rearrangement as demonstrated by the example **296** → **297** (Scheme 5.38).[148b] The azidoallene **271b** was produced analogously. Both allenyl azides did not cyclize to the corresponding triazafulvenes as it could be expected if the effects of substituents on the rates of ring closure (k_2) **254** → **259** (see Table 5.2) are extrapolated to the electron-withdrawing phenylsulfonyl group. Irradiation of **297** in dichloromethane at low temperature caused migration of the phenylsulfonyl group and formation of the nitrile **298**.[190] This result is quite different to that of the photolysis of allenyl azides **271a,b**, which led to the 2-methylene-2*H*-azirines **299a,b**.[190–192] The yields of these highly strained heterocycles were limited to ≤60% by a photochemical secondary reaction, which afforded the fragments **300** and **301**. The methyleneazirines **299** proved

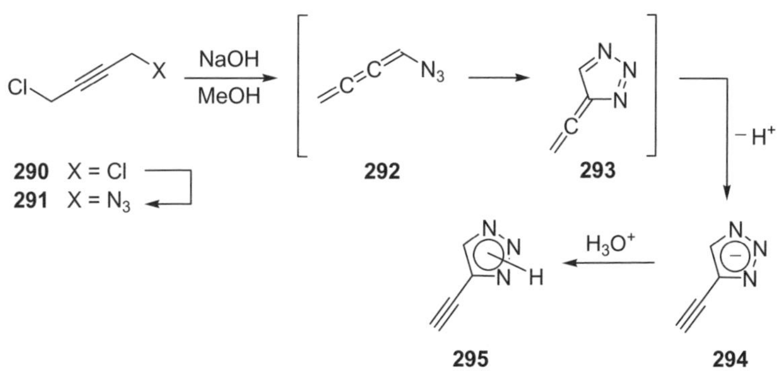

Scheme 5.37 *Synthesis of 4-ethynyl-1,2,3-triazole via azidobutatriene*[22]

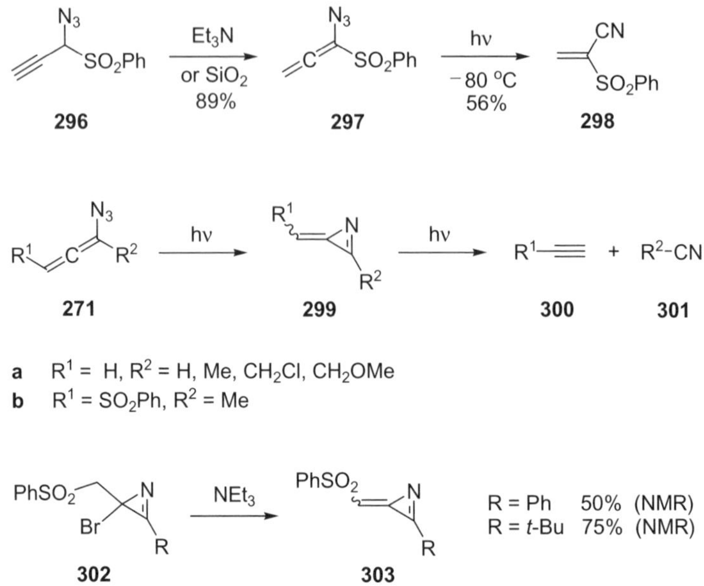

Scheme 5.38 *Photolysis of allenyl azides and generation of 2-methylene-2H-azirines*[148b,190–192]

to be the first heterocyclic triafulvenes. However, similar compounds have been synthesized by the elimination reaction **302** → **303** quite recently.[190]

5.6 Generation of Ethynyl Azides

Already in 1910, Forster and Newman studied the addition of bromine at vinyl azide (**1**) in order to prepare ethynyl azide by elimination of two molecules of hydrogen bromide (see Scheme 5.28).[1a] Since the 1950s, many groups have tried to generate 1-azidoalk-1-ynes, but these species are a mystery even till today.[193] Several attempts at preparing the title compounds were unsuccessful or led to unwanted products.[194] For example, the reaction of (phenylethynyl)sodium or alkyl-1-ynyllithiums and tosyl azide yielded only 1,2,3-triazole derivatives instead of the target compounds.[195] Even in the case of generating 1-azido-2-phenylethyne (**305**) in situ, the sequential chemistry of such a short-lived intermediate is still unclear (Scheme 5.39). Thus, the starting material **304** is transformed under a variety of conditions and in high yield into *cis*- and *trans*-dicyanostilbenes **308**, which are formally dimers of the carbene **306**.[196] However, the known[197] trapping products of **306** including the cyclopropane framework cannot be detected when **304** was treated in the presence of cyclohexene or tolane. Therefore, the formation of **308** from **304** was explained *via* the intermediates **305** and **309** by Boyer and Selvarajan.[196a] But alternative routes leading from **304** to **308** without the intermediary 1-azidoalkyne **305** were also discussed.[196b,198] Hassner and Isbister postulated the 2*H*-azirine **307**, which should give rise to the final product **308**.[196b] Quite recently, it has been shown that irradiation of **304**

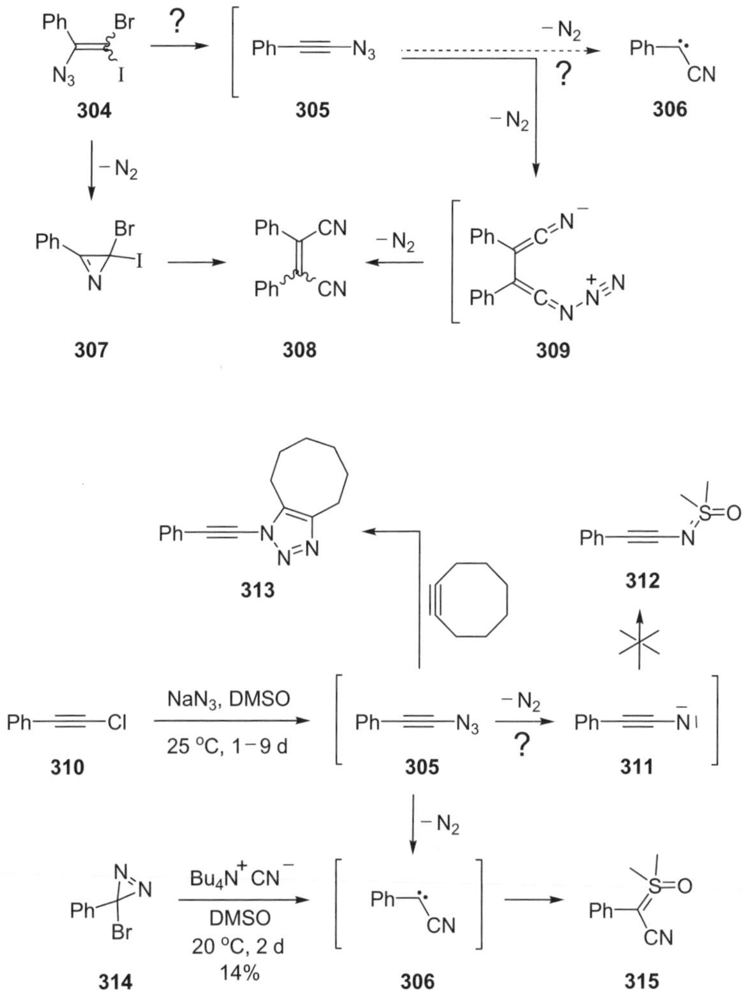

Scheme 5.39 *Attempts to generate and to trap 1-azido-2-phenylethyne*[157,193,196,198,199,201,202]

in dichloromethane at −70 °C induced the complete transformation to the highly unstable heterocycle **307** generated in 90% yield and characterized by its NMR data.[157] Heating of the solution of **307** at 85 °C afforded nearly quantitatively the nitriles **308**. In the light of these results, it is plausible that the thermal conversion of **304** also proceeded *via* **307** leading to **308** without the intermediate **305**. On the other hand, treatment of the precursor **310** with sodium azide in dimethyl sulfoxide was said to furnish in low yield (5–8%) the sulfoximine **312**, which was explained by formation of short-lived azide **305** followed by loss of dinitrogen and trapping of the nitrene **311** by the solvent.[198,199] Several other products but no stilbene derivative **308** were obtained from the reaction of **310**. The compound **312** seems to be a logical interception product of **305** and **311**, respectively, because the synthesis of similar sulfoximines from azides and dimethyl sulfoxide is a well-known process.[200] In the case of **311**, however, Auer and coworkers demonstrated in an *ab initio* study that the nitrene does not correspond to a local minimum of energy and that cleavage of dinitrogen from **305** should generate the carbene **306**.[201] When the experiment starting with **310** was repeated quite recently, it has been shown that the real product **315** (10% yield), resulting *via* **306**, was erroneously taken for **312**.[202] Furthermore, **315** was also prepared from the substrate **314** and tetrabutylammonium cyanide in dimethyl sulfoxide, most probably *via* 3-cyano-3-phenyl-3*H*-diazirine and its decay product **306**.[157] Other sulfoxonium ylides similar to **315** were synthesized previously by storing or heating diazo compounds in dimethyl sulfoxide.[203] The carbene trapping product **315** was not detected if the transformation **304** → **308** was performed in the solvent dimethyl sulfoxide.[157] This outcome excluded the intermediates **305** and **306** in the formation of **308**. When alkyne **310** was treated with cyclooctyne and lithium azide in dimethylformamide, the triazole derivative **313** was isolated in 7.3% yield. Several control experiments were performed to exclude, for example, the generation of **313** from **310** and 4,5,6,7,8,9-hexahydrocycloocta-1,2,3-triazole possibly produced from lithium azide and cyclooctyne. Thus, the interception product **313** is currently the most plausible evidence for short-lived ethynyl azides.[157]

The geometrical structure and the heat of formation have been calculated for 1-azido-2-nitroethyne, which should correspond to an energy minimum and, accordingly, should be able to exist.[204] In an *ab initio* study, the energy barriers were compared for the cleavage of dinitrogen from 1-azidoalkynes bearing different functional groups.[201] It was shown that the parent compound as well as **305** and silyl-substituted ethynyl azides are relatively stable whereas azidoacetylenes with donor groups, especially such as amino or ethylsulfanyl, should split off dinitrogen very easily. Nevertheless, the reaction of 1-chloro-2-alkylsulfanylethyne with sodium azide in dimethyl sulfoxide was claimed to afford the isolable 1-alkylsulfanyl-2-azidoethynes.[205] However, it was proved that the ^{13}C NMR data reported for these products were not compatible with the assumed structures.[201] Moreover, several attempts to reproduce the generation of these 1-azidoalk-1-ynes were unsuccessful.[157,202]

5.7 Conclusion

Although this review is by no means comprehensive, it should give an impression of the great number of feasible syntheses of the title compounds and their possibilities of reactions. The unique combination of C,C double or triple bonds and the azido group often allows not only the common reactions of these parts but also specific transformations into a variety of very different products. Much attention was paid to the results of the last 20 years, but it was tried to mention all important facts about the chemistry of the title compounds including also their history.

At the latest during the last two decades, it turned out that vinyl azides are not only compounds for experts in organic chemistry but also very useful and general tools in synthetic chemistry. Although these azides are known for about hundred years, efficient approaches, developed in the last 45 years, were necessary to supply the starting materials, which have been utilized to show the versatility of vinyl azides. This is demonstrated by reactions induced by thermolysis, photolysis, cycloaddition, or attack by electrophiles or nucleophiles leading, for example, to several kinds of important nitrogen heterocycles and especially to natural products and biologically active compounds.

Most probably, the rapid electrocyclic ring closure of allenyl azides was responsible for the fact that the chemistry of these azides was discovered so lately. This cyclization of azidoallenes allowed to create cascade reactions for the synthesis of functionalized NH-1,2,3-triazoles.

Since 1910, many groups have tried to generate ethynyl azides, but these species have proved to be the Yetis within the azides even today. However, latest results have shown that 1-azidoalk-1-ynes can be trapped by 1,3-dipolar cycloaddition although they lost dinitrogen very rapidly to produce cyanocarbenes.

Acknowledgment

The author thanks Dr N. Ramezanian and Dr J. R. Fotsing for assistance with the manuscript.

References

[1] (a) M.O. Forster, S.H. Newman, *J. Chem. Soc.* **1910**, *97*, 2570–9. (b) M.O. Forster, S.H. Newman, *J. Chem. Soc.* **1911**, *99*, 1277–82.
[2] R.H. Wiley, J. Moffat, *J. Org. Chem.* **1957**, *22*, 995–6.
[3] K. Fries, P. Ochwat, *Ber. Dtsch. Chem. Ges.* **1923**, *56*, 1291–304.
[4] L.F. Fieser, J.L. Hartwell, *J. Am. Chem. Soc.* **1935**, *57*, 1482–4.
[5] Review on azidoquinones: H.W. Moore, *Chem. Soc. Rev.* **1973**, *2*, 415–55.
[6] T.E. Stevens, W.D. Emmons, *J. Am. Chem. Soc.* **1958**, *80*, 338–41.
[7] (a) A.N. Nesmeyanov, M.I. Rybinskaya, T.G. Kelekhsaeva, *J. Org. Chem. USSR (Engl. Transl.)* **1968**, *4*, 897–904. (b) M.I. Rybinskaya, A.N. Nesmeyanov, N.K. Kochetkov, *Russ. Chem. Rev. (Engl. Transl.)* **1969**, *38*, 433–55.
[8] F. Köhler, dissertation, Chemnitz University of Technology (Germany), **2002**.
[9] Recent articles including synthesis of vinyl azides by nucleophilic substitution: (a) G.G. Furin, Y.V. Gatilov, I.Y. Bagryanskay, E.L. Zhuzhgov, *J. Fluorine Chem.* **2001**, *110*, 21–4. (b) M.S. Ozer, S. Thiébaut, C. Gérardin-Charbonnier, C. Selve, *Synth. Commun.* **1998**, *28*, 2429–41. (c) S. Cosgun, M. Özer, F. Hamdoune, *et al.*, *J. Fluorine Chem.* **2001**, *107*, 375–86. (d) N.A. Anisimova, N.G. Makarova, G.A. Berkova, V.M. Berestovitskaya, *Russ. J. Gen. Chem.* **2006**, *76*, 1545–9. (e) J. Finnerty, U. Mitschke, C. Wentrup, *J. Org. Chem.* **2002**, *67*, 1084–92.
[10] J. Jonas, C. Mazal, Z. Rappoport, *J. Phys. Org. Chem.* **1994**, *7*, 652–4.
[11] V.G. Ostroverkhov, E.A. Shilov, *Ukrain. Khim. Zhur.* **1957**, *23*, 615–22; *Chem. Abstr.* **1958**, *52*, 7828d.
[12] (a) U. Türck, H. Behringer, *Chem. Ber.* **1965**, *98*, 3020–4. (b) G. L'abbé, J.-P. Dekerk, P. Van Stappen, *Bull. Soc. Chim. Belg.* **1981**, *90*, 1073–4. (c) H. Hopf, N. Krause, *Tetrahedron Lett.* **1986**, *27*, 6177–80. (d) B.A. Trofimov, A.G. Mal'kina, R.N. Kudyakova, *Russ. J. Org. Chem.* **1993**, *30*, 1616–9.
[13] (a) F. Palacios, D. Aparicio, J.M. de los Santos, I. Perez de Heredia, G. Rubiales, *Org. Prep. Proced. Int.* **1995**, *27*, 171–8. (b) M.A. Arnold, S.G. Durón, D.Y. Gin, *J. Am. Chem. Soc.* **2005**, *127*, 6924–5. (c) M.A. Arnold, K.A. Day, S.G. Durón, D.Y. Gin, *J. Am. Chem. Soc.* **2006**, *128*, 13255–60.
[14] (a) L. Birkofer, A. Ritter, *Angew. Chem.* **1965**, *77*, 414–26; *Angew. Chem. Int. Ed. Engl.* **1965**, *4*, 417–29. (b) L. Birkofer, P. Wegner, *Chem. Ber.* **1966**, *99*, 2512–7.

[15] (a) M. Ochiai, M. Kunishima, K. Fuji, Y. Nagao, *J. Org. Chem.* **1988**, *53*, 6144–5. (b) A. Degl'Innocenti, A. Capperucci, G. Reginato, A. Mordini, A. Ricci, *Tetrahedron Lett.* **1992**, *33*, 1507–8.
[16] (a) A.N. Nesmeyanov, M.I. Rybinskaya, *J. Org. Chem. USSR (Engl. Transl.)* **1966**, *2*, 2041–5. (b) H. Priebe, *Acta Chem. Scand.* **1987**, *B41*, 640–5. (c) T. Kitamura, P.J. Stang, *Tetrahedron Lett.* **1988**, *29*, 1887–90. (d) M. Haddach, R. Pastor, J.G. Riess, *Tetrahedron Lett.* **1990**, *31*, 1989–90. (e) M. Haddach, R. Pastor, J.G. Riess, *Tetrahedron* **1993**, *49*, 4627–38. (f) K. Kuramochi, H. Watanabe, T. Kitahara, *Synlett* **2000**, 397–9.
[17] (a) A.N. Nesmeyanov, M.I. Rybinskaya, *Dokl. Akad. Nauk SSSR* **1964**, *158*, 408–10; *Chem. Abstr.* **1964**, *61*, 14664h. (b) Y. Tanaka, S.R. Velen, S.I. Miller, *Tetrahedron* **1973**, *29*, 3271–83. (c) E.J. Trybulski, L. Benjamin, S. Vitone, A. Walser, R.I. Fryer, *J. Med. Chem.* **1983**, *26*, 367–72. (d) T.L. Gilchrist, G.E. Gymer, *Adv. Heterocycl. Chem.* **1974**, *16*, 33–85.
[18] (a) O. Dimroth, G. Fester, *Ber. Dtsch. Chem. Ges.* **1910**, *43*, 2219–23. (b) J.C. Sheehan, C.A. Robinson, *J. Am. Chem. Soc.* **1951**, *73*, 1207–10. (c) L.W. Hartzel, F.R. Benson, *J. Am. Chem. Soc.* **1954**, *76*, 667–70.
[19] (a) R. Hüttel, *Ber. Dtsch. Chem. Ges.* **1941**, *74*, 1680–7. (b) J.C. Sheehan, C.A. Robinson, *J. Am. Chem. Soc.* **1949**, *71*, 1436–40.
[20] C. Berndt, diploma thesis, Chemnitz University of Technology (Germany), **2008**.
[21] (a) J. Kalcher, W.M.F. Fabian, *Theor. Chem. Acc.* **2003**, *109*, 195–9. (b) A. Peña-Gallego, J. Rodriguez-Otero, E.M. Cabaleiro-Lago, *Eur. J. Org. Chem.* **2005**, 3228–32.
[22] K. Banert, *Chem. Ber.* **1989**, *122*, 1175–8.
[23] (a) V.O. Kuz'min, S.G. Fridman, *Mem. Inst. Chem. Ukrain. Acad. Sci.* **1935**, *2*, 55–64; *Chem. Abstr.* **1937**, *31*, 4660^{5}. (b) V.O. Kuz'min, M.I. Zemlyans'kiĭ, *Mem. Inst. Chem. Ukrain. Acad. Sci.* **1935**, *2*, 183–9; *Chem. Abstr.* **1937**, *31*, 3467^{1}. (c) V.O. Kuz'min, M.I. Zemlyans'kiĭ, *Mem. Inst. Chem. Ukrain. Acad. Sci.* **1935**, *2*, 191–3; *Chem. Abstr.* **1937**, *31*, 3467^{2}. (d) S.G. Fridman, *Mem. Inst. Chem. Ukrain. Acad. Sci.* **1936**, *3*, 587–604; *Chem. Abstr.* **1937**, *31*, 7861^{4}.
[24] (a) T.L. Gilchrist, R. Mendonça, *Synlett* **2000**, 1843–5. (b) A.S. Timén, P. Somfai, *J. Org. Chem.* **2003**, *68*, 9958–63.
[25] (a) P.G. Owston, R. Peters, P.A. Tasker, *J. Chem. Res. (S)* **1985**, 352–3. (b) P.G. Owston, R. Peters, P.A. Tasker, *J. Chem. Res. (M)* **1985**, 3686–94. (c) M.J. Alves, T.L. Gilchrist, *Tetrahedron Lett.* **1998**, *39*, 7579–82. (d) R.M. Moriarty, J.S. Khosrowshahi, *Tetrahedron Lett.* **1986**, *27*, 2809–12.
[26] (a) M.J. Alves, J.F. Bickley, T.L. Gilchrist, *J. Chem. Soc., Perkin Trans. 1* **1999**, 1399–401. (b) Y.S.P. Álvares, M.J. Alves, N.G. Azoia, J.F. Bickley, T.L. Gilchrist, *J. Chem. Soc., Perkin Trans. 1* **2002**, 1911–9.
[27] A. Hassner, G. L'abbé, M.J. Miller, *J. Am. Chem. Soc.* **1971**, *93*, 981–5.
[28] (a) P. Ykman, G. Mathys, G. L'abbé, G. Smets, *J. Org. Chem.* **1972**, *37*, 3213–6. (b) M.-a. Kakimoto, M. Kai, K. Kondo, *Chem. Lett.* **1982**, 525–6. (c) T. Patonay, J. Jekő, É. Rimán, *Synth. Commun.* **2002**, *32*, 2403–15.
[29] (a) J.A. VanAllen, W.J. Priest, A.S. Marshall, G.A. Reynolds, *J. Org. Chem.* **1968**, *33*, 1100–2. (b) T. Sasaki, K. Kanematsu, M. Murata, *Tetrahedron* **1973**, *29*, 529–32.
[30] (a) E.J. Trybulski, R.I. Fryer, E. Reeder, S. Vitone, L. Todaro, *J. Org. Chem.* **1986**, *51*, 2191–202. (b) M. Alajarin, R.-Á. Orenes, Á. Vidal, A. Pastor, *Synthesis* **2003**, 49–52.
[31] (a) A.G. Hortmann, D.A. Robertson, B.K. Gillard, *J. Org. Chem.* **1972**, *37*, 322–4. (b) T. Sakai, I. Kawabata, T. Kishimoto, T. Ema, M. Utaka, *J. Org. Chem.* **1997**, *62*, 4906–7.
[32] (a) G. Smolinsky, *J. Am. Chem. Soc.* **1961**, *83*, 4483–4. (b) G. Smolinsky, *J. Org. Chem.* **1962**, *27*, 3557–9.
[33] M.S.F. Lie Ken Jie, M.S. Alam, *Chem. Phys. Lipids* **2001**, *111*, 29–35.
[34] Review: K. Banert, *Synthesis* **2007**, 3431–46.
[35] (a) G. Smolinsky, C.A. Pryde in *The Chemistry of the Azido Group*, S. Patai, ed.; John Wiley & Sons, Inc., New York, **1971**, pp. 555–85. (b) G. L'abbé, A. Hassner, *Angew. Chem.* **1971**, *83*, 103–9; *Angew. Chem. Int. Ed. Engl.* **1971**, *10*, 98–104. (c) A. Hassner in *Azides and Nitrenes*, E.F.V. Scriven, ed.; Academic Press, Orlando, **1984**, pp. 35–94. (d) K. Banert in

Houben–Weyl, Methoden Org. Chem., 4th ed., Vol. *E 15* (eds.: H. Kropf, E. Schaumann), Thieme, Stuttgart, **1993**, pp. 818–75. (e) S.J. Collier in *Science of Synthesis*, Vol. *33* (ed.: G.A. Molander), Thieme, Stuttgart, **2006**, pp. 541–63.

[36] A. Hassner, L.A. Levy, *J. Am. Chem. Soc.* **1965**, *87*, 4203–4.

[37] Review: A. Hassner, *Acc. Chem. Res.* **1971**, *4*, 9–16.

[38] (a) F.W. Fowler, A. Hassner, L.A. Levy, *J. Am. Chem. Soc.* **1967**, *89*, 2077–82. (b) A. Hassner, F.W. Fowler, *J. Org. Chem.* **1968**, *33*, 2686–91.

[39] J. Schweng, E. Zbiral, *Liebigs Ann. Chem.* **1978**, 1089–95.

[40] K. Banert, B. Meier, *Angew. Chem.* **2006**, *118*, 4120–3; *Angew. Chem. Int. Ed.* **2006**, *45*, 4015–9.

[41] (a) A.M. Salunkhe, P.V. Ramachandran, H.C. Brown, *Tetrahedron Lett.* **1999**, *40*, 1433–6. (b) A. Padwa, T. Stengel, *Arkivoc* **2005**, (*v*), 21–32. (c) W. Zhao, E.M. Carreira, *Chem. Eur. J.* **2006**, *12*, 7254–63. (d) N.C. Srivastav, T. Manning, D.Y. Kunimoto, R. Kumar, *Bioorg. Med. Chem.* **2007**, *15*, 2045–53.

[42] R. Kumar, M. Nath, D.L.J. Tyrrell, *J. Med. Chem.* **2002**, *45*, 2032–40.

[43] (a) A. Hassner, F. Boerwinkle, *J. Am. Chem. Soc.* **1968**, *90*, 216–8. (b) A. Hassner, F. Boerwinkle, *Tetrahedron Lett.* **1969**, 3309–12.

[44] (a) M. Curini, F. Epifano, M.C. Marcotullio, O. Rosati, *Tetrahedron Lett.* **2002**, *43*, 1201–3. (b) V. Nair, T.G. George, V. Sheeba, A. Augustine, L. Balagopal, L.G. Nair, *Synlett* **2000**, 1597–8.

[45] (a) A. Kirschning, M.A. Hashem, H. Monenschein, L. Rose, K.-U. Schning, *J. Org. Chem.* **1999**, *64*, 6522–6. (b) A. Kirschning, H. Monenschein, C. Schmeck, *Angew. Chem.* **1999**, *111*, 2720–2; *Angew. Chem. Int. Ed.* **1999**, *38*, 2594–6. (c) J. Barluenga, M. Álvarez-Pérez, F.J. Fañanás, J.M. González, *Adv. Synth. Catal.* **2001**, *343*, 335–7.

[46] (a) G.R. Harvey, K.W. Ratts, *J. Org. Chem.* **1966**, *31*, 3907–10. (b) G.R. Harvey (Monsanto Co.) U.S. 3471523, **1969**; *Chem. Abstr.* **1970**, *72*, 3252f.

[47] Review on acceptor-substituted allenes: K. Banert, J. Lehmann in *Modern Allene Chemistry*, N. Krause, A.S.K. Hashmi, eds.; Wiley-VCH, Weinheim, **2004**, pp. 359–424.

[48] A. Melzer, dissertation, Chemnitz University of Technology (Germany), **2001**.

[49] R.A. Abramovitch, M. Konieczny, W. Pennington, S. Kanamathareddy, M. Vedachalam, *J. Chem. Soc., Chem. Commun.* **1990**, 269–70.

[50] (a) D.O. Spry, A.R. Bhala, *Heterocycles* **1986**, *24*, 1799–806. (b) X. Huang, R. Shen, T. Zhang, *J. Org. Chem.* **2007**, *72*, 1534–7.

[51] J.R. Fotsing, dissertation, Chemnitz University of Technology (Germany), **2004**.

[52] J.R. Fotsing, K. Banert, *Synthesis* **2006**, 261–72.

[53] K. Banert, F. Köhler, K. Kowski, B. Meier, B. Müller, P. Rademacher, *Chem. Eur. J.* **2002**, *8*, 5089–93.

[54] A. Padwa, T.J. Blacklock, P.H.J. Carlsen, M. Pulwer, *J. Org. Chem.* **1979**, *44*, 3281–7.

[55] I. Scharf, dissertation, Chemnitz University of Technology (Germany), **2009**.

[56] W. Fendel, dissertation, Chemnitz University of Technology (Germany), **1997**.

[57] G. Smolinsky, C.A. Pryde, *J. Org. Chem.* **1968**, *33*, 2411–6.

[58] (a) C.J. Moody, J.G. Ward, *J. Chem. Soc., Perkin Trans. 1* **1984**, 2903–9. (b) J.-C. Marié, C. Courillon, M. Malacria, *Arkivoc* **2007** (*v*), 277–92. (c) A. Perosa, M. Selva, P. Tundo, *J. Chem. Soc., Perkin Trans. 2* **2002**, 1033–7.

[59] G. Drefahl, K. Ponsold, G. Schubert, *J. Prakt. Chem.* **1969**, *311*, 919–24.

[60] (a) C.J. Moody, M. Pass, C.W. Rees, G. Tojo, *J. Chem. Soc., Chem. Commun.* **1986**, 1062–3. (b) C.J. Moody, A.L. Beck, W.J. Coates, *Tetrahedron Lett.* **1989**, *30*, 4017–8.

[61] K. Banert, M. Hagedorn, C. Liedtke, A. Melzer, C. Schöffler, *Eur. J. Org. Chem.* **2000**, 257–67.

[62] H. Hemetsberger, D. Knittel, H. Weidmann, *Monatsh. Chem.* **1969**, *100*, 1599–603.

[63] D. Knittel, *Synthesis* **1985**, 186–8.

[64] (a) H. Hemetsberger, I. Spira, W. Schönfelder, *J. Chem. Res.* (S) **1977**, 247. (b) H. Hemetsberger, I. Spira, W. Schönfelder, *J. Chem. Res. (M)* **1977**, 2701–19. (c) C.J. Moody, C.W. Rees, J.A.R. Rodrigues, S. Chung Tsoi, *J. Chem. Res. (S)* **1985**, 238–9. (d) P. Molina, E. Aller, M.A. Lorenzo, *Synthesis* **1993**, 1239–42. (e) L.C. Meurer, P.E. Finke, S.G. Mills,

et al., *Bioorg. Med. Chem. Lett.* **2005**, *15*, 645–51 and 1755. (f) G.A. Pinna, G. Loriga, G. Murineddu, *et al.*, *Chem. Pharm. Bull.* **2001**, *49*, 1406–11.
[65] (a) G. Túrós, A. Csámpai, T. Lovász, A. Györfi, H. Wamhoff, P. Sohár, *Eur. J. Org. Chem.* **2002**, 3801–6. (b) Z.-F. Xu, M.-W. Ding, *Chin. J. Appl. Chem.* **2003**, *20*, 198–200. (c) Z.-F. Xu, M.-W. Ding, *Huaxue Shiji* **2003**, *25*, 291–5.
[66] (a) K. Kondo, S. Morohoshi, M. Mitsuhashi, Y. Murakami, *Chem. Pharm. Bull.* **1999**, *47*, 1227–31. (b) L. Rodriguez-Salvador, E. Zaballos-Garcia, E. Gonzales-Rosende, M.L. Testa, J. Sepulveda-Arques, R.A. Jones, *Tetrahedron* **2000**, *56*, 4511–4. (c) A. Shafiee, J. Shahbazi Mojarrad, M.A. Jalili, H.R. Adhami, F. Hadizadeh, *J. Heterocycl. Chem.* **2002**, *39*, 367–73.
[67] (a) D. Knittel, H. Hemetsberger, H. Weidmann, *Monatsh. Chem.* **1970**, *101*, 157–60. (b) H. Hemetsberger, D. Knittel, *Monatsh. Chem.* **1972**, *103*, 194–204.
[68] (a) H. Hemetsberger, D. Knittel, *Monatsh. Chem.* **1972**, *103*, 205–9. (b) C. Vogel, P. Delavier, *Tetrahedron Lett.* **1989**, *30*, 1789–92.
[69] (a) T. Patonay, R.V. Hoffman, *J. Org. Chem.* **1995**, *60*, 2368–77. (b) M.M. Sá, G.P. Silveira, A.J. Bortoluzzi, A. Padwa, *Tetrahedron* **2003**, *59*, 5441–7.
[70] H. Bretschneider, H. Hörmann, *Monatsh. Chem.* **1953**, *84*, 1033–42.
[71] (a) T. Patonay, J. Jekö, É. Juhász-Tóth, *Eur. J. Org. Chem.* **2008**, 1441–8. (b) É. Juhász-Tóth, T. Patonay, *Eur. J. Org. Chem.* **2002**, 3055–64.
[72] M. Rens, L. Ghosez, *Tetrahedron Lett.* **1970**, 3765–8.
[73] Review: H. Heimgartner, *Angew. Chem.* **1991**, *103*, 271–97; *Angew. Chem. Int. Ed. Engl.* **1991**, *30*, 238–64.
[74] Recent applications: (a) B. Iliev, A. Linden, H. Heimgartner, *Helv. Chim. Acta* **2003**, *86*, 3215–34. (b) P. Köttgen, A. Linden, H. Heimgartner, *Helv. Chim. Acta* **2006**, *89*, 731–46. (c) K.A. Brun, A. Linden, H. Heimgartner, *Helv. Chim. Acta* **2002**, *85*, 3422–43.
[75] (a) M. Henriet, M. Houtekie, B. Techy, R. Touillaux, L. Ghosez, *Tetrahedron Lett.* **1980**, *21*, 223–6. (b) C. Bernard, L. Ghosez, *J. Chem. Soc., Chem. Commun.* **1980**, 940–1.
[76] A. Gagneux, S. Winstein, W.G. Young, *J. Am. Chem. Soc.* **1960**, *82*, 5956–7.
[77] K. Banert, *Umlagerungen organischer Azide. Reaktionsmechanismen und Anwendungen in der Synthese*, Schäuble Verlag, Rheinfelden (Germany), **1993**.
[78] H. Priebe, *Angew. Chem.* **1984**, *96*, 728–9; *Angew. Chem. Int. Ed. Engl.* **1984**, *23*, 736–8.
[79] K. Banert, *Angew. Chem.* **1985**, *97*, 231–2; *Angew. Chem. Int. Ed. Engl.* **1985**, *24*, 216–7.
[80] (a) K. Banert, W. Fendel, J. Schlott, *Angew. Chem.* **1998**, *110*, 3488–91; *Angew. Chem. Int. Ed.* **1998**, *37*, 3289–92. (b) V.K. Brel, *Synthesis* **2007**, 2674–80. (c) V.K. Brel', E.V. Abramkin, *Russ. J. Gen. Chem.* **1994**, *64*, 1764–8.
[81] A. Bohle, diploma thesis, Chemnitz University of Technology (Germany), **2007**.
[82] C.J. Nielsen, P. Klaeboe, H. Priebe, *J. Mol. Struct.* **1986**, *147*, 217–29.
[83] K. Banert, *Tetrahedron Lett.* **1985**, *26*, 5261–4.
[84] K. Banert, *Chem. Ber.* **1987**, *120*, 1891–6.
[85] K. Banert, M. Hagedorn, *Angew. Chem.* **1989**, *101*, 1710–1; *Angew. Chem. Int. Ed. Engl.* **1989**, *28*, 1675–6.
[86] K. Banert, *Liebigs Ann./Recueil* **1997**, 2005–18.
[87] C. Schöffler, dissertation, Chemnitz University of Technology (Germany), **2000**.
[88] K. Banert, J. Schlott, *Tetrahedron* **2000**, *56*, 5413–9.
[89] K. Banert in *Houben–Weyl, Methoden Org. Chem.*, 4th ed., Vol. *E 15* (eds.: H. Kropf, E. Schaumann), Thieme, Stuttgart, **1993**, pp. 1344–7.
[90] K. Nishiyama, M. Oba, A. Watanabe, *Tetrahedron* **1987**, *43*, 693–700.
[91] E. Öhler, S. Kanzler, *Liebigs Ann. Chem.* **1994**, 867–76.
[92] T.-s. Chou, S.-J. Lee, M.-L. Peng, D.-J. Sun, S.-S.P. Chou, *J. Org. Chem.* **1988**, *53*, 3027–31.
[93] (a) R.V. Hoffman, B.S. Severns, *J. Org. Chem.* **1996**, *61*, 5567–73. (b) S. Mangelinckx, N. De Kimpe, *Synlett* **2006**, 369–74.
[94] W.L. Stepp (Dept. Air Force), US 3217017, **1965**; *Chem. Abstr.* **1966**, *64*, 1896f.
[95] B. Müller, dissertation, Chemnitz University of Technology (Germany), **2002**.

[96] (a) T.M.V.D. Pinho e Melo, A.M.d'A. Rocha Gonsalves, C.S.J. Lopes, T.L. Gilchrist, *Tetrahedron Lett.* **1999**, *40*, 789–92. (b) T.M.V.D. Pinho e Melo, C.S.J. Lopes, A.L. Cardoso, A.M.d'A. Rocha Gonsalves, *Tetrahedron* **2001**, *57*, 6203–8. (c) A. Gómez-Zavaglia, A. Kaczor, A.L. Cardoso, T.M.V.D. Pinho e Melo, R. Fausto, *J. Phys. Chem. A* **2006**, *110*, 8081–92.

[97] (a) C.R. Alonso-Cruz, A.R. Kennedy, M.S. Rodriguez, E. Suárez, *Tetrahedron Lett.* **2007**, *48*, 7207–10. (b) C.R. Alonso-Cruz, A.R. Kennedy, M.S. Rodriguez, E. Suárez, *J. Org. Chem.* **2008**, *73*, 4116–22.

[98] C.R. Alonso-Cruz, A.R. Kennedy, M.S. Rodriguez, E. Suárez, *Org. Lett.* **2003**, *5*, 3729–32.

[99] S. Riela, C. Aprile, M. Gruttadauria, P. Lo Meo, R. Noto, *Molecules* **2005**, *10*, 383–93.

[100] (a) J.N. Denis, J. Vicens, A. Krief, *Tetrahedron Lett.* **1979**, 2697–700. (b) A. Hassner, A.S. Amarasekara, *Tetrahedron Lett.* **1987**, *28*, 5185–8. (c) M. Tingoli, M. Tiecco, D. Chianelli, R. Balducci, A. Temperini, *J. Org. Chem.* **1991**, *56*, 6809–13. (d) J.-F. Huot, F. Outurquin, C. Paulmier, *Chem. Lett.* **1991**, 1957–60. (e) A.A. Sherman, L.O. Kononov, A.S. Shashkov, G.V. Zatonsky, N.E. Nifant'ev, *Mendeleev Commun.* **1998**, 9–12.

[101] (a) S. Tomoda, Y. Matsumoto, Y. Takeuchi, Y. Nomura, *Bull. Chem. Soc. Jpn.* **1986**, *59*, 3283–4. (b) T.K. Chakraborty, G.V. Reddy, *Tetrahedron Lett.* **1990**, *31*, 1335–8.

[102] (a) M. Mizuno, T. Shioiri, *Tetrahedron Lett.* **1999**, *40*, 7105–8. (b) V. Nair, T.G. George, *Tetrahedron Lett.* **2000**, *41*, 3199–201. (c) M.-Y. Chang, C.-Y. Lin, P.-P. Sun, *J. Chin. Chem. Soc.* **2005**, *52*, 1061–7.

[103] (a) L.E. Overman, M.J. Sharp, *J. Am. Chem. Soc.* **1988**, *110*, 612–4 and 5934. (b) V.J. Majo, P.T. Perumal, *Tetrahedron Lett.* **1997**, *38*, 6889–92.

[104] (a) W. Zhu, D. Ma, *Chem. Commun.* **2004**, 888–9. (b) Q. Cai, W. Zhu, H. Zhang, Y. Zhang, D. Ma, *Synthesis* **2005**, 496–9.

[105] Y. Masuda, M. Murata, M. Ikeda, S. Watanabe, *J. Chem. Soc., Perkin Trans. 1* **1998**, 1013–4.

[106] (a) R.R. Sauers, S.D. Van Arnum, *Tetrahedron Lett.* **1987**, *28*, 5797–800. (b) R.R. Sauers, S.D. Van Arnum, *Phosphorus Sulfur Silicon Relat. Elem.* **2003**, *178*, 2169–81. (c) C.-Z. Tao, X. Cui, J. Li, A.-X. Liu, L. Liu, Q.-X. Guo, *Tetrahedron Lett.* **2007**, *48*, 3525–9. (d) K. Ilg, H. Werner, *Chem. Eur. J.* **2002**, *8*, 2812–20. (e) C. Shin, Y. Oh, J.H. Cha, A.N. Pae, H. Choo, Y.S. Cho, *Tetrahedron* **2007**, *63*, 2182–90.

[107] (a) G. L'abbé, *Angew. Chem.* **1975**, *87*, 831–8; *Angew. Chem. Int. Ed. Engl.* **1975**, *14*, 775–82. (b) G. L'abbé, *New Synthetic Methods* **1979**, *5*, 1–24.

[108] (a) S. Bräse, C. Gil, K. Knepper, V. Zimmermann, *Angew. Chem.* **2005**, *117*, 5320–74; *Angew. Chem. Int. Ed.* **2005**, *44*, 5188–240. (b) E.F.V. Scriven, K. Turnbull, *Chem. Rev.* **1988**, *88*, 297–368. (c) E.F.V. Scriven (ed.), *Azides and Nitrenes: Reactivity and Utility*, Academic Press, Orlando (USA), **1984**. (d) S. Patai (ed.), *The Chemistry of the Azido Group*, Wiley-Interscience, London, **1971**.

[109] P. Klaeboe, C.J. Nielsen, H. Priebe, S.H. Schei, C.E. Sjøgren, *J. Mol. Struct.* **1986**, *141*, 161–72.

[110] (a) H.M. Badawi, *J. Mol. Struct. (Theochem)* **2002**, *579*, 11–9. (b) C.J. Nielsen, C.E. Sjøgren, *J. Mol. Struct. (Theochem)* **1987**, *150*, 361–79.

[111] G. L'abbé, *Chem. Ind. (London)* **1971**, 278.

[112] A.N. Thakore, J. Buchshriber, A.C. Oehlschlager, *Can. J. Chem.* **1973**, *51*, 2406–14.

[113] Reviews on 2*H*-azirines: (a) J. Backes in *Houben–Weyl, Vol. E 16c* (ed.: D. Klamann), Thieme, Stuttgart, **1992**, pp. 321–69. (b) V. Nair in *The Chemistry of Heterocyclic Compounds, Small-Ring Heterocycles, Vol. 42, Part 1* (ed.: A. Hassner), John Wiley & Sons, Inc., New York, **1983**, pp. 215–332. (c) W.H. Pearson, B.W. Lian, S.C. Bergmeier in *Comprehensive Heterocyclic Chemistry II, Vol 1A* (ed.: A. Padwa), Pergamon, New York, **1996**, pp. 1–60. (d) F. Palacios, A.N. Ochoa de Retana, E. Martinez de Marigorta, J.M. de los Santos, *Eur. J. Org. Chem.* **2001**, 2401–14. (e) T.L. Gilchrist, *Aldrichimica Acta* **2001**, *34*, 51–5. (f) K.M.L. Rai, A. Hassner in *Advances in Strained and Interesting Organic Molecules, Vol. 8* (ed.: B. Halton), Jai, Greenwich, **2000**, pp. 187–257.

[114] For the ring strain in 2*H*-azirines, see: E.-U. Würthwein, T. Hergenröther, H. Quast, *Eur. J. Org. Chem.* **2002**, 1750–5.

[115] Å.S. Timén, E. Risberg, P. Somfai, *Tetrahedron Lett.* **2003**, *44*, 5339–41.
[116] P.N.D. Singh, C.L. Carter, A.D. Gudmundsdóttir, *Tetrahedron Lett.* **2003**, *44*, 6763–5.
[117] (a) W. Bauer, K. Hafner, *Angew. Chem.* **1969**, *81*, 787–8; *Angew. Chem. Int. Ed. Engl.* **1969**, *8*, 772–3. (b) K. Isomura, M. Okada, H. Taniguchi, *Tetrahedron Lett.* **1969**, 4073–6.
[118] First isolation of the natural product azirinomycin: (a) E.O. Stapley, D. Hendlin, M. Jackson, A.K. Miller, S. Hernandez, J.M. Mata, *J. Antibiot.* **1971**, *24*, 42–7. (b) T.W. Miller, E.W. Tristram, F.J. Wolf, *J. Antibiot.* **1971**, *24*, 48–50.
[119] K. Banert, C. Berndt, unpublished results, Chemnitz University of Technology (Germany), **2008**.
[120] (a) J.R. Fotsing, M. Hagedorn, K. Banert, *Tetrahedron* **2005**, *61*, 8904–9. (b) K. Banert, J.R. Fotsing, M. Hagedorn, H.P. Reisenauer, G. Maier, *Tetrahedron* **2008**, *64*, 5645–8.
[121] K. Banert, F. Köhler, B. Meier, *Tetrahedron Lett.* **2003**, *44*, 3781–3.
[122] M.J. Alves, P.M.T. Ferreira, H.L.S. Maia, L.S. Monteiro, T.L. Gilchrist, *Tetrahedron Lett.* **2000**, *41*, 4991–5.
[123] (a) M.J. Alves, T.L. Gilchrist, *J. Chem. Soc., Perkin Trans. 1* **1998**, 299–303. (b) M.J. Alves, N.G. Azoia, J.F. Bickley, A.G. Fortes, T.L. Gilchrist, R. Mendonça, *J. Chem. Soc., Perkin Trans. 1* **2001**, 2969–76. (c) M.J. Alves, A.G. Fortes, F.T. Costa, V.C.M. Duarte, *Tetrahedron* **2007**, *63*, 11167–73. (d) M.J. Alves, A.G. Fortes, F.T. Costa, *Tetrahedron* **2006**, *62*, 3095–102.
[124] Recent examples: (a) E.V. Sadanandan, S.K. Pillai, M.V. Lakshmikantham, A.D. Billimoria, J.S. Culpepper, M.P. Cava, *J. Org. Chem.* **1995**, *60*, 1800–5. (b) P.M. Fresneda, P. Molina, J.A. Bleda, *Tetrahedron* **2001**, *57*, 2355–63. (c) F. Hong, J. Zaidi, B. Cusack, E. Richelson, *Bioorg. Med. Chem.* **2002**, *10*, 3849–58. (d) R.A. Tapia, Y. Prieto, F. Pautet, *et al.*, *Bioorg. Med. Chem.* **2003**, *11*, 3407–12. (e) I. Borza, S. Kolok, A. Gere, *et al.*, *Bioorg. Med. Chem. Lett.* **2003**, *13*, 3859–61. (f) D.J. Bentley, J. Fairhurst, P.T. Gallagher, A.K. Manteuffel, C.J. Moody, J.L. Pinder, *Org. Biomol. Chem.* **2004**, *2*, 701–8. (g) T. Fryatt, H.I. Pettersson, W.T. Gardipee, *et al.*, *Bioorg. Med. Chem.* **2004**, *12*, 1667–87. (h) D. Coowar, J. Bouissac, M. Hanbali, M. Paschaki, E. Mohier, B. Luu, *J. Med. Chem.* **2004**, *47*, 6270–82. (i) S.M. Lee, R. Jeon, *Arch. Pharm. Res.* **2005**, *28*, 1219–23. (j) N.H. Al-Said, K.Q. Shawakfeh, W.N. Abdullah, *Molecules* **2005**, *10*, 1446–57. (k) G.C. Condie, M.F. Channon, A.J. Ivory, N. Kumar, D. StC. Black, *Tetrahedron* **2005**, *61*, 4989–5004. (l) A. Tsotinis, M. Gerasimopoulou, M. Vlachou, D. Moreau, C. Roussakis, *Lett. Drug Des. Discov.* **2006**, *3*, 14–6. (m) L.F. Tietze, F. Major, *Eur. J. Org. Chem.* **2006**, 2314–21. (n) P. Vital, P.-O. Norrby, D. Tanner, *Synlett* **2006**, 3140–4. (o) P. Vital, D. Tanner, *Org. Biomol. Chem.* **2006**, *4*, 4292–8. (p) X. Dong, Z. Zhang, R. Wen, J. Shen, X. Shen, H. Jiang, *Bioorg. Med. Chem. Lett.* **2006**, *16*, 5913–6. (q) M.S. Tichenor, J.D. Trzupek, D.B. Kastrinsky, F. Shiga, I. Hwang, D.L. Boger, *J. Am. Chem. Soc.* **2006**, *128*, 15683–96. (r) I. Borza, É. Bozó, G. Barta-Szalai, *et al.*, *J. Med. Chem.* **2007**, *50*, 901–14. (s) M.A. Colucci, P. Reigan, D. Siegel, A. Chilloux, D. Ross, C.J. Moody, *J. Med. Chem.* **2007**, *50*, 5780–9.
[125] (a) J.M. Chezal, E. Moreau, G. Delmas, *et al.*, *J. Org. Chem.* **2001**, *66*, 6576–84. (b) P. Molina, P.M. Fresneda, S. Delgado, *J. Org. Chem.* **2003**, *68*, 489–99. (c) G.C. Condie, J. Bergman, *J. Heterocycl. Chem.* **2004**, *41*, 531–40. (d) P.J. Roy, C. Dufresne, N. Lachance, *et al.*, *Synthesis* **2005**, 2751–7. (e) T. Lomberget, S. Radix, R. Barret, *Synlett* **2005**, 2080–2.
[126] (a) G.B. Jones, J.E. Mathews, *Tetrahedron* **1997**, *53*, 14599–614. (b) S. Mayer, J.-Y. Mérour, B. Joseph, G. Guillaumet, *Eur. J. Org. Chem.* **2002**, 1646–53. (c) S. Selvi, S.-C. Pu, Y.-M. Cheng, J.-M. Fang, P.-T. Chou, *J. Org. Chem.* **2004**, *69*, 6674–8. (d) A.W. Grubbs, G.D. Artman, III, R.M. Williams, *Tetrahedron Lett.* **2005**, *46*, 9013–6. (e) A. Tsotinis, M. Vlachou, S. Zouroudis, *et al.*, *Lett. Drug Des. Discov.* **2005**, *2*, 189–92 and 428. (f) J.S. Sawyer, D.W. Beight, E.C.R. Smith, *et al.*, *J. Med. Chem.* **2005**, *48*, 893–6. (g) J.S. Daniels, R. Espina, K. Cao, *et al.*, *Chem. Res. Toxicol.* **2007**, *20*, 1709–17.
[127] (a) M. Welch, R.S. Phillips, *Heterocycl. Commun.* **1999**, *5*, 305–10. (b) M.K. Bratenko, M.V. Vovk, V.A. Chornous, N.V. Mel'nichenko, *Russ. J. Org. Chem.* **1999**, *35*, 1812–4. (c) K.L. Milkiewicz, D.J. Parks, T. Lu, *Tetrahedron Lett.* **2003**, *44*, 4257–60. (d) V.N. Yarovenko, S.L. Semenov, I.V. Zavarzin, A.V. Ignatenko, M.M. Krayushkin, *Russ. Chem. Bull.* **2003**, *52*, 451–6. (e) A. Zarghi, A.H. Ebrahimabadi, F. Hassanzadeh, M.R. Heydari, A. Shafiee,

Boll. Chim. Farmac. **2003**, *142*, 251–4. (f) P.R. Kumar, S. Raju, P.S. Goud, *et al.*, *Bioorg. Med. Chem.* **2004**, *12*, 1221–30. (g) P. Gajdos, J. Miklovic, A. Krutosikova, *Chem. Heterocycl. Compd.* **2006**, *42*, 719–25. (h) O. Arad, J. Morros, X. Batllori, J. Teixidó, S. Nonell, J.I. Borrell, *Org. Lett.* **2006**, *8*, 847–50.

[128] B.J. Stokes, H. Dong, B.E. Leslie, A.L. Pumphrey, T.G. Driver, *J. Am. Chem. Soc.* **2007**, *129*, 7500–1.

[129] (a) T.L. Gilchrist, C.W. Rees, J.A.R. Rodrigues, *J. Chem. Soc., Chem. Commun.* **1979**, 627–8. (b) A. Tsotinis, A. Eleutheriades, K.A. Hough, K. Davidson, D. Sugden, *Bioorg. Chem.* **2007**, *35*, 189–204. (c) J.K. MacLeod, A. Ward, A.C. Willis, *Aust. J. Chem.* **1998**, *51*, 177–87.

[130] H. Dong, M. Shen, J.E. Redford, B.J. Stokes, A.L. Pumphrey, T.G. Driver, *Org. Lett.* **2007**, *9*, 5191–4.

[131] (a) S. Chiba, Y.-F. Wang, G. Lapointe, K. Narasaka, *Org. Lett.* **2008**, *10*, 313–6. (b) M.J. Alves, T.L. Gilchrist, J.H. Sousa, *J. Chem. Soc., Perkin Trans. 1* **1999**, 1305–10.

[132] (a) K.M.L. Rai, A. Hassner in *Comprehensive Heterocyclic Chemistry II, Vol. 1A* (ed.: A. Padwa), Elsevier, Oxford, **1996**, pp. 61–96, and references therein. (b) J. Laue, G. Seitz, *Liebigs Ann.* **1996**, 645–8. (c) M. Drögemüller, R. Jautelat, E. Winterfeldt, *Angew. Chem.* **1996**, *108*, 1669–71; *Angew. Chem. Int. Ed. Engl.* **1996**, *35*, 1572–4. (d) M. Drögemüller, T. Flessner, R. Jautelat, U. Scholz, E. Winterfeldt, *Eur. J. Org. Chem.* **1998**, 2811–31. (e) E. Haak, E. Winterfeldt, *Synlett* **2004**, 1414–8. (f) Y. Tamura, Y. Yoshimura, T. Nishimura, S. Kato, Y. Kita, *Tetrahedron Lett.* **1973**, *14*, 351–4. (g) Y. Tamura, S. Kato, Y. Yoshimura, T. Nishimura, Y. Kita, *Chem. Pharm. Bull.* **1974**, *22*, 1291–6. (h) S. Senda, K. Hirota, T. Asao, K. Maruhashi, *J. Am. Chem. Soc.* **1978**, *100*, 7661–4. (i) D.S. Pearce, M.J. Locke, H.W. Moore, *J. Am. Chem. Soc.* **1975**, *97*, 6181–6. (j) M. Zaidlewicz, I.G. Uzarewicz, *Heteroat. Chem.* **1993**, *4*, 73–7. (k) A. Hassner, F.W. Fowler, *J. Am. Chem. Soc.* **1968**, *90*, 2869–75.

[133] (a) A. Hassner, F.W. Fowler, *Tetrahedron Lett.* **1967**, 1545–8. (b) B. Rose, D. Schollmeyer, H. Meier, *Liebigs Ann.* **1997**, 409–12.

[134] (a) A.L. Logothetis, *J. Org. Chem.* **1964**, *29*, 3049–52. (b) V. Nair, *J. Org. Chem.* **1968**, *33*, 2121–3 and 4316. (c) T.C. Gallagher, R.C. Storr, *Tetrahedron Lett.* **1981**, *22*, 2909–12. (d) T.M.V.D. Pinho e Melo, A.L. Cardoso, C.S.B. Gomes, A.M.d'A. Rocha Gonsalves, *Tetrahedron Lett.* **2003**, *44*, 6313–5.

[135] K. Banert, E. Penk, unpublished results, Chemnitz University of Technology (Germany), **2009**.

[136] T.M.V.D. Pinho e Melo, C.S.J. Lopes, A.M.d'A. Rocha Gonsalves, *Tetrahedron Lett.* **2000**, *41*, 7217–20.

[137] T.M.V.D. Pinho e Melo, A.L. Cardoso, A.M.d'A. Rocha Gonsalves, *Tetrahedron* **2003**, *59*, 2345–51.

[138] T.M.V.D. Pinho e Melo, C.S.J. Lopes, A.M.d'A. Rocha Gonsalves, R.C. Storr, *Synthesis* **2002**, 605–8.

[139] S. Lopes, C.M. Nunes, R. Fausto, T.M.V.D. Pinho e Melo, *J. Mol. Struct.* **2009**, *919*, 47–53.

[140] A. Hassner, N.H. Wiegand, H.E. Gottlieb, *J. Org. Chem.* **1986**, *51*, 3176–80.

[141] C. Mazal, J. Jonas, Z. Žák, *Tetrahedron* **2002**, *58*, 2729–33.

[142] A. Hassner, A.S. Miller, M.J. Haddadin, *J. Org. Chem.* **1972**, *37*, 2682–5.

[143] K. Banert, *Chem. Ber.* **1989**, *122*, 123–8.

[144] A. Hassner, D. Tang, J. Keogh, *J. Org. Chem.* **1976**, *41*, 2102–4.

[145] A. Hassner, D.J. Anderson, R.H. Reuss, *Tetrahedron Lett.* **1977**, 2463–6.

[146] A. Kuhtz, diploma thesis, Chemnitz University of Technology (Germany), **2004**.

[147] K. Banert, F. Köhler, *Angew. Chem.* **2001**, *113*, 173–6; *Angew. Chem. Int. Ed.* **2001**, *40*, 174–7.

[148] (a) K. Banert, Y.-H. Joo, T. Rüffer, B. Walfort, H. Lang, *Angew. Chem.* **2007**, *119*, 1187–90; *Angew. Chem. Int. Ed.* **2007**, *46*, 1168–71. (b) J.R. Fotsing, K. Banert, *Eur. J. Org. Chem.* **2005**, 3704–14.

[149] K. Banert, *Angew. Chem.* **1987**, *99*, 932–4; *Angew. Chem. Int. Ed. Engl.* **1987**, *26*, 879–85.

[150] G. L'abbé, G. Mathys, *J. Heterocycl. Chem.* **1974**, *11*, 613–4.
[151] G. L'abbé, G. Mathys, *J. Org. Chem.* **1974**, *39*, 1221–5.
[152] A. Padwa, A. Ku, H. Ku, A. Mazzu, *J. Org. Chem.* **1978**, *43*, 66–72.
[153] A. Padwa, A. Ku, H. Ku, A. Mazzu, *Tetrahedron Lett.* **1977**, 551–4.
[154] (a) A.B. Levy, A. Hassner, *J. Am. Chem. Soc.* **1971**, *93*, 2051–3. (b) J. Harnisch, G. Szeimies, *Chem. Ber.* **1979**, *112*, 3914–33.
[155] E.P. Kyba, D.C. Alexander, *Tetrahedron Lett.* **1976**, 4563–4.
[156] A. Hassner, A.B. Levy, *J. Am. Chem. Soc.* **1971**, *93*, 5469–74.
[157] K. Banert, J. Wutke, unpublished results, Chemnitz University of Technology (Germany), **2009**.
[158] (a) A. Hassner, B.A. Belinka, Jr., *J. Am. Chem. Soc.* **1980**, *102*, 6185–6. (b) A. Hassner, P. Munger, B.A. Belinka, Jr., *Tetrahedron Lett.* **1982**, *23*, 699–702.
[159] (a) G. L'abbé, A. Hassner, *J. Heterocycl. Chem.* **1970**, *7*, 361–6. (b) G. L'abbé, G. Mathys, S. Toppet, *J. Org. Chem.* **1975**, *40*, 1549–52.
[160] G. L'abbé, G. Mathys, S. Toppet, *Chem. Ind. (London)* **1975**, 278–9.
[161] A. Hassner, B.A. Belinka, Jr., M. Haber, P. Munger, *Tetrahedron Lett.* **1981**, *20*, 1863–6.
[162] (a) A. Arques, P. Molina, D. Auñón, *et al.*, *J. Organomet. Chem.* **2000**, *598*, 329–38. (b) E. Ciganek, *J. Org. Chem.* **1970**, *35*, 3631–6. (c) A. Tárraga, P. Molina, D. Curiel, *ARKIVOC* **2002**, (*v*), 85–91.
[163] (a) P. Molina, P.M. Fresneda, P. Almendros, *Tetrahedron* **1993**, *49*, 1223–36. (b) P. Molina, A. Arques, A. Alfas, *J. Org. Chem.* **1993**, *58*, 5264–70. (c) P. Molina, P. Almendros, P.M. Fresneda, *Tetrahedron* **1994**, *50*, 2241–54. (d) P. Molina, E. Aller, A. López-Lázaro, M. Alajarín, A. Lorenzo, *Tetrahedron Lett.* **1994**, *35*, 3817–20. (e) O. Chavignon, J.C. Teulade, D. Roche, *et al.*, *J. Org. Chem.* **1994**, *59*, 6413–8. (f) M.W. Ding, J. Zhu, S.F. Shi, X.P. Liu, *Chin. Chem. Lett.* **2002**, *13*, 942–4. (g) M.-W. Ding, Y. Sun, S.-J. Yang, X.-P. Liu, Z.-J. Liu, *Synth. Commun.* **2003**, *33*, 1651–8.
[164] (a) P. Molina, P. Almendros, P.M. Fresneda, *Tetrahedron Lett.* **1993**, *34*, 4701–4. (b) P. Molina, S. Garcia-Zafra, P.M. Fresneda, *Synlett* **1995**, 43–5. (c) G. Guanti, R. Riva, *Tetrahedron: Asymmetry* **2001**, *12*, 1185–200. (d) F. Palacios, E. Herrán, C. Alonso, G. Rubiales, *Tetrahedron* **2006**, *62*, 7661–6. (e) D. Corona, E. Diaz, Á. Guzmán, C.K. Jankowski, *Spectrochimica Acta Part A* **2005**, *61*, 2788–95. (f) F. Palacios, E. Herrán, C. Alonso, *et al.*, *J. Org. Chem.* **2006**, *71*, 6020–30. (g) F. Palacios, E. Herrán, G. Rubiales, C. Alonso, *Tetrahedron* **2007**, *63*, 5669–76.
[165] (a) P.M. Fresneda, P. Molina, M.A. Sanz, *Synlett* **2000**, 1190–2. (b) P.M. Fresneda, J.A. Bleda, M.A. Sanz, P. Molina, *Synlett* **2007**, 1541–4.
[166] J.A.R. Rodrigues, G.C. Leiva, J.D.F. de Sousa, *Tetrahedron Lett.* **1995**, *36*, 59–62.
[167] P.M. Fresneda, P. Molina, *Synlett* **2004**, 1–17.
[168] (a) Y. Sun, M.-W. Ding, *Synth. Commun.* **2005**, *35*, 41–7. (b) Y. Sun, L.-P. Gao, M.-W. Ding, *Synth. Commun.* **2006**, *36*, 1185–91.
[169] P.C. Montevecchi, M.L. Navacchia, P. Spagnolo, *J. Org. Chem.* **1997**, *62*, 5846–8.
[170] (a) J. Thiem, D. Springer, *Carbohydr. Res.* **1985**, *136*, 325–34. (b) T.R. Burke, Jr., M.S. Smyth, M. Nomizu, A. Otaka, P.P. Roller, *J. Org. Chem.* **1993**, *58*, 1336–40. (c) T.R. Burke, Jr., M.S. Smyth, A. Otaka, P.P. Roller, *Tetrahedron Lett.* **1993**, *34*, 4125–8. (d) S. Chetyrkina, K. Estieu-Gionnet, G. Laïn, M. Bayle, G. Déléris, *Tetrahedron Lett.* **2000**, *41*, 1923–6.
[171] (a) M.P. Cabal, R.S. Coleman, S.J. Danishefsky, *J. Am. Chem. Soc.* **1990**, *112*, 3253–5. (b) H.W. Moore, H.R. Shelden, D.F. Shellhamer, *J. Org. Chem.* **1969**, *34*, 1999–2001. (c) J.D. Hobson, J.R. Malpass, *J. Chem. Soc. C* **1969**, 1499–503. (d) J.D. Hobson, J.R. Malpass, *J. Chem. Soc. C* **1967**, 1645–8.
[172] I. Scharf, diploma thesis, Chemnitz University of Technology (Germany), **2002**.
[173] K. Banert, S. Grimme, R. Herges, *et al.*, *Chem. Eur. J.* **2006**, *12*, 7467–81.
[174] T.C. Gallagher, R.C. Storr, *Tetrahedron Lett.* **1981**, *22*, 2905–8.
[175] (a) G. L'abbé, M. Mahy, M. Bollyn, G. Germain, G. Scheefer, *Bull. Soc. Chim. Belg.* **1983**, *92*, 881–91. (b) G. L'abbé, *Bull. Soc. Chim. Belg.* **1984**, *93*, 579–92. (c) A. Hassner, J. Keogh, *J. Org. Chem.* **1986**, *51*, 2767–70.

[176] V.J. Shiner, Jr., J.S. Humphrey, Jr., *J. Am. Chem. Soc.* **1967**, *89*, 622–30.
[177] K. Banert, *Chem. Ber.* **1989**, *122*, 911–8.
[178] M.G. Baldwin, K.E. Johnson, J.A. Lovinger, C.O. Parker, *J. Polymer Sci. Part B* **1967**, *5*, 803–6.
[179] (a) L.A. Burke, G. Leroy, M.T. Nguyen, M. Sana, *J. Am. Chem. Soc.* **1978**, *100*, 3668–74. (b) T. Yamabe, M. Kaminoyama, T. Minato, K. Hori, K. Isomura, H. Taniguchi, *Tetrahedron* **1984**, *40*, 2095–9. (c) K. Fukushima, H. Iwahashi, *Heterocycles* **2005**, *65*, 2605–18.
[180] (a) G. L'abbé, G. Mathys, *J. Org. Chem.* **1974**, *39*, 1778–80. (b) H. Bock, R. Dammel, S. Aygen, *J. Am. Chem. Soc.* **1983**, *105*, 7681–5. (c) K. Isomura, K. Takehara, M. Ichiki, H. Taniguchi, *Kitakyushu Kogyo Koto Senmon Gakko Kenkyu Hokoku* **1998**, *31*, 103–10.
[181] (a) R.W. Saalfrank, E. Ackermann, M. Fischer, U. Wirth, *Chem. Ber.* **1987**, *120*, 2003–6. (b) R.W. Saalfrank, U. Wirth, C.-J. Lurz, *J. Org. Chem.* **1989**, *54*, 4356–9.
[182] R. Huisgen, *Angew. Chem.* **1980**, *92*, 979–1005; *Angew. Chem. Int. Ed. Engl.* **1980**, *19*, 947–73.
[183] (a) J.S. Meek, J.S. Fowler, *J. Am. Chem. Soc.* **1967**, *89*, 1967. (b) J.S. Meek, J.S. Fowler, *J. Org. Chem.* **1968**, *33*, 985–91. (c) N.S. Zefirov, N.K. Chapovskaya, *J. Org. Chem. USSR (Engl. Transl.)* **1968**, *4*, 1252. (d) A.N. Nesmeyanov, M.I. Rybinskaya, *Dokl. Akad. Nauk SSSR* **1966**, *166*, 1362–5. (e) A.N. Nesmeyanov, M.I. Rybinskaya, *Dokl. Akad. Nauk SSSR* **1966**, *170*, 600–3. (f) A.N. Nesmeyanov, M.I. Rybinskaya, *Zh. Org. Khim.* **1966**, *2*, 2081–6.
[184] Reviews on the synthesis of 1,2,3-triazoles: (a) V.P. Krivopalov, O.P. Shkurko, *Russ. Chem. Rev.* **2005**, *74*, 339–79. (b) H. Dehne in *Houben–Weyl*, 4th ed., Vol. *E 8d* (ed.: E. Schaumann), Thieme, Stuttgart, **1994**, pp. 305–405. (c) A.C. Tomé in *Houben–Weyl, Science of Synthesis*, 5th ed., Vol. *13* (eds.: R.C. Storr, T.L. Gilchrist), Thieme, Stuttgart, **2004**, pp. 415–601.
[185] K. Banert, *Chem. Ber.* **1989**, *122*, 1963–7.
[186] K. Banert, *Targets in Heterocyclic Systems* **2000**, *3*, 1–32.
[187] (a) T. Harrison, A.P. Owens, B.J. Williams, *et al.*, *J. Med. Chem.* **2001**, *44*, 4296–9. (b) A.P. Owens, Merck Sharp and Dohme Limited, UK, WO 9629317, **1996**; *Chem. Abstr.* **1997**, *126*, 8122. (c) R. Baker, J. Elliot, G.I. Stevenson, C.J. Swain, Merck Sharp and Dohme Limited, WO 9701553, **1997**; *Chem. Abstr.* **1997**, *126*, 171610. (d) R. Baker, J. Elliot, G.I. Stevenson, C.J. Swain, Merck Sharp and Dohme Limited, WO 9701554, **1997**; *Chem. Abstr.* **1997**, *126*, 171611. (e) J.D. Moseley, C.J. Swain, B.J. Williams, Merck Sharp and Dohme Limited, UK, GB 2302689, **1997**; *Chem. Abstr.* **1997**, *126*, 277501.
[188] J.C. Loren, K.B. Sharpless, *Synthesis* **2005**, 1514–20.
[189] A. Ihle, dissertation, Chemnitz University of Technology (Germany), **2006**.
[190] J.R. Fotsing, K. Banert, *Eur. J. Org. Chem.* **2006**, 3617–25.
[191] K. Banert, M. Hagedorn, *Angew. Chem.* **1990**, *102*, 90–2; *Angew. Chem. Int. Ed. Engl.* **1990**, *29*, 103–5.
[192] K. Banert, M. Hagedorn, E. Knözinger, A. Becker, E.-U. Würthwein, *J. Am. Chem. Soc.* **1994**, *116*, 60–2.
[193] (a) K. Banert in *Houben–Weyl, Science of Synthesis*, 5th ed., Vol. *24* (ed.: A. de Meijere), Thieme, Stuttgart, **2006**, pp. 1061–72. (b) K.I. Booker-Milburn in *Comprehensive Organic Functional Group Tranformations*, Vol. *2* (eds.: A.R. Katritzky, O. Meth-Cohn, C.W. Rees), Pergamon, Oxford, **1995**, pp. 1054–5. (c) G. Himbert in *Houben–Weyl*, 4th ed., Vol. *E 15* (eds.: H. Kropf, E. Schaumann), Thieme, Stuttgart, **1993**, p. 3455.
[194] (a) J.H. Boyer, C.H. Mack, N. Goebel, L.R. Morgan, Jr., *J. Org. Chem.* **1958**, *23*, 1051–3. (b) J.H. Boyer, R. Selvarajan, *Tetrahedron Lett.* **1969**, 47–50. (c) V.A. Garibina, A.A. Leonov, A.V. Dogadina, B.I. Ionin, A.A. Petrov, *J. Gen. Chem. USSR (Engl. Transl.)* **1985**, *55*, 1771–81.
[195] (a) R. Helwig, M. Hanack, *Chem. Ber.* **1985**, *118*, 1008–21. (b) E. Robson, J.M. Tedder, B. Webster, *J. Chem. Soc.* **1963**, 1863–5.
[196] (a) J.H. Boyer, R. Selvarajan, *J. Am. Chem. Soc.* **1969**, *91*, 6122–6. (b) A. Hassner, R.J. Isbister, *J. Am. Chem. Soc.* **1969**, *91*, 6126–8.
[197] (a) R. Breslow, *J. Am. Chem. Soc.* **1957**, *79*, 5318. (b) R. Breslow, C. Yuan, *J. Am. Chem. Soc.* **1958**, *80*, 5991–4. (c) P.C. Petrellis, H. Dietrich, E. Meyer, G.W. Griffin, *J. Am. Chem. Soc.* **1967**, *89*, 1967–9.

[198] K. Yamabe, R. Tanaka, *Sasebo Kogyo Koto Senmon Gakko Kenkyu Hokoku* **1985**, 22, 119–23; *Chem. Abstr.* **1987**, *106*, 49680t.
[199] R. Tanaka, K. Yamabe, *J. Chem. Soc., Chem. Commun.* **1983**, 329–30.
[200] A.J. Mancuso, D. Swern, *Synthesis* **1981**, 165–85.
[201] E. Prochnow, A.A. Auer, K. Banert, *J. Phys. Chem. A* **2007**, *111*, 9945–51.
[202] K. Banert, M. Hagedorn, unpublished results, Chemnitz University of Technology (Germany), **2007**.
[203] F. Dost, J. Gosselck, *Tetrahedron Lett.* **1970**, 5091–3.
[204] P. Politzer, P. Lane, P. Sjoberg, M.E. Grice, H. Shechter, *Struct. Chem.* **1995**, *6*, 217–23.
[205] (a) S.G. D'yachkova, E.A. Nikitina, N.K. Gusarova, M.L. Al'pert, B.A. Trofimov, *Russ. Chem. Bull.* **2001**, *50*, 751–2. (b) S.G. D'yachkova, E.A. Nikitina, N.K. Gusarova, A.I. Albanov, B.A. Trofimov, *Russ. J. Gen. Chem. (Engl. Transl.)* **2003**, *73*, 782–5.

6

Small Rings by Azide Chemistry

Thomas L. Gilchrist[1] *and Maria José Alves*[2]

[1]*Cunningham Drive, Wirral, CH63 0JX, UK;* [2]*Departamento de Química, Campus Gualtar, Universidade do Minho, P-4710057 Braga, Portugal*

6.1 Introduction

Azides are important for the synthesis of two classes of small ring nitrogen heterocycles: 2*H*-azirines and aziridines. The synthetic methods fall into two broad categories: (1) those in which organic azides are the direct precursors and (2) those in which azides are used to provide intermediates that are then converted into heterocycles. We aimed to provide examples of both types. Routes that lead to 2*H*-azirines and those that lead to aziridines are described in Sections 6.2 and 6.3. Azides have been used less frequently for the preparation of other small nitrogen heterocycles but some examples are given in Sections 6.4 and 6.5.

We became interested in this area of chemistry because we wished to prepare some new and highly electrophilic 2*H*-azirines with potential for use as dienophiles in the Diels–Alder reaction. Vinyl azides appeared to be the most promising precursors. Previously there had been only one report of the cycloaddition of 2*H*-azirines to a simple diene (cyclopentadiene)[1] although highly activated dienes such as tetraphenylcyclopentadienone and 1,3-diphenylisobenzofuran had been used to intercept some transient 2*H*-azirines.[2,3] Our investigations led to the preparation of several new 2*H*-azirines. Cycloaddition reactions with these provided access to some novel fused-ring aziridines.[4] An outline of the results is included in Sections 6.2 and 6.3.

6.2 2*H*-Azirines

Of the methods known for the preparation of 2*H*-azirines **1**, the most widely applicable is that starting from vinyl azides **2** (Scheme 6.1).[5,6] The most common method of

Organic Azides: Syntheses and Applications Edited by Stefan Bräse and Klaus Banert

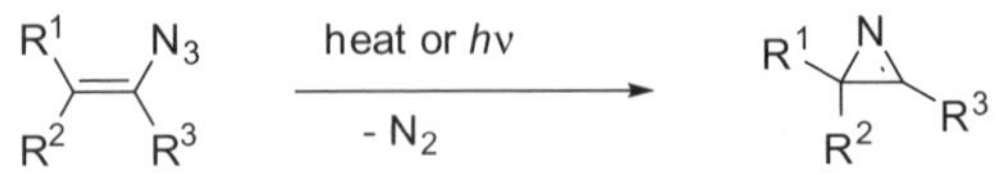

Scheme 6.1 *2H-Azirines from vinyl azides*

Table 6.1 *Examples of the conversion of vinyl azides into 2H-azirines*

R^1	R^2	R^3	Conditions	Yield (%)	Ref.
H	H	Ph	*hv*, pentane	94	7
H	H	Ph	CH_2Cl_2, 150 °C[a]	95	9
H	H	CO_2Bn	CH_2Cl_2, 150 °C[a]	85	9
H	H	CO_2Bn	PhMe, 110 °C[a]		10
H	H	^tBu	*hv*, pentane	81	7
H	H	$CO_2{}^tBu$	PhMe, 110 °C[a]		11
H	Ph	N(Me)Ph	DMF, rt[b]	50	12
Me	Me	NEt_2	Et_2O, rt[b]	94	13
CF_3	CF_3	OMe	Aq. diglyme, 25 °C[b]	11	14
F	CF_3	F	$C_2H_2Cl_4$, 25–40 °C[b]	25	15
H	Ph	COPh	CCl_4, 77 °C	63	16
H	$P(O)(OEt)_2$	$CH{=}CH_2$	PhMe, 110 °C	80	17
Me	Me	$CH_2P(O)(OEt)_2$	*hv*, MeCN	96	18
Me	CO_2Et	COPh	PhMe, 110 °C	68	19
H	2,6-$C_6H_3Cl_2$	CO_2Et	PhMe, 110 °C	82	20
Cl	CO_2Et	CO_2Et	Heptane, 98 °C	98	21
Me	H	Ph	Microwave	85	8

[a] Azirine not isolated but used in solution.
[b] Vinyl azide generated *in situ*.

decomposition is thermolysis in an inert solvent such as toluene or heptane, but other procedures include photolysis[7] and microwave irradiation in the absence of a solvent.[8] Representative examples are given in Table 6.1.

As indicated in Table 6.1 some 2*H*-azirines are isolated from one-pot reactions in which the vinyl azide precursors are produced *in situ*. For example, 3-dialkylamino-2*H*-azirines **4** have been prepared from tertiary amides **3**; the vinyl azide intermediates are generated by successive chlorination and reaction with sodium azide (Scheme 6.2).[12,22] The conversion of these vinyl azides into azirines takes place at room temperature and it is clear that the nature of the substituents on the double bond of vinyl azides influences their temperature of decomposition. Fluoro substituted vinyl azides also decompose at room temperature.[15]

Related procedures have been used to generate a series of chiral aminoazirines derived from amino acids. These azirines have then been used as reagents in peptide synthesis.[23–25] An example is provided by the synthesis of a homoproline-derived azirine **5** (Scheme 6.3).[23]

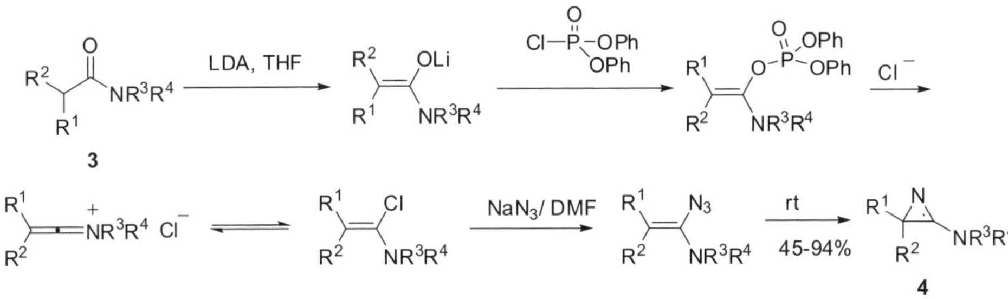

Scheme 6.2 *3-Dialkylamino-2H-azirines from tertiary amides*[12,22]

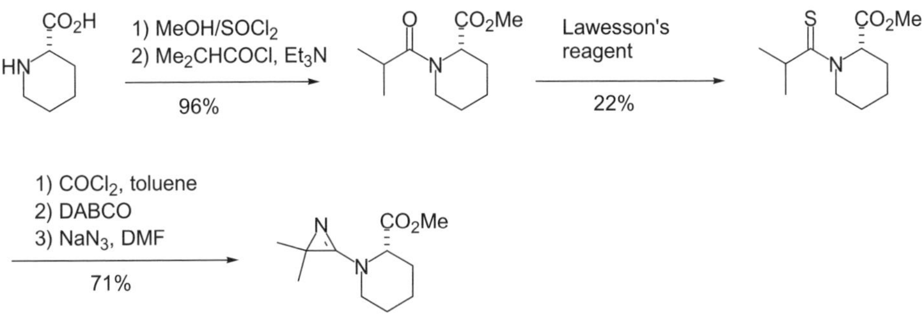

5

Scheme 6.3 *Synthesis of a chiral 3-dialkylamino-2H-azirine*[23]

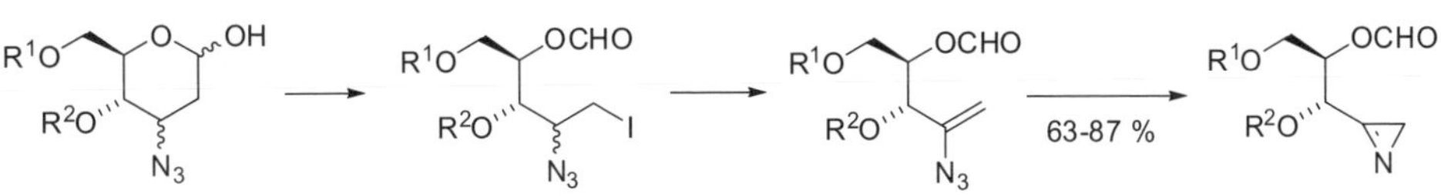

6 7 8 9

Scheme 6.4 *2H-Azirines from 3-azido-2,3-dideoxyhexanopyranoses*[27]

There are also examples of azirine syntheses from vinyl azides that are produced by less conventional routes.[26,27] Such a sequence is illustrated in Scheme 6.4[27] 3-Azido-2,3-dideoxyhexanopyranoses **6** were oxidized in the presence of iodine to hydroxyl radicals that fragmented, by cleavage of a C–C bond, to give iodoazides **7**. These were then converted in good yields into the vinyl azides **8** and, from them, into the azirines **9**.

1,4-Diazidobutadienes **11** have been generated by electrocyclic ring opening of 3,4-diazidocyclobutenes **10**.[26] The diazidobutadienes were photolyzed to give, among other products, bi-2*H*-azirin-2-yls **12** (Scheme 6.5) but these compounds are not isolable and were detected spectroscopically.

Most 2*H*-azirines are susceptible to ring opening by nucleophiles and some are thermally unstable, so the conditions under which they are generated can be critical to success.

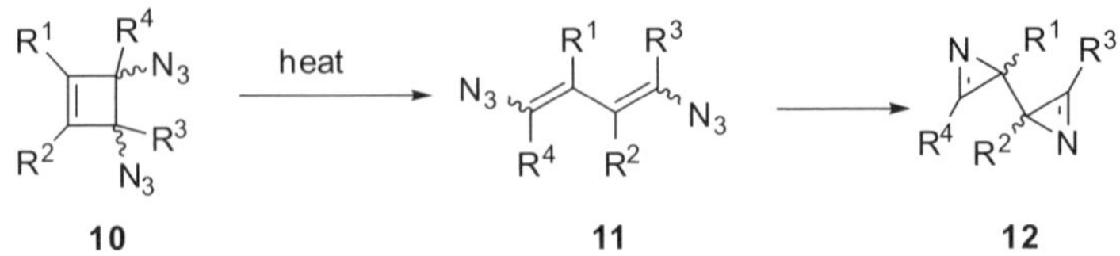

Scheme 6.5 *Bi-2H-azirinyls from 3,4-diazidocyclobutenes*[26]

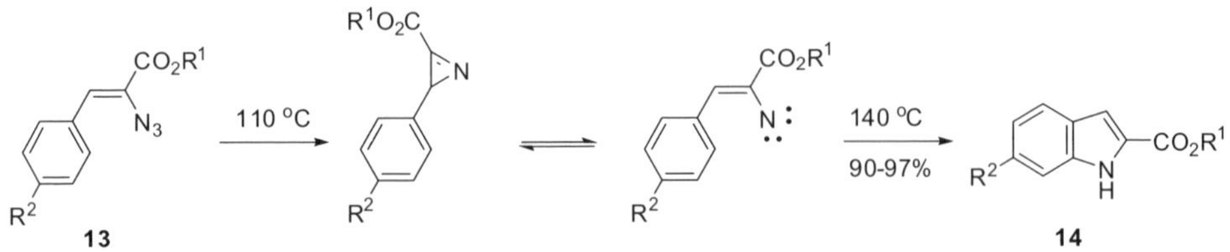

Scheme 6.6 *Conversion of alkyl 2-azidocinnamates into indole-2-carboxylic esters*[28]

For example, 2-arylazirines **13** can rearrange to indoles **14** when heated in xylene (Scheme 6.6) and the direct thermal conversion of alkyl 2-azidocinnamates into indoles has been exploited as a route to indole-2-carboxylic acid esters.[28,29] The temperature has to be controlled when these vinyl azides were decomposed in solution in order to isolate 2*H*-azirines.[30] Dilute solutions are also usually preferable in order to avoid bimolecular reactions. Photolysis is sometimes preferable to thermolysis for the generation of thermally unstable azirines.[18]

Hassner and co-workers have suggested empirical rules for determining which vinyl azides produce azirines on thermolysis and which predominantly give other products.[16] Vinyl azides of type **2** in which R^3 is hydrogen or a ketocarbonyl group predominantly give nitriles or other heterocycles, although, azirines can sometimes be detected as intermediates if the reaction conditions are carefully controlled. For example, both 3-phenylazirine and 3-*tert*-butylazirine have been intercepted in a Diels–Alder reaction with 1,3-diphenylisobenzofuran.[3] We have found that 3-*tert*-butoxycarbonyl-2*H*-azirine and other 3-alkoxycarbonylazirines can be generated cleanly in boiling toluene from alkyl 2-azidoacrylates, but attempts to isolate them lead to extensive decomposition. These azirines have been successfully intercepted, without isolation, by 1,3-dienes. These 'one pot' syntheses of bicyclic aziridines are described in Section 6.3. Indirect evidence for the formation of 2*H*-azirine intermediates has also been obtained by interception with nucleophiles, as in the case of the photolysis of 4-azido-1,2-dihydronaphthalene **15** in methanol (Scheme 6.7).[7] Five- and six-membered cyclic 3-azidoenones do not give azirines on decomposition, although on the basis of a theoretical study it has been suggested that azirines may be intermediates in the formation of some final products.[31] Transient bicyclic azirines are also implicated as intermediates in the ring expansion reactions of some aryl azides.[32]

There are a few examples of indirect syntheses of 2*H*-azirines from unsaturated azides. The photolysis of azidoallenes has provided a route to methylene-2*H*-azirines **16**.[33–35] The

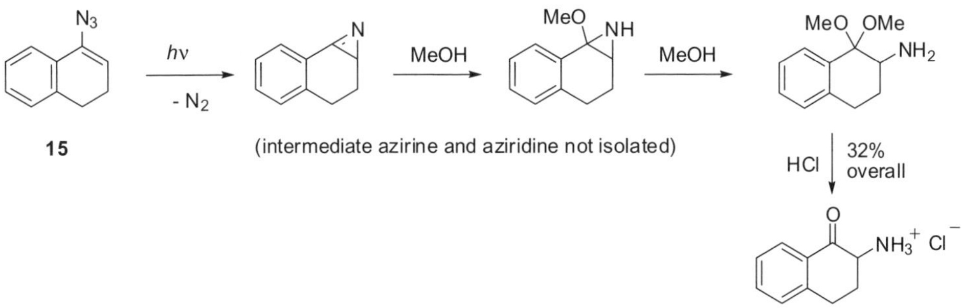

Scheme 6.7 *Interception of a transient 2H-azirine by reaction with methanol*[7]

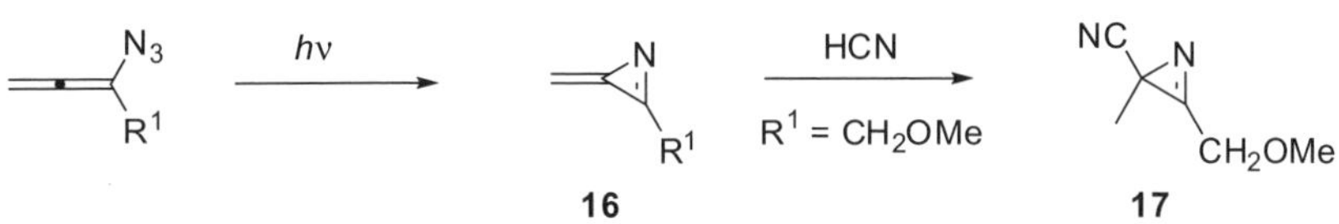

Scheme 6.8 *Methylene-2H-azirines from azidoallenes*[33]

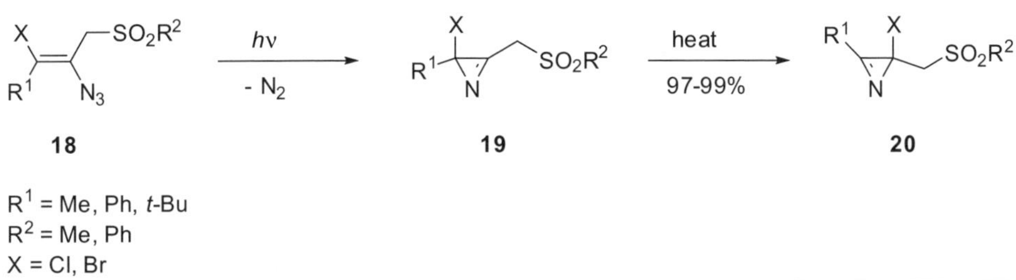

Scheme 6.9 *Thermal isomerization of halo-2H-azirines*[36]

relatively stable 2*H*-azirine **17** was obtained by the interception of the methyleneazirine **16** (R^1 = CH_2OMe) by HCN (Scheme 6.8).[33]

The photolysis of the halo substituted vinyl azides **18** produced the azirines **19**; these were then thermally isomerized to the more stable 2*H*-azirines **20** (Scheme 6.9).[36] This type of isomerization has also been observed with fluoro substituted azirines.[15]

6.3 Aziridines

Many of the methods used for the synthesis of aziridines rely directly or indirectly upon the use of azides.[37,38] The formation of aziridines from organic azides and alkenes can proceed by either one of two well-established mechanisms: (1) the initial loss of nitrogen from the azide to generate a nitrene intermediate that then adds to the double bond or (2) the [3+2] cycloaddition of the azide to the alkene to form a triazoline that then loses nitrogen (Scheme 6.10).[39] In practical terms the distinction is not always clear-cut because

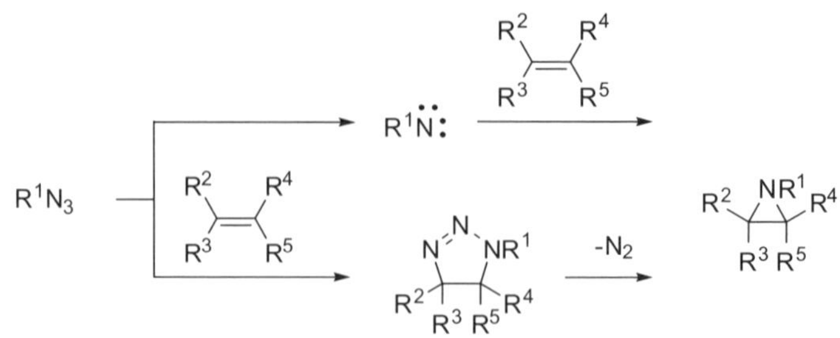

Scheme 6.10 *Alternative mechanisms for the formation of aziridines from alkenes and organic azides*[39]

some triazoline intermediates, especially those produced by intramolecular addition, spontaneously lose nitrogen to produce aziridines. Nevertheless the two routes are described separately below in Sections 6.3.1 and 6.3.2. Reactions of azides and alkenes in the presence of transition metals can also provide a method to synthesize aziridines. These reactions may involve nitrenoid complexes of the metals although the reaction mechanisms have not always been established; we have included examples in Section 6.3.1. We have also given brief descriptions of indirect syntheses of aziridines from precursors that have been produced by the use of inorganic azides; the conversion of oxiranes into azidohydrins and their subsequent ring closure to aziridines is an example of such an indirect route. These methods are presented Section 6.3.3.

6.3.1 Aziridines *via* Nitrene Intermediates

Nitrene intermediates can be generated thermally, photochemically or by the use of some metal catalysts. The temperature at which organic azides decompose varies between about 25 °C and 200 °C depending upon the substituent.[40] Cyanogen azide, acyl azides and aroyl azides decompose at low temperatures because of the relatively low bond order of the N–N_2 bond. Acyl and aroyl azides rearrange spontaneously to isocyanates with the elimination of nitrogen (Curtius rearrangement) and nitrene intermediates cannot be produced thermally from them. There are also some other types of organic azides in which nitrenes are not generated thermally because substituents participate in the elimination of nitrogen; *o*-nitroaryl azides and some other *o*-substituted aryl azides are examples. Among the groups of azides that can produce nitrenes on thermolysis relatively few are efficient sources of aziridines. Other reactions such as hydrogen shifts, reaction with nucleophiles and insertion into C–H bonds often predominate. Useful thermal reactions are generally restricted to those between electron rich alkenes and azides bearing electron withdrawing groups, particularly alkyl azidoformates.

Nitrene addition reactions have somewhat more scope when the reactions are carried out by azide photolysis; for example, the acylnitrene pivaloylnitrene adds in moderate yield and with good stereoselectivity to *cis*-alkenes when it is generated from pivaloyl azide by photolysis in dichloromethane.[41] Similarly *p*-cyanobenzoylnitrene, when generated by photolysis of the azide **21** in the presence of 2,5-dihydrofuran, gives the aziridine **22** in moderate yield (Scheme 6.11).[42] This nitrene and other *p*-substituted benzoylnitrenes also react with C_{60} to give fulleroaziridines such as **23**.[43]

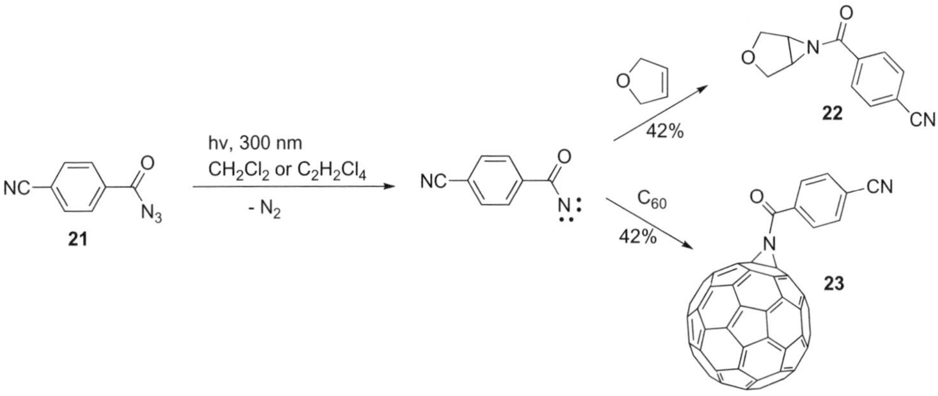

Scheme 6.11 *Cycloaddition reactions of photochemically generated p-cyanobenzoylnitrene*[42,43]

Table 6.2 *Aziridines from Azides R^1N_3 via Nitrenes*

R^1 in R^1N_3	Conditions	Ref.
CO_2Et	100 °C or *hv*, 254 nm	44
$C(OEt){=}NSO_2Me$	80 °C or *hv*, 300 nm	45
$COBu^t$	*hv*, 300 nm, CH_2Cl_2	41
CO–*p*–C_6H_4–CN	*hv*, 300 nm	42,43
C_6F_5	*hv*, 300 nm	46
Ferrocenyl	80 °C or *hv*	47

An outline of azides that have been shown to give aziridines via nitrenes either thermally or photochemically and in synthetically useful yields is given in Table 6.2.

A further restriction on the synthetic utility of the nitrene addition reaction is its unpredictable stereochemistry in reactions with disubstituted alkenes such as *cis*- and *trans*-but-2-ene. Nitrenes can exist in a singlet or triplet state. For most nitrenes the triplet (diradical) state is the ground state. Nitrenes that are generated thermally or by direct photolysis are initially in the singlet state and their (concerted) addition to alkenes is stereospecific. If the alkenes are relatively unreactive the nitrene can convert into its ground triplet state either partially or completely before addition. The resultant aziridines are produced with varying degrees of stereoselectivity because the addition of the triplet nitrene is a stepwise process. The triplet species can also be produced directly by photosensitized addition.[44]

Alkoxycarbonylnitrenes have been the most widely used in synthesis. They add to a range of simple alkenes and to enol esters and other electron rich alkenes. For example, photochemically generated ethoxycarbonylnitrene adds to enol acetates in high yield but the resulting aziridines are very unstable.[48] The nitrene also adds to vinylsilanes in good yield.[49] Furthermore aziridine diesters are generated from some α,β-unsaturated esters.[50] Even octafluoronaphthalene reacts with ethyl azidoformate to give an aziridine (although

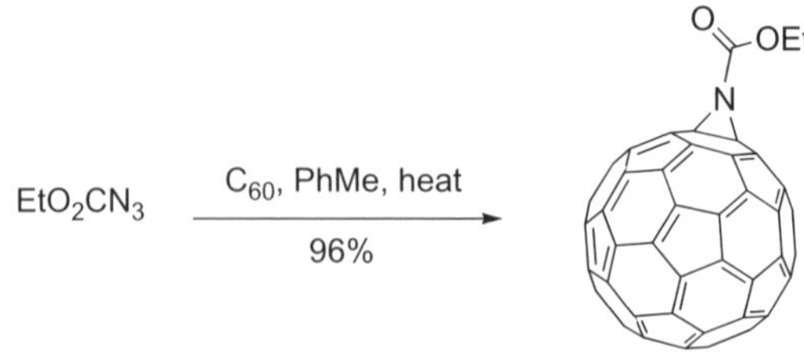

24

Scheme 6.12 *The addition of ethoxycarbonylnitrene to C_{60}*[53]

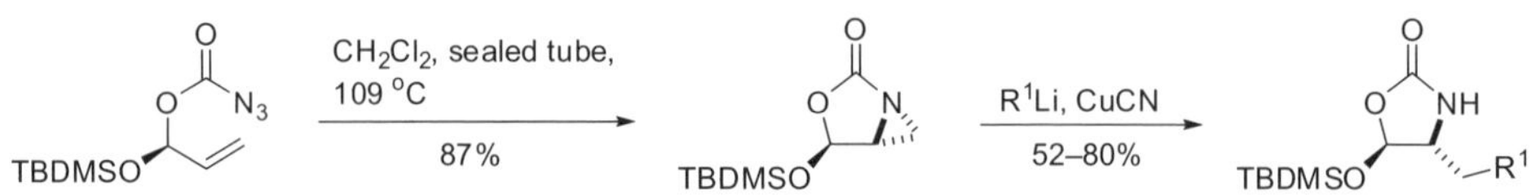

25

Scheme 6.13 *An intramolecular nitrene addition as part of a synthesis of (–)-bestatin derivatives*[56]

this might be *via* a triazoline intermediate).[51] An aziridine was isolated in good yield from the reaction of ethoxycarbonylnitrene with 1,2-dimethylcyclobutene, but the corresponding reaction with 1-methylcyclopropene led only to an acyclic product[52]. A fulleroaziridine **24** was isolated in high yield (based on C_{60} consumed) from the reaction of ethyl azidoformate and C_{60} in boiling toluene (Scheme 6.12).[53] The related but bulky nitrene generated thermally from 1,3,5-tri-*tert*-butylphenyl azidoformate has also been shown to add to C_{60}.[54]

Some intramolecular cycloaddition reactions of alkyl azidoformates are particularly successful and have useful applications in synthesis.[55,56] An example is a route to derivatives of (–)-bestatin, an aminopeptidase inhibitor, in which the key steps are the formation of the bicyclic aziridine **25** followed by ring opening (Scheme 6.13).[56]

The preparation of aziridines from alkenes *via* nitrene-metal complexes has increased it's importance as a method in recent years, especially because of the opportunities it offers to produce enantiopure aziridines.[57] Sulfonylnitrenoid complexes are the most common; metal catalyzed aziridinations using sulfonyl azides have been known since the 1960s.[58] Many catalyzed reactions have been carried out using *N*-sulfonyliminoiodinanes ($RSO_2N{=}IPh$), and these are usually more successful than those with sulfonyl azides. However sulfonyl azides offer the advantage that no iodobenzene is formed as a byproduct.[59] Toluene-*p*-sulfonyl azide (tosyl azide, TsN_3) and *p*-nitrobenzenesulfonyl azide (nosyl azide, NsN_3) have both been used successfully, NsN_3 having the advantage that the arenesulfonyl group is more easily removed from the products.[60,61] An example of asymmetric aziridination is the photochemical reaction of styrene with tosyl azide in the presence of the chiral diimine **26**-$CuPF_6$ complex which led to the formation of the aziridine **27** with

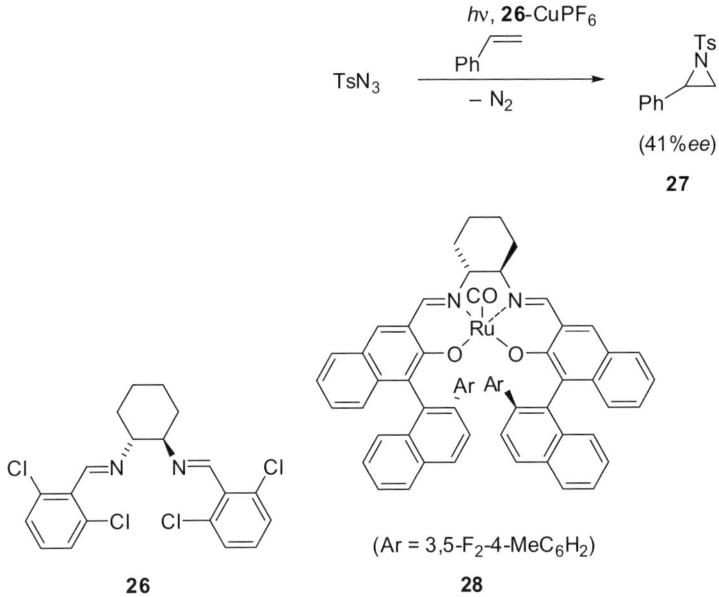

Scheme 6.14 *Catalytic asymmetric aziridination using chiral diimine ligands*[61,62]

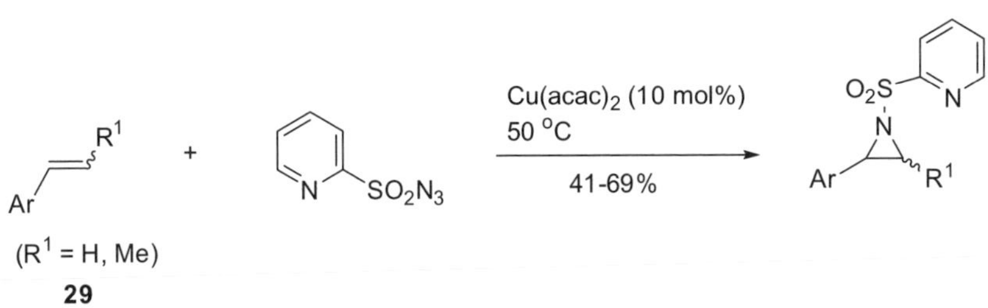

Scheme 6.15 *Copper–catalyzed aziridination of styrenes by pyridine-2-sulfonyl azide*[63]

an *ee* of 41% (Scheme 6.14).[62] More recently the method has been refined: a more robust and much more efficient catalyst is the fluorinated (OC)Ru(salen) complex **28**.[61] This has enabled styrene and substituted styrenes to be converted into arenesulfonylaziridines in high yield and with excellent enantioselectivity at room temperature using tosyl azide or nosyl azide as the reagents.

A comparative study of the aziridination of styrene using a variety of arenesulfonyl azides and a $Cu(acac)_2$ catalyst has shown that pyridine-2-sulfonyl azide and related substituted pyridines are particularly efficient.[63] It seems likely that the nitrogen atom of the pyridine ring coordinates to the copper ion and drives the formation of an internally stabilized nitrenoid intermediate. The method has been used to achieve aziridination of a range of substituted styrenes **29** in good yield and without the need for the alkene to be present in large excess (Scheme 6.15).

Diphenylphosphoryl azide **30** is another useful reagent for catalyzed aziridination.[64] Several substituted styrenes have been converted into the corresponding *N*-phosphory-

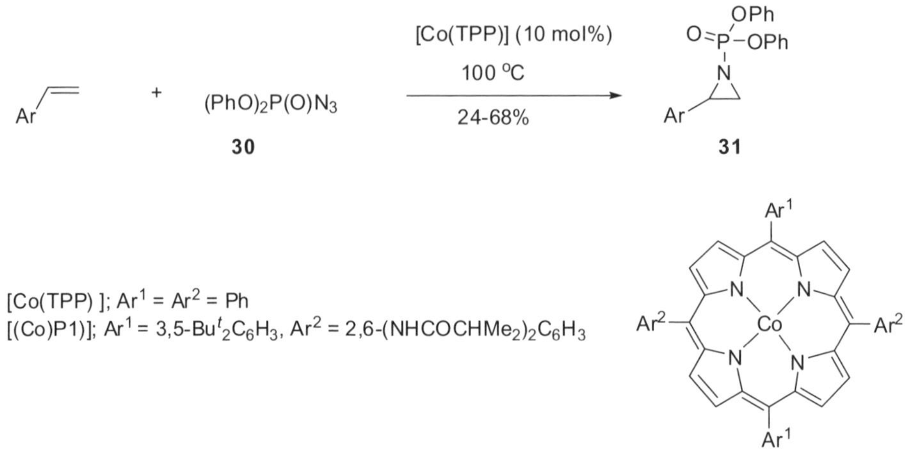

Scheme 6.16 *Cobalt(II) porphyrin catalyzed aziridination of styrenes*[64,65]

lated aziridines **31** in good yield in the presence of cobalt(II) tetraphenylporphyrin [Co(TPP)] as catalyst (Scheme 6.16). Cobalt ions appear to be essential for this conversion. The diphenylphosphoryl substituent is easily removed from the aziridines. A modified cobalt(II) porphyrin complex (designated [Co(P1)]; see Scheme 6.16) was designed to allow hydrogen bonding interactions in the nitrenoid intermediate. This proved to be an excellent catalyst for the conversion of styrenes into *N*-arensulfonylaziridines in high yield.[65] Cobalt(II) tetraphenylporphyrin also catalyzes the decomposition of ethyl azidoformate but it promotes the formation of *O*-ethyl carbamate by hydrogen abstraction instead of cycloaddition.[66]

6.3.2 Aziridines *via* Triazolines

The 1,3-dipolar addition of azides to alkenes is a well-established method for the synthesis of 4,5-dihydro-1,2,3-triazoles (triazolines). The relationship between the rate of the cycloaddition reaction and the nature of the alkene was investigated in detail for azidobenzene by Sustmann and Trill.[67] Essentially they demonstrated that the reaction is faster with both electron rich and electron poor alkenes than it is with ethene. In general the reaction is most favourable with a combination of an electron deficient azide and an electron rich alkene or with an electron rich azide and an electrophilic alkene, although steric effects can also influence the rate with particular alkenes. The product triazolines are often not very stable and there are several routes by which they can be converted into more stable species. These include elimination or oxidation to (aromatic) 1,2,3-triazoles and decomposition with the loss of nitrogen. Aziridines are among the products that can be formed when nitrogen is lost. In principle aziridines can be generated either through diradical or a zwitterionic intermediate (Scheme 6.17). The decomposition can be brought about both thermally and photochemically.

The rate at which nitrogen is lost from triazolines is very dependent upon substituents. For example, *N*-cyanotriazolines decompose at temperatures between 0 °C and 35 °C,

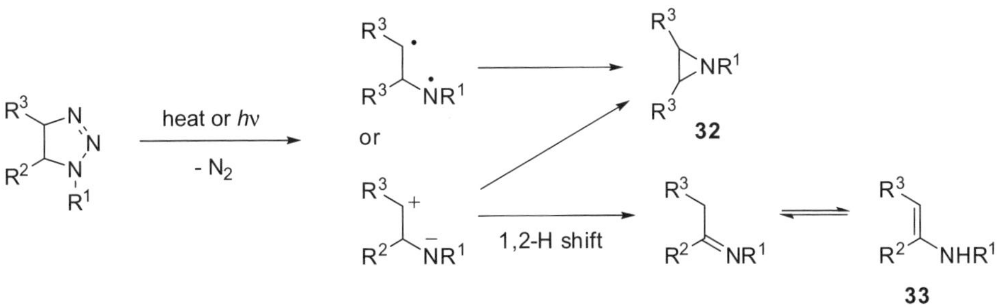

Scheme 6.17 Aziridines and imines by elimination of nitrogen from triazolines

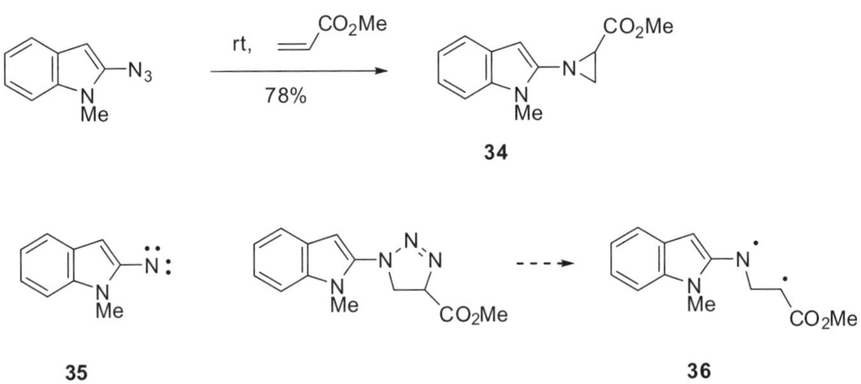

Scheme 6.18 Formation of an aziridine from 2-azido-1-methylindole and methyl acrylate, and possible reaction intermediates[69]

leading to the formation of mixtures of *N*-cyanoaziridines **32** (R^1 = CN) and alkylidene-cyanamides **33** (R^1 = CN).[68] In reactions wherein aziridines are isolated but triazolines are neither isolated nor detected it can be difficult to distinguish between nitrene and triazoline mechanisms. An example of this is the reaction of 2-azido-1-methylindole and methyl acrylate at room temperature which led directly to the isolation of the aziridine **34** in 78% yield.[69] The authors initially favoured a mechanism in which a nitrene intermediate **35** adds to methyl acrylate but the rate of loss of nitrogen from a possible triazoline intermediate would also be enhanced by the nature of the substituents in an intermediate diradical **36** (Scheme 6.18). Several similar reactions were later interpreted in terms of a triazoline mechanism.[70,71] Even with an established nitrene precursor such as ethyl azidoformate, thermal cycloaddition to give a triazoline can compete with nitrene addition to an alkene.[72]

Thermolysis of methyl 1-phenyltriazoline-4-carboxylate **37** is also very efficient: the aziridine **38** is isolated in 95% yield after the triazoline is heated in toluene at 85 °C. In contrast the triazoline **37** is converted at room temperature into diazoester **39** by triethylamine (Scheme 6.19).[73]

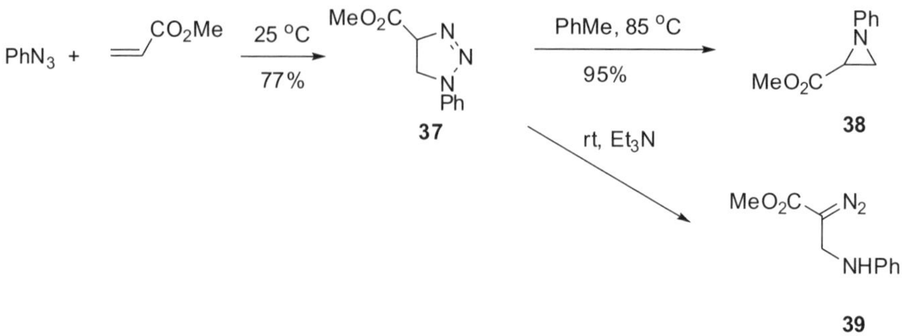

Scheme 6.19 *Thermolysis of methyl 1-phenyltriazoline-4-carboxylate*[73]

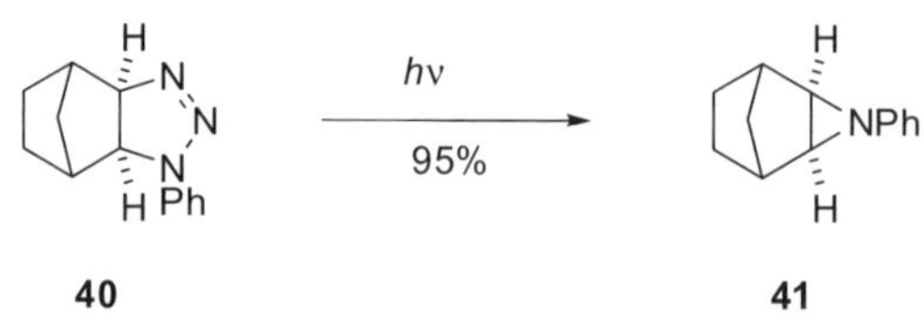

Scheme 6.20 *Photolysis of a triazoline*[74]

Photolysis is sometimes superior to thermolysis as a means of converting triazolines into aziridines. The cycloadduct **40** formed from azidobenzene and norbornene provides an example; it is cleanly photolyzed to give the aziridine **41** (Scheme 6.20) but its thermal decomposition leads to a mixture of products.[74]

In contrast to the nitrene cycloadditions shown in Schemes 6.11 and 6.12, C_{60} reacts with aryl azides at room temperature to give isolable triazolines. These extrude nitrogen on photolysis to give the corresponding fulleroaziridines whereas thermal decomposition leads predominantly to ring opened products.[75]

Some applications of the triazoline route to aziridines from alkenes and azides are described below.

6.3.2.1 Combinations of Nucleophilic Alkenes and Electron Deficient Azides

Vinyl ethers and other oxygen substituted alkenes readily undergo 1,3-dipolar cycloaddition reactions with organic azides but in most cases the products are rapidly converted into 1,2,3-triazoles by elimination. There are, however, a few examples of elimination of nitrogen from the triazolines to give aziridines.[38] The silyl ketene acetal **42** was shown to react with fluoroalkanesulfonyl azides **43** at 0 °C to give isolable aziridines **44** in moderate yield (45–48%) along with α-amino esters **45**.[76] A zwitterionic intermediate is suggested as the precursor of both products (Scheme 6.21). Although the formation of the esters **45** requires a hydrolytic step the ratio of the products **44** and **45** is not affected by the use of a moist solvent, indicating that the amino esters are not formed exclusively by cleavage of the aziridines.

A short synthetic sequence to a benzazocenone **46**, a potential building block for mitomycenoids and related antitumour compounds, makes use of an intramolecular addition

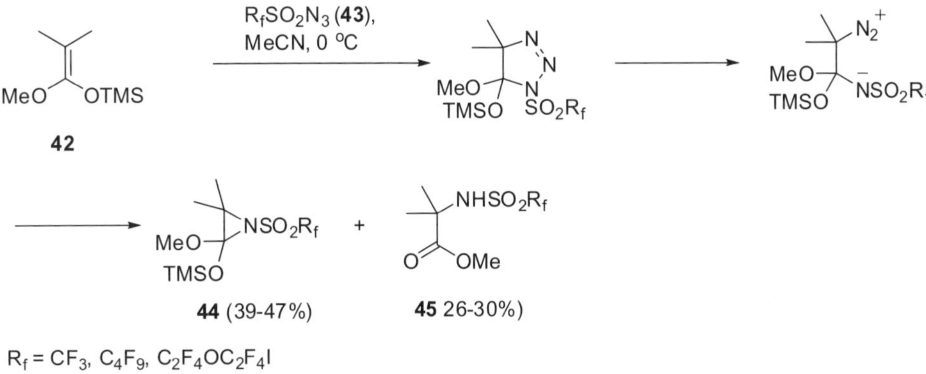

Scheme 6.21 *Aziridines from a silylketene acetal*[76]

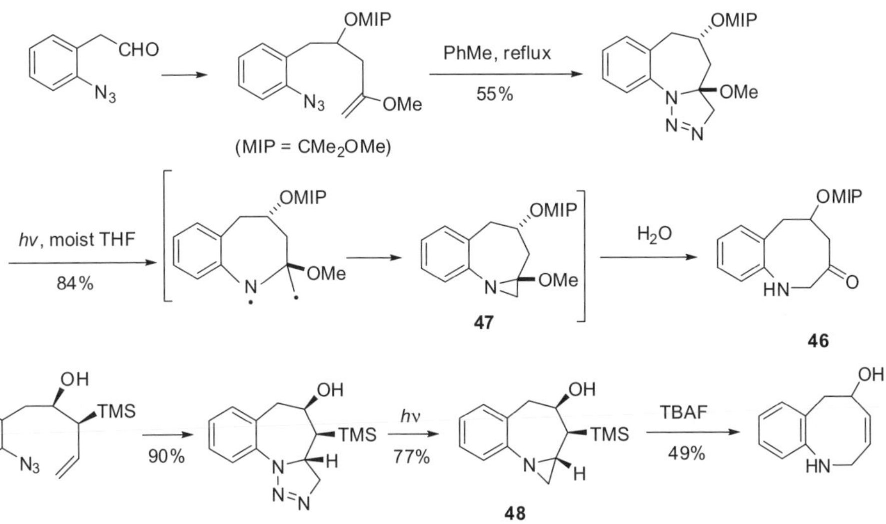

Scheme 6.22 *Routes to benzazocenes by intramolecular cycloaddition of azides*[77,78]

of an aryl azide to a vinyl ether (Scheme 6.22). Its photolysis leads to the generation of a bicyclic aziridine **47** but this is hydrolyzed in situ to the benzazocenone **46**.[77] A related reaction sequence but with an allylsilane as the internal component leads to an isolable aziridine intermediate **48**.[78]

6.3.2.2 Combinations of Electrophilic Alkenes and Nucleophilic Azides

1,3-Dipolar cycloaddition reactions of unactivated azides to electrophilic alkenes are common; examples were given earlier in Schemes 6.18 and 6.19. Another example is the addition of benzyl azide to cyclopentenone, which, in the presence of a Lewis acid (TMSOTf), gave a mixture of the enaminone **49** and the aziridine **50** in good overall yield

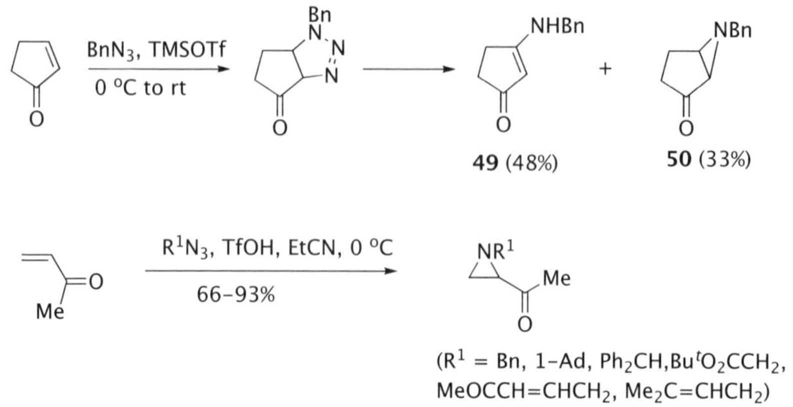

Scheme 6.23 *Catalyzed reactions of azides with enones*[79,80]

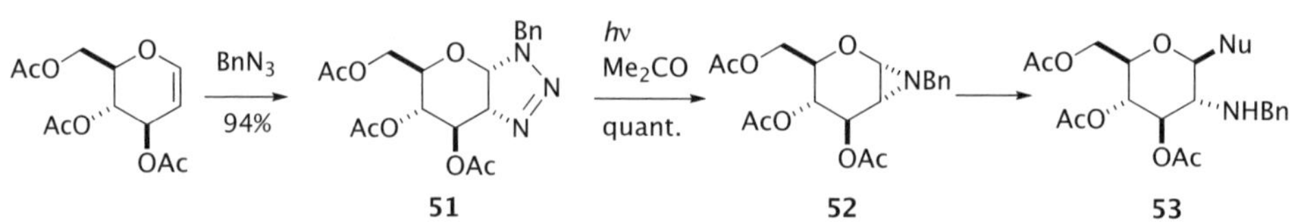

Scheme 6.24 *Aminoglucosides from tri-O-acetyl-D-glucal*[81]

(Scheme 6.23) (see also Chapter 7).[79] Under these conditions the intermediate triazoline is not isolable; the Lewis acid appears to have the dual role of activating the enone and facilitating the decomposition of the triazoline. Triflic acid is a very efficient catalyst for promoting the formation of aziridines from azides and electron deficient alkenes such as but-3-en-2-one (Scheme 6.23).[80] The acid appears to activate the alkene to nucleophilic attack by the azide; it is not necessary to invoke triazolines as intermediates although they may be involved.

Enol ethers are usually classified as nucleophilic alkenes but glycals with electron-withdrawing groups are sufficiently activated to undergo cycloaddition to nucleophilic azides such as benzyl azide. For example, tri-*O*-acetyl-D-glucal and benzyl azide reacted together when heated in trimethyl orthoformate to give the isolable triazoline **51**. This was quantitatively converted by photolysis in acetone into aziridine **52**, from which aminoglucosides **53** were obtained by nucleophilic ring opening (Scheme 6.24). The role of the solvent in the cycloaddition step is crucial to avoid aromatization of the triazoline **51**; the authors suggest that trimethyl orthoformate succeeds because it is not nucleophilic and can act as an acid scavenger.[81]

Some of the most synthetically useful reactions are intramolecular cycloadditions. An intramolecular cyclopentenone cycloaddition led to the formation of the tricyclic aziridine **54**, an intermediate for the preparation of the alkaloid (±)-cephalotaxine, in good yield (Scheme 6.25).[82] The intermediate triazoline was not detected. A related synthesis of the aziridine **55** was used as a step in a synthesis of the alkaloid (±)-aspidospermidine.[83]

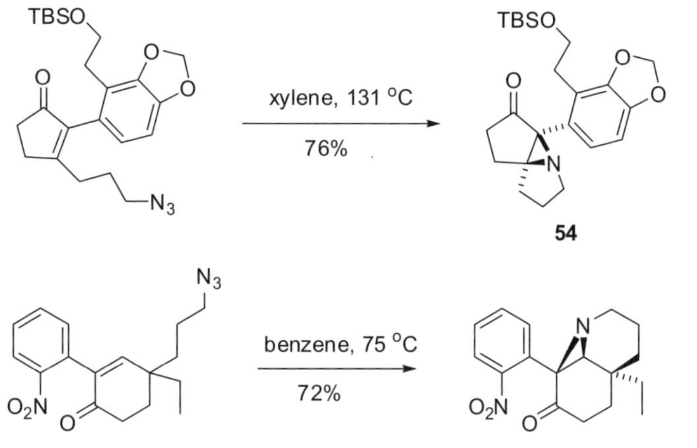

54

55

Scheme 6.25 *Intramolecular addition of alkyl azides to enones*[82,83]

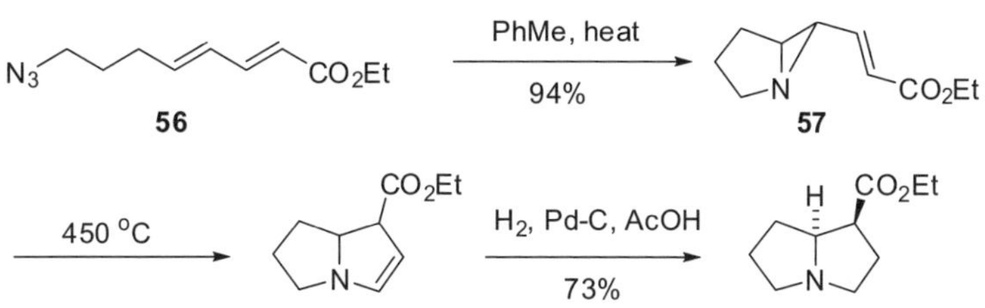

56

57

58 (major isomer)

Scheme 6.26 *A route to the pyrrolizidine ring system*[84]

Intramolecular 1,3-dipolar cycloadditions of azides have also been investigated as a route to the pyrrolizidine ring system.[84] For example, ethyl 8-azidoocta-2,4-dienoate (**56**) was converted in high yield into the labile vinylaziridine **57** when it was heated under reflux in toluene. In other experiments it was demonstrated that activation of one of the double bonds of the diene was essential in order to achieve efficient intramolecular cycloaddition. The vinylaziridine **57** was then converted into the pyrrolizidine **58** by flash pyrolysis followed by catalytic hydrogenation of the product (Scheme 6.26).

6.3.3 Aziridines from Epoxides or 1,2-Diols

The ring opening of epoxides by inorganic azides is the initial step of a general synthetic route to aziridines. Because enantiopure epoxides are readily available by Sharpless epoxidation and other methods, their ring opening by azide ions provides one of the best approaches to the synthesis of enantiopure aziridines.[85,86] The stereochemistry of the aziridines can be reliably predicted on the basis of the mechanisms of the steps involved.

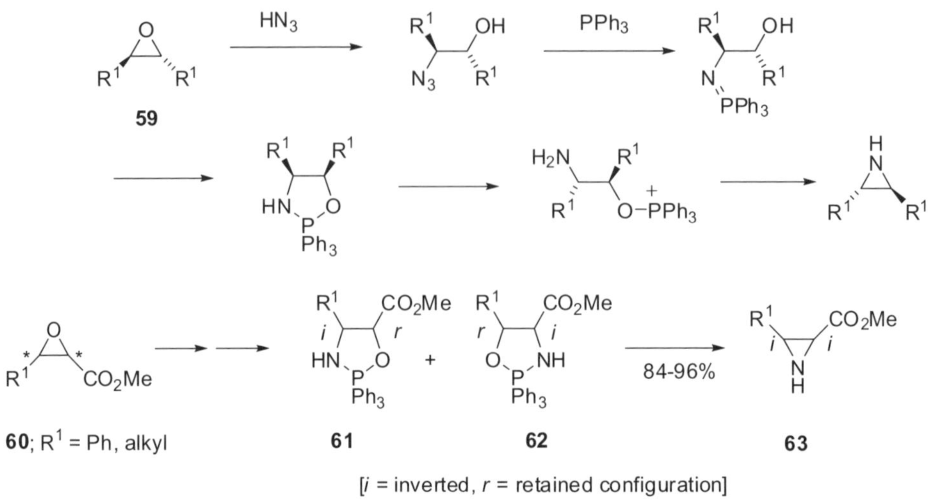

Scheme 6.27 *Enantioselective conversion of chiral epoxides into aziridines*[85,86,88,90]

This is illustrated for a symmetrically 2,3-disubstituted *trans*-epoxide **59** in Scheme 6.27. Initial ring opening by azide ion occurs with inversion of configuration. The resulting azidoalcohol is reductively cyclized by reaction with triphenylphosphine. The cyclization step also proceeds with inversion of configuration as shown. This was used to convert the two enantiomers of diethyl oxirane-2,3-dicarboxylate (**59**, R^1 = CO_2Et and its *cis* epimer) into the corresponding aziridines.[87–89] Even when the substituents are different on the epoxide and the ring opening is not regioselective, both centres are inverted in the aziridine. Thus the enantiopure glycidic esters **60** were converted into the aziridines **63** in good overall yield and with high *ee* through the intermediate 1,3,2-oxazphospholidines **61** and **62**.[90]

Monosubstituted epoxides are normally opened by sodium azide at the unsubstituted carbon atom but in the presence of a Lewis acid this preference can be overridden. Thus styrene oxide **64** is opened predominantly at the higher substituted position by sodium azide in the presence of lithium tetrafluoroborate.[91] Diethylaluminium azide has been used for the opening of trisubstituted epoxides and the predominant products are those derived from attack on the tertiary carbon atom.[92] For epoxyalcohols the reagent $Ti(O\textit{i}Pr)_2(N_3)_2$ has been used to achieve regioselective ring opening. An example of its use is the conversion of epoxyalcohol **65** into azidoalcohols **66** and **67** (Scheme 6.28).[93,94]

Enantiopure aziridines can also be obtained from chiral 1,2-diols. These are converted into cyclic sulfates **68**,[95–97] or cyclic sulfites **69**.[98] These compounds are also cleaved by reaction with lithium azide or sodium azide to azidoalcohols which can then be converted into aziridines as above. The sulfates **68** can be prepared in one pot and in high yield from the diols by reaction with thionyl chloride followed by oxidation with sodium periodate and ruthenium trichloride.[95] Sulfites **69** are available from diols simply by reaction with thionyl chloride. An example of the use of a cyclic sulfate is the preparation of the aziridine **71** in good yield and with high *ee* from the chiral sulfate **70** (Scheme

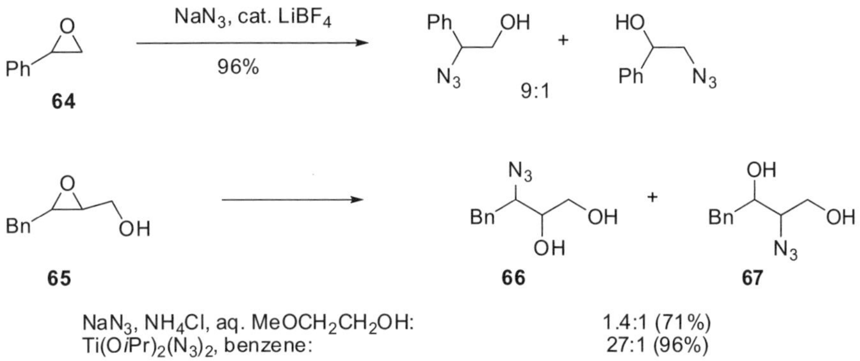

Scheme 6.28 *Regioselective opening of epoxides*[91,93]

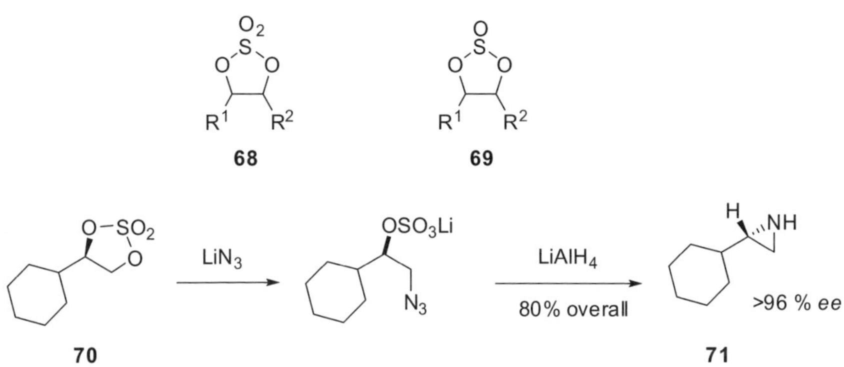

Scheme 6.29 *The use of cyclic sulfates and sulfites for the preparation of aziridines*[96,98]

6.29).[96] Other groups, such as methanesulfonyl,[99] can be used to activate the hydroxyl group of azidoalcohols before reductive cyclization to aziridines.

6.3.4 Aziridines from Vinyl Azides *via* 2*H*-Azirines

Aziridines can be prepared by nucleophilic addition to isolable 2*H*-azirines. This is an important method for the synthesis of aziridines bearing a heteroatom substituent on position 2 or 3.[38] Here we limit the examples to reactions in which vinyl azides are the starting materials and the intermediate 2*H*-azirines derived from them are not isolated, usually because they are unstable.

A simple intramolecular nucleophilic addition is illustrated in Scheme 6.30.[100] Normally alkyl 2-azidocinnamates are converted into alkyl indole-2-carboxylates when they are heated in solution (Scheme 6.6). The azidoester **72** gave the corresponding indole in very low yield, however. The major product was the aziridine **74** which is formed by interception of the intermediate 2*H*-azirine **73** by the internal nucleophile.

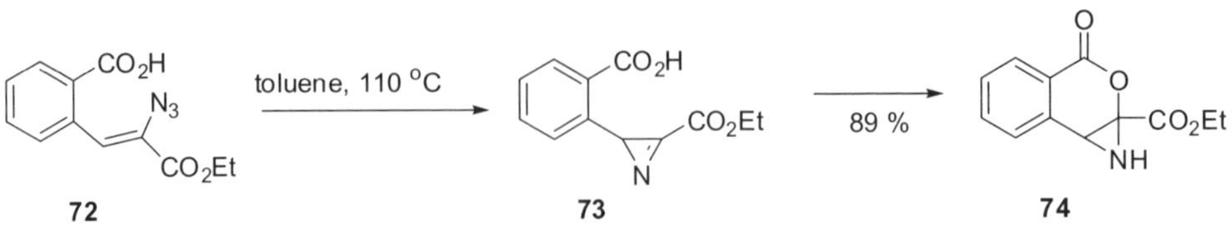

Scheme 6.30 *Interception of a 2H-azirine by an internal nucleophile*[100]

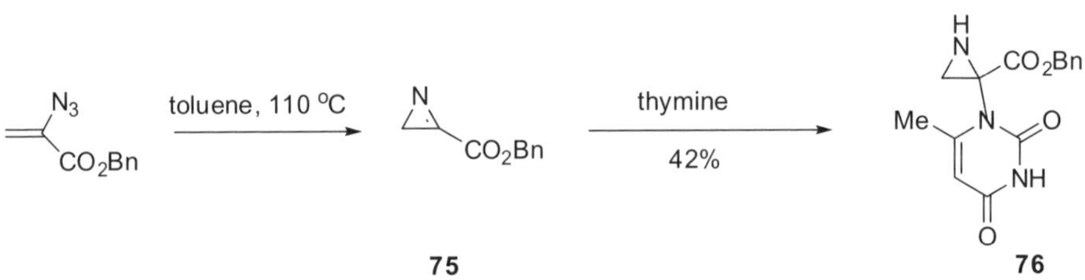

Scheme 6.31 *Reaction of benzyl 2H-azirine-3-carboxylate with thymine*[10]

Nitrogen nucleophiles also add readily to 2*H*-azirines but the resulting aziridines are usually very unstable because of the electron donating effect of the nitrogen substituent. Aromatic nitrogen heterocycles are an exception because the reaction products do not have a localized lone pair. The unstable azirine **75** was intercepted in situ by a series of purine and pyrimidine bases to give isolable aziridines; for example, the aziridine **76** was obtained from a reaction with thymine (Scheme 6.31).[10]

The reactivity of 2*H*-azirines such as **75** also offers the opportunity to carry out cycloaddition reactions with simple nucleophilic 1,3-dienes. We found that the azirines could be generated in toluene or heptane solution from the vinyl azides and the solutions could then be used in Diels–Alder reactions with dienes.[4,11] The cycloaddition reactions were carried out at room temperature and the resulting aziridines were isolated by chromatography. The reactions are regioselective and *endo* selective. For example, the bicyclic aziridine **78** was isolated in good yield from the reaction of 1-methoxybutadiene with *tert*-butyl 2*H*-azirine-3-carboxylate **77** (Scheme 6.32).[11] Similar adducts were isolated from reactions of the azirines **75** and **77** with cyclopentadiene, cyclohexadiene, 1-acetoxybutadiene and other dienes. The enantiopure diene **79** gave a single adduct, the aziridine **80**.[101,102] On the other hand the chiral azirine **81** showed little diastereoselectivity in its reactions with dienes.[103,104] The Diels–Alder selectivity is improved with chiral esters, either with **81**, or others in the presence of Lewis acids.[105] Chiral Lewis acid catalysts have also been used to promote the asymmetric Diels–Alder reaction but diastereoselectivities are moderate.[105]

The reaction of the azirine **75** with furan proved exceptional in that it was slow at room temperature and it eventually gave only the *exo* cycloadduct **82** (Scheme 6.33).[101] This is consistent with other Diels–Alder reactions of furan from which the thermodynamic product, the *exo* adduct, is isolated because the cycloaddition is reversible.

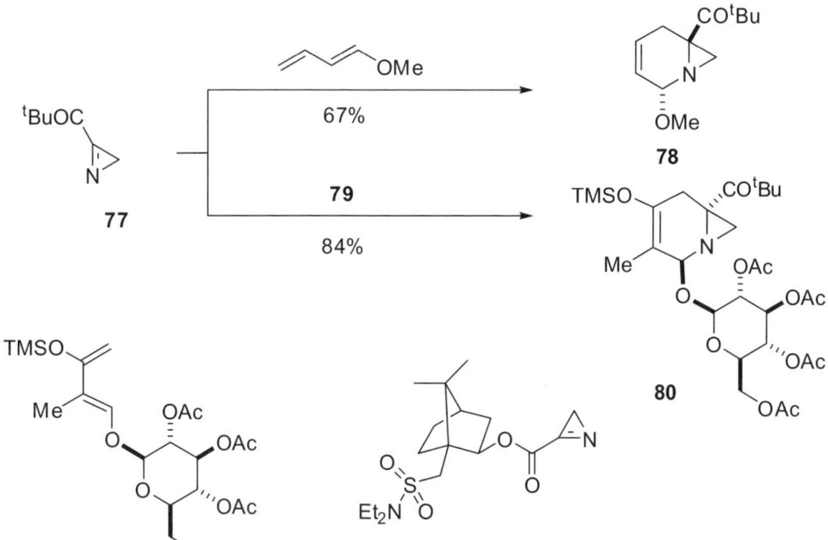

Scheme 6.32 *Diels–Alder reactions of alkyl 2H-azirine-3-carboxylates*[11,103]

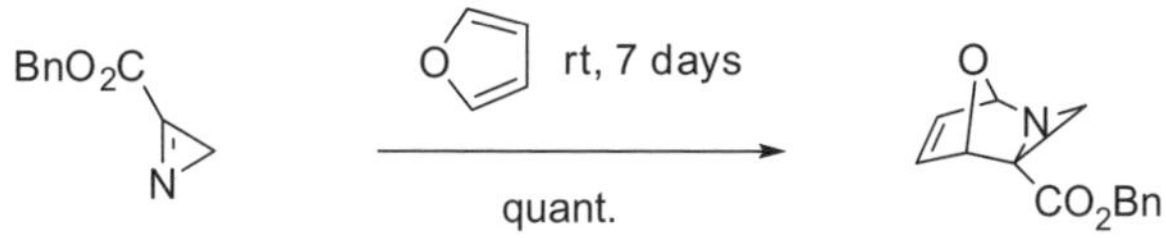

Scheme 6.33 *Cycloaddition of benzyl 2H-azirine-3-carboxylate to furan*[101]

6.4 Triaziridines

The only other three-membered heterocycles that are available through azide chemistry are triaziridines, and these only with very special structural features. The addition of nitrenes or nitrene equivalents to azo compounds normally leads to the formation of azimines **83**, the open chain tautomers of triaziridines **84**.[106] In order to circumvent the formation of azimines Prinzbach and co-workers designed a series of compounds that bear an azide function in close proximity to the π face of an azoalkane.[107] These compounds were then irradiated in dilute solution to give triaziridines among other products.[108] For example, the azide **85** gave the triaziridine **86** (25%), a crystalline compound that could be purified by sublimation (Scheme 6.34). The cage compounds **87** (X = CH, CMe, P and P=O) were prepared in an analogous way.

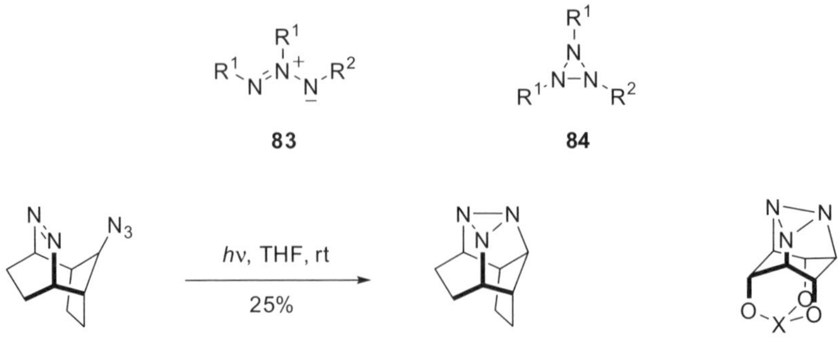

83 84 85 86 87

Scheme 6.34 *Triaziridines*[108]

6.5 Azetidinones

There are a few methods for the synthesis of azetidin-2-ones that have organic azides as precursors. In these reactions the nitrogen atom of the four membered ring system is not derived from the azide function, however. For example, 4-azidopyrrol-2(5*H*)-ones **88** are converted into azetidinones **90** *via* zwitterionic intermediates **89** (Scheme 6.35).[31,109]

Simple azetidine-2-ones **92** have also been obtained in moderate yield, along with carbamate esters **93**, from the boron trifluoride catalyzed reaction of benzyl azides with a cyclopropanone acetal **91**.[110] The complex **94** has been suggested as an intermediate in the ring expansion reaction (Scheme 6.36).

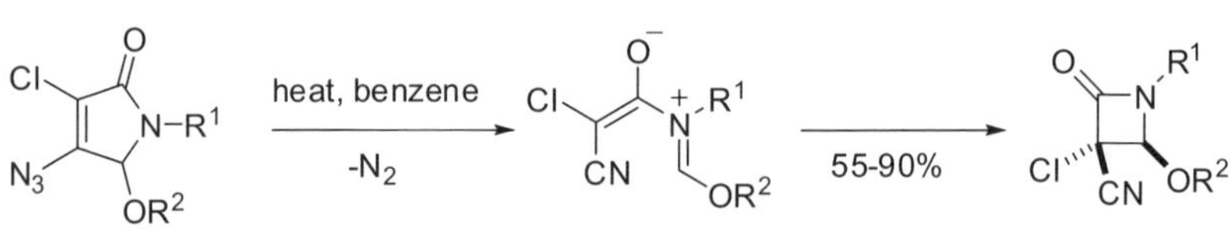

88 89 90

Scheme 6.35 *Azetidin-2-ones from 4-azidopyrrol-2(5H)-ones*[109]

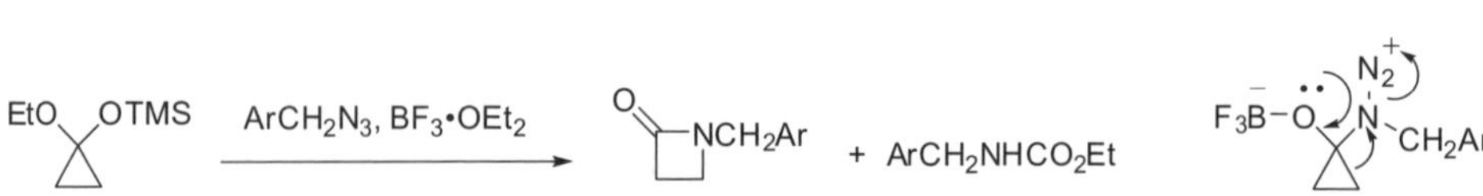

91 92 (27-45%) 93 (13-60%) 94

Scheme 6.36 *Azetidin-2-ones from a cyclopropanone acetal*[110]

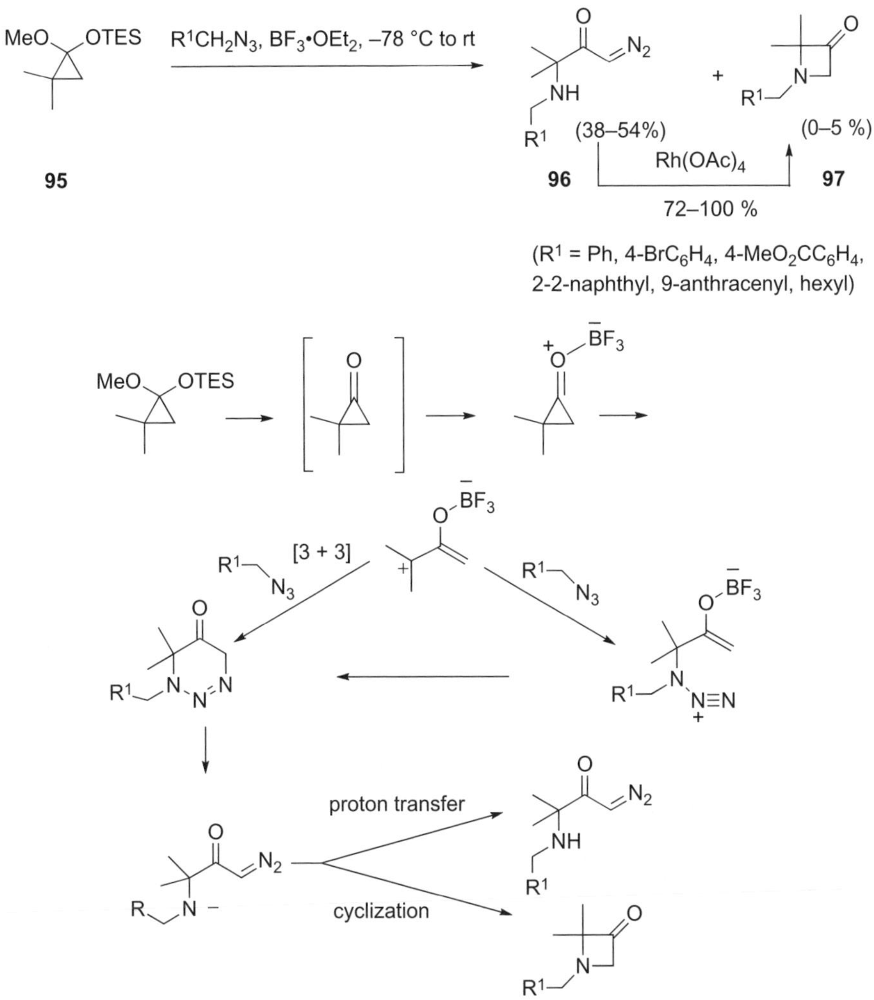

Scheme 6.37 *Azetidin-3-ones from the cyclopropanone acetal* ***95***[111]

Azetidin-3-ones are also available from a related reaction with the cyclopropanone acetal **95**.[111] In this reaction the major products are diazoketones **96** but these can be cyclized in good yield to azetidin-3-ones **97** by reaction with dirhodium tetraacetate. The mechanism proposed by the authors for the reaction is shown in Scheme 6.37.

References

[1] H. Hemetsberger, D. Knittel, *Monatsh. Chem.* **1972**, *103*, 205.
[2] D.J. Anderson, A. Hassner, *J. Org. Chem.* **1973**, *38*, 2565.
[3] A. Hassner, D.J. Anderson, *J. Org. Chem.* **1974**, *39*, 2031.
[4] T.L. Gilchrist, *Aldrichimica Acta* **2001**, *34*, 51.

[5] F. Palacios, A.M.O. de Retana, E.M. de Marigorta, J.M. de los Santos, *Eur. J. Org. Chem.* **2001**, 2401.
[6] F. Palacios, A.M.O. de Retana, E.M. de Marigorta, J.M. de los Santos, *Org. Prep. Proced. Int.* **2002**, *34*, 219.
[7] A. Hassner, F.W. Fowler, *J. Am. Chem. Soc.* **1968**, *90*, 2869.
[8] P.N.D. Singh, C.L. Carter, A.D. Gudmundsdottir, *Tetrahedron Lett.* **2003**, *44*, 6763.
[9] A.S. Timen, E. Risberg, P. Somfai, *Tetrahedron Lett.* **2003**, *44*, 5339.
[10] T.L. Gilchrist, R. Mendonça, *Synlett* **2000**, 1843.
[11] M.J. Alves, T.L. Gilchrist, *Tetrahedron Lett.* **1998**, *39*, 7579.
[12] J.M. Villalgordo, H. Heimgartner, *Helv. Chim. Acta* **1992**, *75*, 1866.
[13] M. Rens, L. Ghosez, *Tetrahedron Lett.* **1970**, 3765.
[14] C.G. Krespan, *J. Org. Chem.* **1969**, *34*, 1278.
[15] C.S. Cleaver, C.G. Krespan, *J. Am. Chem. Soc.* **1965**, *87*, 3716.
[16] A. Hassner, N.H. Wiegand, H.E. Gottlieb, *J. Org. Chem.* **1986**, *51*, 3176.
[17] V.K. Brel, *Synthesis* **2007**, 2674.
[18] R.A. Abramovitch, M. Konieczny, W. Pennington, S. Kanamathareddy, M. Vedachalam, *Chem. Commun.* **1990**, 269.
[19] T. Patonay, J. Jeko, E. Juhasz-Toth, *Eur. J. Org. Chem.* **2008**, 1441.
[20] T.L. Gilchrist, C.W. Rees, J.A.R. Rodrigues, *Chem. Commun.* **1979**, 627.
[21] T. Melo, A. Gonsalves, C.S.J. Lopes, T.L. Gilchrist, *Tetrahedron Lett.* **1999**, *40*, 789.
[22] J.M. Villalgordo, H. Heimgartner, *Helv. Chim. Acta* **1993**, *76*, 2830.
[23] R.A. Breitenmoser, H. Heimgartner, *Helv. Chim. Acta* **2002**, *85*, 885.
[24] S. Stamm, A. Linden, H. Heimgartner, *Helv. Chim. Acta* **2003**, *86*, 1371.
[25] S. Stamm, H. Heimgartner, *Helv. Chim. Acta* **2006**, *89*, 1841.
[26] K. Banert, S. Grimme, R. Herges, *et al.*, *Chem. Eur. J.* **2006**, *12*, 7467.
[27] C.R. Alonso-Cruz, A.R. Kennedy, M.S. Rodriguez, E. Suárez, *Org. Lett.* **2003**, *5*, 3729.
[28] H. Hemetsberger, D. Knittel, *Monatsh. Chem.* **1972**, *103*, 194.
[29] R.E. Bolton, C.J. Moody, C.W. Rees, G. Tojo, *J. Chem. Soc., Perkin Trans. 1* **1987**, 931.
[30] H. Hemetsberger, D. Knittel, H. Weidmann, *Monatsh. Chem.* **1970**, *101*, 161.
[31] D. Suárez, T.L. Sordo, *J. Am. Chem. Soc.* **1997**, *119*, 10291.
[32] E.F.V. Scriven, K. Turnbull, *Chem. Rev.* **1988**, *88*, 297.
[33] K. Banert, M. Hagedorn, *Angew. Chem. Int. Ed.* **1990**, *29*, 103.
[34] K. Banert, M. Hagedorn, E. Knözinger, A. Becker, E.-U. Würthwein, *J. Am. Chem. Soc.* **1994**, *116*, 60.
[35] J.R. Fotsing, K. Banert, *Eur. J. Org. Chem.* **2006**, 3617.
[36] J.R. Fotsing, K. Banert, *Synthesis* **2006**, 261.
[37] J.B. Sweeney, in *Aziridines and Epoxides in Organic Synthesis* (Ed.: A. K. Yudin), Wiley-VCH, Weinheim, **2006**, p. 117.
[38] G.S. Singh, M. D'Hooghe, N. De Kimpe, *Chem. Rev.* **2007**, *107*, 2080.
[39] S. Bräse, C. Gil, K. Knepper, V. Zimmermann, *Angew. Chem. Int. Ed.* **2005**, *44*, 5188.
[40] G. L'abbé, *Chem. Rev.* **1969**, *69*, 345.
[41] G.R. Felt, W. Lwowski, *J. Org. Chem.* **1976**, *41*, 96.
[42] K. Buck, D. Jacobi, Y. Plögert, W. Abraham, *J. Prakt. Chem.* **1994**, *336*, 678.
[43] J. Averdung, J. Mattay, D. Jacobi, W. Abraham, *Tetrahedron* **1995**, *51*, 2543.
[44] J.S. McConaghy, W. Lwowski, *J. Am. Chem. Soc.* **1967**, *89*, 4450.
[45] A. Subbaraj, O.S. Rao, W. Lwowski, *J. Org. Chem.* **1989**, *54*, 3945.
[46] R.A. Abramovitch, S.R. Challand, Y. Yamada, *J. Org. Chem.* **1975**, *40*, 1541.
[47] R.A. Abramovitch, C.I. Azogu, R.G. Sutherland, *Chem. Commun.* **1971**, 134.
[48] J.F.W. Keana, S.B. Keana, D. Beetham, *J. Org. Chem.* **1967**, *32*, 3057.
[49] A.R. Bassindale, P.A. Kyle, M.C. Soobramanien, P.G. Taylor, *J. Chem. Soc., Perkin Trans. 1* **2000**, *7*, 1173.
[50] M.P. Sammes, A. Rahman, *J. Chem. Soc., Perkin Trans. 1* **1972**, 344.
[51] M.G. Barlow, G.M. Harrison, R.N. Haszeldine, W.D. Morton, P. Shawluckman, M.D. Ward, *J. Chem. Soc., Perkin Trans. 1* **1982**, 2101.
[52] J.N. Labows, D. Swern, *Tetrahedron Lett.* **1971**, 4523.

[53] T. Ishida, K. Tanaka, T. Nogami, *Chem. Lett.* **1994**, 561.
[54] M.R. Banks, J.I.G. Cadogan, I. Gosney, P.K.G. Hodgson, P.R.R. Langridgesmith, D.W.H. Rankin, *Chem. Commun.* **1994**, 1365.
[55] S.C. Bergmeier, D.M. Stanchina, *J. Org. Chem.* **1997**, *62*, 4449.
[56] S.C. Bergmeier, D.M. Stanchina, *J. Org. Chem.* **1999**, *64*, 2852.
[57] P. Müller, C. Fruit, *Chem. Rev.* **2003**, *103*, 2905.
[58] H. Kwart, A.A. Khan, *J. Am. Chem. Soc.* **1967**, *89*, 1951.
[59] S. Cenini, E. Gallo, A. Caselli, F. Ragaini, S. Fantauzzi, C. Piangiolino, *Coord. Chem. Rev.* **2006**, *250*, 1234.
[60] T. Fukuyama, C.K. Jow, M. Cheung, *Tetrahedron Lett.* **1995**, *36*, 6373.
[61] K. Omura, T. Uchida, R. Irie, T. Katsuki, *Chem. Commun.* **2004**, 2060.
[62] Z. Li, R.W. Quan, E.N. Jacobsen, *J. Am. Chem. Soc.* **1995**, *117*, 5889.
[63] H. Han, S.B. Park, S.K. Kim, S.B. Chang, *J. Org. Chem.* **2008**, *73*, 2862.
[64] G.Y. Gao, J.E. Jones, R. Vyas, J.D. Harden, X.P. Zhang, *J. Org. Chem.* **2006**, *71*, 6655.
[65] J.V. Ruppel, J.E. Jones, C.A. Huff, R.M. Kamble, Y. Chen, X.P. Zhang, *Org. Lett.* **2008**, *10*, 1995.
[66] M. Mitani, M. Takayama, K. Koyama, *J. Org. Chem.* **1981**, *46*, 2226.
[67] R. Sustmann, H. Trill, *Angew. Chem, Int. Ed.* **1972**, *11*, 838.
[68] M.E. Hermes, F.D. Marsh, *J. Org. Chem.* **1972**, *37*, 2969.
[69] E. Foresti, P. Spagnolo, P. Zanirato, *J. Chem. Soc., Perkin Trans. 1* **1989**, 1354.
[70] P. Zanirato, *J. Chem. Soc., Perkin Trans. 1* **1991**, 2789.
[71] S. Gronowitz, P. Zanirato, *J. Chem. Soc., Perkin Trans. 2* **1994**, 1815.
[72] P.P. Nicholas, *J. Org. Chem.* **1975**, *40*, 3396.
[73] G. Szeimies, R. Huisgen, *Chem. Ber.* **1966**, *99*, 491.
[74] R. Huisgen, L. Möbius, G. Müller, H. Stangl, G. Szeimies, J.M. Vernon, *Chem. Ber.* **1965**, *98*, 3992.
[75] J. Averdung, J. Mattay, *Tetrahedron* **1996**, *52*, 5407.
[76] Y. Xu, S.Z. Zhu, *Tetrahedron* **2001**, *57*, 669.
[77] M.A. Ciufolini, M.Y. Chen, D.P. Lovett, M.V. Deaton, *Tetrahedron Lett.* **1997**, *38*, 4355.
[78] R. Ducray, N. Cramer, M.A. Ciufolini, *Tetrahedron Lett.* **2001**, *42*, 9175.
[79] D.S. Reddy, W.R. Judd, J. Aube, *Org. Lett.* **2003**, *5*, 3899.
[80] J.M. Mahoney, C.R. Smith, J.N. Johnston, *J. Am. Chem. Soc.* **2005**, *127*, 1354.
[81] R.S. Dahl, N.S. Finney, *J. Am. Chem. Soc.* **2004**, *126*, 8356.
[82] G.A. Molander, M. Hiersemann, *Tetrahedron Lett.* **1997**, *38*, 4347.
[83] M.G. Banwell, D.W. Lupton, *Org. Biomol. Chem.* **2005**, *3*, 213.
[84] T. Hudlicky, J.O. Frazier, G. Seoane, *et al.*, *J. Am. Chem. Soc.* **1986**, *108*, 3755.
[85] D. Tanner, *Angew. Chem. Int. Ed.* **1994**, *33*, 599.
[86] H.M.I. Osborn, J. Sweeney, *Tetrahedron Asymmetry* **1997**, *8*, 1693.
[87] S. Saito, N. Bunya, M. Inaba, T. Moriwake, S. Torii, *Tetrahedron Lett.* **1985**, *26*, 5309.
[88] D. Tanner, C. Birgersson, H.K. Dhaliwal, *Tetrahedron Lett.* **1990**, *31*, 1903.
[89] A. Breuning, R. Vicik, T. Schirmeister, *Tetrahedron Asymmetry* **2003**, *14*, 3301.
[90] J. Legters, L. Thijs, B. Zwanenburg, *Tetrahedron Lett.* **1989**, *30*, 4881.
[91] F. Kazemi, A.R. Kiasat, S. Ebrahimi, *Synth. Commun.* **2003**, *33*, 999.
[92] C.E. Davis, J.L. Bailey, J.W. Lockner, R.M. Coates, *J. Org. Chem.* **2003**, *68*, 75.
[93] M. Caron, P.R. Carlier, K.B. Sharpless, *J. Org. Chem.* **1988**, *53*, 5185.
[94] X. Ginesta, M. Pasto, M.A. Pericas, A. Riera, *Org. Lett.* **2003**, *5*, 3001.
[95] Y. Gao, K.B. Sharpless, *J. Am. Chem. Soc.* **1988**, *110*, 7538.
[96] B.B. Lohray, Y. Gao, K.B. Sharpless, *Tetrahedron Lett.* **1989**, *30*, 2623.
[97] G.V. Shustov, A.V. Kachanov, V.A. Korneev, R.G. Kostyanovsky, A. Rauk, *J. Am. Chem. Soc.* **1993**, *115*, 10267.
[98] B.B. Lohray, J.R. Ahuja, *Chem. Commun.* **1991**, 95.
[99] D. Tanner, H.M. He, *Tetrahedron* **1992**, *48*, 6079.
[100] D.M.B. Hickey, A.R. Mackenzie, C.J. Moody, C.W. Rees, *J. Chem. Soc., Perkin Trans. 1*, **1987**, 921.

[101] M.J. Alves, N.G. Azoia, J.F. Bickley, A.G. Fortes, T.L. Gilchrist, R. Mendonca, *J. Chem. Soc., Perkin Trans. 1* **2001**, 2969.

[102] M.J. Alves, I.G. Almeida, A.G. Fortes, A.P. Freitas, *Tetrahedron Lett.* **2003**, *44*, 6561.

[103] Y.S.P. Alvares, M.J. Alves, N.G. Azoia, J.F. Bickley, T.L. Gilchrist, *J. Chem. Soc., Perkin Trans. 1* **2002**, 1911.

[104] M.J. Alves, J.F. Bickley, T.L. Gilchrist, *J. Chem. Soc., Perkin Trans. 1* **1999**, 1399.

[105] A.S. Timen, P. Somfai, *J. Org. Chem.* **2003**, *68*, 9958.

[106] C. Leuenberger, L. Hoesch, A.S. Dreiding, *Helv. Chim. Acta* **1982**, *65*, 217.

[107] W. Marterer, O. Klingler, R. Thiergardt, L. Knothe, H. Prinzbach, *Chem. Ber.* **1991**, *124*, 609.

[108] W. Marterer, O. Klingler, R. Thiergardt, E. Beckmann, H. Fritz, H. Prinzbach, *Chem. Ber.* **1991**, *124*, 621.

[109] H.W. Moore, L. Hernandez, D.M. Kunert, F. Mercer, A. Sing, *J. Am. Chem. Soc.* **1981**, *103*, 1769.

[110] S. Grecian, P. Desai, C. Mossman, J.L. Poutsma, J. Aubé, *J. Org. Chem.* **2007**, *72*, 9439.

[111] P. Desai, J. Aubé, *Org. Lett.* **2000**, *2*, 1657.

7

Schmidt Rearrangement Reactions with Alkyl Azides

Scott Grecian[1] *and Jeffrey Aubé*[2]

[1]*Lacamas Laboratories, 3625 North Suttle Road, Portland, Oregon 97217, USA;* [2]*Department of Medicinal Chemistry, University of Kansas, School of Pharmacy, Malott Hall, 1251 Wescoe Hall Drive, Room 4070, Lawrence, KS 66045-7582, USA*

7.1 Introduction and Early Attempts (1940–60)

The sobriquet 'Schmidt reaction' refers to a number of related processes in which hydrazoic acid or an alkyl azide reacts with an electrophile.[1–6] The product obtained will depend on the reaction partner (carboxylic acids generate amines, aldehydes give nitriles and formamides, and ketones provide amides, respectively) but all of these reactions have certain features in common. They are the initial nucleophilic addition of the azide component to an electrophile (usually promoted by acid) eventually followed by a rearrangement step driven by the energetically favorable loss of nitrogen. The classical Schmidt reaction of hydrazoic acid with ketones is generally considered to involve an intermediate dehydration step.[7,8] Despite the diversity of reactions that comprise the Schmidt family, its single most common variant is shown in Scheme 7.1. This overall process can be regarded as an insertion of NH between the carbonyl and an α alkyl group.[9]

The development of this version of the Schmidt reaction predates the experience of anyone working in organic chemistry today (Schmidt's original report was in 1924) but the extension of the concept to nucleophilic alkyl azides is a decidedly contemporary development. For example, if one compares the two reactions shown in Scheme 7.2, it is clear that the direct insertion of an alkyl azide into cyclohexanone allows for the introduction of additional nitrogen atom substitution and, in the case of an alkyl azide tethered to a carbonyl substrate, the possibility of an intramolecular reaction. The goal of this chapter

Organic Azides: Syntheses and Applications Edited by Stefan Bräse and Klaus Banert

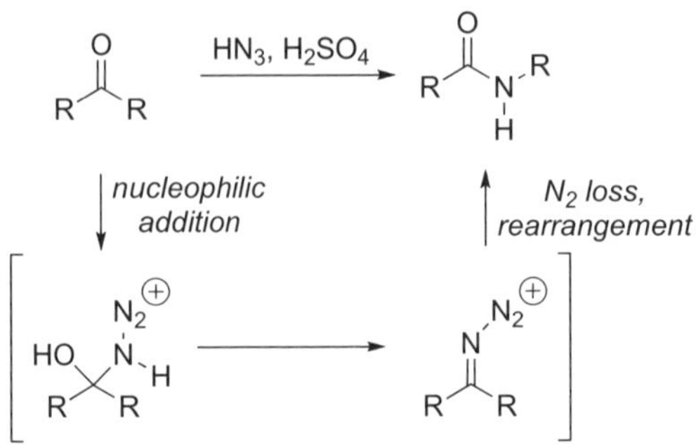

Scheme 7.1 *Generalized Schmidt reaction mechanism*

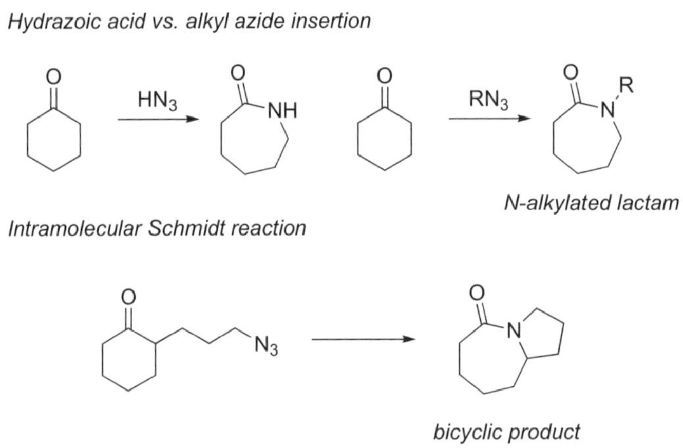

Scheme 7.2 *Comparison between classical (hydrazoic acid-mediated) Schmidt reactions and those involving alkyl azides*

is to survey the history, recent developments, and synthetic utility of the Schmidt chemistry of azides.

The similarity of an alkyl azide to hydrazoic acid is sufficiently obvious that it could not escape the attention of the early architects of the Schmidt reaction. However, the first efforts to engage alkyl azides in Schmidt reactions were unsuccessful. In the 1940s, Briggs and Smith independently reported that no amide was obtained when they tried to react methyl azide with acetophenone, with only unspecified 'decomposition' of the azide noted.[10,11] A few years later, Joseph Boyer and coworkers registered a limited success, showing that certain alkyl azides reacted with aromatic aldehydes to give amides in generally low yields (Scheme 7.3).[12] Moreover, they found that aromatic aldehydes, when treated with β- or γ-hydroxy azides in the presence of sulfuric acid, provided oxazoline products in good yield.[13] At the time, the latter reaction was proposed to occur by direct azide attack onto the carbonyl; the differences in yield were ascribed to greater acid stability of the starting hydroxyalkyl azides thanks to hydrogen bonding (mechanism not

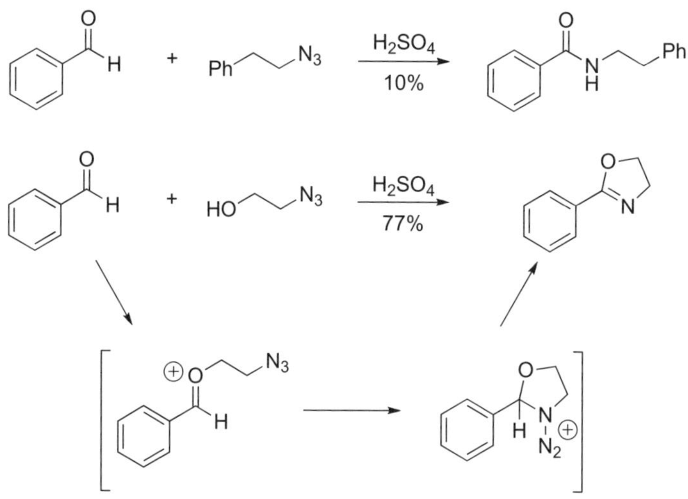

Scheme 7.3 *Early examples of some successful reactions of azides and aldehydes*

shown). However, hindsight helped workers nearly fifty years later[14] to formulate a more satisfying explanation involving initial oxygen attack to form an oxonium ion. The formation of this oxonium ion renders the addition of azide intramolecular, which is then followed by rearrangement.

There were very few reports of Schmidt reactions involving alkyl azides for almost 40 years after Boyer's papers appeared. Some papers described sequences that resembled the bona fide Schmidt reaction in that azides ultimately afforded lactams, but were thermally enacted and mechanistically distinct from the classical Schmidt reaction (see Section 7.5). In the early 1990s, a series of disclosures finally established synthetically useful versions of the Schmidt reaction using various kinds of alkyl azides as the key substrates. The following discussion of this chemistry will concentrate first on those reactions in which the electrophilic partner is a carbonyl group or carbonyl equivalent. Later sections will describe processes in which an alkyl azide attacks a carbocation derived from an alkene, alcohol, or a related precursor. Finally, the ways in which these reactions have been used to construct natural products or other compounds of interest will conclude this chapter.

7.2 Schmidt Reactions of Alkyl Azides with Carbonyl Compounds

7.2.1 Intramolecular Reactions

The first synthetically useful Schmidt reactions of alkyl azides with ketones were intramolecular. Thus, it was shown that azido-tethered ketones, when treated with Brønsted or Lewis acids in CH_2Cl_2 at room temperature, were converted to lactams in good to excellent yields (Table 7.1).[15] Mono- and bicyclic ketones are attractive substrates for the preparation of fused bi- and tricyclic lactams and the range of ring sizes accommodated

Table 7.1 *Intramolecular Schmidt reactions of cyclic azidoketones*[16]

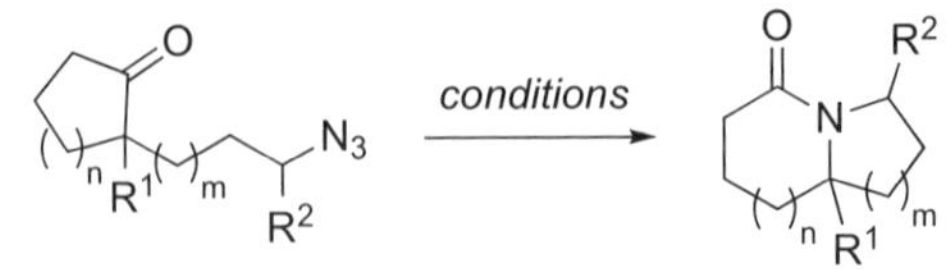

Entry	Ketone	Lactam	Conditions	Yield (%)
1	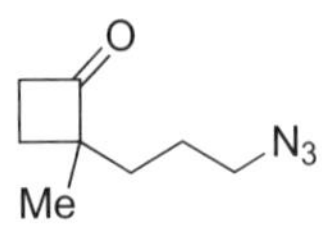	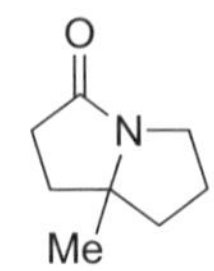	TFA, 25 min	66
2	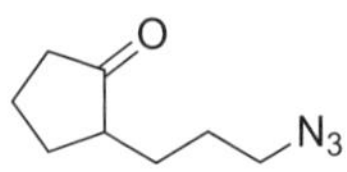	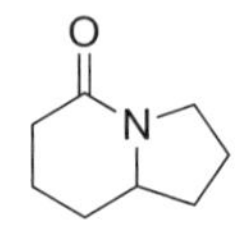	TFA, 40 min	83
3	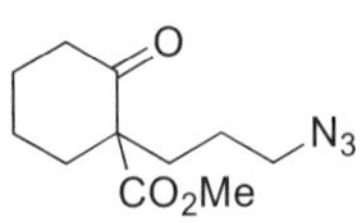	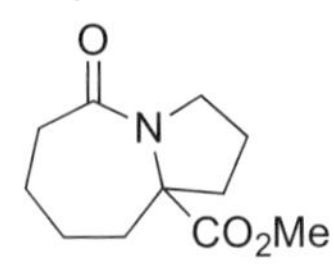	TFA, 16 h	66
4	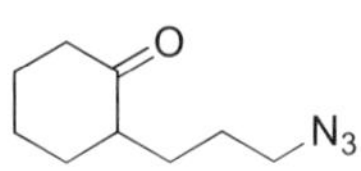	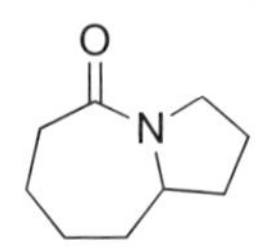	TFA, 3.5 h	85
5	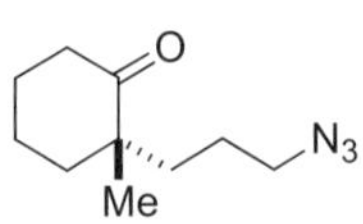 96% er	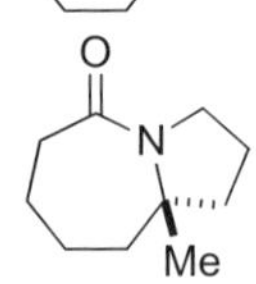 95% er	TFA, 3.5 h	87
6	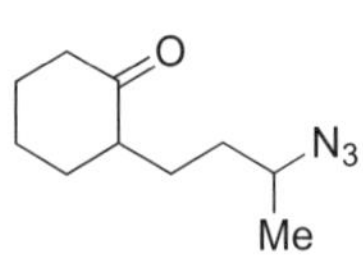	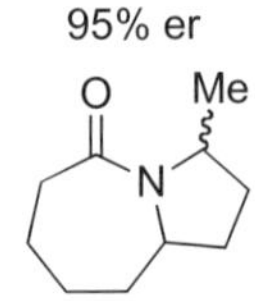	TFA, 15 min	74
7	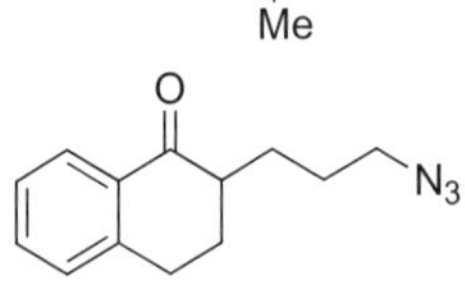	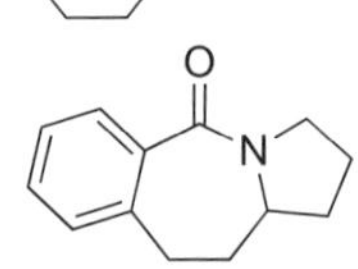	TFA, 1 h	91
8	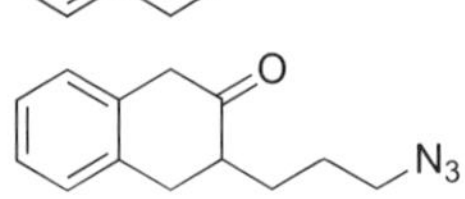	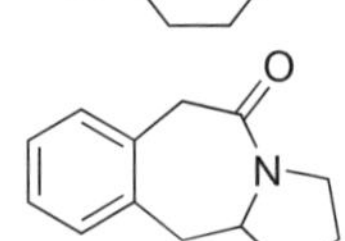	TfOH, 2 days	45
9	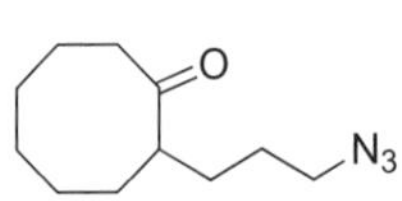	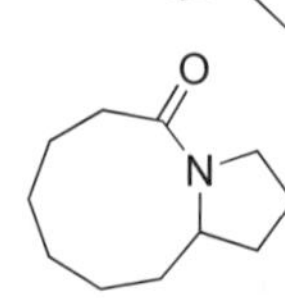	TFA, 16 h	96
10	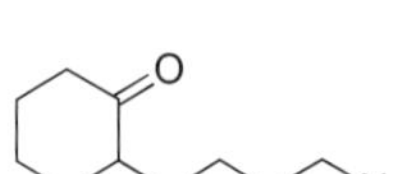	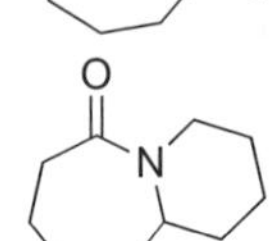	$TiCl_4$, 16 h	91

in the starting alicyclic ketone span 4 to 12 atoms.[16] When electron-withdrawing substituents were attached to the migrating carbon, the rate of rearrangement was considerably reduced (cf. entries 3 and 4). These reactions are stereospecific in that the stereogenic migrating carbon is transferred with retention of configuration (entry 5). This intramolecular version of the Schmidt reaction requires that the ketone and azide groups be separated by four or five atoms, although the latter require potent Lewis acid activation to provide the corresponding lactams and are slower (cf. entries 4 and 10).

The success of the intramolecular Schmidt reaction was ascribed to the relative ease of the initial nucleophilic addition step when compared to the intermolecular variants tried by Briggs, Smith, and Boyer. In the case of δ-substituted azido ketones, the reaction proceeds through a particularly favorable six-membered tetrahedral intermediate. Keto azides attached with a tether lengthened by a single methylene group need to go through the seven-membered ring version of this intermediate; such reactions are less favorable and require the use of more demanding Lewis acid conditions. In either case, migration of bond **a** gives rise to fused lactams with concomitant loss of N_2. Alternatively, migration of bond **b** would provide a corresponding bridged lactam; this modality was not observed until much later (Scheme 7.4; see also Section 7.9). It is noteworthy that the success of these reactions finally put to rest the urban legend that Schmidt reactions of alkyl azides were not possible because they were unable to undergo the dehydration step typically associated with hydrazoic acid chemistry (see Scheme 7.1).

The observation that electron-withdrawing groups attached to the migrating carbon (e.g. CO_2Me, aryl) impede these reactions is also consistent with the mechanistic viewpoint shown in Scheme 7.3 and not easily reconciled with an alternative involving acid-promoted decomposition to a highly reactive nitrene, which would then undergo C–C insertion. The high reactivity of such a nitrene would make its formation rate limiting; thus the overall rate should be relatively insensitive to structural changes about the migrating carbon. Indeed, accumulating experimental and computational results continue to be consistent with the rate-limiting formation of a tetrahedral intermediate. In fact, the alignment of the N–N bond of diazo cation in these intermediates is regularly observed to direct the regioselectivity of the insertion event.

Acyclic ketones and aldehydes having 3 or 4 carbon atoms between the azido and carbonyl group also undergo rearrangements to afford *N*-substituted lactams when treated

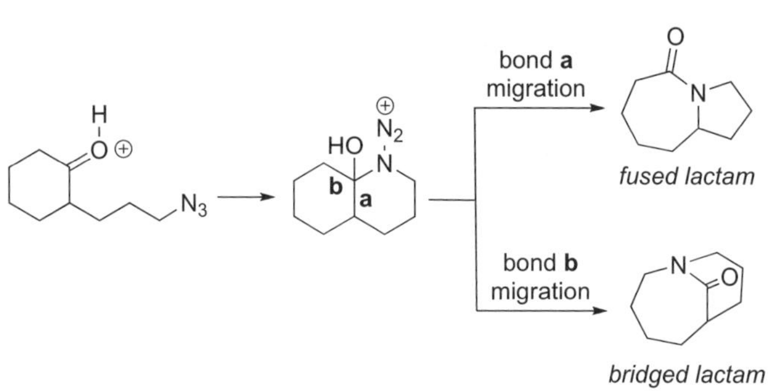

Scheme 7.4 *Regiochemical possibilities for the intramolecular Schmidt reaction of a 2-alkylated cyclohexanone*

with either stoichiometric TFA or $TiCl_4$.[17] The need for a full equivalent of protic or Lewis acid arises from the greater Lewis basicity of the amide product compared to the ketone (i.e. the reaction is subject to product inhibition). When ketones were employed in these examples, the on-tether carbonyl substituent $C(R^2)_2$ preferentially migrated, giving amides **1** (Table 7.2). When azido aldehydes were treated with protic or Lewis acids, significant amounts of lactams **2** were also isolated, arising from formal migration of the aldehyde proton. The product distribution depended heavily on the tether length of the starting materials: reactions of substrates with two-carbon tethers failed, three-carbon tethers gave lactams **2** exclusively, and four- or five-carbon tethers gave only amides **1**. It was reasoned that a combination of minimized steric repulsion and orbital alignment of the diazo cation with the migrating C–C bond in the tetrahedral intermediate determined which pathway predominated (see example given in Scheme 7.5; only two of the possible intermediate conformations are shown).

Latent ketone equivalents can also undergo Schmidt rearrangements. For example, treatment of ketal **3** with TFA at room temperature overnight, concentration, and treat-

Table 7.2 *Intramolecular Schmidt reactions of acyclic ketones and aldehydes*

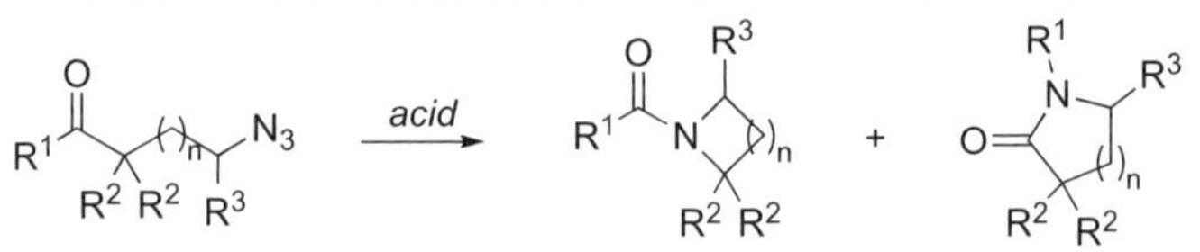

Entry	n	R^1	R^2	R^3	Acid	Yield (%)	
						1	**2**
1	0	H	H	C_6H_{13}	TFA or $TiCl_4$	–	–
2	1	H	C_7H_7	H	$TiCl_4$	–	94
3	1	CH_3	H	H	TFA or $TiCl_4$	–	–
4	2	H	C_7H_7	H	TFA	68–81	0–8
5	2	CH_3	H	H	TFA	75	–
6	2	Ph	H	H	TFA	77	–
8	3	H	C_7H_7	H	TFA	71	–

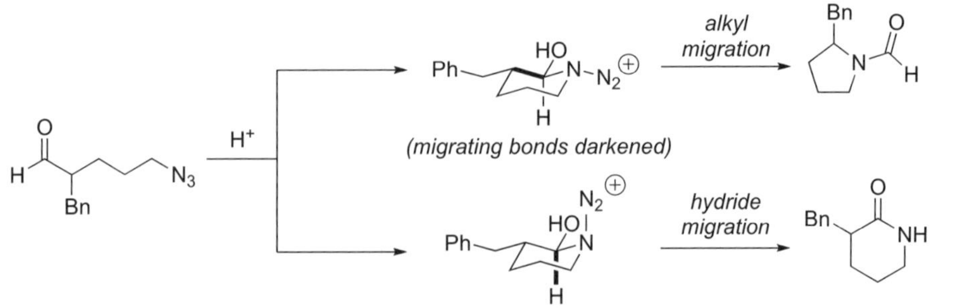

Scheme 7.5 *Rearrangement pathways of a δ-azidoaldehyde*

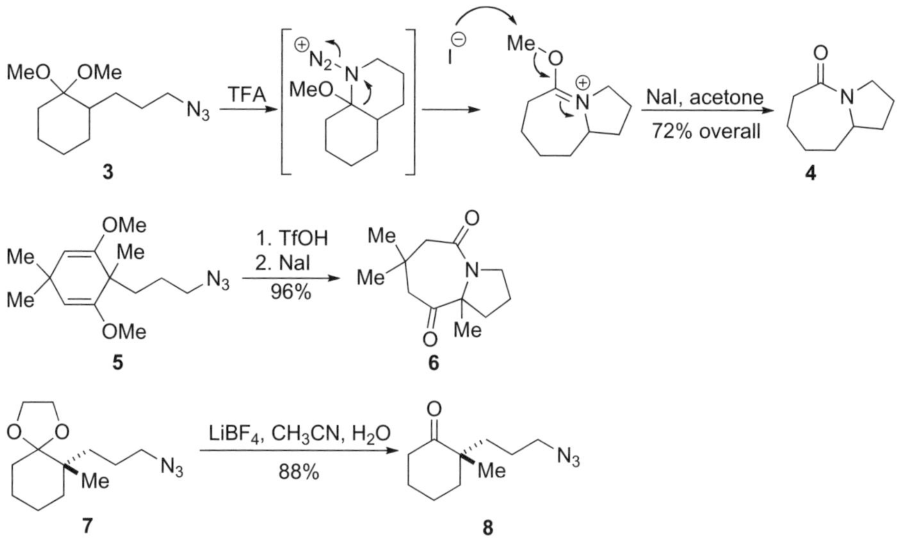

Scheme 7.6 *Two reactions and one non-reaction of some azido ketals*

ment of the resulting crude residue with NaI in acetone furnished lactam **4** in 77% yield (Scheme 7.6).[18] It was also found that enol ethers, such as the cyclohexadienone bis-enol ether **5**, could be converted into the corresponding lactams efficiently. In this case, the unreacted ketone was unmasked in the product **6**. Though in principle these processes could involve initial formation of the ketone prior to tetrahedral intermediate formation, no lactam was observed by TLC prior to the addition of NaI, suggesting that this was not the case. In some cases, it may be advantageous to deprotect a masked ketone in the presence of a tethered azido group without initiating the Schmidt reaction. This was proved possible by treatment with wet $LiBF_4$ in CH_3CN (i.e., for the conversion of **7** to **8**).

Although the intramolecular Schmidt reaction has proved to encompass considerable scope, with certain substrates side reactions predominate. In particular, benzyl azides are prone to acid-promoted rearrangement to generate an iminium intermediate, which then may be captured by an appended ketone.[19] Again, the mode of reactivity was found to depend on the tether length separating the azido and keto groups (Scheme 7.7). When four carbon atoms were present between reactive groups, Schmidt products such as **9** were routinely isolated. However, compounds with longer tethers favored the Mannich pathway, and products such as **10** were the only tractable products. A Schmidt reaction from a ketone bearing a longer tether would require relatively awkward 7- or 8-membered tetrahedral intermediates. The fact that Mannich bases form in these reactions suggests that the competing Schmidt reaction intermediate provides ample time for the phenyl group to take part in a C $\rightarrow$ N migration.

7.2.2 Intermolecular Reactions

Following the introduction of the intramolecular azido-Schmidt reaction, it was reported that potent Lewis acids such as $TiCl_4$ were found to facilitate certain intermolecular

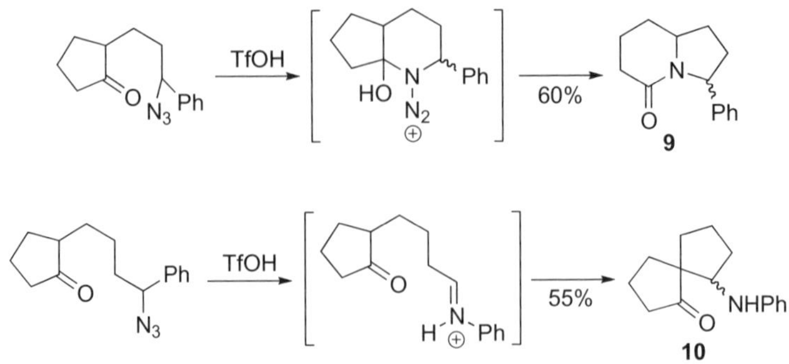

Scheme 7.7 Competition between Schmidt and Mannich pathways

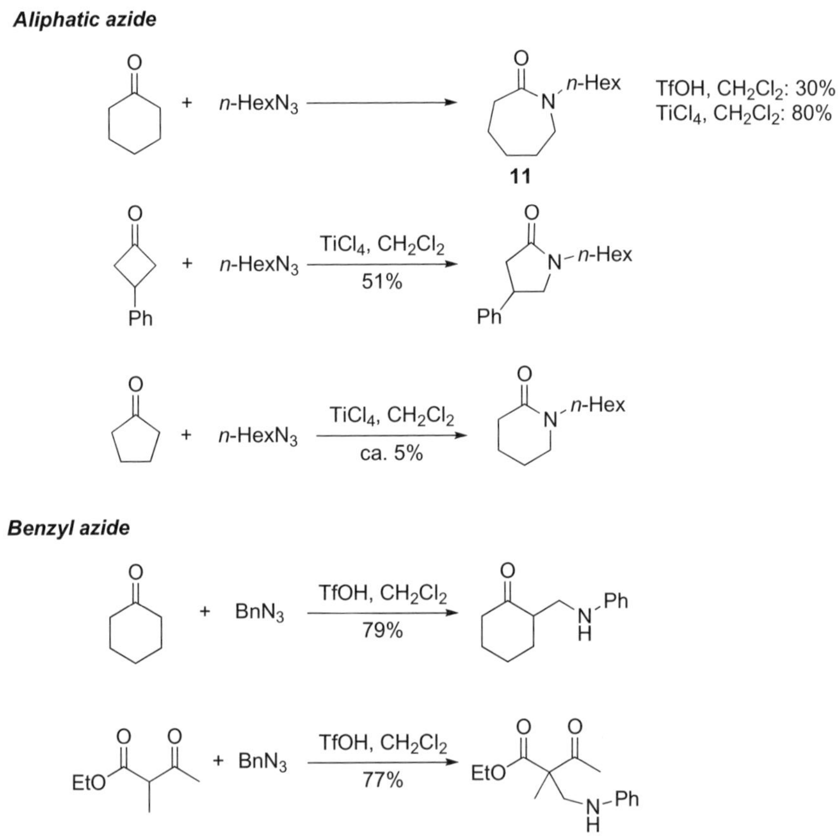

Scheme 7.8 Schmidt versus Mannich pathways in intermolecular reactions

versions of the reaction (Scheme 7.8).[20,21] However, only the strongest protic acids are useful in this context. Thus, treatment with TFA does not afford appreciable yields of compound **11** but TfOH does promote its formation in modest yield (30%).[22] The intermolecular reaction is also relatively finicky. Sterically unhindered cyclohexanones, 3-phenylcyclobutanone, and bicyclic ketones worked well, while cyclopentanone gave

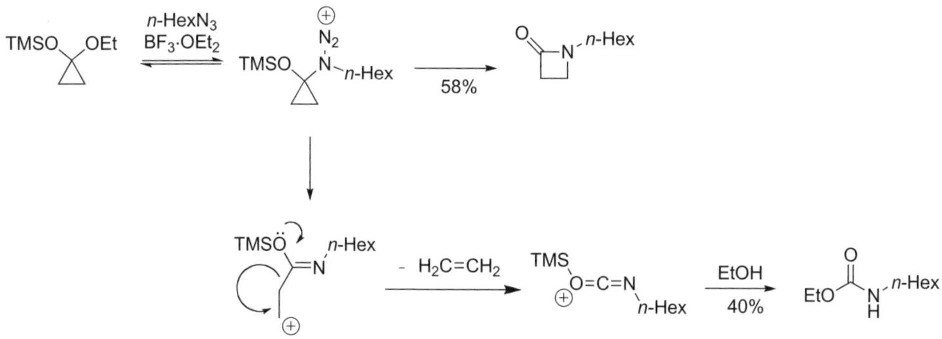

Scheme 7.9 *Reactions of cyclopropanone with alkyl azides*

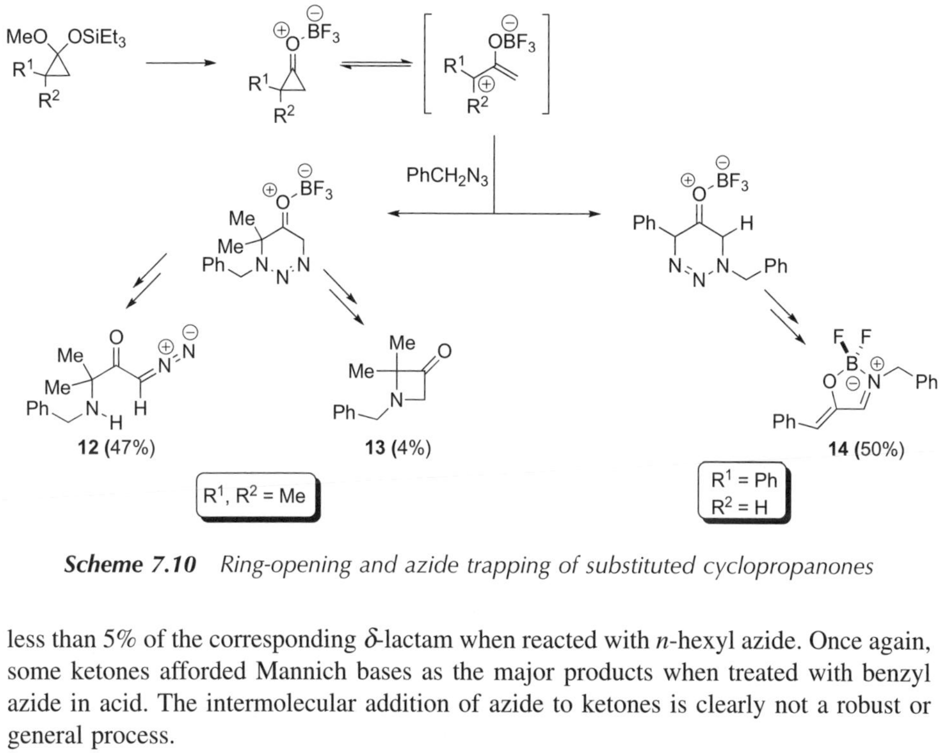

Scheme 7.10 *Ring-opening and azide trapping of substituted cyclopropanones*

less than 5% of the corresponding δ-lactam when reacted with *n*-hexyl azide. Once again, some ketones afforded Mannich bases as the major products when treated with benzyl azide in acid. The intermolecular addition of azide to ketones is clearly not a robust or general process.

The highly ring strained cyclopropanone (which is conveniently stored and used as a mixed ketal) also undergoes Schmidt chemistry to afford *N*-substituted β-lactams along with ethyl carbamates in about a 1 : 1 ratio (Scheme 7.9).[23,24] Presumably, the β-lactam is formed *via* the typical ring expansion mechanism described above. Ethyl carbamates are more mechanistically intriguing, requiring the formal loss of ethylene and N_2 followed by recombination of a silylated isocyanate with ethanol.

The chemistry of substituted cyclopropanones is even more engagingly complicated. For example, 2,2-dimethylcyclopropanone acetals were shown to react with alkyl azides in the presence of $BF_3{\cdot}OEt_2$ to afford α-amino-α'-diazomethylketones like **12** along with a small amount of 3-azetidinones **13** (Scheme 7.10). Interestingly, aryl-substituted cyclopropanones provide neither of these species but instead gave rise to [1.2.3]oxaborazole products such as **14**. It was proposed that substituted cyclopropanones undergo acid-promoted C_2–C_3 ring opening to give an oxyallyl cation (see Section 7.3 for additional examples of alkyl azides reacting with allylic and oxyallylic cations). The regiochemistry of azide addition to this intermediate along with the presence of an available proton positioned to allow elimination of N_2 determine whether α-amino-α'-diazomethylketones (no proton available) or [1.2.3]oxaborazoles (proton available for elimination) are formed.

7.2.3 Reactions of Hydroxyalkyl Azides

As shown in the previous section, intermolecular reactions of alkyl azides with ketones have limited practical utility due to their restricted substrate scope, competing side reactions, and the harsh conditions required for reaction. These deficits can be blamed on the unfavorable initial addition of azide to ketone to form the tetrahedral intermediate, which is slow in the absence of intramolecularity. To overcome these limitations, an *in situ* tethering technique was developed in which a hydroxyalkyl azide was used instead of an alkyl azide (Table 7.3).[22,25] Experimentally, this involves combining a ketone and hydroxyalkyl azide in CH_2Cl_2 in the presence of $BF_3{\cdot}OEt_2$ or $TiCl_4$ followed by basic hydrolysis (aq. KOH or $NaHCO_3$). Cyclic ketones of varying ring size were well tolerated in such reactions, as was variation in functionality and electronic character. Even some acyclic ketones such as acetone led to good yields of the *N*-hydroxyalkyl amide products. Both 2-azidoethanol and 3-azidopropanol were especially efficient partners for these reactions, while longer chain-length hydroxyalkyl azides gave lower yields (not shown).

Mechanistically, these reactions are related to the oxazoline synthesis shown in Scheme 7.3. Both involve the initial formation of an oxonium ion followed by the now-intramolecular addition of the appended azide to this cation (Scheme 7.11). In the present case, migration with loss of N_2 occurs to afford an iminium ether salt (e.g. **16**), which can be isolated by precipitation with cold THF. Treatment of the iminium salt with aqueous $NaHCO_3$ then delivers an *N*-hydroxyalkyl lactam.

Iminium ethers prepared from medium-ring or acyclic ketones can undergo the ring expansion process described above but are also eligible to afford lactones through an alternate mechanism (Scheme 7.12).[26,27] In some cases, a modest level of product control is possible by variation of reaction conditions. For example, when a strong base such as KOH (pH ca. 14) was used to hydrolyze the iminium ether derived from cycloheptanone, lactam **17** was solely formed by one of two possible mechanisms: (1) direct attack at the distal carbon of the hydroxyalkyl azide (indicated in the scheme by an asterisk) or (2) attack at the cationic center, followed by collapse of the tetrahedral intermediate. Labeling experiments suggest that pathway (2) is the more likely.[28] Alternatively, the transient formation of a tetrahedral intermediate is required for the formation of lactone **18** (control experiments ruled out direct conversion of **17** to **18** under these strongly basic conditions). *N*-Protonation of the aminal (pK_a ca. 11) was proposed to occur when aqueous $NaHCO_3$ (pH ca. 9) was used, allowing for cleavage of the corresponding C–N bond and affording **18** as the major product.

It has been shown that in addition to iminium ether hydrolysis, other heteroatom and *C*-based nucleophiles may be incorporated into the lactam products.[28,29] Through the addition of nucleophiles in a solution of DMF at ambient temperature, a range of lactams

Table 7.3 *Reactions of hydroxyalkyl azides with ketones*

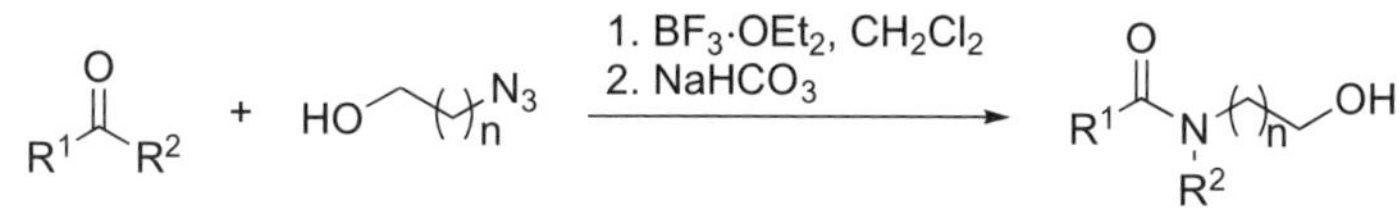

15a, *n* = *1*
15b, *n* = *2*

Entry	Ketone	Hydroxyalkyl azide	Lactam	Yield (%)
1	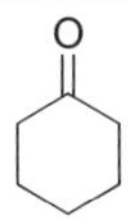	**15a**	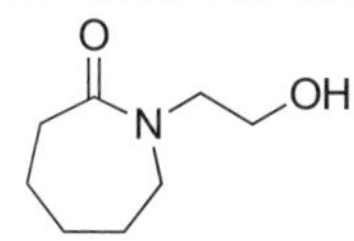	90
2	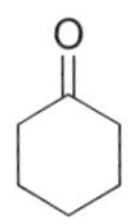	**15b**	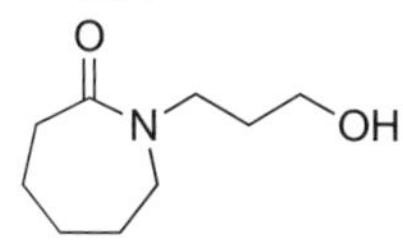	98
3	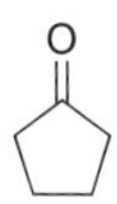	**15a**	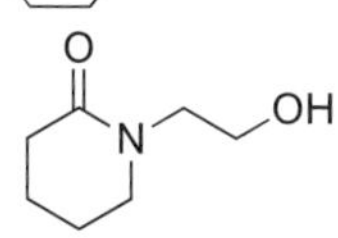	96
4	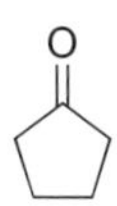	**15b**	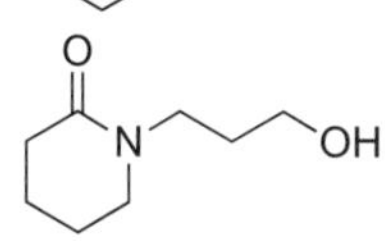	98
5	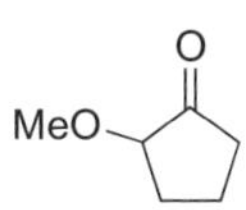	**15b**	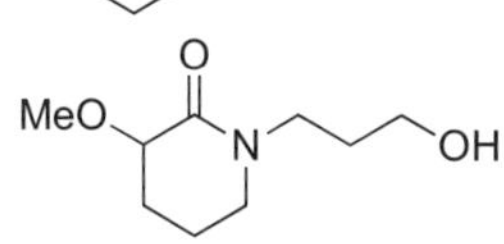	51
6	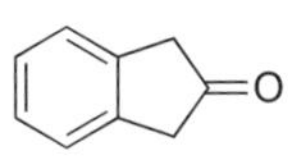	**15b**	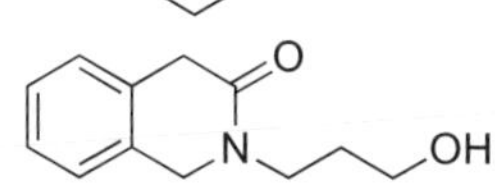	80
7	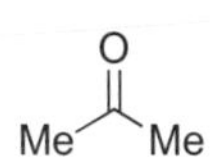	**15b**	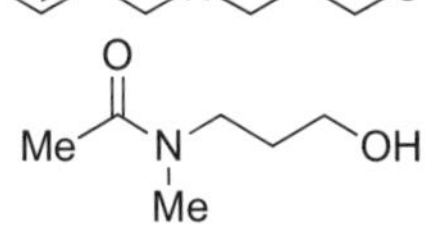	88
8	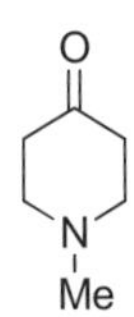	**15b**	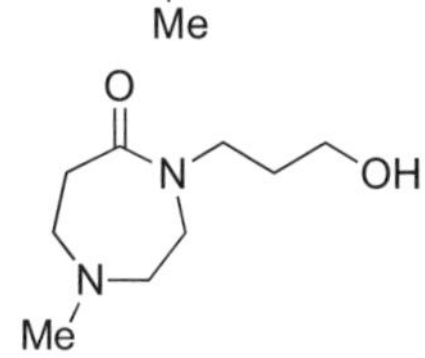	73

bearing terminal functional groups including ethers, halides, azides, and sulfides were readily prepared (Table 7.4). It was also found that mild reduction ($NaBH_4$) of iminium ethers to the tertiary amine was possible, whereas reduction of the corresponding lactams typically required $LiAlH_4$ or borane.

When unsymmetrical ketones undergo *N*-insertion reactions, two regioisomeric lactam products are possible. The regioselectivity of Lewis acid-mediated reactions of

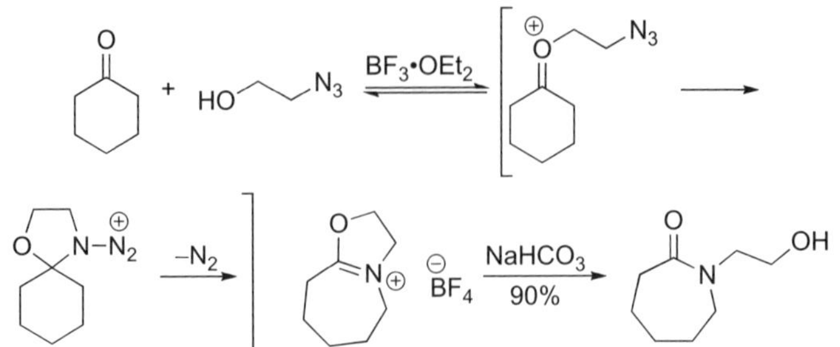

16

Scheme 7.11 Mechanistic overview of the Schmidt reaction of hydroxyalkyl azides with ketones

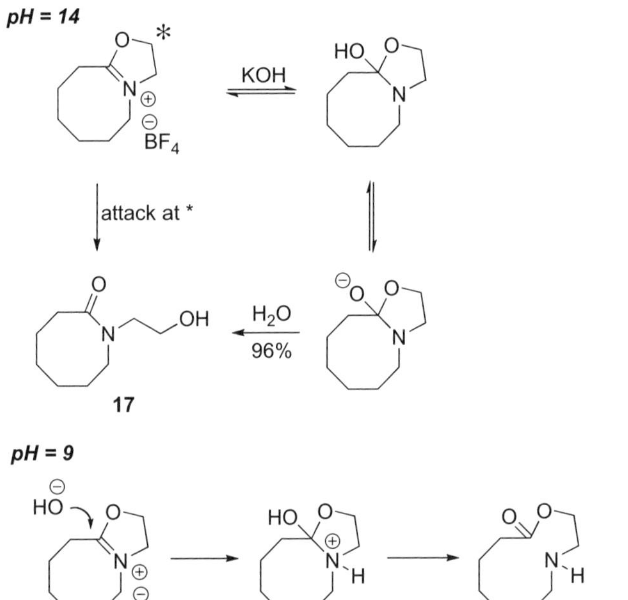

43% **18** (+ 27% **17**)

Scheme 7.12 *Effect of pH on mode of iminium ether hydrolysis*

hydroxyalkyl azides with α-substituted ketones was examined.[30] It was found that when cycloalkanones bearing α-methyl or ethyl substituents were treated with 1,2-azidoethanol or 1,3-azidopropanol in the presence of $BF_3{\cdot}OEt_2$, the less substituted carbon migrated preferentially (Scheme 7.13). When appended with bulkier alkyl groups at the α-position, migration of the more heavily substituted carbon was favored. Finally, when substituted with inductively electron-withdrawing groups (e.g. OMe, Ph), migration of the less-substituted group occurred.

When 4-substituted cyclohexanones such as 4-*tert*-butylcyclohexanone are reacted with chiral, enantioenriched 1,2- or 1,3-hydroxyalkyl azides, asymmetric ring expansion occurs

Table 7.4 *Nucleophilic additions to iminium ethers*

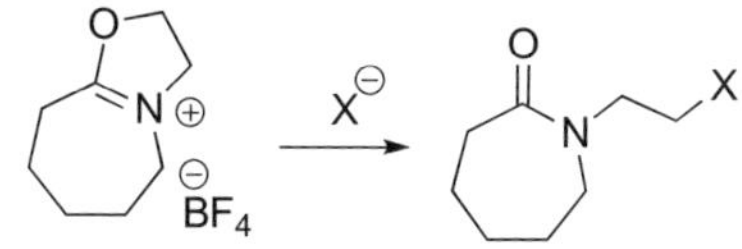

Entry	Nucleophile	X	Yield (%)
1	$(n\text{-Bu})_4NI$	I	55
2	NaCN	CN	82
3	NaSPh	SPh	95
4	NaN_3	N_3	85

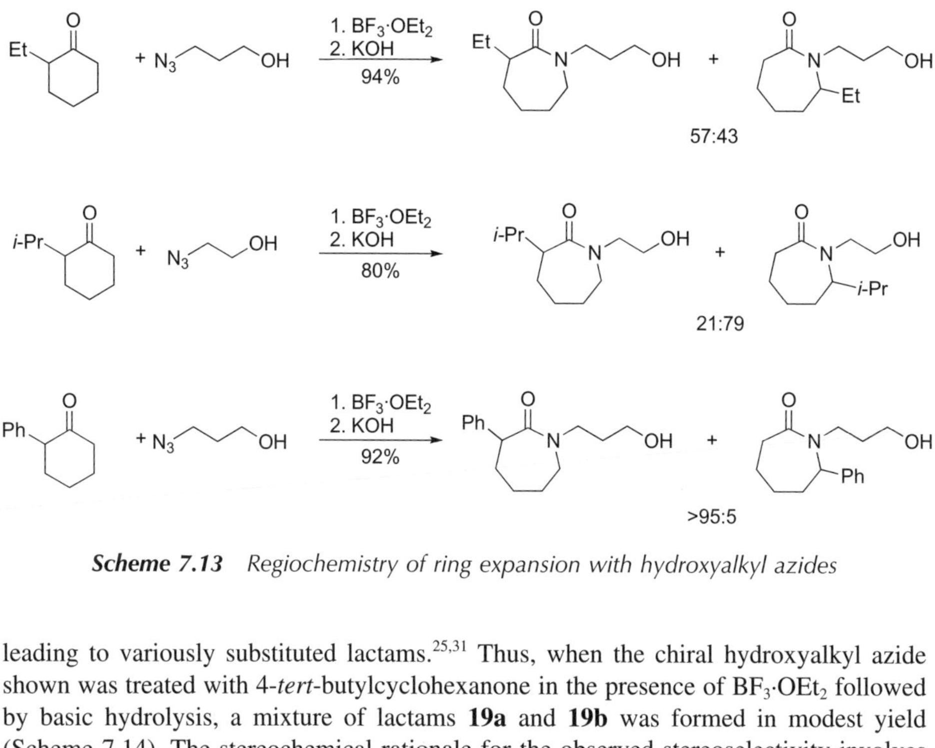

Scheme 7.13 *Regiochemistry of ring expansion with hydroxyalkyl azides*

leading to variously substituted lactams.[25,31] Thus, when the chiral hydroxyalkyl azide shown was treated with 4-*tert*-butylcyclohexanone in the presence of $BF_3{\cdot}OEt_2$ followed by basic hydrolysis, a mixture of lactams **19a** and **19b** was formed in modest yield (Scheme 7.14). The stereochemical rationale for the observed stereoselectivity involves equatorial addition of azide to the oxonium ion (supported by ab initio calculations).[32] Rapid inversion allows the diazonium ion to align itself with either bond for the migration event. Recent computational studies along with a growing body of experimental evidence suggest that attractive cation–π interactions between the diazonium cation and the phenyl group may influence the stereoselectivity of these reactions (see Section 7.9 for another application of this effect).[33]

When (*R*)-3-azido-1-phenylpropan-1-ol was reacted with 4-*tert*-butylcyclohexanone ($BF_3{\cdot}OEt_2$ then KOH), diastereomeric lactams were formed quantitatively in a 95 : 5 ratio

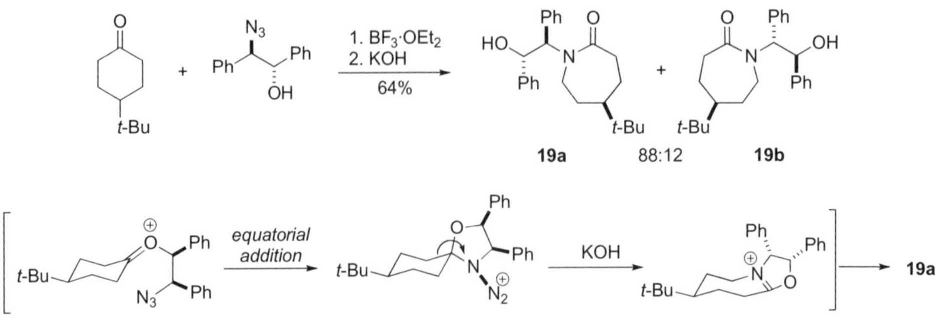

Scheme 7.14 *Asymmetric Schmidt reaction*

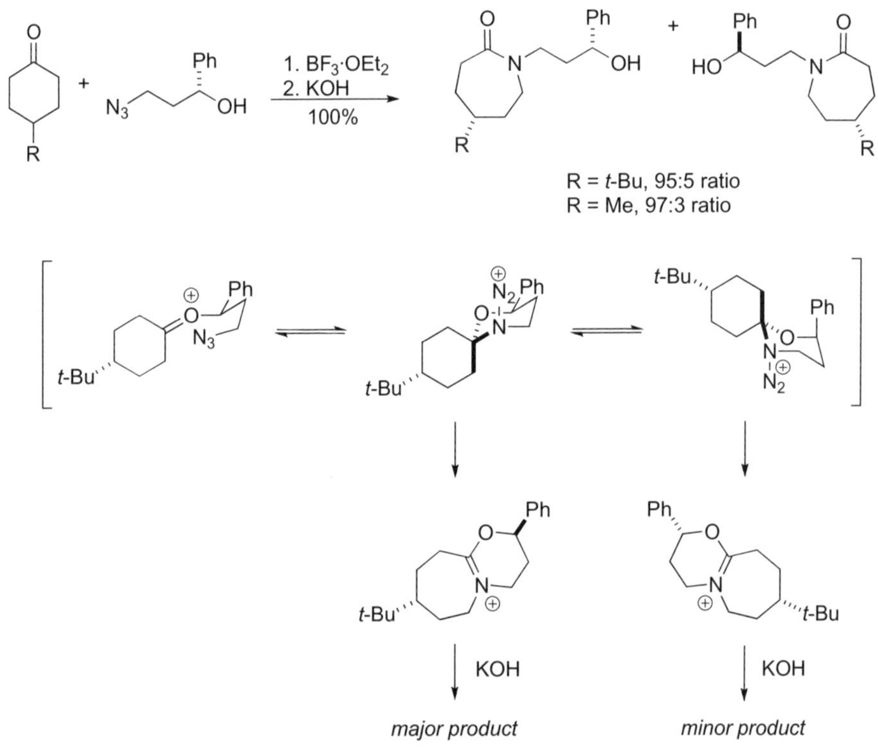

Scheme 7.15 *Asymmetric Schmidt reactions with a monosubstituted hydroxyalkyl azide*

(Scheme 7.15).[31,34,35] Similarly high diastereoselectivity was also observed for 4-methylcyclohexanone and a variety of substitution patterns were tolerated about the ketone. The key stereochemical feature at work in these reactions is the conformation of the azide-containing tether. Minimization of diaxial interaction in the two chairlike intermediates probably accounts for the high diastereoselectivity of these processes (favoring

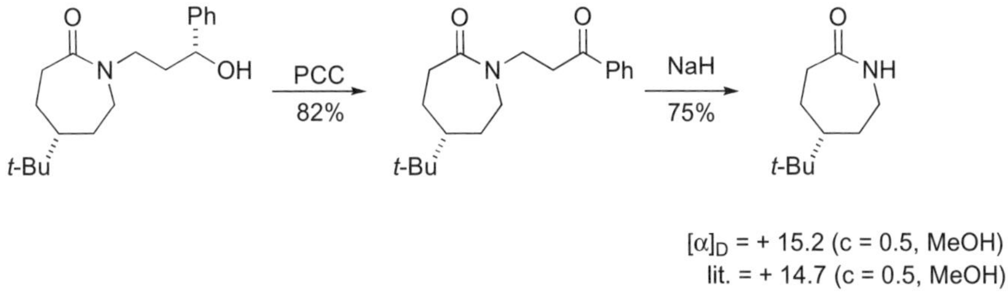

Scheme 7.16 *Removal of chiral tether*

the intermediate containing an equatorial phenyl group).[36] Enantioenriched unsubstituted lactams were procured through oxidation of the alcohol followed by a base-assisted retro-Michael addition (Scheme 7.16).

The diastereoselectivities of asymmetric Schmidt reactions using certain aryl-containing hydroxyalkyl azides could not be explained on the basis of minimized steric interactions alone. It was proposed that their stereochemical course was affected by attractive non-bonded interactions of an aromatic group with a diazonium cation in the reaction intermediates.[33,37] In general, the reaction stereoselectivity of these sequences is controlled by the equilibrium between **A** and **B** (Scheme 7.17). The observed stereoselectivity was attributed to (1) equatorial addition of azide to a common oxonium ion to form both **A** and **B**, (2) rearrangement through migration of a methylene group antiperiplanar to the leaving N_2^+ group, and (3) comparable rates of reaction of **A** and **B** (meaning that the observed stereoselectivity depends only on the relative stability of these intermediates).

When azide **20** was reacted with 4-*tert*-butylcyclohexanone under the usual conditions, lactams **21a** and **21b** were formed in a 64:36 ratio (Scheme 7.18). This was initially surprising because the analogous *i*-Pr and Me-derivatives were much more selective, even though the A value for phenyl is larger than that for either *i*-Pr or Me. The relatively poor stereoselectivity was attributed to the stabilization of the intermediate leading to the minor product **21b** by a cation–π interaction. When azide **22** was used, the stereoselectivity was reversed and much higher. In this case, the steric interaction of the methyl group with an ortho-proton on the phenyl group increases the attractive cation–π interactions by additionally favoring a conformer in which the phenyl group's π system is directed toward the N_2^+ leaving group. These ideas were supported by structure–selectivity relationship studies and ab initio calculations.

It was also found that a heteroatom-containing hydroxyalkyl azide may also exert an electrostatic effect on the stereochemical course of azide-mediated ring expansions.[38] When (*R*)-3-azido-2-methoxypropan-1-ol was reacted with 4-*tert*-butylcyclohexanone under Lewis acid promoted ring expansion conditions, a highly stereoselective reaction occurred to provide a 7:93 mixture of lactams **24a** and **24b**, respectively (Scheme 7.19). In this case, the intermediate containing the methoxy group in a *cis*-1,3-diaxial orientation vis-à-vis the positively charged leaving group correlated with the major product of the reaction.

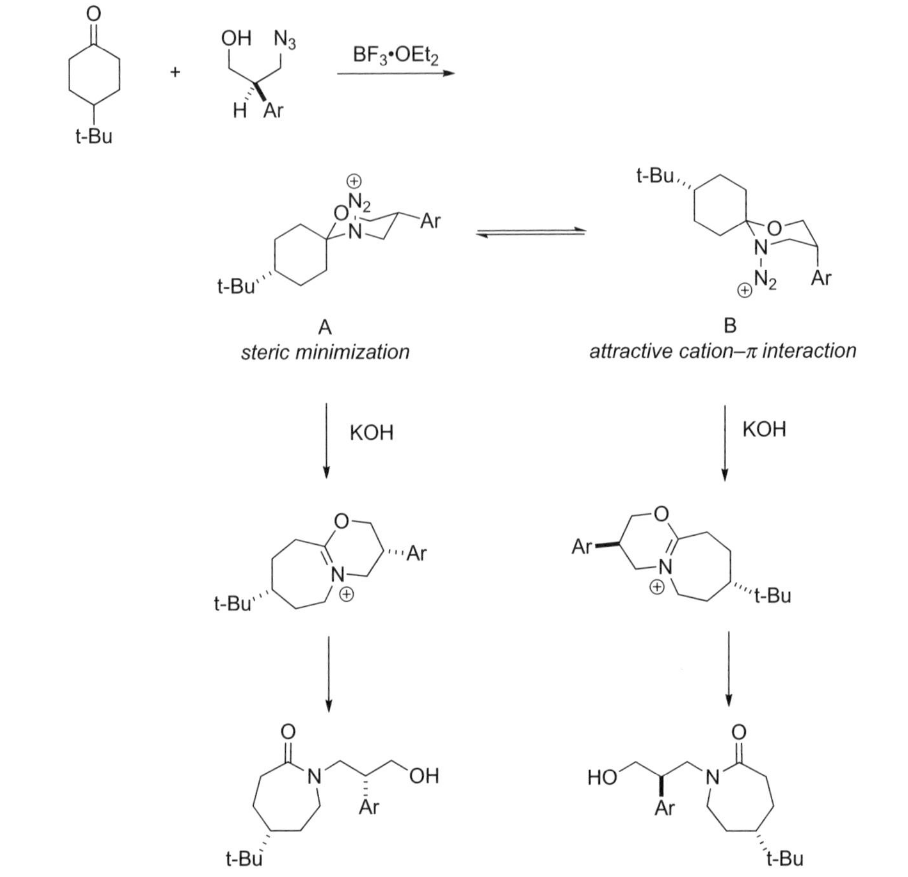

Scheme 7.17 *Steric interactions vs. cation–π interactions in azide-mediated ring expansions*

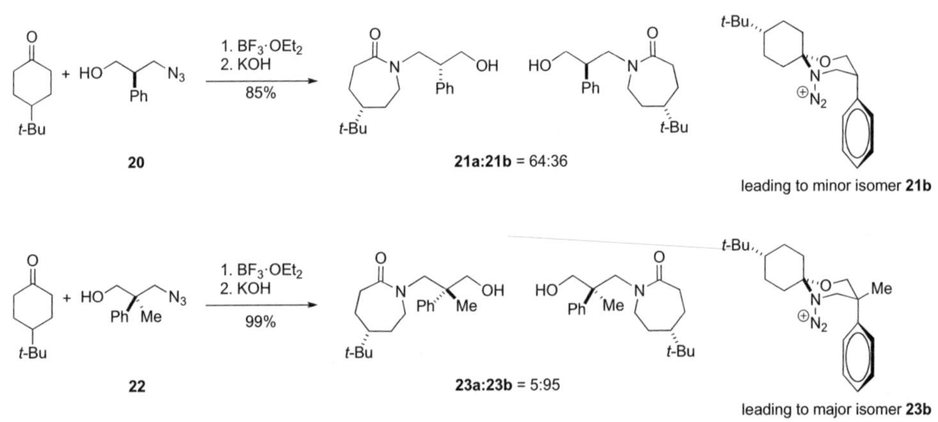

Scheme 7.18 *Role of cation–π interactions in asymmetric Schmidt reactions*

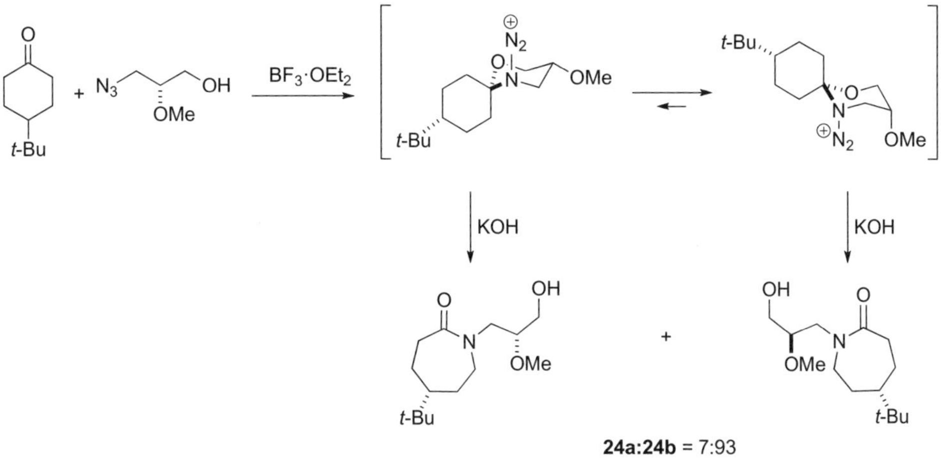

Scheme 7.19 *Cation–n attractive interactions leading to highly diastereoselective ring expansion*

7.3 Schmidt Reactions of Alkyl Azides with Carbocations

The introduction of Schmidt reactions of cations derived from alcohols or olefins with azides was contemporary to that of the carbonyl chemistry described in the previous section. The dean of the cation/azide school of Schmidt chemistry was William Pearson, then working at the University of Michigan. In 1992, the Pearson group disclosed that the treatment of azido alkene **25** with trifluoromethanesulfonic acid (TfOH) and then base afforded a bicyclic enamine in 71% yield (Scheme 7.20).[39,40] Under these strongly acidic conditions, protonation of the alkene gave a cation that was then attacked by the proximal nitrogen of the appended azide. This was followed by rearrangement of the resulting aminodiazonium ion and elimination to deliver neutral product.

A number of azido-tethered alkenes and azido-tethered tertiary alcohols similarly generate rearrangement products when treated with TfOH or $SnCl_4$ followed by a basic or reductive workup.[41] Enamines are obtained when a basic workup is employed whereas the addition of a reducing agent provides tertiary amines (Scheme 7.21); in many cases,

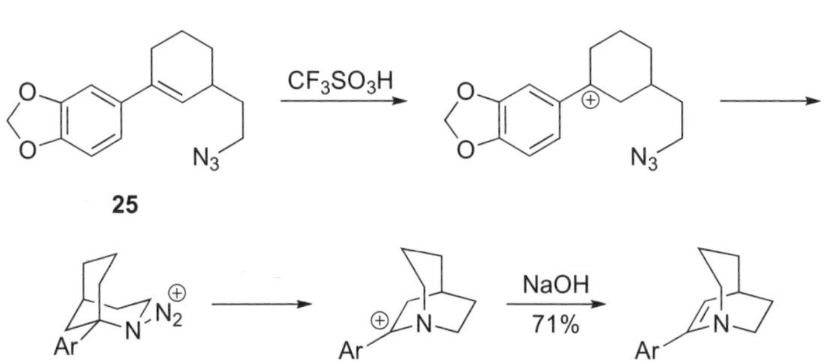

Scheme 7.20 *Intramolecular Schmidt reaction of a cation derived from an alkene*

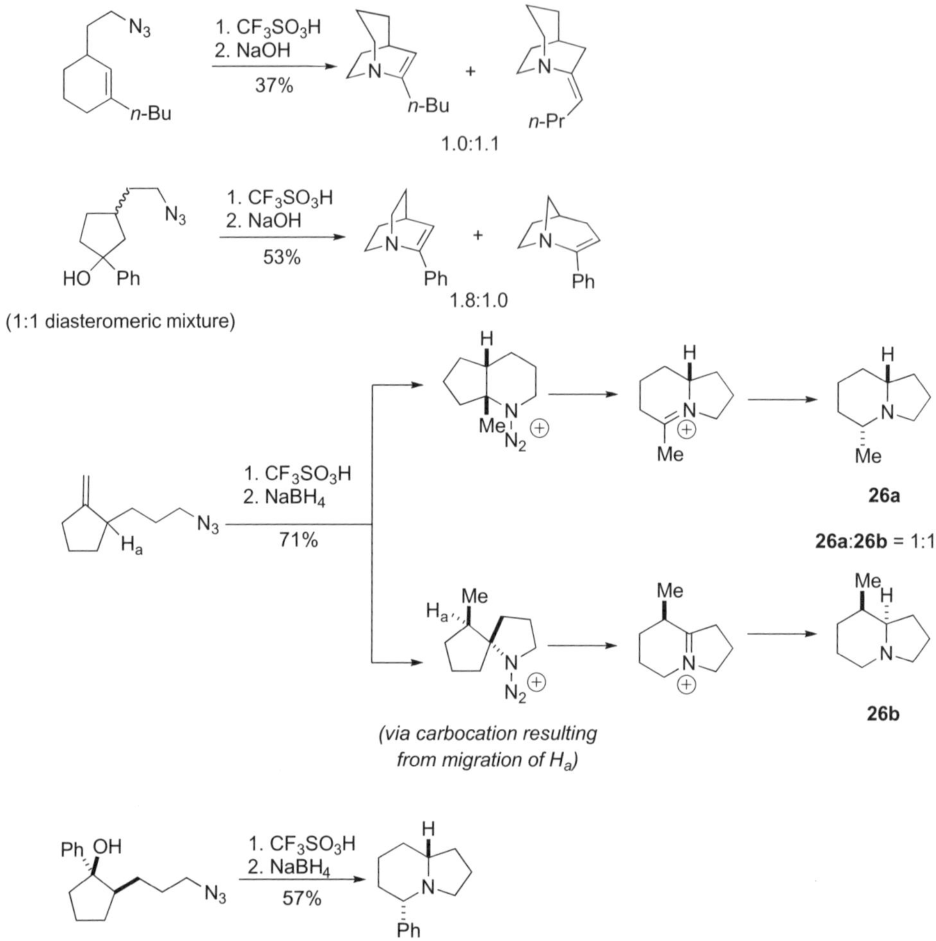

Scheme 7.21 *Acid promoted rearrangement of azido alkenes and azido alcohols. Only a few of the various possible routes to* ***26a*** *and* ***26b*** *are shown; the original paper should be consulted for a full mechanistic description*[41]

mixtures of products were observed. Several interesting observations were noted in these studies including rapid rearrangement of the initially formed carbocation to another more stable cation prior to capture by the pendant azide, which was established through deuterium labeling studies (and accounts for the formation of **26b**). The origin of the regioselectivity in these processes is not immediately obvious. It is believed that, like their carbonyl counterparts, these processes involve a diazonium ion intermediate that can in some cases attain antiperiplanar alignment to more than one bond. Computational and experimental evidence suggests that three factors determine which bond migrates: (1) configuration of the diazonium group in the reactive conformer, (2) the inherent migratory aptitude of the migrating group, and (3) the stability of the newly developed positive charge at the origin of the migrating carbon.

Molina and coworkers likewise reported that treatment of 4-aryl-3-butenyl azides with TfOH yielded 1-pyrrolines in modest yields (Scheme 7.22).[42] These processes surely involve initial olefin protonation, subsequent azide addition to the resulting cation, and finally hydride migration (or proton elimination). Hydride migration was exclusively favored over the migration of other possible groups. Additionally, a rough correlation between the electronic stabilization of the cation and yield was observed.

Pearson and coworkers also developed intermolecular reactions of azides with carbocations. Thus, benzylic or tertiary alcohols, when treated with alkyl azides in the presence of $SnCl_4$ or TfOH followed by $NaBH_4$-mediated reduction, provided tertiary or secondary amines.[43] Both cyclic and acyclic alcohols were compatible with these reactions, though mixtures of constitutionally isomeric products often resulted (Scheme 7.23).

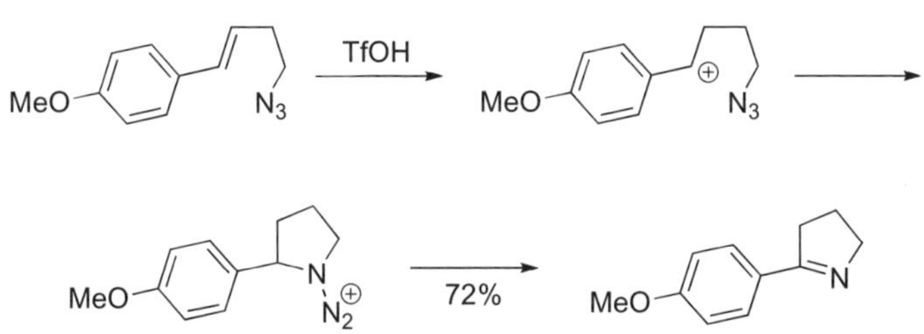

Scheme 7.22 *Intermolecular Schmidt reaction leading to a 1-pyrroline*

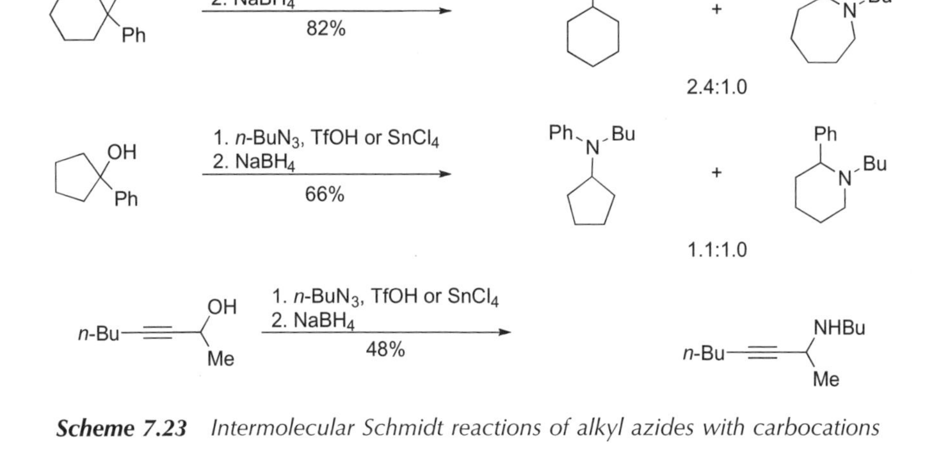

Scheme 7.23 *Intermolecular Schmidt reactions of alkyl azides with carbocations*

It is tempting to think that allylic cations would behave similarly to other stabilized cations in their reactions with alkyl azides.[44,45] In practice what happens is invariably an initial formal [3+3] cycloaddition of azide to the allylic cation, which is followed by a migration event (typically of hydride) or trapping by a nucleophile. Pearson and coworkers found that when sulfonylindole **27** was treated with $SnCl_4$ at −78 °C, followed by basic workup, triazoline adducts **28** were obtained as mixture of chloride epimers (Scheme 7.24).[46] When *N*-alkyl indoles (e.g. **29**) were subjected to Lewis acidic conditions, triazines such as **30** were obtained as the sole products. In the former case involving an *N*-sulfonyl indole, a [3+2] cycloaddition pathway explains the product, whereas the *N*-alkyl indoles examined underwent [3+3] cycloaddition.

West and coworkers showed $BF_3{\cdot}OEt_2$ treatment of certain azide containing 1,4-dien-3-ones at −78 °C followed by warming to 0 °C and subsequent exposure to air, formed peroxy-bridged indolizidinones along with traces of dihydropyridones (Scheme 7.25).[47] To account for these products, these workers suggested an interrupted Nazarov reaction mechanism wherein azide adds to an oxyallyl cation in a [3+3] cycloaddition, followed by loss of N_2 to generate a tetrahydrotriazine intermediate (a stepwise mechanism, not shown, is also possible). The Nazarov reaction is a conrotatory cyclization leading to a five-membered ring. Loss of N_2 affords a zwitterionic intermediate that reacts with $^3O_2$ to form peroxyindolizines **32a,b** as a mixture of diastereomers. A small amount of dihydropyridone **31** is ascribed to a 1,5-H shift of the same zwitterionic intermediate.

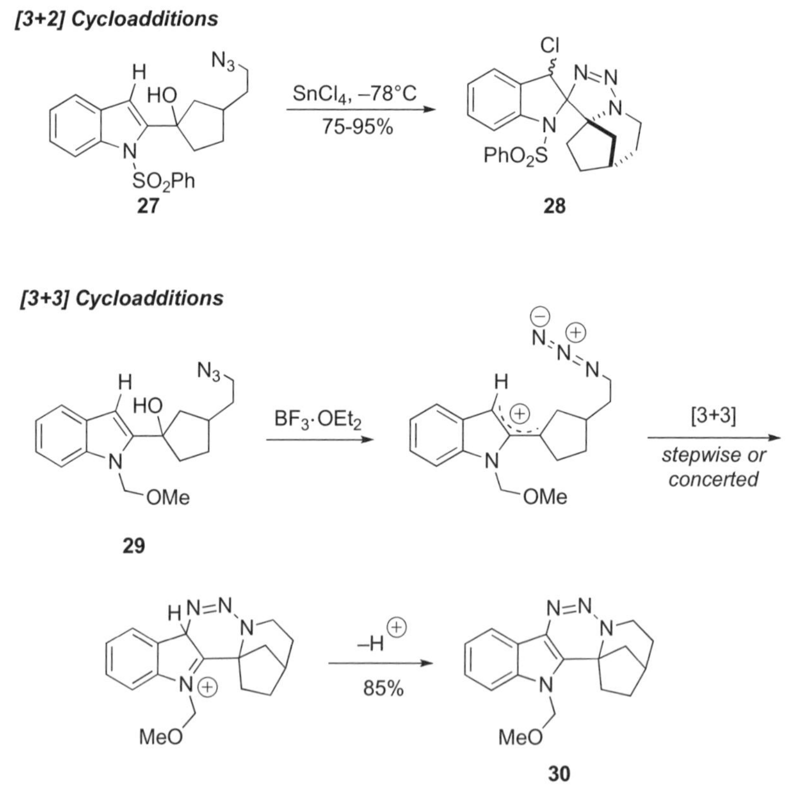

Scheme 7.24 *Intramolecular cycloadditions of azido-tethered allylic cations*

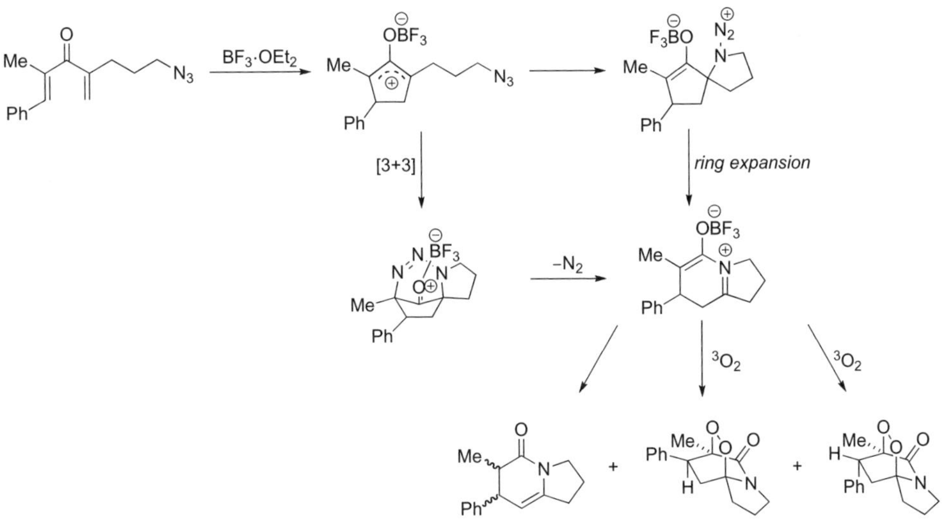

31 32a 32b
72% (1.0:5.0:5.0)

Scheme 7.25 *Intramolecular reactions of azides with oxyallyl cations*

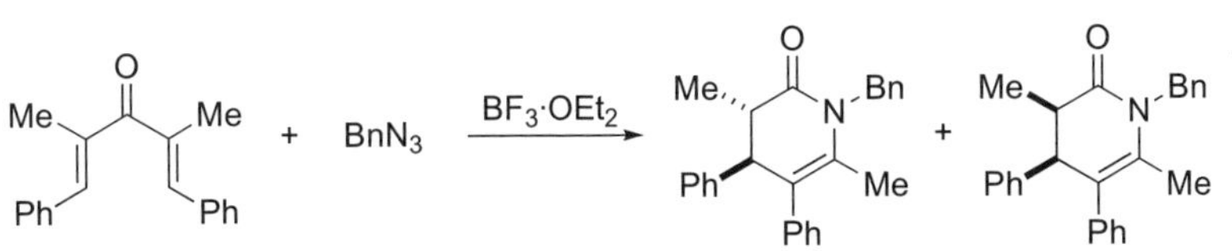

85%, 2:1

Scheme 7.26 *Intermolecular azide trapping of a dienone*

Following their initial discovery of interrupted Nazarov reactions, West and coworkers reported that intermolecular azide trapping of oxyallyl cation intermediates was possible, delivering dihyropyridones in good yields with modest to high diastereoselectivity (Scheme 7.26).[48] Although suprafacial 1,5-H transfer is consistent with the formation of the *trans*-products, the formation of substantial amounts of *cis*-isomers may be explained by stereoselective protonation of the boron enolate intermediate (not shown). The lack of molecular oxygen incorporation in these examples may be due to increased rate of proton transfer compared to the more rigid bicyclic systems when tethered azido dienones were used.

7.4 Metal-mediated Schmidt Reactions of Alkyl Azides with Alkenes and Alkynes

The lack of regioselectivity, the limitation that tertiary or benzylic alcohols must be used as precursors to Schmidt reactions with carbocations, and the harsh Lewis acidic conditions required for initiation prompted Pearson and coworkers to explore an alternative strategy for effecting such reactions. In this regard, they reported that mercury trifluoromethanesulfonate [$Hg(OTf)_2$] can act as a facilitator of intramolecular Schmidt reactions of azido alkenes.[49] The advantages of this method of activation are much improved regioselectivity, a greater range of substitution pattern, and the tolerance of acid-sensitive functionality in the target molecules (Table 7.5). The likely mechanism for these processes involves initial coordination of Hg^{2+} to the alkene (Scheme 7.27). Attack of the tethered azide on to the resulting mercuronium ion, followed by migration (emboldened bond) produces an iminium ion intermediate which persists until reduction, at which point it leads to the amine products.

Table 7.5 *$Hg(OTf)_2$-promoted Schmidt reactions with $NaBH_4$ workup*

Entry	Azide	Product	Yield (%)
1	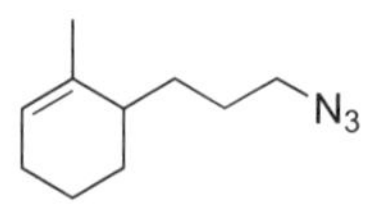	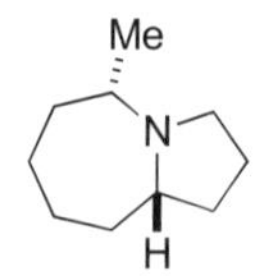	31
2	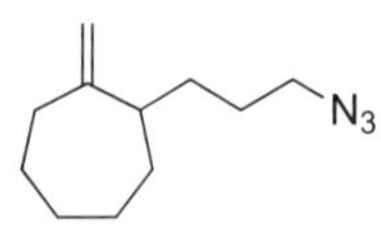	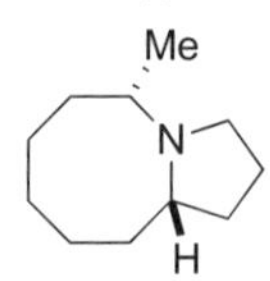	31
3		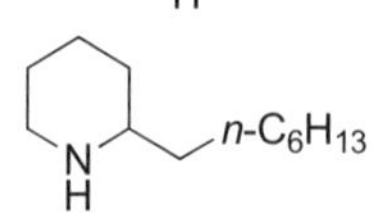	73
4	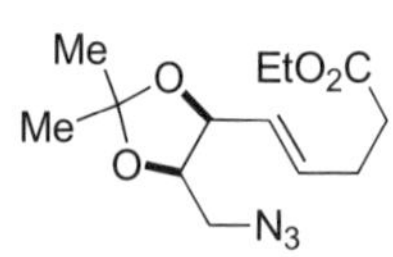	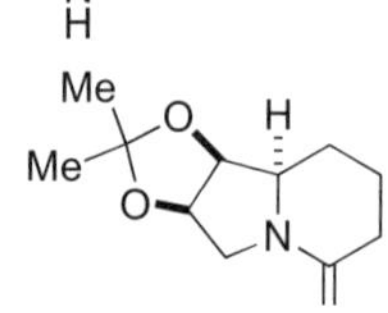	34

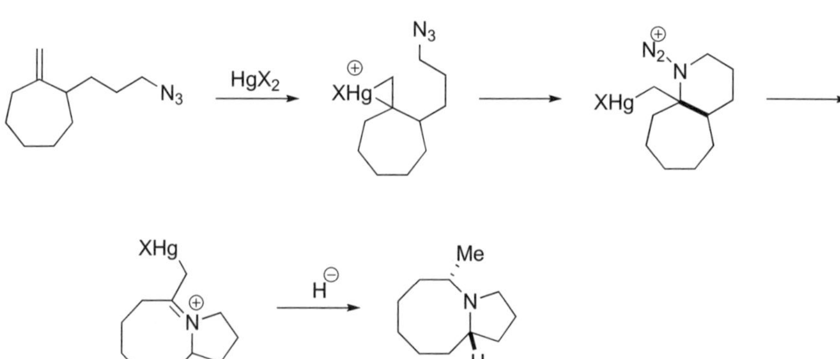

Scheme 7.27 *Mechanism of Hg(II)-promoted Schmidt reactions*

Toste and coworkers have shown that intramolecular Schmidt reactions of acetylenic azides were possible through Au(I) catalysis.[50] Thus, treatment of a series of acetylenic azides with (dppm)Au_2Cl_2 provides pyrroles in good yields (Table 7.6). These reactions were tolerant of varying substitution and chemoselective: when a 1,5-enyne was present, the desired cyclization still proceeded, albeit with reduced efficiency (entry 5).

The proposed mechanism involves initial Au(I) activation of the alkyne followed by nucleophilic addition to the alkyne by azide (Scheme 7.28). This is followed by loss of N_2 to furnish a cationic intermediate, which is stabilized by electron donation from Au(I). A formal 1,2-hydrogen shift regenerates the Au(I) catalyst and the pyrrole after tautomerization. A nitrene intermediate was deemed unlikely because Au(I) did not decompose azides that were not homopropargylic. Notably, Au(I) effectively acts as both a π-acid and electron donor; an appropriate ligand environment is required to tune the electronic character of Au(I) for reaction success.

Table 7.6 *Au(I)-promoted acetylenic Schmidt reactions*

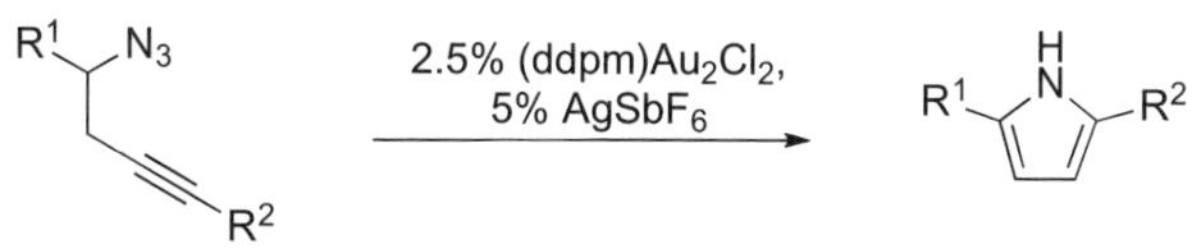

Entry	Azide	Pyrrole	Yield (%)
1	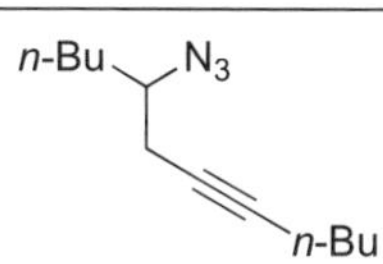	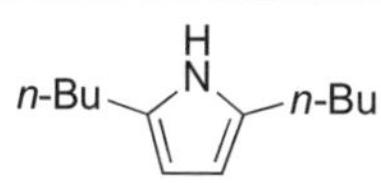	82
2	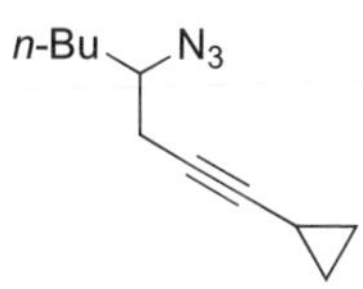	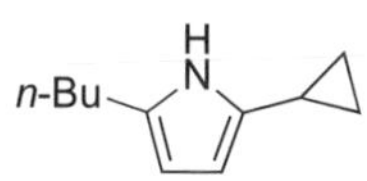	78
3	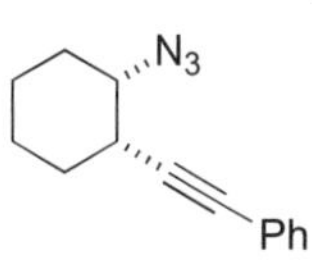	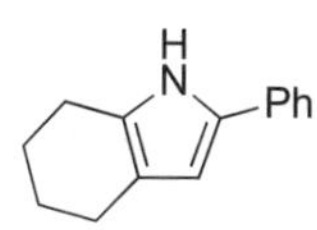	73
4	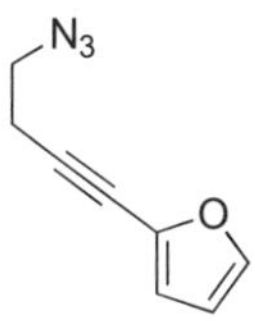	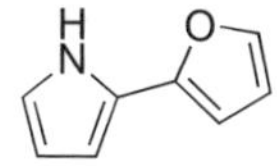	61
5	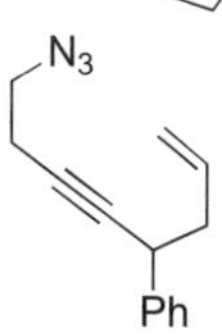	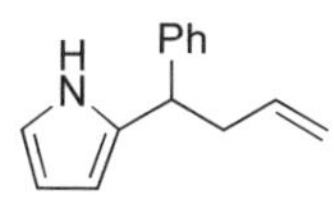	41

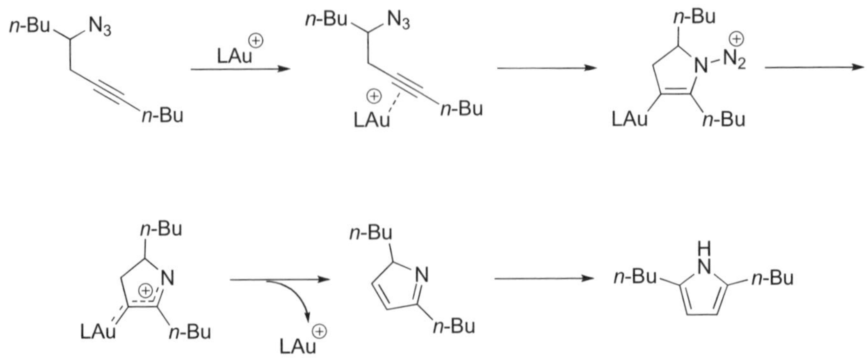

Scheme 7.28 Mechanism of Au(I)-catalyzed Schmidt reactions

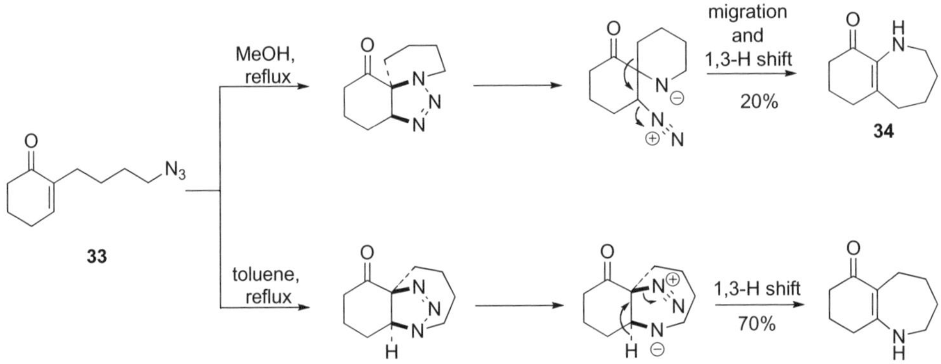

Scheme 7.29 Mechanisms proposed to account for the formation of enaminones ***34*** *and* ***35***

7.5 Reactions of Alkyl Azides with α,β-Unsaturated Ketones

Unlike their saturated counterparts, enones do not generally undergo simple ring expansion chemistry. Instead, they react with azides to undergo Lewis acid-assisted 1,3-dipolar cycloaddition reactions followed by either ring contraction or aziridine formation. This kind of transmogrification was first discovered in a thermal setting (Scheme 7.29). Thus, Sha and coworkers found that when azido-tethered enone **33** was heated in refluxing MeOH, enamine **34** was formed.[51] Apparently, initial 1,3-dipolar cycloaddition provides an unstable triazoline, which fragments, undergoes ring adjustment to afford an intermediate that finally undergoes a 1,3-H shift. When the solvent was switched to toluene and **33** was heated to reflux, enaminone **35** was isolated in good yield instead of the isomeric **34**. Which pathway is operative is determined by the regiochemistry of the initial 1,3-dipolar cycloaddition event and this in turn is affected by solvent.

Table 7.7 *Reactions of benzyl azide with enones in the presence of TMSOTf*

Entry	Enone	Products (yield %)
1	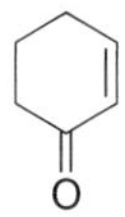	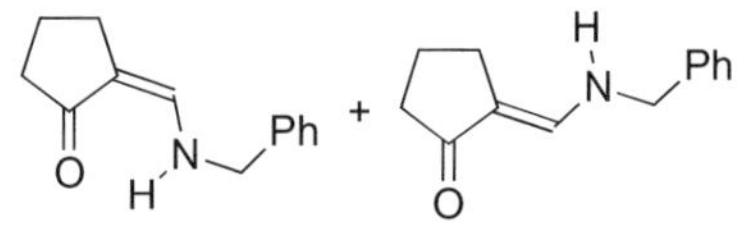 (3:2, 93%)
2	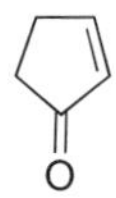	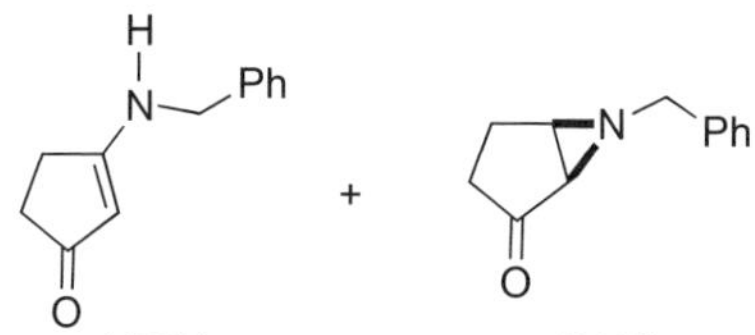 (48%) (33%)
3	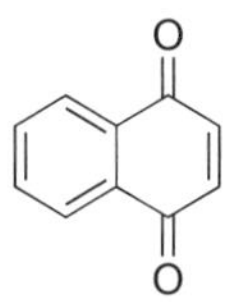	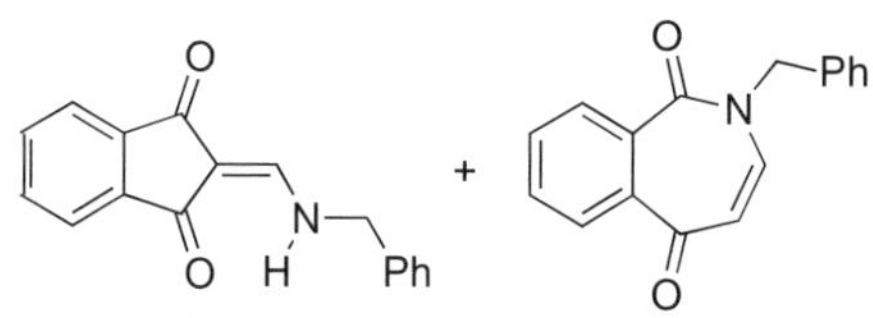 (49%) (14%)
4	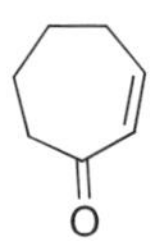	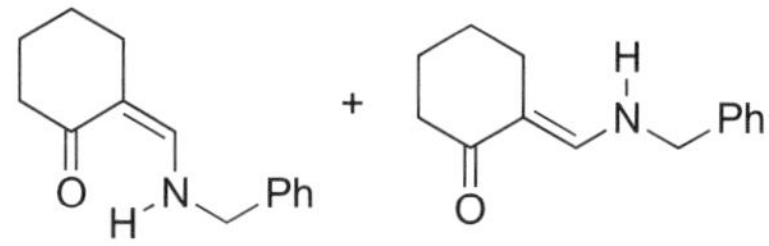 (9:1, 78%)
5	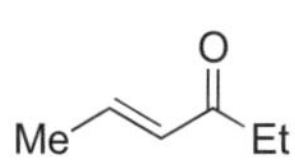	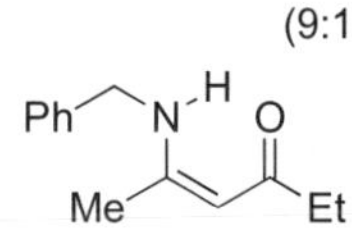 (49%)

Alkyl azides react intermolecularly with 2-cyclohexenones in the presence of TMSOTf to provide exocyclic enaminone products through ring-contraction (Table 7.7, entry 1).[52] On the other hand, 2-cyclopentenone did not undergo ring contraction but provided a mixture of endocyclic enaminone and aziridine instead. This was explained by considering the substantial ring strain that would develop on going from a 5-membered to a 4-membered ring. Of all of the enone partners surveyed, only one provided any ring-expanded lactam and even then only as the minor product (entry 3).

Mechanistically, these processes involve Lewis-acid activation of the enone, subsequent 1,3-dipolar cycloaddition of the enone with azide, and ring opening of the unstable triazoline (Scheme 7.30). In the case of exocyclic enaminone formation, antiperiplanar arrangement of the methylene group (path a) with the diazonium ion facilitated ring contraction; a 1,3-H shift ultimately provided product *via* path a. On the other hand, migration of an axially oriented R group led to the observed endocyclic enaminones (path b).

Johnston and coworkers have explored other dimensions of azide/enone reactivity. Thus, aziridines were obtained in moderate to excellent yields when acyclic enones were

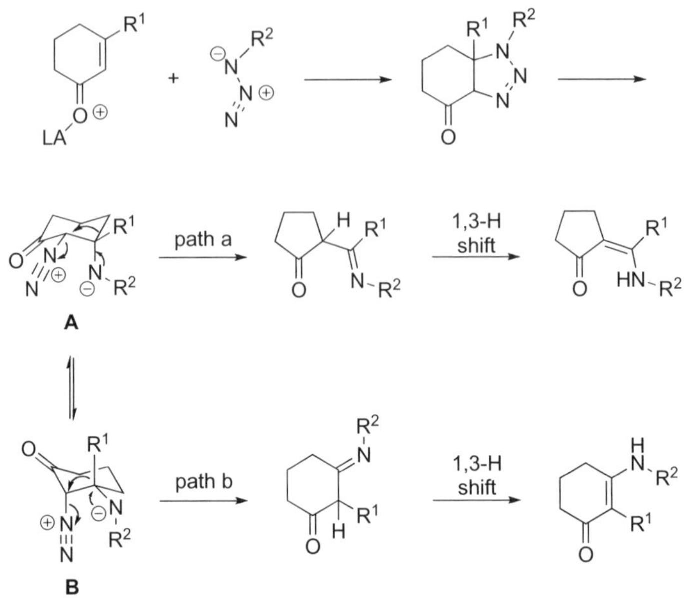

Scheme 7.30 Proposed mechanisms to account for the formation of enaminones

treated with alkyl azides in the presence of TfOH at 0 °C (Scheme 7.31).[53] When such reactions were performed with an unsaturated *N*-acrylol-carbamate, formal *anti*-aminohydroxylation products were isolated. An attractive feature of these reactions is that they were highly diastereoselective. The isolation of a triazoline and resubmission to the reaction conditions did not lead to the expected oxazolidinone products. This surprising result, along with substantial enone substituent effects in crossover experiments, led the authors to suggest that a concerted mechanism might instead account for product formation (not shown).

7.6 Reactions of Alkyl Azides with Epoxides

Epoxides are another viable electrophilic partner for alkyl azides. Baskaran and coworkers found that azido-tethered epoxides afforded bicyclic, tertiary amines bearing a hydroxymethyl substituent upon treatment with a Lewis acid followed by reduction.[54] This one-pot procedure was applicable to bicyclic azido epoxides with varying ring sizes, affording the corresponding bicyclic amines in preparatively useful yields (Table 7.8). These reactions probably involve the initial Lewis acid-assisted formation of a cation followed by azide attack. Migration and concomitant loss of N_2 then generates an iminium ion, which persists until stereoselective reduction by hydride (Scheme 7.32).

Murphy and coworkers have also reported rearrangement reactions of azido-tethered epoxides in which the azide tether was attached directly to a phenyl-substituted epoxide.[55] Thus, when they treated epoxide **36** with $SnCl_4$ at 0 °C in THF, ketone **37** was isolated

Acid-promoted aziridinations

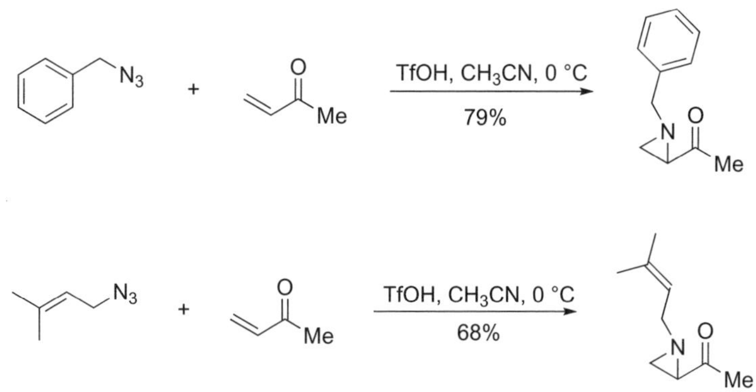

Formal anti-aminohydroxylations

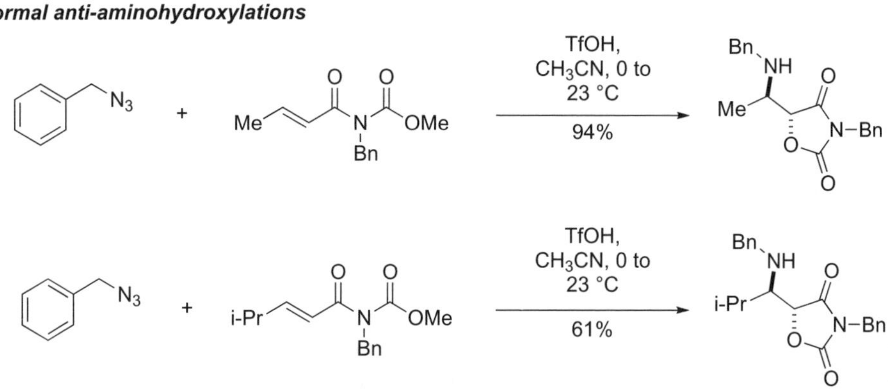

Scheme 7.31 *Acid-promoted aziridination and formal aminohydroxylation of electron-deficient olefins*

Table 7.8 *Reactions of alkyl azides with epoxides in the presence of $EtAlCl_2$ followed by $NaBH_4$ workup*

Entry	Epoxide	Product	Yield (%)
1	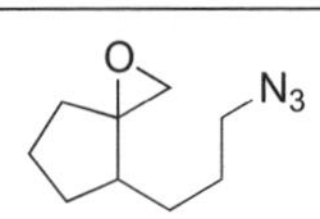	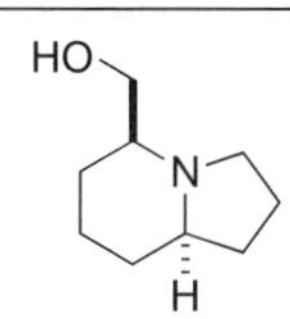	63
2	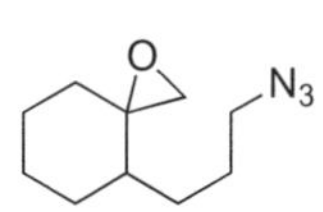	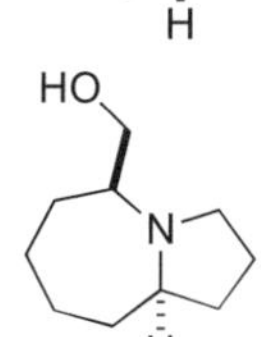	42
3	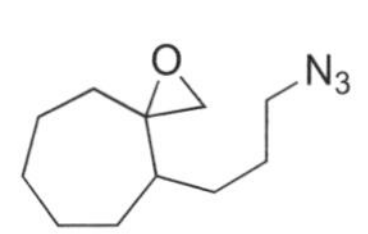	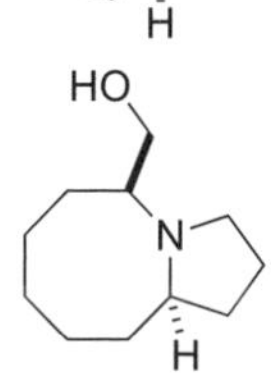	47

in modest yield (Scheme 7.33). When only one aryl group was attached to the epoxide, they were able to intercept the iminium ion intermediate by $NaBH_4$ reduction, thus isolating pyrrolidine products as their *O*-mesylates. When epoxide **38** was treated with $BF_3{\cdot}OEt_2$, followed by aqueous workup, tricyclic ketone **39** was secured in excellent yield. In this case, neighboring group participation by the aryl group occurs after initial attack at the benzylic position by azide. The resulting aziridine is then cleaved, restoring aromaticity. Finally, migration of the methylene group occurred and upon workup, ketone **39** was obtained.

7.7 Combined Schmidt Rearrangement Cascade Reactions

Domino reactions, which combine sequential transformations in a single pot, can allow the rapid development of complex products from simple starting materials.[56–58] One design feature required for successful domino process involves control over the order of reaction events. Since Bronsted acids or Lewis acids are required for the initiation of nearly all azido-Schmidt reactions, combining such reactions with other acid-accelerated processes presents a logical starting point for assembling Schmidt-centric tandem reactions.

Initial attempts to utilize the Schmidt reactions of azides and ketones focused on combining that process with the ubiquitous Diels–Alder cycloaddition reaction. The first example of a combined Diels–Alder/Schmidt reaction was made in the context of a formal total synthesis of (±)-stenine,[59] which led to a detailed examination of the scope and the development of the overall sequence.[60] Since intermolecular Schmidt reactions of ketones are much less facile than their intramolecular counterparts, the prior unification of ketone and azide through a Diels–Alder cycloaddition can facilitate an intramolecular Schmidt reaction. Thus, when an azide-containing diene **40** was combined with 2-cyclopentenone in the presence of $MeAlCl_2$, tricyclic lactam **41** was obtained in modest yield (Scheme 7.34). Another strategy employed was to inhibit the intramolecular reaction of an azido and keto group within the same molecule through the incorporation of enone separating the two groups. When **42** was treated with butadiene in the presence of $MeAlCl_2$, bicyclic amide **43** was formed in moderate yield. Better yields were obtained when electron-rich dienes were used, with siloxy dienes being of particular utility because they reveal a ketone suitable for downstream manipulation upon workup (e.g., **44** → **45**).

Tu and coworkers demonstrated a tandem semi-pinacol/Schmidt rearrangement sequence beginning with certain azido-tethered epoxides.[61] When epoxide **46** was

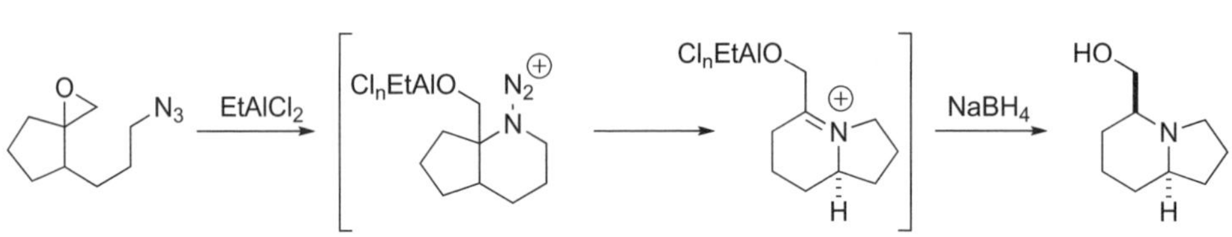

Scheme 7.32 *Mechanism of epoxide opening by azides*

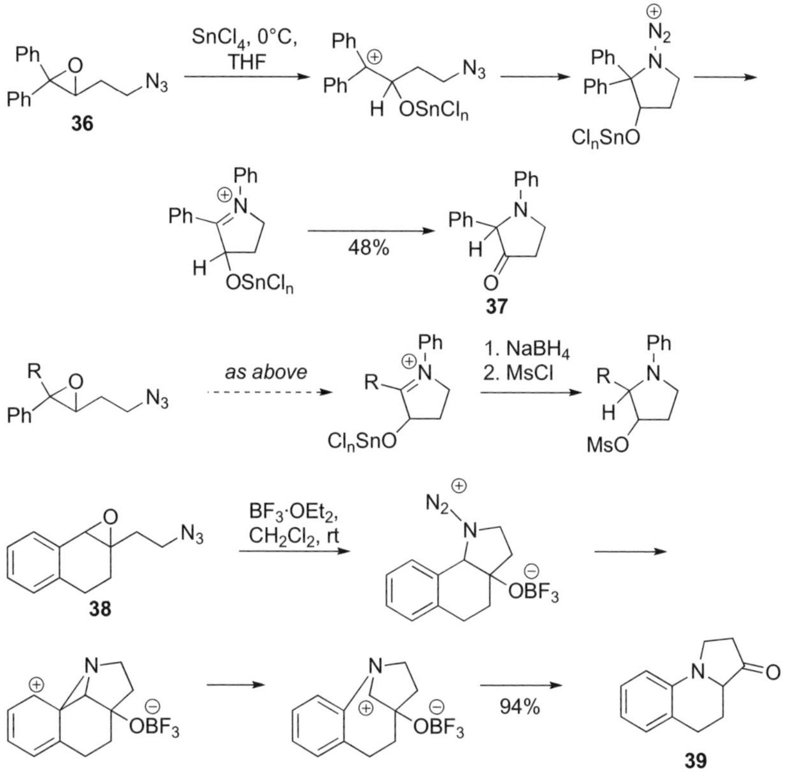

Scheme 7.33 *Rearrangements of azido-tethered phenyl-substituted epoxides*

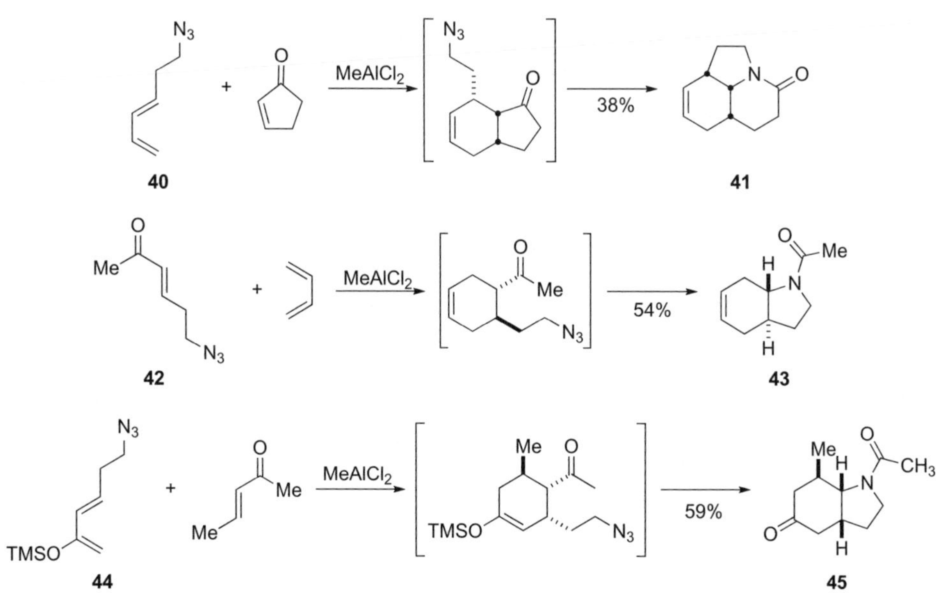

Scheme 7.34 *Domino Diels-Alder/Schmidt reactions*

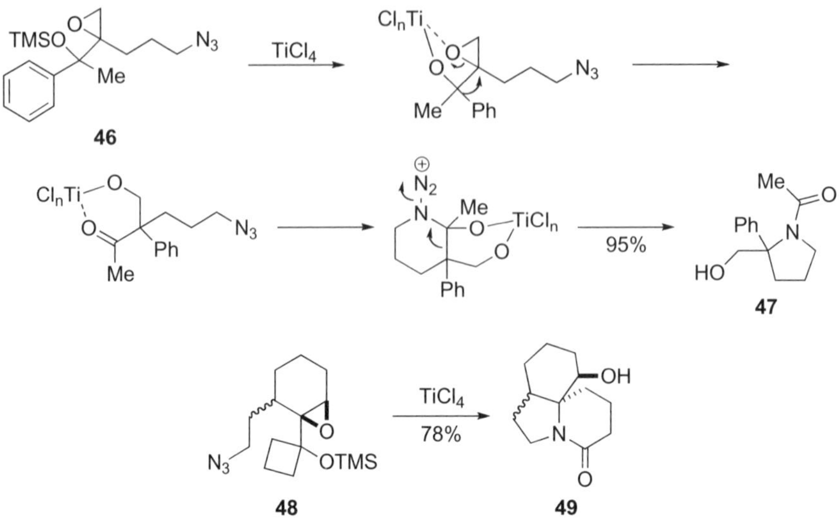

Scheme 7.35 *Tandem semipinacol/Schmidt rearrangements*

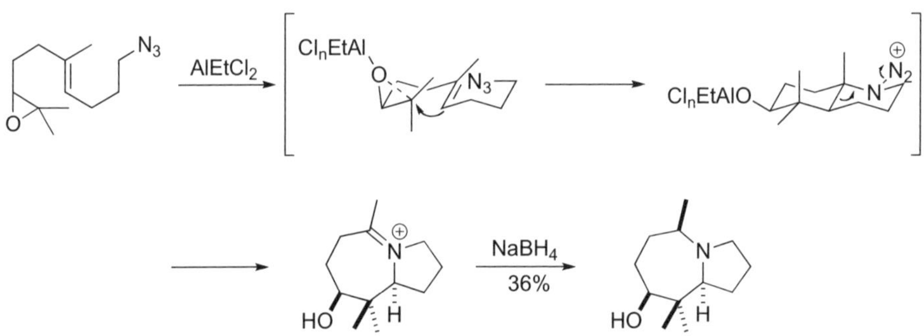

Scheme 7.36 *Cationic cyclization/rearrangement cascade*

subjected to the action of $TiCl_4$ followed by aqueous workup, amide **47** was obtained in excellent yield (Scheme 7.35). In this case, Lewis acid activation of the epoxide was followed by migration of the phenyl group, azide addition to form a tetrahedral intermediate, and finally ring contraction with expulsion of N_2. The ability to efficiently construct tricyclic lactams in a single pot (e.g., **48** → **49**) represents a powerful extension of this work.

The Baskaran group demonstrated that a cationic cyclization Schmidt rearrangement sequence could be executed using an olefinic epoxide bearing an appropriately positioned azide (Scheme 7.36). Thus, treatment of the substrate shown with $EtAlCl_2$ led to a classically inspired epoxide-triggered cationic cyclization wherein the resulting cation could be intercepted by azide.[54] Migration and loss of N_2 occurred spontaneously to furnish an iminium ion, which was subjected to a reductive workup to afford a bicyclic amine with good stereocontrol.

7.8 Schmidt Rearrangements in the Total Synthesis of Natural Products

Schmidt reactions of every stripe provide access to nitrogenous heterocycles. In particular, azide-mediated Schmidt chemistry, especially the intramolecular Schmidt reactions of carbonyls and cations, lead very nicely into polycyclic skeleta that appear in a wide variety of alkaloids. In addition to this, the plethora of methods for the installation of azide and the fact that azides readily withstand a variety of conditions (and thereby can be carried forward without change over a number of synthetic steps) have spurred researchers to examine the controlled execution of azide rearrangements in increasingly challenging total synthesis endeavors.[62] The following examples are presented to illustrate issues of strategy, stereocontrol, and regiocontrol and are roughly organized in terms of electrophilic system employed and increasing chemical complexity.

The first application of the intramolecular Schmidt reaction of alkyl azides with ketones was reported by a group led by Desmaële and d'Angelo in 1993.[63] These researchers prepared the ring system of the homoerythrina alkaloids using a sequence involving alkylation followed to by an intramolecular Michael reaction to afford an ester (Scheme 7.37). Conversion of the ester group to the azide was followed by an efficient intramolecular Schmidt reaction enacted by dissolution in TFA to afford the desired ring system.

A total synthesis of (–)-indolizidine 209B (**50**), isolated from amphibian sources and a blocker of nicotinic receptor channels, was reported in 1993.[64] Starting from ketal **51**, which was synthesized from pulegone in 4 steps,[65] a 4-step sequence provided azide **52** in excellent overall yield (Scheme 7.38). Unmasking of the ketal with $LiBF_4$ followed by dissolution of the resulting ketone in TFA promoted the intramolecular Schmidt reaction uneventfully, providing **53**. The stereoselective incorporation of the side chain needed for the completion of the total synthesis was achieved through a one-pot Grignard addition-reduction sequence. Thus, treating **53** with $C_5H_{11}MgBr$ afforded an iminium ion that was reduced with $NaBH_4$ to afford the target alkaloid. The diastereoselectivity of reduction was attributed to axial attack of hydride onto the iminium ion, proposed to react through the conformation shown.[66]

Although of only modest biological interest, lasubine II (**54**), has attracted considerable interest from organic chemists for the validation of new methodologies for alkaloid total synthesis. A total synthesis of this target revealed a limitation of the intramolecular

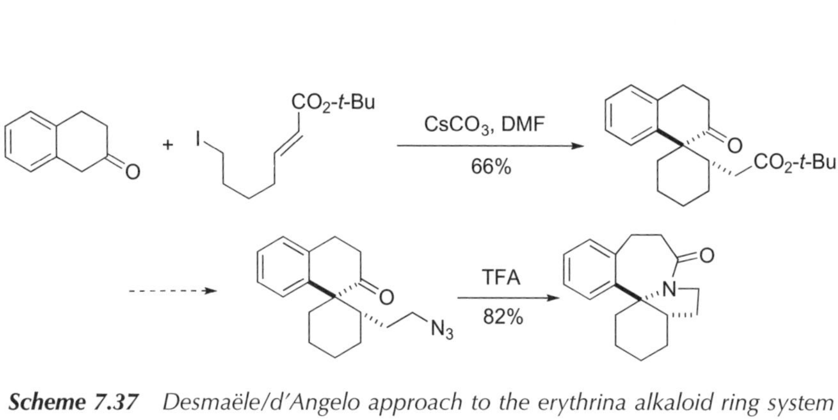

Scheme 7.37 *Desmaële/d'Angelo approach to the erythrina alkaloid ring system*

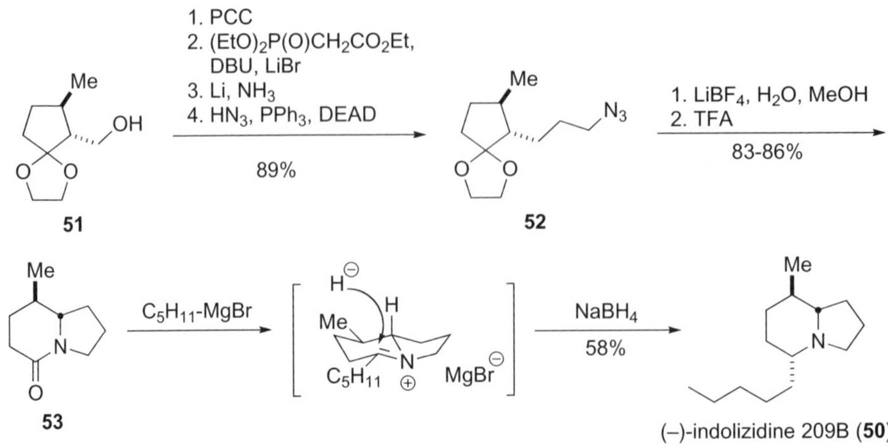

Scheme 7.38 *Total synthesis of (–)-indolizidine 209B*

Schmidt reactions as it pertains to quinolizidine ring systems and led to the development of a workaround involving the photochemical rearrangement of nitrones.[67]

The synthesis began with a TIPS-protected cyclopentenone **55** (a chemotype frequently encountered in prostaglandin synthesis), which was converted to alcohol **56** in four steps (Scheme 7.39). After conversion to the azide, several attempts to carry out the intramolecular Schmidt reaction were made, with $TiCl_4$ in refluxing methylene chloride ultimately identified as the only viable conditions. Under these circumstances, a ca. 1 : 1 mixture of lactams **57a** and **57b** were isolated in moderate yield; it was shown through control experiments that enolization and epimerization α to the carbonyl occurred in competition to ring expansion under these vigorous reaction conditions. To circumvent this problem, a modified approach involving nitrone rearrangement was adopted. Preparation of nitrone **58** was accomplished as depicted and upon photolysis, **57a** was formed along with a small amount of **57b** (dr $\geq$ 13 : 1). The formal total synthesis was then completed through the previously known intermediate **59**.[68]

Early work established the dependence of Schmidt reaction efficiency on tether length between the carbonyl and azide; four carbons turning out to be optimal.[15,16] This point was used to synthetic advantage in a total, asymmetric synthesis of aspidospermidine (**60**)[69,70] that drew much inspiration from Stork's total synthesis of this classic target.[71] Thus, racemic 2-ethylcyclopentanone was converted into ketal **61** in enantioenriched form in 9 steps and 19% overall yield (Scheme 7.40). Compound **61** was treated with $TiCl_4$ with the expectation of accomplishing a Schmidt reaction of the azide with the free ketone. However, the azide exclusively attacked the ketal instead to provide **62**, a bridged *ortho*-aminal. It was conjectured that the Schmidt reaction with the available ketone was disfavored due to the exo stereochemical orientation of the azide-containing side chain vis-à-vis the bicyclic ring system. In contrast, when the ketal was unmasked using $LiBF_4$ and then reacted under similar conditions, the resulting diketone exclusively afforded lactam

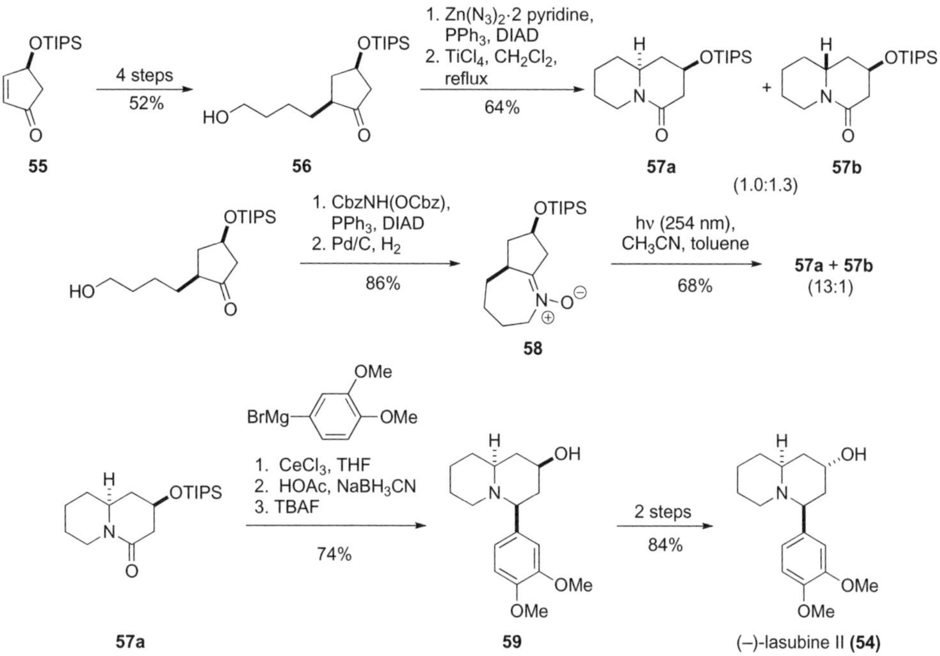

Scheme 7.39 *Formal synthesis of (–)-lasubine II*

64, derived from reaction of the ketone in the most favored 1,5 relationship to the azide instead of the alternative 1,4-oriented ketone. It was ultimately shown that the diketone had epimerized in the course of deprotection, affording **63**. Only in this isomer could the azide group reach the carbonyl group and engage in a Schmidt reaction to give **64**. Lactam **64** was converted to amine **65** and subsequently subjected to a regioselective Fischer indole synthesis protocol that gave mostly the desired regioisomer (the isomer corresponding to aspidospermidine was accompanied by 13% of the regioisomer, which is not shown in the scheme). The synthesis was completed as originally established by Stork and Dolfini.[71]

(–)-Sparteine (**66**) is widely used as a chiral ligand for a variety of asymmetric transformations. Although the (+)-enantiomer is naturally occurring, it is not easily obtained from natural sources. This was once considered a serious drawback of many sparteine-promoted asymmetric reactions, but it is now largely a moot point due to the development of efficient sparteine surrogates, principally by O'Brien and colleagues.[72,73] The total synthesis began with the interesting, useful, and C_2-symmetric ketone **67**, which was readily derived from norbornadiene using a doubly asymmetric hydrosilylation reaction originally reported by Hayashi (Scheme 7.41).[74,75] The most direct route from this compound to sparteine would involve two Schmidt reactions, possibly simultaneous, that would provide a bislactam precursor. Although, a single intramolecular Schmidt reaction could be carried out, it was not possible to carry out a second Schmidt reaction on alkylated lactams related to **68**.[76] Ultimately, a sequential rearrangement approach was taken which involved $TiCl_4$-induced Schmidt rearrangement to give **68**.[77] Nitrone **69** was then

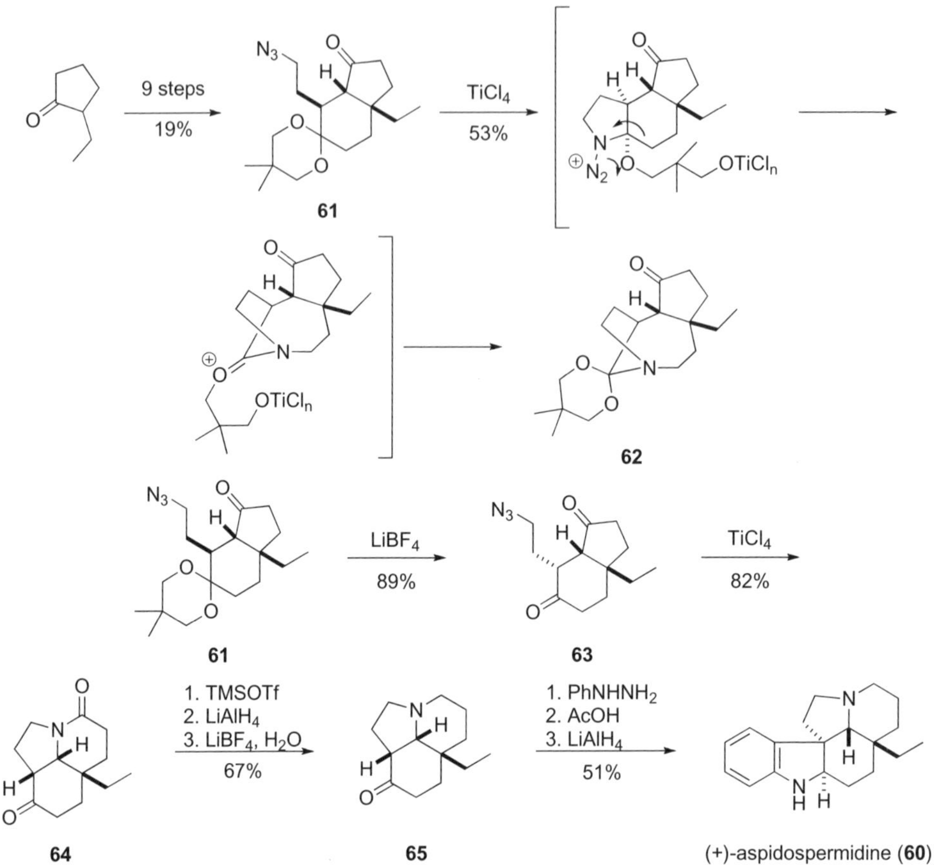

Scheme 7.40 *Total synthesis of (+)-aspidospermidine*

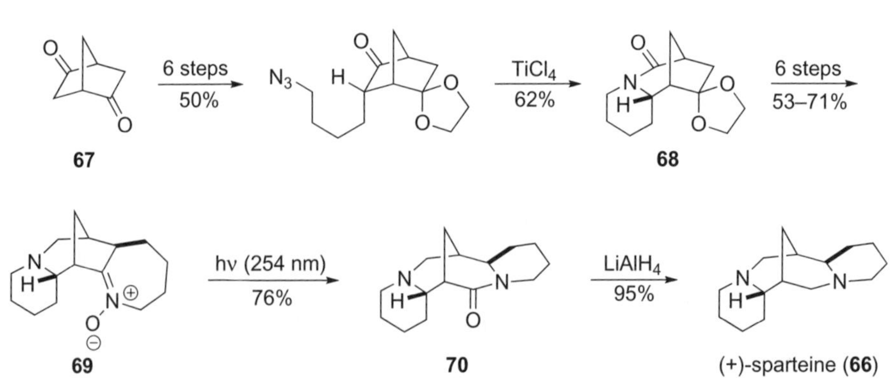

Scheme 7.41 *Total synthesis of (+)-sparteine featuring two types of ring expansions*

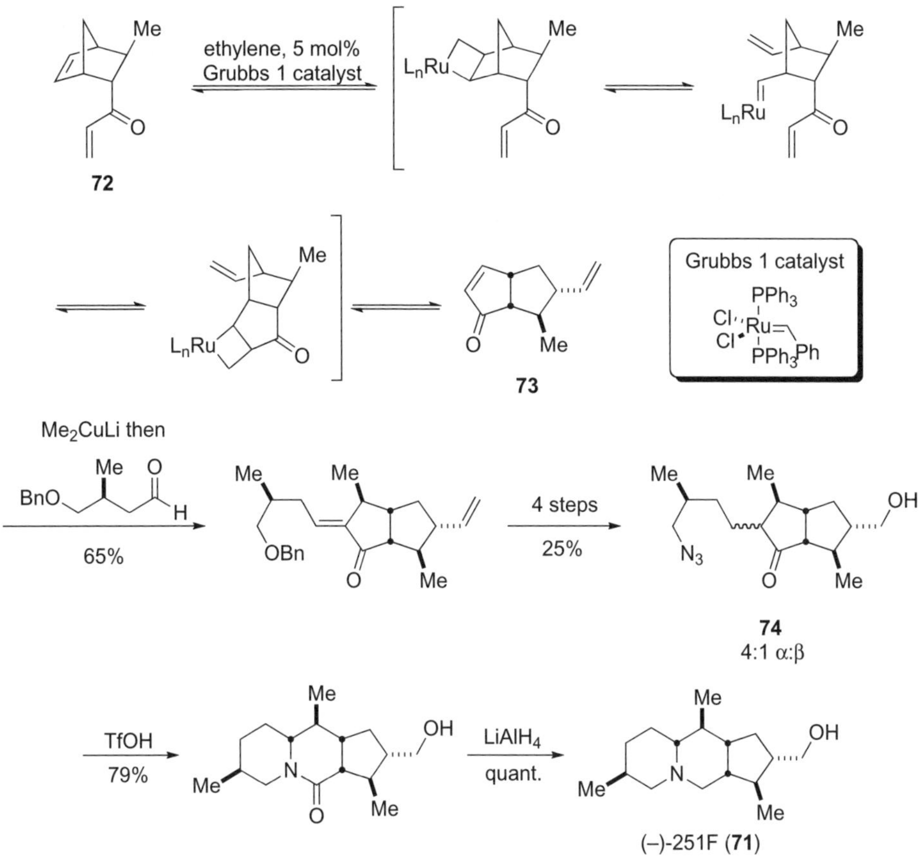

Scheme 7.42 *Total synthesis of alkaloid (–)-251F*

secured through a six-step sequence and photolyzed to afford lactam **70**. LAH reduction of this lactam then afforded (+)-sparteine. This synthesis, along with that of lasubine (Scheme 7.39, above) established such nitrone rearrangements as useful alternatives to the intramolecular Schmidt reaction in ornery cases.[78]

Quinolizidine-containing alkaloid 251F (**71**) was extracted from the skin of a Colombian dendrobatid frog; although many such alkaloids have potent biological activity, the pharmacological profile of this particular agent was unknown (and remains) due to the paucity of material isolated (only 300 μg of this material was isolated).[79] A synthesis of (–)-251F involving a late stage intramolecular nitrogen ring expansion reaction was reported, although the most interesting element of this synthesis is not the Schmidt reaction but rather the means by which the precursor ketone was prepared (Scheme 7.42).[80,81] Thus, ketone **72** (itself prepared using a straightforward Diels–Alder reaction of cyclopentadiene) underwent skeletal rearrangement when treated with Grubbs's first-generation metathesis catalyst in an approach inspired by that group's synthesis of capnellene.[82] This complex reaction involves both ring-opening and ring-closing metathesis steps (only the most direct pathway leading to the desired **73** is shown); related reactions have been increasingly used by other researchers in other total synthesis endeavors.[83] The remainder of the synthesis entailed substrate-directed reactions leading to azide **74**, which was converted to lactam by acid treatment and finally provided the target alkaloid after carbonyl removal.

Two total syntheses of the *Stemona* alkaloid stenine that utilize the intramolecular Schmidt reaction have been reported. In the first, an intramolecular Diels–Alder/intramolecular Schmidt rearrangement was carried out to achieve the rapid formation of the BCD ring system of stenine (Scheme 7.43).[59] To this end, triene **76** (*E*:*Z*=85:15) was synthesized by a straightforward but lengthy route (13 steps) and subjected to $MeAlCl_2$, thus providing a mixture of three lactam products. The major lactam (**77a**), which possesses the stereochemistry of the natural product, was converted to an advanced intermediate toward stenine previously reported by Hart.[84,85] A key stereochemistry-determining step in this sequence was an iodolactonization affording **78** and the subsequent radical-mediated allylation of this intermediate.

The reaction of trienyl azide **76** was noteworthy as the first reported example of a combined Diels–Alder/Schmidt reaction of any type (see Section 7.7 above). In addition, the isolation of compound **77b** represented the first known observation of a bridged lactam from an intramolecular Schmidt reaction (in contrast to the formation of bridged *ortho*-amide **62** shown in Scheme 7.40). Mechanistically, an endo Diels–Alder leads to **77a** and **77b** (box at bottom of Scheme 7.43), while an exo transition state leads to **77c** (transition state not shown). The formation of fused or bridged lactam depends on antiperiplanar migration of carbon from the equatorially or axially disposed N_2^+ groups in the intermediates shown.

An intermolecular Diels–Alder/Schmidt approach was deployed for an efficient second-generation synthesis of (±)-stenine (Scheme 7.44).[86] Thus, the union of diene **79** and 2-cyclohexenone *via* the venerable Diels–Alder reaction provided the context for an intramolecular Schmidt reaction. Surprisingly, it was found that the use of $SnCl_4$ led to product resulting from an exo Diels–Alder adduct **80a** as the major product (3:1 ratio). This compound was readily converted to the natural product target using standard chemistry. This 9-step route to (±)-stenine proceeded in 14% overall yield, a significant improvement in terms of efficiency over the first-generation synthesis. A key feature that permitted this improvement was the relatively simple preparation of the substrate needed for the intermolecular Diels–Alder reaction as opposed to the previously used intramolecular variant. Also helpful was the direct installation of the ethyl group early in the sequence (obviating the need to remove a single methylene group from an allyl group in the late stages of the synthesis).

A related synthesis of (±)-neostenine (**81**), an isomer of (±)-stenine with reported antitussive properties,[87] was accomplished by effecting a switch in diastereoselectivity of the key step (Scheme 7.45).[88] Thus, when diene **79** and 2-cyclohexenone were combined in the presence of $BF_3{\cdot}OEt_2$, isomeric lactams **80a** and **82** were formed as mixture of ethyl diastereomers along with a small amount of Diels–Alder adduct. The ethyl stereoisomers completely converged onto a single isomer upon treatment with NaOMe in MeOH and subsequent processing afforded lactone **83**. The synthesis required installation of a methyl group on the more hindered convex face of this ring system, which was accomplished by the multistep sequence shown. Chemoselective reduction of **84** then completed the

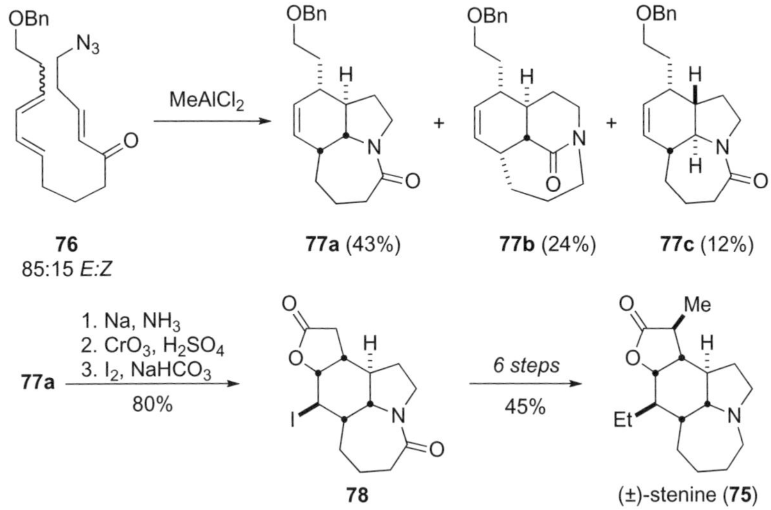

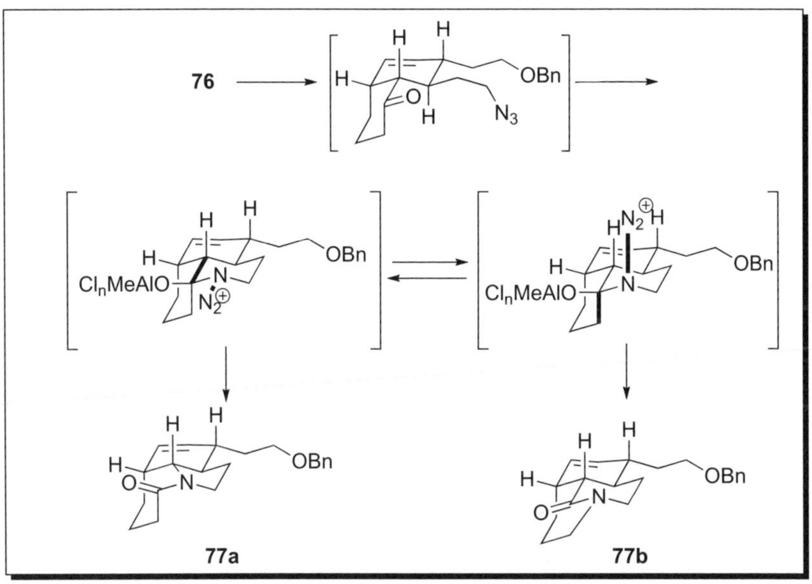

Scheme 7.43 *First generation Diels–Alder Schmidt approach to (±)-stenine and mechanism for formation of lactams **77a** and **77b**. Migrating groups are indicated by darkened bonds*

synthesis of (±)-neostenine, the structure of which was verified by X-ray crystallographic analysis of the synthetic material.

Pearson and coworkers exploited cationic azide capture as a key step in their formal synthesis of (±)-gephyrotoxin **85** (Scheme 7.46).[89] Treatment of azide **86** with TfOH caused rearrangement to isomeric iminium ions. After reduction, exchange of the bromide with acetate, and another reduction, gave alcohols **87** and **88**. The major isomer **87** had previously been converted into (±)-gephyrotoxin (**85**) by Kishi.[90] The formal

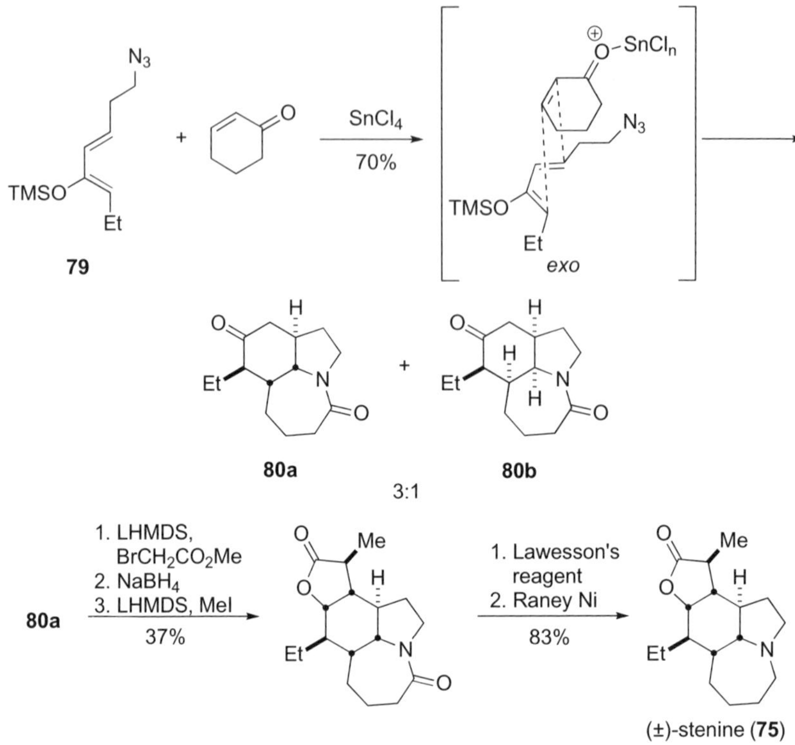

Scheme 7.44 *Second-generation approach to (±)-stenine*

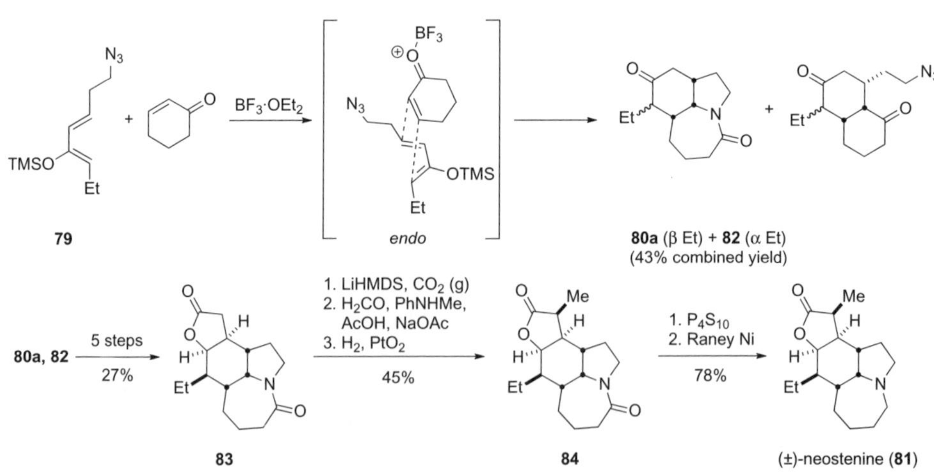

Scheme 7.45 *Total synthesis of (±)-neostenine (**81**)*

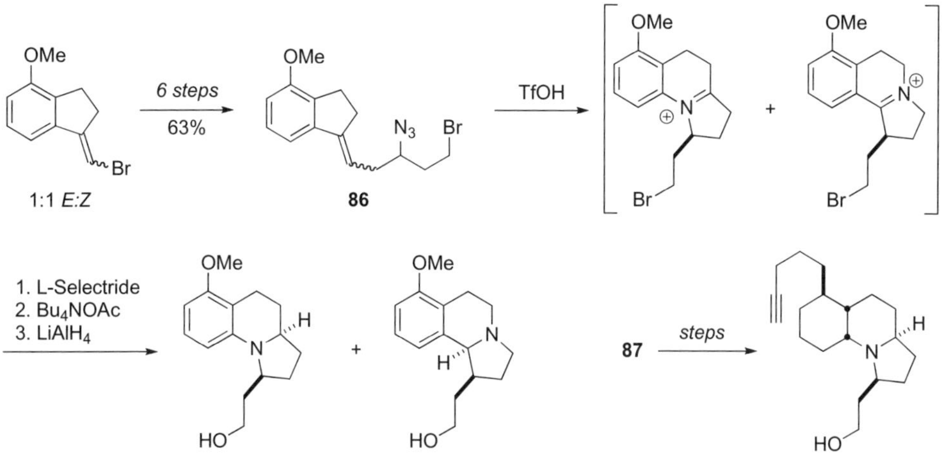

Scheme 7.46 *Formal total synthesis of (±)-gephyrotoxin*

synthesis proceeded in 22% overall yield and 9 steps from commercially available 4-methoxy-1-indanone.

Following up on their development of a Lewis acid-initiated semipinacol/Schmidt rearrangement, Tu and coworkers applied such a sequence to the total synthesis of (±)-stemonamine (**89**), which has been used for centuries in China and Japan as an insecticide and to treat respiratory disease (Scheme 7.47).[91] A simple, known azido enone was converted in 3 steps to the tandem semipinacol/Schmidt precursor **90**, which was treated with $TiCl_4$ in CH_2Cl_2, enacting smooth rearrangement in a highly diastereoselective fashion to **91**. Formation of the cyclopentenone ring was then carried out in a 5-step sequence to afford a 1 : 1 mixture of diastereomers **92a,b**. Separation of these isomers was straightforward and **92a** was converted into the target (±)-stemonamine **89** using the sequence shown.

7.9 Schmidt Rearrangements of Alkyl Azides in the Synthesis of Interesting Non-natural Products

More than one rearrangement pathway is possible from the tetrahedral intermediate of an intramolecular Schmidt reaction of cyclic ketones. The vast majority of lactams isolated *via* this chemistry arise from migration of the carbon to which the azidoalkyl chain is appended, affording fused products. However, it has been increasingly recognized that bridged lactams can also form and can even predominate in special cases (the first example of this was encountered in the first-generation total synthesis of stenine, shown in Scheme 7.43). Bridged lactams where the nitrogen atom is at a bridging position are theoretically interesting because they contain an amide bond that is twisted out of plane. Molecules with this feature display unusual spectral characteristics and reactivity that has classically been ascribed to loss of amide bond resonance or strain.[92,93]

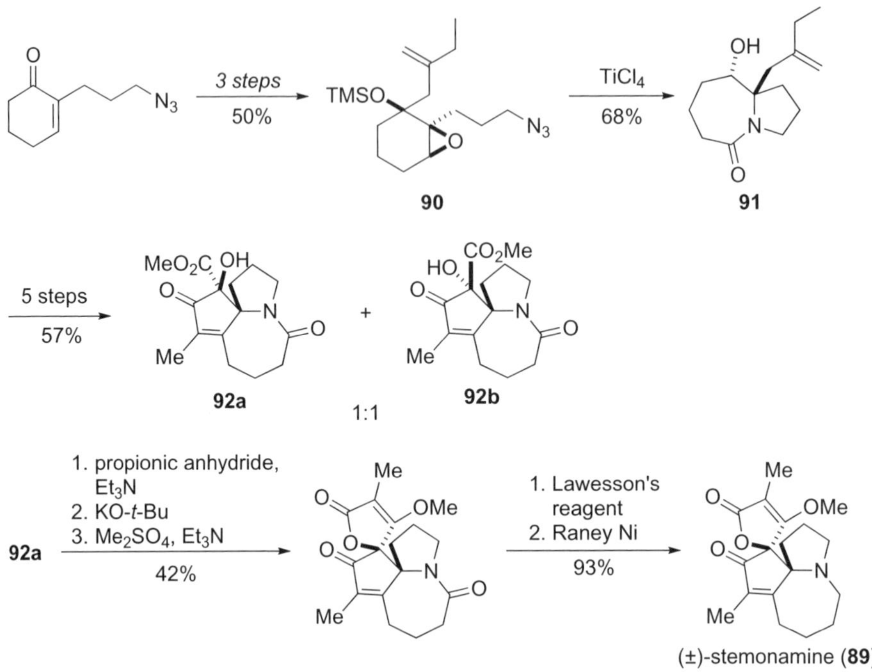

Scheme 7.47 *Total synthesis of (±)-stemonamine*

A series of ketones **93a–d** was synthesized and subjected to Lewis acidic conditions (Table 7.9).[94] When **93a** was subjected to the action of $MeAlCl_2$, **94a** was formed exclusively in high yield. However, when a 4-*tert*-butyl group was introduced into the starting ketones, bridged lactams **95b–d** were observed. The placement of an aryl group α to the ketone (cf. entries 2-4) dramatically increased the yield of bridged lactam products and more electron-rich aryl groups gave the highest conversion to bridged lactams **95**.

Without the *tert*-butyl group present, the reactive conformation leading to lactam **94a** appears to place the phenyl group in an axial position; an antiperiplanar arrangement of the diazonium ion and the methylene group required for migration leads to the fused compound **94a** (Scheme 7.48). For compounds containing a *tert*-butyl group but lacking an α–phenyl substituent, a similar reactive conformation occurs leading mostly to **94b**, with a small amount of bridged lactam **95b** also being observed. However, when both the diazonium ion and the aryl group are in a 1,3-diaxial relationship, a reactive conformation in which the leaving N_2^+ group is axial can be stabilized by a cation–π effect due to the phenyl group in a 1,3-diaxial relationship (see also Scheme 7.17). This leads to mostly bridged lactams **95c** and **95d**. Follow-up work has focused on the unusual reactivity of these compounds.[95]

The release of dinitrogen in Schmidt rearrangements provides a powerful driving force for the construction of strained bridged lactam products. Tani and Stoltz took advantage of this in a landmark synthesis of 2-quinuclidonium tetrafluoroborate **97** (Scheme 7.49).[96] In this work, an azide **96** bearing an azidoalkyl group placed in a position non-adjacent to the reacting ketone was treated with HBF_4. In this way, a mixture of quinuclidinium salt **97** and the isomeric lactam **98** were formed; note that a fused lactam is not possible

Table 7.9 *Synthesis of fused and bridged lactams*

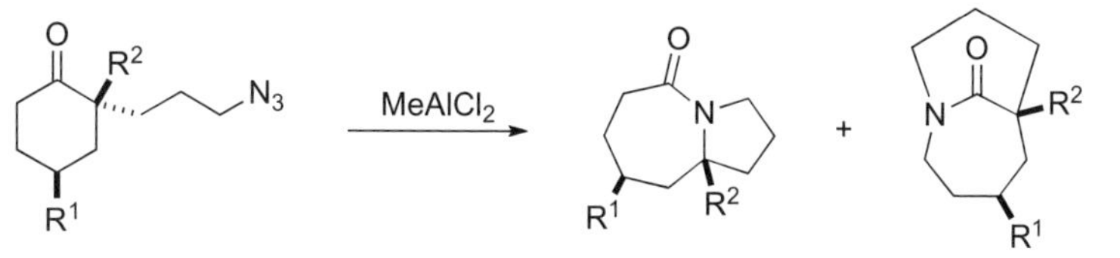

93a-d **94a-d** **95a-d**

Entry	Compound	R^1	R^2	Yield (%)	
				94	**95**
1	**a**	H	Ph	96	0
2	**b**	*t*-Bu	H	57	17
3	**c**	*t*-Bu	Ph	20	51
4	**d**	*t*-Bu	*p*-(MeO)C_6H_4	10	65

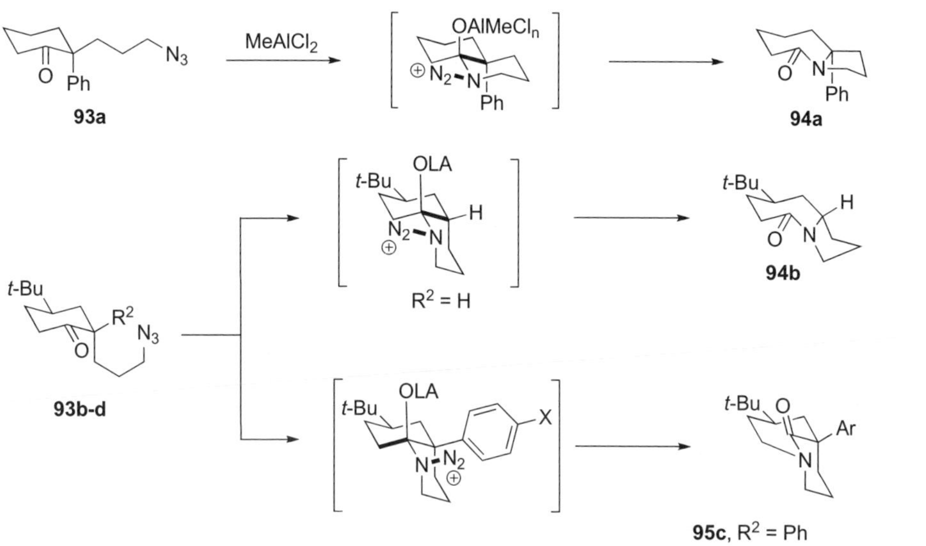

Scheme 7.48 *Reactive conformations leading to fused and bridged lactams*

from this kind of substrate. Ultimately, recrystallization of this mixture produced crystalline quinuclidinium tetrafluoroborate **97**, the structure of which was confirmed by X-ray crystallographic analysis.[97]

The use of Schmidt rearrangement chemistry toward indolizidine-containing compounds by Pearson and coworkers was extended to include the synthesis of several dopamine analogs and non-opiate antinociceptive agents (two examples are illustrated in Scheme 7.50).[98] These compounds bear some resemblance to more standard opioids like apomorphine, butaclamol, or isobutaclamol, and illustrate a platform for the potential synthesis of additional analogues.

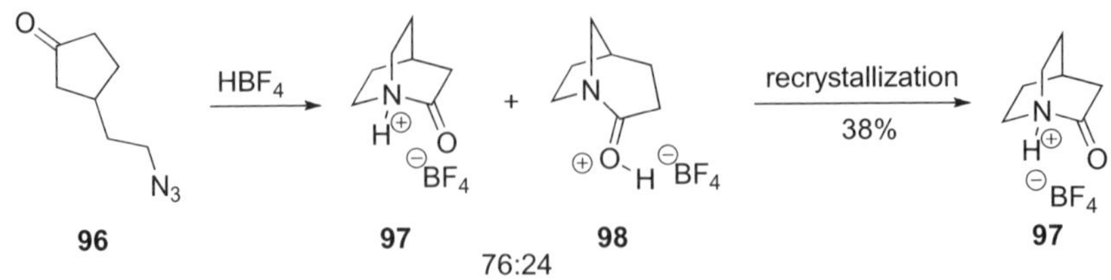

Scheme 7.49 Synthesis of quinuclidinium tetrafluoroborate

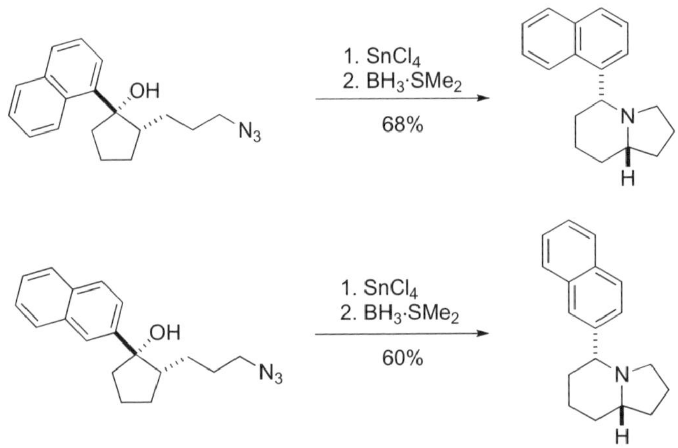

Scheme 7.50 Schmidt reaction route to dopamine analogs

7.10 Schmidt Rearrangements of Hydroxyalkyl Azides toward Biologically Relevant Compounds

The Schmidt reaction of hydroxyalkyl azides with ketones has been applied to the synthesis of *β*-turn mimics (Scheme 7.51).[99] Toward this end, symmetrical pyranone **99** was reacted with a chiral hydroxyalkyl azide in the presence of $BF_3{\cdot}OEt_2$; subsequent hydrolytic workup gave a 9 : 1 mixture of **100a,b**. Reductive removal of the phenylhydroxyethyl side chain, amide hydrolysis, esterification, and peptide coupling gave a peptide hybrid that was processed to its cyclic analogue using standard chemistry. NMR and X-ray crystallographic analysis of **101** suggested a solution structure consistent with type I *β*-turn mimicry.

A modular synthesis of a series of *γ*-turn mimics was enabled by the Lewis acid-mediated reactions of enantiopure hydroxyalkyl azides with piperidinones. This involved the preparation of enantiopure azides **103**, the side chains of which corresponded to those in naturally occurring side amino acids. These azides were combined with piperidinones **102** in the presence of $BF_3{\cdot}OEt_2$ or TfOH, and the resulting iminium ethers hydrolyzed to the corresponding lactams **104** (Table 7.10).

This chemistry was extended to the synthesis of libraries of substituted *γ*-turn-inspired molecules by taking advantage of the ability to add nucleophiles to the intermediate

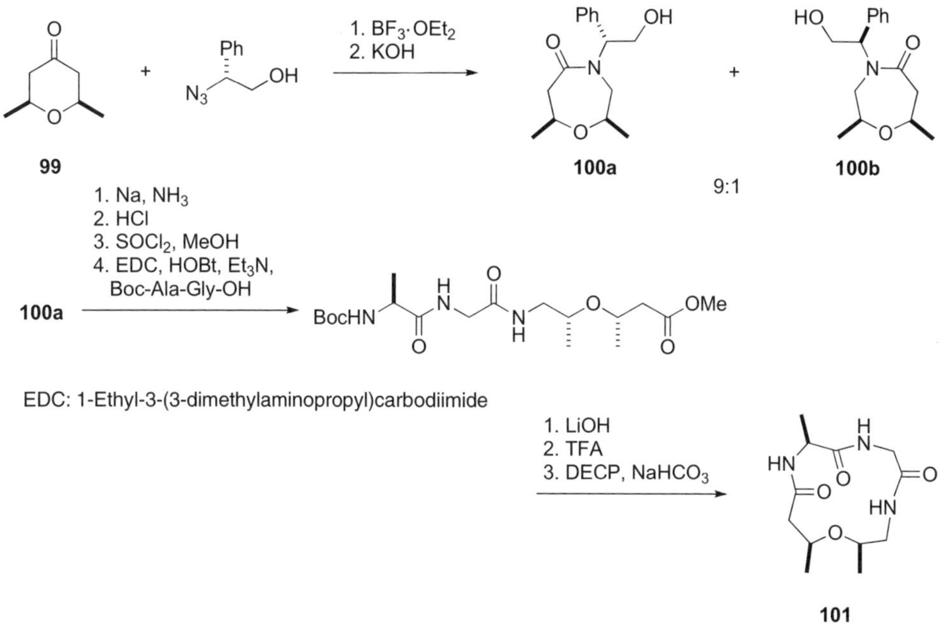

Scheme 7.51 *Synthesis of a β-turn peptidomimetic*

Table 7.10 *Syntheses of γ-turn mimics*

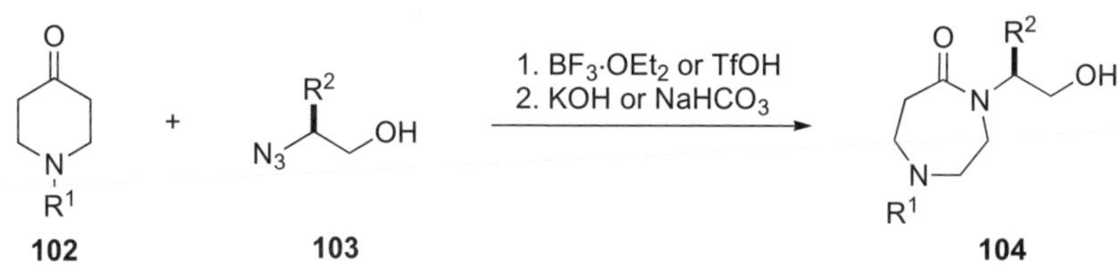

Entry	Compounds	R^1	R^2	Yield (%)
1	**a**	Bn	$-CHMe_2$	89
2	**b**	Bn	$-CH_2CHMe_2$	87
3	**c**	Bn	$-CO_2Bn$	62
4	**d**	Fmoc	$-CHMe_2$	63
5	**e**	Cbz	$-CH_2CHMe_2$	52
6	**f**	GlyCbz	$-CH_2CHMe_2$	79

iminium ethers formed in this kind of ring expansion reaction (Scheme 7.52; cf. Table 7.4 in Section 7.2).[100] The example shown utilized a sublibrary of differently substituted thiophenols in the ring-opening step. By modestly substituting R^1, R^2, and R^3 with four different groups each, libraries containing up to 64 variations of Markush structure **105** were readily generated.

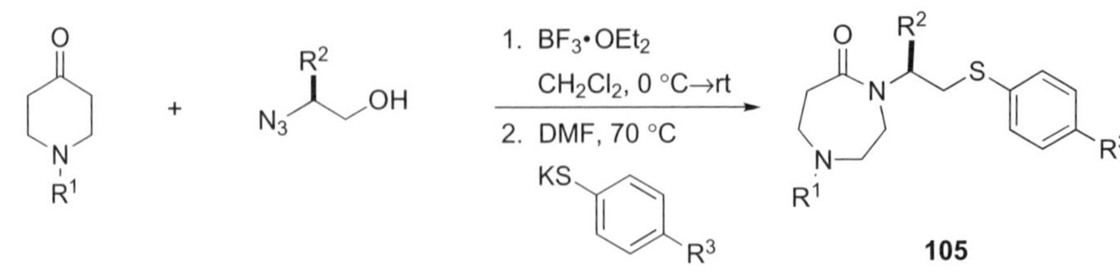

105

Scheme 7.52 *Three-component parallel synthesis of lactam libraries*

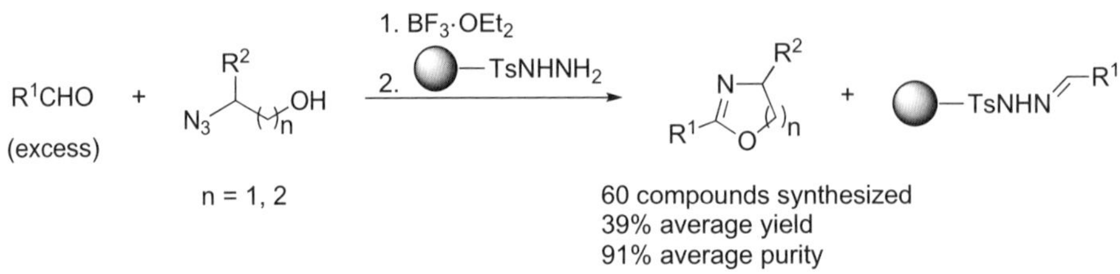

Scheme 7.53 *Parallel synthesis of oxazolines and 1,3-dihydrooxazines*

An additional library was synthesized using the same reaction of hydroxyalkyl azides with aldehydes originally reported by Boyer in 1955 (Scheme 7.3, Section 7.1).[12] Thus, a library of oxazolines and dihydrooxazines was synthesized from aromatic aldehydes and azides by treatment with $BF_3{\cdot}OEt_2$ followed by capture of the excess aldehyde (or azide, not shown) with a resin bound scavenger (Scheme 7.53).[101]

7.11 Final Comments

The Schmidt reaction of hydrazoic acid has been a mainstay of heterocyclic organic chemistry for nearly a hundred years. In the past decade-plus, it has been increasingly appreciated that alkyl azides participate a rich nucleophilic chemistry of their own. Perhaps the most surprising element of this discovery was that it took so long to come to fruition. However, there were perfectly valid reasons that giants in the field, such as Briggs and Smith, failed in their early attempts. Chief among these is the fact that N_3-containing compounds are poor nucleophiles, poor enough that while hydrazoic acid reacts with aldehydes, ketones, and acids, while under protic acid conditions methyl azide reacts only with aldehydes (and even then, just barely). One possible explanation is based on the size difference of a proton vs. an alkyl group, although this point has not been rigorously proved. Thus, it was not until the early 1990s that the dual enablers of the azido Schmidt reaction, strong Lewis acids and intramolecularity, were found to vastly expand the Schmidt repertoire to include alkyl azides.

The ramifications of this discovery are still being worked out, but it is already clear that the use of azides in ring expansion chemistry allows for an additional element of control (regiochemical, stereochemical) and in some cases facilitates reactions that would be difficult at best using hydrazoic acid. What is evident is that the future will see additional forward momentum in the development of this fascinating functional group.

Acknowledgments

The authors acknowledge the US National Institutes of Health for the support of their programs in alkyl azide chemistry (GM49093 and P50GM069663) and Jennifer Treece and Angelica Meyer for assistance with the manuscript.

References

[1] R.A. Abramovich, E.P. Kyba, in *The Chemistry of the Azido Group* (ed.: S. Patai), John Wiley & Sons, Ltd, London, **1971**, pp. 221–329.
[2] D.V. Banthorpe, in *The Chemistry of the Azido Group* (ed.: S. Patai), John Wiley & Sons, Ltd, London, **1971**, pp. 397–440.
[3] E.P. Kyba, in *Azides and Nitrenes: Reactivity and Utility* (ed.: E.F.V. Scriven), Academic, Orlando, **1984**, pp. 2–34.
[4] P.A.S. Smith, in *Molecular Rearrangements, Vol. 1* (ed.: P. de Mayo), John Wiley & Sons, Inc., New York, **1963**, pp. 457–591.
[5] S. Uyeo, *Pure Appl. Chem.* **1963**, *7*, 269–83.
[6] H. Wolff, *Organic Reactions* **1946**, *3*, 307–36.
[7] P.A.S. Smith, E.P. Antoniadu, *Tetrahedron* **1960**, *9*, 210–29.
[8] L.E. Fikes, H. Shechter, *J. Org. Chem.* **1979**, *44*, 741–4.
[9] M.B. Smith, J. March, *March's Advanced Organic Chemistry: Reactions, Mechanisms, and Structure*, 5th ed., John Wiley & Sons, Inc., New York, **2001**.
[10] L.H. Briggs, G.C. De Ath, S.R. Ellis, *J. Chem. Soc.* **1942**, 61–3.
[11] P.A.S. Smith, *J. Am. Chem. Soc* **1948**, *70*, 320–3.
[12] J.H. Boyer, J. Hamer, *J. Am. Chem. Soc.* **1955**, *77*, 951–4.
[13] J.H. Boyer, F.C. Canter, J. Hamer, R.K. Putney, *J. Am. Chem. Soc.* **1956**, *78*, 325–7.
[14] J.G. Badiang, J. Aubé, *J.Org. Chem.* **1996**, *61*, 2484–7.
[15] J. Aubé, G.L. Milligan, *J. Am. Chem. Soc.* **1991**, *113*, 8965–6.
[16] G.L. Milligan, C.J. Mossman, J. Aubé, *J. Am. Chem. Soc.* **1995**, *117*, 10449–59.
[17] H.-L. Lee, J. Aubé, *Tetrahedron* **2007**, *63*, 9007–15.
[18] C.J. Mossman, J. Aubé, *Tetrahedron* **1996**, *52*, 3403–8.
[19] A. Wrobleski, J. Aubé, *J. Org. Chem.* **2001**, *66*, 886–9.
[20] J. Aubé, G.L. Milligan, C.J. Mossman, *J. Org. Chem.* **1992**, *57*, 1635–7.
[21] P. Desai, K. Schildknegt, K. Agrios, A.C.J. Mossman, G.L. Milligan, J. Aubé, *J. Am. Chem. Soc.* **2000**, *122*, 7226–32.
[22] V. Gracias, K.E. Frank, G.L. Milligan, J. Aubé, *Tetrahedron* **1997**, *53*, 16241–52.
[23] P. Desai, J. Aubé, *Org. Lett.* **2000**, *2*, 1657–9.
[24] S. Grecian, P. Desai, C.J. Mossman, J.L. Poutsma, J. Aubé, *J. Org. Chem.* **2007**, *72*, 9439–47.
[25] V. Gracias, G.L. Milligan, J. Aubé, *J. Am. Chem. Soc* **1995**, *117*, 8047–8.
[26] J.E. Forsee, J. Aubé, *J. Org. Chem.* **1999**, *64*, 4381–5.
[27] J.E. Forsee, B.T. Smith, K.E. Frank, J. Aubé, *Synlett* **1998**, 1258–60.
[28] E. Fenster, B.T. Smith, V. Gracias, G.L. Milligan, J. Aubé, *J. Org. Chem.* **2008**, *73*, 201–5.
[29] V. Gracias, G.L. Milligan, J. Aubé, *J. Org. Chem.* **1996**, *61*, 10–11.
[30] B. Smith, V. Gracias, J. Aubé, *J. Org. Chem.* **2000**, *65*, 3771–4.
[31] K. Sahasrabudhe, V. Gracias, K. Furness, *et al.*, *J. Am. Chem. Soc.* **2003**, *125*, 7914–22.

[32] N.D. Hewlett, J. Aubé, J.L. Radkiewicz-Poutsma, *J. Org. Chem.* **2004**, *69*, 3439–46.
[33] C.E. Katz, T. Ribelin, D.G. English, *et al.*, *J. Org. Chem.* **2008**, *73*, 3318–27.
[34] K. Furness, J. Aubé, *Org. Lett.* **1999**, *1*, 495–7.
[35] T.P. Ribelin, J. Aubé, *Nat. Protocols* **2008**, *3*, 137–43.
[36] R.W. Hoffmann, *Chem. Rev.* **1989**, *89*, 1841–60.
[37] C.E. Katz, J. Aubé, *J. Am. Chem. Soc.* **2003**, *125*, 13948–9.
[38] T. Ribelin, C.E. Katz, D. Winthrow, *et al.*, *Angew Chem., Int. Ed.* **2008**, *47*, 6233–5.
[39] W.H. Pearson, J.M. Schkeryantz, *Tetrahedron Lett.* **1992**, *33*, 5291–4.
[40] W.H. Pearson, *J. Heterocyclic Chem.* **1996**, *33*, 1489–96.
[41] W.H. Pearson, R. Walavalkar, J.M. Schkeryantz, W.-K. Fang, J.D. Blickensdorf, *J. Am. Chem. Soc.* **1993**, *115*, 10183–94.
[42] P. Molina, J. Alcántra, C. López-Leonardo, *Synlett* **1995**, 363–4.
[43] W.H. Pearson, W.-k. Fang, *J. Org. Chem.* **1995**, *60*, 4960–1.
[44] A.G. Schultz, M. Macielag, M. Plummer, *J. Org. Chem.* **1988**, *53*, 391–5.
[45] A.G. Schultz, S.O. Myong, S. Puig, *Tetrahedron Lett.* **1984**, *25*, 1011–4.
[46] W.H. Pearson, W.-K. Fang, J.W. Kampf, *J. Org. Chem.* **1994**, *59*, 2682–4.
[47] A. Rostami, Y. Wang, A.M. Arif, R. McDonald, F.G. West, *Org. Lett.* **2007**, *9*, 703–6.
[48] D. Song, A. Rostami, F.G. West, *J. Am. Chem. Soc.* **2007**, *129*, 12019–22.
[49] W.H. Pearson, D. Hutta, W.-K. Fang, *J. Org. Chem.* **2000**, *65*, 8326–32.
[50] D.J. Gorin, N.R. Davis, D.F. Toste, *J. Am. Chem. Soc.* **2005**, *127*, 11260–1.
[51] C.-K. Sha, S.-L. Ouyang, D.-Y. Hsieh, R.-C. Chang, S.-C. Chang, *J. Org. Chem.* **1986**, *51*, 1490–4.
[52] D.S. Reddy, W.R. Judd, J. Aubé, *Org. Lett.* **2003**, *5*, 3899–902.
[53] J.M. Mahoney, C.R. Smith, J.N. Johnston, *J. Am. Chem. Soc.* **2005**, *127*, 1354–5.
[54] G. Reddy, P.B. Varghese, S. Baskaran, *Org. Lett.* **2003**, *5*, 583–5.
[55] S. Lang, A.R. Kennedy, J.A. Murphy, A.H. Hayne, *Org. Lett.* **2003**, *5*, 3655–8.
[56] T.-L. Ho, *Tandem Reactions in Organic Synthesis*, John Wiley & Sons, Inc., New York, **1992**.
[57] L.F. Tietze, *Chem. Rev.* **1996**, *96*, 115–36.
[58] L.F. Tietze, A. Modi, *Med. Res. Rev.* **2000**, *20*, 304–22.
[59] J.E. Golden, J. Aubé, *Angew. Chem. Int. Ed.* **2002**, *41*, 4316–8.
[60] Y. Zeng, D.S. Reddy, E. Hirt, J. Aubé, *Org. Lett.* **2004**, *6*, 4993–5.
[61] P. Gu, Y.-M. Zhao, Y.Q. Tu, Y. Ma, F. Zhang, *Org. Lett.* **2006**, *8*, 5271–3.
[62] E. Nyfeler, P. Renaud, *Chimia* **2006**, *60*, 276–84.
[63] M.-A. Le Dréau, D. Desmaële, F. Dumas, J. d'Angelo, *J. Org. Chem.* **1993**, *58*, 2933–5.
[64] J. Aubé, P.S. Rafferty, G.L. Milligan, *Heterocycles* **1993**, *35*, 1141–7.
[65] J.N. Marx, L.R. Norman, *J. Org. Chem.* **1975**, *40*, 1602–6.
[66] P. Deslongchamps, *Stereoelectronic Effects in Organic Chemistry*, Pergamon, Oxford, **1983**.
[67] V. Gracias, Y. Zeng, P. Desai, J. Aubé, *Org. Lett.* **2003**, *5*, 4999–5001.
[68] D. Ma, W. Zhu, *Org. Lett.* **2001**, *3*, 3927–9.
[69] R. Iyengar, K. Schildknegt, J. Aubé, *Org. Lett.* **2000**, *2*, 1625–7.
[70] R. Iyengar, K. Schildknegt, M. Morton, J. Aubé, *J. Org. Chem.* **2005**, *70*, 10645–52.
[71] G. Stork, J.E. Dolfini, *J. Am. Chem. Soc.* **1963**, *85*, 2872–3.
[72] M.J. Dearden, C.R. Firkin, J.-P.R. Hermet, P. O'Brien, *J. Am. Chem. Soc.* **2002**, *124*, 11870–1.
[73] J.-P.R. Hermet, D.W. Porter, M.J. Dearden, *et al.*, *Org. Biomol. Chem.* **2003**, *1*, 3977–88.
[74] R. Kuwano, M. Sawamura, J. ShiraI, M. Takahashi, Y. Ito, *Tetrahedron Lett.* **1995**, *36*, 5239–42.
[75] A. Berkessel, M. Schroeder, C.A. Sklorz, *et al.*, *J. Org. Chem.* **2004**, *69*, 3050–6.
[76] J.A. Wendt, J. Aubé, *Tetrahedron Lett.* **1996**, *37*, 1531–4.
[77] B.T. Smith, J.A. Wendt, J. Aubé, *Org. Lett.* **2002**, *4*, 2577–9.
[78] Y. Zeng, B.T. Smith, J. Hershberger, J. Aubé, *J. Org. Chem.* **2003**, *68*, 8065–7.
[79] T.F. Spande, H.M. Garraffo, H.J.C. Yeh, Q.L. Pu, L.K. Pannell, J.W. Daly, *J. Nat. Prod.* **1992**, *55*, 707–22.
[80] A. Wrobleski, K. Sahasrabudhe, J. Aubé, *J. Am. Chem. Soc.* **2002**, *124*, 9974–5.

[81] A. Wrobleski, K. Sahasrabudhe, J. Aubé, *J. Am. Chem. Soc.* **2004**, *126*, 5475–81.
[82] J.R. Stille, R.H. Grubbs, *J. Am. Chem. Soc.* **1986**, *108*, 855–6.
[83] N. Holub, S. Blechert, *Chem. Asian J.* **2007**, *2*, 1064–82.
[84] C.-Y. Chen, D.J. Hart, *J. Org. Chem.* **1990**, *55*, 6236–40.
[85] C.-Y. Chen, D.J. Hart, *J. Org. Chem.* **1993**, *58*, 3840–9.
[86] Y. Zeng, J. Aubé, *J. Am. Chem. Soc.* **2005**, *127*, 15712–3.
[87] H.-s. Chung, P.-m. Hon, G. Lin, P. P.-h. But, H. Dong, *Planta Medica* **2003**, *69*, 914–20.
[88] K.J. Frankowski, J.E. Golden, Y. Zeng, Y. Lei, J. Aubé, *J. Am. Chem. Soc.* **2008**, *130*, 6018–24.
[89] W.H. Pearson, W.-K. Fang, *J. Org. Chem.* **2000**, *65*, 7158–74.
[90] R. Fujimoto, Y. Kishi, J.F. Blount, *J. Am. Chem. Soc.* **1980**, *102*, 7154–6.
[91] M.-Y. Zhao, P. Gu, Y.-Q. Tu, C.-A. Fan, Q. Zhang, *Org. Lett.* **2008**, *10*, 1763–6.
[92] A. Greenberg, in *Structure and Reactivity* (eds.: J.F. Liebman, A. Greenberg), VCH, New York, **1988**, pp. 139–78.
[93] A. Greenberg, in *The Amide Linkage: Structural Significance in Chemistry, Biochemistry, and Materials Science* (eds.: A. Greenberg, C.M. Breneman, J.F. Liebman), Wiley–Interscience, New York, **2000**, pp. 47–83.
[94] L. Yao, J. Aubé, *J. Am. Chem. Soc.* **2007**, *129*, 2766–7.
[95] Y. Lei, A.D. Wrobleski, J.E. Golden, D.R. Powell, J. Aubé, *J. Am. Chem. Soc.* **2005**, *127*, 4552–3.
[96] K. Tani, B.M. Stoltz, *Nature* **2006**, *441*, 731–4.
[97] T. Ly, M. Krout, D.K. Pham, K. Tani, B.M. Stoltz, R.R. Julian, *J. Am. Chem. Soc.* **2007**, *129*, 1864–5.
[98] W.H. Pearson, B.M. Gallagher, *Tetrahedron* **1996**, *52*, 12039–48.
[99] D.S. Reddy, D. Vander Velde, J. Aubé, *J. Org. Chem.* **2004**, *69*, 1716–9.
[100] E. Fenster, D.K. Rayabarapu, M. Zhang, *et al.*, *J. Comb. Chem.* **2008**, *10*, 230–4.
[101] P. Chaudhry, F. Schoenen, B. Neuenswander, G. Lushington, J. Aubé, *J. Comb. Chem.* **2007**, *9*, 473–6.

8

Radical Chemistry with Azides

Ciril Jimeno and Philippe Renaud

University of Bern, Department of Chemistry and Biochemistry, Freiestrasse 3, CH-3012 Bern, Switzerland

8.1 Introduction

Organic azides represent a very interesting class of compounds since they are precursor of a wide range of nitrogen containing organic molecules.[1] Moreover, they possess a unique reactivity pattern orthogonal to many other organic functional groups. This assessment is also true in the field of radical chemistry. For instance, azides can be considered as protected amino groups that are perfectly tolerated in C-C bond forming reactions. This is illustrated by the synthesis of (–)-batzelladine D recently disclosed by P.A. Evans.[2] In this synthesis, a pyrrolo[1,2f]pyrimidine is built via a 5-*exo* trig radical cyclization. The two azido substituents present in the molecule are not reduced under the reaction conditions used to run the cyclization (tributyltin hydride and triethylborane in the presence of air at room temperature) (Scheme 8.1).

The stability of azides in radical reactions is further illustrated by the radical fragmentation of carbohydrate anomeric alkoxyl radicals bearing an azido substituent at position 3 reported by Suarez (Scheme 8.2).[3,4] After fragmentation, the 2-azido-substituted radical is iodinated and no β-fragmentation of the azide is observed.

However, under suitable conditions, azides react with a variety of radicals and this is the basis of several useful synthetic procedures for the formation of carbon–nitrogen bonds. For instance, synthesis of azides by radical addition of an azidyl radical to alkenes (Scheme 8.3a) and by reaction of an alkyl radical with an azidating reagent (Scheme 8.3b) will be presented. The reduction of azides leading to aminyl radicals (Scheme 8.3c) and the addition of alkyl radicals to alkyl azides (Scheme 8.3d) will also be discussed.

Organic Azides: Syntheses and Applications Edited by Stefan Bräse and Klaus Banert

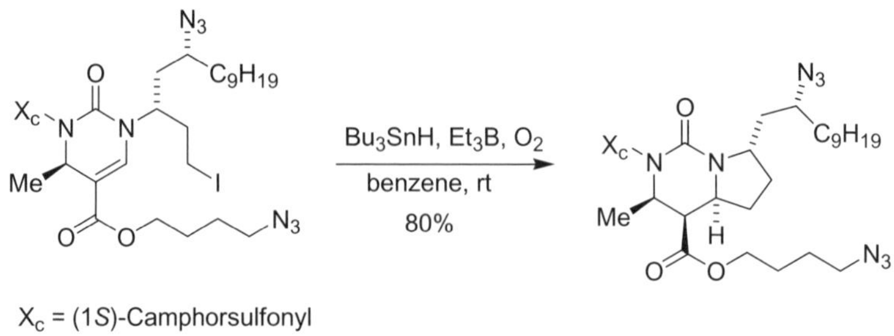

Scheme 8.1

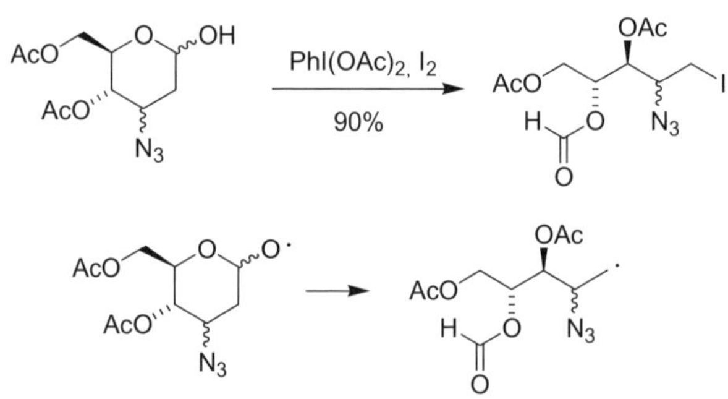

Scheme 8.2

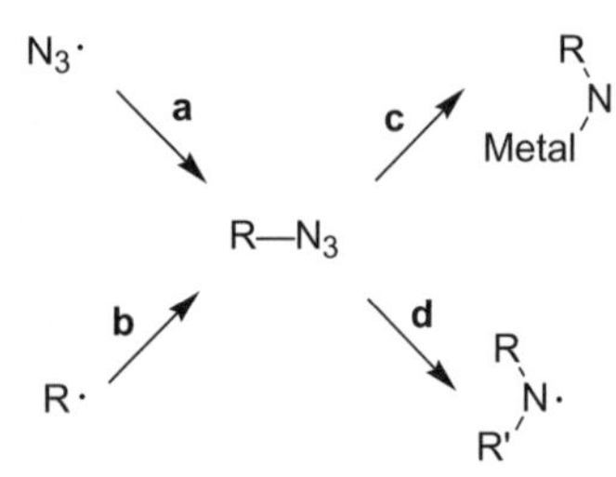

Scheme 8.3

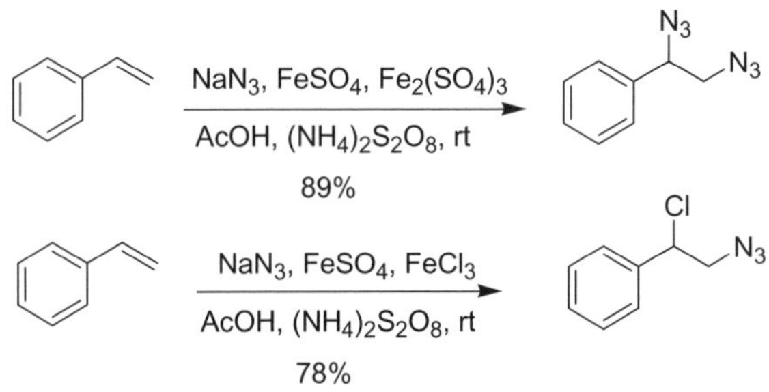

Scheme 8.4

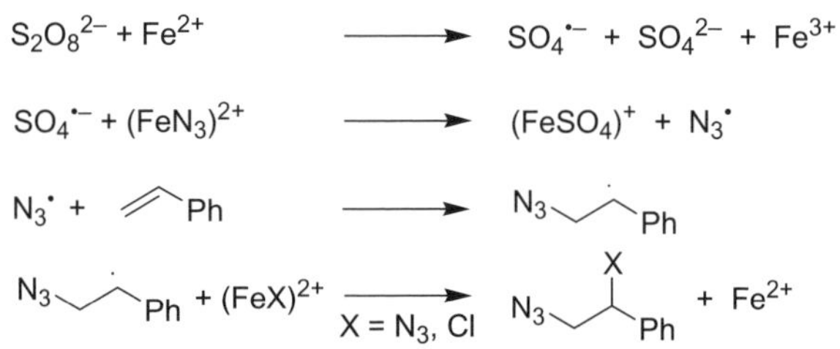

Scheme 8.5

8.2 Addition of the Azidyl Radical onto Alkenes

The use of the azidyl radical in organic synthesis offers the opportunity to functionalize olefins into the corresponding alkylazides, which are equivalent forms of the ubiquitous amino group. The reaction with olefins is especially interesting due to the electrophilic character of the azidyl radical.[5]

8.2.1 Metal Generated Azidyl Radicals

The generation of the azidyl radical by oxidation of the azide anion was accomplished by Minisci *et al.* using Fe^{2+} salts and *tert*-butyl hydroperoxide.[6] Further developments in the same group led to the use of a Fe^{2+}/Fe^{3+} system and several oxidizing agents such as H_2O_2, permanganate, or Ce^{4+} salts.[7,8] In this way, the diazidation of olefins is achieved under mild reaction conditions. Indeed, iron salts allow to control the generation of the azidyl radical and to sustain a radical chain process. Selective diazidation or choroazidation of styrene are performed by using $Fe_2(SO_4)_3$ or $FeCl_3$, respectively, in the presence of ammonium peroxydisulfate (Scheme 8.4).[9,10]

A reaction mechanism involving the formation of the azidyl radical from an iron(III) azide was proposed (Scheme 8.5). Only catalytic amounts of iron(II) salts are actually needed to sustain the radical chain. Moreover, other types of functionalization are possible depending on the iron(III) salt used.

A practical procedure for the generation of the azidyl radical using cerium ammonium nitrate (CAN) was developed somewhat later.[11] This method is taking advantage of the

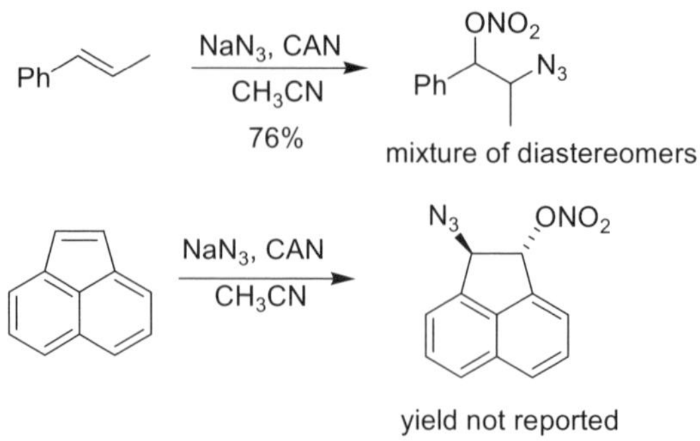

Scheme 8.6

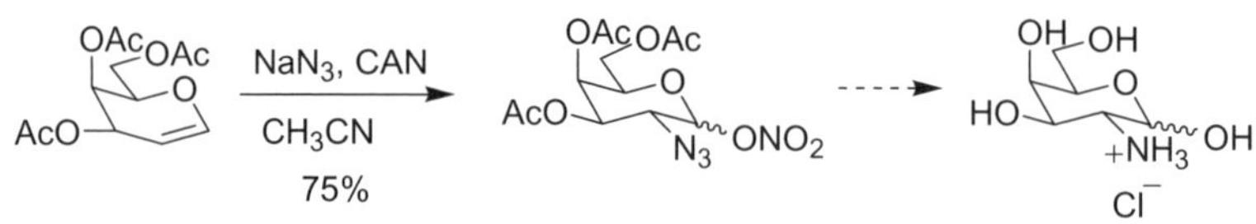

Scheme 8.7

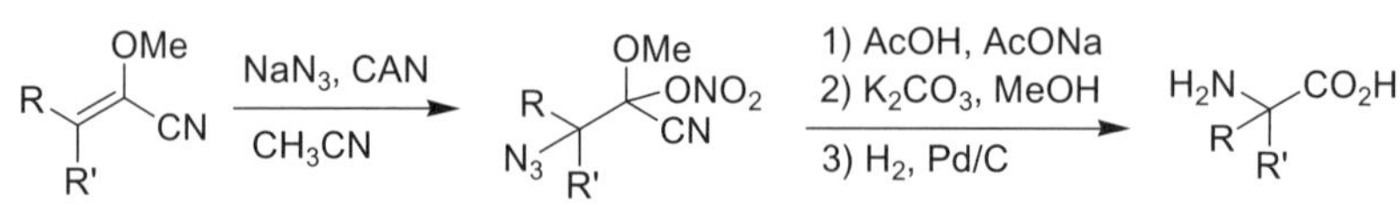

Scheme 8.8

old reaction between CAN and sodium azide that yields quantitatively dinitrogen.[12] An efficient nitroazidation of olefins was achieved (Scheme 8.6). For instance, nitroazidation of *trans*-α-methylstyrene affords a mixture of *syn* and *anti* isomers in good yield. However, a single diastereomer (*trans*) is obtained when the reaction is performed on the cyclic acenaphthylene.

The NaN_3/CAN procedure is very popular in carbohydrate chemistry to prepare 2-aminosugars (Scheme 8.7).[13,14] The azide is introduced stereoselectively in equatorial orientation (*galacto* configuration). Several authors took advantage of this chemistry for the synthesis of other amino sugars.[15–19]

α-Methoxy acrylonitriles are readily prepared by olefination of aldehydes and ketones with Wittig reagents. Their nitroazidation using NaN_3/CAN yields the corresponding β-azidonitriles. Hydrolysis under mild conditions gives the α-azido carboxylic acids that can be subsequently hydrogenated to α-amino acids (Scheme 8.8).[20]

Magnus group used the NaN_3/CAN combination to achieve the α-azidation of silyl enol ethers. This reagent allows to prepare the corresponding α-azido ketones in good yields (Scheme 8.9).[21] Only azidation products are obtained, and no product arising from an hypothetical nitration is built. This method has been used for the synthesis of (–)-barreazines A and B.[22]

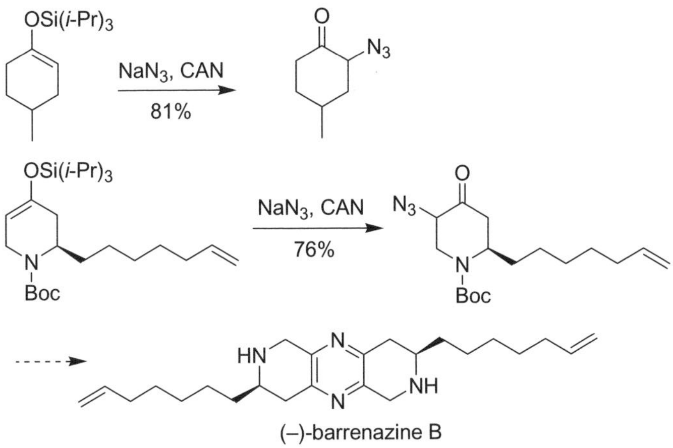

Scheme 8.9

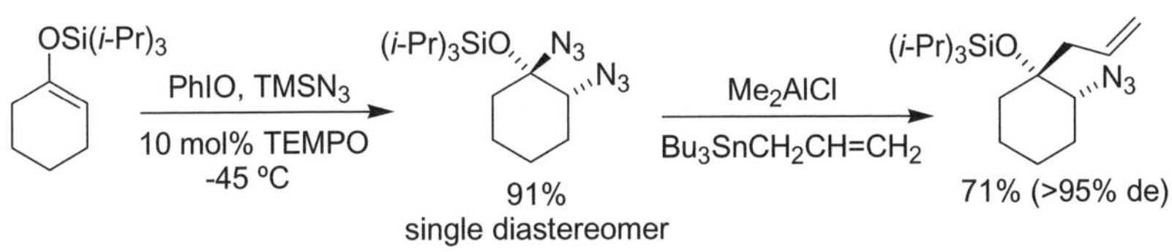

Scheme 8.10

Other metals are also used to azidate olefins or their allylic positions. Pb(IV) salts in combination with $TMSN_3$ are particularly useful for these purposes. Although the reaction mechanism is unknown, the formation of radical pairs was proposed. In most examples, though, ionic mechanisms are invoked, and some experiments clearly indicated the absence of free radicals.[23–25] Manganese(III) salts in the presence of a large excess of sodium azide are also use for the diazidation of olefins in good yields.[26,27] Even though the formation of azidyl radicals was not discarded, the authors favored a ligand-transfer oxidation from an azido-Mn(III) complex (see also Section 8.3.1.4).

8.2.2 Azidation using Hypervalent Iodine Compounds

Hypervalent iodine compounds represent a useful source of azidyl radicals. For instance, a mixture of iodosyl benzene (PhIO), acetic acid, and sodium azide is employed for the diazidation of several olefins. A carbocationic mechanism has been proposed but azidyl radicals are likely responsible for the observed reactivity.[28,29]

The diazidation of silyl enol ethers was also achieved by Magnus.[30] The reactivity of the PhIO/$TMSN_3$ mixture is extremely temperature dependent. Catalytic amounts of TEMPO activate the diazidation process (Scheme 8.10).

Under the optimized reaction conditions, diazidation of cyclic silyl enol ethers can be performed in good yields and generally high diastereoselectivities.[21,30] The isolated diazides also are useful intermediates in the way to more elaborated compounds. For instance,

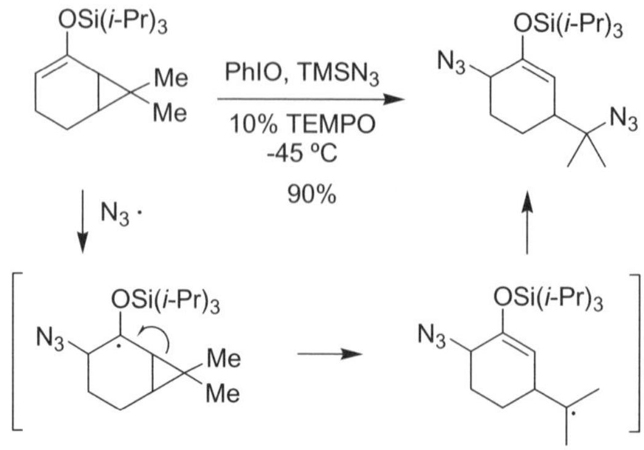

Scheme 8.11

the azide at C1 could be efficiently substituted by carbon nucleophiles in the presence of Me_2AlCl, with retention of configuration, generally in high diastereoselectivity, and without affecting the azide at C2 (Scheme 8.10).[21,30]

Proof of the involvement of azidyl radicals in this process was obtained by using a silyl enol ether with a fused cyclopropyl ring (Scheme 8.11). The rearranged product isolated in high yield clearly suggests the participation of radical intermediates.

Tingoli reported that a $PhI(OAc)_2/NaN_3$ mixture can be used in conjunction with diphenyl diselenide to afford the azidoselenation of olefins (Scheme 8.12). Many different olefins were bifunctionalized in this way. The regiochemistry of the adducts as well as the reaction of radical probes such as *β*-pinene support strongly the formation of free radicals.[31,32]

8.2.3 Halogen Azides as a Source of Azidyl Radicals

The seminal work of Hassner has shown that halogen azides react with alkenes both in ionic and radical pathways leading to opposite regiochemistry (Scheme 8.13).[33,34] The nature of the reagent and the polarity of the media play a fundamental role. Iodine azide reacts via an ionic pathway whereas the two mechanisms may operate for bromine and chlorine azides. Apolar solvents such as pentane and irradiation of the reaction mixture favor the radical process.

Chlorine azide is used for the radical azidation of glycals under UV radiation, whereas an ionic mechanism operates in the dark.[35] As for the NaN_3/CAN system, high equatorial selectivity for the azide at C2 is obtained, whereas the selectivity at the anomeric carbon was poorer (Scheme 8.14).

Halogen azides add also to *α,β*-unsaturated carbonyl compounds[36] and allenes,[37] often following ionic pathways. In the case of allenes, for example, IN_3 behaves exclusively as an ionic reagent, whereas BrN_3 reacts via an ionic mechanism at −65 °C and a radical one at room temperature. In the later case, however, unstable bisadducts are obtained that often lead to explosive decompositions.[37] An interesting alternative to halogen azides was developed by combining $PhIO/TMSN_3$ with tetraethylammonium halides.[38] Although the exact nature of the reagent formed is unknown, it is likely that halogen azides are slowly generated *in situ*, thus overcoming their inherent unstability.

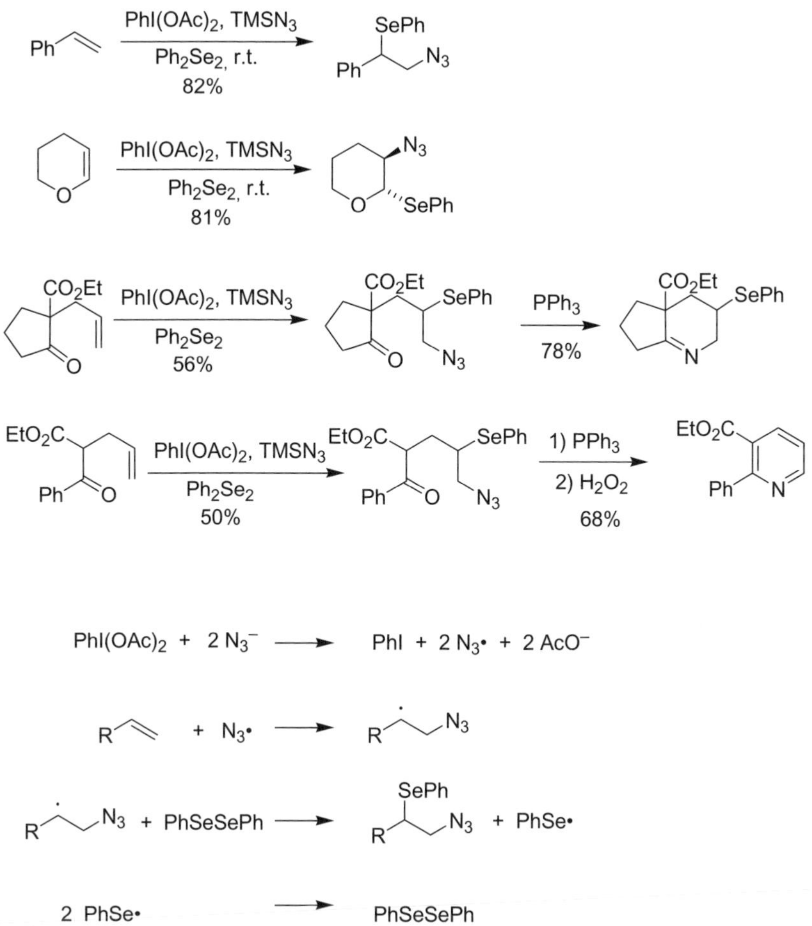

Scheme 8.12

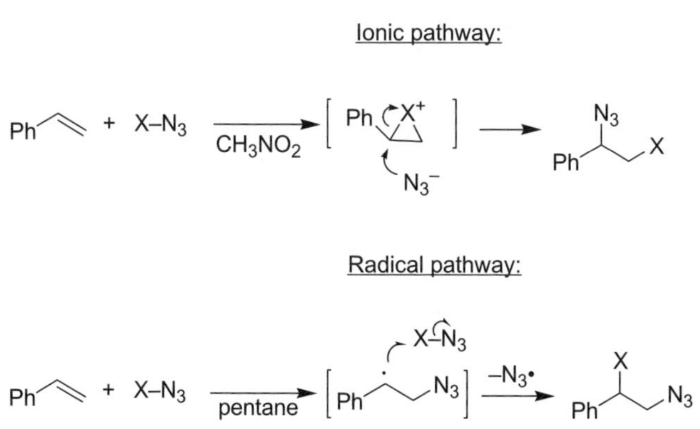

Scheme 8.13

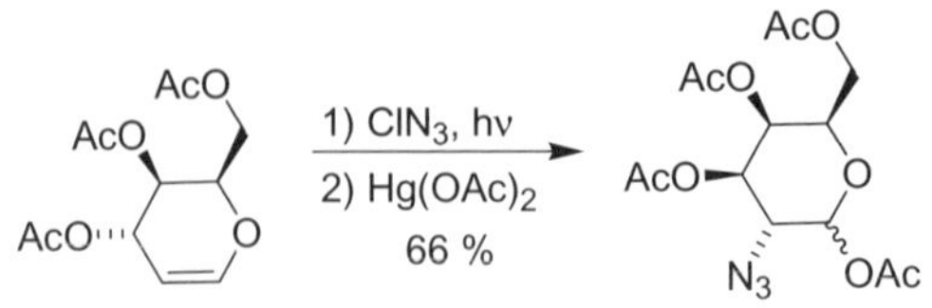

Scheme 8.14

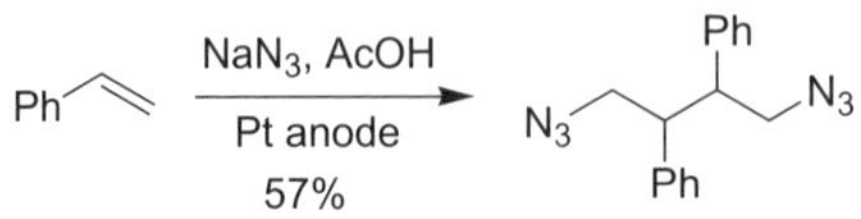

Scheme 8.15

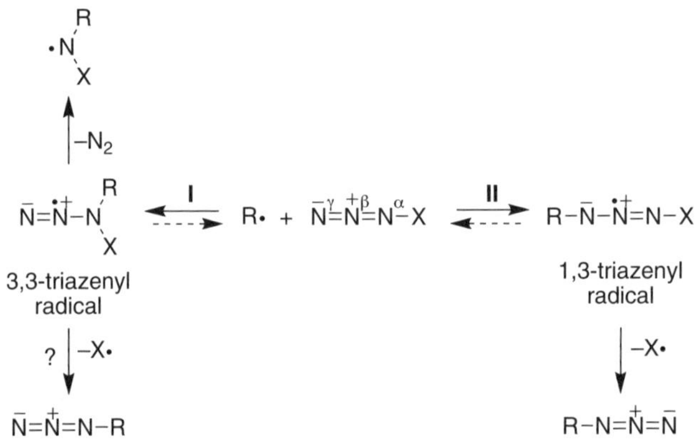

Scheme 8.16

8.2.4 Electrochemically Generated Azidyl Radicals

The relatively low standard potential of the $N_3\cdot/N_3^-$ redox couple in aqueous solution (+1.33 ± 0.01 V vs. NHE) is even more reduced in organic solvents, and therefore the generation of azidyl radicals by electrochemical methods is perfectly feasible.[5] The multigram scale dimerization of styrene represents an early synthetic application of the electrochemical process (Scheme 8.15).[39,40] However, the scope of this reaction is so far limited since other substrates give poor yields and/or significant amounts of by-products.

8.3 Azidation of Carbon Centered Radicals

The addition of a carbon-centered radical to a nitrogen-containing trap is not a very common process. Organic azides of type X–N_3 have also been investigated as radical trap. They can undergo homolytic addition at either the inner (N^α, path I) or the terminal (N^γ, path II) nitrogen to give a 3,3-triazenyl or a 1,3-triazenyl radical, respectively (Scheme 8.16).[41] The 3,3,-triazenyl radical (path I) evolves presumably by rapid loss of nitrogen to furnish an aminyl radical (this behavior will be discussed under point 5). Fragmentation

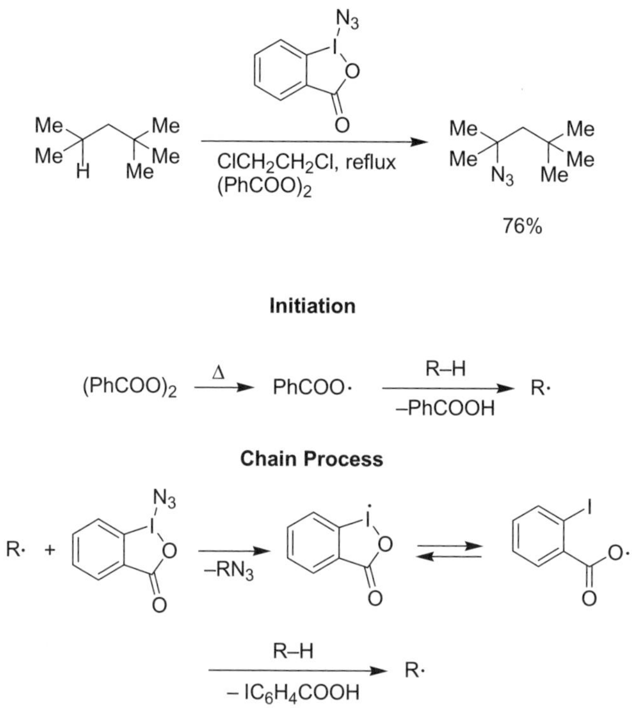

Scheme 8.17

of X· to deliver the azide R–N_3 cannot be excluded. On the other hand, the 1,3-triazenyl radical (path II) may fragment to liberate the radical X· and R–N_3. This last reaction corresponds to an azidation of the radical R·.

8.3.1 Radical Azidation

8.3.1.1 Azidation with Iodine Derivatives

Zhdankin reported the preparation of stable azidoiodinanes.[42] These compounds are good azidating agents for various organic substrates and they can be used for direct azidation of hydrocarbons at high temperature in the presence of a radical initiator. For example, 2,2,4-trimethylpentane reacts with 1-azido-1,2-benziodoxole-3-(1*H*)-one in refluxing 1,2-dichloroethane in the presence of benzoylperoxide to afford the corresponding azide in 76% yield (Scheme 8.17). The proposed mechanism involves hydrogen atom abstraction by the 2-iodobenzoyl radical followed by azidation of the alkyl radical by the azidoiodinane. This elegant chain process proceeds with moderate to good yields at the secondary and tertiary positions of several different alkanes.

Substitution of activated hydrogen atoms by azido groups has been investigated recently by Bols. Initially, reactions were rung with IN_3 and good results were obtained for azidation of benzyl ethers[43–45] (Scheme 8.18). Interestingly, the α-azidobenzyl ethers react efficiently with alkyl and aryl Grignard reagent to afford the corresponding α-alkylbenzyl or diarylmethyl ethers in good yields.[44,45]

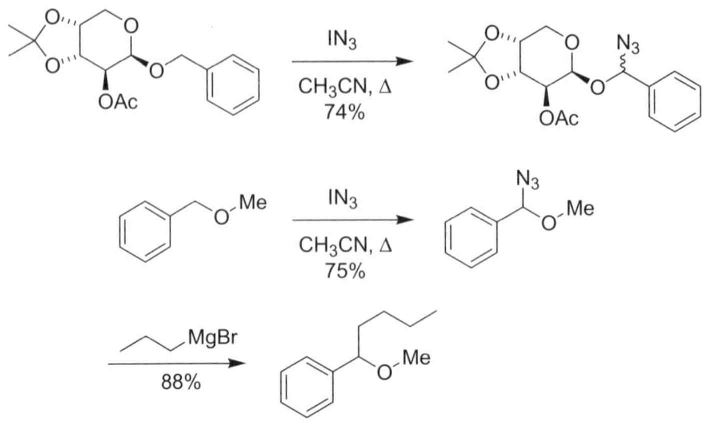

Scheme 8.18

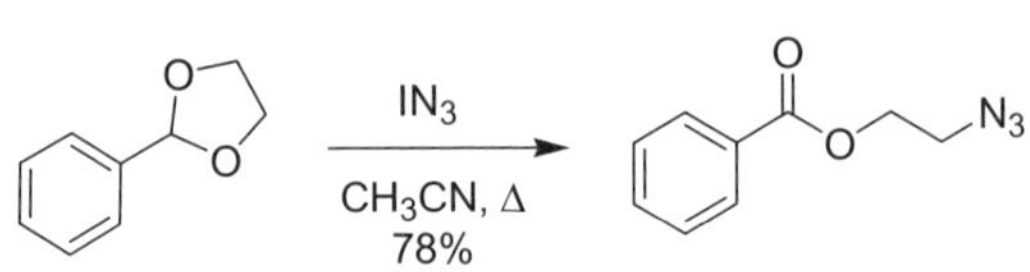

Scheme 8.19

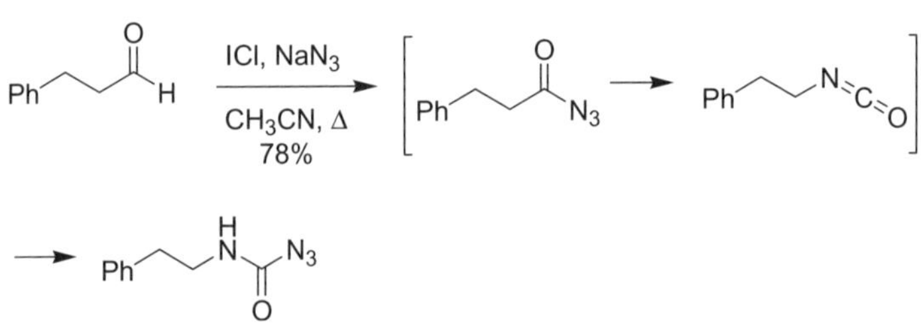

Scheme 8.20

Cyclic benzylidene acetals have also been found to react with IN_3 to provide ring-opened product via an azido-Hanessian reaction (Scheme 8.19).[46]

The same reagent allows to convert aldehydes into acyl azides.[47] If the reaction is run in the presence of NaN_3, the intermediate acyl azide is converted into a carbamoyl azide via Curtius rearrangement (Scheme 8.20).

Recently, Bols has shown that the hazardous IN_3 can be replaced by $PhI(N_3)_2$ generated in situ from $PhI(OAc)_2$ and $TMSN_3$.[48,49] This reagent leads to higher reaction rate, higher yield in several cases, and safer reaction conditions. IN_3 has been used successfully to the reactions described in Scheme 8.18–8.20. Both reagents (IN_3 and $PhI(N_3)_2$) afford very similar product distributions rationalized by closely related mechanisms involving a common hydrogen atom abstraction step by the azidyl radical (Scheme 8.21). The

Reaction with IN_3:

$$N_3\bullet + H\text{–}R \longrightarrow R\bullet + H\text{–}N_3$$

$$R\bullet + N_3\text{–}I \longrightarrow R\text{–}N_3 + I\bullet$$

$$I\bullet + I\text{–}N_3 \longrightarrow I_2 + N_3\bullet$$

Reaction with $PhI(N_3)_2$:

$$N_3\bullet + H\text{–}R \longrightarrow R\bullet + H\text{–}N_3$$

$$R\bullet + PhI(N_3)_2 \longrightarrow R\text{–}N_3 + PhI + N_3\bullet$$

Scheme 8.21

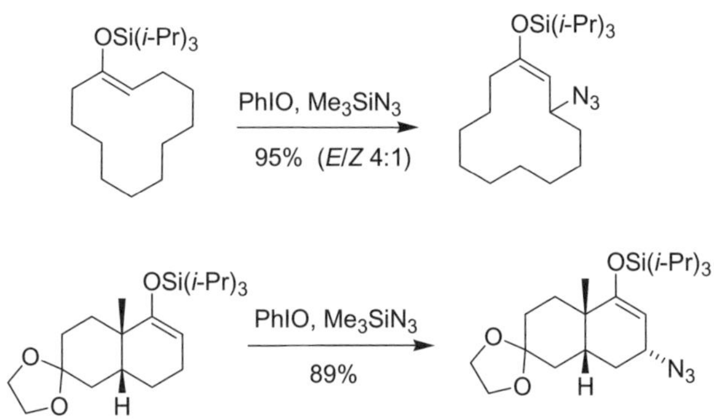

Scheme 8.22

propagation of the chain process is provided by azidation of the carbon-centered radical with IN_3 or $PhI(N_3)_2$.

Magnus developed also a combination of iodosyl benzene and trimethylsilyl azide to azidate the β-carbon of silyl enol ethers (Scheme 8.22).[30,50] The total synthesis of lycorane *amaryllidaceae* alkaloids takes advantage of this process.[51] A tentative ionic mechanism is proposed for this process. However, involvement of radicals cannot be totally excluded.[52]

Under the same reaction conditions, anilines[53] and amides[54,55] lead to the isolation of C–H substituted products at carbon atoms adjacent to nitrogen (Scheme 8.23). These azidoamines/azidoamides are precursor of iminium/acycliminium ions that are useful intermediates for C–C bond formation.

8.3.1.2 Azidation with Sulfonyl Azides

In an early experiment, Abramovitch and Breslow observed the formation of traces of alkyl azides during Curtius-type rearrangement of sulfonyl azides.[56–58] This formation was rationalized by a direct reaction between alkyl radicals and sulfonyl azides. Roberts examined the reaction of aryl and alkyl sulfonyl azides with allylstannanes in order to

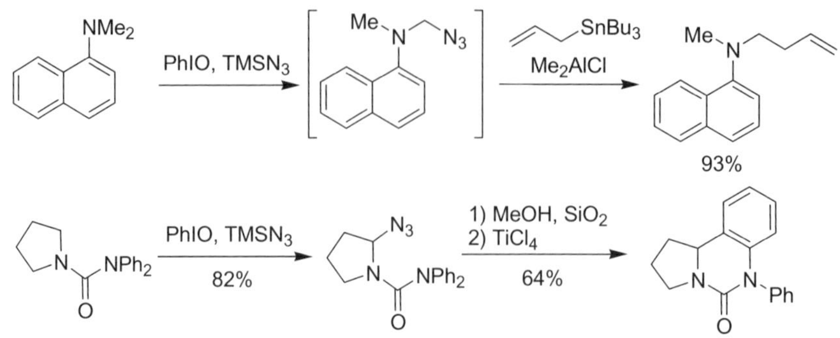

Scheme 8.23

develop an homolytic allylation reaction at nitrogen.[59] In this process, the formation of allylsulfones and stannyl azides represent the major side reactions. These products arise from the addition of the tin radical at the terminal nitrogen followed by fragmentation of a sulfonyl radical. These initial observations led to an optimized procedure for the radical azidation of iodides with ethanesulfonyl azide (Scheme 8.24).[60,61] Ethanesulfonyl azide, easily prepared from ethanesulfonyl chloride and sodium azide, is a stable liquid that can be heated at 100 °C without decomposition. The crucial step of this process is the addition of the alkyl radical at the *N*-terminal position of the azido moiety to give a 1,3-triazenyl radical that fragments to liberate the corresponding alkyl azide and the ethanesulfonyl radical (see pathway I in Scheme 8.16).[62] After sulfur dioxide extrusion, the ethyl radical can propagate the chain by an iodine atom transfer process. The radical azidation with ethanesulfonyl azide and DLP works well with a variety of secondary and tertiary alkyl iodides (see Scheme 8.24 for typical examples). Primary alkyl iodides are not efficiently converted into azides. This inefficiency is caused by the nearly thermoneutral iodine atom transfer between the ethyl radical and the primary alkyl iodide as well as by the lower nucleophilicity of primary alkyl radicals relative to secondary and tertiary ones.

Xanthates are also suitable precursors for radical group transfer reactions. For instance, the anomeric xanthate prepared from α-bromo-2-desoxyglucose and potassium *O*-ethyl xanthate gives the corresponding anomeric azide as a single α-anomer in 74% yield (Scheme 8.25). Interestingly, the preparation of such α-anomeric azides is much more difficult than the β-isomers when classical nucleophilic substitution approaches are employed. Moreover, these anomeric azides are useful intermediates for the preparation of biologically important *N*-linked glycoconjugates.

Porter reported the first decarboxylative azidation of thiohydroxamates (Barton PTOC esters).[63] Moderate to good yields of the desired azides were obtained together with the *S*-pyridyl derivatives resulting from the direct trapping of the radical by the Barton ester. Interestingly, these reactions proved to be diastereoselective as demonstrated by the first example depicted in Scheme 8.12. The formation of the *S*-pyridyl derivatives (Barton rearrangement) limits severely the scope of this process. The use of a new type of thiohydroxamate esters (MPDOC esters) allows to overcome this difficulty (Scheme 8.26).[64] These esters are more stable than the classical Barton esters and less prone to rearrange

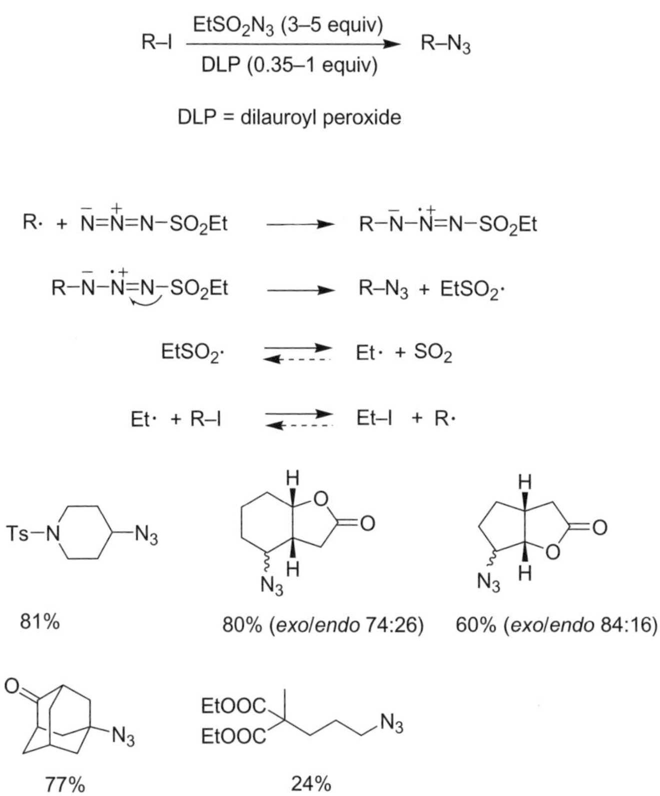

Scheme 8.24

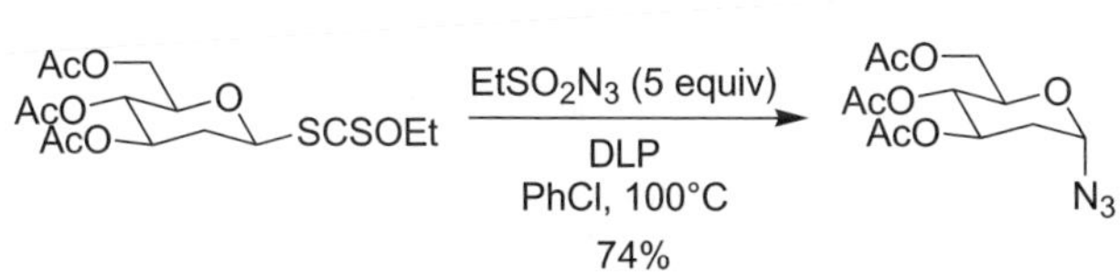

Scheme 8.25

under radical conditions. In the case of α-alkoxy and α-amino acids, optimal results are obtained with the even more stable MMDOC esters developed recently by Kim.

In order to broaden the scope of the reaction, ethanesulfonyl azide was replaced by benzenesulfonyl azide and in the presence of hexabutylditin as chain transfer reagent.[61] Because of the instability of the phenyl radical, the intermediate benzenesulfonyl radical does not liberate SO_2. The proposed chain reaction is described in Scheme 8.27. Under these reaction conditions, the azidation reaction is clean and fast (≤4 h) and purification of the products is easier than in the procedure involving $EtSO_2N_3$. Cyclization reactions are efficiently performed (Scheme 8.27).

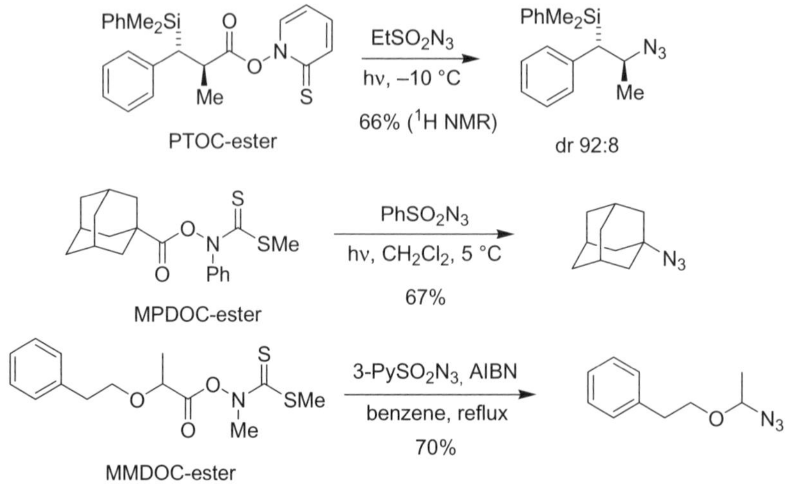

Scheme 8.26

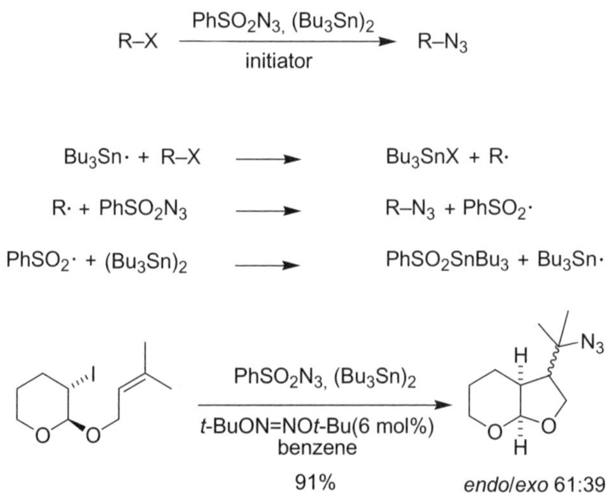

Scheme 8.27

The intermolecular addition of carbon-centered radicals to unactivated alkenes followed by azidation (a formal carboazidation of alkenes) has been reported.[65,66] A one-pot procedure similar to the one used for intramolecular reactions gives good results (Scheme 8.28). Slow addition of benzenesulfonyl azide is not necessary because this electrophilic reagent does not react with the initial electrophilic or ambiphilic radicals. Excellent results are obtained with α-iodo and α-xanthate esters. α-Bromoacetates give also satisfactory results. The carboazidation process allows to prepare pyrrolidinone derivatives in a straightforward manner (Scheme 8.28, bottom example). A tin-free version of this reaction using triethylborane instead of hexabutylditin has also been reported.[67,68]

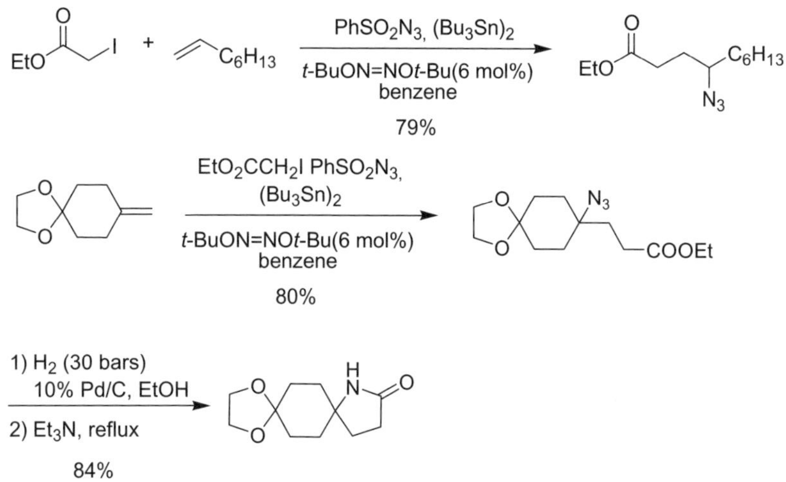

Scheme 8.28

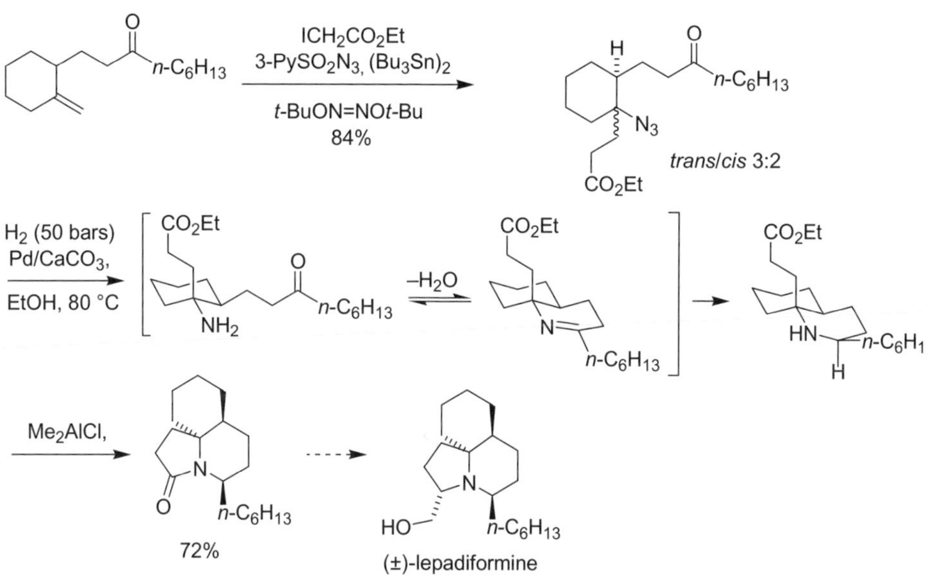

Scheme 8.29

The carboazidation has been applied in natural product synthesis. For instance, lepadiformine[69] (Scheme 8.29) and hyacinthacine A_1[70] (Scheme 8.30) have been prepared via carboazidation of a methylenecyclohexane derivative and a chiral allylsilane, respectively (Scheme 8.29 and 8.30). In this last case, a remarkable diastereoselectivity was observed in the carboazidation process.[71,72] Both reactions are based on the use of 3-pyridylsulfonyl azide instead of benzenesulfonylazide to facilitate the purification of the intermediate azides.[73]

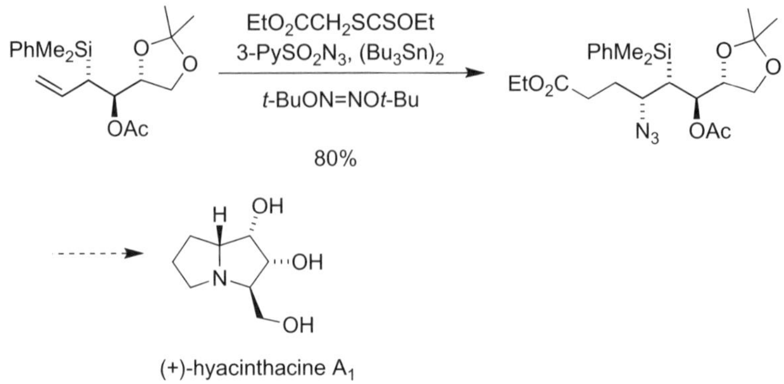

Scheme 8.30

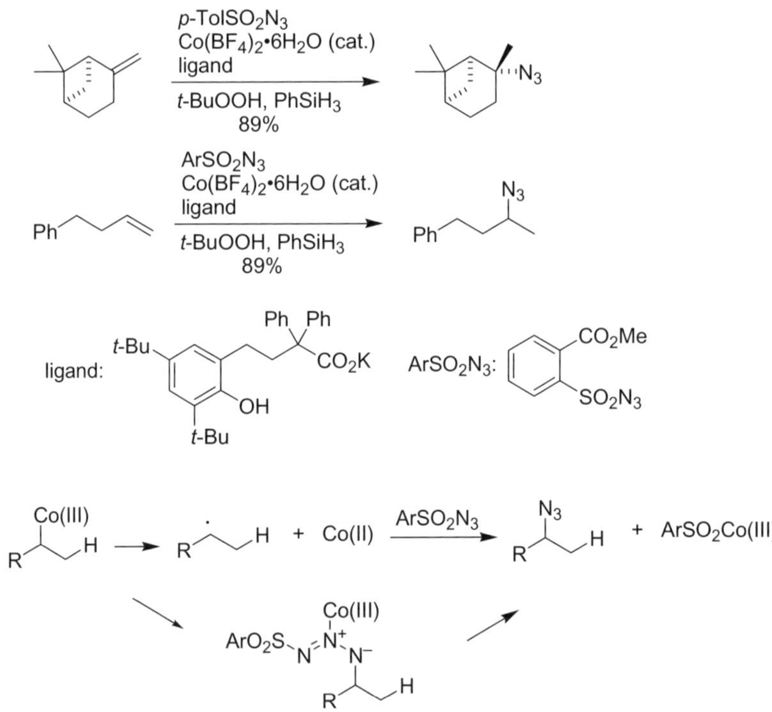

Scheme 8.31

8.3.1.3 Cobalt-catalyzed Hydroazidation of Olefins

Carreira has developed an interesting Co-catalyzed hydroazidation of alkenes (Scheme 8.31) (see also Chapter 4).[74,75] This reaction employed sulfonyl azides as nitrogen source. It allows a direct access to secondary and tertiary azides. The mechanism of this reaction is not fully understood and radical intermediates may be involved.

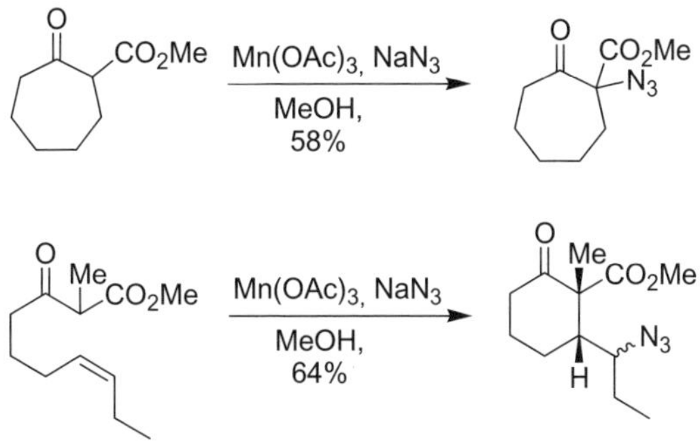

Scheme 8.32

8.3.1.4 Azidation with Mn(III)-NaN_3

Snider has reported that the radicals formed in Mn(III)-based oxidative free-radical cyclizations of *β*-keto esters and malonate esters can be trapped with sodium azide and Mn(III) to give cyclic and bicyclic azides in moderate to good yield (Scheme 8.32).[76] The mechanism involves presumably an azide transfer from a Mn(III)-azide species as proposed by Fristad for the diazidation of alkenes.[26]

8.3.2 Radical Additions to Alkyl and Aryl Azides

Kim reported efficient radical cyclizations involving azides as radical traps (Scheme 8.33).[77] This intramolecular amination process is efficient and proceeds via a 3,3-triazenyl radical that readily eliminates nitrogen to afford an aminyl radical. The use of $(TMS)_3SiH$ gives better results than Bu_3SnH since azides are relatively inert toward tris(trimethylsilyl) silyl radicals.[77,78] However, tin hydride can be used with alkyl iodides since the iodine atom abstraction by the stannyl radical is faster than its addition to the azide.

Murphy has investigated domino radical cyclizations of iodoaryl azides.[79] This reaction found several applications in the total synthesis of natural product, as for instance Murphy's synthesis of aspidospermidine (Scheme 8.34),[80,81] vindoline,[82] horsfiline,[83,84] and coerulescine.[83]

The preparation of indoles from azidophenylacetylene derivatives has been reported by Montevecchi (Scheme 8.35).[85] In this approach, the initial alkenyl radical is generated by a reversible addition of a thiyl radical onto an alkyne.

8.4 Aminyl and Amidyl Radicals via Reduction of Azides

The reduction of alkylazides by single electron transfer (SET) cleanly furnishes the alkylaminyl radicals plus molecular nitrogen. A variety of metal reagents and photochemical methods exist. The reaction can also be performed with organic radicals, which add to the azide and release dinitrogen, too (see also Section 8.3.2). The resulting aminyl radical can be simply reduced to the corresponding amine or can further react with radical traps (olefins) to form C–N bonds.

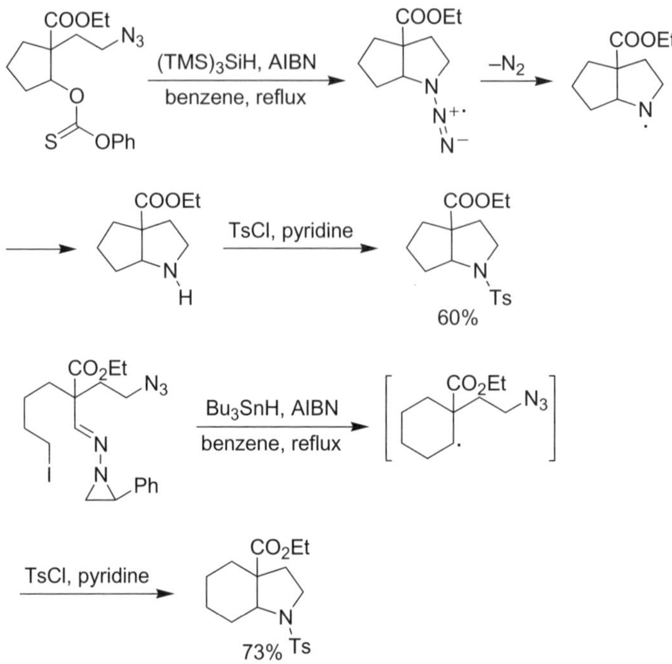

Scheme 8.33

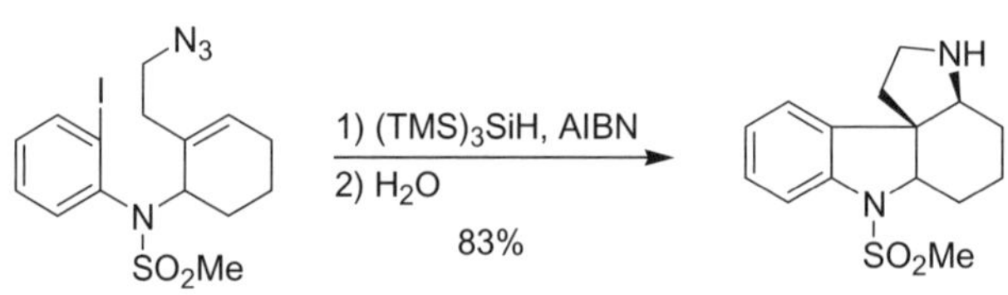

Scheme 8.34

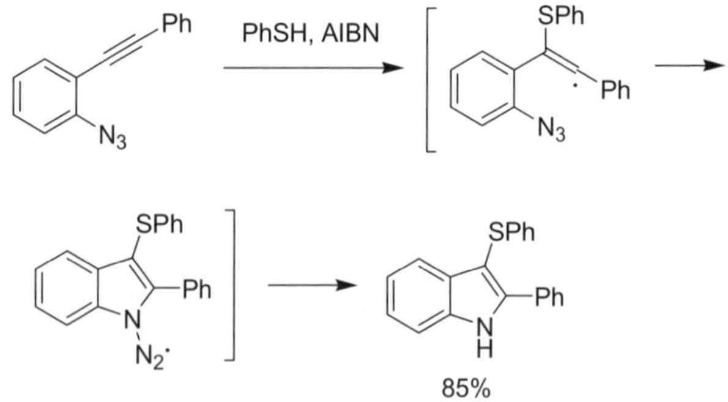

Scheme 8.35

8.4.1 Photo- and Electrochemical Reductions of Organic Azides to Amines

Initial studies focused on the photolysis of organic azides (acyl azides) in isopropanol to furnish the corresponding amide.[86] Similar results are obtained by using a radical initiator.[87] The reaction is triggered by addition of the 2-hydroxypropyl radical to the acylazide. The triazenyl radical intermediate undergo a 1,5-hydrogen transfer coupled to the loss of nitrogen and formation of acetone. The benzoylaminyl radical is finally reduced by isopropanol thus propagating the chain reaction (Scheme 8.36). Sulfonyl azides are also successfully reduced under photolysis in isopropanol.[88]

8.4.2 Reduction of Organic Azides with Metals

This reaction also works presumably through the intermediacy of aminyl radicals, as inferred from the nitrogen evolution that accompanies the process. The use of water, alcohols, or some kind of proton donor is essential. When Cu(0) is used, typical by-products of oxidation of alcohols are often observed, which also support a radical mechanism. However, the formation of nitrenes cannot be discarded.[89–91] More selective reaction conditions involve zinc dust in the presence of acetic acid, which have found widespread use in the literature. An early example of application implied the reduction of geranylazide to geranylamine (Scheme 8.37).[91,92] This method has been also successfully applied in carbohydrate chemistry.[93,94]

Fürstner reported that organic azides influence the zinc-induced ring-opening reactions of deoxyhalogenosugars.[95] This effect is equally effective intra- and intermolecularly (Scheme 8.38) and it is assumed that the presence of an alky azides favors a radical mediated reduction over the formation of an organozinc species.

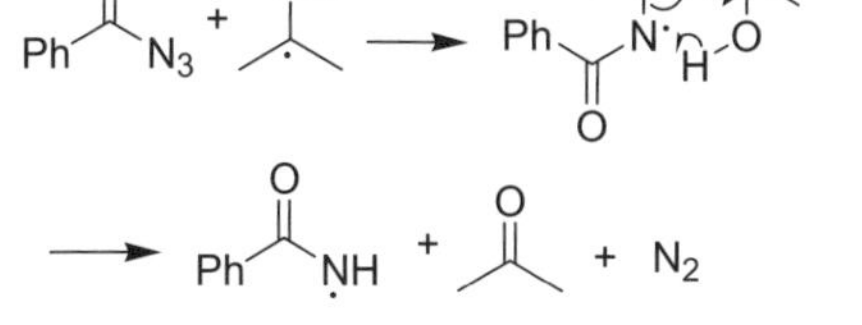

Scheme 8.36

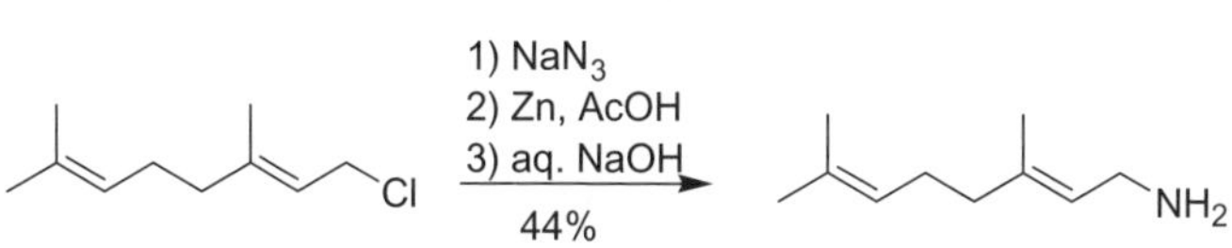

Scheme 8.37

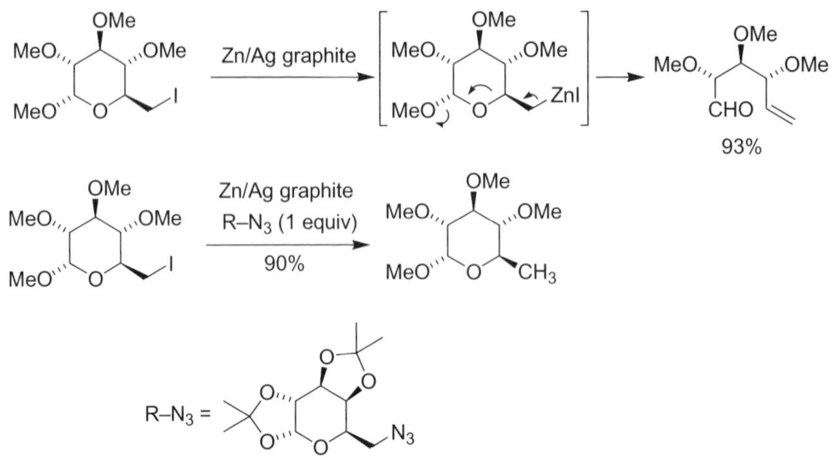

Scheme 8.38

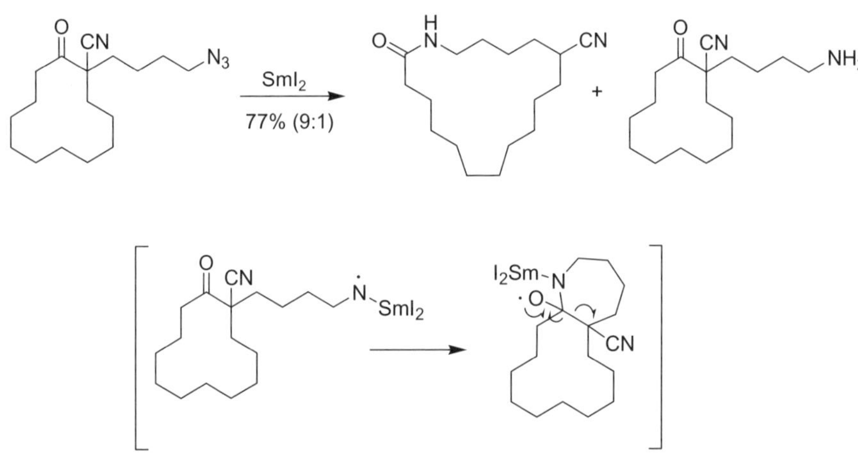

Scheme 8.39

Mg(0) or Ca(0) in methanol,[96] as well as In(0)/NH_4Cl[97] are reported to reduce azides under mild conditions. Electrochemical reduction can be achieved easily too. For example, phenylazide was reduced quantitatively to aniline using a Pt electrode and hexafluoroisopropanol as a proton source.[98]

8.4.3 Reduction of Organic Azides with SmI_2

Some low valent metal salts are good reagents for the reduction of azides through a SET mechanism. $CrCl_2$ in acidic aqueous solutions was initially used.[91,99] However, samarium diiodide turns out to be a more convenient reagent under neutral conditions in organic solvents.[100] Simultaneous reduction-ring expansion of some azidoketones allows to prepare macrolactams (Scheme 8.39).[101]

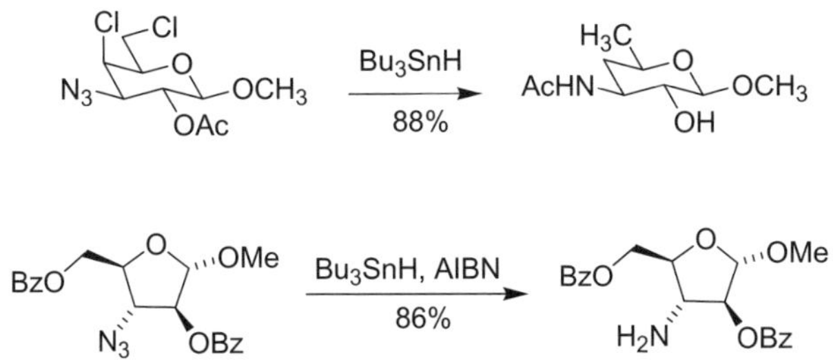

Scheme 8.40

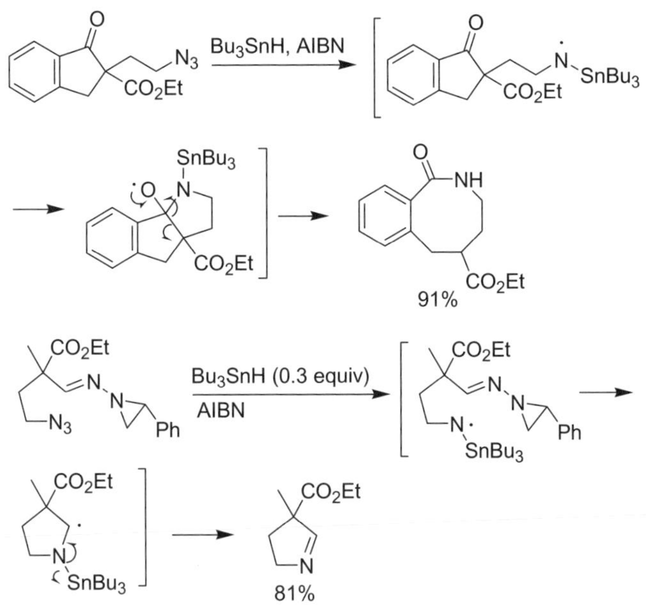

Scheme 8.41

8.4.4 Radical Reactions of Organic Azides with Tributyltin Hydride

Tributyltin hydride has been commonly used to perform the reduction of azides.[102] This transformation includes the formation of an *N*-alkylstannanaminyl radical intermediate. Radical initiators such as AIBN are known to facilitate the reaction.[103–105] The preparation of azidosugars from the corresponding azides is well documented as illustrated by the reaction of Redlich involving azide reduction, dechlorination, and acetyl transfer (Scheme 8.40, top)[106] and by the reaction of Mikhailopulo (Scheme 8.40, bottom).

The ability of the intermediate aminyl radicals to rearrange was exploited by several authors.[107] For instance, macrocyclic lactams are obtained in excellent yields via a ring expansion process developed by Kim (Scheme 8.41, top).[108] Preparation of cyclic imines via a tin hydride promoted radical cyclization onto an hydrazone is also possible (Scheme

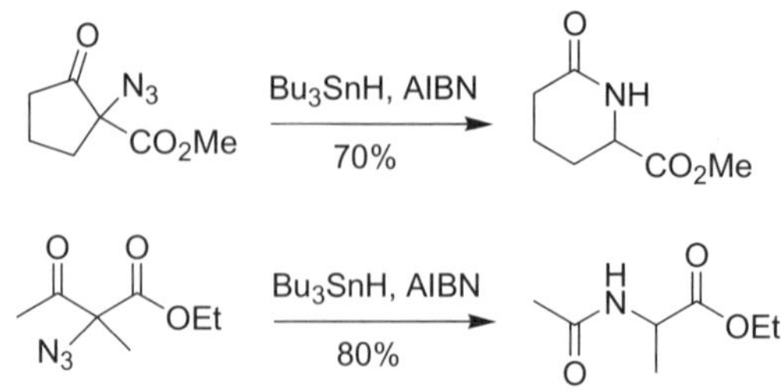

Scheme 8.42

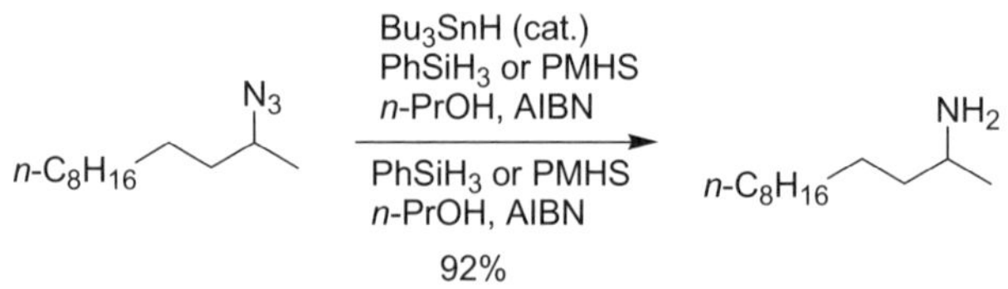

Scheme 8.43

8.41, bottom).[77] Radical cyclization of tributylstannylaminyl radicals generated from azides to carbonyl compounds,[109] nitriles and alkenes[110] are reported. Such radicals give also efficient 1,5-hydrogen transfers leading to C-centered radicals.[110]

A one-atom ring-expansion of α-azidoketones allows to prepare lactams according to Benati (Scheme 42, top).[111] The same reaction can be used for the conversion of α-azido-α-methylacetylacetate to α-aminoacid under radical reduction with Bu_3SnH (Scheme 8.42, bottom).[112]

Due to the toxicity of tin reagents, Fu developed a method using catalytic amounts of Bu_3SnH that is regenerated by using phenylsilane or polymethylhydrosiloxane (PMHS) in alcoholic solvents (Scheme 8.43).[113] Tributylgermanium hydride can substitute Bu_3SnH in some reaction such as the reduction of arylazides to anilines.[114]

8.4.5 Radical Reductions of Organic Azides with Silanes

Spagnolo and co-workers discovered that thiols catalyze the reduction of azides in the presence of triethylsilane.[115] Arylazides are reduced almost quantitatively to anilines (Scheme 8.44). Aliphatic azides were more reluctant to reduction.

The quest for tin-free reductive methods led the same authors to develop the use of indium(III) hydride Cl_2InH, generated in situ from triethylsilane and $InCl_3$.[116] Aromatic and aliphatic azides as well as sulfonyl azides and acyl azides are reduced in moderate to excellent yield to the corresponding amines and amides. Azido nitriles are efficiently converted to the pyrrolidin-2-imines (Scheme 8.45).

Reductions in aqueous media can be carried out using $(Me_3Si)_3SiH$ with a radical initiator. An amphiphilic thiol has to be used with water soluble substrates (Scheme 8.46).[117]

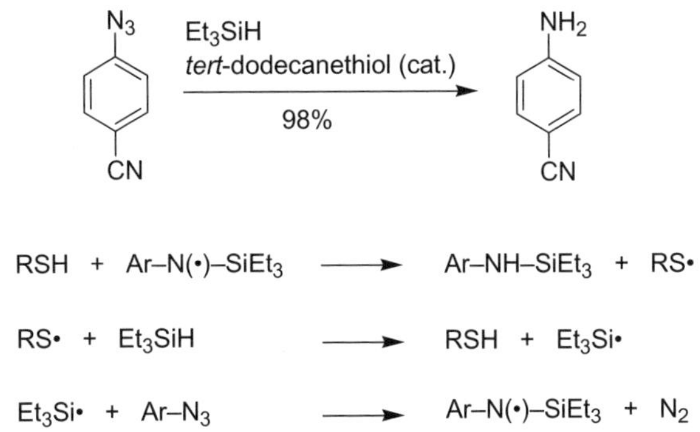

Scheme 8.44

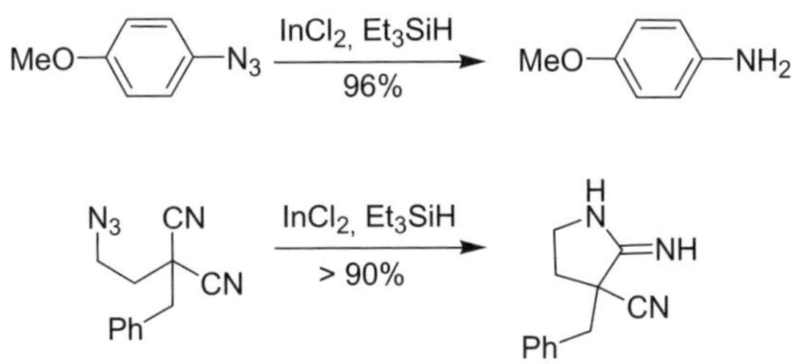

Scheme 8.45

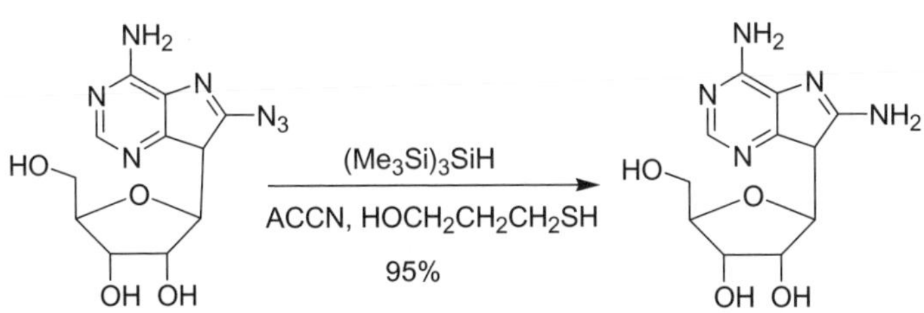

ACCN = 1,1'-azobis(cyclohexanecarbonitrile)

Scheme 8.46

8.4.6 Radical Reactions of Organic Azides with $FeCl_2$

Bach *et al.* reported the use of iron(II) chloride to perform intramolecular aminochlorinations starting from allyloxycarbonylazides. Trimethylsilyl chloride is used as a stoechiometric additive in the presence of a catalytic amounts of iron(II) chloride (10 mol%). Two examples are given in Scheme 8.47 and the proposed radical mechanism is described in Scheme 8.48.[118–120]

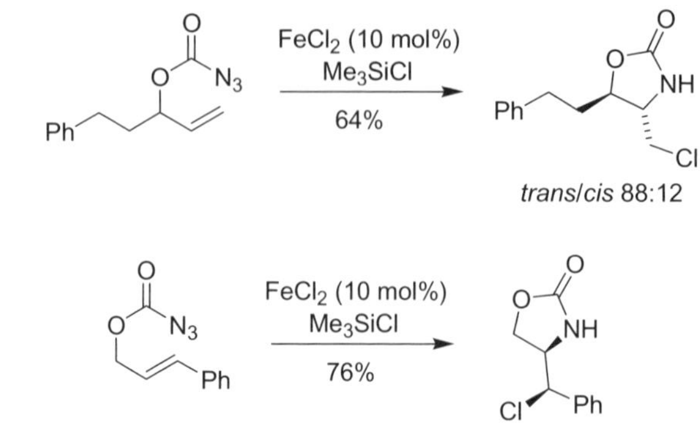

Scheme 8.47

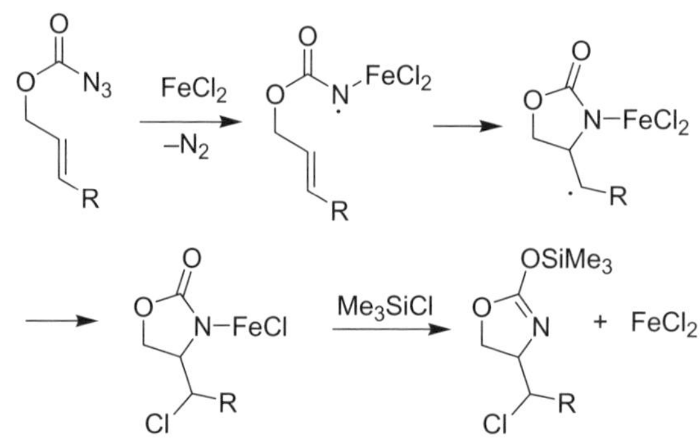

Scheme 8.48

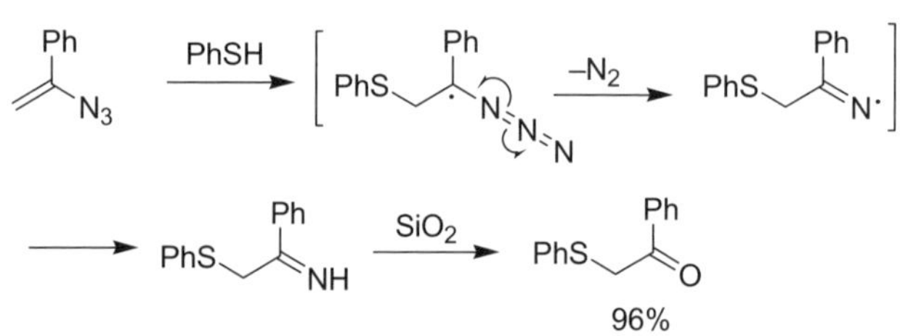

Scheme 8.49

8.5 Fragmentation Reaction of α-Azidoalkyl Radicals

Montevecchi has reported that addition of thiyl radicals onto vinyl azide affords iminyl radicals via N_2-extrusion of the intermediate α-azidoalkyl radical.[121] Under these reducing condition, the iminyl radical is reduced to the imine and finally, after chromatography, the corresponding ketone is isolated in excellent yield (Scheme 8.49).

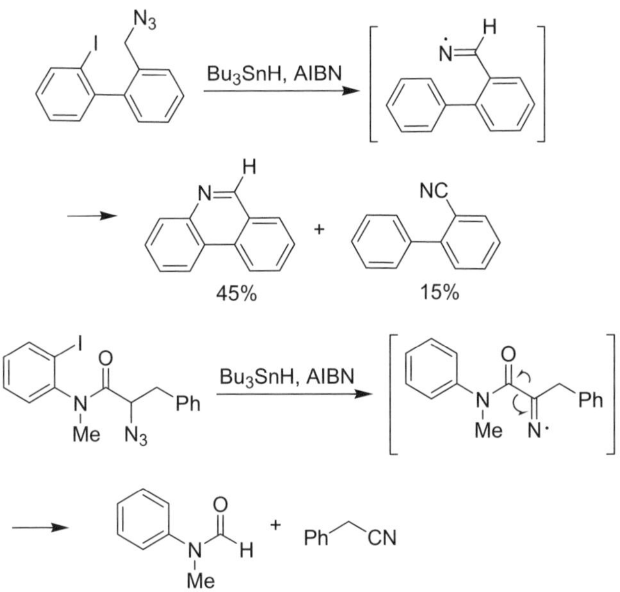

Scheme 8.50

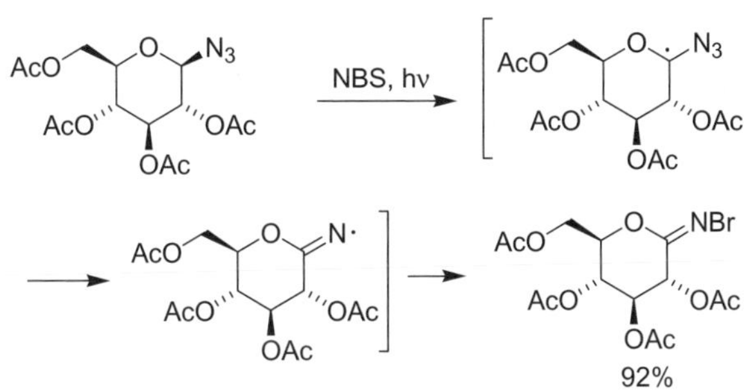

Scheme 8.51

Leardini and Spagnolo have investigated the generation of iminyl radicals from azides via hydrogen atom transfer in the presence of tin hydride.[122] Depending on the nature of the iminyl radicals, homolytic aromatic substitution was observed as well as conversion to the corresponding nitrile (Scheme 8.50, top). *β*-Fragmentation of carbamoyl radicals has been observed as major reaction pathway with *α*-azidoamides (Scheme 8.50, bottom).

Bromoimines have been prepared from glycosyl azides by treatment with N-bromosuccinimide under irradiation.[123,124] The reaction involves abstraction of the anomeric hydrogen atom leading to an anomeric radicals that undergo nitrogen extrusion and bromination (Scheme 8.51).

The radical fragmentation of *β*-hydroxy azides has been investigated by Suarez.[125] For example, treatment of an azidoalcohol derived from limonene with diacetoxyiodobenzene

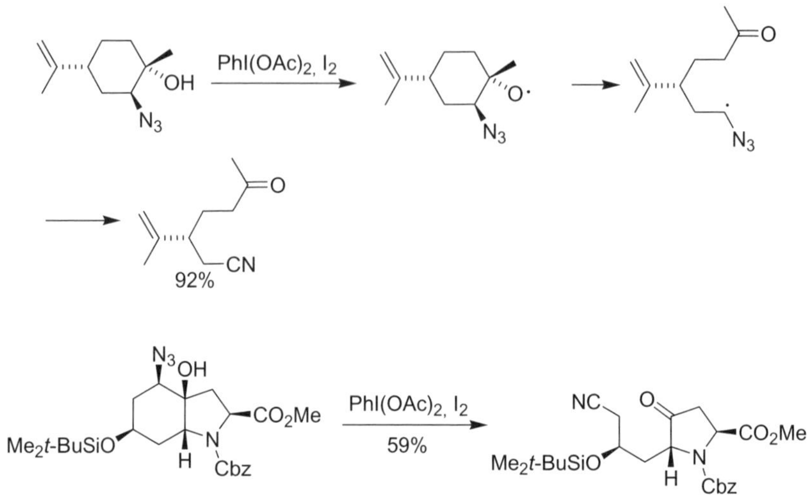

Scheme 8.52

and iodine (Suarez reagent) affords an acyclic nitrile via β-fragmentation of the alkoxyl radical followed by N_2 elimination and oxidation (Scheme 8.52, top). Similarly, Wipf has observed that a bicyclic β-azidoalcohol fragments when treated with Suarez reagent to give a monocylic pyrrolidine.[126] Transannular fragmentation of the intermediate alkoxyl radical to form the stabilized α-aminoradical is not observed demonstrating the strong directing ability of the azide (Scheme 8.52, bottom).

8.6 Conclusions

The radical chemistry of azides represents a rich field of investigation that has led to some very useful synthetic procedures. The diversity of radical processes involving azides is truly remarkable. Indeed, azides are stable under many radical reaction conditions and can therefore be considered as protected amines. Due to their stability, several highly effective radical mediated processes to introduce an azide moiety into saturated and unsaturated molecules have been reported. However, under favorable conditions, azides act as efficient radical traps. This reactivity allows for instance the formation of C–N bonds as well as the generation of aminyl radicals (after loss of dinitrogen).[127] The radical chemistry of azides provides an entry into a particularly diverse chemistry that deserves to be further investigated.

References

[1] S. Bräse, C. Gil, K. Knepper, V. Zimmermann, *Angew. Chem. Int. Ed.* **2005**, *44*, 5188.
[2] P. Evans, J. Qin, J. Robinson, B. Bazin, *Angew. Chem. Int. Ed.* **2007**, *46*, 7417.

[3] C.R. Alonso-Cruz, A.R. Kennedy, M.S. Rodriguez, E. Suarez, *Org. Lett.* **2003**, *5*, 3729.
[4] C.R. Alonso-Cruz, A.R. Kennedy, M.S. Rodríguez, E. Suárez, *J. Org. Chem.* **2008**, *73*, 4116.
[5] M.S. Workentin, B.D. Wagner, J. Lusztyk, D.D.M. Wayner, *J. Am. Chem. Soc.* **1995**, *117*, 119.
[6] F. Minisci, R. Galli, U. Pallini, *Gazz. Chim. Ital.* **1961**, *90*, 1023.
[7] F. Minisci, R. Galli, *Tetrahedron Lett.* **1962**, *3*, 533.
[8] F. Minisci, R. Galli, M. Cecere, *Gazz. Chim. Ital.* **1964**, *94*, 67.
[9] F. Minisci, R. Galli, M. Cecere, *Chim. Ind.* **1966**, *48*, 347.
[10] R. Galli, V. Malatesta, *Org. Prep. Proc. Int.* **1971**, *3*, 227.
[11] W.S. Trahanov, M.D. Robbins, *J. Am. Chem. Soc.* **1971**, *93*, 5256.
[12] E. Sommer, H. Pincas, *Chem. Ber.* **1915**, *48*.
[13] R.U. Lemieux, R.M. Ratcliffe, *Can. J. Chem.* **1979**, *57*, 1244.
[14] J. Lehmann, W. Reutter, D. Schoning, *Chem. Ber.* **1979**, *112*, 1470.
[15] H. Paulsen, P. Matschulat, *Liebigs Ann. Chem.* **1991**, 487.
[16] A. Marra, F. Gaufeny, P. Sinay, *Tetrahedron* **1991**, *47*, 5149.
[17] C.R. Bertozzi, M.D. Bednarski, *Tetrahedron Lett.* **1992**, *33*, 3109.
[18] W.K. Berlin, W.S. Zhang, T.Y. Shen, *Tetrahedron* **1991**, *47*, 1.
[19] P. Smid, W.P.A. Jörning, A.M.G. van Duuren, G.J.P.H. Boons, G.A. van der Marel, J.H. van Boom, *J. Carbohydrate Chem.* **1992**, *11*, 849.
[20] D.L.J. Clive, N. Etkin, *Tetrahedron Lett.* **1994**, *35*, 2459.
[21] P. Magnus, L. Barth, *Tetrahedron Lett.* **1992**, *33*, 2777.
[22] M. Martinez, L. Sarandeses, J. Sestelo, *Tetrahedron Lett.* **2007**, *48*, 8536.
[23] E. Zbiral, *Synthesis* **1972**, 285.
[24] H. Hugl, E. Zbiral, *Tetrahedron* **1973**, *29*, 753.
[25] H. Hugl, E. Zbiral, *Tetrahedron* **1973**, *29*, 759.
[26] W.E. Fristad, T.A. Brandvold, J.R. Peterson, S.R. Thompson, *J. Org. Chem.* **1985**, *50*, 3647.
[27] B.B. Snider, H. Lin, *Synth. Commun.* **1998**, *28*, 1913.
[28] R.M. Moriarty, J.S. Khosrowshahi, *Tetrahedron Lett.* **1986**, *27*, 2809.
[29] R.M. Moriarty, J.S. Khosrowshahi, *Synth. Commun.* **1987**, *17*, 89.
[30] P. Magnus, M.B. Roe, C. Hulme, *J. Chem. Soc., Chem. Commun.* **1995**, 263.
[31] M. Tingoli, M. Tiecco, D. Chianelli, R. Balducci, A. Temperini, *J. Org. Chem.* **1991**, *56*, 6809.
[32] M. Tingoli, M. Tiecco, L. Testaferri, R. Andrenacci, R. Balducci, *J. Org. Chem.* **1993**, *58*, 6097.
[33] A. Hassner, *Acc. Chem. Res.* **1971**, *4*, 9.
[34] A. Hassner, F.P. Boerwink, A.B. Levy, *J. Am. Chem. Soc.* **1970**, *92*, 4879.
[35] N.V. Bovin, S.E. Zurabyan, A.Y. Khorlin, *Carbohydrate Res.* **1981**, *98*, 25.
[36] R.C. Cambie, J.L. Jurlina, P.S. Rutledge, B.E. Swedlund, P.D. Woodgate, *J. Chem. Soc., Perkin Trans.* **1982**, *1*, 327.
[37] A. Hassner, J. Keogh, *J. Org. Chem.* **1986**, *51*, 2767.
[38] A. Kirschning, M.A. Hashem, H. Monenschein, L. Rose, K.U. Schoning, *J. Org. Chem.* **1999**, *64*, 6522.
[39] H. Schäfer, *Angew. Chem. Int. Ed.* **1970**, *9*, 158.
[40] H J. Schäfer, *Angew. Chem. Int. Ed.* **1981**, *20*, 911.
[41] B.P. Roberts, J.N. Winter, *J. Chem. Soc., Perkin Trans.* **1979**, *2*, 1353.
[42] V.V. Zhdankin, A.P. Krasutsky, C.J. Kuehl, *et al.*, *J. Am. Chem. Soc.* **1996**, *118*, 5192.
[43] C. Viuf, M. Bols, *Angew. Chem. Int. Ed.* **2001**, *40*, 623.
[44] H. Pedersen, S. Sinning, A. Bulow, O. Wiborg, L. Falborgc, M. Bols, *Org. Biomol. Chem.* **2004**, *2*, 2861.
[45] M. Baruah, M. Bols, *J. Chem. Soc., Perkin Trans.* **2002**, *1*, 509.
[46] M. Baruah, M. Bols, *Synlett* **2002**, 1111.
[47] L. Marinescu, J. Thinggaard, I.B. Thomsen, M. Bols, *J. Org. Chem.* **2003**, *68*, 9453.
[48] C.M. Pedersen, L.G. Marinescu, M. Bols, *Org. Biomol. Chem.* **2005**, *3*, 816.
[49] L.G. Marinescu, C.M. Pedersen, M. Bols, *Tetrahedron* **2005**, *61*, 123.
[50] P. Magnus, J. Lacour, *J. Am. Chem. Soc.* **1992**, *114*, 767.

[51] P. Magnus, J. M. Bailey, M. J. Porter, *Tetrahedron* **1999**, *55*, 13927.
[52] P. Magnus, J. Lacour, P. A. Evans, M. B. Roe, C. Hulme, *J. Am. Chem. Soc.* **1996**, *118*, 3406.
[53] P. Magnus, J. Lacour, W. Weber, *J. Am. Chem. Soc.* **1993**, *115*, 9347.
[54] P. Magnus, C. Hulme, *Tetrahedron Lett.* **1994**, *35*, 8097.
[55] P. Magnus, C. Hulme, W. Weber, *J. Am. Chem. Soc.* **1994**, *116*, 4501.
[56] M.F. Sloan, W.B. Renfrow, D.S. Breslow, *Tetrahedron Lett.* **1964**, 2905.
[57] R.A. Abramovitch, W.D. Holcomb, *J. Chem. Soc., Chem. Commun.* **1969**, 1298.
[58] D.S. Breslow, M.F. Sloan, N.R. Newburg, W.B. Renfrow, *J. Am. Chem. Soc.* **1969**, *91*, 2273.
[59] H.S. Dang, B.P. Roberts, *J. Chem. Soc., Perkin Trans.* **1996**, *1*, 1493.
[60] C. Ollivier, P. Renaud, *J. Am. Chem. Soc.* **2000**, *122*, 6496.
[61] C. Ollivier, P. Renaud, *J. Am. Chem. Soc.* **2001**, *123*, 4717.
[62] D. Masterson, J. Shackleford, *Synlett* **2007**, 1302.
[63] D.S. Masterson, N.A. Porter, *Org. Lett.* **2002**, *4*, 4253.
[64] E. Nyfeler, P. Renaud, *Org. Lett.* **2008**, *10*, 985.
[65] P. Panchaud, C. Ollivier, P. Renaud, S. Zigmantas, *J. Org. Chem.* **2004**, *69*, 2755.
[66] P. Renaud, C. Ollivier, P. Panchaud, *Angew. Chem. Int. Ed.* **2002**, *41*, 3460.
[67] P. Panchaud, P. Renaud, *J. Org. Chem.* **2004**, *69*, 3205.
[68] P. Panchaud, P. Renaud, *Chimia* **2004**, *58*, 232.
[69] P. Schar, P. Renaud, *Org. Lett.* **2006**, *8*, 1569.
[70] L. Chabaud, Y. Landais, P. Renaud, *Org. Lett.* **2005**, *7*, 2587.
[71] L. Chabaud, Y. Landais, P. Renaud, *Org. Lett.* **2002**, *4*, 4257.
[72] L. Chabaud, Y. Landais, P. Renaud, *et al.*, *Chem., Eur. J.* **2008**, *14*, 2744.
[73] P. Panchaud, P. Renaud, *Adv. Synth. Catal.* **2004**, *346*, 925.
[74] J. Waser, H. Nambu, E.M. Carreira, *J. Am. Chem. Soc.* **2005**, *127*, 8294.
[75] J. Waser, B. Gaspar, H. Nambu, E.M. Carreira, *J. Am. Chem. Soc.* **2006**, *128*, 11693.
[76] B.B. Snider, J. R. Duvall, *Org. Lett.* **2004**, *6*, 1265.
[77] S. Kim, G.H. Joe, J.Y. Do, *J. Am. Chem. Soc.* **1994**, *116*, 5521.
[78] M. Santagostino, J.D. Kilburn, *Tetrahedron Lett.* **1995**, *36*, 1365.
[79] M. Kizil, J.A. Murphy, *J. Chem. Soc., Chem. Commun.* **1995**, 1409.
[80] M. Kizil, B. Patro, O. Callaghan, J.A. Murphy, H.B. Hursthouse, D. Hibbs, *J. Org. Chem.* **1999**, *64*, 7856.
[81] B. Patro, J.A. Murphy, *Org. Lett.* **2000**, *2*, 3599.
[82] S.Z. Zhou, S. Bommezijn, J.A. Murphy, *Org. Lett.* **2002**, *4*, 443.
[83] D.E. Lizos, J.A. Murphy, *Org. Biomol. Chem.* **2003**, *1*, 117.
[84] D. Lizos, R. Tripoli, J.A. Murphy, *Chem. Commun.* **2001**, 2732.
[85] P.C. Montevecchi, M.L. Navacchia, P. Spagnolo, *Eur. J. Org. Chem.* **1998**, 1219.
[86] L. Horner, G. Bauer, J. Dörges, *Chem. Ber.* **1965**, *98*, 2631.
[87] L. Horner, G. Bauer, *Tetrahedron Lett.* **1966**, 3573.
[88] M.T. Reagan, A. Nickon, *J. Am. Chem. Soc.* **1968**, *90*, 4096.
[89] H. Kwart, A.A. Kahn, *J. Am. Chem. Soc.* **1967**, *89*, 1950.
[90] H. Kwart, A.A. Kahn, *J. Am. Chem. Soc.* **1967**, *89*, 1951.
[91] D.N. Kirk, M.A. Wilson, *J. Chem. Soc., Chem. Commun.* **1970**, 64.
[92] D.A. Sutton, *J. Chem. Soc.* **1944**, 306.
[93] A. Hasegawa, M. Kiso, *Carbohydrate Res.* **1975**, *44*, 121.
[94] Y. Takeda, S. Horito, *Carbohydrate Res.* **2005**, *340*, 211.
[95] A. Furstner, J. Baumgartner, D.N. Jumbam, *J. Chem. Soc., Perkin Trans.* **1993**, *1*, 131.
[96] S.N. Maiti, P. Spevak, A.V.N. Reddy, *Synth. Commun.* **1988**, *18*.
[97] G. Vidyasagar Reddy, G. Venkat Rao, D.S. Iyengar, *Tetrahedron Lett.* **1999**, *40*, 3937.
[98] D.E. Herbranson, M.D. Hawley, *J. Org. Chem.* **1990**, *55*, 4297.
[99] E.J. Kaufmann, R.C. Thompson, *J. Am. Chem. Soc.* **1977**, *99*, 1824.
[100] M.H.C. Goulaouic-Dubois, *Tetrahedron Lett.* **1995**, *36*, 7427.
[101] C. Goulaouic-Dubois, A. Guggisberg, M. Hesse, *Tetrahedron* **1995**, *51*, 12035.
[102] M. Frankel, D. Wagner, D. Gertner, A. Zilkha, *J. Organometal. Chem.* **1967**, *7*, 518.
[103] N.E. Poopeiko, T.I. Pricota, I.A. Mikhailopulo, *Synlett* **1991**, 342.

[104] H.H. Wasserman, R.K. Brunner, J.D. Buynak, C.G. Carter, T. Oku, R.R.P., *J. Am. Chem. Soc.* **1985**, *107*, 519.
[105] D.B. Werz, A. Adibekian, P.H. Seeberger, *Eur. J. Org. Chem.* **2007**, 1976.
[106] H. Redlich, W. Roy, *Liebigs Ann. Chem.* **1981**, 1215.
[107] Y. Maeda, K. U. Ingold, *J. Am. Chem. Soc.* **1980**, *102*, 328.
[108] S. Kim, G.H. Joe, J.Y. Do, *J. Am. Chem. Soc.* **1993**, *115*, 3328.
[109] S. Kim, K.S. Yoon, S.S. Kim, H.S. Seo, *Tetrahedron* **1995**, *51*, 8437.
[110] S. Kim, K.M. Yeon, K.S. Yoon, *Tetrahedron Lett.* **1997**, *38*, 3919.
[111] L. Benati, D. Nanni, C. Sangiorgi, P. Spagnolo, *J. Org. Chem.* **1999**, *64*, 7836.
[112] L. Benati, G. Bencivenni, R. Leardini, *et al.*, *J. Org. Chem.* **2005**, *70*, 3046.
[113] D.S. Hays, G.C. Fu, *J. Org. Chem.* **1998**, *63*, 2796.
[114] L. Benati, G. Bencivenni, R. Leardini, *et al.*, *J. Org. Chem.* **2006**, *71*, 434.
[115] L. Benati, G. Bencivenni, R. Leardini, *et al.*, *J. Org. Chem.* **2006**, *71*, 5822.
[116] L. Benati, G. Bencivenni, R. Leardini, *et al.*, *Org. Lett.* **2006**, *8*, 2499.
[117] A. Postigo, S. Kopsov, C. Ferreri, C. Chatgilialoglu, *Org. Lett.* **2007**, *9*, 5159.
[118] T. Bach, B. Schlummer, K. Harms, *Chem. Commun.* **2000**, 287.
[119] T. Bach, B. Schlummer, K. Harms, *Synlett* **2000**, 1330.
[120] T. Bach, B. Schlummer, K. Harms, *Chem. Eur. J.* **2001**, *7*, 2581.
[121] P.C. Montevecchi, M.L. Navacchia, P. Spagnolo, *J. Org. Chem.* **1997**, *62*, 5846.
[122] G. Bencivenni, T. Lanza, R. Leardini, *et al.*, *J. Org. Chem.* **2008**, *73*, 4721.
[123] J.P. Praly, L. Somsak, S.H. Mahmoud, Z. Elkharraf, G. Descotes, I. Farkas, *J. Carbohydrate Chem.* **1992**, *11*, 201.
[124] J.P. Praly, C. Distefano, L. Somsak, G. Descotes, *J. Chem. Soc., Chem. Commun.* **1992**, 200.
[125] R. Hernandez, E.I. Leon, P. Moreno, E. Suarez, *J. Org. Chem.* **1997**, *62*, 8974.
[126] P. Wipf, D. A. Mareska, *Tetrahedron Lett.* **2000**, *41*, 4723.
[127] M. Minozzi, D. Nanni, P. Spagnolo, Review article on the conversion of azides to amiayl radicals. *Chem. Eur. J.* **2009**, *15*, 7830.

9

Cycloaddition Reactions with Azides: An Overview

Christine Schilling, Nicole Jung and Stefan Bräse

Institute of Organic Chemistry, Karlsruhe Institute of Technology (KIT), Fritz-Haber-Weg 6, Karlsruhe, D-76131, Germany

9.1 Huisgen 1,3-dipolar cycloaddition

The history of 1,3-dipoles began in the nineteenth century, when Curtius[1] reported on the diazoacetic ester. A few years later, Buchner *et al.*[2] successfully performed the first 1,3-dipolar cycloaddition of the diazoacetic ester with α,β-unsaturated esters. Despite the fact that, over the subsequent years, different 1,3-dipoles have been discovered, only a few have been generally effective in organic synthesis, e.g. the well-known Diels–Alder reaction.[3]

In 1963, Huisgen *et al.* published a systematic study on the concerted 1,3-dipolar cycloaddition, based on previous results from Smith and coworkers about the 1,3-addition of diazoalkanes, ozone and azides,[4] as well as kinetic studies on the mechanism.[5] The addition of a 1,3-dipole **1** (*a-b-c*), possessing ambivalent electrophilic as well as nucleophilic activity to a multiple bond system like the dipolarophile **2** (*d-e*), leads to a remarkably wide variety of five-membered heterocyclic compounds.[6]

Nowadays a broad range of different 1,3-dipoles, ozone, azides[7–10] and diazoalkanes on the one hand as well as dipoles like nitrones, nitro compounds, carbonyl ylides, nitrile oxides, nitrile imines and ylides on the other hand, are well-established. The addition of these 1,3-dipoles to an alkene is one of the most frequently used cycloaddition reactions in organic synthesis.[3,6]

Beyond the concerted reaction mechanism proposed by Huisgen (Scheme 9.1, mechanism 1), a different mechanism via a singlet diradical intermediate was discussed in the

Organic Azides: Syntheses and Applications Edited by Stefan Bräse and Klaus Banert

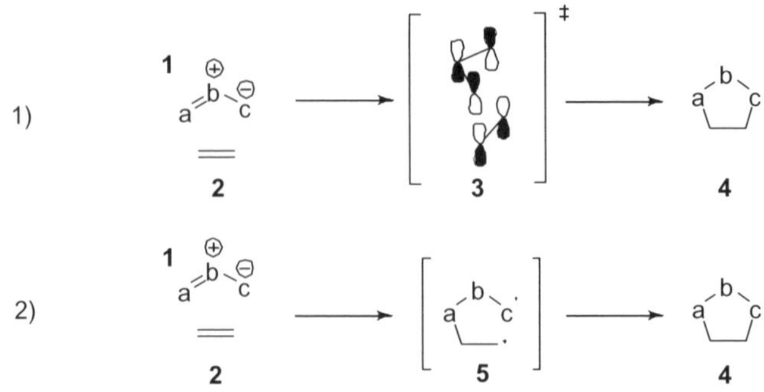

Scheme 9.1 *Discussed mechanism of 1,3-dipolar cycloaddition reaction: (1) concerted; (2) diradical mechanism*[3,6]

1960s.[11] However the diradical mechanism (Scheme 9.1, mechanism 2) was disproved (Scheme 9.1).

To date, the Huisgen 1,3-dipolar cycloaddition is described as nonconcerted when catalysts, processing via metallacycle intermediates, are used, leading to substituted heterocycles[12] in excellent selectivity (for specific examples, see Sections 9.2 and 9.5). The outstanding discovery of the Cu(I)-catalyzed azide-alkyne cycloaddition by Meldal and Sharpless particularly improved the rates as well as the regioselectivity of the reaction (see subsequent chapter) and is now the most commonly used 1,3-dipolar cycloaddition in organic synthesis.[13,14]

9.2 Copper-Catalyzed Azide-Alkyne Cycloaddition (CuAAC)

9.2.1 General Aspects of the CuAAC Reaction

The Huisgen 1,3-dipolar cycloaddition to triazoles can be performed under copper-catalysis and is then known as Copper-Catalyzed Azide-Alkyne Cycloaddition (CuAAC).[15,16]

As the Huisgen reaction is very slow in forming triazoles at room temperature, the use of a copper catalyst accelerates the cycloaddition up to approximately 10^7 times and enables reactions in aqueous systems. Moreover, the advanced cycloaddition offers several supplementary advantages that are characteristic of Click chemistry. While former uncatalyzed transformations to triazoles always yielded a mixture of the 1,4- and 1,5-regioisomers, the CuAAC produces only 1,4-triazoles. These 1,4-disubstituted-1,2,3-triazoles are produced with excellent yields and purities (for other applications see also ref.[17]). Furthermore, the formation of triazoles is nearly independent of the substituents of both reaction partners. Neither substituents with sterical nor electronical effects have great influence on the outcome of the CuAAC reaction.[18]

The active copper species in these transformations is Cu(I), but Cu(0) and Cu(II)-species may be used as well, depending on the chemical and biological environment and the required reaction conditions. Cu(II) salts are the catalysts that are most often selected for Click chemistry, as they can be easily reduced *in situ* to Cu(I) in the presence of reducing agents. The latter have to be added in great excess to guarantee that no Cu(II) remains in solution, which would lead to side products of the Click reaction. The reducing agent of choice in aqueous systems is ascorbate,[19–22] while approaches in organic solvents can be performed with hydrazine.[23]

Cu(I) salts are added less frequently because of the instability of the Cu(I)-species under aerobic conditions. To prevent oxidation of the catalyst, the Click reactions either have to be performed under anaerobic conditions or a stabilizing ligand has to be added to the reaction mixture. Several of these ligands are known to be able to prolong the life span of the oxidation state of the Cu(I)-catalyst, thereby allowing reactions under air. The most famous of these is the tris(triazolylmethyl)amine ligand (TBTA) investigated by Fokin *et al.*,[24] but there are other examples of polydentate ligands accelerating triazole formation as demonstrated by Matyjaszewski *et al.*[23] Besides the compatibility of the Click reaction with aerobic conditions, these ligands, discussed in Section 9.3.1 of this chapter, effect a rate-accelerating property of the triazole formation. In organic solvents, the salts CuBr and CuI can be substituted for $[Cu(CH_3CN)_4]PF_6$,[19] $(EtO)_3P{\cdot}CuI$,[25] $[Cu(PPh_3)_3]Br$[26,27], which show enhanced solubility in non-aqueous systems. Beyond these, Cu(I) was added in the form of a solid-supported catalyst[28,29] and has been used in the form of Cu(I) zeolithes[30,31] and Al_2O_3 nanoparticles.[32]

Cu(0) catalyzes Click reactions indirectly *via* the formation of Cu(I) by comproportionation of Cu(II) and Cu(0). The essential Cu(II) species can be added in the form of copper salts, but it is not mandatory due to the presence of traces of copper oxides and carbonates on the metal surface. Although this method benefits from very low copper contamination, high selectivity and the isolation of pure triazoles in high yields, the Cu(0)-catalyzed Click reaction is disadvantageous because of prolonged reaction times.

9.2.2 Mechanism of the CuAAC Reaction

The proposed mechanism of the CuAAC reaction outlined in Scheme 9.2 is based on several investigations over the last decade but has not yet been completely proven. Some details, especially those concerning the complexation of the Cu(I)-species and the origin of selectivity of the cycloaddition, are still unknown. The first step of the CuAAC reaction is the formation of Cu(I) acetylides from Cu(I) **6** and alkynes **7** through the addition of a base. As the CuAAC reaction has shown to have a second order dependence on the copper species, Cu(I) acetylides in the form of different bridged-aggregates are discussed as important intermediates **8**.[20] Step B of the reaction cycle describes the coordination of azides to the Cu(I) acetylide species and the formation of complex **10**, which can be presented in form **A** or **B**. While structure **A** refers to the coordination of the acetylide and the azide to just one copper atom, structure **B** illustrates that a coordination of the starting materials to different copper atoms is feasible as well.[20,33,34] The transition state **10B** is often the preferred means of both six membered-variants, because a complex with two Cu(I) species in the transition state may explain the absolute regioselectivity of the reaction. After the bond formation between the nitrogen in position 3 (referring to the numbering of the triazole product) and the acetylene moiety, transition states **10A** and **10B** can be proposed. They form Cu(I) triazolides **12** and undergo proteolysis to give 1,4-disubstituted 1,2,3-triazoles **13** in combination with the recovery of the copper catalyst **6**.[19,35,36]

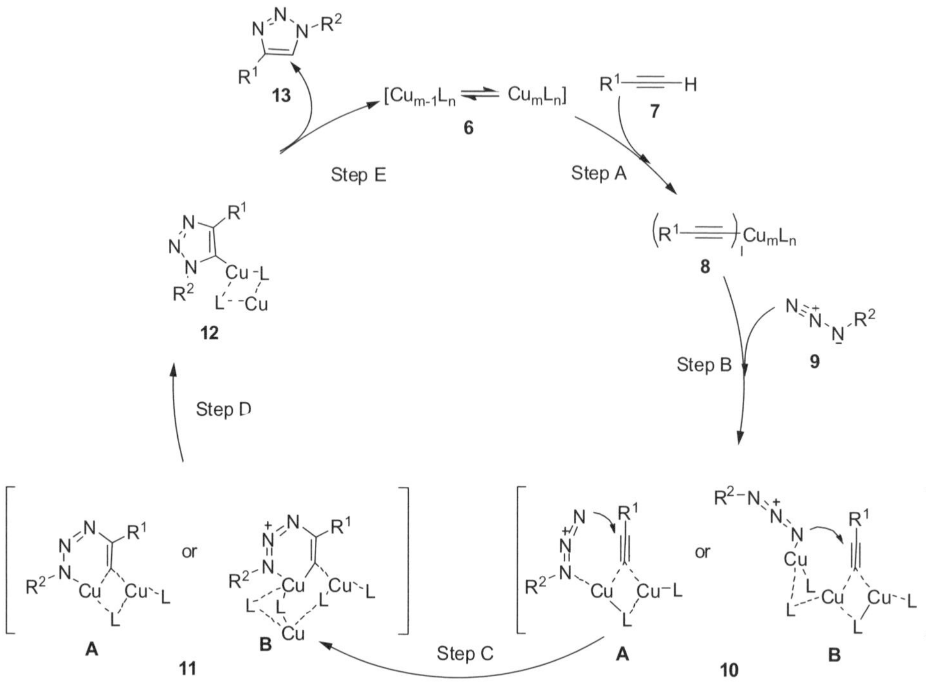

Scheme 9.2 *Proposed mechanism for CuAAC reaction*[13]

9.3 Acceleration of the Click Reaction[37,38]

9.3.1 Addition of Ligands

As mentioned in Section 9.2, several ligands can be added to stabilize the Cu(I) species and thereby facilitate Click reactions without an inert atmosphere by maintaining a high concentration of the catalytically active copper species. Furthermore, ligands can accelerate cycloaddition reactions drastically by chelating the catalyst. The origin of these accelerating effects and the catalytic process are not yet fully understood, but several observations lead to the assumption that the ligands influence the reaction rate by affecting the equilibrium distribution between the copper clusters in solution and the equilibration rate among the clusters.[13] The most active ligands are summarized in Figure 9.1. Aliphatic ligands of type **15** and **22** are generally more active then pyridine-containing ligands **14** or **17**. The denticity of the chelating ligand is also crucial. While tridentate ligands accelerate the reaction rate dramatically, the addition of tetradentate ligands that saturate the coordination of Cu(I) have been found to have a slightly weaker effect on the cycloaddition.[23,24] Former investigations[23] presumed that the tetradentate ligands interfere with the essential coordination of the alkyne to the copper center and therefore lower the rate of reaction. A more recent study of TBTA complexes showed that, at least concerning the precursor of the active catalyst, the Cu(I) species does not coordinate with the tertiary amine of TBTA and that a dinuclear coordination complex is formed with a triazole unit bridging two Cu-centers through the medial and proximal nitrogens.[39] Similar observa-

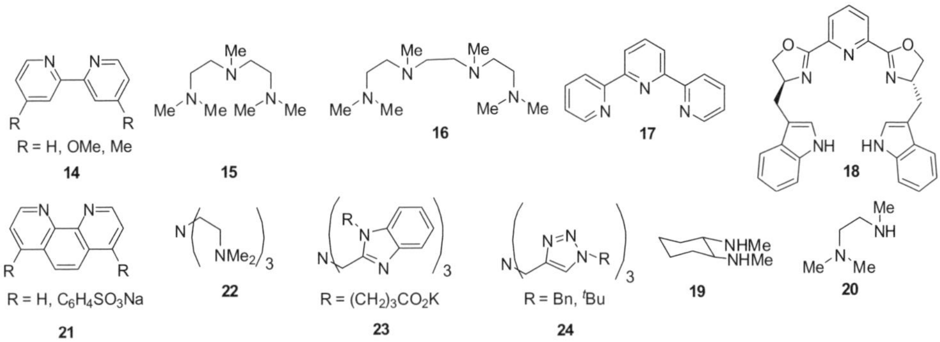

Figure 9.1 *Ligands with rate accelerating effect on the CuAAC reaction*

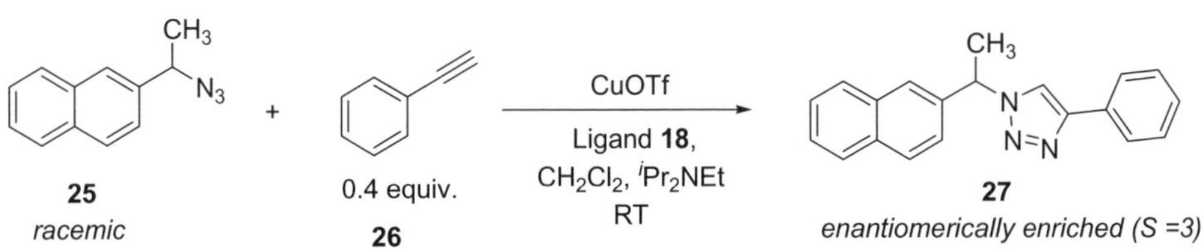

Scheme 9.3 *Kinetic resolution using Click chemistry*[54]

tions of dinuclear and tetranuclear copper complexes that are involved in ligand-free cycloadditions confirm the importance of these species and their rate-accelerating influence.[34]

Up to now, the two most active catalysts are TBTA[20,40–42] and bathophenanthrolinedisulfonic acid.[43–47] Considering the aforementioned tasks of the ligands, the TBTA catalyst is often preferred because of its potential for stabilization.[43] TBTA and TTA (tristriazolylamine)[44–46] belong to the group of tertiary amine catalysts that are surrounded by three identical triazoles. Those tertiary amines are known for imidazolyl (**23**)[42–47] and also for pyridine[47] and alkyl-based substituents.[23]

Beyond those rather complicated amine ligands, other common bases are applied as ligands as well.[48] Lutidine,[49] pyridine[50] and amino acids like histidine[47,48] and proline[51–53] have been tested as rate-accelerating additives.

Ligand addition to the CuAAC reaction can affect kinetic resolution by using chiral precursors and ligands. The first example for asymmetric kinetic resolution of the azide-alkyne cycloaddition was given by Fokin and Finn, who tested several copper complexes of the bis(oxazolinyl)pyridine (pybox) family in combination with the racemic mixture of the compound **25** (Scheme 9.3).[54] Kinetic resolution could be observed up to a maximum enantiomer selectivity factor of three using ligand **18**.

9.3.2 Addition of Base

While the addition of a base in the reaction mixture does not make much difference to the transformations under Cu(II), its presence under Cu(I) salt addition is much more significant. In the latter case, the success of the triazole formation depends on the presence of amine bases, even if ligands such as TBTA are added, because heterocyclic nitrogen compounds do not deliver enough basicity. Alternatively, a similar Cu(I)-cluster-activating effect can be achieved through ultrasonication in the absence of a base.[20,55]

9.4 Copper-free Click Chemistry

The azide-alkyne cycloaddition is highly relevant for biological applications. However, *in vivo* applications are limited by the fact of the toxicity of copper ions for living organisms.

The great demand of metal-free Click reactions for *in vivo* studies has been an immense challenge. Some researchers at the Scripps Research Institute used the resulting slow kinetics in the absence of copper ions for the target-guided synthesis of enzyme inhibitors.[56] Here, the regiospecificity was induced through the binding pocket of the enzyme. Thus, though metal-free strategies have been developed, the reactions are not regioselective, very slow and require higher temperatures in the absence of transition metals.

Bertozzi and coworkers[57–59] suggested some highly strained cyclooctynes (18 kcal/mol of ring strain) to lower the activation barrier without the use of metal catalysts discussed in previous works of Wittig *et al.*[60] The first generation of cyclooctynes (Figure 9.2) showed an increasing reactivity to the [3 + 2]-cycloaddition in the row of oxo-, fluoro- and difluoro-substituents in the α-position of the triple bond. Thereafter, the difluorocyclooctynes were tested for imaging glycans in live cells because of their inertness in a biological environment. This led to the discovery of the second generation, expanding the variety of substrates useful for biologists.

Electron-withdrawing groups tend to lower the LUMO of the alkyne and thus increase the interaction with the HOMO of the azide. The kinetics are comparable to the Cu-catalyzed Click reaction; however, cycloadditions with these compounds are not regioselective. Specific DFT-measurements of the transition state were taken and previously published by Houk and coworkers.[61]

Although cyclooctynes enable copper-free Click reactions, they are not compatible with bioconjugations in aqueous media, because of their hydrophobic character. Thus, Bertozzi and coworkers uncovered a new method for the synthesis of a biocompatible cyclooctyne (6,7-dimethoxyazacyclooct-4-yne = DIMAC)[58] to detect azide-labeled biomolecules *via* copper-free Click chemistry. The first conjugations to biotin as well as cell-surface glycans[62] were arranged.

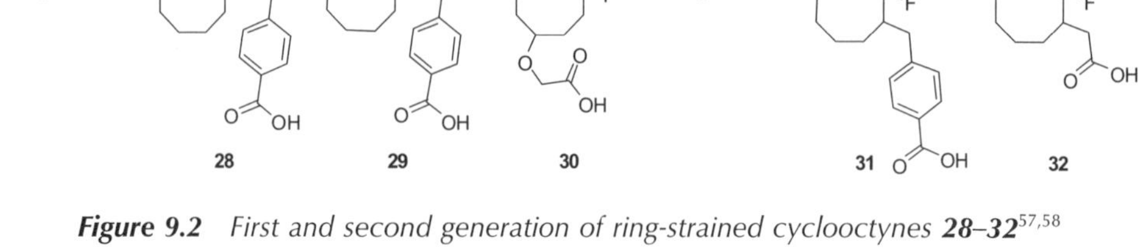

Figure 9.2 *First and second generation of ring-strained cyclooctynes* ***28–32***[57,58]

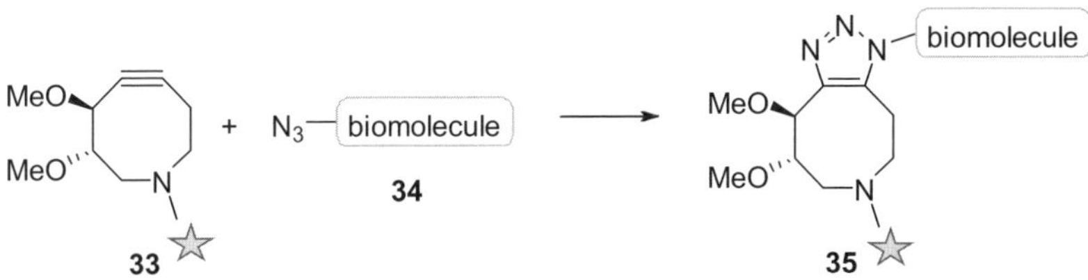

33 35

Scheme 9.4 *Biolabeling* via *copper-free Click chemistry with 6,7-dimethoxyazacyclooct-4-yne (**33**)*[58]

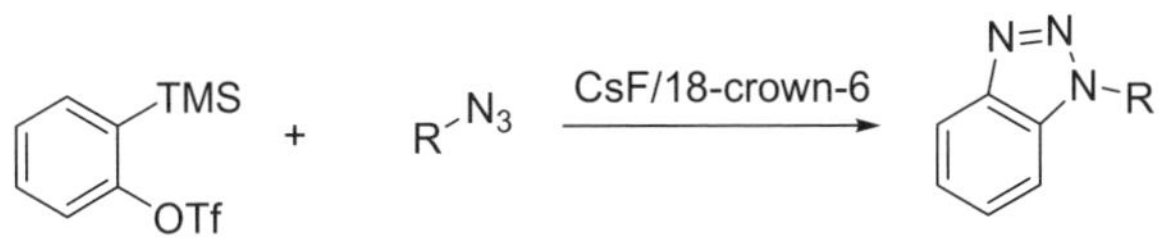

36 37 38

Scheme 9.5 *Click chemistry with arynes – a fluoride-promoted* o*-elimination of triflate **36** with functionalized azides **37***[65]

In the last years, other substrates like oxanorbornadienes by Cornelissen *et al.*[63] as well as dibenzocyclooctynes by Boon *et al.*[64] were tested as (high) potential compounds for copper-free azide-alkyne bioconjugations.

Arynes can be added under mild conditions and rapid reaction times to azides through fluoride-promoted *o*-elimination. The addition of arynes to azides has been already shown by Huisgen and Wittig, respectively. Recently, *O*-(trimethylsilyl)phenyltriflate and various fluoride sources in combination with a complementary crown ether were used for the [3 + 2]-cycloaddition with functionalized azides (Scheme 9.5).[65]

9.5 Ruthenium-Catalyzed Azide-Alkyne Cycloaddition (RuAAC)

The Ruthenium-Catalyzed Azide-Alkyne Cycloaddition (RuAAC) complements the well-established CuAAC reaction, as the formation of 1,5-substituted triazoles (instead of 1,4-substituted 1,2,3-triazoles) can be achieved with high regioselectivity. In contrast to the CuAAC reaction, triazoles that are synthesized *via* RuAAC reaction can be formed from terminal as well as internal alkynes. This offers the possibility of the formation of fully-substituted triazoles.

There are other characteristic properties that have to be considered concerning the RuAAC reaction:

1. Many aprotic solvents like THF, dioxane, toluene or DMF can be used for RuAAC, but protic solvents such as MeOH or ^{i}PrOH result in reduced yields and the formation of side products.
2. Present knowledge suggests that the RuAAC is not sensitive to reactions in the presence of atmospheric oxygen.
3. Reactions can be carried out between room temperature and 110 °C.

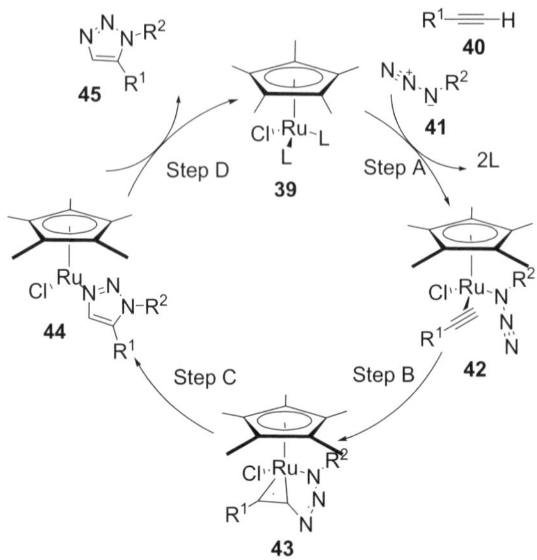

Scheme 9.6 Proposed mechanism for the RuAAC-reaction[69]

The mechanism of the RuAAC reaction has been investigated by several groups and is summarized in Scheme 9.6.[66–68] The proposed catalytic cycle includes the formation of the catalytically active species [Cp*RuCl] and the formal substitution of the spectator ligands by the alkyne **40** and the azide **41** to give complex **42**. After oxidative coupling of the alkyne and the azide, the intermediate species **43** undergoes reductive elimination and releases the aromatic triazole product **45**.

The choice of Ru(II)-catalyst is of crucial importance for the success of the RuAAC reaction. Up until now, Ru(II)-catalysts bearing a η^5-pentamethylcyclopentadienyl ligand are the only catalytic systems that show high selectivity as well as excellent yields. It has been suggested that the presence of the electron rich Cp*-ligand (stabilizing the higher formal oxidation state of the metal center) is irreplaceable within the Ru-catalyst. Cp*-containing catalysts like $[Cp^*RuCl]_4$, $Cp^*RuCl(PPh_3)_2$, Cp*RuCl(COD), Cp*RuCl(NBD) show special reactivity and selectivity. This high activity of the Cp*RuCl-catalysts may also originate in the lability of the bystander ligands in those complexes that enable the formation of the intermediate **42** and the sterically-demanding nature of Cp*, facilitating the reductive elimination in the catalytic cycle.

With respect to the synthetic availability and stability, $Cp^*RuCl(PPh_3)_2$ and Cp*RuCl(COD) are the most frequently studied complexes. The latter offers advantages because of the more labile properties of its ligands and the enhanced activity of the catalyst, allowing for reaction of internal alkynes and aryl azides even at room temperature.[69]

The success and reaction rate of the RuAAC reaction are nearly unaffected by the substituents of the alkyne, but strongly dependent on the nature of the azides. Cycloadditions of primary azides result in high yields, whereas secondary azides react more slowly and produce lower yields. Tertiary azides are hardly clickable *via* the RuAAC procedure. The regioselectivity of the RuAAC reaction is excellent in cases where the alkyne substituent bears a hydrogen-bond-donor functionality that directs the participating

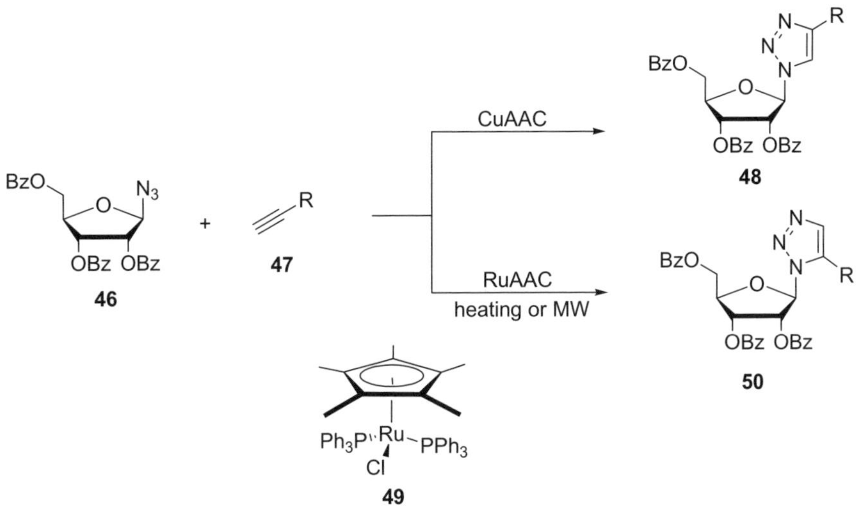

Scheme 9.7 *Preparation of 1,4- and 1,5-disubstituted triazoles* via *CuAAC- or RuAAC-reaction*[70]

components to yield triazoles with the hydrogen-donor substituent in position 5. In cases with substituents without hydrogen-donor functionality, the regioselectivity is directed by electronic and sterical effects. The new bond in the metallacycle intermediate **43** is formed between the more nucleophilic carbon of the alkyne and the nitrogen in position 3 (the more electronegative carbon assumes position 4 in the target triazole **45**) (Scheme 9.6).

The combination of CuAAC and RuAAC offers the possibility of synthesizing triazoles with diverse residues and facilitates the alternation of the regioselectivity. As such, a broad spectrum of substrates is available and used, for example, in the derivatizion of nucleosides through the synthesis of triazole analogs. Agrofoglio *et al.*[70] synthesized a small library of 1,4- and 1,5-disubstituted triazolo derivatives and compared the yielded ratio under Cu(I)- as well as Ru(II)-catalysis (with microwave irradiation) (Scheme 9.7).

The RuAAC reaction has been applied to the synthesis of a variety of other substrates, for instance, in the formation of peptide bond surrogates,[71] 1-protected 5-amido 1,2,3-triazoles[72] and it has been used for the replacement of the lactone moiety in naturally-occurring lignans.[73] Just as a rate-accelerating effect can be demonstrated in the CuAAC reaction, the RuAAC reaction can be undertaken under microwave irradiation as well.[74]

9.6 Use of Other Metals for the Cycloaddition of Azides and Alkynes: Ni(II), Pt(II), Pd(II)

To date, there is only one investigation concerning metals beyond ruthenium and copper that may catalyze triazole formation through an AAC reaction. Matyjaszewski *et al.* reported on the addition of palladium, platinum and nickel to azides and alkynes in comparison to the same model system without any catalyst. The rate-accelerating effect of the metals could only be detected in the case of the addition of $PtCl_2$ and $PdCl_2$ in combination with PM-DETA (**15**) (Figure 9.1) as a ligand. The accelerating effects monitored were rather low, especially in the case of nickel, so that further investigation has yet to follow.[23]

9.7 Cycloaddition Reactions with Azides for the Synthesis of Tetrazoles

9.7.1 Intermolecular Approaches

Tetrazoles with a similar structure to triazoles tolerate various chemical environments. They are stable under oxidizing and reducing conditions as well as strongly acidic and basic media. Thus, this class of heterocycles presently plays a crucial role in the field of coordination chemistry,[75] material science application and medicinal chemistry.[76,77] The pharmacokinetic potential of tetrazoles, frequently used as metabolically stable surrogates of carboxylic acids, makes the synthesis of this nitrogen-rich heterocycle particularly fascinating.

In 1932, a method for the synthesis of tetrazoles *via* the reaction of hydrazoic acid (HN_3) with organic cyanides was first reported. Despite many disadvantages such as toxicity, harsh reaction conditions as well as expensive reagents, the aforementioned or similar procedures led to the direct formation of tetrazole rings through the Huisgen 1,3-dipolar cycloaddition.[75] Against all odds, only nitriles activated by strong electron-withdrawing groups can be effectively used as dipolarophiles.[77]

The first examples of a direct 1,5-substituents tetrazole synthesis *via* an intermolecular [2 + 3]-cycloaddition reaction of organic azides (based on the results for a similar synthesis of triazoles) were developed by Sharpless *et al.* in 2002 using sulfonyl or acyl cyanides (Scheme 9.8).[76,78] In both cases, the Click reaction involved simple heating of the organic azides (hindered, aliphatic or aryl azides) and the nitriles into a homogenous liquid at higher reaction temperature (80–100 °C), in which no further purification was necessary.

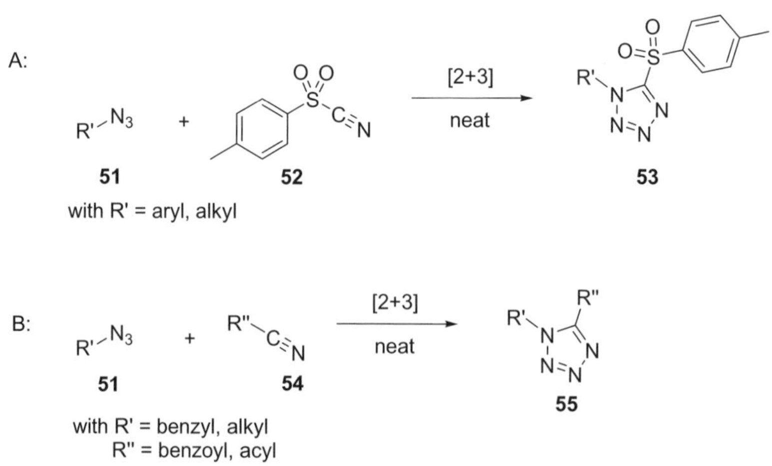

Scheme 9.8 *[2 + 3] Dipolar cycloaddition for the synthesis of 1,5-disubstituted sulfonyl tetrazoles (A) and acyl tetrazoles (B)*[76,78]

In the case of acyltetrazoles (**55**), a slight excess (1.5 equiv.) of nitrile was used for the full conversion of the cycloaddition.

The 5-substitutents of the resulting sulfonyl (**53**) and acyl tetrazoles (**55**) are known to be replaceable with a wide range of O-, N- and C-nucleophiles through an addition-elimination pathway, leading to a variety of 1,5-disubstituted tetrazoles hardly accessible by a direct Huisgen azide-nitrile cycloaddition.[79,80]

Thus, Dondoni *et al.*[77] synthesized a new class of tetrazole-tethered C-glycosyl α-amino acids using serine and cysteine as nucleophiles based on Demko's and Sharpless' aforementioned Click azide-sulfonyl cyanide cycloaddition/sulfonyl substitution route (Scheme 9.8).

A further synthesis of acyltetrazoles was reported in 2007, in which acyl cyanides (10 equiv.), bearing a hydrogen at the α-position, were transformed into tetrazoles in higher yield and purity in the presence of $ZnBr_2$ as a catalyst (1 equiv.) at room temperature.[81]

In the same year, a successful Lewis acid catalysis was developed by Sedelmeier *et al.*[82] The group synthesized 5-substituted tetrazoles in a direct conversion using dialkylaluminium azides (**57**), which are inexpensive, soluble in organic solvents and non-toxic (Scheme 9.9). The proposed mechanism for the 1,3-dipolar cycloaddition (Scheme 9.9A) suggests Lewis acid properties of the aluminium center activating the nitriles in the azide addition. Different 5-substituted tetrazoles (**63**) were obtained in excellent yields after a simple workup procedure (Scheme 9.9B). However, the reaction temperature varied between –40 °C and 120 °C, depending on the reactivity of the substrates.

Several other approaches are known to produce tetrazoles *via* cycloaddition with azides. Fang *et al.* investigated the synthesis of tetrazoles by a one-pot tandem reaction.[83] It was shown that alcohols and aldehydes can be transferred into the corresponding nitriles, which undergo successive cycloaddition through the addition of azides (Scheme 9.10).

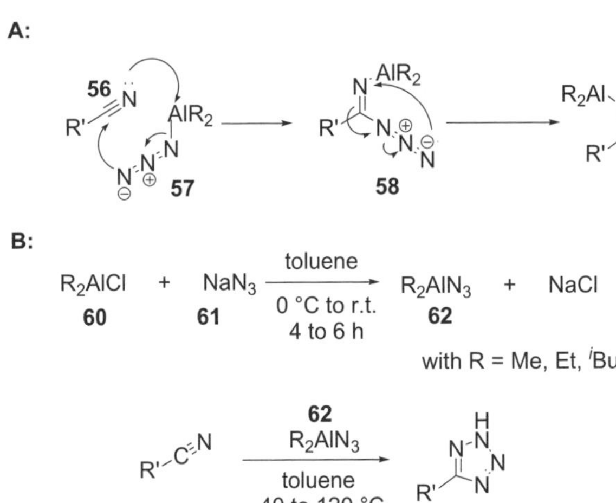

Scheme 9.9 *Proposed mechanism (A) and the [2 + 3]-cycloaddition route to tetrazoles (B)*[82]

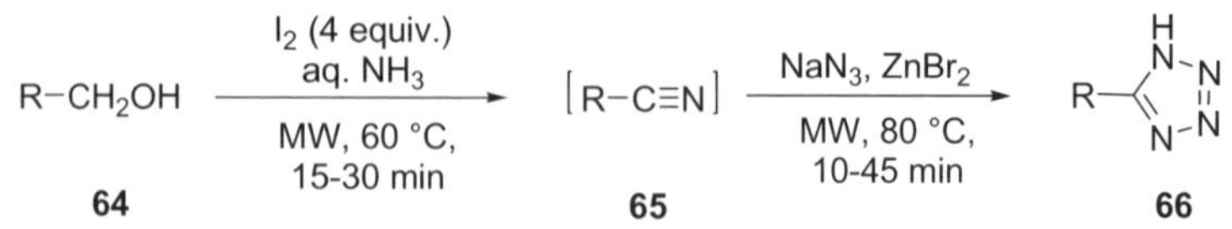

Scheme 9.10 Tetrazole synthesis by a one-pot domino reaction[83]

9.7.2 Intermolecular Approaches

Intermolecular cycloadditions of nitriles with azides are a useful tool for the synthesis of heterocycles such as tetrazolopyridines or tetrazolopiperazines. Couty *et al.* performed the intermolecular addition of nitriles as well as alkynes to azides, yielding triazoles and tetrazoles respectively.[84] Starting with *N*-cyanomethyl amino alcohols, the hydroxyl functionality was initially chlorinated and substituted by sodium azide. The resulting compound containing a nitrile as well as an azide group was heated for one hour at 150 °C, producing the target tetrazolopiperazines in 70–84% yield (Scheme 9.11). A similar method for the synthesis of the heterocycle-formation – starting from acyclic precursors – was suggested by Fleet *et al.*, who synthesized tetrazoles of manno- and rhamnofuranoses through the cycloaddition of azides to nitriles.[84]

Other intramolecular conversions show the formation of tetrazolo[1,5-α]pyridines through the conversion of pyridine *N*-oxides into the corresponding 2-azidopyridines. These azides exist in equilibrium with the tetrazoles, resulting from the cycloaddition of the azide with the pyridine ring. The ratio of the equilibrium is dependent on the substitution of the heterocycle; however the tetrazole usually dominates the azide species and can be converted into several other derivatives via hydrogenation, alkylation or arylation.[85]

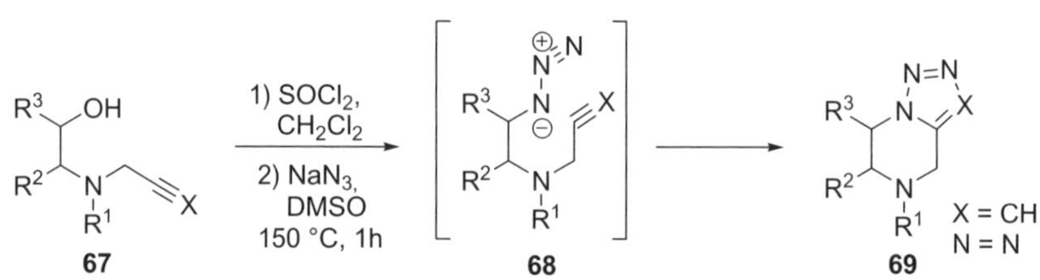

Scheme 9.11 Synthesis of tetrazolopiperazines and triazolopiperazines by cycloaddition of azides to nitriles[84]

9.8 Click Chemistry for the Synthesis of Dihydrotriazoles

The activation enthalpies for the cycloaddition of alkynes and alkenes are similar; however, since triazoles are more stable (30-40 kcal/mol) than dihydrotriazole products, the cycloaddition is very exothermic and therefore irreversible.

Houk *et al.*[86] studied substituent effects in the 1,3-dipolar cycloaddition of azides with alkenes and alkynes on the basis of reactant concentrations, in order to explore the equilibrium constant and examine the reversibility or irreversibility of these reactions. They affirmed a lower barrier for the cycloaddition between electron-deficient azides such as formyl- and methanesulfonylazides and electron-rich alkenes and conjugated alkenes (Scheme 9.12).

Reactions of bicyclo-[2.2.1]hept-5-ene-2,3-dicarboximides with aryl azides led to dihydrotriazoles with an *exo* orientation of the nitrogen-rich cycle (Alder rule) and an *endo* orientation of the imide (Scheme 9.13). Kas'yan *et al.*[87] found out that both the electron-donor as well as electron-acceptor substituents in the azide molecules accelerate the reaction.

In 1987, Buchanan *et al.* published a new route to chiral hydroxypyrrolidines *via* intramolecular 1,3-cycloaddition starting from 2,3-*O*-isopropylidene-*D*-erythrose. They confirmed that dihydrotriazoles act as a stable intermediate after the cycloaddition of the azide derivative, but only in the case of the *E*-isomer. However, the analytical data of the resulting diazo ester, based on both isomers (*E*- and *Z*-isomer), was similar after treatment with sodium ethoxide.

In the field of dihydrotriazoles, some results were previously published by Quast *et al.* on the photoextrusion of nitrogen out of dihydrotriazole[88] as well as dihydrotetrazole[89] derivatives. Their dihydrotriazole synthesis was based on Trost's and Pearson's 1,3-dipolar cycloaddition from the lithium enolate of methylisobutyrate with alkyl azides at a lower temperature (–78 °C).[90]

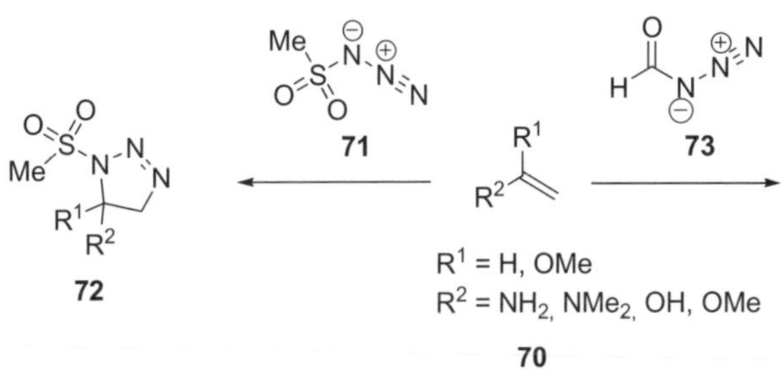

Scheme 9.12 *1,3-Dipolar cycloaddition of methanesulfonyl- (**71**) and formylazide (**73**) with enamines and enols **70***[86]

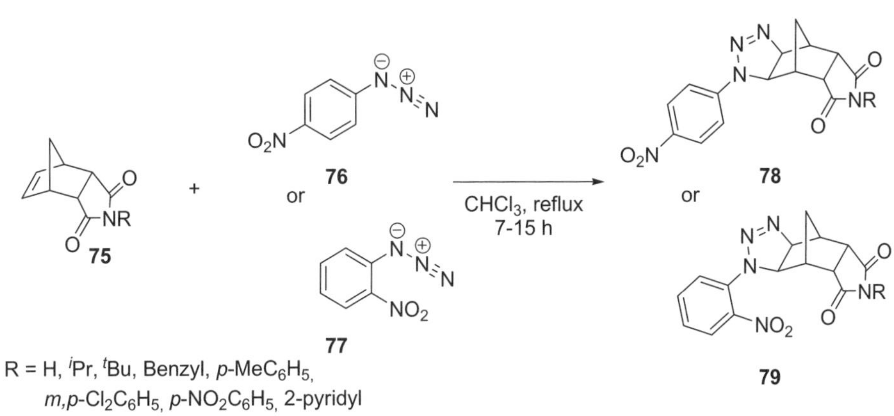

Scheme 9.13 *1,3-Dipolar cycloaddition of bicyclo-[2.2.1]hept-5-ene-2,3-dicarboximides with aromatic azides following the Alder rules*[87]

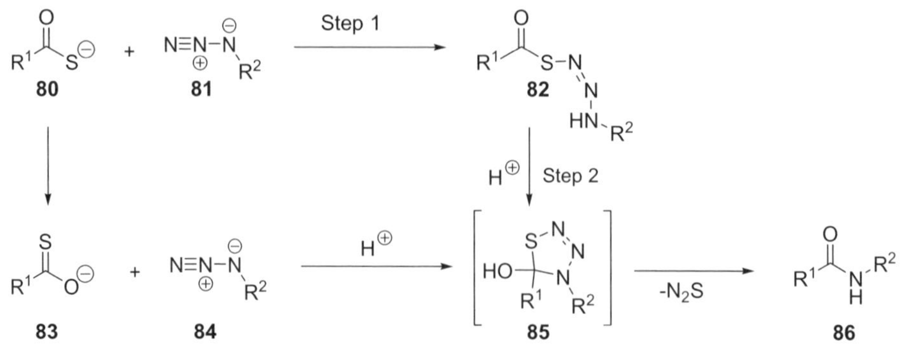

Scheme 9.14 Pathways for reaction of electron-rich and -deficient azides with thio acids[91]

9.9 Cycloaddition Reactions with Azides to Give Thiatriazoles

Cycloaddition reactions of azides with sulfur-containing dipolarophiles are less important and more seldomly used than the aforementioned Click reactions or the synthesis of tetrazoles, which may originate at least partly in the relatively low stability of the target compounds. As the Huisgen reaction, as well as the formation of tetrazoles can be performed at higher temperatures, the synthesis of thiatriazoles depends strongly on the availability of successful reactions at room temperature under mild conditions. The thiatriazole synthesis gains synthetic relevance through the secondary products that can be accessed *via* thermolysis. Williams *et al.* demonstrated that the heating of thiatriazoles **85** (Scheme 9.14) can yield amides (**86**) and that therefore, the cycloaddition of azides and thio acids is a useful tool that complements synthetic protocols such as the Staudinger reaction.[91]

The addition of azides to thio acids (**80**) is successful for electron-rich as well as for electron-deficient azides, but just as in case of the latter ones, a cycloaddition mechanism could be confirmed. Scheme 9.14 presents the supposed mechanism for both examples, referring to a two-step path for the reaction of electron-rich azides consisting of the addition of the azide to the thio acid and subsequent cyclization *via* nucleophilic attack on the carbonyl group (Step 2).

Azides can undergo cycloaddition reactions with sulfonyl isocyanates as well. L'Abbé *et al.* showed the formation of 4-sulfonyl-tetrazolin-5-ones that underwent thermolytically-induced decomposition and produced carbodiimides with the formation of N_2.[92]

References

[1] T. Curtius, *Ber. Dtsch. Chem. Ges.* **1883**, *16*, 2230–1.
[2] E. Buchner, *Ber. Dtsch. Chem. Ges.* **1888**, *21*, 2637–47.
[3] K.V. Gothelf, K.A. Jørgensen, *Chem. Rev.* **1998**, *98*, 863–909.
[4] L.I. Smith, *Chem. Rev.* **1938**, *23*, 193–285.
[5] R. Huisgen, G. Szeimies, L. Möbius, *Chem. Ber.* **1967**, *100*, 2494–2507.
[6] R. Huisgen, *Angew. Chem., Int. Ed.* **1963**, *2*, 565–632.
[7] S. Bräse, D. Keck, *SOS*, Thieme, Stuttgart **2007**, *31b*, 1827–43.

[8] V. Zimmermann, B. Lesch, S. Bräse, *SOS*, Thieme, Stuttgart **2009**, *41*, in preparation.
[9] S. Bräse, T. Muller, *SOS*, Thieme, Stuttgart **2007**, *31b*, 1845–72.
[10] F. Avemaria, V. Zimmermann, S. Bräse, *Synlett* **2004**, 1163–5.
[11] K.N. Houk, J. Gonzáles, Y. Li, *Acc. Chem. Res.* **1995**, *28*, 81–90.
[12] S. Bräse, A. Friedrich, M. Gartner, T. Grab, T. Schröder, *Topics in Heterocycl.* Springer, Berlin **2008**.
[13] M. Meldal, C.W. Tornøe, *Chem. Rev.* **2008**, *108*, 2952–3015.
[14] C.W. Tornøe, C. Christensen, M. Meldal, *J. Org. Chem.* **2002**, *67*, 3057–64.
[15] S. Bräse, C. Gil, K. Knepper, V. Zimmermann, *Angew. Chem., Int. Ed.* **2005**, *44*, 5188–5240.
[16] C.I. Schilling, S. Bräse, *Org. Biomol. Chem.* **2007**, *5*, 3586–8.
[17] T. Schröder, M. Gartner, T. Grab, S. Bräse, *Org. Biomol. Chem.* **2007**, *5*, 2767–9.
[18] P. Wu, V.V. Fokin, *Aldrichim. Acta* **2007**, *40*, 1, 7–17.
[19] V.V. Rostovtsev, L.G. Green, V.V. Fokin, K.B. Sharpless, *Angew. Chem., Int. Ed.* **2002**, *41*, 2596–9.
[20] V.O. Rodionov, V.V. Fokin, M.G. Finn, *Angew. Chem., Int. Ed.* **2005**, *44*, 2210–15.
[21] B.H.M. Kuijpers, S. Groothuys, A.R. Keereweer, *et al.*, *Org. Lett.* **2004**, *6*, 3123–6.
[22] D.A. Ossipov, J. Hilborn, *Macromol.* **2006**, *39*, 1709–18.
[23] P.L. Golas, N.V. Tsarevsky, B.S. Sumerlin, K. Matyjaszewski, *Macromol.* **2006**, *39*, 6451–7.
[24] T.R. Chan, R. Hilgraf, K.B. Sharpless, V.V. Fokin, *Org. Lett.* **2004**, *6*, 2853–5.
[25] F. Perez-Balderas, M. Ortega-Munoz, J. Morales-Sanfrutos, *et al.*, *Org. Lett.* **2003**, *5*, 1951–4.
[26] M. Malkoch, K. Schleicher, E. Drockenmuller, *et al.*, *Macromol.* **2005**, *38*, 3663–78.
[27] P. Wu, A.K. Feldman, A.K. Nugent, *et al.*, *Angew. Chem., Int. Ed.* **2004**, *43*, 3928–32.
[28] C. Girard, E. Onen, M. Aufort, S. Beauvière, E. Samson, J. Herscovici, *Org. Lett.* **2006**, *8*, 1689–92.
[29] R. Guezguez, K. Bougrin, K. El Akri, R. Benhida, *Tetrahedron Lett.* **2006**, *47*, 4807–11.
[30] S. Chassaing, A.S.S. Sido, A. Alix, M. Kumarraja, P. Pale, J. Sommer, *Chem. Eur. J.* **2008**, *14*, 6713–21.
[31] S. Chassaing, M. Kumarraja, A.S.S. Sido, P. Pale, J. Sommer, *Org. Lett.* **2007**, *9*, 883–6.
[32] M.L. Kantam, V.S. Jaya, B. Sreedhar, M.M. Rao, B. M.J. Choudary, *Mol. Catal.* **2006**, *256*, 273–7.
[33] V.D. Bock, H. Hiemstra, J.H. Van Maarseveen, *Eur. J. Org. Chem.* **2006**, 51–68.
[34] B.F. Straub, *Chem. Commun.* **2007**, 3868–70.
[35] M. Ahlquist, V.V. Fokin, *Organometallics* **2007**, *26*, 4389–91.
[36] F. Himo, T. Lovell, R. Hilgraf, *et al.*, *J. Am. Chem. Soc.* **2005**, *127*, 210–16.
[37] A. Baron, Y. Bleriot, M. Sollogoub, B. Vauzeilles, *Org. Biomol. Chem.* **2008**, *6*, 1898–1901.
[38] I. Geci, V.V. Filichev, E.B. Pedersen, *Chem Eur. J.* **2007**, *13*, 6379–86.
[39] P.S. Donelly, S.D. Zanatta, S.C. Zammit, J.M. White, S.J. Williams, *Chem. Commun.* **2008**, 2459–61.
[40] Q. Wang, T.R. Chan, R. Hilgraf, *et al.*, *J. Am. Chem. Soc.* **2003**, *125*, 3192–3.
[41] S.I. van Kasteren, H.B. Kramer, H.H. Jensen, *et al.*, *Nature*, **2007**, *446*, 1105–9.
[42] V.O. Rodionov, S.I. Presolski, S. Gardinier, Y.H. Lim, G. Finn, *J. Am. Chem. Soc.* **2007**, *129*, 12696–12704.
[43] W.G. Lewis, F.G. Magallon, V.V. Fokin, M.G.J. Finn, *J. Am. Chem. Soc.* **2004**, *126*, 9152–3.
[44] X.L. Sun, C.L. Stabler, C.S. Cazalis, E.L. Chaikof, *Bioconjugate Chem.* **2006**, *17*, 52–7.
[45] W.H. Zhan, H.N. Barnhill, K. Sivakumar, H. Tian, Q. Wang, *Tetrahedron Lett.* **2005**, *46*, 1691–5.
[46] F. Tian, M.L. Tsao, P.G. Schultz, *J. Am. Chem. Soc.* **2004**, *126*, 15962–3.
[47] V.O. Rodionov, S.I. Presolski, D.D. Diaz, V.V. Fokin, M.G. Finn, *J. Am. Chem. Soc.* **2007**, *129*, 12705–12.
[48] K. Tanaka, C. Kageyama, K. Fukase, *Tetrahedron Lett.* **2007**, *48*, 6475–9.

[49] J.H. Van Maarseveen, W.S. Horne, M.R. Ghadiri, *Org. Lett.* **2005,** *7*, 4503–6.
[50] H.N. Gopi, K.C. Tirupula, S. Baxter, S. Ajith, I.M. Chaiken, *ChemMedChem* **2006**, *1*, 54–7.
[51] W.M. Xu, X. Huang, E.J. Tang, *Comb. Chem.* **2005**, *7*, 726–33.
[52] A.K. Feldman, B. Colasson, V.V. Fokin, *Org. Lett.* **2004**, *6*, 3897–9.
[53] X. Zhang, R.P. Hsung, L. You, *Org. Biomol. Chem.* **2006**, *4*, 2679–82.
[54] J.C. Meng, V.V. Fokin, M.G. Finn, *Tetrahedron Lett.* **2005**, *46*, 4543–6.
[55] B. Sreedhar, P.S. Reddy, *Synth. Commun.* **2007**, *37*, 805–12.
[56] M. Whiting, J. Muldoon, Y.C. Lin, *et al.*, *Angew. Chem., Int. Ed.* **2006**, *45*, 1435–9.
[57] J.A. Codelli, J.M. Baskin, N.J. Agard, C.R. Bertozzi, *J. Am. Chem. Soc.* **2008**, *130*, 11486–93.
[58] E.M. Sletten, C.R. Bertozzi, *Org. Lett.* **2008**, *10*, 3097–9.
[59] J.-F. Lutz, *Angew. Chem., Int. Ed.* **2008**, *47*, 2182–4.
[60] G. Wittig, A. Krebs, *Chem. Ber.* **1961**, *94*, 3260–75.
[61] D.H. Ess, G.O. Jones, K.N. Houk, *Org. Lett.* **2008**, *10*, 1633–6.
[62] J.M Baskin, J.A. Prescher, S.T. Laughlin, *et al.*, *Proc. Natl. Acad. Sci. U.S.A.* **2007**, *104*, 16793–7.
[63] S.S. van Berkel, A.J. Dirks, M.F. Debets, *et al.*, *ChemBioChem* **2007**, *8*, 1504–8.
[64] X. Ning, J. Guo, M.A. Wolfert, G.-J. Boons, *Angew. Chem., Int. Ed.* **2008**, *47*, 2253–5.
[65] L. Campbell-Verduyn, P.H. Elsinga, L. Mirfeizi, R.A. Dierckx, B.L. Feringa, *Org. Biomol. Chem.* **2008**, *6*, 3461–3.
[66] C.-T. Zhang, X. Zhang, F.-L. Qing, *Tetrahedron Lett.* **2008**, *49*, 3927–30.
[67] M.M. Majireck, S.M. Weinreb, *J. Org. Chem.* **2006**, *71*, 8680–3.
[68] L. Zhang, X. Chen, P. Xue, *et al.*, *J. Am. Chem. Soc.* **2005**, *127*, 15998–9.
[69] B.C. Boren, S. Narayan, L.K. Rasmussen, *et al.*, *J. Am. Chem. Soc.* **2008**, *130*, 8923–30.
[70] U. Pradere, V. Roy, T.R. McBrayer, R.F. Schinazi, L.A. Agrofoglio, *Tetrahedron* **2008**, *64*, 9044–51.
[71] A. Tam, U. Arnold, M.B. Soellner, R.T. Raines, *J. Am. Chem. Soc.* **2007**, *129*, 12670–1.
[72] S. Oppilliart, G. Mousseau, L. Zhang, *et al.*, *Tetrahedron* **2007**, *63*, 8094–8.
[73] D. Imperio, T. Pirali, U. Galli, *et al.*, *Bioorg. Med. Chem.* **2007**, *15*, 6748–57.
[74] L.K. Rasmussen, B.C. Boren, V.V. Fokin, *Org. Lett.* **2007**, *9*, 5337–9.
[75] P.N. Gaponik, S.V. Voitekhovich, O.A. Ivashkevich, *Russ. J. Chem. Rev.* **2006**, *75*, 507–39.
[76] Z.P. Demko, K B. Sharpless, *Angew. Chem. Int. Ed.* **2002,** *41*, 2110–13.
[77] M. Aldhoun, A. Massi, A. Dondoni, *J. Org. Chem.* **2008**, in press.
[78] P.Z. Demko, K.B. Sharpless, *Angew. Chem. Int. Ed.* **2002**, *41*, 2113–16.
[79] M.A. Gol'tsberg, G.I. Koldobskii, *Russ. J. Org. Chem.* **1996**, *32*, 1194.
[80] A.P. Koreneva, G.I. Koldobskii, *Russ. J. Org. Chem.* **2000**, *36*, 1698–9.
[81] I.F. Clemencon, B. Ganem, *Tetrahedron* **2007**, *63*, 8665–9.
[82] V. Aureggi, G. Sedelmeier, *Angew. Chem., Int. Ed.* **2007**, *46*, 8440–4.
[83] J. Fleet, J.-J. Shie, J.-M. Fang, *J. Org. Chem.* **2007**, *72*, 3141–4.
[84] B. Davis, T.W. Brandstetter, C. Smith, L. Hackett, B.G. Winchester, G.W.J. Fleet, *Tetrahedron Lett.* **1995**, *36*, 7507–10.
[85] J.M. Keith, *J. Org. Chem.* **2006**, *71*, 9540–3.
[86] G.O. Jones, K.N. Houk, *J. Org. Chem.* **2008**, *73*, 1333–42.
[87] I.N. Tarabara, A.O. Kas'yan, M.Y. Yaravoi, S.V. Shishkina, O.V. Shishkin, L.I. Kas'yan, *Russ. J. Org. Chem.* **2004**, *40*, 992–8.
[88] H. Quast, L. Bieber, *Angew. Chem., Int. Ed.* **1975**, *14*, 428–9.
[89] H. Quast, B. Seiferling, *Tetrahedron Lett.* **1982**, *23*, 4681–4.
[90] B.M. Trost, W.H. Pearson, *J. Am. Chem. Soc.* **1981**, *103*, 2483–5.
[91] R.V. Kolakowski, N. Shangguan, R.R. Sauers, L.J. Williams, *J. Am. Chem. Soc.* **2006**, *128*, 5695–5702.
[92] E. van Loock, J.-M. Vandensavel, G. L'Abbé, G. Smets, *J. Org. Chem.* **1973**, *38*, 2916–17.

10

Dipolar Cycloaddition Reactions in Peptide Chemistry

Christian Wenzel Tornøe[1] *and Morten Meldal*[2]

[1]*H. Lundbeck A/S, Ottiliavej 9, DK-2500 Valby, Denmark;* [2]*Carlsberg Laboratory, Gamle Carlsberg Vej 10, DK-2500 Valby, Denmark*

10.1 Introduction

Dipolar cycloaddition reactions (DCRs) are unique reactions when considering their versatility, their atom economy and the highly functionalized heterocyclic products they provide. The significant contributions by Rolf Huisgen has led to the description of a 1,3-dipolar cycloaddition reaction, where a dipole (formally a zwitterionic molecule) adds to a dipolarophile (an alkene or alkyne) to form a five-membered heterocyclic ring.[1,2] Two σ-bonds are formed in this concerted reaction where bond breaking and bond formation occurs simultaneously and in a stereospecific manner, which also can be described by frontier molecular orbital theory as the highest occupied molecular orbital of one component reacting with the lowest unoccupied molecular orbital of the other component.

The field of dipolar cycloaddition reactions with azides in peptide chemistry has developed rapidly since Tornøe and Meldal first described the copper(I)-catalyzed cycloaddition between azides and peptide-linked terminal alkynes to exclusively form the 1,4-substituted [1,2,3]-peptidotriazoles.[3] This reaction was later termed CuAAC (copper-catalyzed azide-alkyne cycloaddition) or simply the 'click' reaction, and has been used extensively in all areas of chemistry, biology and material sciences due to its mild reaction conditions, high functional group tolerance, generally high yields and exclusive formation of 1,4-substituted [1,2,3]-triazoles. The mechanistic aspects of the CuAAC leading to the 1,4-substitution are complicated and have been discussed elsewhere.[4–6]

Organic Azides: Syntheses and Applications Edited by Stefan Bräse and Klaus Banert

Even though the CuAAC is a rather new reaction, more than 800 publications on CuAAC 'click' chemistry has been published (May 2008), and it has been extensively reviewed.[5–11] The focus of this chapter will be on azides in 1,3-dipolar cycloaddition reactions, mainly catalyzed by transition metals, in peptide chemistry. Protein ligation and protein modification by dipolar cycloaddition reactions has been reviewed and will not be included.[5] Angell and Burgess[12] published an excellent review on peptidomimetics generated by CuAAC in early 2007 with a thorough overview of the field and since then more than twenty new publications describing dipolar cycloaddition reactions in peptide chemistry have appeared.

This review is intended to cover all reactions and technology in peptide chemistry concerning dipolar cycloaddition reactions of azides, and the collection of literature was concluded on May 15 2008. The authors would like to apologize in advance if any references are missing.

10.2 Amino Acid Derivatives by DCR

Glycosyl amino acids for synthesis of glycopeptide mimetics have frequently been obtained through DCRs. Building blocks with heterocyclic-linked *C*-glycosyl amino acids and their incorporation into novel glycopeptides were described by Dondoni *et al.*[13] Two regioisomeric *C*-glycosyl triazole alanines **1** were obtained by thermal 1,3-dipolar cycloaddition between ethynyl *C*-glycoside and azido-functionalized alanine. Kuijpers *et al.* presented an efficient synthesis of α- and β-triazole-linked glycosyl-amino acids and glycopeptides by 'clicking' together either azidosugars and acetylenic amino acids (to produce **2**) or with equal efficiency acetylenic glycosides and azide-containing amino acids (to produce **3**).[14] The preparation of β-amino acid building blocks suitable for solid-phase synthesis was reported by Ziegler *et al.* where compounds derived from aspartic acid (see Figure 10.1) were reacted with either propargyl or azido glycosides.[15] Copper(I)-catalysis using the $P(OEt)_3{\cdot}CuI$-complex and microwave heating (80 °C for 30 minutes) afforded protected triazolyl-linked glycosylated β-amino acids (**4** and **5** in Figure 10.1) in good to high yields (65–98%). To limit the formation of side-products and product decomposition in the cycloaddition step, microwave heating ensured short reaction times.

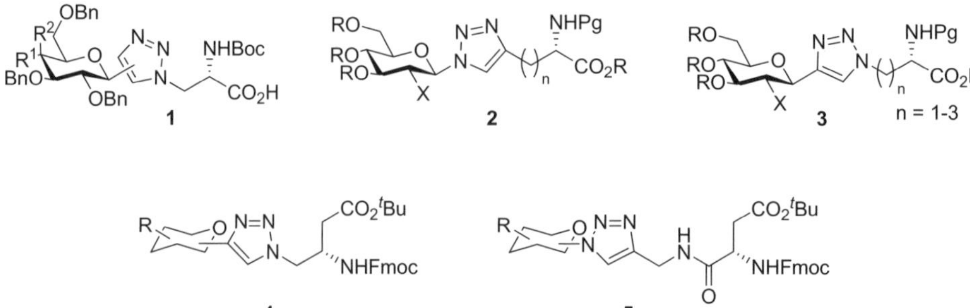

Figure 10.1 Glycoconjugated amino acids prepared by dipolar cycloaddition reactions. Pg, protecting group

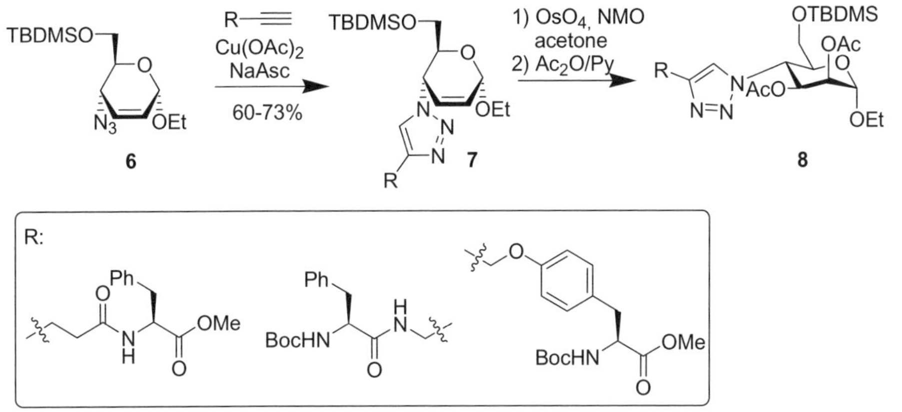

Scheme 10.1 *Synthesis of mannopyranoside and glucopyranoside derivatives of amino acids*

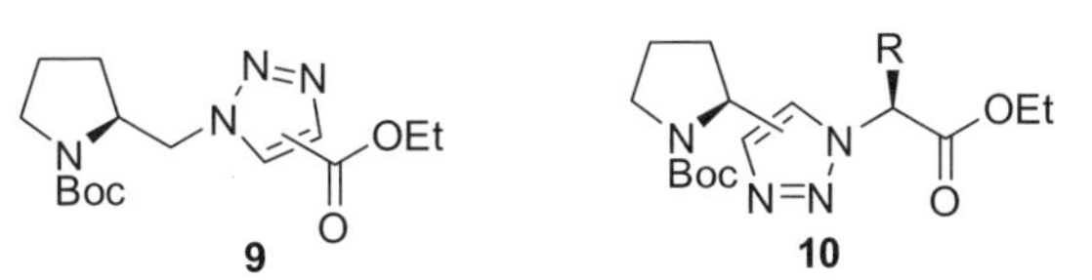

Figure 10.2 *Dipeptide mimics containing a 1,4- or 1,5-substituted triazole were produced by a thermal or copper(I)-catalyzed 1,3-dipolar cycloaddition reaction*

The group of Srivastava expanded the range of triazole-linked glycosyl amino acids by preparing unsaturated triazole-derivatives **7** in good yields (60–73%, see Scheme 10.1) using copper(II) sulfate and sodium ascorbate (NaAsc) to catalyze the cycloaddition reaction between **6** and alkyne-substituted amino acids.[16] The products were subjected to osmium-catalyzed dihydroxylation to afford the corresponding mannopyranoside derivatives **8**. Reactions of α- and β-glucopyranosyl azide with the same alkynes provided triazolyl-glucopyranosyl amino acids directly in 40–84% yield (not shown).

DCR has also been used to prepare amino acid and dipeptide mimetics. The synthesis of Pro-Gly and Pro-Xaa dipeptide mimetics (**9** and **10**, respectively, where Xaa is an unspecified amino acid) was reported by Gmeiner *et al.* and the triazolyldipeptides were assembled using a 1,3-dipolar cycloaddition (thermal or Cu^I-catalyzed) to provide 1,4- and 1,5-substituted triazoles (see Figure 10.2).[17] The conformational behavior of the resulting peptides was analyzed by NMR and FTIR and the results suggested that the cis/trans ratios could be tuned by the triazole and its substitution.

Tetrazole analogues of amino acids have been prepared from Z-protected amino acids,[18] and Z-protected amino nitriles.[19] More recently, the group of Sureshbabu *et al.* extended this formal [3 + 2] cycloaddition between a nitrile-containing protected α-amino acid **11** and azide to also include Fmoc-amino acids and prepared β- and γ-tetrazolyl α-amino acids **12** for incorporation into peptides (see Scheme 10.2).[20] The carboxamide side-chains of asparagine and glutamine were dehydrated and converted to the tetrazole by treatment with sodium azide and zinc bromide to afford tetrazole-isosteres of aspartic acid and

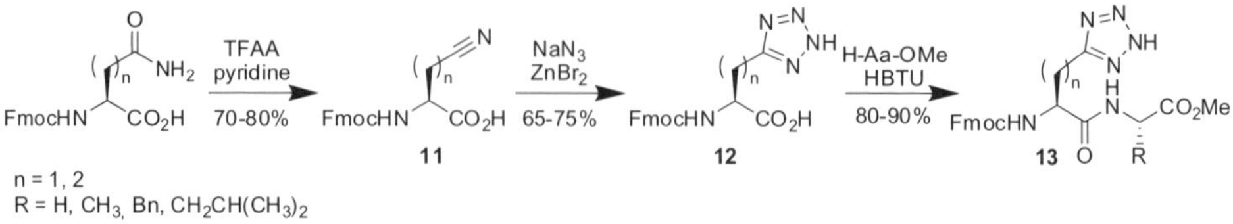

Scheme 10.2 *Tetrazole-analogues of aspartic acid and glutamic acid were prepared from nitrile containing Fmoc-amino acids*

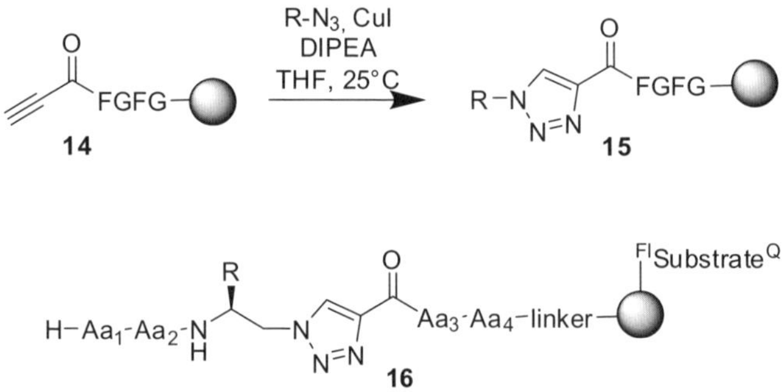

Scheme 10.3 *The first reported synthesis of peptidotriazoles by CuAAC. Fl, fluorophore (2-aminobenzoic acid); Q, quencher (3-nitrotyrosine)*

glutamic acid. These were employed in peptide coupling to afford the dipeptide **13** in high yield.

10.3 Peptide Backbone Modifications by DCR

CuAAC was pioneered within solid-phase peptide synthesis and it was demonstrated that the Cu(I)-catalysis was compatible with standard Fmoc-based solid-phase peptide synthesis protocols and protecting groups.[3] Furthermore, the diversity of azides used (α-azido acids, aryl, alkyl, and sugar-derived azides) in the cycloaddition reaction indicated a large potential for this reaction. A full paper followed where the regioselectivity of the CuAAC 'click' reaction was investigated and proven to be completely selective for the 1,4-substituted [1,2,3]-triazole using 2D NMR.[21] Immobilized alkynes (for example **14**, Scheme 10.3) were subjected to CuAAC and efficiently converted to the corresponding triazoles **15** except when the sterically hindered 2-azido-2,2-diphenylacetic acid or the electron-deficient trimethylsilyl azide were used. It was also demonstrated that peptide synthesis could be continued when using amino azides, thus allowing for the preparation of [1,2,3]-triazoles incorporated in the peptide backbone. Various solvents (acetonitrile, dichloromethane, tetrahydrofuran, *N,N*-dimethylformamide and *N,N*-diisopropylethylamine) were shown to be compatible with the CuAAC (typically performed at 25 °C). Shortly after, the Sharpless group published their very efficient conditions for the aqueous CuAAC 'click'

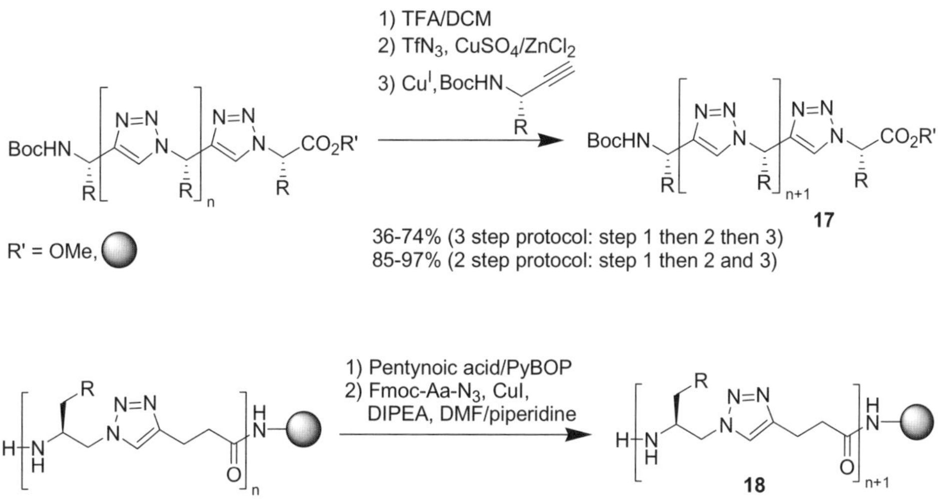

Scheme 10.4 *Synthesis of peptidotriazole-oligomers by CuAAC. PyBOP, 1*H*-benzotriazole-1-yl-oxy-tris-pyrrolidino-phosphonium hexafluorophosphate*

reaction ($CuSO_4$ and sodium ascorbate in water/^{t}BuOH).[22] A combinatorial library of putative inhibitors of the cysteine protease *Leishmania mexicana* CPB2.8ΔCTE, was generated by the split-and-mix protocol on the hydrophilic $PEGA_{1900}$-resin to afford approximately 450.000 peptidotriazoles (**16** in Scheme 10.3), utilizing the efficient CuAAC 'click' reaction to form a central triazole from resin-bound alkyne and β-amino azides.[23] On-bead screening was possible since the synthesis design afforded a one-bead two-compounds construct containing both inhibitory peptide and a fluorophore/quencher-labeled FRET (fluorescence resonance energy transfer) peptide substrate. Active peptidotriazole inhibitors were identified after enzyme incubation and cleavage of substrate by selection of dark beads where no substrate hydrolysis had occurred using automated beadsorting.[24] Twenty-three peptidotriazoles were resynthesized on solid support and isolated in 30–89% yield. These were shown to inhibit the protease with K_i's from 0.076 μM to 1000 μM.

Angelo and Arora described the synthesis of 1,4-substituted triazole oligomers **17** that displayed amino acid side chains on a non-peptidic triazole backbone.[25] The elegant oligomer synthesis consisted of 3 steps: Boc-deprotection, diazotransfer to an amino acid and a Cu(I)-catalyzed 1,3-DCR with a Boc-protected substituted propargylamine derived from amino acids. The cycle yield ranged from 36–74% (3 step protocol, Scheme 10.4). The authors then optimized their triazolamer (triazole oligomer) synthesis and reported the one-pot sequential diazotransfer followed by an *in situ* copper(I)-catalyzed cycloaddition to amino acid-derived alkynes (2 step protocol, Scheme 10.4).[26] $ZnCl_2$ was found to be the most efficient soluble diazotransfer catalyst on solid-phase, but not compatible with Boc-based solution-phase triazolamer synthesis due to Boc-cleavage. The isolated yield of a specific triazole trimer was significantly improved from the previously reported 3 step protocol[25] (28% yield) to 62% by the new one-pot sequential procedure. Zhang and Fan investigated conditions carefully and developed a protocol for the solid-phase synthesis of peptidotriazoles containing multiple triazole moieties by utilizing a modified

CuAAC with CuI, ascorbic acid and 20% piperidine in DMF.[27] These conditions ensured dissolution and stabilization of the active copper(I)-species and simultaneously removed the Fmoc-group of the resulting triazole to afford oligomeric peptidotriazoles **18** with very high purity.

Aucagne and Leigh described the chemoselective derivatization of alkyne-substituted tripeptides by CuAAC 'click' chemistry followed by in-situ deprotection of the TMS-alkyne with silver hexafluorophosphate and then a second CuAAC reaction (see Scheme 10.5).[28] Compound **19** was functionalized with azido-dipeptides **20** to afford bis-triazole pseudopeptides **21** in good purity and yield.

The Raines group published an elegant paper on the replacement of an amide bond with a 1,5-substituted [1,2,3]-triazole **22** and incorporation of this dipeptide into bovine pancreatic ribonuclease (RNase A) by semisynthetic methods and showed that the melting temperature (T_m) and catalytic activity of the resulting RNase A variants were retained.[29] The triazolyl dipeptide was prepared by a Ru(II)-catalyzed alkyne-azide cycloaddition[30] which afforded 1,5-substituted triazoles, selectively (see Scheme 10.6).

Triazole-substituted piperazine amides mimicking β-turn structures were reported by the Burgess group.[31] Fifteen piperazine amides were prepared and isolated in high yields (**23**, 71–96%, Scheme 10.7) and used in couplings with dichlorotriazine derivatives (tagged with fluorescein or an alkyne) to afford combinatorial libraries **24**, which were

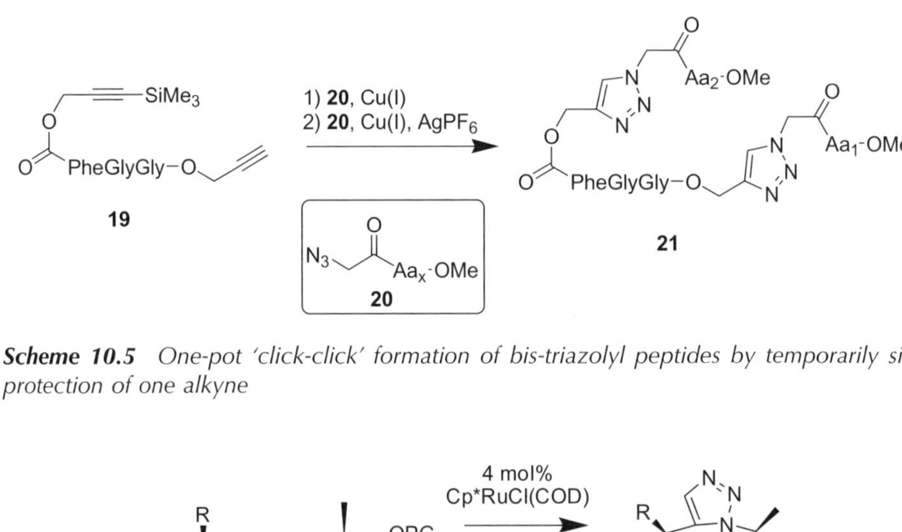

Scheme 10.5 One-pot 'click-click' formation of bis-triazolyl peptides by temporarily silyl protection of one alkyne

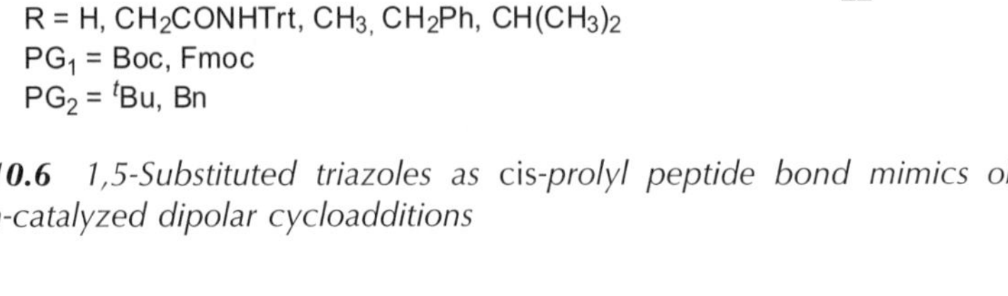

R = H, $CH_2CONHTrt$, CH_3, CH_2Ph, $CH(CH_3)_2$
PG_1 = Boc, Fmoc
PG_2 = ^{t}Bu, Bn

Scheme 10.6 1,5-Substituted triazoles as cis-prolyl peptide bond mimics obtained by ruthenium-catalyzed dipolar cycloadditions

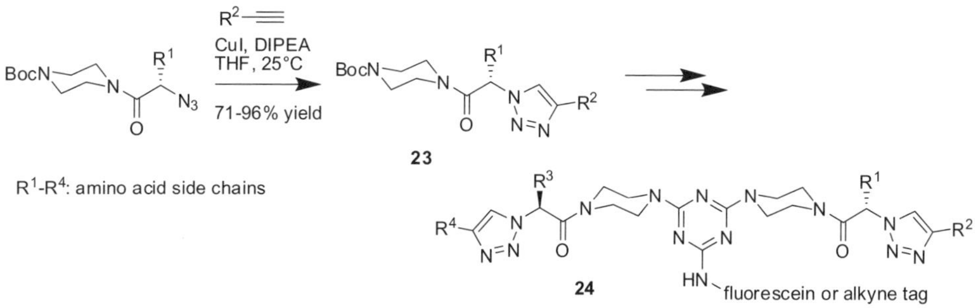

Scheme 10.7 *β-turn mimics produced from piperazine amides and dichlorotriazines*

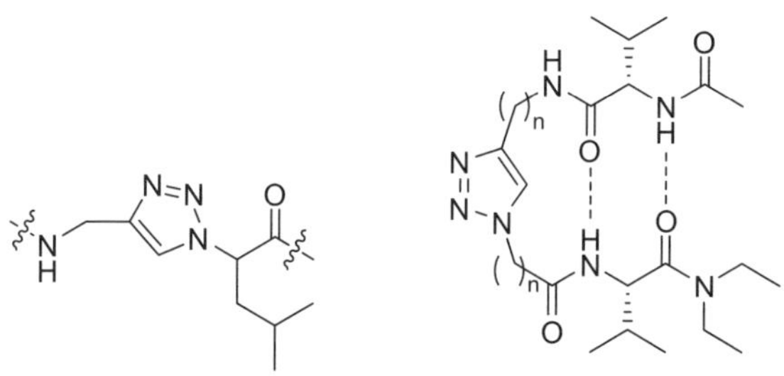

Figure 10.3 *Copper(I)-catalyzed azide-alkyne cycloadditions were used to incorporate triazoles into α-helical structures* **25** *and β-turn mimics* **26**, *respectively, and the secondary structures were spectroscopically verified*

screened for their binding to the TrkA, TrkC and p75 receptor. Four compounds displaying ~2-fold selectivity for TrkA were identified, all containing a threonine/lysine type of side chains (R^1/R^2).

The group of Ghadiri used the 'click' reaction between L-azido leucine and Fmoc-propargylamine to assemble an ε^2-amino acid in 91% yield.[32] This amino acid (**25** in Figure 10.3) replaced a dipeptide fragment in a known α-helical peptide structure and three peptide sequences were prepared using solid-phase peptide synthesis and shown to retain their α-helical structure. Oh and Guan prepared four β-turn mimics by clicking together an azide-derived peptide with an alkyne-fragment to afford triazolyl-linked peptides (**26**, see Figure 10.3) and used proton NMR and FTIR to establish the optimal length of the linkers.[33] It was demonstrated with FTIR that intramolecular hydrogen bonding was predominant thus indicating that a stable β-turn mimic was obtained, when n = 3.

Danishefsky *et al.* prepared azido-substituted glycosyl amino acids which where clicked onto alkyne-labeled peptides in solution catalyzed by nanosized copper(0) to afford hybrid molecules **27** that could have potential use as anticancer vaccines (illustrated in Scheme 10.8).[34] An immune response may be elicited by using appropriate carbohydrate antigens and the authors demonstrated the linking concept with three different antigens.

Fukase *et al.* discovered that non-basic histidine residues accelerated the CuAAC by coordinating and stabilizing the Cu(I)-species, similar to triethylamine.[35] This non-basic

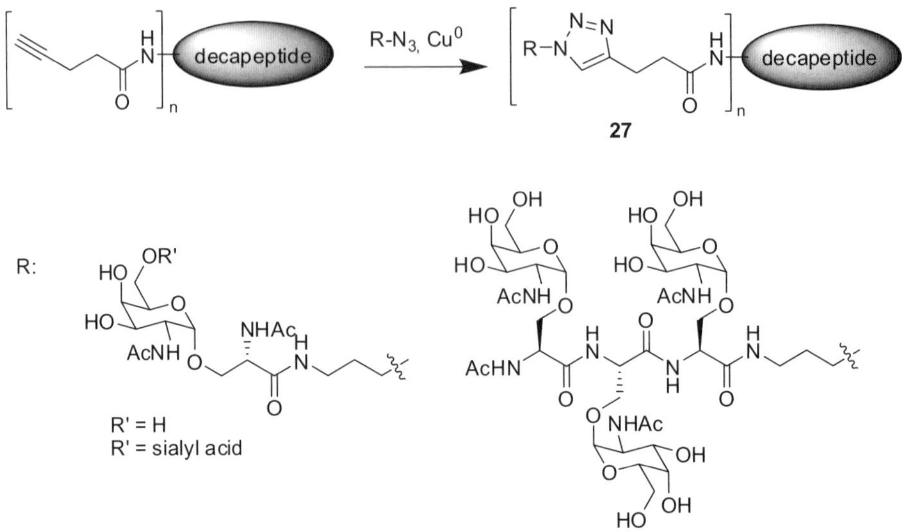

Scheme 10.8 Copper(I)-catalyzed conjugation of antigens and peptides

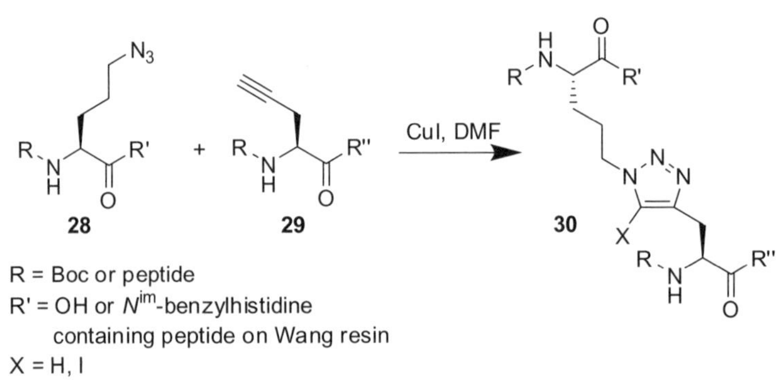

Scheme 10.9 Peptide-conjugation through triazole formation catalyzed by histidine-coordinated copper(I)

'click' reaction was exemplified by the conjugation of azide-derivatized ornithine residues **28** with propargylglycinyl peptides **29**, either in solution or immobilized on a Wang resin (see Scheme 10.9). The desired triazole-conjugates **30** were obtained in high yields, though occasionally as a mixture of 1,4-substituted triazole and the 5-iodo-1,4-substituted triazole.

10.4 Other Peptide Modifications by DCR

In order to introduce reactive azides/alkynes in peptides, the Rutjes group utilized alcalase mediated dipeptide synthesis with amino acid amides and either acetylenic or azido amino

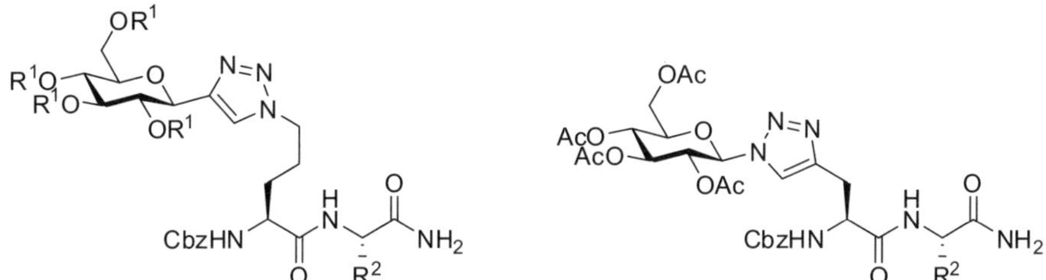

31
R^1 = Ac, Bn
R^2 = H, Bn

32
R^2 = H, Bn

Figure 10.4 *Examples of glycodipeptides produced by enzyme catalyzed peptide coupling and 'click' chemistry*

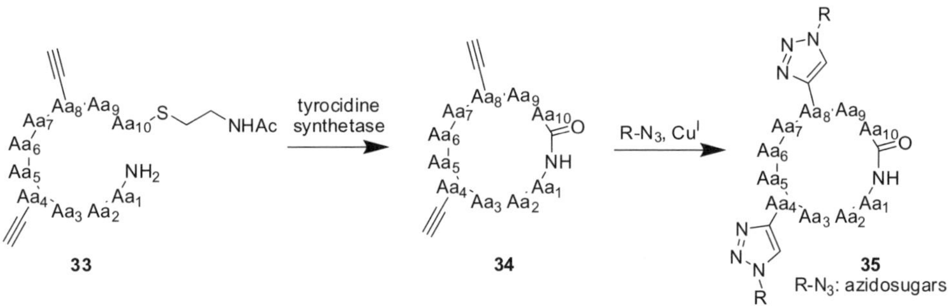

33 **34** **35**

Scheme 10.10 *Enzyme-mediated cyclization of alkyne-substituted peptide thioethers and subsequent CuAAC glycoconjugation*

acid methyl esters.[36] The resulting dipeptides were efficiently clicked with azido- or alkyne-substituted glucose to form triazolyl glycodipeptides **31** and **32** in high yields (62 to >99%, see Figure 10.4).

Lin and Walsh used part of the tyrocidine synthetase to cyclize decapeptide *N*-acetyl cysteamine thioethers **33** containing 1-3 propargylglycinyl residues to form macrocyclic peptides **34** (17 examples).[37] These peptides were glycoconjugated using azidolabeled sugars by copper(I)-catalysis (see Scheme 10.10) and two of the resulting glycopeptides **35** showed equipotent minimal inhibitory concentration of bacteria growth compared to the antibiotic peptide tyrocidine, but much improved minimal hemolytic concentration.

Wang *et al.* described the assembly of large oligomannose clusters onto a cyclic TASP (template-assembled synthetic proteins) decapeptide template[38] by 'click' chemistry and studied the resulting epitope mimic for their binding to HIV-neutralizing antibody 2G12 by surface plasmon resonance (see Scheme 10.11).[39] Four lysine residues, acylated with propiolic acid, where subjected to CuAAC with oligomannosyl azides followed by derivatization of two other lysines to form **36**. This intermediate was 'clicked' with alkynylated T-helper peptides to afford **37** in 70% yield and showed affinity for 2G12 but no *in-vivo* data was presented.

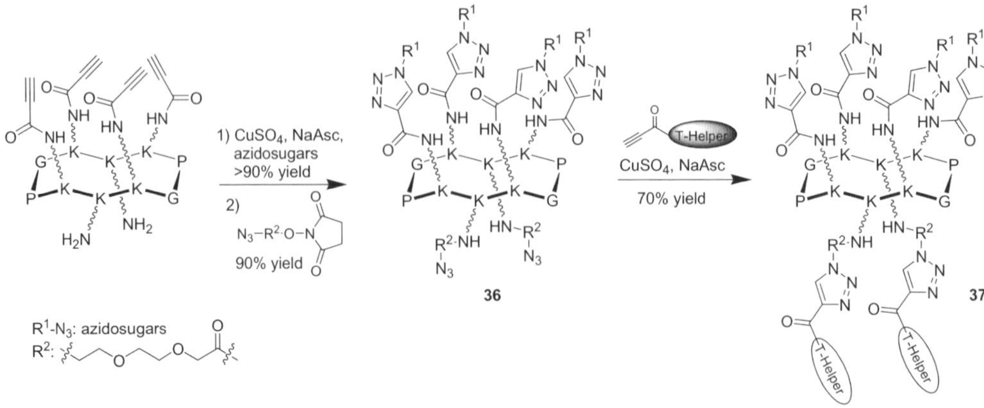

Scheme 10.11 *Formation of epitope mimics by sequential 'click' chemistry*

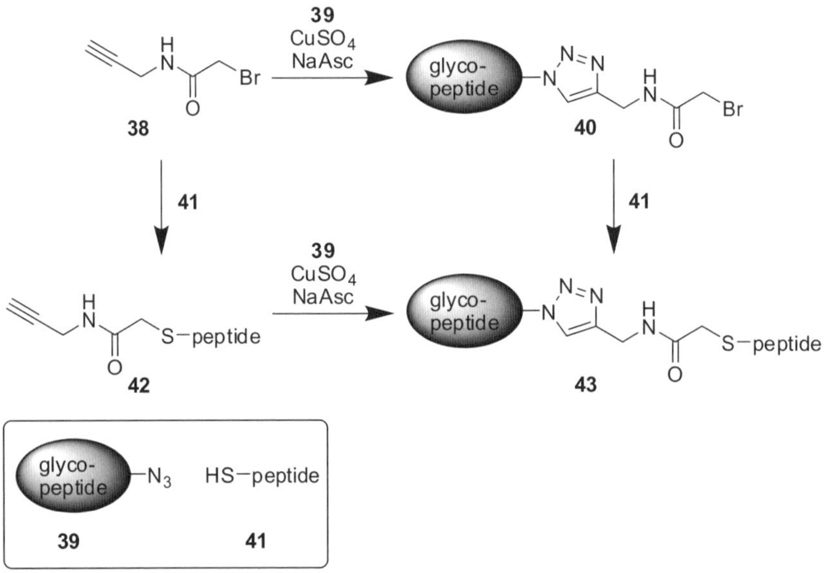

Scheme 10.12 *Methods for linking a peptide with a glycopeptide using the CuAAC 'click' reaction*

Macmillan described methods for linking a glycopeptide covalently to a peptide employing copper(I)-catalyzed formation of a [1,2,3]-triazole.[40] Two synthetic routes towards **43** (see Scheme 10.12) were employed: CuAAC 'click' chemistry between 1) *N*-(propargyl)-bromoacetamide (**38**) and **39** followed by alkylation of free cysteines or 2) alkynylated cysteine residues **42** and an azido-derived glycopeptide **39**. No application of the products **43** was described.

Taillefumier *et al.* were the first to report on cyclic *β*-peptoids and described macrocycles comprised of *N*-propargyl-*β*-alanine oligomers **44**.[41] The cyclic tetramer was

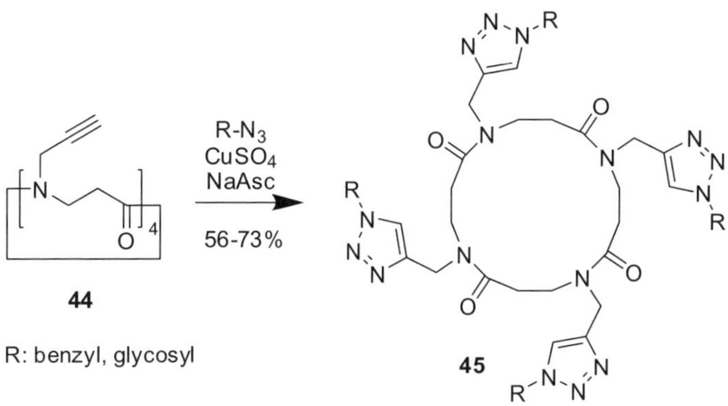

Scheme 10.13 *Derivatization of cyclic β-peptoids using 'click' chemistry*

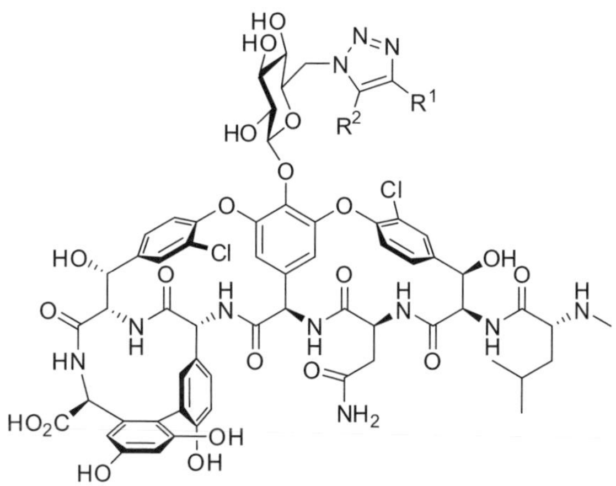

46 (20-90% yield)
23 examples (R^1 = various substituents, R^2 = H)
1 example (R^1 = R^2 = CO_2H)

Figure 10.5 *Improved antibacterial activity was observed after triazole-functionalization of vancomycin*

further derivatized by 'click' chemistry resulting in benzyl- or glycosyl-triazole functionalization as illustrated in Scheme 10.13 to afford **45** in good yield (73% and 56%, respectively).

The group of Thorson described a cycloaddition library **46** between 24 diverse alkynes and an azido-substituted vancomycin.[42] The copper(I)-catalyzed conjugation between the alkynes and the azido-sugar moiety of vancomycin proceeded in 20–90% yield (see Figure 10.5). Two compounds (R^1 = *n*-heptyl, R^2 = H and R^1 = $(CH_2)_8CO_2H$, R^2 = H) showed improved antibacterial activity against three clinically relevant strains compared to vancomycin itself.

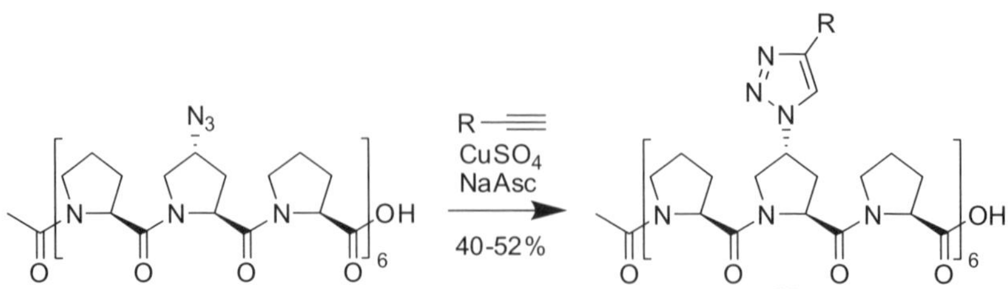

Scheme 10.14 *Derivatization of polyprolines by CuAAC*

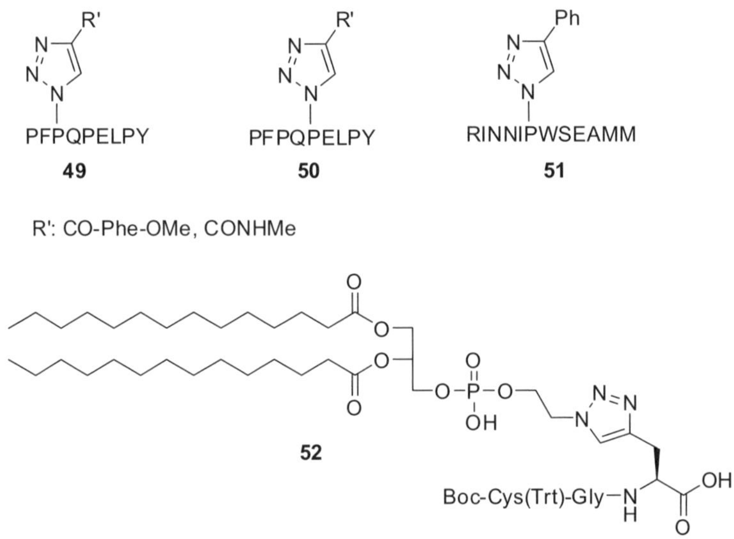

Figure 10.6 *Triazole-derivatized gluten peptides* ***49*** *and* ***50****, an HIV-1 fusion inhibitor* ***51*** *and a lipopeptide* ***52***

Wennemers *et al.* reported the derivatization of oligoproline Ac-[Pro-(4R)Azp-Pro]$_6$-OH **47** by subjecting the azidoprolines (Azp) to a copper(I)-catalyzed 1,3-dipolar cycloaddition reaction (Scheme 10.14).[43] The oligopeptide **47** showed a clear preference for the polyproline II (PPII) helix both before and after derivatization with various alkynes (**47** and **48**, respectively).

The group of Overkleeft used the crystal structure of a peptide/HLA-DQ2 complex to identify two proline residues in the peptide that allowed for functionalization with the aim of improving the affinity towards HLA-DQ2.[44] Proline 3 and 5 in the epitope (PFPQPELPY) derived from a gluten peptide, were replaced with (4*S*)- or (4*R*)-4-azido-L-proline on solid-phase and functionalized using CuAAC with either *N*-methyl-2-propynamide or *N*-propynoyl-L-phenylalanine methylester. The binding affinity of the resulting peptides (**49** and **50** in Figure 10.6) to HLA-DQ2 was determined by a competition assay resulting in IC$_{50}$s ranging between 1 μM and 29 μM (similar to the unsubstituted peptide, $IC_{50} = 17\,\mu M$). Further characterization of one triazolyl-peptide revealed that it did not elicit a T-cell response on its own, but was able to block T-cell proliferation in the presence of the natural gluten peptide, albeit at low concentrations (around 200–400 μM). The group of Chaiken demonstrated that triazolyl-peptide conjugates could inhibit the binding of HIV-1 envelope glycoprotein (gp120) to CD4, a prerequisite for HIV-1 infection.[45] They started with a micromolar peptide fusion inhibitor and replaced a proline with (4S)-4-azido-proline which was 'clicked' with various alkynes (20 examples). The most potent fusion inhibitor **51** ($IC_{50} = 22\,nM$) showed a 50-fold improvement in blocking the gp120-CD4 binding compared to the underivatized peptide. The group of Moroder investigated the *C*-terminal lipidation of peptides and proteins by copper(I)-catalyzed 1,3-dipolar cycloaddition reaction between a propargylglycinyl residue and phosphatidylethanol azide.[46] The resulting lipopeptide **52** was elongated with a tetrapeptide-thioester by native chemical ligation and finally derivatized with a fluorescent label to study its uptake and distribution in HeLa cells, and equilibrium was established within 30 minutes.

Linking two peptide chains by dipolar cycloaddition reactions was explored by Eichler *et al.* who used the CuAAC between different immobilized tetrapeptides acylated with propiolic acid and an azidoacetylated pentapeptide, to obtain assembled peptides in variable yields (19–88%, not shown).[47] Furthermore, they described an elegant strategy towards scaffolded triazolyl-peptides **54** using the cyclic scaffold **53** with two orthogonally protected lysines and a nitrophenyl alanine to sequential 'click' different azidopeptides onto the scaffold after selective removal of the protecting groups as illustrated in Scheme 10.15.

The pseudodipeptide **55** (see Scheme 10.16) was regarded as a conformationally restricted Xaa-Pro dipeptide and Scolastico *et al.* functionalized the azido-group by copper(I)-catalyzed dipolar cycloaddition with an alkynylated sugar, biotin or fluorescein to afford substituted triazolyl-Xaa-Pro-dipeptide mimics **56** in good to excellent yields.[48]

Mash *et al.* described an azido-functionalized mixture of hexaols which were reacted with alkynylated ligands for the human melanocortin receptor 4 (hMC4R) using copper(I)-catalysis to form melanocyte-stimulating hormone-ligands **57** (in Figure 10.7) linked to

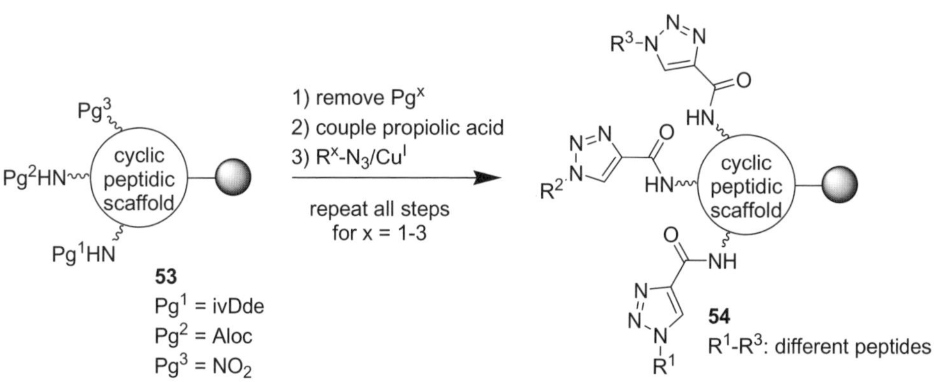

Scheme 10.15 *Scaffolded triazolyl-peptides by sequential 'click' chemistry. ivDde, 1-(4,4-dimethyl-2,6-dioxo-cyclohexylidene)-3-methyl-butyl*

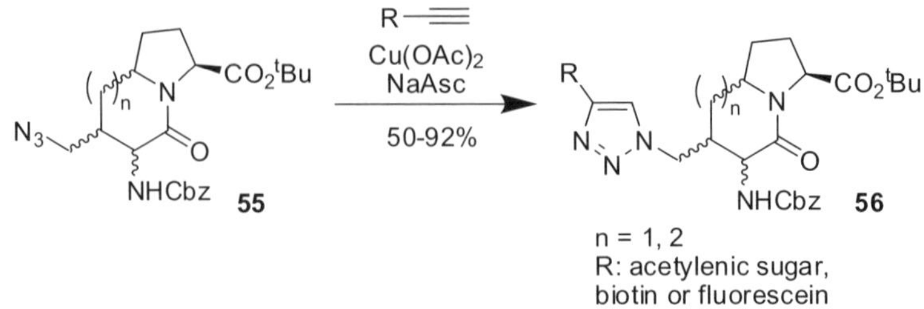

Scheme 10.16 *Copper(I)-catalyzed functionalization of a pseudodipeptide*

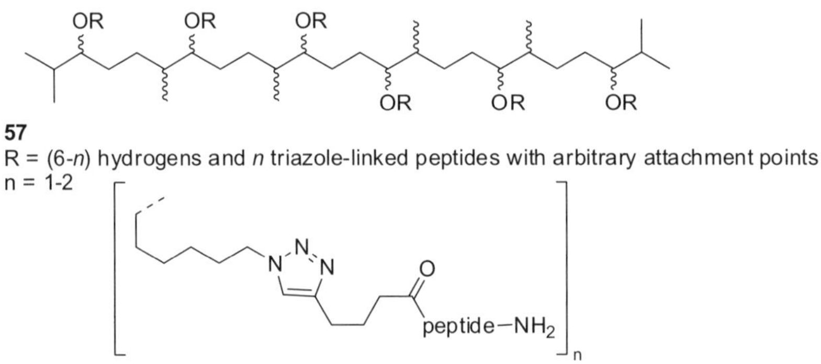

Figure 10.7 *Triazolyl-peptide ligands covalently linked to a squalene derived hexaol*

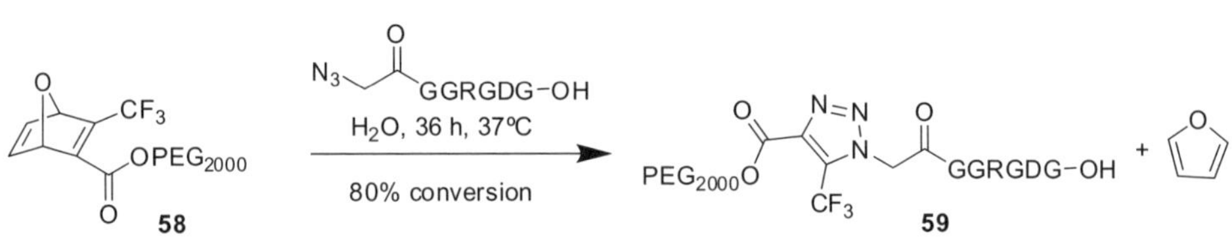

Scheme 10.17 *1,3-Dipolar cycloaddition followed by a retro Diels-Alder reaction afforded a 1,4,5-substituted triazole from an azido-peptide and a substituted oxanorbornadiene*

a squalene-derived scaffold through a [1,2,3]-triazole.[49] The resulting diastereomeric and regioisomeric mixtures were characterized in a competitive binding assay and shown to retain their affinity for the hMC4R except for the divalent molecule (n = 2), which did not show cooperativity by the two peptide ligands (probably due to a slow off-rate at the receptor).

The group of Rutjes reported on the low temperature, copper(I)-free cycloaddition between an azide and an electron deficient and strained double bond (for example oxanorbornadiene **58** in Scheme 10.17) at 25–37 °C to form a triazoline, followed by a spontaneous retro Diels-Alder reaction that expelled furan to produce a 1,4,5-substituted triazole (**59**).[50] The copper-free cycloaddition-retro Diels-Alder was anticipated to be compatible

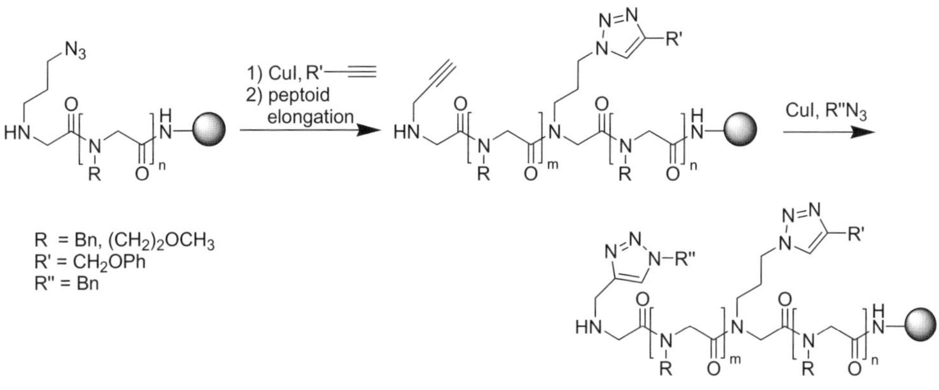

Scheme 10.18 *N-substituted glycine-oligomers functionalized by 'click' chemistry*

with *in* vivo ligation reactions, and this was exemplified by using an oxanorbornadiene functionalized with poly(ethylene) glycol (PEG) at 37 °C to afford a PEG-triazole-linked peptide **59**. A similar technique involving constrained cycloalkynes was developed by Bertozzi's group for protein ligation reactions.[51]

A sequential peptoid elongation and copper(I)-catalyzed alkyne-azide cycloaddition was developed by the group of Kirshenbaum to produce functionalized peptoid oligomers.[52,53] *N*-substituted glycine oligomers were assembled with a terminal alkyne or azide residue and both underwent cycloaddition with high conversion (>95%) to afford triazolyl peptoids **60** (see Scheme 10.18). A bifunctional peptoid hexamer **61** was also prepared by this strategy containing a ferrocene and estradiol group with the aim of generating a platform of functionalized peptoids that could be used as biosensors (illustrated in Figure 10.8).

17α-Ethynyl-estradiol was coupled to azide-substituted peptoid oligomers and the resulting estradiol-peptoid conjugates **62** were shown to displace 17β-estradiol (E2) at the estrogen receptor in a competitive binding experiment (albeit with >25-fold reduced potency compared to E2).[54]

The solid-phase synthesis of adenyl-peptide conjugates was described by Filippov *et al.* utilizing 'click' chemistry to attach the azide-derived 2-alkoxy-8-hydroxy adenine **63** to a major histocompatibility complex (MHC) class I epitope.[55] Fmoc-chemistry was used to assemble the alkyne-conjugated peptide followed by CuAAC with **63** and it was shown that one of the adenyl peptide conjugates could enhance the T-cell response compared to a mixture of **63** and the unmodified peptide (see Scheme 10.19).

Cellular delivery of peptide nucleic acids (PNA) or modified oligonucleotides is hampered by poor penetration and therefore a new conjugation strategy between cell-penetrating peptides and DNA/PNA/thioacetamido nucleic acid (TANA) was presented by Kumar *et al.*[56] The authors prepared azido- and alkyne-functionalized derivatives and subjected them to a copper(I)-catalyzed cycloaddition to form novel conjugates (**64** and **65** in Figure 10.9) between peptides and DNA/PNA/TANA with high purity. Torrence *et al.* demonstrated that by 'clicking' together an azido-substituted 2′,5′-oligoadenylate

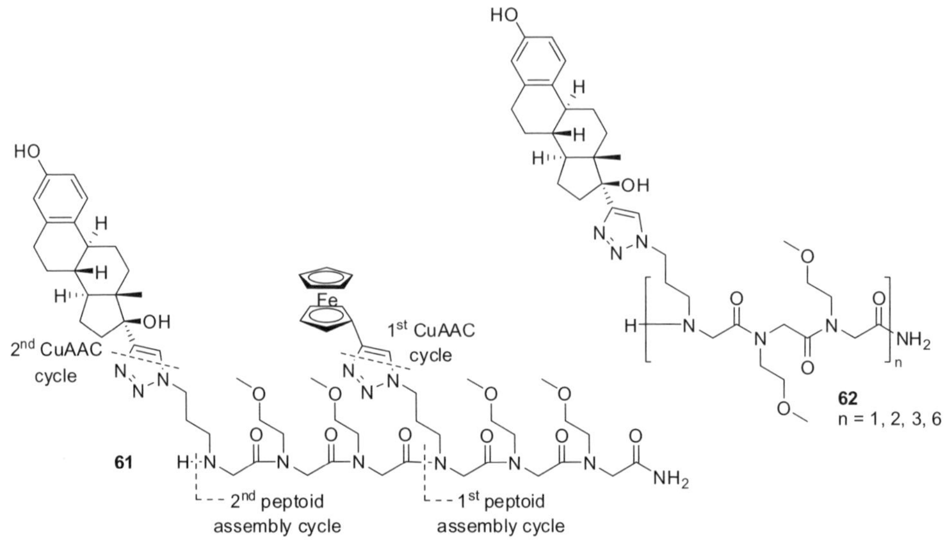

Figure 10.8 *Estradiol-functionalized peptoids as potential biosensors*

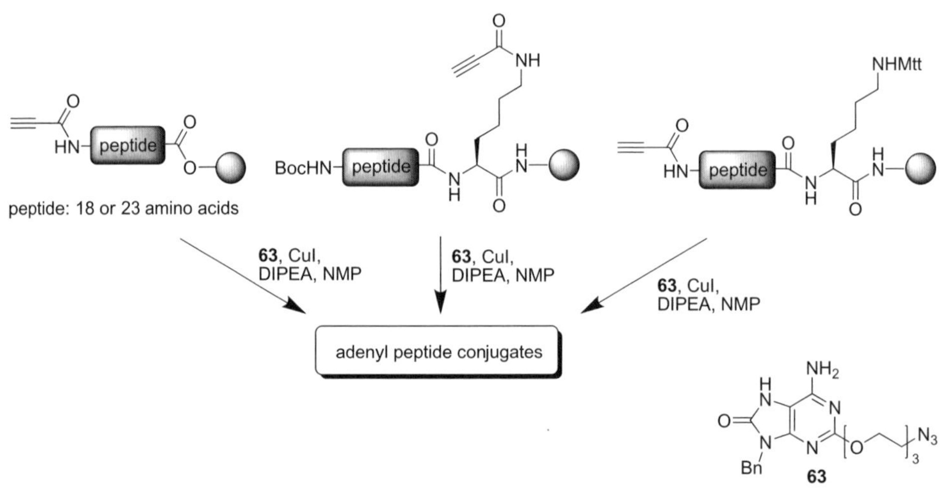

Scheme 10.19 *Copper(I)-catalyzed formation of adenyl peptide conjugates*

tetramer with an alkyne-labeled dodecapeptide the resulting chimera **66** was able to activate human RNase L that effects cellular RNA degradation.[57] A similar chimera, linked by a sulfhydryl-ether instead of the triazole, was assembled by sulfhydryl attack on an α-chloroacetylated peptide and the chimera was shown by confocal microscopy to be effectively taken up into intact cells. However, the cellular uptake of the triazole chimera **66** was not determined.

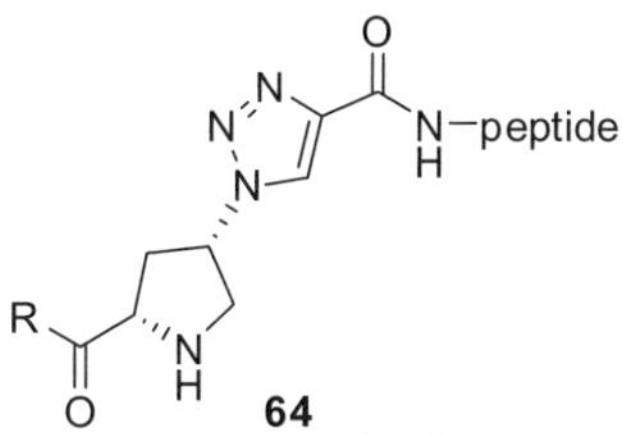

64
R = OH, PNA, TANA

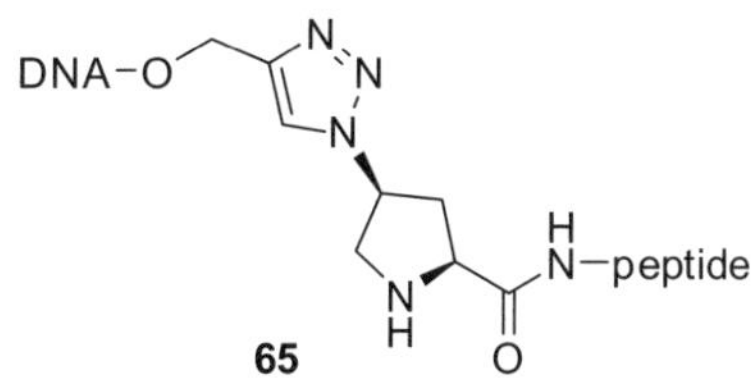

65

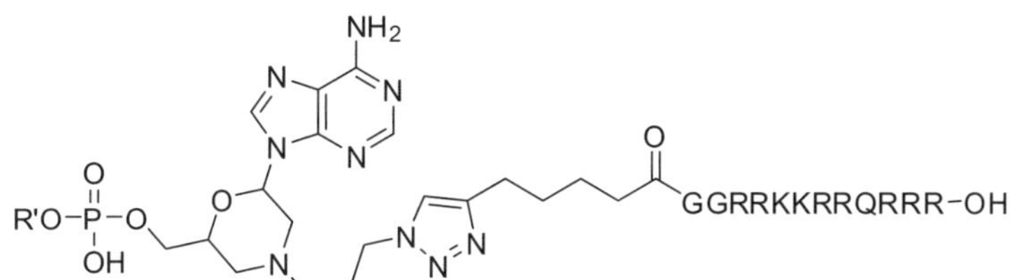

66
R' = H or p5'A2'p5'A2'p5'A

Figure 10.9 *Triazole-based chimeras between oligonucleotides and peptides. p5'A2'p5'A2'p5'A, an oligoadenylate*

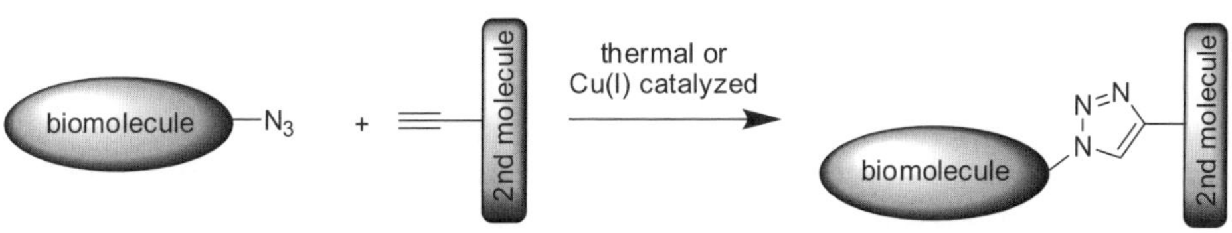

Scheme 10.20 *Conjugation reactions of a biomolecule by 1,3-dipolar cycloaddition*

Ju and Seo reported thermal and copper(I)-catalyzed 1,3-dipolar cycloaddition between azide-labeled DNA and a fluorescent molecule or alkyne modified glass surface (illustrated in Scheme 10.20).[58]

The Yao group has published two reports on triazole libraries generated by 'click' chemistry from alkyne-derivatized warhead molecules and various azides to afford hydroxamates **67**, aldehydes **68** and vinyl sulfones **69** as illustrated in Figure 10.10.[59,60] The hydroxamates (96 compounds) were derivatized with hydrophobic groups on the triazole and used to obtain an inhibition fingerprint of several matrix metalloproteases, such as MMP-7, thermolysin and collagenase by *in situ* screening. Low micromolar inhibitors of MMP-7 were identified by this method and shown to have some selectivity (10-35 fold) towards thermolysin and collagenase. Libraries **68** and **69** (384 compounds) were also generated by CuAAC (384-well format) and screened *in situ* for caspase inhibition which resulted in the identification of low micromolar inhibitors of caspase-3 and -7 (both reversible and irreversible inhibitors). The 'click' assembly of 12 hydroxamate probes **70** was subsequently reported and the products were used to profile various matrix metalloproteases.[61] An azide, containing a photoreactive group (benzophenone) and a

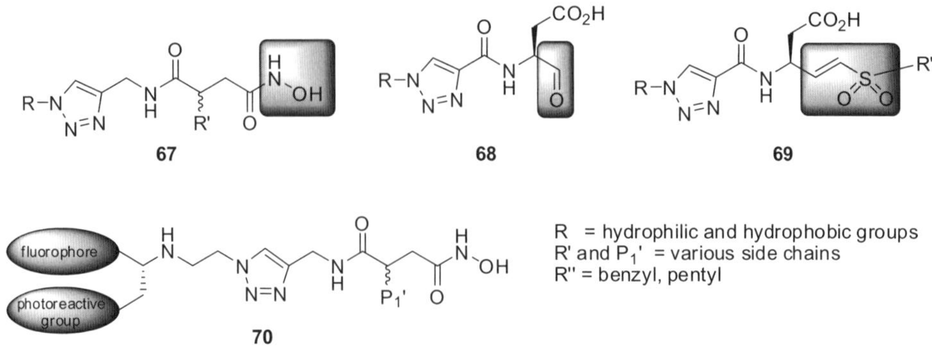

Figure 10.10 *Libraries of peptide-like structures containing warheads used to study the activity fingerprint of several proteases*

fluorophore (rhodamine), underwent copper(I)-catalyzed cycloaddition with alkyne-derivatized hydroxamate warheads with full conversion to form compounds **70**, which were used in gel-based fingerprinting of several MMPs, carbonic anhydrase and anthrax lethal factor.

10.5 Macrocyclization by DCR

The cyclodimerization of an 11- and 19-mer peptide by CuAAC was reported by Finn *et al.* providing selectively rings with 76 and 124 atoms, respectively.[62] These results spurred a detailed investigation into the mechanistic aspects of the cyclodimerization and indirectly of the CuAAC, and it was concluded that this reaction was effective for producing large cyclic dimers (for example **71** in Scheme 10.21).

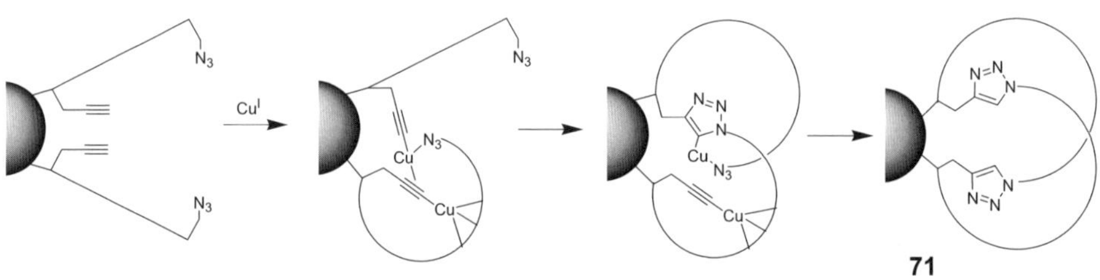

Scheme 10.21 *Cyclodimerization of peptide chains by CuAAC*

Angell and Burgess described the synthesis of β-turn mimics by intramolecular copper(I)-catalyzed cycloaddition of dipeptide **72** in Scheme 10.22.[63] However, they noted the prevalence to form dimeric cycloaddition products **74** instead of the desired macrocycle **73**, and yields were low due to poor solubility of both products. Molecular dynamics and NMR were used to identify plausible type I and II β-turn conformations for **73**, however, the circular dichroism spectrum did not correspond to typical type I and II β-turn structures and this discrepancy was not discussed further. Burke Jr. *et al.* reported the preparation

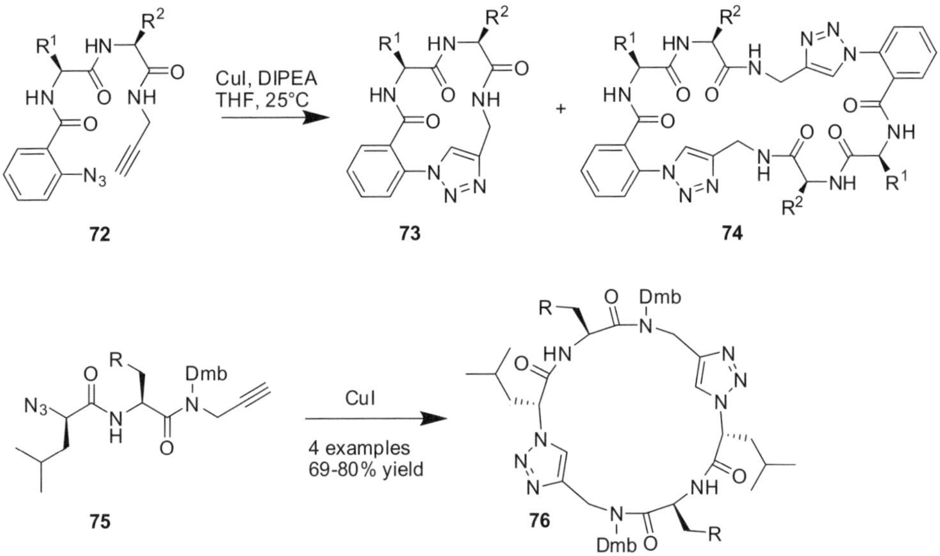

Scheme 10.22 *Copper(I)-catalyzed cyclization affording both monomeric and dimeric macrocycles. Dmb, 2,4-dimethoxybenzyl*

of an azide- and alkyne-functionalized tetrapeptide which, when subjected to copper(I)-catalysis, afforded the expected monomeric cyclization product at 1 mM substrate concentration while at 2 mM the macrocyclic dimer was formed as the major product.[64] Both macrocycles were tested for their Grb SH2 domain-binding inhibition and they displayed binding affinities of 0.23 μM and 0.0018 μM, respectively. Ghadiri *et al.* described the synthesis of C_2 symmetric cyclic triazolyl-peptides **76** by subjecting alkynylated azido-dipeptides **75** to a copper(I)-catalyzed domino dimerization macrocyclization thus forming two triazoles in a 20-membered ring.[65,66] By alternating the stereochemistry of the two amino acids involved in the cycloaddition it was possible to alter the ratio between cyclic dimerization and trimerization. The unprotected cyclic pseudopeptides suffered from very low solubility so a backbone protecting group was incorporated to ease purification.

The group of van Maarseveen investigated the conformational properties of several amide bonds in the cyclopeptide cyclo[ProTyrProVal], which is a potent tyrosinase inhibitor, and replaced amide bonds with 1,4-substituted triazoles by CuAAC.[67,68] Peptides **77** and **79** were synthesized in solution and cyclized under forcing conditions to provide the cyclo-triazolylpeptides (**78** and **80** in Scheme 10.23) and both compounds retained the biological activity found in the parent cyclopeptide.

Inhibition of the signal transducers and activators of transcription 3 (STAT3) dimerization is believed to be a novel approach to treat cancers where constitual activation of STAT3 is found. Wang *et al.* described the synthesis of a conformationally constrained peptidomimetic (**82** in Scheme 10.24) that was 3-fold more potent that the linear peptide **81** in blocking the STAT3 dimerization.[69] The intramolecular copper(I)-catalyzed cycloaddition was performed on the tripeptide, Boc-Lys(N_3)-Thr(^{t}Bu)-Pra-OMe, in 80% yield and subsequently elongated to produce **82**.

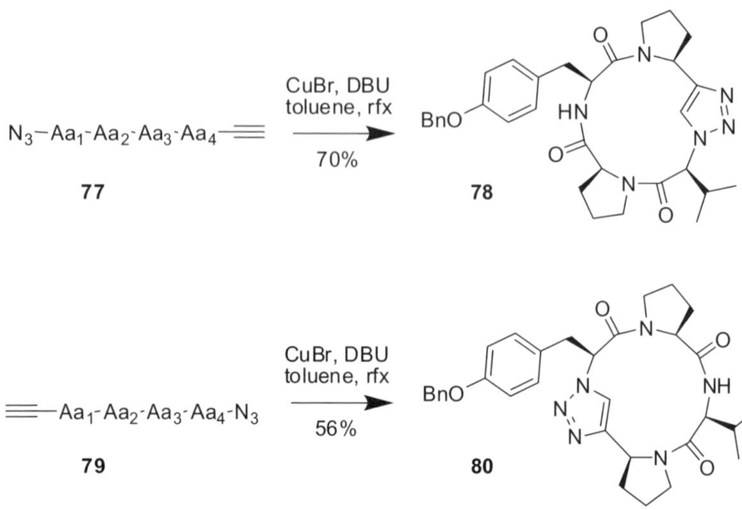

Scheme 10.23 *Copper(I)-catalyzed macrocyclization affording tyrosinase inhibitors **78** and **80***

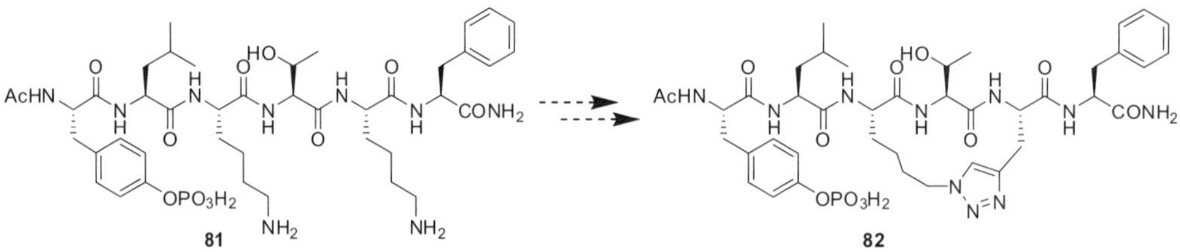

Scheme 10.24 *Linear and macrocyclic inhibitors of STAT3 dimerization*

The first cyclic peptidomimetic (**83** in Figure 10.11) containing a triazole mimicking a disulfide bond was prepared by intramolecular copper(I)-catalyzed cycloaddition between a propargylglycine residue and 3-azido-alanine on a PEG-based polymer.[70] The cycloaddition performed equally well with and without the peptide protecting groups and the desired cyclic peptide was isolated in 76% yield after purification. In a similar fashion, Inguimbert *et al.* performed on-resin peptide cyclization of VEGFR1 mimics using CuAAC and isolated cyclomonomeric products **84**.[71] Linear peptides were assembled on Rink amide MHBA resin and CuAAC afforded the cyclic triazole-linked peptides in decent crude purity and yield (50–70%); however, RP-HPLC purification performed poorly and only 4–9% of pure (>95%) cyclic peptide **84** could be isolated. The cyclization-strategy did unfortunately not improve the inhibitory activity of **84** for displacing biotinylated VEGF at the VEGFR1 compared to the linear analogues.

Billing and Nilsson reported the cyclodimerization of a bifunctional molecule **85** displaying both an azido- and an alkyne-group.[72] A propynoyl-dipeptide was coupled to an azido-aminoglucopyranoside and cyclodimerized by CuAAC to afford **86** in decent yields (33–64%, Scheme 10.25). The products were believed to be artificial receptor prototypes. A similar strategy was used to assemble macrocycle **87** containing two amino acid resi-

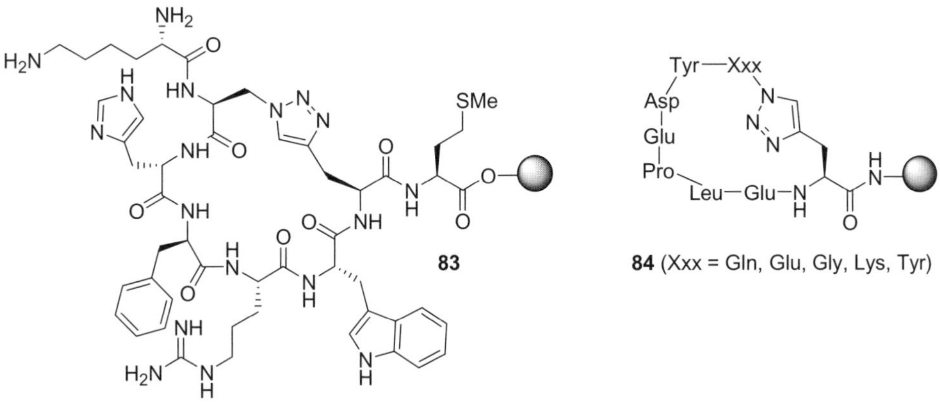

83

84 (Xxx = Gln, Glu, Gly, Lys, Tyr)

Figure 10.11 *Macrocyclization of peptides by 'click' chemistry between propargylic acid residues and an azide-functionalized amino acid*

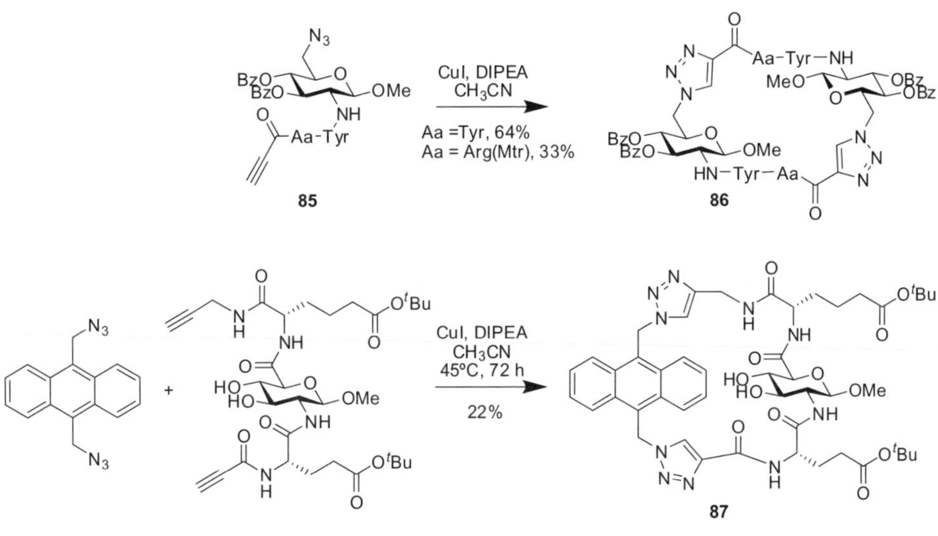

85

86

87

Scheme 10.25 *Bis-triazole formation affording peptidic molecular receptors*

dues, a glucosamine derivative and anthracene.[73] The fluorescent macrocycle, envisioned to be an example of a simple molecular receptor, was constructed by copper(I)-catalyzed formation of a bis-triazole and isolated in low yield (22%) due to its amphiphilic nature.

10.6 Dendrimers and Polymers

The group of Liskamp reported the development of an efficient microwave-assisted CuAAC between various azido-functionalized peptides and di/tetra/octa- and hexadecavalent alkyne dendrimers (**88** in Scheme 10.26).[74] The resulting peptide cycloadducts **89**

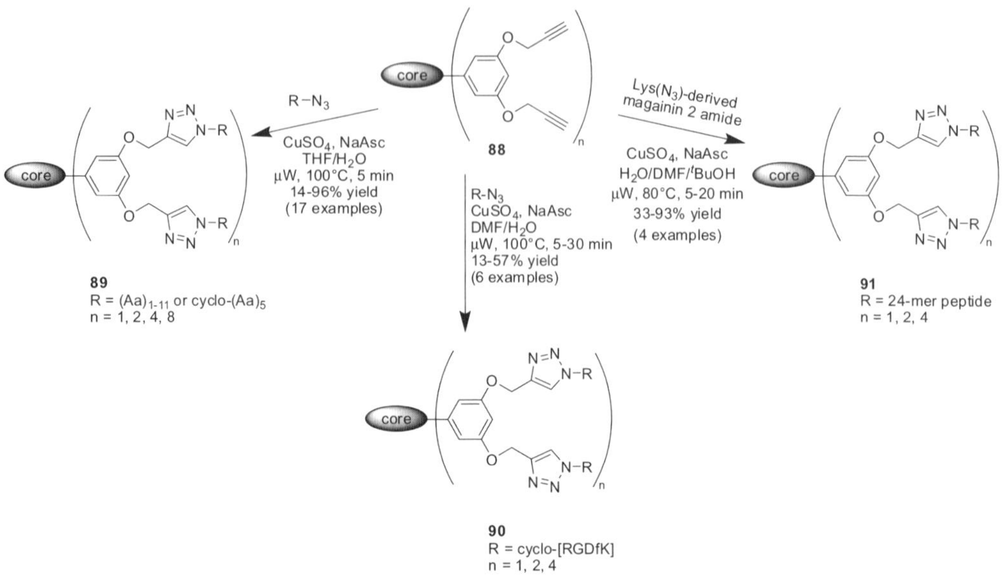

Scheme 10.26 *Dendrimers functionalized with triazolyl-peptides by microwave-assisted CuAAC*

were formed in 14–96% yield; however, poor solubility was observed for some products, which may explain the lower yields. A similar paper described the assembly of $\alpha_v\beta_3$ integrin-directed multivalent peptides by combining cyclo[Arg-Gly-Asp-D-Phe-Lys(N_3)] with **88** to form dendrimers **90** which were taken up *in vivo* by tumor cells expressing $\alpha_v\beta_3$ integrin.[75] Another report described 'click' chemistry of the alkyne-substituted dendrimer **88** with the antimicrobial peptide, magainin 2 amide, which was *C*- or *N*-terminally modified with Lys(N_3), to afford multivalent magainin 2 dendrimers **91**, which were examined for their membrane pore forming properties.[76] The tetravalent and octavalent dendrimers were approximately 100-fold more active than the parent magainin 2 peptide when measured in leakage experiments from large unilamellar vesicles comprised of dioleoylphosphatidylcholine.

The calixarene scaffold is ideal for displaying multiple units of a biomolecule and the CuAAC between alkynes and azido-derived calixarenes is a powerful tool for exploring the multivalency potential of diverse ligands. This concept has been demonstrated by Bew *et al.*, who showed that azido-calix[4]arenes were efficiently derivatized on the upper rim using 'click' chemistry with alkyne-derived α-amino acids, dipeptides and sugars to provide triazolyl-calix[4]arenes **92** as illustrated in Figure 10.12.[77] The Sharma group described a triazole-containing dendrimer **93** which was based on aspartic acid and lysine.[78] The dual surface dendrimer **93** was assembled by CuAAC between alkyne- or azido-derived aspartic acid, lysine or cystine-residues to afford dendrimers in good yield (65–70%) with potential use in drug delivery or gene therapy. Rutjes *et al.* prepared azido-functionalized polystyrene by reacting polystyrene-bromide with trimethylsilyl azide/tetrabutylammonium fluoride and used this in the 'click' reaction with an alkyne-

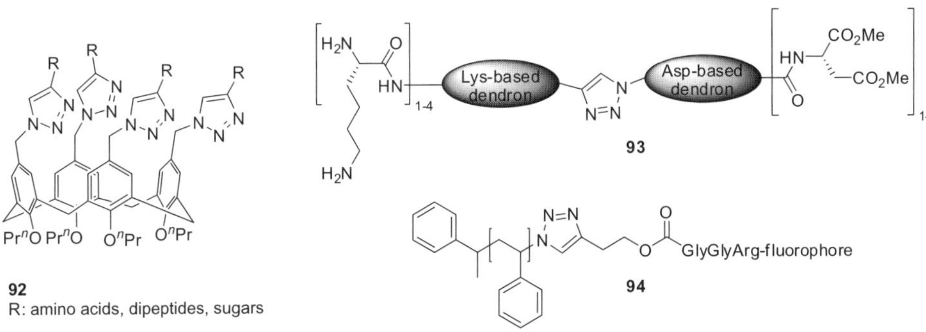

92
R: amino acids, dipeptides, sugars

93

94

Figure 10.12 *Peptidic dendrimers and polymers produced by CuAAC*

modified tripeptide to generate an amphiphile **94** comprised of a hydrophobic polystyrene-core decorated with cationic arginine residues.[79] The amphiphilic nature of the reactants posed a solubility challenge for the CuAAC reaction but full conversion was obtained in THF after 36 h at 35 °C.

10.7 Isotopic Labeling by DCR

95

Scheme 10.27 *Radiolabeling of biomolecules by copper(I)-catalyzed azide-alkyne cycloadditions*

Only few methods exists today for ^{18}F-labeling of peptides (typically acylation and reductive amination) so Årstad and Glaser developed methods for radiolabeling peptides and small molecules with fluorine-18.[80] A radiolabeled azide was reacted with an alkyne-derivatized peptide, using copper(I) catalysis, to afford a ^{18}F-triazole-linked peptide **95** useful for *in vivo* imaging (see Scheme 10.27).

The authors later described a labeling system with 2-[^{18}F]-fluoroethyl azide.[81] This was clicked onto a pentapeptide capped with propiolic acid to produce a fluorine-18 labeled triazolylpeptide (**96** in Scheme 10.28) in high radiochemical yield (92%). The CuAAC was also selected for the [^{18}F]-labeling of neurotensin(8–13) (NT(8-13)) by the Wuest group.[82] NT(8–13) was azido-functionalized to afford **97**, which was reacted with 4-[^{18}F] fluoro-*N*-(prop-2-ynyl)benzamide using excess copper(II) sulfate and sodium ascorbate in an aqueous buffer. The [^{18}F]-labeled NT(8–13) **98** was isolated in 66% radiochemical yield. However, the *in-vitro* binding of **98** (IC_{50} = 66 nM) was reduced >100-fold compared to NT(8–13) (IC_{50} = 0.4 nM) due to the *N*-terminal triazole-modification. [^{18}F]-Fluoroalkynes were used to selectively fluorine-18 label various *N*-(3-azidopropionyl)-

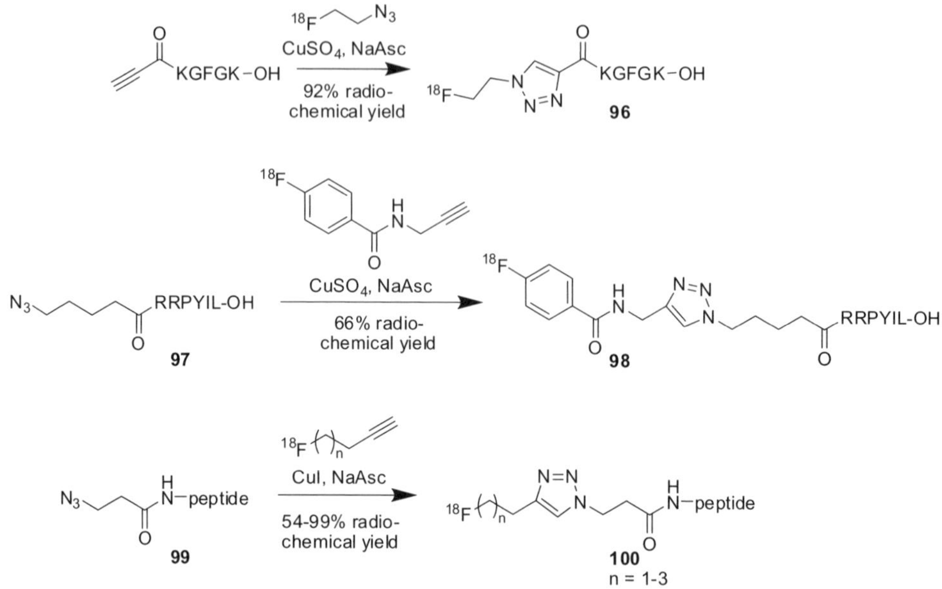

Scheme 10.28 Fluorine-18 labeling of peptides and biomolecules by copper(I)-catalyzed azide-alkyne cycloadditions

peptides **99** by an optimized Cu(I)-catalyzed cycloaddition to form [^{18}F]-triazolylpeptides **100**.[83] A large excess of CuI and azidopeptide were used to decrease the reaction time to 10 minutes in order to limit the radiodecay, and products were isolated in yields of 54–99%.

10.8 Perspective

A large number of peptidomimetics are currently entering clinical trials, typically protease inhibitors[84] and anti-cancer agents,[85] which emphasizes the importance of developing reactions which efficiently modify peptides or peptide-like structures to increase their drug-like properties. Azides are important dipoles in 1,3-dipolar cycloaddition reactions and react with dipolarophiles such as alkynes and nitriles, to afford [1,2,3]-triazoles and tetrazoles, respectively.

Copper-catalyzed azide-alkyne cycloadditions have become increasingly popular due to their almost quantitative formation of 1,4-substituted triazoles, regioselectively, and the remarkable functional group tolerance, which is important when dealing with peptides or peptidomimetics. The majority of publications on dipolar cycloaddition reactions in peptide chemistry has focused on the CuAAC and reported peptide bond isosteres, side-chain functionalization, glycoconjugation, macrocyclization and isotopic labeling of peptides. We will most likely see an increasing number of applications where peptides are modified by dipolar cycloadditions in the future.

References

[1] R. Huisgen, *Angew. Chem. Int. Ed. Engl.* **1963**, *2*, 565–632.
[2] R. Huisgen, *Angew. Chem. Int. Ed. Engl.* **1963**, *2*, 633–96.
[3] C.W. Tornøe, M. Meldal, in Peptides: The wave of the future, M. Lebl, R.A. Houghten, eds.; Kluwer Academic Publishers, Dordrecht, **2001**, pp. 263–4.
[4] V.A. Rodionov, V.V. Fokin, M.G. Finn, *Angew. Chem. Int. Ed. Engl.* **2005**, *44*, 2210–15.
[5] M. Meldal, C.W. Tornøe, *Chem. Rev.* **2008**, *108*, 2952–3015.
[6] V.D. Bock, H. Hiemstra, J.H. van Maarseveen, *Eur. J. Org. Chem.* **2006**, 51–68.
[7] H.C. Kolb, K.B. Sharpless, *Drug Discovery Today* **2003**, *8*, 1128–37.
[8] P. Wu, V.V. Fokin, *Aldrich Chim. Acta* **2007**, *40*, 7–17.
[9] M.V. Gil, M.J. Arévalo, Ó. López, *Synthesis* **2007**, 1589–1620.
[10] J.E. Moses, A.D. Moorhouse, *Chem. Soc. Rev.* **2007**, *36*, 1249–62.
[11] J.-F. Lutz, *Angew. Chem. Int. Ed. Engl.* **2007**, *46*, 1018–25.
[12] Y. Angell, K. Burgess, *Chem. Soc. Rev.* **2007**, *36*, 1674–89.
[13] A. Dondoni, P.P. Giovannini, A. Massi, *Org. Lett.* **2004**, *6*, 2929–32.
[14] B.H.M. Kuijpers, S. Groothuys, A.R. Keereweer, *et al.*, *Org. Lett.* **2004**, *6*, 3123–6.
[15] N. Pietrzik, C. Schips, T. Ziegler, *Synthesis* **2008**, 519–26.
[16] J.V. dos Anjos, D. Sinou, S.J. de Melo, R.M. Srivastava, *Synthesis* **2007**, 2647–52.
[17] A. Paul, H. Bittermann, P. Gmeiner, *Tetrahedron* **2006**, *62*, 8919–27.
[18] Z.P. Demko, K.B. Sharpless, *Org. Lett.* **2002**, *4*, 2525–7.
[19] Z. Grzonka, E. Rekowska, B. Liberek, *Tetrahedron* **1971**, *27*, 2317–22.
[20] V.V. Sureshbabu, R. Venkataramanarao, S.A. Naik, G. Chennakrishnareddy, *Tetrahedron Lett.* **2007**, *48*, 7038–41.
[21] C.W. Tornøe, C. Christensen, M. Meldal, *J. Org. Chem.* **2002**, *67*, 3057–64.
[22] V.V. Rostovtsev, L.G. Green, V.V. Fokin, K.B. Sharpless, *Angew. Chem. Int. Ed. Engl.* **2002**, *41*, 2596–9.
[23] C.W. Tornøe, S.J. Sanderson, J.C. Mottram, G.H. Coombs, M. Meldal, *J. Comb. Chem.* **2004**, *6*, 312–24.
[24] C. Christensen, T. Groth, C.B. Schiødt, N.T. Foged, M. Meldal, *QSAR & Comb. Sci.* **2003**, *22*, 737–44.
[25] N.G. Angelo, P.S. Arora, *J. Am. Chem. Soc.* **2005**, *127*, 17134–5.
[26] N.G. Angelo, P.S. Arora, *J. Org. Chem.* **2007**, *72*, 7963–7.
[27] Z. Zhang, E. Fan, *Tetrahedron Lett.* **2006**, *47*, 665–9.
[28] V. Aucagne, D.A. Leigh, *Org. Lett.* **2006**, *8*, 4505–7.
[29] A. Tam, U. Arnold, M.B. Soellner, R.T. Raines, *J. Am. Chem. Soc.* **2007**, *129*, 12670–1.
[30] L. Zhang, X. Chen, P. Xue, *et al.*, *J. Am. Chem. Soc.* **2005**, *127*, 15998–9.
[31] Y. Angell, D. Chen, F. Brahimi, H.U. Saragovi, K. Burgess, *J. Am. Chem. Soc.* **2008**, *130*, 556–65.
[32] W.S. Horne, M.K. Yadav, D. Stout, M.R. Ghadiri, *J. Am. Chem. Soc.* **2004**, *126*, 15366–7.
[33] K. Oh, Z. Guan, *Chem. Commun.* **2006**, 3069–71.
[34] Q. Wan, J. Chen, G. Chen, S. J. Danishefsky, *J. Org. Chem.* **2006**, *71*, 8244–9.
[35] K. Tanaka, C. Kageyama, K. Fukase, *Tetrahedron Lett.* **2007**, *48*, 6475–9.
[36] S. Groothuys, B.H.M. Kuijpers, P.J.L.M. Quaedflieg, *et al.*, *Synthesis* **2006**, 3146–52.
[37] H. Lin, C.T. Walsh, *J. Am. Chem. Soc.* **2004**, *126*, 13998-14003.
[38] G. Tuchscherer, C. Servis, G. Corradin, U. Blum, J. Rivier, M. Mutter, *Protein Sci.* **1992**, *1*, 1377–86.
[39] J. Wang, H. Li, G. Zou, L.-X. Wang, *Org. Biomol. Chem.* **2007**, *5*, 1529–40.
[40] D. Macmillan (ULC Business PLC). PCT Int. Appl. WO 2008001109, *Chem. Abstr.* **2008**, *148*, 121969.
[41] O. Roy, S. Faure, V. Thery, C. Didierjean, C. Taillefumier, *Org. Lett.* **2008**, *10*, 921–4.
[42] X. Fu, C. Albermann, C. Zhang, J.S. Thorson, *Org. Lett.* **2005**, *7*, 1513–15.
[43] M. Kümin, L.-S. Sonntag, H. Wennemers, *J. Am. Chem. Soc.* **2007**, *129*, 466–7.
[44] V.V. Kapoerchan, M. Wiesner, M. Overhand, *et al.*, *Bioorg. Med. Chem.* **2008**, *16*, 2053–62.

[45] H.N. Gopi, K.C. Tirupula, S. Baxter, S. Ajith, I.M. Chaiken, *Chem. Med. Chem.* **2006**, *1*, 54–7.
[46] H.-J. Musiol, S. Dong, M. Kaiser, *et al.*, *Chem. Bio. Chem.* **2005**, *6*, 625–8.
[47] R. Franke, C. Doll, J. Eichler, *Tetrahedron Lett.* **2005**, *46*, 4479–82.
[48] D. Arosio, M. Bertoli, L. Manzoni, C. Scolastico, *Tetrahedron Lett.* **2006**, *47*, 3697–3700.
[49] B. Jagadish, R. Sankaranarayanan, L. Xu, *et al.*, *Bioorg. Med. Chem. Lett.* **2007**, *17*, 3310–13.
[50] S.S. van Berkel, A.J. Dirks, M.F. Debets, *et al.*, *Chem. Bio. Chem.* **2007**, *8*, 1504–8.
[51] N.J. Agard, J.A. Prescher, C.R. Bertozzi, *J. Am. Chem. Soc.* **2004**, *126*, 15046–7.
[52] H. Jang, A. Fafarman, J.M. Holub, K. Kirshenbaum, *Org. Lett.* **2005**, *7*, 1951–4.
[53] J.M. Holub, H. Jang, K. Kirshenbaum, *Org. Biomol. Chem.* **2006**, *4*, 1497–1502.
[54] J.M. Holub, M.J. Garabedian, K. Kirshenbaum, *QSAR & Comb. Sci.* **2007**, *26*, 1175–80.
[55] J.J. Weterings, S. Khan, G.J. van der Heden, *et al.*, *Bioorg. Med. Chem. Lett.* **2006**, *16*, 3258–61.
[56] K. Gogoi, M.V. Mane, S.S. Kunte, V.A. Kumar, *Nucleic Acids Res.* **2007**, *35*, e139.
[57] L. Zhou, C.S. Thakur, R.J. Molinaro, *et al.*, *Bioorg. Med. Chem.* **2006**, *14*, 7862–74.
[58] J. Ju and T.S. Seo (Columbio University). PCT Int. Appl. WO 2004055160, *Chem. Abstr.* **2004**, *141*, 85159.
[59] J. Wang, M. Uttamchandani, J. Li, M. Hu, S.Q. Yao, *Org. Lett.* **2006**, *8*, 3821–4.
[60] S.L. Ng, P.-Y. Yang, K.Y.T. Chen, R. Srinivasan, S.Q. Yao, *Org. Biomol. Chem.* **2008**, *6*, 844–7.
[61] J. Wang, M. Uttamchandani, J. Li, M. Hu, S.Q. Yao, *Chem. Commun.* **2006**, 3783–5.
[62] S. Punna, J. Kuzelka, Q. Wang, M.G. Finn, *Angew. Chem. Int. Ed. Engl.* **2005**, *44*, 2215–20.
[63] Y. Angell, K. Burgess, *J. Org. Chem.* **2005**, *70*, 9595–8.
[64] W.J. Choi, Z.-D. Shi, K.M. Worthy, *et al.*, *Bioorg. Med. Chem. Lett.* **2006**, *16*, 5265–9.
[65] W.S. Horne, C.D. Stout, M.R. Ghadiri, *J. Am. Chem. Soc.* **2003**, *125*, 9372–6.
[66] J.H. van Maarseveen, W.S. Horne, M.R. Ghadiri, *Org. Lett.* **2005**, *7*, 4503–6.
[67] V.D. Bock, R. Perciaccante, T.P. Jansen, H. Hiemstra, J.H. van Maarseveen, *Org. Lett.* **2006**, *8*, 919–22.
[68] V.D. Bock, D. Speijer, H. Hiemstra, J.H. van Maarseveen, *Org. Biomol. Chem.* **2007**, *5*, 971–5.
[69] J. Chen, Z. Nikolovska-Coleska, C.-Y. Yang, *et al.*, *Bioorg. Med. Chem. Lett.* **2007**, *17*, 3939–42.
[70] M. Roice, I. Johannsen, M. Meldal, *QSAR & Comb. Sci.* **2004**, *23*, 663–73.
[71] V. Goncalves, B. Gautier, A. Regazzetti, *et al.*, *Bioorg. Med. Chem. Lett.* **2007**, *17*, 5590–4.
[72] J.F. Billing, U.J. Nilsson, *J. Org. Chem.* **2005**, *70*, 4847–50.
[73] J.K.M. Ågren, J.F. Billing, H.E. Grundberg, U.J. Nilsson, *Synthesis* **2006**, 3141–5.
[74] D.T.S. Rijkers, G.W. van Esse, R. Merkx, *et al.*, *Chem. Commun.* **2005**, 4581–3.
[75] I. Dijkgraaf, A.Y. Rijnders, A. Soede, *et al.*, *Org. Biomol. Chem.* **2007**, *5*, 935–44.
[76] C.J. Arnusch, H. Branderhorst, B. de Kruijff, R.M.J. Liskamp, E. Breukink, R.J. Pieters, *Biochemistry* **2007**, *46*, 13437–42.
[77] S.P. Bew, R.A. Brimange, N. L'Hermite, S.V. Sharma, *Org. Lett.* **2007**, *9*, 3713–16.
[78] V. Haridas, K. Lal, Y.K. Sharma, *Tetrahedron Lett.* **2007**, *48*, 4719–22.
[79] A.J. Dirks, S.S. van Berkel, N.S. Hatzakis, *et al.*, *Chem. Commun.* **2005**, 4172–4.
[80] E. Årstad and M. Glaser (Hammersmith Imanet Limited). PCT Int. Appl. WO 2006067376, *Chem. Abstr.* **2006**, *145*, 110205.
[81] M. Glaser, E. Årstad, *Bioconjugate Chem.* **2007**, *18*, 989–93.
[82] T. Ramenda, R. Bergmann, F. Wuest, *Lett. Drug Design Disc.* **2007**, *4*, 279–85.
[83] J. Marik, J.L. Sutcliffe, *Tetrahedron Lett.* **2006**, *47*, 6681–4.
[84] G. Fear, S. Komarnytsky, I. Raskin, *Pharmacol. Ther.* **2007**, *113*, 354–68.
[85] C. Avendano, J.C. Menendez, *Clin. Transl. Oncol.* **2007**, *9*, 563–70.

11

Photochemistry of Azides: The Azide/Nitrene Interface

Nina Gritsan[1] *and Matthew Platz*[2]

[1]*Institute of Chemical Kinetics and Combustion of Siberian Branch of Russian Academy of Sciences, Novosibirsk, Russia;* [2]*Chemistry Department, the Ohio State University, Columbus, Ohio, USA*

11.1 Introduction

Organic azides are widely used in synthetic organic chemistry.[1–4] An important application of the photochemistry of organic azides is the photoaffinity labeling of biopolymers. This technique was invented by Singh *et al.*,[5] and adapted for use with azides by Bayley and Knowles.[6] For example, aryl azide based photoaffinity labeling has been employed to obtain information on the higher order structure of RNA and RNA-protein complexes.[7] For many years azide photochemistry was used by industrial scientists in the field of lithography.[8] Materials chemists use azide photochemistry in the formation of electrically conducting polymers[9] and for surface modification and functionalization.[10]

It is commonly believed[1–3,11–15] that photolysis and thermolysis of the organic azides leads mainly to the dissociation of N–N bond with formation of molecular nitrogen and nitrenes, as first proposed by Tiemann in 1891.[16] Nitrenes, species containing neutral, monovalent nitrogen atoms, are typically very reactive and short-lived intermediates. Azides form bonds to many elements,[1–3,11–15] and, consequently, many types of nitrenes are known or can be imagined. Nitrenes and other reactive intermediates can be involved in many types of reactions which results in complex mixture of possible products. Physical organic chemists seek to understand the role of nitrenes and other intermediates in azide photochemistry and how the structures of intermediates control their reactivity.[1,2,11–15,17–20]

Organic Azides: Syntheses and Applications Edited by Stefan Bräse and Klaus Banert

Thus, the diverse applications of organic azides and the complicated nature of their photochemistry attract significant interest of the scientific community. There has been recent, dramatic progress in mechanistic understanding of the photochemistry of organic azides as a result of the application of modern spectroscopic techniques and high level *ab initio* molecular orbital (MO) calculations.[14,15,17–20] The goal of this chapter is to present a modern view of the mechanism of organic azide photochemistry. The most attention will be paid to the direct observations of the reactive intermediates and to the study of their reactions.

11.2 Photochemistry of Hydrazoic Acid (HN_3)

Photodecomposition of the simplest azide, hydrazoic acid (HN_3), yields the parent nitrene, imidogen (NH).[21–26] This nitrene can also be produced by thermolysis[27] and multiphoton dissociation[28,29] of HN_3. Photolysis of HN_3 in the gas phase with 248, 266, 283 and 308 nm light generates NH almost exclusively in the lowest singlet state ($^1\Delta$).[22–26] The N_3 fragment and H atom were also observed as primary products of HN_3 photodissociation at 266, 248 and 193 nm[30] with quantum yields 0.04, 0.14 and 0.2, respectively.[30c] Formation of NH in different excited states was observed upon photolysis of HN_3 with light of wavelength shorter than 240 nm.[21,22,25] For instance, NH in the $X\ ^3\Sigma^-$, $a^1\Delta$, $b^1\Sigma^+$, $A^3\Pi$, and $c^1\Pi$ states were formed by UV photolysis of HN_3 at 193 nm, at 300 K with quantum yields ≤0.0019, 0.4, 0.017, 0.00015 and 0.00061, respectively.[25]

NH in its lowest singlet state ($a^1\Delta$) inserts readily into paraffin CH-bonds, abstracts hydrogen atoms from hydrocarbons and undergoes relaxation to the ground triplet state.[31] For example, the ratio of these channels is 0.6 : 0.1 : 03 in the case of reaction of ^{1}NH with ethane (Scheme 11.1).[31b]

Reactions of the ground state, ^{3}NH ($X\ ^3\Sigma^-$), play an important role in combustion processes.[32] Triplet NH reacts with molecular hydrogen, water and CO_2.[33] Modern theoretical study[34] demonstrates that reactions with H_2 and H_2O proceed via hydrogen atom abstraction. The ^{3}NH abstracts hydrogen atoms from starting material, hydrazoic acid,[35] and from hydrocarbons to form aminyl ($NH_2\cdot$) and alkyl radicals[36,37] in spite of the fact that some reactions are endothermic, depending on the alkane.[37] Absolute rate constants for many of these reactions have been measured in the gas phase.[36,37] Triplet NH also reacts with alkenes via formation of an intermediate triplet diradical which then decomposes into several reaction channels.[38]

The major products of the gas phase reaction of ^{3}NH with molecular oxygen are $NO^\cdot$ and $OH^\cdot$ radicals.[39,40] It was proposed on the basis of quantum chemical calculations,[40,41] that the primary product is iminoperoxide (HNOO) which undergoes 1,3-hydrogen shift

$$NH(\Delta^1) + C_2H_6 \longrightarrow C_2H_5NH_2^* \quad \text{(insertion, ~60\%)}$$

$$NH(\Delta^1) + C_2H_6 \longrightarrow {}^\bullet C_2H_5 + {}^\bullet NH_2 \quad \text{(abstraction, ~10\%)}$$

$$NH(\Delta^1) + C_2H_6 \longrightarrow NH(^3\Sigma^-) + C_2H_6 \quad \text{(relaxation, ~30\%)}$$

Scheme 11.1 *Reactions of NH in the lowest singlet state ($a^1\Delta$) with ethane*[31b]

Scheme 11.2 *Mechanism of the gas phase reaction of* ^{3}*NH with molecular oxygen*[39–41]

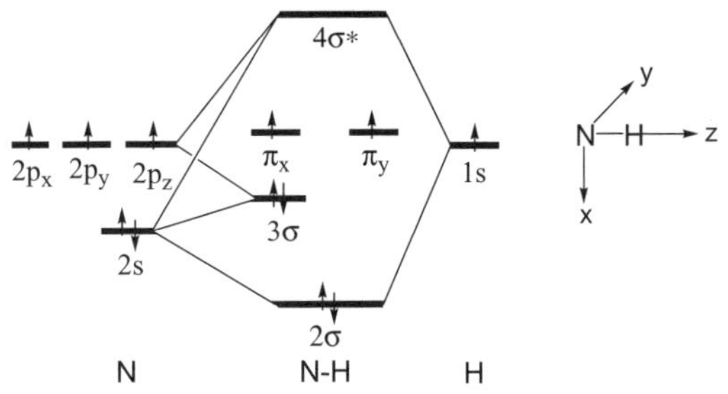

Figure 11.1 *Molecular orbitals of nitrene NH, the 1σ orbital, which is not shown, is the 1s AO on nitrogen*

yielding hydroperoxinitrene (HOON). The latter undergoes fast dissociation to the NO˙ and OH˙ radicals (Scheme 11.2).

The photochemistry of HN_3 in the argon and nitrogen matrices was studied for the first time by J. Pimentel *et al.* and the IR spectra of triplet nitrene NH and radical NH_2 were recorded.[42] A series of studies of matrix isolated NH (in ground triplet $^3\Sigma$ and excited singlet $^1\Delta$ states) and its deutero-substitute analogue (ND) were later performed using UV and luminescence spectroscopy.[43–47] The spectroscopy and relaxation of the lowest excited singlet state of NH / ND ($^1\Delta$) were studied in detail in Ne, Ar, Kr and Xe matrices.[46,47]

The heavy atom host accelerates intersystem crossing in either the excited state of HN_3 or ^{1}NH, which led to good yields of ^{3}NH in Xe.[48] Matrix isolated ^{3}NH reacts with CO to form NHCO[49] and with molecular oxygen, an excellent triplet nitrene trap, to form *trans*-HNOO which was characterized by IR spectroscopy.[48] The EPR spectrum of triplet imidogen immobilized in a cryogenic matrix has not yet been observed. One negative attempt was reported in Ar, Kr and Xe matrices in 1967.[50] The zero-field splitting parameter for triplet NH ($|D/hc| = 1.863\,cm^{-1}$)[51] was obtained in the gas phase using laser magnetic-resonance spectroscopy.

Systematic mechanistic studies of the photochemistry of HN_3 in solution and the chemistry of NH with hydrocarbons, particularly *cis*- and *trans*- alkenes have not been performed.

In the context of this chapter, the electronic structure and spectroscopy of the simplest nitrene, NH, are very important because they will be useful in the analysis of more complicated nitrenes. The electronic structure of NH can be understood on the basis of elementary molecular orbital considerations (Figure 11.1).

In NH, two valence molecular orbitals, corresponding to the N–H bond (2σ) and the lone pair on nitrogen (3σ), are occupied by electron pairs. Two more valence electrons must be distributed between two, degenerate, non-bonding molecular orbitals (NBMOs), π_x, and π_y, which consist of the $2p_x$ and $2p_y$ AOs on nitrogen (Figure 11.1). The three

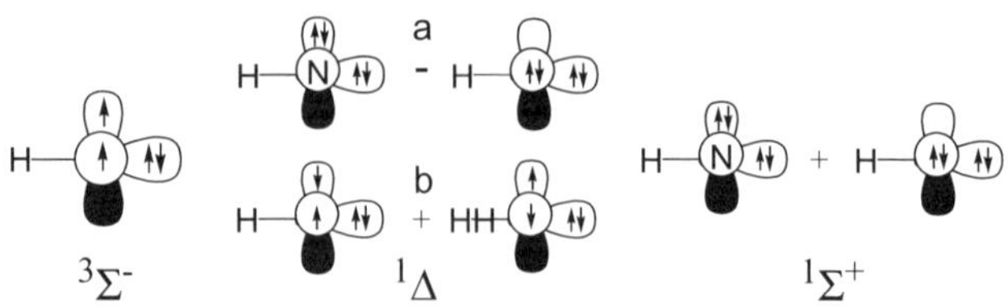

Figure 11.2 Schematic depictions of one spin component of the lowest triplet state ($^3\Sigma^-$), of the (a) 'closed-shell' and (b) 'open-shell' components of the lowest singlet state ($^1\Delta$) and of the second singlet excited state ($^1\Sigma^+$) of NH

lowest electronic states of NH – $^3\Sigma^-$, $^1\Delta$, and $^1\Sigma^+$ (Figure 11.2) – all arise from the electronic configurations in which the two electrons are distributed between these two NBMOs. The Pauli exclusion principle prevents electrons with the same spin from simultaneously appearing in the same region of space. Thus, the triplet has the lowest Coulombic repulsion energy of all of the low-lying states; hence, it is the ground state of NH ($^3\Sigma^-$, Figure 11.2).

The 'closed-shell' component of $^1\Delta$ is a linear combination (with a minus sign) of two configurations in which the two non-bonding electrons occupy the same 2p orbital; whereas, in the 'open-shell' component one electron occupies each of the 2p AOs. The two components of a $^1\Delta$ state (Figure 2) may look different, but symmetry considerations reveal shows that they are degenerate.

The third electronic state of NH, $^1\Sigma^+$, is a linear combination of the same configurations as in the $^1\Delta$ state, but with a positive sign (Figure 11.2). The motions of the non-bonding electrons are 'anti-correlated' in the $^1\Sigma^+$ state, so they have a higher Coulombic repulsion energy than in $^1\Delta$. This is the reason why $^1\Sigma^+$ is a higher energy electronic state than $^1\Delta$.

The experimental absorption spectra of NH in the lowest triplet ($X\ ^3\Sigma^-$) and singlet ($a^1\Delta$) states have similar bands in the near UV region with maxima at 336 and 324 nm, respectively.[21,52,53] Both transitions are associated with the electron promotion from a 3σ (lone pair) orbital to singly occupied π-orbitals (Figure 11.1).[14] The singlet and triplet absorptions of imidogen are very similar because the same orbitals are involved in the excitation of each spin state. As the transitions are localized on the nitrogen atom, these bands are characteristic of nitrenes in general and will appear in the same spectral region in alkyl and arylnitrenes.

The singlet-triplet splitting of NH was determined experimentally by spectroscopy of neutral NH[53,54] and by negative ion photoelectron spectroscopy (PES) of the NH anion.[55] In the latter experiment, the anion NH is prepared in the gas phase and exposed to monochromatic UV-laser light. This leads to ejection of photoelectrons whose kinetic energies are analyzed. The imidogen anion (NH) can ionize to form either the singlet or triplet nitrene, thus, the difference in the kinetic energies of the photoelectrons leading to ^{1}NH and ^{3}NH is simply the singlet-triplet splitting of NH. A value of 1.561 eV (36 kcal/mol) for the singlet-triplet splitting (ΔE_{ST}) in NH was obtained very accurately from the spectroscopic data.[53–55]

11.3 Photochemistry of Alkyl Azides

The photolysis of alkyl azides at room temperature cleanly forms imines as products.[56–60] In general, light and heat induced decomposition of alkyl azide does not produce alkyl nitrenes, which can be intercepted in respectable yields with a bimolecular trap.[56,57] For example, Moriarty and Reardon studied photolysis of *n*-butyl, *n*-amyl, 4-heptyl, 2-phenylethyl, 3-phenylpropyl, 4-phenylbutyl, phenylmethyl, 1-phenylbutyl, 1,1-diphenylethyl, and cyclohexyl azides.[58] The primary products of reaction were found to be nitrogen and imines derived from hydrogen, alkyl or aryl migration to nitrogen. Hydrogen atom was found to migrate up to five times faster than an *n*-alkyl group. For cyclohexyl azide only α-hydrogen migration occurs.[58] In the case of phenylmethyl azide, the phenyl/hydrogen migration aptitudes is equal to unity.[58] The ethyl/methyl migration aptitudes were determined to be in the range 1.0–1.4 for photolysis of a series of 2-substituted 2-butyl azides, where the 2-substituent was an aryl, *n*-propyl or $Ph(CH_2)_n$ (n = 1–3) groups.[59]

Later,[60] Kyba and Abramovitch studied the photolysis of nine *sec*- and *tert*-alkyl azides (Scheme 11.3) in detail and measured the migratory rations.

It was found that the range of migratory aptitude does not deviate greatly from unity and photolysis of alkyl azides bearing pendant aryl groups (**1a,b**) does not lead to intramolecular trapping of a nitrene (Scheme 11.4). On the basis of these data, it was proposed that singlet excited alkyl azides eliminate nitrogen with concomitant rearrangement to form imine products, without the intervention of a nitrene intermediate.[60]

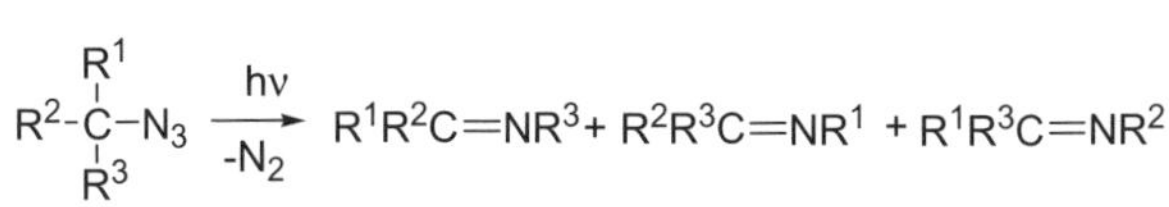

R^1 = Ph, 2-PhC_6H_4, $PhCH_2CH_2$, t-Bu, *n*-Pr, *n*-Am
R^2 = Ph, Me, *n*-Pr, *n*-Am
R^3 = Ph, Me, H, *n*-Am

Scheme 11.3 *Photo-rearrangement of* tert-*alkyl azides*

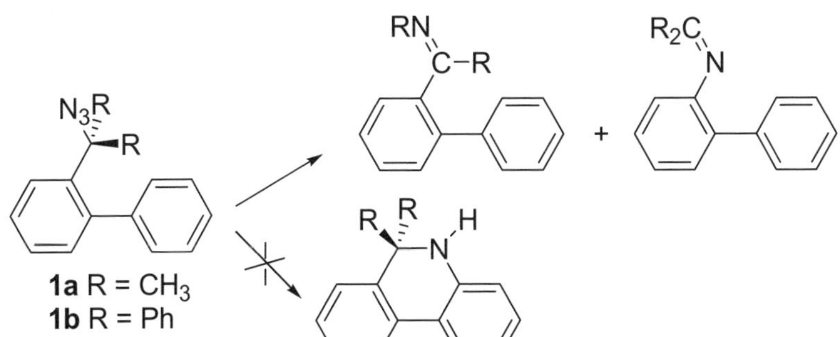

1a R = CH_3
1b R = Ph

Scheme 11.4 *Photo-rearrangement of di-methyl and di-phenyl derivatives of 2-biphenylmethyl azide*[60]

The quantum yields of photolysis of a series of tertiary alkyl azides (**2a-d**) were measured and found to be in the range of 0.27–0.53.[59]

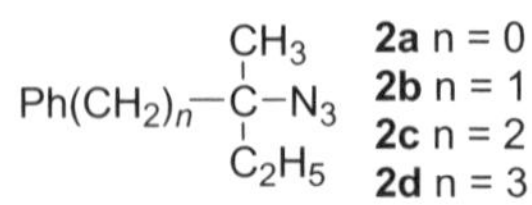

Scheme 11.4a

There are only few examples of photolysis of alkyl azides, which result in any process other than rearrangement to an imine.[56,61,62] Photolysis of highly fluorinated azide **3** in cyclohexane gave amide **5** in 18% yield after a hydrolytic workup, implicating a nitrene C–H insertion product **4**.[61]

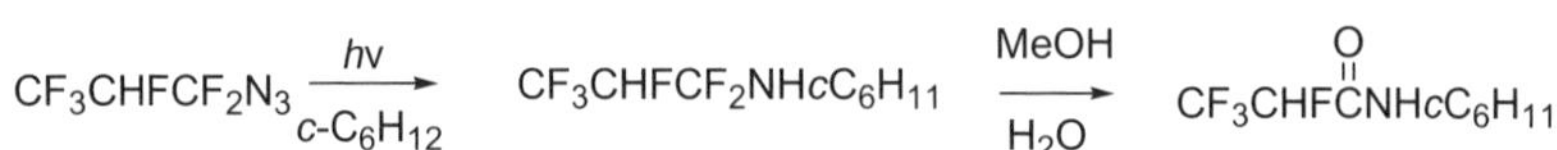

3 4 5

Scheme 11.5 *Phototransformation of fluorosubstituted* n-*propyl azide*[61]

An intramolecular cyclization via a formal C–H insertion as a minor process was observed upon photolysis of a steroidal azide.[62] Unique features of the photochemistry of α-azidoacetophenones and β-azidopropiophenones will be discussed later in this section.

Since there is no evidence of nitrene intermediates in the solution photochemistry of alkyl azides, attempts were made to detect triplet alkylnitrenes in matrix photochemical experiments. Photolysis of CH_3N_3 or CD_3N_3 at cryogenic temperatures in Ar, N_2 and CO_2 matrices fails to produce an IR spectrum attributable to triplet methylnitrene.[63,64] The IR spectrum of imine CH_2=NH (or CD_2=ND) is observed instead.

In subsequent studies, emphasis was given to studies of the matrix photochemistry of tertiary alkyl azides. Dunkin and Thomson studied the photochemistry of *tert*-butyl azide (**6**) in an N_2 matrix at 12 K.[65] Using IR spectroscopy they detected the formation of only one product – imine **7**.

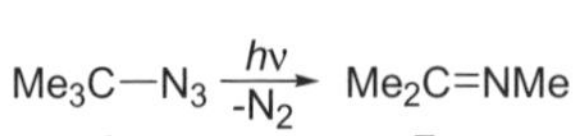

6 7

Scheme 11.6 *Matrix photochemistry of* tert-*butyl azide*[65]

The formation of strained bridgehead imines was observed upon photolysis of a series of matrix-isolated bridgehead azides, namely, adamantyl, bicyclo[2.2.2]octyl, bicyclo[3.2.1]heptyl and norbornyl azides.[66–70] For example, the photochemistry of matrix-isolated 1-azidonorbornane (**8**) was studied using monochromatic irradiation, IR, UV and ESR spectroscopy, and trapping with methanol and CO.[70] Three types of imines (*E*- and *Z*-isomers of **9** and **10**) and traces of triplet nitrene **11** were detected (Scheme 11.7). Imines **9** and **10** are light-sensitive, undergo interconversion and decompose to form unidentified products.

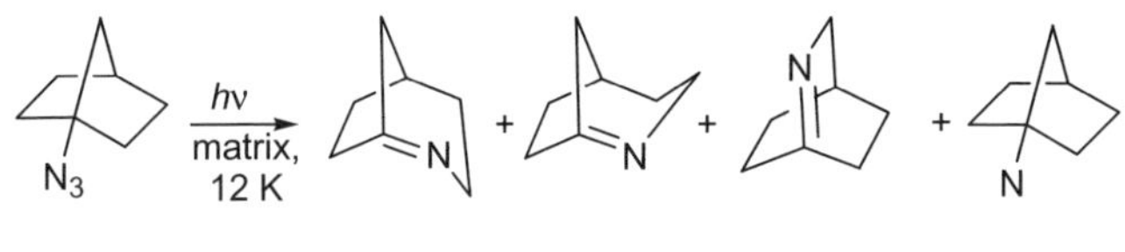

Scheme 11.7 *Matrix photochemistry of 1-azidonorbornane (**8**)*[70]

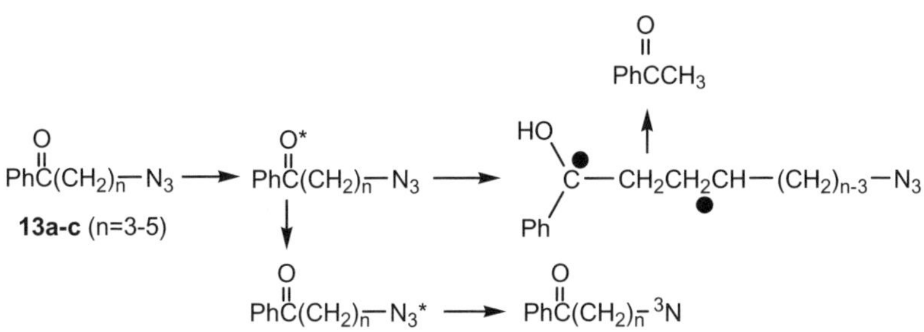

Scheme 11.8 *The primary processes in the photochemistry of α-benzoyl-ω-azidoalkanes*[73]

Triplet nitrene **11** was identified based on the ESR signal observed at 8124 G (E = 0, $|D/hc| = 1.65\,cm^{-1}$), the narrow band in the electronic absorption spectrum at 298 nm, and photochemical trapping with CO in Ar at 36 K.[70]

The first experimental evidence for the existence of alkylnitrenes was provided by ESR spectroscopy in glassy matrixes.[71] Very weak ESR signals were detected upon direct irradiation.[71] Sensitized photolysis of alkyl azides was found to be the most appropriate procedure to generate triplet alkylnitrenes in glassy matrixes.[71,72]

Later,[73] triplet alkylnitrenes **12a-c** were generated by intramolecular sensitization in solution. The photochemistry of three α-benzoyl-ω-azidoalkanes ($PhCO(CH_2)_nN_3$ (**13a-c**, n = 3–5) was studied and two competitive processes were found to proceed from the triplet ketone (Scheme 11.8): energy transfer to azide to yield triplet alkyl nitrene and γ-hydrogen abstraction to yield Norrish type II products. Photolysis of azides with a longer methylene chain (**13b,c**) yields mainly acetophenone, a product of Norrish type II reaction. The major products of the azide **13a** photolysis were substituted pyrrole and pyrroline derived from triplet nitrene **12a**.[73]

Recently, the Gudmundsdottir group performed a detailed study of the intramolecular sensitization of a number of α-azidoacetophenones (**14**) and β-azidopropiophenones (**15**) in solution (using product analysis and laser flash photolysis techniques)[74,75] and in argon matrices.[75] Compounds **16** and **17** are the major products in solution and their yields are in the range of 50 up to >99% depending on the substituent in the aryl group (Schemes 11.9 and 11.10). Donating substituents (R = *p*-OMe, *p*-SMe) lead to the growth of this yield. The mechanisms of product formation were proposed but not sufficiently substantiated.[74,75]

Transient absorption bands with maxima at ~320 nm were detected in the laser flash photolysis of **14** and **15** in solution and assigned to triplet alkylnitrenes, which were found

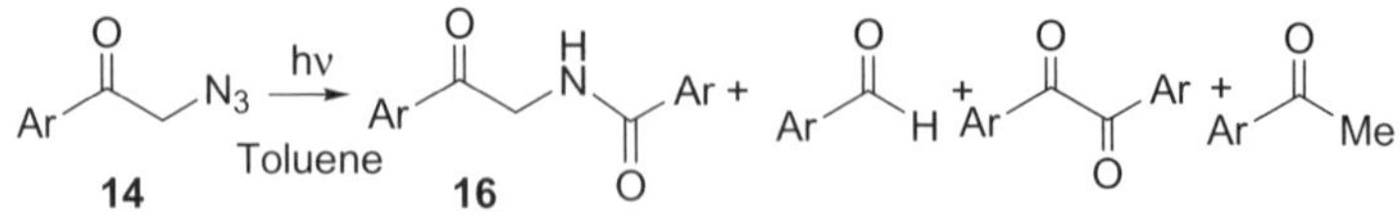

Scheme 11.9 *Products of the α-azidoacetophenones (**14**) photolysis*[74]

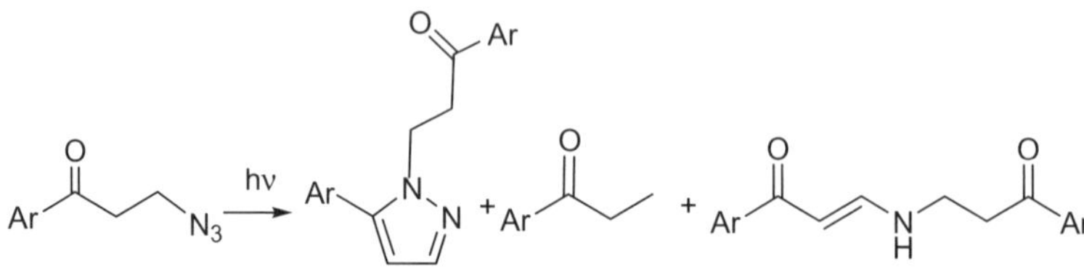

Scheme 11.10 *Products of the β-azidopropiophenones (**15**) photolysis*[75b]

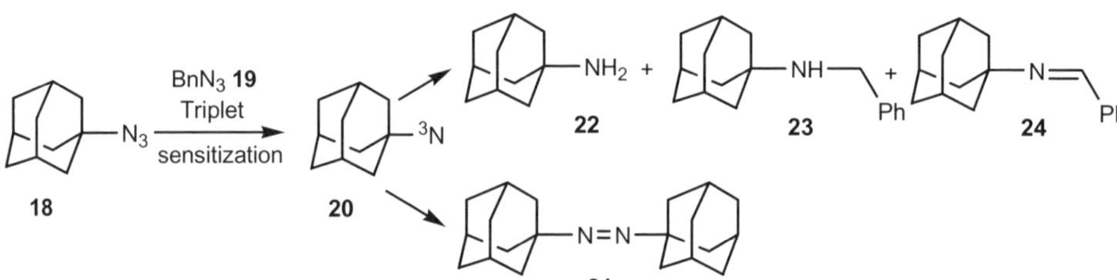

Scheme 11.11 *Triplet-sensitized photolysis of 1-azidoadamantane*[77]

to be long-lived with a lifetime of tens of milliseconds. The rate constant of their reaction with oxygen was estimated[74b,75b] to be 3–5 × $10^4 M^{-1} s^{-1}$. Corresponding nitro-compounds were isolated after photolysis of **14** and **15** in solutions saturated with oxygen.

Lewis and Saunders studied intermolecular sensitization of a series of alkyl azides.[76] It was found that the triplet energies of alkyl azides are 75–80 kcal/mol and the quantum yields for azide disappearance with appropriate triplet sensitizers approach unity. Recently,[77] the Gudmundsdottir group performed product analysis upon triplet sensitization of 1-azidoadamantane (**18**) and benzyl azide (**19**). It was found that triplet nitrene **20** undergoes dimerization yielding aza-adamantane **21** and hydrogen abstraction from the solvent yielding the following products (**22–24**, Scheme 11.11). Product distribution depends on the sensitizer and solvent nature. Formation of products typical[78] of direct irradiation of **18** (**25** and **26**) in a low yield have also been observed[77] (Scheme 11.12). The triplet benzyl nitrenes participate mainly in the reaction of hydrogen abstraction from the solvent (toluene).[77]

It was mentioned previously (Scheme 11.5), that photolysis of highly fluorinated azide **3** yields an intermolecular nitrene C–H insertion product. This implies a relatively long lifetime of fluorinated singlet nitrenes and the possibility of their relaxation to a ground triplet state. Indeed,[79] formation of triplet CF_3N was observed upon photolysis of CF_3N_3

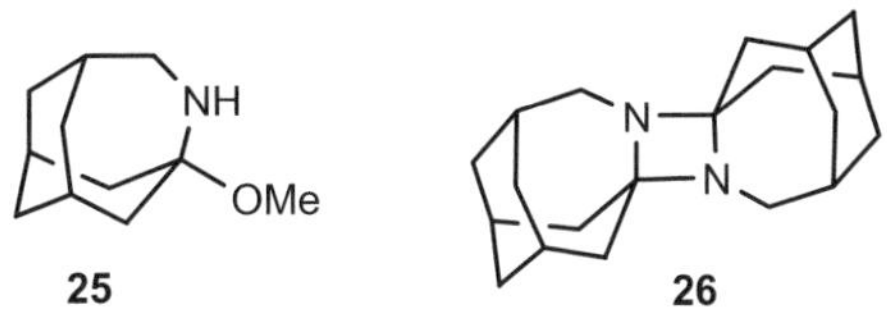

25 **26**

Scheme 11.12 *Products of direct irradiation of 1-azidoadamantane*[78]

in an Ar matrix and in polycrystalline pentane at cryogenic temperatures. A persistent ESR spectrum of triplet CF_3N in pentane at 6–10 K ($|D/hc| = 1.736\,cm^{-1}$) is very similar to that of triplet NH ($|D/hc| = 1.863\,cm^{-1}$)[51] and CH_3–N ($|D/hc| = 1.720\,cm^{-1}$).[80] The electronic absorption spectrum of $^3CF_3N$ was also detected in an Argon matrix at 14 K.

Note, that the absorption[81–84] and emission[83–89] spectra of triplet methylnitrene ^{3}MeN are well known. The 0–0 transition in the absorption spectrum of ^{3}MeN was found to be at 316.9 nm in an N_2 matrix[82] and at 314.3 nm in the gas phase,[83] similar to the spectrum of the parent NH (336 nm).[21,52] The singlet methylnitrene ^{1}MeN has never been produced as a trappable species and its detection by femtosecond spectroscopy has failed.[90] Nevertheless, negative ion photoelectron spectroscopy of MeN anion demonstrates that ^{3}MeN is lower in energy than the singlet by 1.352 ± 0.011 eV (31.2 ± 0.3 kcal/mol).[91] The features assigned to singlet nitrene in the photoelectron spectrum of MeN^- were interpreted as belonging to a resonance, rather than to a true minimum on the singlet MeN potential energy surface.[91]

Therefore, no experimental data is available which indicates formation of singlet alkylnitrenes as discrete intermediates upon photolysis of alkyl azides with the exception of perfluorinated alkylnitrenes. There is also doubt that ^{1}MeN is a true intermediate, namely, a species which is characterized by a minimum on the potential energy surface (PES). To investigate this issue a series of quantum chemical calculations were performed.[92–95]

According to the early calculations,[92] there is no minimum on the PES that corresponds to singlet MeN. As these calculations were performed at low level of theory, they were repeated in the 1990s.[93,95] The most recent studies, performed using the CAS/MP2[94] and CASSCF/CASPT2[95] techniques, predict a very shallow minimum for singlet MeN in the $^1A'$ state. The rearrangement of ^{1}MeN to CH_2=NH was predicted to be very exothermic ($\Delta H = -83$ kcal/mol) with a barrier 0.5–3.8 kcal/mol depending on the level of theory.[94,95] The thermal decomposition of MeN_3 was also predicted to occur in two steps, via a singlet nitrene intermediate.[94]

Thus, the predicted reaction mechanism is very dependant on the method employed in the calculations. The long wavelength transition of singlet CH_3N in the $^1A'$ state was calculated[14] at 287 nm with oscillator strength $f = 5 \times 10^{-3}$. Therefore, spectroscopic detection with pico- or femtosecond time-resolution should be performed to solve this problem.

11.4 Photochemistry of Vinyl Azides

The simplest vinyl azide, H_2C=CH–N_3, has been known for about 100 years. However, vinyl azides became an important and synthetically useful class of organic compounds only in the late 1960s.[96] Until recently,[97] photolysis and thermolysis of vinyl azides have been the main methods for the synthesis of azirines – higly strained nitrogen-containing heterocycles.

Smolinsky and Pryde[98] first observed azirine formation, together with small amount of ketenimin, by gas-phase pyrolysis of α-aryl substituted vinyl azides.

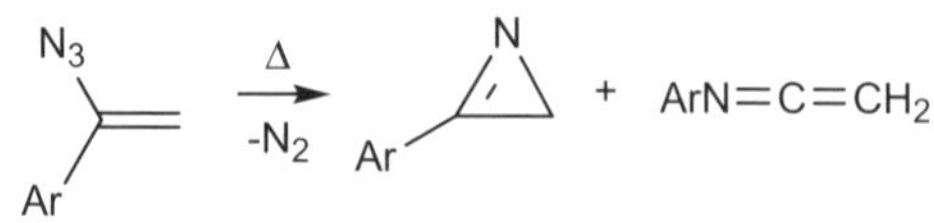

Scheme 11.13 Pyrolysis of α-aryl substituted vinyl azides[98]

Hassner and Fowler[99] first discovered that photolysis of α-substituted vinyl azides produce 2-mono- or 2,3-disubstituted-1-azirines with a large chemical yield (80–90%).

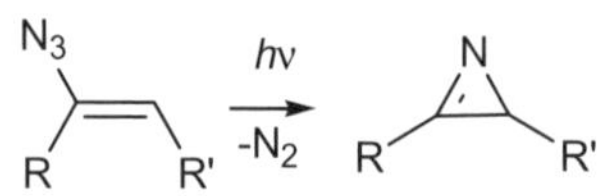

Scheme 11.14 Photolysis of α-substituted vinyl azides[99]

Isolable 1-azirines were formed upon photolysis of α-unsubstituted (R=H) vinyl azides only at low temperature and underwent further decomposition upon heating.[100]

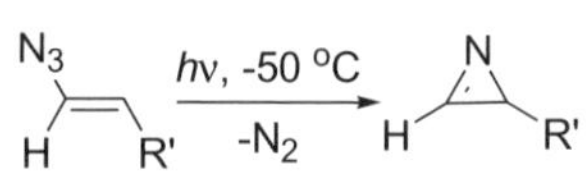

Scheme 11.15 Low temperature photolysis of α-unsubstituted vinyl azides[100]

The formation of 1-azirines (**27**) along with ketenimines (**28**) upon photolysis and thermolysis of vinyl azides was explained by the intermediacy of singlet vinylnitrenes (^{1}**29**).[101] Ketenimins can serve as a precursor to nitriles (**30**) if R′=H.

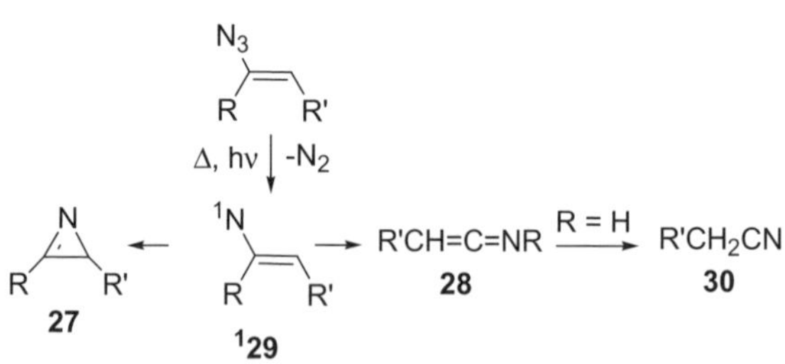

Scheme 11.16 Step-wise mechanism of vinyl azide photolysis and thermolysis[101]

However, singlet vinylnitrenes have never been observed by time resolved spectroscopy and triplet vinylnitrenes have not been observed by either time resolved or matrix spectroscopy. The formation of an azirine was observed upon photolysis of α-azidostyrene in an argon matrix at cryogenic temperature (8 K), but even under these conditions nitrene

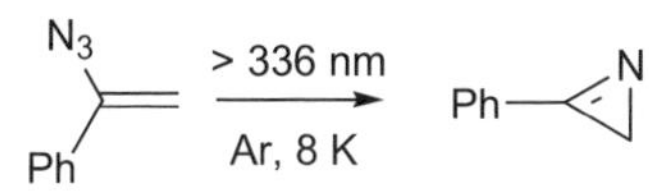

Scheme 11.17 *Photolysis of α-azidostyrene in an argon matrix at 8 K*[102]

species were not detected.[102] Therefore, the concerted formation of azirines (without intervention of a singlet nitrene) upon photolysis and thermolysis of vinyl azides was considered more reasonable.[96,103]

The situation with vinylnitrenes is analogous to methylnitrene and it is not clear if either of these singlet nitrenes are true reactive intermediates with finite lifetimes. Quantum chemical calculations can help to explain the properties of vinylnitrenes. All calculations reported in the literature have been concerned with only the simplest vinyl azide and vinylnitrene, CH_2=CH–N_3 and CH_2=CH–N (**29a**), respectively.[104–106] Early theoretical studies on the vinyl azide to azirine transformation employed semi-empirical calculations,[104a] or *ab initio* calculations performed at relatively low levels of theory,[104b,c] and have focused only on the closed-shell singlet excited state ($^1A'$) of vinylnitrene. We will discuss only the latest calculations.[105,106]

Cramer and Parasuk[105] have performed very accurate calculations of the electronic structure and energies of the lowest states of **29a**. Single-point calculations at the CASSCF(4,4) geometry were carried out at the MRCI and CASPT2 levels using cc-pVDZ and cc-pVTZ basis sets. Calculations predict **29a** to have a $^3A''$ ground state and the lowest open-shell singlet ($^1A''$) and closed-shell singlet ($^1A'$) states lie 15 and 40 kcal/mol higher in energy, respectively. The C=C bond length (1.346 Å) in $^1A'$ state is typical for the carbon-carbon double bond. It is lengthened to 1.391 Å in the $^3A''$ state and to 1.461 Å in the $^1A''$ state. The lowest singlet state of **29a** is open-shell and resembles a 1,3-biradical.

The ΔE_{ST} is much lower in **29a** (15 kcal/mol)[105] than in NH (36 kcal/mol)[53–55] or CH_3N (31.2 kcal/mol)[91] because the C=C substituent allows the π electron in the $^1A''$ state to become localized in a region of space that is disjoint from the region of space that is occupied by the σ electron. As will be discussed in Section 11.6 for the same reason, the lowest singlet state in phenylnitrene is also the open-shell $^1A_2$ state of **29a**.

According to the (4,4) CASSCF/6-31G* calculations, the nitrene **^{1}29a** in the $^1A''$ state can close to the azirine without any barrier and this state was found to be the transition state for interchange of the enantiotopic pair of hydrogens in 2*H*-azirine (**27a**).[106] Therefore, if a barrier does exist, it is probably very small. This conclusion, based on the results of calculations, is wholly consistent with the fact noted above, that the triplet and singlet vinyl nitrenes have escaped detection. However, further experimental studies, using very fast laser flash photolysis techniques, along with higher level *ab initio* calculations are certainly warranted.

11.5 Photochemistry of Carbonyl Azides and Azide Esters

The thermal rearrangement of carbonyl azides (Curtius rearrangement)[107] giving isocyanates in quantitative yields at 60–80 °C has been known over the years and has been studied in detail.[108–115] It is now generally accepted[14,108,113] and supported by modern quantum chemical calculations[114,115] that loss of nitrogen and migration of R are concerted processes in the Curtius rearrangement of carbonyl azides.

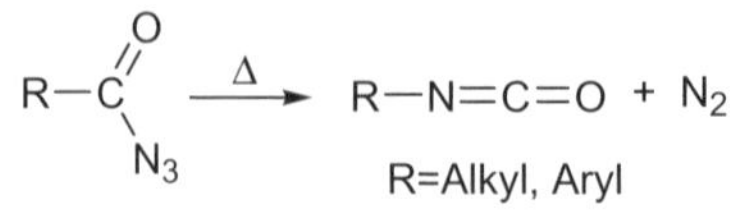

Scheme 11.18 *Thermal decomposition of carbonyl azides*[108]

In turn, photolysis of carbonyl azides gives rise to two types of reactions. The photo-Curtius rearrangement proceeds to form isocyanate. In addition, bimolecular trapping products, typical of the reactions of singlet carbonylnitrenes, are also observed.[108–110,115–127]

The mechanisms of photo-Curtius rearrangements have long been debated.[108–110,119–133] It has been shown that the yield of isocyanates, formed upon photolysis of a series of carbonyl azides (R–CO–N_3, R = *t*-Butyl, Aryl), remains constant in the presence and in the absence of nitrene traps.[111,112,117,118] For example, the yield of isocyanate **33a** produced on photolysis of pivaloyl azide (R = *tr*-Butyl, **31a**) in methylene chloride (an inert solvent) is 40%. Photolysis of **31a** in cyclohexene leads to a 45% yield of aziridine adduct **34a** and a 41% yield of isocyanate **33a**. Trapping the nitrene does not depress the yield of isocyanate. Hence, isocyanate **33a** and adduct **34a** cannot be derived from the same reactive intermediate, but instead the isocyanate must be formed from the excited state of the azide or from the electronically or thermally excited nitrene.

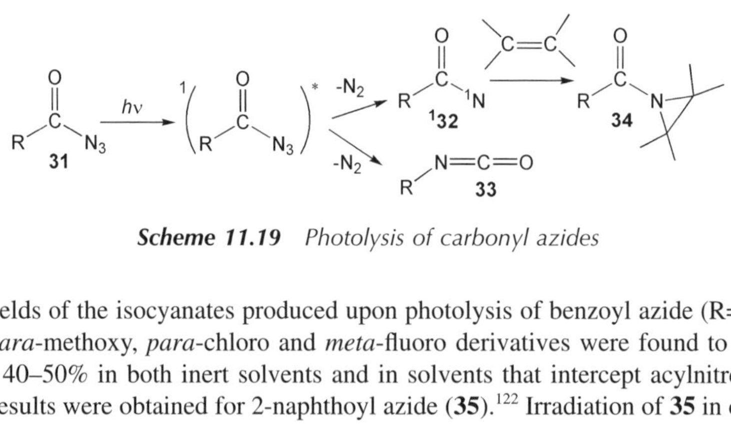

Scheme 11.19 *Photolysis of carbonyl azides*

The yields of the isocyanates produced upon photolysis of benzoyl azide (R=Ph, **31b**) and its *para*-methoxy, *para*-chloro and *meta*-fluoro derivatives were found to be in the range of 40–50% in both inert solvents and in solvents that intercept acylnitrenes.[117,118] Similar results were obtained for 2-naphthoyl azide (**35**).[122] Irradiation of **35** in cyclohexane at room temperature produces isocyanate (**36**, ~45%), *N*-cyclohexyl-2-naphthamide – the product of 2-naphthoylnitrene (**37**) insertion (~50%), and a trace (<1%) of 2-naphthamide (**38**).[122] Therefore, it was concluded that carbonylnitrenes (R–CO–N) do not rearrange to isocyanates (R–N=C=O) at a rate that is competitive with their capture by trapping agents.[108,111,115,118,122]

The Schuster group comprehensively studied the photochemistry of **35**[122] and substituted benzoyl azides[123–125] in order to determine the multiplicity of the ground state of aroylnitrenes. First, they demonstrated[122] that both direct and triplet-sensitized photolysis generates products characteristic of the reactions of ^{1}**37**. Earlier,[120,121] similar results had been obtained for benzoyl azide (R=Ph, **31b**). The direct and triplet sensitized photolysis of **31b** produced the same trapping products and these products were characteristic of a singlet nitrene ^{1}**32b**.

Second, photolysis of **35** in cyclohexane solution containing either *cis*- or *trans*-4-methyl-2-pentene (0.2 M) forms aziridines (**40**) with complete (>98%) retention of olefin stereochemistry.[122] Based on the Skell-Woodworth hypothesis developed for carbenes,[134] they concluded, that aziridines are formed in the reaction of singlet 2-naphthoylnitrenes (^{1}**37**).[122] Besides, extending the nitrene lifetime by diluting the concentration of the trapping agent (0.01 M) does not lead to relaxation of a putative excited singlet nitrene to its putative lower energy triplet state.[122]

Finally, ESR signals attributable to triplet nitrene ^{3}**37** were not observed after irradiation of **35** in fluorolube at 77 K. A nitrene-like triplet ESR spectrum was not detected either after photolysis of benzoyl azide **31b** in glassy matrices.[71,72b]

Therefore, the experimental data are most consistent with a singlet ground state of carbonyl nitrenes. Triplet carbonylnitrenes were not detected in either chemical trapping or spectroscopic experiments. The explanation of these results was available only recently.[135–137]

To understand the reason for the singlet multiplicity of the ground state of carbonyl azides, the singlet-triplet splitting for benzoyl- (**32b**) and 2-naphthoylnitrenes (**37**) were calculated at the B3LYP/6-31G* level of theory.[135] The triplet states were still computed to be lower in energy by about 5.0 and 4.5 kcal/mol for **32b** and **37**, respectively. However, these values are much lower than the well known ΔE_{ST} values for NH (36 kcal/mol)[53–55] and CH_3N (31.2 ± 0.3 kcal/mol).[91] The significant stabilization of the singlet state relative to the triplet state in aroyl nitrenes is attributed to the bonding interaction between the nitrogen and oxygen atoms.[135–137] Due to this bonding, the structure of the singlet species resembles that of a cyclic oxazirine,[135] although the calculated N–O distance (~1.76 Å) is much longer than a normal N–O single bond (about 1.5 Å in strained rings).[138]

Later,[115,136,137] it was demonstrated by calculation of ΔE_{ST} for formyl- and acetylnitrenes at very high levels of theory (CCSD(T) or CBS-QB3), that simple DFT calculations overestimate ΔE_{ST} by about 9 kcal/mol. Therefore, aroylnitrenes are indeed species predicted to have a singlet ground state.

Having understood the nature of the singlet species, Pritchina and colleagues[136,137] studied the photolysis of benzoyl azide **31b** in argon matrix at cryogenic temperature in order to detect this species directly by spectroscopic methods. It was shown that photolysis (254 nm) of matrix isolated **31b** affords at least two products. One of these products has an IR spectrum characteristic of isocyanate **33b**. Another product (λ_{max} = ~300 nm) is transformed to **33b** upon further irradiation at 313 nm. The electronic absorption spectrum of the latter product as well as its IR spectrum coincide well with the calculated spectra of singlet species ^{1}**32b** with a structure resembling that of cyclic oxazirine (Figures 11.3 and 11.4).[136,137]

Results of our investigations[136,137] were reproduced and the formation of a small amount of cyanate **41** was also detected.[139] Thus, the photolysis of **31b** could be described by Scheme 11.20.

Kinetics of the reactions of singlet species ^{1}**32b** in solution at room temperature were studied using time-resolved IR spectroscopy (TRIR)[136,140] and nanosecond laser flash photolysis.[115,141] The absolute rate constants of bimolecular reactions of ^{1}**32b** with

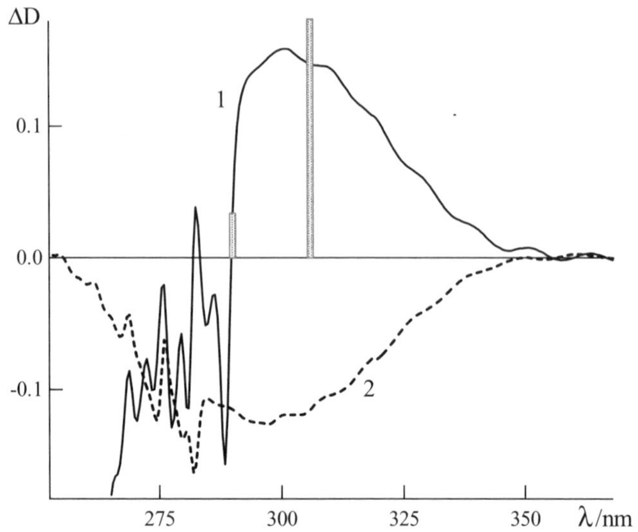

Figure 11.3 *Difference electronic absorption spectra recorded upon irradiation of **31b** at 254 nm for 2 min in argon matrix at 12 K (1) and the sample after further irradiation at 313 nm for 8 min (2).[136] The positions and relative intensities of the absorption bands calculated for species **^{1}32b** at the CASSCF/CASPT2 level are indicated by vertical bars*

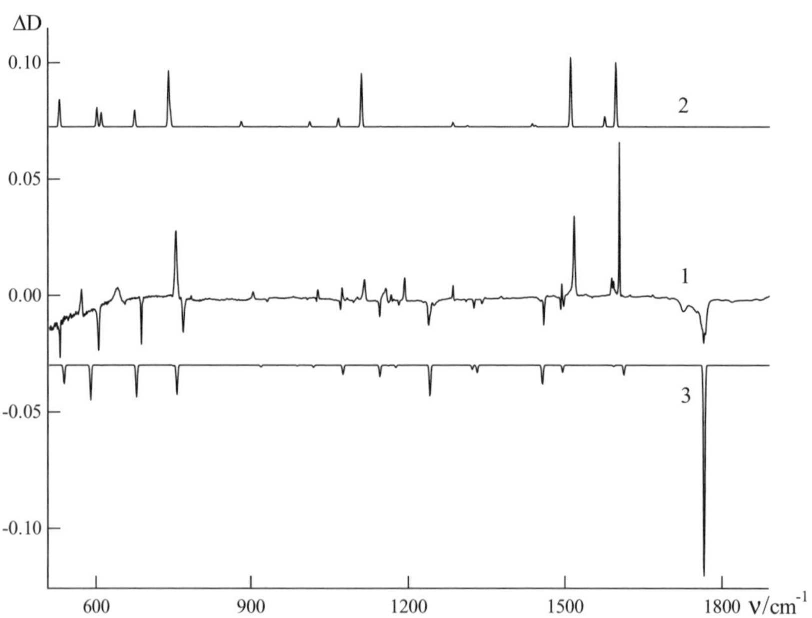

Figure 11.4 *Changes in the IR spectrum (1) recorded after additional irradiation at 313 nm (8 min) of the sample of **31b** irradiated at 254 nm (2 min).[137] Calculated IR spectra of isocyanate **33b** (2) and singlet species **^{1}32b** (3)*

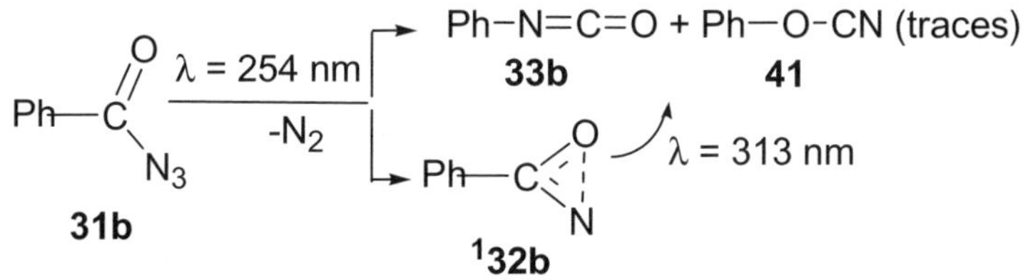

Scheme 11.20 *Photochemistry of benzoyl azide in argon matrix at 12 K*[136,137,139]

acetonitrile, methanol, water, cyclohexene, a series of alkenes and some nucleophilic anions were measured.[115,141] A very small negative activation energy was determined for reaction of **^{1}32b** with 1-hexene. This result is in good agreement with theoretical predictions.[115]

11.5.1 Photochemistry of Azide Esters

In contrast to carbonyl azides, photolysis and thermolysis of azidoformates (RO–CO–N_3, R=Alkyl, Aryl) yield mainly products derived from capture of the nitrenes (RO–CO–N).[108,142–146] Carbethoxy azide **42** has been studied most extensively. Formation of products characteristic of reactions of carbethoxynitrene **^{1}43** have been observed by thermolysis and photolysis of azide **42**[143,146] and by α-elimination of arysulfonate ion from *N*-(*p*-nitrobenzenesulfonyloxy) urethane.[143,145] The reaction of **^{1}43** with *cis* and *trans*-4-methyl-2-pentene was studied as a function of alkene concentration. At large alkene concentrations, aziridination is stereospecific, but upon dilution of the alkene, the stereospecificity is lost.[143] The triplet nitrene **^{3}43** also reacts with the olefins, but non-stereospecifically, presumably through intermediate biradical formation (Scheme 11.21).[107,143] These results are completely analogous to studies of carbenes in which a stereospecific singlet intermediate is produced initially, and subsequently relaxes to a less selective, lower energy triplet intermediate.[147]

Insertion of **^{1}43** into the CH bond of alkanes and the OH bond of alcohols, addition of **^{1}43** to acetylenes, and its reaction with benzene, followed by azepine formation, are all well documented.[108]

Similar results (formation of nitrene derived products and the absence of isocyanate and products of its transformations) were obtained upon photolysis of (4-acetylphenoxy) carbonyl azide.[122,123]

The triplet ground state of nitrene **43** was proven by ESR spectroscopy ($|D| = 1.603\,cm^{-1}$, $|E| = 0.0215\,cm^{-1}$).[72b] A similar ESR spectrum ($|D| = 1.65\,cm^{-1}$, $|E| = 0.024\,cm^{-1}$) was recorded for the triplet (4-acetylphenoxy)carbonylnitrene.[122,123]

Recently, the photochemistry of azide **42** was studied by laser flash photolysis (λ_{ex} = 266 nm) techniques in Freon-113 ($CF_2ClCFCl_2$) at room temperature.[148] The formation of at least two intermediates, viz., triplet nitrene **^{3}43** (λ_{max} = 400 nm, lifetime 1.5 μs) and ethoxycarbonyl radical **44** (λ_{max} = 333 nm, lifetime 0.4 μs), was observed (Scheme 11.22). The singlet nitrene **^{1}43** was deduced to have a lifetime between 2 and 10 ns in Freon-113 at ambient temperature. The kinetics of reactions of **^{3}43** with tetramethylethylene and triethylsilane were also measured.[148]

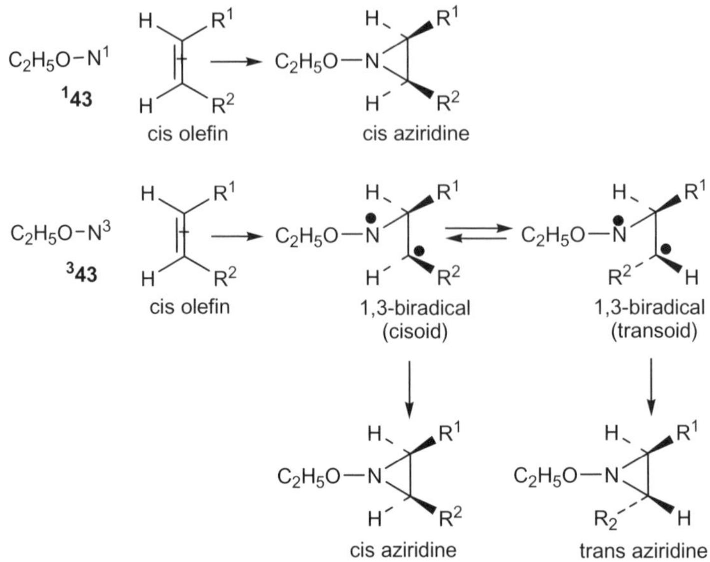

Scheme 11.21 *The stereospecific and non-stereospecific addition of carbethoxynitrenes **43** to alkenes*[143]

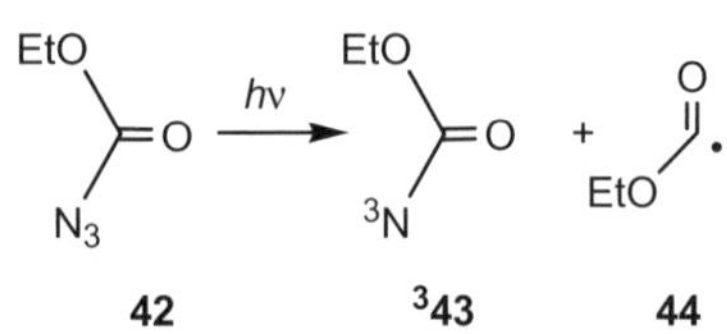

Scheme 11.22 *Laser flash photolysis of carbethoxy azide **42** in Freon-113 at ambient temperature*[148]

The photochemistry of carbomethoxy azide (**45**) and its deuterated derivative was studied in inert gas matrices at 4 and 10 K.[149,150] The IR spectrum of the reaction products shows characteristic lines assigned to methoxyisocyanate (**46**) and the products of its further phototransformations, viz., formaldehyde and isocyanic acid.[149] Triplet carbmethoxynitrene was not detected,[149,150] apparently due to the photochemical transformation of the latter to **46**.

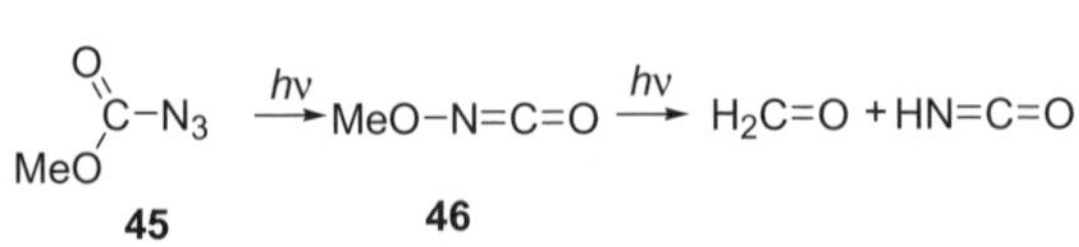

Scheme 11.23 *Photolysis of carbmethoxy azide in neon matrix at 4 K*[149]

To understand the difference in the properties of carbonylnitrenes (R–CO–N) and nitrenoformates (RO–CO–N), the structures and ΔE_{ST} for simplest nitrenes (R=H and Me) were analyzed at the CCSD(T) and CBS-QB3 levels of theory.[115,137] According to the calculations both nitrenoformates (CH_3O–CO–N and HO–CO–N) have triplet ground states, whereas carbonylnitrenes (CH_3–CO–N and H–CO–N) are singlet ground state species. At this level of theory, the singlet excited states of nitrenoformates have the same nature as the ground singlet states of carbonylnitrenes, viz., they have a structure resembling that of a cyclic oxazirine. The difference in the ΔE_{ST} between carbonylnitrenes and nitrenoformates was attributed to the smaller bonding interaction between the nitrogen and oxygen atoms in the latter case.[115,137]

11.6 Photochemistry of Phenyl Azide and Its Simple Derivatives

The photochemistry of phenyl azide and its simple derivatives have received the most attention in the literature. The results of early studies were summarized in a number of reviews.[1,2,11–13,151–153] Over the last decade, modern time-resolved spectroscopic techniques and high level *ab initio* calculations have been successfully applied and reveal the detailed description of aryl azide photochemistry. This progress was analyzed in recent reviews.[14,15,17–20] Femtosecond time resolved methods have been recently employed to study the primary photophysical and photochemical processes upon excitation of aryl azides.[154–161] The precise details by which aryl azide excited states decompose to produce singlet arylnitrenes and how rapidly the seminal nitrenes lose heat to solvent and undergo unimolecular transformations were detailed. As a result of the application of modern experimental and theoretical techniques, phenylnitrene (PhN) – the primary intermediate of phenyl azide photolysis, is now one of the best characterized of all known organic nitrenes.[14,15,17–20,157–159]

In this section we will briefly consider the most important early results which created the basis for the interpretation of the more recent studies. The largest part of this section will be devoted to the experimental and theoretical discoveries of the last decade.

11.6.1 Photochemistry of Phenyl Azide

Photolysis and pyrolysis of phenyl azide (**47**) and most of its derivatives in hydrocarbons leads to polymeric tars instead of diagnostic insertion products and aziridines.[162] A fortunate exception is the formation of azepines in the presence of primary and secondary amines. The Huisgen group[163] was the first to observe that thermolysis of **47** in the presence of aniline leads to extrusion of molecular nitrogen and formation of azepine **48a** (Scheme 11.24). Eight years later, Doering and Odum[164] demonstrated that azepine **48b** is formed with a high yield (≥70%) upon photolysis of **47** in diethylamine (Scheme 11.24). Aside from azepines, the formation of *ortho*-substituted aniline **49** (in a yield of 39%) was discovered upon photolysis of **47** in ethanethiol (Scheme 11.24).[165]

The azirine **50** and/or ketenimine (1,2-didehydroazepine) **51** were proposed as the trappable reactive intermediates produced upon photolysis of **47** in solution (Scheme 11.25).[162–165]

The high dilution of solutions of phenyl azide suppresses polymer formation and azobenzene forms instead.[166,167] This indicates that singlet intermediates (**50** and/or **51**) serve as a reservoir for triplet phenylnitrenes (^{3}**52**), which either undergo dimerization or react

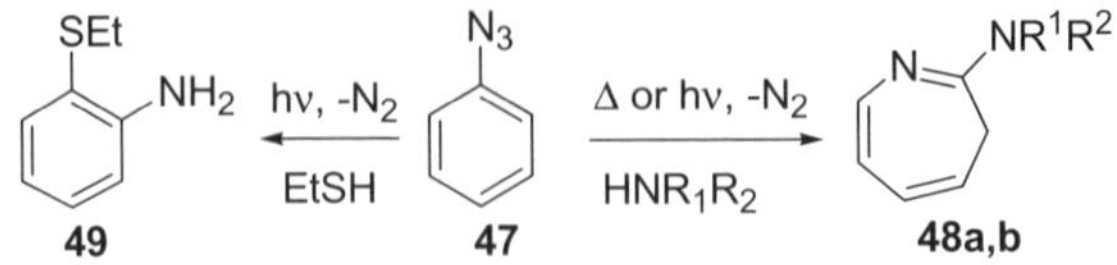

Scheme 11.24 *Adduct formation upon photolysis and pyrolysis of phenyl azide*[163–165]

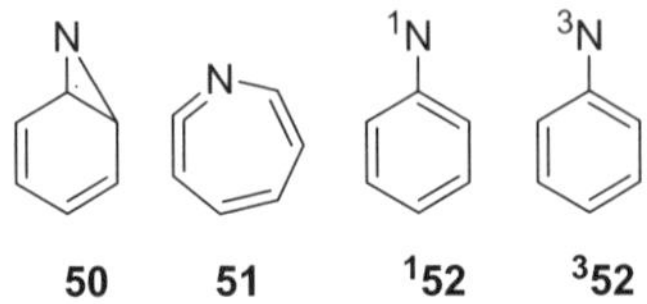

Scheme 11.25 *Structure of intermediates of phenyl azide photolysis*

with azide **47** to give azobenzene. These reactions have never been directly monitored for ^{3}**52** by time-resolved techniques. However, dimerization of substituted triplet phenylnitrenes (*para*-nitro[167] and 2,4,6-tribromo[168]), as well as polycyclic 1-naphthyl-,[169,170] 1-anthranyl-[169] and 1-pyrenylnitrenes,[170,171] were studied by laser flash photolysis techniques. The decay of triplet arylnitrenes and/or formation of corresponding azo-compounds were found to obey second-order kinetics with rate constants in the range of $0.55–2.1 \times 10^9\,M^{-1}\,s^{-1}$ in benzene at room temperature.[168–171]

Later, it was demonstrated that photolysis of the dilute hydrocarbon solutions ($<10^{-4}$ M) of simple derivatives of **47** in the presence of oxygen gives the corresponding nitro- and nitrosobenzenes with a yield of ~80%.[13,172,173] The latter are also the products of triplet arylnitrenes reactions.[13,172–175]

Formation of triplet phenylnitrene (^{3}**52**) was detected by EPR spectroscopy after photolysis of **47** in glassy matrices at 77 K.[176] The temperature dependence of the EPR signal demonstrated that the triplet state is the ground state of phenylnitrene.[176] Shortly thereafter, Reiser's group[177] recorded the UV-Vis spectrum of ^{3}**52** in a glassy matrix. Later it was found that ^{3}**52** is extremely light sensitive and that upon photoexcitation at 77 K, it rapidly isomerizes to the isomeric ketenimine **51**.[168] Figure 11.5 shows the spectrum of ^{3}**52** recorded in EPA at 77 K.

In 1978, Chapman and LeRoux detected the formation of ketenimine **51** using matrix isolation IR spectroscopy.[102] Irradiation of **47** in an argon matrix at 8 K with light of $\lambda > 360$ nm (or $\lambda > 216$ nm) led to the formation of a product giving an intense IR band at 1895 cm^{-1} characteristic of a heterocumulene structure (–N=C=C–). The ketenimine **51** was unstable upon further irradiation ($\lambda > 360$ nm); however, the formation of azirine **50** was not observed.[102] The later spectroscopic studies in matrices[178–180] demonstrated the formation of both **51** and ^{3}**52** with a characteristic triplet EPR spectrum ($|D/hc| = 1.027$ cm^{-1}, $|E/hc| = 0$ cm^{-1}).[179] The ratio of ^{3}**52** and **51** in the initial steps of irradiation was found to be ~4 : 1 at 334 nm excitation.[179] Irradiation of **47** in argon matrix at 280 nm preferentially gave **51**.[180] The shorter wavelength excitation (254 nm) of **47** in an argon matrix yields ^{3}**52** and **51** in the ration of ~1 : 2.[181]

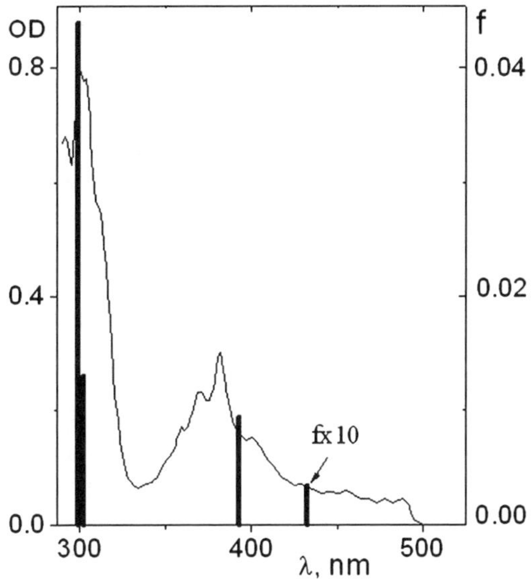

Figure 11.5 *The difference absorption spectrum of* ^{3}***52*** *in EPA glass at 77 K.*[168] *The computed positions and oscillator strengths (f, right-hand axis) of the absorption bands of* ^{3}***52*** *are depicted as solid vertical lines.*[197] *Reprinted with permission from ref.*[197] *Copyright 1999 ACS Publications*

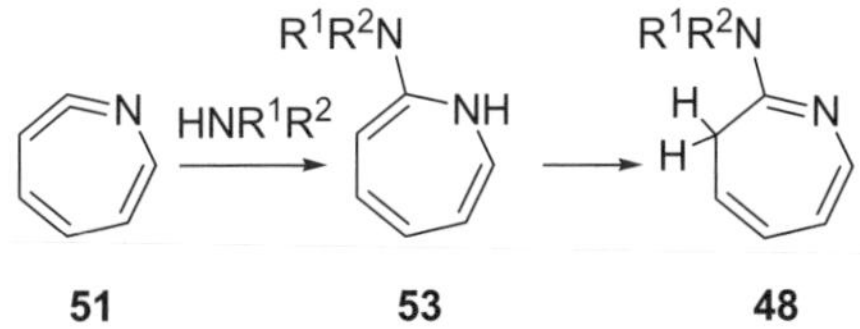

Scheme 11.26 *Reaction of ketenimine with primary and secondary amines*

The solution phase photochemistry of phenyl azide **47** is temperature dependent.[168] Photolysis of **47** in the presence of diethylamine at ambient temperature yields azepine **48b**. Lowering the temperature suppresses the yield of **48b** and below 160 K, azobenzene, the product of triplet nitrene dimerization, is produced. Thus, high temperature favors reactions of singlet state intermediates whilst low temperatures favor reactions associated with triplet phenylnitrene.

The formation of **51** in solution was recorded by laser flash photolysis techniques (broad band at ~350 nm),[168,171] and this assignment was unambiguously proved by time-resolved IR spectroscopy (TRIR).[182,183] It was also established that ketenimine **51** is the species trapped by nucleophiles in solution.[183] The tautomerization of the primary trapping product, 1H-azepine (**53**), to the final 3H-azepine (**48**) was also studied in detail in the 1970's (Scheme 11.26).[184]

Recently,[185] the low temperature ^{13}C NMR and IR spectra of **51** incorporated into a hemicarcerand were reported. The lifetime of **51** in the inner phase of a hemicarcerand was found to be 32 min at 194 K. Encapsulation also dramatically increased the lifetime of the triplet nitrene **^{3}52** ($\tau \approx$ 75 days at 194 K).[185b]

The photochemistry of **47** was also studied in the gas phase.[186,187] Some groups reported that gas-phase photolysis of **47** produced the absorption and emission spectra of **^{3}52**.[186] Cullin and co-workers[187] demonstrated that the spectra observed were actually that of the cyanocyclopentadienyl radical. UV-photolysis of **47** produces singlet phenylnitrene (**^{1}52**) with excess vibrational energy, which in the gas phase can not be shed by collisions with solvent molecules. Thus, hot **^{1}52** explores the C_6H_5N surface and eventually finds the global minimum, cyanocyclopentadiene (**54**), which can shed its excess energy by losing a hydrogen atom to form the cyanocyclopentadienyl radical (**55**). These results are in excellent agreement with Wentrup's gas-phase pyrolysis studies.[188] To date, phenylnitrene has not been detected in the gas-phase.

Thus, in the late 1980s a series of intermediates produced by the photolysis of phenyl azide had been directly observed (**^{3}52** in matrices and low temperature glasses, **51** in matrices and liquids, and **55** in the gas-phase). However, the results obtained in solution and inert gas matrices differ substantially from those obtained in low temperature glasses. In glasses, triplet nitrene **^{3}52** is the major product, whereas ketenimine **51** is the major product in solution at ambient temperature and often in the inert gas matrices at ~10 K. Formation of **51** in inert gas matrices was explained by the slow vibrational relaxation of the hot singlet nitrene **^{1}52** in these matrices, which competes with fast isomerization to **51**.[168,180]

At that time, the primary intermediate, singlet phenylnitrene **^{1}52**, had still escaped direct detection. The benzazirine **50** had not been detected either, although the formation of **49** in ethanthiol (Scheme 11.24) was an indication of its intermediacy in the phototransformation of **47**. A series of reviews were published in 1992,[12,13,153] which economically explained much of the photochemistry of phenyl azide (Scheme 11.27).

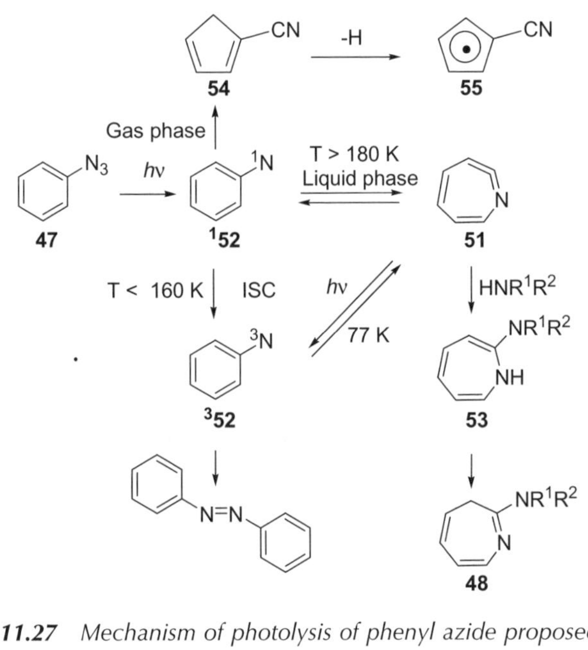

Scheme 11.27 *Mechanism of photolysis of phenyl azide proposed in 1992*

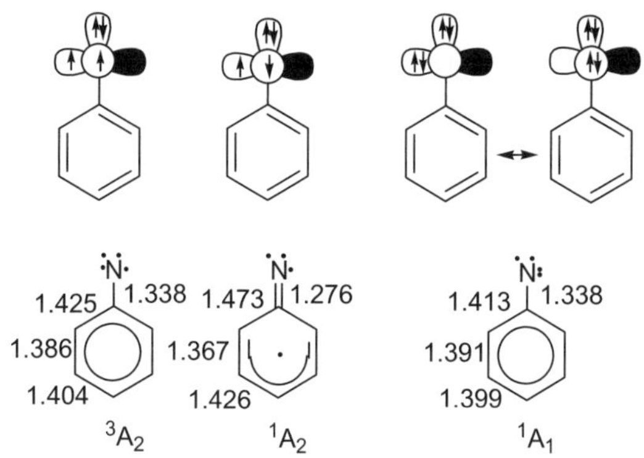

Figure 11.6 *Orbital occupancies for nonbonding electrons and CASSCF(8,8)/6-31G(d) optimized geometries of the lowest triplet and singlet states of phenylnitrene* ***52****. Reprinted with permission from ref.*[20] *Copyright 2006 ACS Publications*

At the same year, two computational studies clarifying the electronic structure of nitrene **52** were published.[189,190] In phenylnitrene **52**, a lone pair of electrons occupies a hybrid orbital, rich in 2s character and the two non-bonding electrons both occupy pure 2p orbitals. One of these is a p-π orbital, and the other a p orbital on nitrogen that lies in the plane of the benzene ring. The near-degeneracy of the two 2p orbitals gives rise to four low-lying spin states – a triplet ($^3A_2$), an open-shell singlet ($^1A_2$), and two closed-shell singlets ($^1A_1$). The orbital occupancies and CASSCF(8,8)/6-31G* geometries of the three lowest spin states are shown in Figure 11.6.

In the $^3A_2$ and $^1A_2$ states, the 2p-π orbital and the in-plane 2p orbital on N are both singly occupied. The two $^1A_1$ states of **52** are a mixture of two dominant configurations – one in which the in-plane p orbital on N is doubly occupied and the 2p-π orbital is empty. The latter is slightly lower in energy than the configuration in which these orbital occupancies are reversed.[106,189,190] The two $^1A_1$ states differ only by the sign of this linear combination. In both the $^3A_2$ and $^1A_1$ states the C–N bond is relatively long, and the phenyl ring shows little bond-length alternation (Figure 11.6). In the $^1A_2$ state, however, strong delocalization of the electron in the nitrogen p-π orbital into the aromatic ring leads to a very short C–N bond (1.276 Å).[106,189,190] This delocalization confines the electron in the π-orbital and the electron of opposite-spin in the in-plane 2p AO on nitrogen to different regions of space, thus minimizing their mutual Coulombic repulsion energy.[106] This is the reason for the strong stabilization of the $^1A_2$ state relative to the $^1A_1$ state of **52**.

High levels of theory predict that the lowest singlet state ($^1A_2$) of **¹52** lies about 18 kcal/mol higher in energy than the triplet ground state ($^3A_2$),[106,189–192] in excellent agreement with the experimental results obtained by photoelectron (18 ± 2 kcal/mol)[193] and electron detachment (18.3 ± 0.7 kcal/mol)[194] spectroscopy. The second singlet state of **52**, $^1A_1$, lies about 30 kcal/mol above the ground triplet state.[106,189–193]

In 1997 the primary intermediate – singlet phenylnitrene **¹52**, was finally directly detected by LFP techniques. Gritsan, Yuzawa and Platz[195] and the Wirz group[196] simultaneously reported the observation of singlet phenylnitrene (**¹52**) with λ_{max} at about 350 nm and a lifetime of ≈1 ns at ambient temperature. More accurate measurements[197]

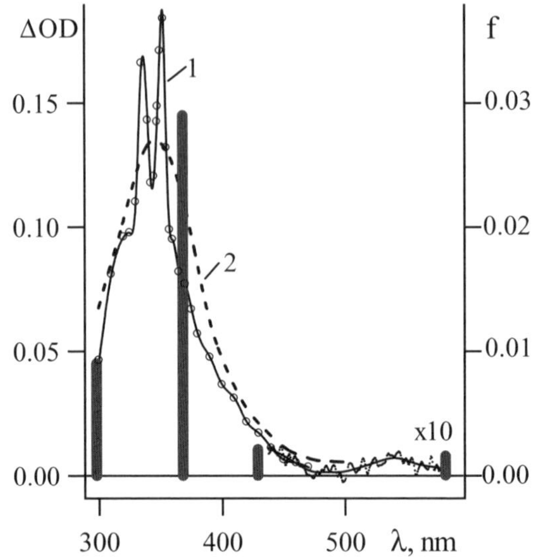

*Figure 11.7 Transient absorption spectra recorded after laser excitation (266 nm, 35 ps) of phenyl azide **47** in pentane: after 2 ns at 233 K (1) and after 10 ns at 295 K (2). The computed positions and oscillator strengths (f, right-hand axis) of the absorption bands of ^{1}**52** are depicted as solid vertical lines*

revealed the structure of this band (336 and 352 nm) and also a very low intensity band at 540 nm (Figure 11.7). Figure 11.7 also demonstrates the electronic absorption spectrum of ketenimine **51** (spectrum 2), recorded on a nanosecond time scale at ambient temperature.[195]

The assignment of the transient absorption in Figure 11.7 to the ^{1}**52** was supported by the calculation of its electronic absorption spectrum. Indeed, the spectrum of ^{1}**52** ($^1A^2$) calculated at the CASPT2 level is in good agreement with the transient spectrum of Figure 11.7. The only intense absorption band in the spectrum of ^{1}**52** is localized around 350 nm (Figure 11.7). According to the calculations the main configuration involved in this transition consists of excitation of an electron from the lone pair orbital (n_z) on nitrogen to the singly occupied nitrogen 2p orbital that lies in the molecular plane (p_y). A similar type of transition contributes to the intense absorption band of ^{3}**52** at ~300 nm[197] (Figure 11.5). The spectra of the simplest nitrenes ^3NH (λ_{max} = 336 nm)[21,52] and $^3NCH_3$ (λ_{max} = 316.9 nm)[82] are also associated with these types of analogous transitions.

The calculations also demonstrate that the electronic absorption spectra of ^{1}**52** and ^{3}**52** are very similar, but that all of the calculated and experimentally detected bands of ^{1}**52** (Figure 11.7) exhibit a red-shift compared to those of ^{3}**52** (Figure 11.5).[197] This is reasonable because both of these species have very similar open-shell electronic configurations ($^3A_2$ and $^1A_2$).

The decay of ^{1}**52** in pentane was monitored at 350 nm over a wide temperature range of 150–270 K. This allowed direct measurement of the rate constants for intersystem crossing (k_{ISC}) and for rearrangement (k_R), and the Arrhenius parameters for the latter.[197]

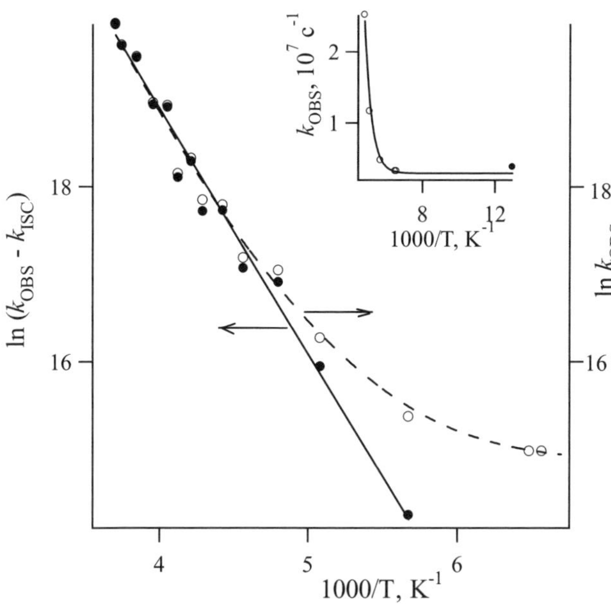

Figure 11.8 *Arrhenius treatment of the k_{OBS} data (open circles) and $k_R = k_{OBS} - k_{ISC}$ (filled circles) for singlet phenylnitrene **¹52** deduced upon assuming that k_{ISC} is independent of temperature. Insert: temperature dependence of k_{OBS} data. Reprinted with permission from ref.[197] Copyright 1999 ACS Publications*

The decay of **¹52** and the growth of the products (**51** and or **³52**) are first order and can be analyzed to yield an observed rate constant, k_{OBS}. The magnitude of k_{OBS} decreases with decreasing temperature until about 170 K, whereupon it reaches a limiting value.[197] Analysis of this temperature dependence (Figure 11.8) gave the rate constant of intersystem crossing ($k_{ISC} = 3.2 \pm 0.3 \times 10^6\,s^{-1}$) and the Arrhenius parameters for the rearrangement of **¹52** ($E_a = 5.6 \pm 0.3$ kcal/mol, $A = 10^{13.1\pm0.3}\,s^{-1}$).[197]

Recently,[198] the LFP of **47** was studied at 77 K where singlet nitrene **¹52** cleanly relaxes to the triplet state **³52**. The rate constant of intersystem crossing at 77 K was found to be $3.8 \pm 0.3 \times 10^6\,s^{-1}$. Thus k_{ISC} for **¹52** is indeed temperature independent. The spectrum of **¹52** at 77 K is similar to that detected in solution (Figure 11.7, spectrum 1).[197,198]

Scheme 11.27 describes the rearrangement of **¹52** to **51** as one-step reaction. However, the computational work of Karney and Borden[106] demonstrates this to be a two-step process involving benzazirine **50**, the species trapped by ethanethiol (Scheme 11.24). The first step, cyclization of **¹52** to the azirine **50**, is predicted to be the rate-determining step (Scheme 11.28). The CASPT2 energetics of rearrangement is depicted in Figure 11.9.

The CASPT2 calculated barrier to cyclization of 9.2 kcal/mol[106] is somewhat higher than the experimental barrier of 5.6 ± 0.3 kcal/mol.[197] The discrepancy between the calculated and experimental barrier heights is due to the general tendency of the CASPT2 method to overstabilize open-shell species (in this case, $^1A_2$-**3**) relative to closed-shell species (in this case, all the other stationary points on the reaction path).[199] For an

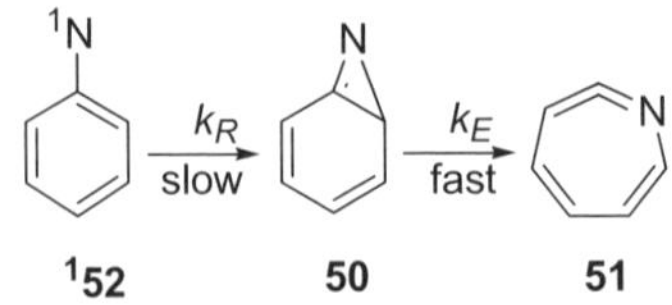

Scheme 11.28 *Two-step rearrangement of singlet phenylnitrene to azepine*

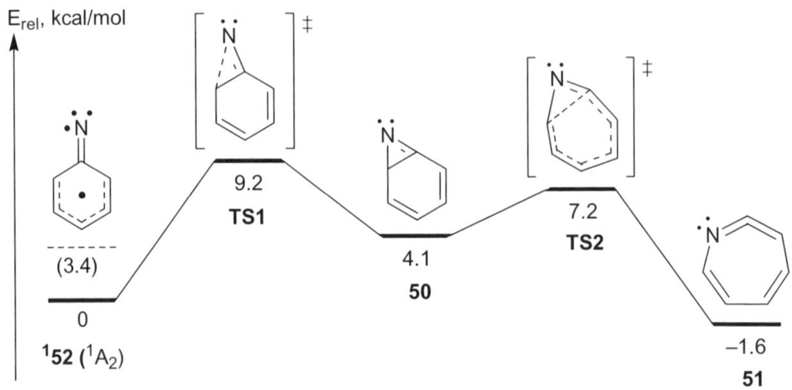

Figure 11.9 *Energetics of the ring expansion of singlet phenylnitrene **¹52** (¹A₂), calculated at the CASPT2/6-311G(2d,p)//CASSCF(8,8)/6-31G(d) level. Reprinted with permission from ref.*[20] *Copyright 2006 ACS Publications*

analogous system, the error was 3.4 kcal/mol.[106] Taking into account this error, the theoretical barrier for the cyclization of **¹52** is predicted to be 5.8 kcal/mol, which is in excellent agreement with the experimental activation energy (5.6 ± 0.3 kcal/mol).[197]

The CASPT2 barrier for the process **50** → **51** is only ca. 3 kcal/mol, and this reaction is calculated to be exothermic by about 6 kcal/mol (Figure 11.9). These computational results are consistent with the failure to detect **50**.[182,183] Nevertheless, although **50** has not been observed spectroscopically, it can be intercepted by ethanethiol (Scheme 11.24).

On this basis, in recent reviews[14,15,17,19,20] the mechanism of phenyl azide photolysis described by Scheme 11.27 was accepted and supplemented by the two-step mechanism of the singlet phenylnitrene rearrangement to azepine **51** (Scheme 11.28).

However, the formation of azepine **51**, upon photolysis of **47**, in inert gas matrices was not fully understood.[102,178–181] Nevertheless, it was reasonably explained by the formation of **51** in the reaction of the hot singlet nitrene, **¹52**$^{\#}$.[168,180] This hypothesis is consistent with the results of recent studies of the photolysis of **47** and a series of its simple derivatives in solution at room temperature by femtosecond transient absorption spectroscopy[154–157] and femtosecond IR spectroscopy.[158]

It was found that the N–N bond cleavage in aryl azides proceeds on a femtosecond time scale (ca. 100–500 fs).[154–157] For example, the times of formation of singlet nitrenes in acetonitrile produced upon excitation at 266 nm are ~100 fs for biphenyl-4-ylnitrene and 280 ± 150 fs for biphenyl-2-ylnitrene[155,156] and the singlet arylnitrenes are formed with an excess of vibrational energy. Moreover, the shorter excitation wavelength results in a greater excess of vibrational energy. Vibrational cooling proceeds on a picosecond time scale (10 ps for singlet biphenyl-4-ylnitrene in acetonitrile[155,156] and 11 ps for 3,5-dichlorobiphenyl-2-ylnitrene[154] in cyclohexane).

The singlet phenylnitrene **^{1}52**, which is formed upon photodissociation of **47**, has a considerable excess of vibrational energy and can easily overcome the potential energy barrier for isomerization. Indeed, studies utilizing femtosecond time-resolved IR spectroscopy[158] have demonstrated that a portion of the total yield of ketenimine **51** is formed on a picosecond time scale. The ketenimine **51** was also formed in the vibrationally hot state. The formation of ketenimine **51** and its vibrational cooling proceed within 10–50 ps. Unfortunately, attempts to separate these two processes and determine the characteristic time of the ketenimine formation failed.[158]

The vibrational cooling of the singlet nitrene **^{1}52** competes with its transformation to ketenimine **51** on a picosecond time scale.[158] The singlet nitrene **^{1}52** in its ground vibrational state also undergoes two-step transformation to the same product **51** with a time constant ~1 ns in pentane at ambient temperature.[195–197]

Due to its short lifetime, the singlet phenylnitrene **^{1}52** has never been trapped chemically, the only exception being protonation with the formation of phenyl nitrenium ion **54**.[200–202] It was predicted theoretically, that **54** has a singlet ground state which is favored by 21.2 kcal/mol over the lowest triplet state.[203] Protonation of **^{1}52** competes with its isomerization to **51** only at low pH $\leq$ 1.[200] McClelland and co-workers performed product analysis and LFP studies of the photolysis of phenyl azide in water at pH = 0–2 and proposed a mechanism of reactions of nitrenium ion **54** (Scheme 11.29).[200,201] The nitrenium ion **54** reacts with water or other nucleophiles yielding substituted anilines.[200,201] Note, that lifetime of **^{1}52** in water (25–50 ps) was estimated to be much shorter than in hydrocarbons (~1 ns).

Recently, the nitrenium ion **54** was directly detected by transient absorption spectroscopy in pure formic acid.[202] The decay of nitrene **^{1}52** (τ = 12 ps) produces **54** with a broad absorption band centered at 500 nm. The lifetime of **54** is 110 ps in pure formic acid.[202]

Taking into account very recent results the mechanism of phenyl azide photolysis in condensed phase could be described as shown in Scheme 11.29.

Recently, the theoretical analysis of the PES along the reaction coordinate corresponding to the elimination of molecular nitrogen has been performed for phenyl azide and a number of its derivatives.[156] These PES sections were calculated for the ground (S_0) and two excited (S_1 and S_2) states. It was predicted,[156] that the first excited singlet state of aryl azides apparently is dissociative. The second excited state (S_2) is a bound state, and its geometry is similar to that of the ground S_0 state. Moreover, the oscillator strength of the $S_0 \rightarrow S_1$ transition is much lower than that of $S_0 \rightarrow S_2$. Thus, the absorption of the UV light leads to the population of the S_2 state. The characteristic time of the formation of the vibrationally hot singlet nitrene (100–500 fs) most probably corresponds to the internal conversion from the S_2 to the dissociative S_1 state.

It was also found,[156] that the PES of the S_0 and S_1 states intersect. This means that the S_0 state of phenyl azide correlates with ground state of the molecular nitrogen and the upper state of singlet nitrene (the closed-shall $^1A_1$ state). In turn, the phenyl azide S_1 state correlates with ground state molecular nitrogen and the lowest state of singlet nitrene (the open-shall $^1A_2$ state). Consequently, the S_1 state of phenyl azide and some of its

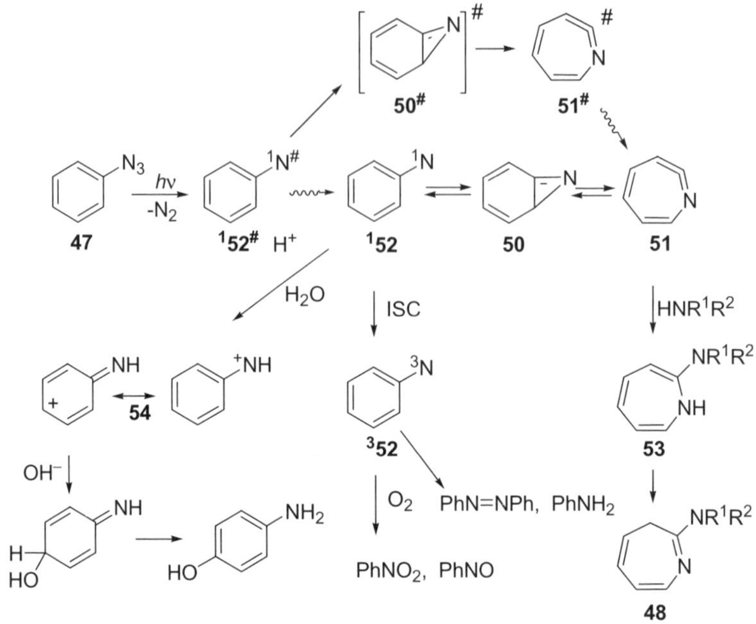

Scheme 11.29 *Full scheme of the condensed phase photochemistry of phenyl azide* ***47***[14,15,17,19,20,158,200–202]

derivatives can undergo very fast relaxation to the ground state through the conical intersection. This explains why, in spite of the very short time of singlet nitrene formation, the quantum yield of the phenyl azide photodissociation is much less than unity. It was measured to be about 0.5 at ambient temperature and in glasses at 77 K.[204–206]

Therefore, the mechanism of phenyl azide photolysis is now understood in detail. Most of the intermediates have been directly detected and their spectroscopy and reactivity have been studied experimentally and analyzed theoretically. It should be noted that great progress has been achieved only in the last decade.

11.6.2 Photochemistry of Simple Derivatives of Phenyl Azide

The photochemistry of simple derivatives of **47** has also been studied in solution using product analysis and time-resolved techniques and in matrices at cryogenic temperatures using spectroscopic methods.[11–13,162] In the last decade, a comprehensive study of the substituent effect on the reactivity of phenylnitrenes have been performed.[154,207–214] The influence of substituents on the spectroscopy and dynamics of singlet phenylnitrene has been reviewed recently.[19,20]

Photolysis of most substituted phenyl azides in hydrocarbons, as in the case of parent **47**, leads to modest yields of identifiable products (azo-benzenes, nitro- and nitroso-benzenes, anilines etc.) along with polymeric tars or resins.[162] Formation of azepines in the presence of primary and secondary amines is also typical of photolysis of the most substituted phenyl azides.[162,183] In some cases, the products of formal bond insertion or

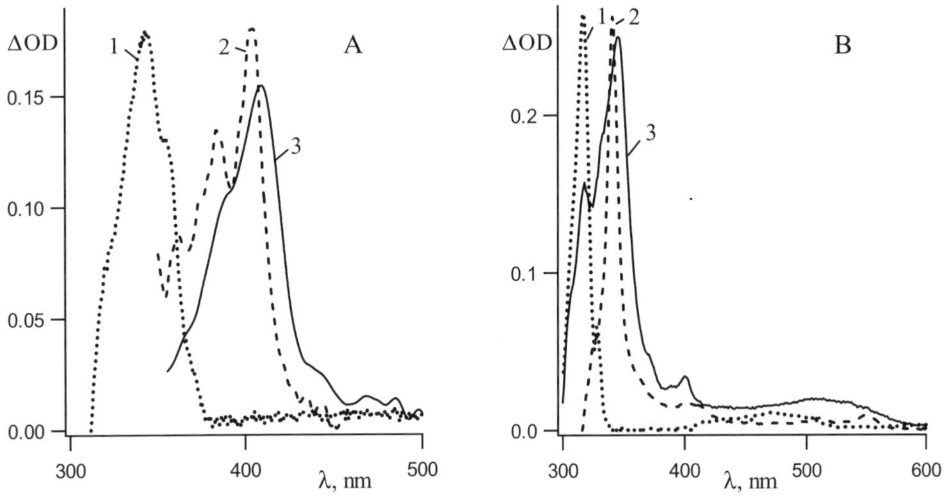

Figure 11.10 *A: Transient absorption spectra recorded after laser excitation (266 nm, 35 ps) of* para*-phenyl (1),* ortho*-dicyano (2) and* ortho*-phenyl (3) substituted phenyl azides in pentane. B: The difference absorption spectra of triplet* para*-phenyl (1),* ortho*-dicyano (2) and* ortho*-phenyl (3) substituted phenylnitrenes in glassy methylcyclohexane at 77 K*

addition to the double bonds were also identified.[12,13,19,162] A number of *ortho*-substituted phenyl azides give clean reactions with high yields of identifiable products of cyclization involving the *ortho*-substituent.[162] The variety of products formed upon the aryl azide photolysis is a result of different reactions of the reactive intermediates – arylnitrenes, azirines and azepines. Therefore, the influence of the substituents on the reactivity of these intermediates will be discussed in this section.

The transient absorption spectra of a series of substituted singlet phenylnitrenes are characterized by an intense absorption band in the near-UV or visible region with maxima at 320–440 nm (Figure 11.10A). Analysis of the data available verifies that the *ortho*-substituents influence the absorption spectra of singlet phenylnitrenes more significantly than do *para*-substituents.[19] Moreover, the shift of the near-UV absorption band of singlet arylnitrenes correlates with the shift of the intense near-UV absorption bands of triplet nitrenes (Figure 11.10A,B).

The values of k_{OBS} in the substituted singlet phenylnitrenes were also measured over a wide temperature range.[154,207–214] As in the case of **^{1}52** (Figure 11.8), the magnitude of k_{OBS} decreases as the temperature decreases, until a limiting value is reached (Figure 11.11). The temperature-independent rate constant, observed at low temperature, was associated with intersystem rate constant – k_{ISC}.

In addition, the k_{ISC} of **^{1}52** and a series of its *ortho*-dialkyl derivatives,[198] as well as *para*-and *ortho*-biphenylnitrenes[154,214] were measured recently in glassy matrices at 77 K. It was demonstrated that the k_{ISC} measured at 77 K and estimated from liquid phase measurements are in very good agreement (Table 11.1). Thus, the value of k_{ISC} is indeed temperature independent and could be estimated from the analysis of the temperature dependence of k_{OBS} (Figure 11.11).

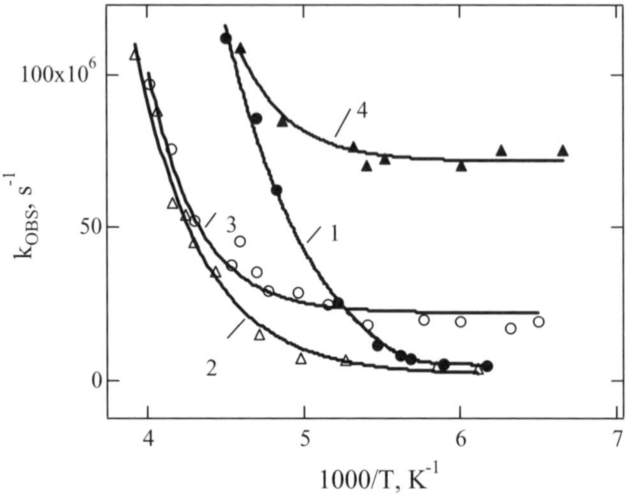

Figure 11.11 *Temperature dependence of the k_{OBS} values of* para-*fluoro (1),* para-*chloro (2),* para-*bromo (3), and* para-*iodo (4) singlet phenylnitrene in pentane. Reprinted with permission from ref.*[208] *Copyright 1999 ACS Publications*

Table 11.1 *Kinetic parameters of* para *substituted singlet aryl nitrenes ($X–C_6H_4–N$) in pentane*

Para-X	τ_{295K}, ns	k_{ISC} ($\times10^6 s^{-1}$)	E_a (kcal/mol)	Log A (s^{-1})	Ref.
H	~1	3.2 ± 0.3	5.6 ± 0.3	13.1 ± 0.3	197
		~3.8[a]			198
CH_3	~1	5.0 ± 0.4	5.8 ± 0.4	13.5 ± 0.2	208
CF_3	1.5	4.6 ±0.8	5.6 ± 0.5	12.9 ± 0.5	208
$C(O)CH_3$	5.0	8 ± 3	5.3 ± 0.3	12.5 ± 0.3	208
F	~0.3	3.5 ± 1.4	5.3 ± 0.3	13.2 ± 0.3	208
Cl	~1	3.9 ± 1.5	6.1 ± 0.3	13.3 ± 0.3	208
Br	~3	17 ± 4	4.0 ± 0.2	11.4 ± 0.2	208
I	b	72 ± 10	b	b	208
OCH_3	<1	>500	b	b	208
CN	8 ± 4	6 ± 2	7.2 ± 0.8	13.5 ± 0.6	212
Ph	15 ± 2	12 ± 1	6.8 ± 0.3	12.7 ± 0.3	214
		9.3 ± 0.4[a]			214
$N(CH_3)_2$[c]	0.12	8300 ± 200	b	b	216
NO_2[d]	<20	>50	b	b	217

[a]Measured at 77 K in 3-methylpentane glassy matrix,
[b]not measured,
[c]in toluene,
[d]n benzene.

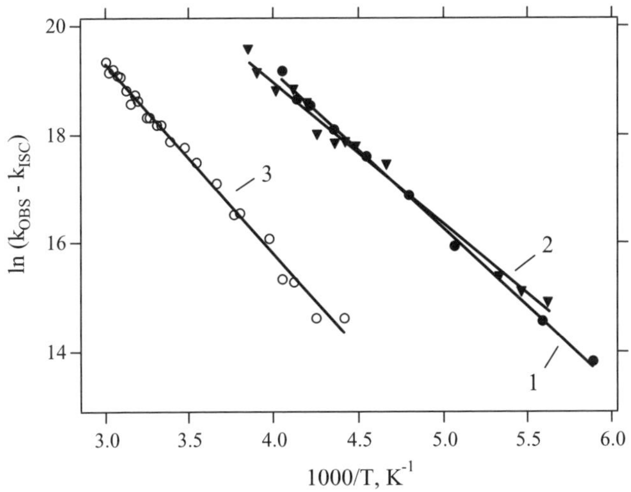

Figure 11.12 *Arrhenius treatment of the $k_R(=k_{OBS} - k_{ISC})$ data for singlet* para-*methyl (1),* ortho-*methyl (2) and* ortho,ortho-*dimethyl- (3) phenylnitrene in pentane. Reprinted with permission from ref.*[209] *Copyright 1999 ACS Publications*

After taking into account that k_{ISC} is temperature independent, plots of ln(k_{OBS}-k_{ISC}) were used to deduce the Arrhenius parameters for cyclization of the substituted singlet arylnitrenes (Figure 11.12, Tables 11.1 and 11.2).

Table 11.1 demonstrates that there is a noticeable heavy atom (Br, I) effect on k_{ISC}. However, the influence of the π donating substituents (OCH_3, $N(CH_3)_2$) is more pronounced. This is consistent with the solution phase photochemistry of *para*-methoxy and *para*-dimethylaminophenyl azides, which largely yield azobenzenes on photolysis.[183] It is interesting to note, that both electron donating and withdrawing substituents accelerate ISC. Noticeable acceleration of ISC (by a factor of 5) was also revealed for *ortho*-alkyl and *ortho,ortho*-dialkyl substituents (Table 11.2).[198,209] Thus, all simple derivatives of phenylnitrene have a k_{ISC} value similar or higher than that of parent ^{1}**52**.

Unfortunately, there is no straightforward explanation for the substituent effect on the ISC rate for arylnitrenes. It was only concluded that the vibrationally-averaged SOC matrix elements are required for a quantitative description of the ISC in arylnitrenes.[215]

Table 11.1 shows also that *para*-substituents have little influence on the rate constant of singlet phenylnitrene rearrangement, k_R. This is not very surprising given that theory predicts that singlet phenylnitrene has an open-shell electronic structure.[106,189–192] Therefore, cyclization of singlet arylnitrenes requires only that the nitrogen bend out of the molecular plane, so that the singly occupied σ non-bonding molecular orbital (NBMO) can interact with the singly occupied π NBMO.[106]Azirine formation is simply the cyclization of a quinoidal 1,3-biradical, which originally has two orthogonal, anti-parallel spins. Thus, large substituent effects are not anticipated.

Note, that similar to the case of **47**, photolysis of a series of *para*- and *meta*-substituted phenyl azides in nitrogen and argon matrices at 12 K yield corresponding ketenimines as major products.[218]

Table 11.2 *Kinetic parameters of ortho-substituted phenylnitrenes in Hydrocarbons*

Substituent	τ_{295K}, ns	k_{ISC} ($\times 10^6\,s^{-1}$)	Log A (s^{-1})	E_a (kcal/mol)	Ref
2-methyl (**59a**)	~1	10 ± 1	12.8 ± 0.3	5.3 ± 0.4	209
2,6-dimethyl (**59b**)	12 ± 1	15 ± 3	13.0 ± 0.3	7.0 ± 0.3	209
2,4,6-trimethyl (**59c**)	8 ± 1	29 ± 3	13.4 ± 0.4	7.3 ± 0.4	209
2,6-diethyl (**59d**)	~9	10 ± 2[a]	12.1 ± 0.5	5.2 ± 0.5	198
2,4,6-tri-*tert*-Bu (**59f**)		6.8	–	–	198
2-fluoro (**62a**)	8 ± 1	3.3 ± 0.5	13.0 ± 0.3	6.7 ± 0.3	213
2,6-difluoro (**62d**)	240	2.4 ± 0.3	11.5 ± 0.5	7.3 ± 0.7	213
2,3,4,5,6,-penta-fluoro (**62e**)	56 ± 4	3.3 ± 1.5	12.8 ± 0.6	7.8 ± 0.6	213
2-cyano	~2	2.8 ± 0.3	12.8 ± 0.3	5.5 ± 0.3	212
2,6-dicyano	~2.3	6.2 ± 0.8	13.5 ± 0.2	6.5 ± 0.4	212
2-pyrimidyl	~13[b]	80 ± 20	–	–	210
2-phenyl	0.016[c]	17 ± 1[a]	–	–	214
2-phenyl-4,6-dichloro	0.26	14 ± 1[a]	11.6 ± 0.2[d]	2.7 ±0.2[d]	154

[a] Measured at 77 K in 3-methylpentane glassy matrix;
[b] 295 K, CH_2Cl_2;
[c] 295 K, CH_3CN;
[d] the effective Arrhenius parameters.

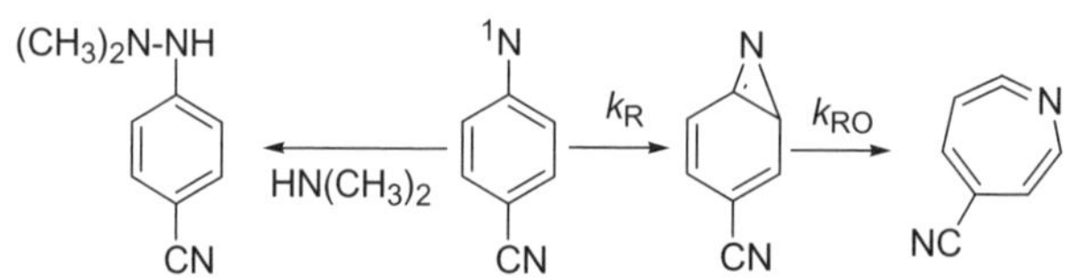

Scheme 11.30 *Reactions of singlet* para-*cyanophenylnitrene*[212,219]

Two *para* substituents, phenyl and cyano, depress k_R and retard the rate of cyclization significantly (Table 11.1). Phenyl and cyano are both radical stabilizing substituents. When attached to the carbon atom *para* to the nitrene nitrogen, these substituents concentrate spin density at this carbon and reduce the spin density at the carbons *ortho* to the nitrene nitrogen. The reduced spin density at carbons *ortho* to the nitrogen atom lowers the rate at which the 1,3-biradical cyclizes. The lifetimes of these singlet nitrenes at ambient temperature are 15 ns (phenyl) and 8 ns (cyano) and the activation barriers to cyclization are 6.8[214] and 7.2 kcal/mol,[212] respectively, compared to 5.6 kcal/mol for parent **¹52**. These results are in quantitative agreement with the CASPT2/6-31G* calculations.[212,214] The longer lifetime of singlet *para*-cyanophenylnitrene (**55**) explains the high yield (>70%) of hydrazine (**58**) observed upon photolysis of *para*-cyanophenyl azide in dimethylamine (Scheme 11.30).[219]

No traces of the 3*H*-azepines were reported among the products of *para*-nitro- and *para*-dimethylaminophenyl azide photolysis in the presence of diethylamine[183] or even in neat dimethylamine.[219a] This agrees with the results of calculations of Cramer and co-authors.[215] In the case of the highly electron-withdrawing NO_2 substituent, the barrier to cyclization was calculated to be 1 kcal/mol higher than that for parent ^{1}**52**. For the highly electron-donating NHMe substituent, this barrier is about 4 kcal/mol higher than that of ^{1}**52**.[215] The predicted reduction of the reactivity, along with much faster ISC (Table 11.1), accounts for the absence of 3H-azepines for these *para*-substituted phenyl azides. Photolysis of *para*-azidoaniline in an argon matrix also yields mainly triplet nitrene.[175] Secondary photolysis of the triplet *para*-aminophenylnitrene gives not only ketenimine, but the corresponding azirine as well.[175]

The influence of *ortho*-substituents on the singlet arylnitrene rearrangement is more pronounced. Thus photolysis of *ortho*-alkyl substituted aryl azides (e.g. *o*-methyl, *o*-ethyl and *o*-isopropyl) in diethylamine affords nucleophilic trapping products that are consistent with initial cyclization of singlet nitrene to the unsubstituted *ortho* carbon only.[220] Murata and Tomioka have observed the tetracyanoethylene trapping of singlet 2,4,6-tri-methylphenylnitrene, as well as of its ring-expansion product.[221] The cyclization of singlet nitrene to the unsubstituted *ortho* carbon only was observed also in the case of *ortho*-fluorophenylnitrene.[222] These results[220–222] demonstrate that steric effects play a role in determining the barrier to ring expansion.

On the other hand, singlet *ortho*-cyano[211,223] and *ortho*-acetylphenylnitrenes[224] undergo cyclization not only away from the substituent, but also toward the cyano or acetyl group. Both steric and electronic effects play important roles in these cases and nearly cancel each other.[212]

For most *ortho*-substituted phenylnitrenes, as well as for *para*-substituted ones,[198,207–214] the cyclization to benzazirine is the rate-determining step of the process of nitrene isomerization to ketenimine (Scheme 11.31), similar to the case of the parent ^{1}**52**. The lifetimes of these singlet nitrenes and Arrhenius parameters for their rearrangement are summarized in Table 11.2.

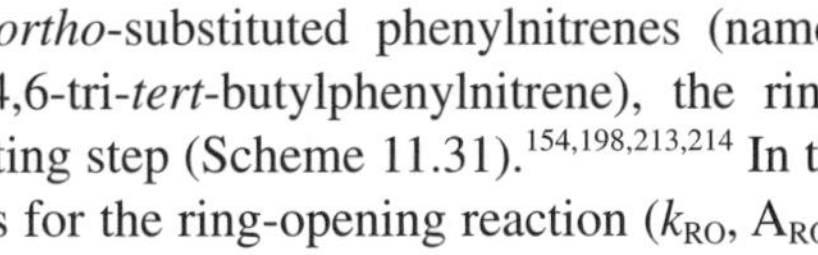

Scheme 11.31 *The rearrangement of substituted phenylnitrenes*

However for a few *ortho*-substituted phenylnitrenes (namely, *ortho*-fluorophenyl-, *ortho*-biphenyl- and 2,4,6-tri-*tert*-butylphenylnitrene), the ring-opening reaction was found to be the rate-limiting step (Scheme 11.31).[154,198,213,214] In the case of these nitrenes, the Arrhenius parameters for the ring-opening reaction (k_{RO}, A_{RO}, E_{RO}) could be obtained (Table 11.3).

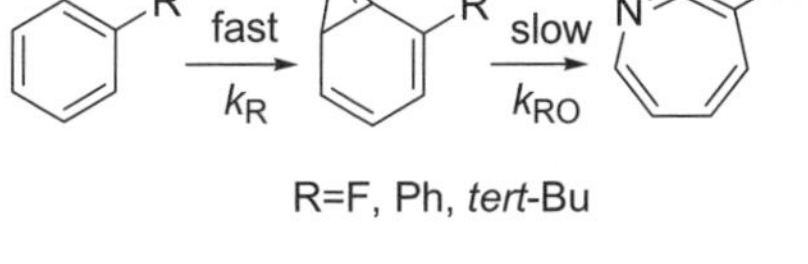

Scheme 11.32 *The rearrangement of some* ortho-*substituted phenylnitrenes*

Table 11.3 *Lifetime of benzazirines, kinetic parameters for their ring-opening reaction in pentane and calculated barriers of this reaction ($\Delta H^{\#}$)*

Substituent	τ_{298} (ns)	log A_{RO} (s^{-1})	E_{RO} (kcal/mol)	$\Delta H^{\#}$ (kcal/mol)	Ref
2,4,6-tri-*tert*-butyl	62 ± 2	12.6 ± 0.2	7.4 ± 0.2	6.3	198
2-fluoro	100 ± 10	13.5 ± 0.4	9.0 ± 0.5	7.0	213
2-phenyl	13 ± 1	12.1 ± 0.1	5.7 ± 0.2	4.7	214
2-phenyl (d_9-analogue)	11 ± 1	12.6 ± 0.1	6.3 ± 0.1	–	214

A detailed kinetic study[209] demonstrated that a single *ortho*-methyl substituent has no influence on the rate of cyclization of the singlet tolylnitrene (**59a**, Table 11.2). In contrast to the case of **59a**, cyclization of di-*ortho*-methyl substituted nitrenes **59b,c** necessarily proceeds towards a carbon bearing a substituent. In the case of **59b,c** the resulting steric effect extends the lifetimes of **59b,c** at ambient temperature to about 10 ns and raises the barrier to cyclization by about 1.5 kcal/mol,[209] in quantitative agreement with the results of CASPT2 calculations of Karney and Borden.[225] Note, that photolysis of *ortho,ortho*-dimethylphenyl azide in a nitrogen matrix at 12 K gives only triplet nitrene.[226]

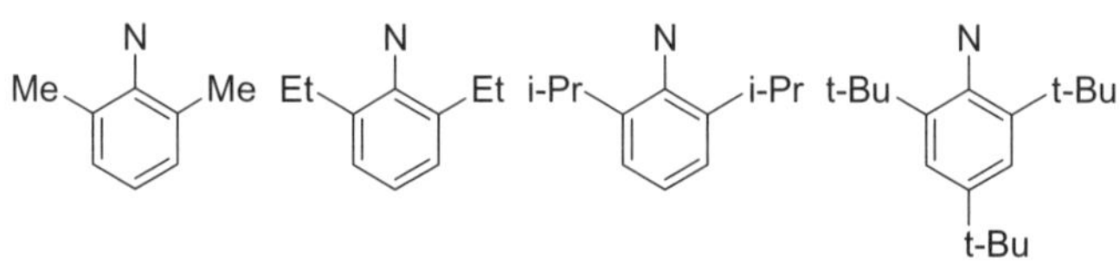

59b **59d** **59e** **59f**

Scheme 11.33 *Structures of the* ortho-*alkyl substituted singlet phenylnitrenes*

Unexpected results were obtained for the rearrangement of singlet arylnitrenes with bulky *ortho*-alkyl substituents (**59d-f**).[198] The lifetimes of the nitrenes **59d** and **59e** were found to be shorter than that of **59b** and singlet nitrene **59f** was not observed in liquid solution due to its very short lifetime. The benzazirine **60f** was detected instead (λ_{max} = 285 nm, Figure 11.13, spectrum 1) and proven to be a precursor of ketenimine **61f** (350 nm, Figure 11.13, spectrum 2). Therefore it was possible to measure the barrier for its ring-opening reaction (Table 11.3).[198]

According to the calculations,[198] the origin of the dramatic drop of the barrier in the case of **59f** is due to the strain released between the nitrogen atom and the alkyl substituent when the nitrogen atom moves away from the substituent during the cyclization (Scheme 11.34).

Theory predicts that the bulky alkyl substituents will alter not only the energy barrier of the first cyclization step but the second ring-expansion step as well.[198] Therefore in the case of **59f**, the rate-determining step is the second ring-expansion reaction as was found experimentally. The calculated barrier (6.3 kcal/mol) is very close to the experimentally determined activation energy of this reaction (E_{RO} = 7.4 ± 0.2 kcal/mol, Table 11.3).[198]

Unlike most arylnitrenes, polyfluorinated arylnitrenes have bountiful bimolecular chemistry (Scheme 11.36).[12,227,228] Therefore, polyfluorinated aryl azides are useful

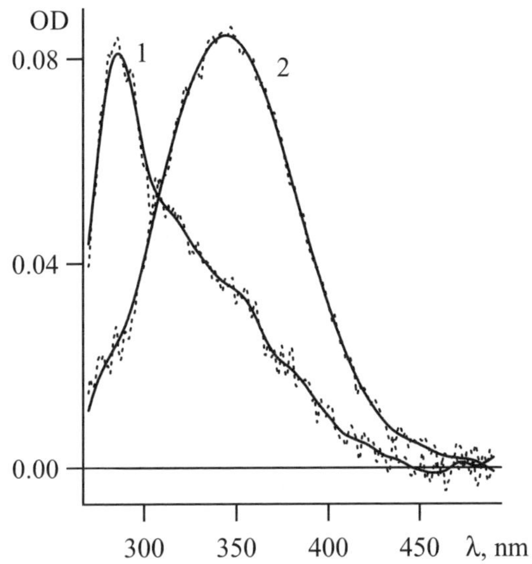

Figure 11.13 *Transient absorption spectra produced upon LFP (266 nm) of 2,4,6-tri-*tert-*butylphenyl azide in pentane at ambient temperature over a window of 10 ns just after the laser pulse (1) and 1 μs after the laser pulse (2). Reprinted with permission from ref.*[20] *Copyright 2006 ACS Publications*

Scheme 11.34 *Effect of the bulky* ortho-tert-*butyl substituent on the reaction of singlet nitrene cyclization*

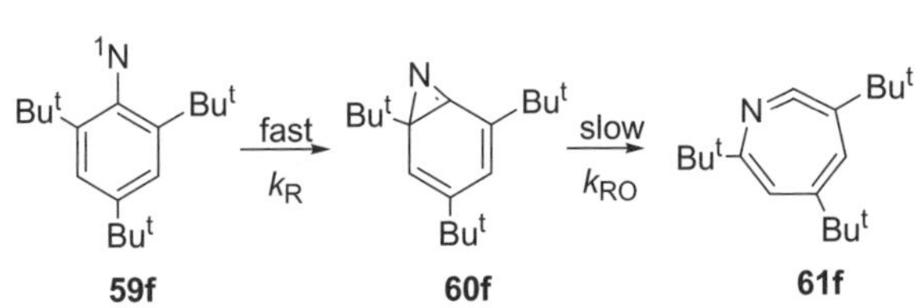

Scheme 11.35 *Two-step rearrangement of the singlet 2,4,6-tri-*tert-*butylphenylnitrene* ***59f***[198]

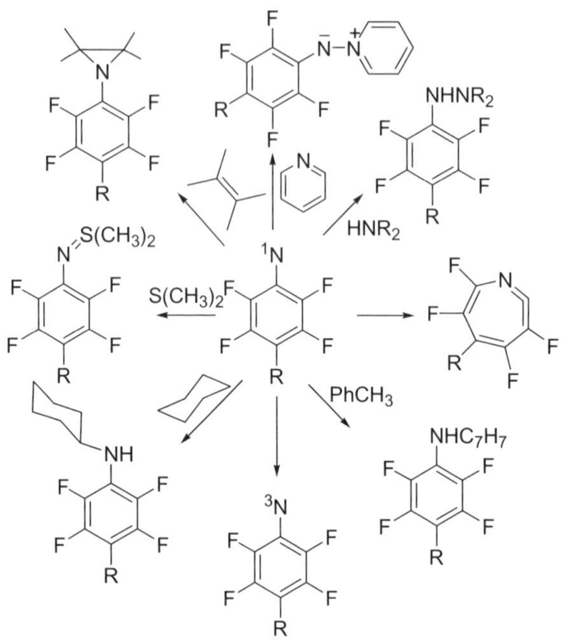

Scheme 11.36 *Products of bimolecular reactions of singlet perfluorophenyl nitrenes*

reagents in synthetic organic chemistry,[229] in photoaffinity labeling,[7,230] and for the covalent modification of polymer surfaces.[9,10]

The effects of the number and positions of fluorine substituents on the ring expansion of phenylnitrene have been extensively investigated by the Platz group.[229,231] They concluded that fluorine substitution at both *ortho* positions is required to inhibit the ring expansion effectively.[12] Similar to the case of *ortho,ortho*-dimethylphenyl azide,[226] photolysis of perfluoro- and *ortho,ortho*-difluorophenyl azides in an nitrogen matrix at 12 K gives only triplet nitrenes.[232–234]

To understand the fluorine effect quantitatively, the kinetics of fluoro substituted phenylnitrenes (**62a-d**, Figure 11.14) was studied[213] and the data were interpreted with the aid of molecular orbital calculations.[213,225]

Although the singlet *ortho*-flurophenylnitrene (**62a**) undergoes cyclization to the unsubstituted *ortho* carbon only,[222] the barrier to this process is larger by ~1 kcal/mol than that of the parent system (Tables 11.1, 11.2). Placement of fluorine substituents at both *ortho* positions (**62d**) raises the barrier to cyclization by about 3 kcal/mol, relative to the unsubstituted system. Both results are consistent with the calculations of Karney and Borden (Figure 1.14).[225]

According to the calculations,[213] the origin of the pronounced influence of *ortho*-fluoro substitution on prolonging the lifetime of singlet arylnitrene **62d** and increasing the activation energy for cyclization is due to a combination of the steric effect and the extraordinary electronegativity of the fluorine atom which reinforce each other.

The nitrene **62a** was the first singlet arylnitrene for which the ring-opening reaction was experimentally found to be the rate-determining step of rearrangement to azepine.[213]

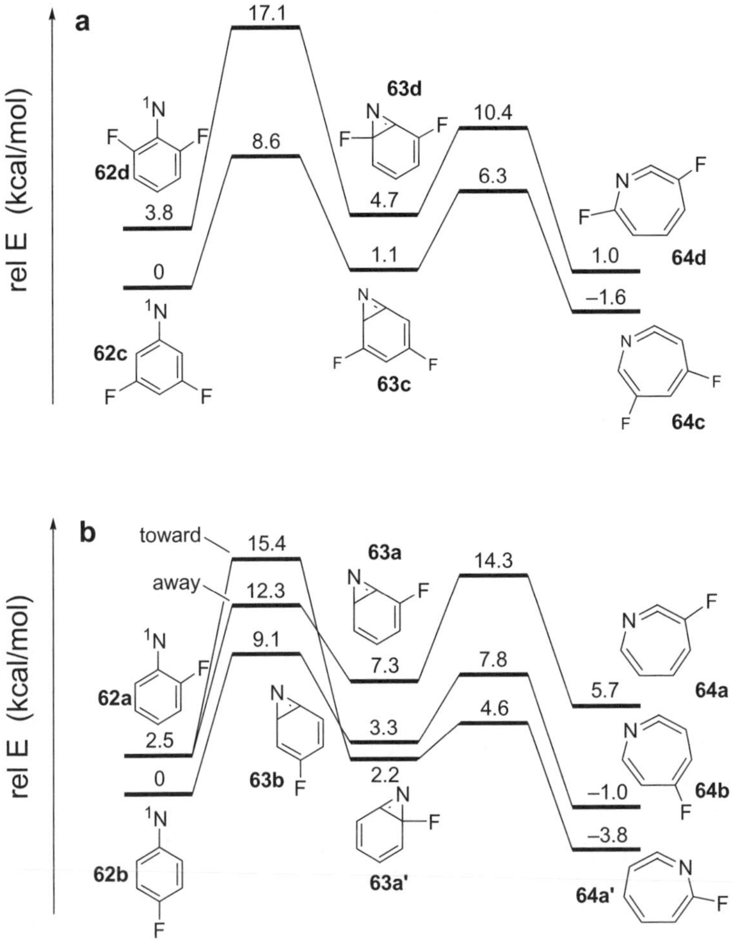

Figure 11.14 *Relative energies (in kcal/mol) of species involved in the ring expansions of singlet fluoro-substituted phenylnitrenes calculated at the CASPT2/cc-pVDZ//CASSCF(8,8)/6-31G* level. (a) Difluorinated phenylnitrenes. (b) Monofluorinated phenylnitrenes. Reprinted with permission from ref.*[213] *Copyright 2001 ACS Publications*

This is consistent with the results of CASSCF/CASPT2 calculations (Figure 11.14).[213] As shown in Figure 11.14, in all cases except the 'away' ring expansion of **62a**, the transition state for the second step of the ring expansion (**63** → **64**) is computed to be lower in energy than that for the first step (**62** → **63**) at the CASPT2 level of theory. The rate constant, k_{RO}, for the ring-opening reaction of **63a** was measured and the Arrhenius parameters were found to be $A_{RO} = 10^{13.5\pm0.4}\,M^{-1}\,s^{-1}$ and $E_{RO} = 9000 \pm 500$ cal/mol (Table 11.3).

The addition of the second fluorine substituent (benzazirine **63d**) decreases the barrier to conversion of azirine **63d** to ketenimine **64d** slightly (Figure 11.14), although the

Table 11.4 *Rate constants of reaction of substituted ketenimines (1,2-didehydroazepines) with diethylamine (DEA) in cyclohexene*[182,183,212]

$XPhN_3$, X	k_{DEA}, $M^{-1}s^{-1}$	$XPhN_3$, X	k_{DEA}, $M^{-1}s^{-1}$	$XPhN_3$, X	k_{DEA}, $M^{-1}s^{-1}$
H	6.5×10^6	*p*-CO_2NMe_2	4.4×10^7	*p*-CN	1.6×10^9
p-Ph	3.6×10^5	*p*-COMe	2.8×10^8	*o*-CN	3.5×10^9
p-SMe	1.6×10^5	*p*-Cl	1.3×10^8	*o,o*-diCN	8.3×10^9
p-OMe	2.5×10^4	*p*-Br	1.7×10^8		
p-CO_2H	3.0×10^7	*p*-I	2.7×10^8		

barrier for **63d** → **64d** is still predicted to be ca. 2.5 kcal/mol higher than the barrier for **50** → **51** at the same level of theory.[106] Note, that formation of the corresponding azirines was observed upon irradiation of triplet perfluoro- and *ortho,ortho*-difluorophenylnitrenes in an argon matrix at 77 K.[233,234]

It was mentioned previously that protonation of phenyl azide ^{1}**52** yields phenylnitrenium ion (Scheme 11.29).[200–202] However, the protonation of ^{1}**52** competes with its isomerization to **51** only at low pH ≤ 1.[200] Surprisingly, the *para*-biphenylnitrene, 2-fluorenylnirene and a series of their derivatives yield nitrenium ions without the added acids.[200,235] The water is the proton-donor in this case. The reactivity of these nitrenium ions has been studied in some detail,[236,237] since the nitrenium ions are proposed to be the DNA-binding intermediates responsible for carcinogenecity of aromatic amines.[236,238]

Along with the substituent effect on the reactivity of singlet phenylnitrenes, the influence of substituents on the reactions of ketenimines with nucleophiles was also studied in detail. As in the case of unsubstituted ketenimine **51**, its simple derivatives could be trapped by nucleophiles in solution.[182–184] The primary products, corresponding 1H-azepines, undergo subsequent isomerization to final products.[162,184] Reaction of ketenimines with primary and secondary amines is the most studied of the reactions with nucleophiles. Rate constants of this reaction with DEA (Table 11.4) were measured for a series of substituted ketenimines using TRIR spectroscopy,[182,183] as well as conventional LFP techniques.[212]

Table 11.4 demonstrates that substituents exert considerable influence on the rate constant of this reaction. The rate constant of reaction with DEA varies over more than 5 orders of magnitude depending on the nature and position of substituents and increases significantly with the electron-withdrawing power of the substituents.

Recently[239] the absolute rate constants of reactions of **51** and a number of its derivatives with typical amino acids, nucleosides and other simple reagents of biological interest were measured in water and HEPES buffer using LFP technique.

It is also known,[183] that ketenimines react with aryl azides, the rate constant of reaction between **51** and **47** is $7.5 \times 10^4 M^{-1} s^{-1}$. At very low concentration of aryl azides, the lifetimes of **51** and of its 5-iodo derivative was measured to be 4–5 ms[183] and 24 ms for the 5-methyl derivative[173] (i.e. $k_{OBS} \approx 40$–$250 s^{-1}$). The latter values represent the rate of irreversible conversion of ketenimines to triplet arylnitrenes. In the absence of nucleophilic agents, photolysis of aryl azides yields typical products of triplet arylnitrene reactions – azo compounds and anilines in the absence of oxygen[12,13,162,166,167,183] and nitro and

Table 11.5 *Rate constant of reaction of triplet* para-*substituted phenynitrenes* ($X\text{-}C_4H_4\text{-}N^3$) *with oxygen at ambient temperature*[172,173,241]

X	H	CH_3	NO_2	NH_2	NH_2
k, $10^6 M^{-1} s^{-1}$	2.4 ± 0.1	1.9 ± 0.2	0.8 ± 0.1	4.5 ± 1.2	8 ± 2
Solvent	CH_3CN	CH_3CN	C_6H_{12}	C_6H_{12}	$PhCH_3$

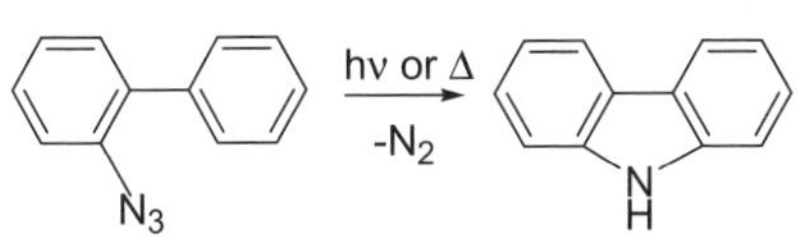

65a **66a**

Scheme 11.37 *Photolysis and thermolysis of* ortho-*biphenyl azide*

nitroso compounds in the presence of oxygen.[13,172–175] The decrease of aryl azide concentration leads to the growth of detectable products.[166,167,172,173]

Reaction of triplet arylnitrenes with oxygen and subsequent reactions of primary intermediates – arylnitrosooxides, are reviewed earlier[13,240] and will not be discussed here. The rate constant of reaction of triplet arylnitrenes with molecular oxygen are presented in Table 11.5. It appeared that the rate constants are substantially lower than the diffusion limit and are in the range $(0.8 - 8) \times 10^6 M^{-1} s^{-1}$. For **^{3}52**, the temperature dependence for this reaction in acetonitrile was measured and the Arrhenius parameters were estimated ($Ea = 4.3 \pm 0.5$ kcal/mol, $A = 10^{9.6\pm0.4} s^{-1}$).[241]

Alkyl, cyano, acetyl and fluoro substituents in the *ortho*-position do not change the mechanism of phenyl azide photochemistry influencing only the rate constants of elementary reactions (k_{ISC}, k_R, k_{RO}, k_{NUC}). At the same time, a number of photochemical and thermal cyclizations involving the *ortho*-substituents are known for *ortho*-substituted phenyl azides.[162] The most interesting, important and well understood reaction of this type is formation of carbazoles **66** on pyrolysis[242] and photolysis.[154,184a,214,243,244] of *ortho*-biphenyl azide **65a** and a series of its derivatives (Scheme 11.37).

Since the early 1970s, the reactive intermediates involved in the transformation of **65** to **66** were studied by trapping,[184a,244] matrix spectroscopy,[169,177] and flash photolysis.[244b,245] Swenton, Ikeler, and Williams[243] demonstrated that carbazole is derived from reaction of a singlet state species, presumably singlet nitrene **^{1}67a**, whereas triplet nitrene **^{3}67a** dimerizes to form azo compound **68**.

It was also demonstrated that, in the presence of DEA, photolysis of azide **65a** leads to the formation of 3H-azepine **71a** (Scheme 11.38), with a concomitant reduction in the yield of carbazole **66a**.[183,244a] The carbazole formation was measured at its absorption maximum (289.4 nm),[244b,245] and a rate constant was found to be $2.2 \times 10^3 s^{-1}$ at 300 K in cyclohexane.[243] This value is about 5–6 orders of magnitude lower than the rate constants of singlet arylnitrenes rearrangement (Tables 11.1 and 11.2). Therefore, the following scheme (Scheme 11.38) could economically describe the early results.[242–245]

Recently the mechanism of carbazole formation upon photolysis of *ortho*-biphenyl azide (**65a**), its deutero- (**65a-d$_9$**) and dichloro (**65b**) derivatives was studied in detail

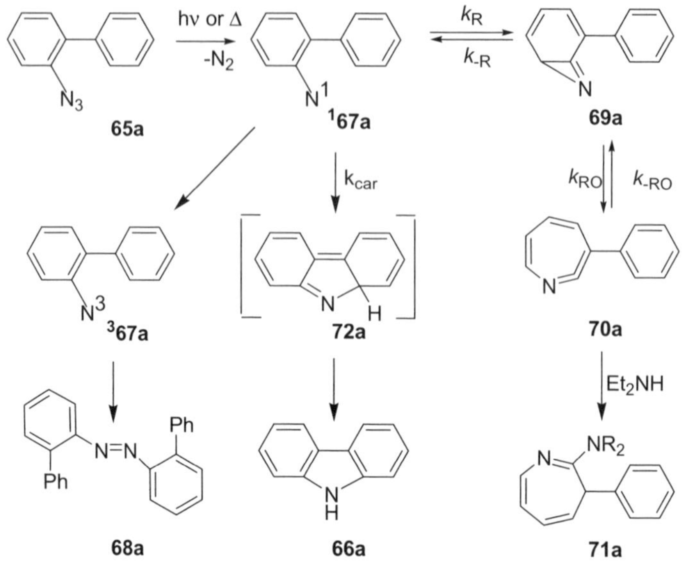

Scheme 11.38 *Mechanism of photolysis of ortho-biphenyl azide proposed based on the early studies*[243–245]

using nanosecond laser flash photolysis,[154,214] time-resolved IR,[158,214] femtosecond transient absorption spectroscopy[154–156] and computational chemistry.[154,156,214]

LFP data obtained at room temperature[214] demonstrate that, in agreement with previous flash photolysis studies,[244b,245] carbazole **66a** is mainly formed on the millisecond time scale in pentane. Moreover, the characteristic ketenimine IR band was detected at $1868\,cm^{-1}$ in CD_3CN. This band appeared faster than the time resolution of the apparatus (~100 ns). The decay of this band was accompanied by the appearance of the carbazole band at $1241\,cm^{-1}$ with a rate constant of $1.0 \pm 0.2 \times 10^4\,s^{-1}$, which is close to the value measured by Sundberg *et al.*[244b] The TRIR experiments thus demonstrate,[214] that ketenimine **70a** does indeed serve as a source of **66a** on the longer time scale, via the mechanism shown in Scheme 11.38.

In addition to the formation of **66a** on the millisecond time scale, discussed previously,[244b,245] some formation of **66a** was detected on the nanosecond time scale as well.[214] The ns growth of carbazole absorption at 290 nm was accompanied by the ns decay of a transient absorption in the visible region between 400 and 500 nm. The time constants for the growth and decay functions are equal to 70 ± 5 ns in pentane at ambient temperature. The precursor of **66a** was assigned to isocarbazole **72a** (Scheme 11.38).[214]

LFP of perdeuterated azide **65a-d₉** at ambient temperature[214] demonstrated a pronounced kinetic isotope effect on the kinetics of carbazole formation on the ns time scale ($k_H/k_D = 3.4 \pm 0.2$), which is consistent with the reaction being the isomerization of isocarbazole **72a** into carbazole **66a** by a 1,5-hydrogen shift. In addition methanol and water were found to accelerate the disappearance of the transient absorption of **72a** produced upon LFP of **65a** in pentane.[214] A reasonable mechanism for this catalysis is shown in Scheme 11.39.

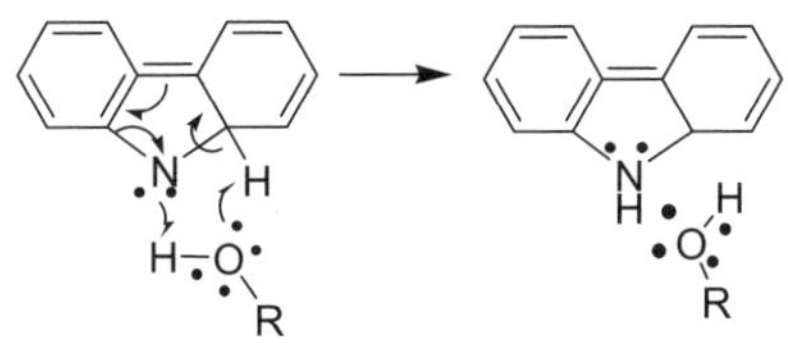

Scheme 11.39 *Catalysis of isocarbazole isomerization by water*[214]

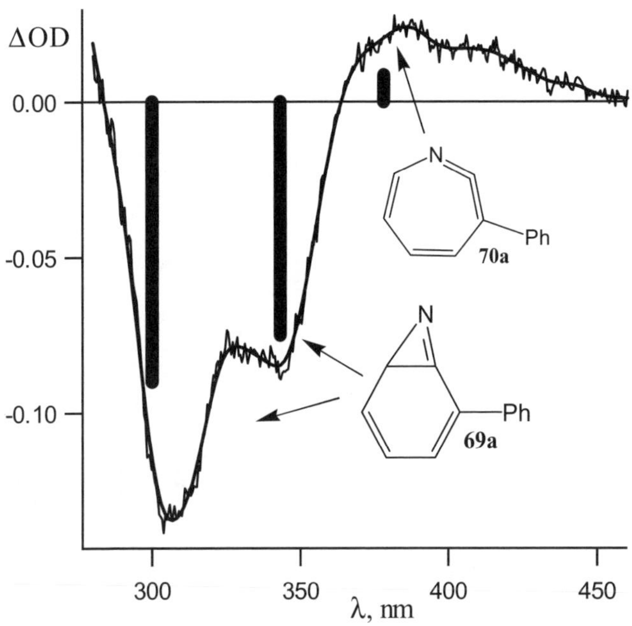

Figure 11.15 *The difference spectrum obtained by LFP of* ortho-*biphenyl azide* **65a** *in pentane at 161 K. The computed positions and relative oscillator strengths of the absorption bands of benzazirine* **69a** *and azepine* **70a** *are depicted as solid vertical lines (negative and positive, respectively)*

Benzazirine **69a** has strong absorptions with maxima at ~305 and 340 nm and was observed upon the LFP of **65a** at low temperature (e.g. 160 K). Decay of **69a** was accompanied by azepine **70a** formation with absorption at 350–400 nm (Figure 11.15).[214]

In Freon-113 at ambient temperature, the lifetime of azirine **69a** is 12 ± 2 ns, which is about 6 times shorter than that for isocarbazole **72a** disappearance and carbazole **66a** formation. The temperature dependence of the observed rate constants for the decay of azirines **69a** and **69a-d₉** were studied and the activation parameters were found to be $E_{RO} = 5.7 \pm 0.1$ kcal/mol and $A_{RO} = 10^{12.1\pm0.1}$ s^{-1} for **69a** and $E_{RO} = 6.3 \pm 0.1$ kcal/mol and $A_{RO} = 10^{12.6\pm0.1}$ s^{-1} for **69a-d$_9$** in satisfactory agreement with DFT calculations (Table 11.3).

At least one additional long lived intermediate absorbing in the range 350–450 nm was observed upon LFP of **65a**.[214,244] According to the DFT calculations,[214] isocarbazole **72a**

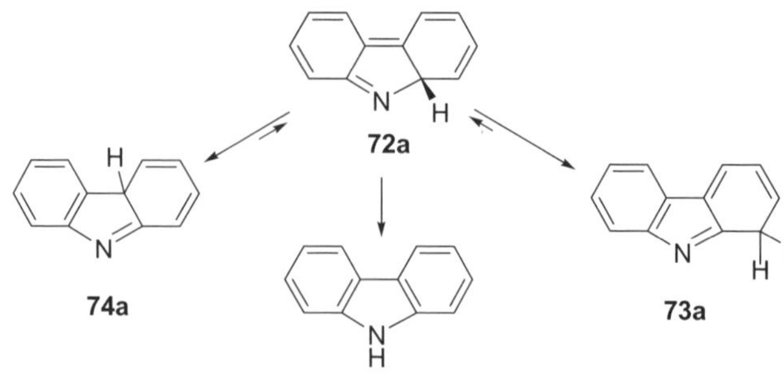

Scheme 11.40 *Mechanism of carbazole formation on a longer time scale*

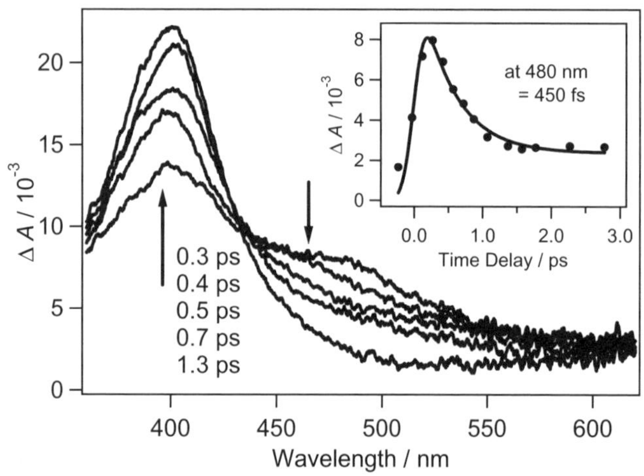

Figure 11.16 *Transient absorption spectra recorded between 0.3–1.3 ps after the laser pulse for* ortho-*biphenylyl azide in acetonitrile. The time dependence of the signal measured at 480 nm is shown as inset. Reprinted with permission from ref.*[156] *Copyright 2006 ACS Publications*

can undergo exothermic 1,5-hydrogen shifts to form not only carbazole **66a**, but isomeric isocarbazoles **73a** and **74a** (Scheme 11.40) as well. Both of these isocarbazoles were predicted to have intense absorption around 360 nm. Presumably, subsequent 1,5-shifts in **73a** and **74a**, reform **72a**, and eventually yield more carbazole **66a**, seconds to minutes after the laser pulse (Scheme 11.40).[214] Therefore the formation of carbazole is not only biphasic, but is most probably triphasic.

The spectrum (λ_{max} = 410 nm) and kinetics of singlet *ortho*-biphenylnitrene ^{1}**67a** were recorded by LFP of **65a** in glassy 3-methylpentane at 77 K. The lifetimes of ^{1}**67a** and ^{1}**67-d$_9$** at 77 K are equal to 59 ± 3 ns and 80 ± 2 ns, respectively.[214] A similar spectrum was detected recently at room temperature using femtosecond transient absorption spectroscopy (Figures 11.16 and 11.17).[155,156]

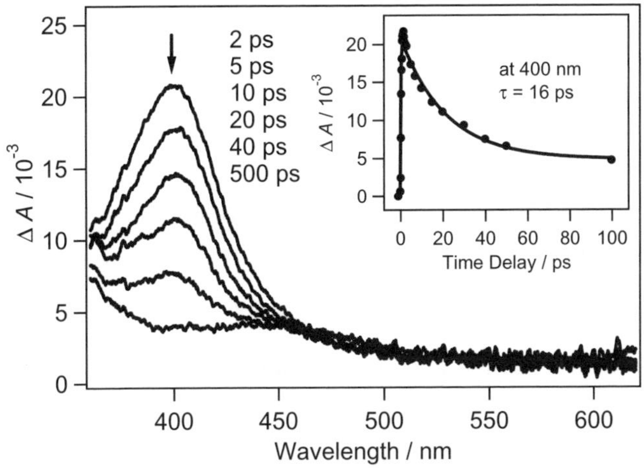

Figure 11.17 *Transient absorption spectra recorded between 2-500 ps for* ortho-*biphenylyl azide in acetonitrile. The time dependences of the signal measured at 400 nm is shown as inset. Reprinted with permission from ref.*[156] *Copyright 2006 ACS Publications*

It was found that the absorption of singlet nitrene **^{1}67a** grows with a time constant 280 ± 150 fs and decays with time constant 16 ± 3 ps. Note, that as in the case of phenyl azide **47**, the quantum yield of photolysis of **65a** is significantly less than unity (about 0.44 at ambient temperature and in glasses at 77 K).[204] Therefore, the S_1 state of **65a**, similar to the case of parent system **47**, undergoes very fast relaxation to the ground state through the conical intersection.

The 16 ps time constant ($k = 6.3 \times 10^{10}\,s^{-1}$) represents the population decay time of singlet nitrene **^{1}67a** by isomerization to isocarbazole **72a** and benzazirine **69a** with the latter process being predominant. Assuming that the pre-exponential factor for cyclization of **^{1}67a** is ~$10^{13}\,s^{-1}$, the activation energy could be estimated as ~3 kcal/mol. This value is in excellent agreement with the (14/14) CASPT2/6-31G*//(14,14)CASSCF/6-31G* calculations (Figure 11.18),[214] if one takes into account the typical underestimation by ~3.4 kcal/mol[106] of the energy of open-shell **^{1}67**. The *ortho*-phenyl group lowers the enthalpy of activation for cyclization, compared to parent phenylnitrene **^{1}52**, by destabilizing singlet nitrene **^{1}67a** sterically, as in the case of *ortho-tert*-butyl substituent (Sheme 11.34).[198] According to the calculations[214] both **69a** and **72a** should be formed from **^{1}67a**, however azirine **69a** should be the kinetically favored product around room temperature in accord with experiment.

It was mentioned above that formation of ketenimine **70a** from azirine **69a** proceeds on a nanosecond time scale ($\tau \sim 12$ ns in Freon 113).[214] However, formation of vibrationally hot **70a** was also detected on a ps time scale in CH_3CN using time resolved IR spectroscopy.[158] Fits to the kinetic traces indicated that ketenimine **70a** is formed with a time constant of ~10 ps and undergoes vibrational cooling with a time constant of ~29 ps. These data demonstrate that as in the case of parent nitrene **^{1}52**, the singlet nitrene **^{1}67a** is born on a fs time scale with excess vibrational energy and isomerizes to vibrationally excited 1,2-didehydroazepine **70a** on a ps time scale.

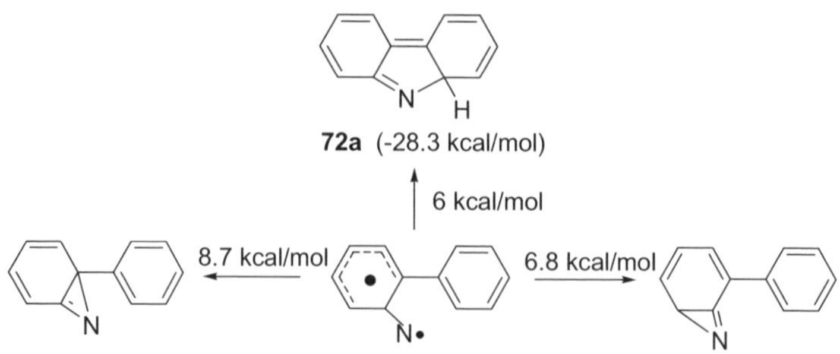

Figure 11.18 Structures of the intermediates formed upon cyclization of singlet ortho-biphenylnitrene (^{1}**67a**), their electronic energies relative to the nitrene ^{1}**67a** (in parenthesis) and the energy differences between the transition states (TSs) and ^{1}**67a**. Reprinted with permission from ref.[20] Copyright 2006 ASC Publications

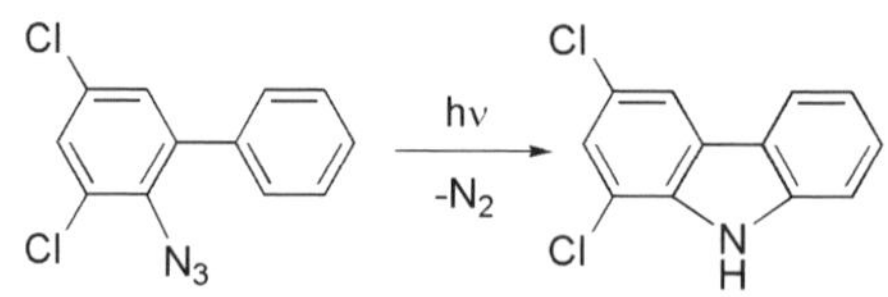

Scheme 11.41 *Photochemistry of 2-azido-3,5-dichlorobiphenyl (**65b**)*

Therefore, Scheme 11.38 represents the oversimplified mechanism of carbazole **66a** formation upon photolysis of *ortho*-biphenyl azide **65a** and should be supplemented with the processes described in Schemes 11.39 and 11.40, as well as by fast formation of **70a** from a hot singlet nitrene ^{1}**67a**.

The chemistry of the unsubstituted *ortho*-biphenyl system is complicated by the fact that the key intermediate of this reaction, ^{1}**67a**, undergoes two cyclization processes at competitive rates, and that azepine formation is reversible. To simplify the chemistry of singlet nitrene and allow straightforward study of the isocarbazole formation, 2-azido-3,5-dichlorobiphenyl (**65b**) was synthesized (Scheme 11.41) and its photochemistry was studied using nano- and picosecond transient absorption spectroscopy.[154]

Indeed, the chlorinated carbazole **66b** is produced predominantly on the nanosecond time scale and only to a minor extent from the corresponding didehydroazepines **70b** and/or **70b′** (Scheme 11.42).[154] The transient absorption in the visible region (Figure 11.19) was assigned to a mixture of two intermediates with maxima at 470 nm, 440 and 425 nm, respectively. One of the intermediates is isocarbazole **72b** (λ_{max} = 320 and 470 nm). Its lifetime in pentane at room temperature is 65 ± 4 ns and is 263 ± 4 ns for perdeuterated analogue, similar to the case of **72a**.[154] The second intermediate (**75b**, λ_{max} = 360, 420 and 440 nm) unfortunately could not be identified.[154]

As in the case of **65a**, the transient absorption spectrum of singlet nitrene ^{1}**67b** was detected at ambient temperature using a fs transient absorption spectroscopy and at 77 K

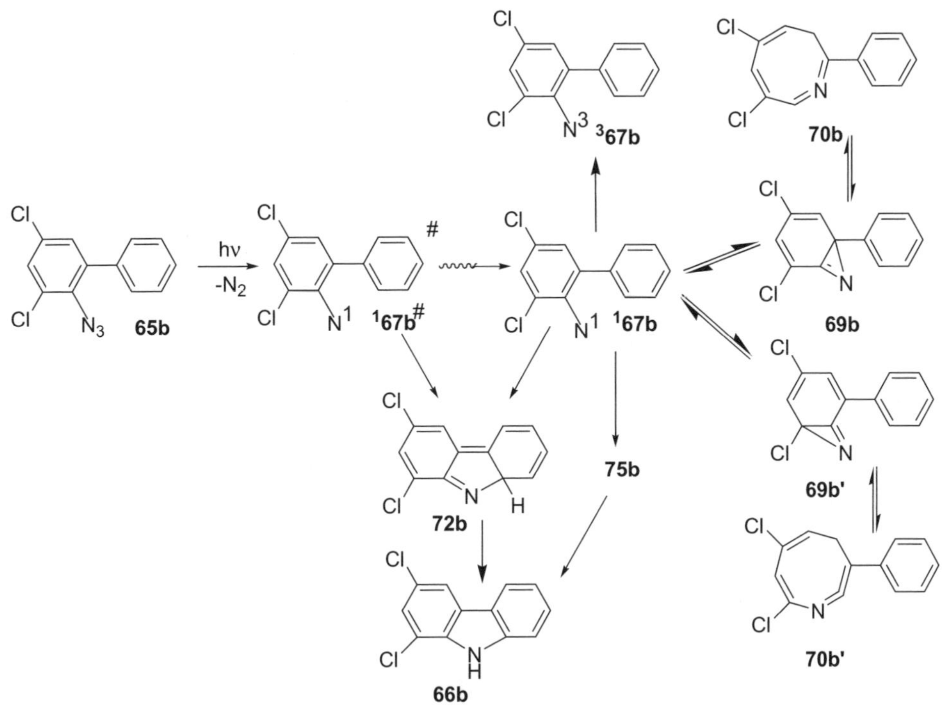

Scheme 11.42 Mechanism of photolysis of 2-azido-3,5-dichlorobiphenyl (**65b**)

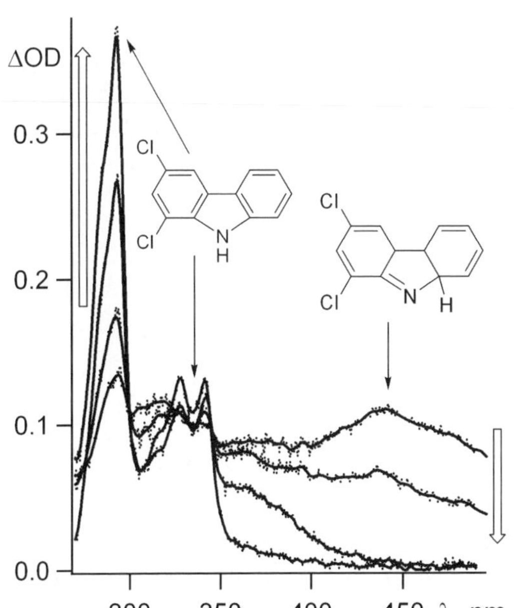

Figure 11.19 *Transient absorption spectra detected over a window of 10 ns following LFP of 2-azido-3,5-dichlorobiphenyl* ***65b*** *in pentane at ambient temperature 15 ns, 60 ns 1 μs and 30 ms after the laser pulse*

by conventional LFP.[154] The spectrum detected in cyclohexane 40 ps after the laser pulse was a mixture of spectra of at least two species – singlet nitrene ^{1}**67b** and isocarbazole **72b**. It was proposed, that a part of isocarbazole **72b** is formed from vibrationally hot singlet nitrene ^{1}**67b**. The time constant for vibrational relaxation of ^{1}**67b** was estimated to be 11 ± 2 ps.

The decay of vibrationally cooled ^{1}**67b** was accompanied by the growth of isocarbazole **72b** with the rate constant $k_{OBS} = 3.8 \pm 0.8 \times 10^{9}\,s^{-1}$ ($\tau = 260 \pm 70$ ps).[154] A considerable acceleration of the singlet nitrene ^{1}**67b** rearrangement was observed in methanol. The rate constant of its decay was found to be $1.6 \pm 0.2 \times 10^{10}\,s^{-1}$ ($\tau = 62 \pm 10$ ps).[154] Nevertheless, this value is about 4 times lower than that for ^{1}**67a** in CH_3CN ($6.3 \times 10^{10}\,s^{-1}$).[155,156] The main contribution to the latter process is the reaction of the azirine formation (^{1}**67a** → **69a**). This process is significantly retarded by *ortho*-chlorine substitution and isocarbazole formation gives the main contribution to the decay of ^{1}**67b**.

The decay of singlet nitrene ^{1}**67b** in hydrocarbon solutions was measured in three different types of experiments and the Arrhenius parameters for the rate constant of ^{1}**67b** rearrangement were estimated to be: $E_a = 2.7 \pm 0.2$ kcal/mol and $A = 10^{11.6\,\pm\,0.2}\,s^{-1}$.[154] The measured activation energy is in perfect agreement with the predicted barrier to isocarbazole formation (~3 kcal/mol).[154,214]

Unexpectedly, the yield of isocarbazole **72b** was found to depend on the energy of the photons used for the excitation of azide **65b**.[154] It drops significantly on going from excitation by a YAG (266 nm) to excimer (308 nm) laser radiation.[154] It is in line with the observation, that in part the isocarbazole **72b** is formed from vibrationally hot singlet nitrene ^{1}**67b**.

Recently,[157] very fast intramolecular cyclization of singlet nitrene was observed upon photolysis of *ortho*-nitrophenyl azide (**76**). It is well known,[246] that pyrolysis and photolysis of **76** leads cleanly to benzofuroxan (**77**). According to the results of recent computational study[247] and early experiments,[246a,b] the pyrolysis of **76** produces **77** by a concerted one-step mechanism. However, photolysis of **76** produces **77** through a stepwise mechanism.[157] Formation of singlet nitrene ^{1}**78** from excited **76** was detected to occur with a time constant ~500 fs. The lifetime of nitrene ^{1}**78** is very short – 8.3 ps, and corresponds to ring-closure reaction rate constant $1.2 \times 10^{11}\,s^{-1}$.

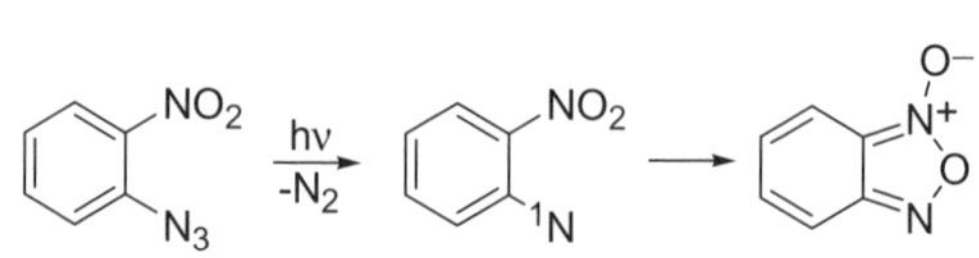

76 ^{1}**78** **77**

Scheme 11.43 *Mechanism of photolysis of* ortho-*nitrophenyl azide*

Therefore, the photochemical cyclizations of *ortho*-substituted aryl azides involving the *ortho*-substituents (such as formations of carbazoles upon photolysis of *ortho*-biphenyl azides and of benzofuroxan upon photolysis of *ortho*-nitrophenyl azide) are found to occur by stepwise mechanism with intermediacy of singlet nitrenes. The cyclizations of singlet nitrenes occur on a picosecond time scale.

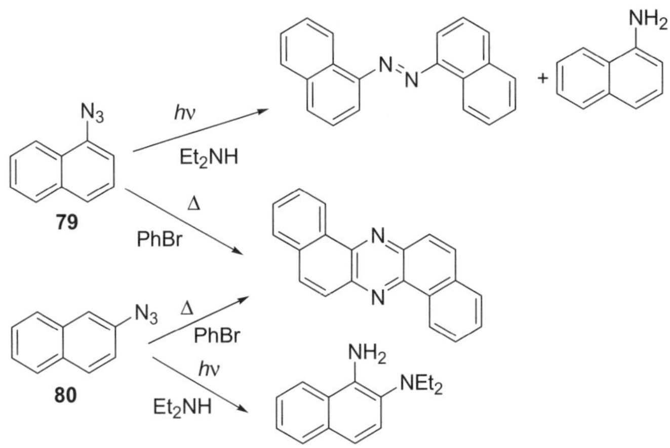

Scheme 11.44 *Photolysis and pyrolysis of 1- and 2-azidonaphthalene*

11.6.3 Photochemistry of Polynuclear Aromatic Azides

Fusion of a benzene ring to another aromatic ring or rings changes the electronic structure of the modified nitrenes and related intermediates sufficiently enough to alter their chemistry, kinetics and thermodynamics. Thus photolysis of polynuclear aromatic azides in the presence of primary and secondary amines leads not to 3H-azepines, but to corresponding diamino-products instead.[162,170,248–250] Nevertheless, in some cases photolysis and thermolysis of polynuclear aromatic azides yields the products (azo-compounds, nitro-compounds etc.) typical of photochemistry of phenyl azides.[11,151,162,169–171,251] For instance, photolysis of polycyclic aromatic azides in glassy matrixes at 77 K gives corresponding arylnitrenes in the ground triplet state, as was demonstrated by the EPR spectroscopy.[170,171,251] In solution, dimerization of triplet polynuclear aromatic nitrenes or their reaction with starting azides produces azo-compounds.[169–171] The photochemical and thermal reactions of polycyclic aromatic azides have been reviewed periodically.[1,11,162,252] Therefore, recent results will be mainly discussed in this chapter.

Much effort has been devoted to the study of the photochemistry of 1- and 2-naphthyl azides (**79** and **80**). The products obtained upon pyrolysis and photolysis of the naphthyl azides were reported in the 1970s and 1980s.[165,248–250,253,254] In 1974, the Suschitzky group[248] discovered that pyrolysis of **79** and **80** in bromobenzene yields a significant amount of dibenzo[a,h]phenazine (Scheme 11.44). The photolytic decomposition of **80** in DEA leads to a diamine product.[248] On the contrary, photolysis of **79** in DEA produces mostly azonaphthalene and aminonaphthalene in low yields[248] along with a very low yield of diamine adduct.[250]

Carroll *et al.* found that the yield of diamine product is sensitive to the photolysis time.[254a] A drastic reduction in the photolysis time leads to a much-improved yield of the diamines. It was also discovered that the yield of diamine products formed upon photolysis of the 1- and 2-naphthyl azides can also be significantly improved by the presence of $(Me_2NCH_2)_2$ (or TMEDA) as co-solvent.[249]

Leyva and Platz demonstrated[250] that reaction temperature plays an important role in the photochemistry of **79**, as with the case of parent phenyl azide **47**. Moderate yields of adducts were observed by simply lowering the temperature of the photolysis of **79** with DEA.

The photoproduct studies mentioned above suggest the intermediacy of azirines in the photochemistry of **79** and **80**. Additional evidence of these intermediates was provided by the observations of adducts in the photolysis of **80** with ethanethiol[254b] and with methanolic methoxide.[253a]

Direct evidence for azirines and didehydroazepines was obtained by Dunkin and Thomson upon UV-irradiation ($\lambda > 330$ nm) of **79** and **80** in nitrogen or argon matrices at 12 K.[255] In the case of **79** the IR band at 1730 cm^{-1} was formed upon initial photolysis and was assigned to tricyclic azirine. The IR bands at 1926 and 1912 cm^{-1}, formed on secondary photolysis of the tricyclic azirine, were attributed to didehydroazepine intermediates. Similarly, photolysis of **80** produces IR bands at 1708, 1723 and 1736 cm^{-1} assigned to corresponding azirines. The IR bands at 1911, 1923 cm^{-1} were formed on secondary photolysis and assigned to azepines. However, a detailed assignment has not been performed.

Recently,[256] the photochemistry of azides **79** and **80** in an argon matrix was reinvestigated and new assignments of the experimental UV-vis and IR spectra of the species observed were presented on the basis of quantum chemical calculations. The primary products were found to be the corresponding triplet nitrenes ^{3}**81** and ^{3}**82**. According to the new assignment,[256] the peaks at 1710–1740 cm^{-1}, observed by Dunkin and Thompson,[255] belong to azirines **83** and **84** (Figures 11.20, 11.21). The peaks at 1910–1930 cm^{-1}, observed on prolonged irradiation, were assigned to azepines **85** and **86**. In addition, evidence was presented for the formation of novel ring expansion products, the cyclic ylides **87** and **88** (Scheme 11.45).

A comprehensive computational study of the potential energy surfaces on the rearrangements leading to ylides **87** and **88** was performed (Figures 11.20 and 11.21).[256] It was predicted that singlet 1-naphthyl nitrene (^{1}**81**) cyclizes selectively at the beta carbon with the formation of azirine **83**. The azirine **90** derived from cyclization at carbon atom 9 is predicted to be very high in energy (Figure 11.20).

Argon matrix photolysis of 1- and 2-naphthyl azides **79** and **80** at 313 nm initially produced the singlet naphthyl nitrenes, ^{1}**81** and ^{1}**82**. Relaxation to the corresponding lower energy, persistent triplet nitrenes ^{3}**81** and ^{3}**82** competes with cyclization to the azirines **83** and **84** which can also be formed photochemically from the triplet nitrenes (Figures 11.20 and 11.21). On prolonged irradiation, the triplet nitrenes ^{3}**81** and ^{3}**82** can be converted to the 7-membered cyclic ketenimines **85** and **86**, respectively, as described earlier by Dunkin and Thomson.[255]

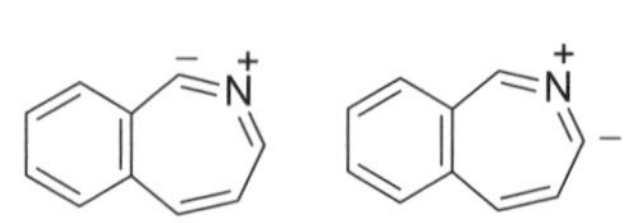

87 **88**

Scheme 11.45 *Structures of cyclic ylides*

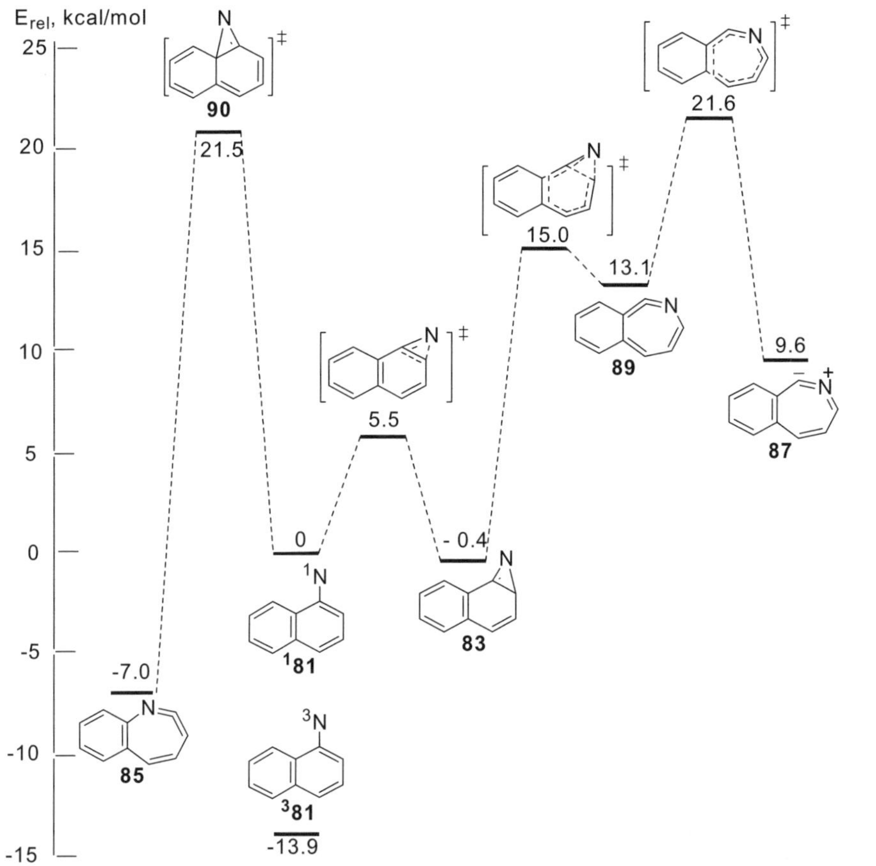

Figure 11.20 The relative energies of valence isomers of 1-naphthylnitrene **¹81** calculated at the CASPT2//CASSCF(12,12)/6-31G* level.[256] All energies include zero-point energy corrections. Note, that **90** is a transition state by DFT, but a shallow minimum by CASSCF

However, instead of the *o*-quinoid ketenimines **89** and **91**, which are the expected primary ring-opening products of azirines **83** and **84**, respectively, the novel bond-shift isomers **87** and **88** were observed, which may be formally regarded as cyclic nitrile ylides. The existence of such ylidic heterocumulenes had been predicted previously.[257]

The photochemistry of naphthyl azides **79** and **80** in solution at ambient temperature has been studied using LFP[170] and TRIR[258] techniques and the femtosecond transient absorption spectroscopy.[156,160,259] LFP study of naphthyl azide photochemistry in glassy solvents at 77 K has also been performed.[258]

In an early ns time resolved study,[170] no transient absorption above 350 nm was observed immediately after pulsed laser excitation of **79**. However, the transient absorption spectrum of triplet nitrene (^{3}**81**, λ_{max} = 370 nm) was observed a few microseconds after the laser pulse. The intensity of the transient absorption of ^{3}**81** increased exponentially with a time constant of 2.8 μs in benzene at ambient temperature. The ^{3}**81** decay monitored within 100 μs of the laser pulse was accompanied by the concurrent formation of

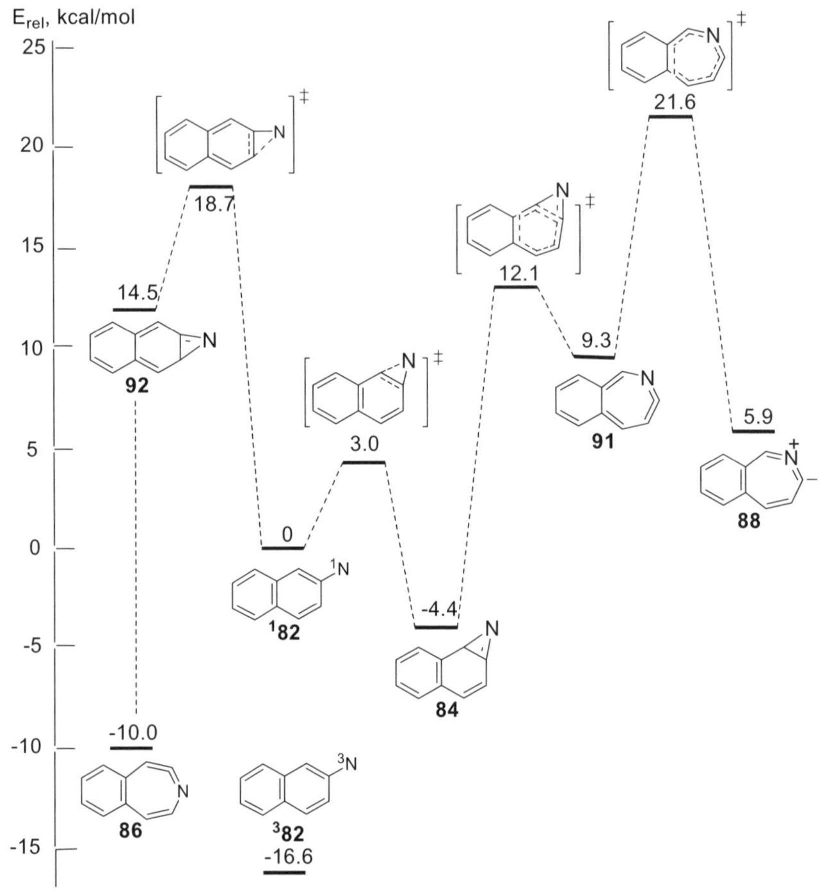

Figure 11.21 *Relative energies of valence isomers of singlet 2-naphthylnitrene* ***¹82*** *calculated at the CASPT2//CASSCF(12,12)/6-31G* level.*[256] *All energies include zero-point energy corrections*

1,1′-azonaphthalene (**93**, λ_{max} = 420 nm), by the second-order reaction with a rate constant of $1 \times 10^9\,M^{-1}\,s^{-1}$. The precursor to **³81** was assigned to tricyclic azirine **83** (Scheme 11.46), a species which serves as a reservoir for singlet 1-naphthylnitrene **¹81**.[170]

The expected azirine **83** with lifetimes of 3.2 ± 0.6 μs was detected by TRIR spectroscopy.[258] The lifetime of **83** was also measured in solvent mixtures of 1 : 3 and 1 : 1 DEA/acetonitrile. The absolute bimolecular rate constant of reaction of **83** with DEA was estimated to be ~$1.4 \times 10^5\,M^{-1}\,s^{-1}$, a value much smaller than that of a diffusion-controlled process. The slow rate of reaction of **83** with DEA is consistent with the reported low yield of DEA photo-adduct observed upon photolysis of **79**.[249]

LFP of **79** in glassy 3-methylpentane at 77 K produces singlet nitrene **¹81**, which is characterized by a structured band in the near-UV region with maxima at 362, 383 and 397 nm in agreement with calculations.[258] At 77 K the singlet nitrene **¹81** cleanly relaxes to the lower energy triplet nitrene **³81** with $k_{ISC} = 1.1 \pm 0.1 \times 10^7\,s^{-1}$.

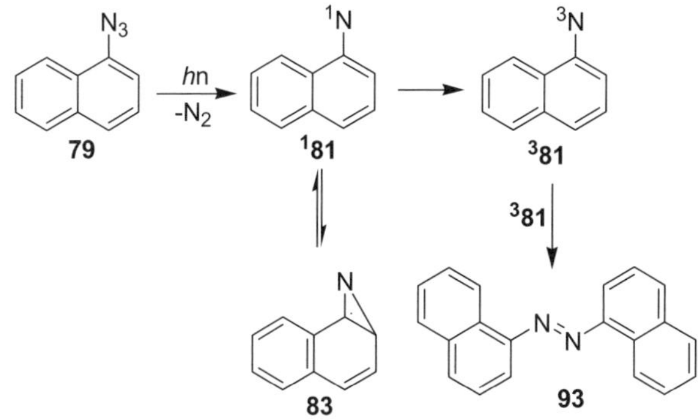

Scheme 11.46 *Photochemistry of 1-naphthyl azide **79** in solution at ambient temperature*

A similar transient absorption spectrum (λ_{max} = 385 nm) was detected at ambient temperature in acetonitrile using ultrafast LFP and assigned to singlet nitrene **¹81**.[156] The absorption of singlet nitrene **¹81** grows with time constant ~730 fs and decays with time constant of 12 ps ($k = 8.3 \times 10^{10}\,s^{-1}$) to form naphthazirine **83**.[156] Assuming that the pre-exponential factor for cyclization of **¹81** is ~$10^{13}\,s^{-1}$, the activation energy could be estimated as ~3 kcal/mol. This value is in reasonable agreement with the CASPT2//CASSCF(12,12)/6-31G* calculations (Figure 11.20),[256] after taking into account the typical underestimation by ~3 kcal/mol[106] of the energy of open-shell **¹81** in the calculation.

Although the singlet nitrene **¹81** is very short-lived, its protonation leading to the formation of 1-naphthylnitrenium cation was observed in 88% formic acid using ultrafast LFP.[160] The rate of formation of this cation was equal to the rate of **¹81** decay (τ = 8.4 ps). The lifetime of 1-naphthylnitrenium ion is 860 ps in 88% formic acid.

According to the RI-CC2/TZVP calculations,[156] the S_2 excited state of 1-naphthyl azide **79** is a bound state with an equilibrium geometry similar to that of the ground S_0 state (Figure 11.22). On the contrary, the S_1 excited state, best characterized as $\pi \rightarrow$ (in plane, π^*, azide) excitation, is dissociative toward the formation of molecular nitrogen and the singlet nitrene **¹81**, although a small barrier (~2 kcal/mol) for N_2 expulsion is predicted. Figure 11.22 demonstrates also an S_0/S_1 crossing when the N–N coordinate is about 1.65 Å. A similar So/S_1 crossing may account for the low quantum yield of photolysis of parent azide **47** and its *ortho*-phenyl derivative **65a**.[204] However, in the case of azide **79** the quantum yield of photolysis is close to unity at ambient temperature and at 77 K.[204]

The calculations also predict[156] that for azide **79**, as well as for azide **47** and biphenyl azides, the oscillator strength of $S_0 \rightarrow S_2$ transition is much larger than that of $S_0 \rightarrow S_1$. Thus UV excitation of these aryl azides is predicted to promote the ground state azide to the S_2 state, which rapidly converts to the dissociative S_1 state. Therefore, the time constant of arylnitrene formation does not relate to the N–N bond cleavage, but instead to the internal conversion from S_2 to S_1 state.

The nanosecond LFP[170] of 2-naphthyl azide **80** gave results similar to that of **79**, except that in this case the transient absorption of triplet nitrene **³82** was not detected and the formation of 2,2′-azonaphthalene was found to be approximately 20 times slower than that of **93**. Naphthazirine **84** was detected in acetonitrile at ambient temperature using

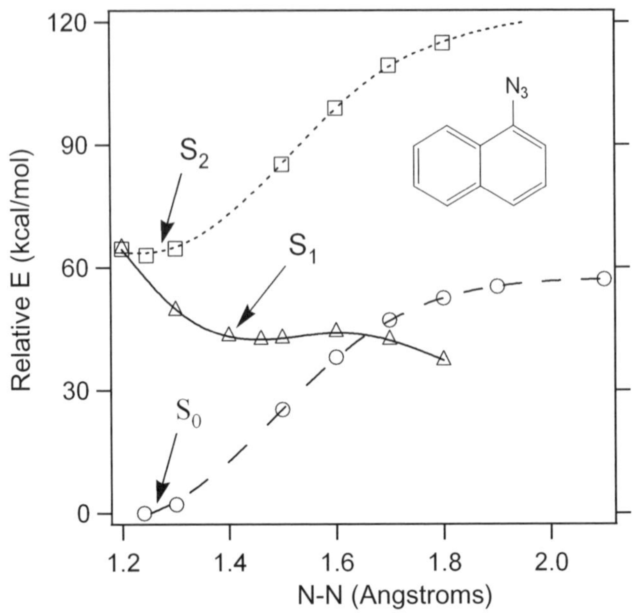

Figure 11.22 *RI-CC2/TZVP fully relaxed potential energy curves for N_2 expulsion in 1-naphthyl azide **79**: ground state (open circles), first excited state (open triangles), and second excited state (open squares).[156] The arrows indicate the N–N bond length of the fully optimized stationary point on each respective energy surface*

TRIR spectroscopy.[258] Its lifetime was determined to be 150 ± 10 μs in excellent agreement with early observation of a slow rate of growth of 2,2′-azonaphthalene.[170] The longer lifetime of **84** (comparing to **83**) is consistent with theoretical calculations as well (Figure 11.21), since **84** is predicted to be 4.4 kcal/mol more stable than singlet nitrene ^{1}**82**, whereas the cyclization of ^{1}**81** to azirine **83** is essentially thermoneutral (Figure 11.20).[256]

The rate constants of azirine **84** decay were determined in the presence of 25% and 50% DEA in acetonitrile at ambient temperature. The absolute bimolecular rate constant of reaction of azide **84** with DEA was estimated as ~$2.5 \times 10^5 M^{-1} s^{-1}$. This rate constant is slightly faster but is still similar to that of azirine **83** ($1.4 \times 10^5 M^{-1} s^{-1}$). Thus, trapping of **84** by DEA is still the dominant process since it has a long lifetime (150 μs) in the absence of the amine trap. This is consistent with the reported high yields of adducts formed upon photolysis of **80** in the presence of secondary amines.[249,248,253,254] A DEA adduct is inefficiently formed upon photolysis of **79** at ambient temperature because of the much shorter lifetime of naphthazirine **83** (~3 μs).[170,258]

In the ultrafast LFP study of **80**, the formation of two intermediates have been observed in acetonitrile at ambient temperature.[259] One of the species, the S_2 excited state of **80** (λ_{max} ~ 350 nm) has a lifetime within the instrument response (300 fs) due to its rapid convertion to the dissociative S_1 state. The second intermediate, singlet nitrene ^{1}**82** (λ_{max} ~ 420 nm) has the shortest lifetime of any singlet arylnitrenes observed to date – 1.8 ps ($k = 5.6 \times 10^{11} s^{-1}$).[259] Assuming that the pre-exponential factor for cyclization of ^{1}**81** is ~$10^{13} s^{-1}$, the activation energy to cyclization could be estimated as ~1.7 kcal/mol.

This value is in reasonable agreement with the CASPT2//CASSCF(12,12)/6-31G* calculations (Figure 11.21),[256] after taking into account the expected underestimation (~3 kcal/mol)[106] of the energy of open-shell **^{1}81** in this calculation.

Unlike **79**, LFP of **80** in a glassy matrix at 77 K failed to produce a detectable transient absorption above 320 nm.[258] On the contrary, the triplet nitrene **^{3}82** is a primary product of 2-naphthyl azide photolysis in argon matrix at 12 K. Note, that triplet nitrene **^{3}82** absorbing at 365 nm was generated and detected by triplet sensitization.[170] To reconcile the observations obtained at 12 and 77 K, it was proposed, that for **^{1}82** $k_R \gg k_{ISC}$ at 77 K, but that $k_{ISC} \gg k_R$ at 12 K. If one assumes that $k_{ISC} = 1.1 \times 10^7\,s^{-1}$ and $A = 10^{13}\,s^{-1}$ for the cyclization reaction, then the barrier to cyclization, E_a, can be bracketed as follows: $0.3 < E_a < 2.1$ kcal/mol. This result is consistent with the above mentioned estimation from fs time-resolved experiments ($E_a \sim 1.7$ kcal/mol) and theoretical predictions (Figure 11.21).[256]

The photochemistry of other polynuclear aromatic azide has been investigated in less detail than that of the naphthyl azides, although some azides have been studied to a certain extent. For instance, photolysis of 1-, 2-, and 9-azidoanthracenes in organic matrices at 77 K yields the corresponding triplet nitrenes, whose electronic absorption spectra[171,177,204,251d,260] and EPR spectra[251a,d] were recorded in the 1960s. The expected azo-compound was formed upon irradiation of 1-azidoanthracene in ethanol.[169] The azo dimer was formed in the reaction of triplet 1-nitrenoanthracene with starting azide with a rate constant $5 \times 10^8\,M^{-1}\,s^{-1}$.[169] The product distributions observed upon photolysis of 2-azidoanthracene in the presence of nucleophiles resemble those formed upon photolysis of 2-azidonaphthalene (**80**) and provide evidence for the intermediacy of both azirine and triplet nitrene intermediates.[253]

The photochemistry of 9-azidoanthracene (**94**) in solution has not been explored by chemical analysis of reaction mixtures. However, the mechanism of photolysis of **94** was investigated recently by laser flash photolysis and computational chemistry.[261] LFP of **94** produces singlet nitrene **^{1}95** with a lifetime of about 20 ns at both ambient temperature and 77 K. Thus the lifetime of **^{1}95** is controlled by intersystem crossing to the ground triplet state **^{3}95**. The rapid rate of ISC of **^{1}95** relative to that of singlet phenylnitrene (**^{1}52**) is in agreement with the calculated value of the singlet-triplet splitting (ΔE_{ST}) of this nitrene.[261] The ΔE_{ST} of **95** was predicted to be 5.3 kcal/mol (CASSCF/CASPT2 procedure),[261] which is much smaller than that for parent **^{1}52** (~18 kcal/mol)[106,189–194] and naphthyl nitrenes **81** (13.9 kcal/mol) and **82** (16.6 kcal/mol).[256] The absence of cyclization of **^{1}95** to form a bridgehead **96** is also in agreement with calculations that indicate that the conversion of **^{1}95** to **96** is endothermic by 23 kcal/mol.[261]

More attention was devoted to the photochemistry of 1-pyrenyl azide (**97**).[170,171,262–264] Irradiation of **97** in deoxygenated benzene solution gives a high yield of 1,1′-azopyrene

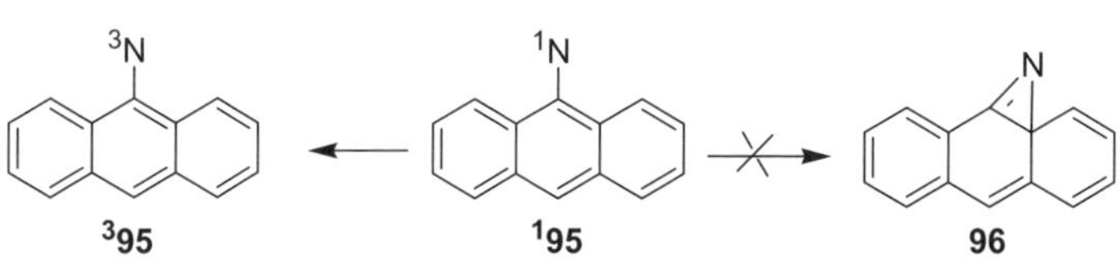

^{3}95 **^{1}95** **96**

Scheme 11.47 *Primary processes in the photochemistry of 9-azidoanthracene **94***

(**98**).[170,171] When **97** was photolyzed in DEA solution with a low-power continuous light source, a large yield (82%) of 1-aminopyrene (**99**) was obtained. High-power laser excitation of **97** gave a large yield of **98** and only a 4% yield of **99**.[170] Triplet sensitization of the decomposition of **97** in benzene containing 1 M DEA at low irradiation power gave **99** in 55% yield.[170]

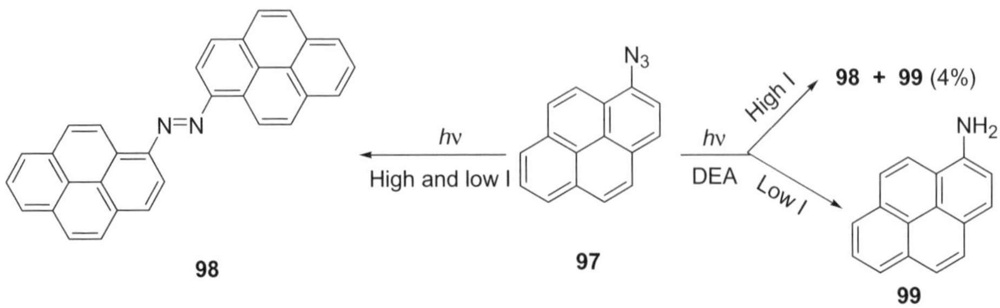

Scheme 11.48 *Products of 1-pyrenyl azide photolysis in the presence and absence of DEA and at high-power laser (High I) and low-power continuous light (Low I) sources*[170,171]

In 1976, Sumitani, Nagakura and Yoshihara demonstrated that LFP of **97** produces a transient species with absorption maximum at 455 nm, which disappeared with a time constant of 22 ns at room temperature and with 34 ns at 77 K and decayed to a species with absorption maximum at 420 nm.[264] The latter species disappeared in a second-order process to give **98**. This spectrum (λ_{max} = 420 nm) was also formed on triplet sensitization and was identical to that of triplet 1-nitrenopyrene (^{3}**100**) observed in a glassy matrix at 77 K. Therefore, the species with the 420 nm absorption maximum was assigned to triplet nitrene ^{3}**100** and its precursor with maximum at 455 nm – to singlet nitrene ^{1}**100**. This assignment is secure after taking into account all recent results on the spectroscopy and dynamics of singlet arylnitrenes,[14,15,19,20] although it was questioned in the 1980s.[170] Note, that singlet 1-pyrenylnitrene (^{1}**100**) was the first singlet arylnitrene directly detected by spectroscopic methods.[264]

Thus the results of a ns time resolved LFP study of 1-pyrenyl azide are very similar to that of 1-azidoanthracene. On the contrary, the photochemistry of 2-pyrenyl azide (**101**, Scheme 11.49) is quite similar to that observed for 2-naphthyl azide **80**.[170,256,258] Irradiation in benzene gives primarily 2,2′-azopyrene (**102**).[170] In the presence of DEA ($HNEt_2$), the product is almost exclusively 1-amino-2-diethylaminopyrene (**103**) and it reaches its maximum value at a very low concentration of DEA (4×10^{-3} M).[170]

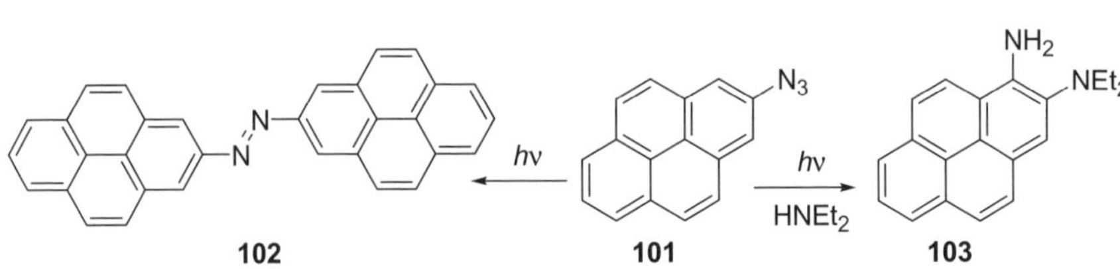

Scheme 11.49 *Photolysis products of 2-pyrenyl azide (**101**)*

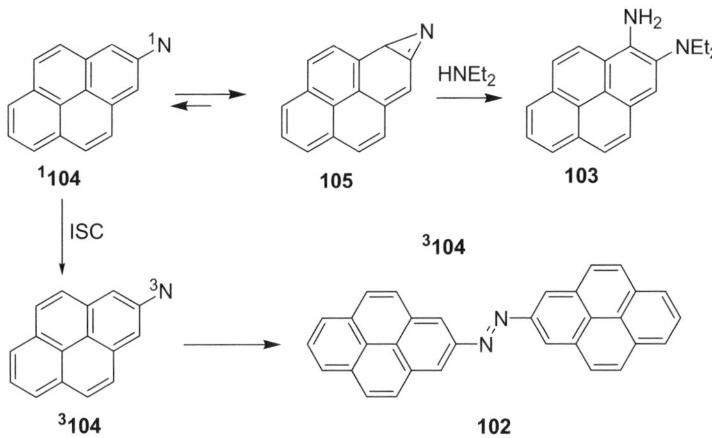

Scheme 11.50 *Proposed mechanism of 2-pyrenyl azide photochemistry*

LFP of **101** produces a transient absorption with maximum at 420 nm.[170] On a longer time scale the absorption of 2,2′-azopyrene (**102**) grows by a second-order process. The 420 nm absorption band was also formed by irradiation of **101** at 77 K in glassy matrix. The EPR spectrum of triplet 2-nitrenepyrene (^{3}**104**) was detected at 77 K and at 4 K.

The same absorption band was observed in DEA as a solvent. It disappeared in a first-order process with a time constant of 2.5 μs with formation of **103**. The authors assigned the 420 nm absorption to triplet 2-nitrenopyrene (^{3}**104**) and proposed that the conversion of singlet nitrene ^{1}**104** to its triplet ground state (^{3}**104**) is reversible.[170] However, the authors expressed some skepticism in this assignment.

Taking into account all recent results and especially the recent data on the ns time resolved LFP study of 2-naphthyl azide (**80**),[258,259] the following mechanism of 2-pyrenyl azide photolysis can be proposed (Scheme 11.50). In our opinion the transient absorption at 420 nm should be assigned to azirine **105**, which reacts with DEA to give **103** and serves as reservoir for triplet nitrene (^{3}**104**), which undergo dimerization or react with azide **101** to give azopyrene **102**.

Therefore, as in the case of parent phenyl azide **47** and its simple derivatives, the photochemistry of polynuclear aromatic azide, especially that of naphthyl azides **79** and **80**, is now well understood. Specifically, the dynamics of the primary photophysical processes as well as the subsequent photochemical steps have been directly investigated using a variety of modern and conventional experimental techniques and computational chemistry. It is clear now, that the difference between the photochemistry of phenyl azide (and its simple derivative) and polynuclear aromatic azide is caused mainly by the difference in the thermodynamics of the singlet nitrene rearrangement to azinine type species.

11.7 Conclusion

Over the last 15 years there has been dramatic progress in our understanding of the chemistry of simple acyl and aryl azides. These advancements are a direct result of the blossoming of computational methods, nano second time resolved spectroscopy and matrix isolation techniques. As we have described in this review the photolysis of azides leads to the extrusion of molecular nitrogen and the release of singlet nitrenes which relax to form a variety of secondary and tertiary intermediates. This field has matured to the point that it is now possible to use nanosecond spectroscopy (and spectroscopic methods with faster or slower time resolution) to directly observe the seminal singlet nitrenes released upon azide photolysis and the intermediates derived from them. It is possible to measure activation barriers to the intramolecular and intermolecular processes that these species undergo. Computational methods complement the experimental techniques by predicting the potential energy surfaces and simulating the UV-Vis and IR spectra of the intermediates of interest. This work has sharpened our intuition. It is possible to use the structure-reactivity data base to qualitatively predict the behavior of the intermediates produced upon photolysis of a simple azide. Thus, this field is close to reaching scientific maturity if it has not done so already.

The new challenge is to develop the same level of understanding of the first 100 ps that follow the excitation of an azide, using femto and picosecond time resolved spectroscopy. The new goal is to understand the nature of the excited state populated by the absorption of light as a function of wavelength and to discover which excited states are dissociative and which are not. The challenge is to develop a map detailing the production of an excited state and how it relaxes to a dissociative state. The singlet nitrene so formed will surely be born with excess vibrational energy. Another goal is to learn if the 'hot' nitrene undergoes chemistry prior to relaxation to a thermalized nitrene and the role of structure, wavelength and solvent on these events. Chemists have long deduced that acyl azide excited states can undergo a concerted Curtius Rearrangement and bypass a nitrene route to isocyanates. The combination of theory and ultrafast time resolved spectroscopy should also give new insights into acyl azide excited state chemistry. A decade from now we hope to provide a review of these explorations and develop an intuition for structure-wavelength-reactivity issues on the ultrafast timescales that rivals what we have developed for the nanosecond realm.

Acknowledgments

Support for this work by the National Science Foundation, the Russian Foundation for Basic Research and the Siberian Branch of Russian Academy of Sciences is gratefully acknowledged.

References

[1] *Nitrenes*; W. Lwowski, ed.; John Wiley & Sons, Inc., New York, **1970**.
[2] *Azides and Nitrenes – Reactivity and Utility*; E.F.V. Scriven, Ed.; Academic Press, New York, **1984**.
[3] (a) *The Chemistry of the Azido Group*, S. Patai, ed.; John Wiley & Sons, Inc., New York, **1971**. (b) *The Chemistry of Halides, Pseudo-Halides and Azides*, Supplement D, S. Patai, Z. Rappoport, eds.; John Wiley & Sons Ltd, Chichester, **1983**. (c) *The Chemistry of Halides, Pseudo-Halides and Azides, Part 1 and 2*, S. Patai, ed.; John Wiley & Sons, Ltd, Chichester, **1995**.

[4] S. Bräse, C. Gil, K. Knepper, V. Zimmermann, *Angew. Chem. Int. Ed.* **2005**, *44*, 5188–5240.
[5] A. Singh, E. R Thornton, F. H. Westheimer, *J. Biol. Chem.* **1962**, *237*, 3006–8.
[6] (a) H. Bayley, J.R. Knowles, *Methods Enzymol.* **1977**, *46*, 69–114. (b) H. Bayley, *Photogenerated Reagents in Biochemistry and Molecular Biology*. Elsevier, New York, **1983**.
[7] (a) J.F. Wang, W.D. Downs, T.R. Cech, *Science* **1993**, *260*, 504–8. (b) J.L. Chen, J.M. Nolan, M.E. Harris, N.R. Paxe, *EMBO J.* **1998**, *17*, 1515–25. (c) R. Pinard, J.E. Heckman, J.M. Burke, *J. Mol. Biol.* **1999**, *287*, 239–51. (d) K.G. Pinney, M.P. Mejia, V.M. Villalobos, *et al.*, *Bioorg. Med. Chem.* **2000**, *8*, 2417–25. (e) K.L. Buchmueller, B.T. Hill, M.S. Platz, K.M. Weeks, *J. Am. Chem. Soc.* **2003**, *125*, 10850–61.
[8] D.S. Breslow, in *Azides and Nitrenes – Reactivity and Utility*; E.F.V. Scriven, ed.; Academic Press, New York, **1984**.
[9] E.W. Meijer, S. Nijhuis, F.C.B.M. Von Vroonhoven, *J. Am. Chem. Soc.* **1988**, *110*, 7209–10.
[10] (a) S.X. Cai, D.R. Glenn, J.F.W. Keana. *J. Org. Chem.* **1992**, *57*, 1299–1304. (b) H. Niino, Y. Koga, A. Yabe, *J. Photochem. Photobiol. A: Chemistry* **1997**, *106*, 9–13. (c) H. Niino, T. Sato, A. Yabe, *Appl. Phys. A* **1999**, *69*, 605–10. (d) M.D. Yan, *React. Func. Polymers* **2000**, *45*, 137–44. (e) C. Henneuse-Boxus, E. Duliere, J. Marchand-Brynaert, *Eur. Polymer J.* **2001**, *37*, 9–18. (f) S.H. Khong, S. Sivaramakrishnan, R.Q. Png, *et al.*, *Adv. Func. Mater.* **2007**, *17*, 2490–9.
[11] E.F.V. Scriven, in *Reactive Intermediates*, R.A. Abramovitch, ed.; Plenum, New York, V. 2, **1982**.
[12] G.B. Schuster, M.S. Platz, *Adv. Photochem.* **1992**, *17*, 69–143.
[13] N.P. Gritsan, E.A. Pritchina, *Russ. Chem. Rev. (Uspekhi Khimii)* **1992**, *61*, 910–39.
[14] N.P. Gritsan, M.S. Platz, W.T. Borden, in *Computational Methods in Photochemistry*; A.G. Kutateladze, ed.; Taylor & Francis, Boca Raton, **2005**.
[15] M.S. Platz, in *Reactive Intermediate Chemistry*, R.A. Moss, M.S. Platz, M. Jons, eds.; Wiley-Interscience, Hoboken, **2004**.
[16] F. Tiemann, *Ber. Dtsch. Chem. Ges.* **1891**, *24*, 4162–7.
[17] W.T. Borden, N.P. Gritsan, C.M. Hadad, *et al.*, *Acc. Chem. Res.* **2000**, *33*, 765–11.
[18] W.L. Karney, W.T. Borden, *Adv. Carbene Chem.* **2001**, *3*, 205–51.
[19] N.P. Gritsan, M.S. Platz, *Adv. Phys. Org. Chem.* **2001**, *36*, 255–304.
[20] N.P. Gritsan, M.S. Platz, *Chem. Rev.* **2006**, *106*, 3844–67.
[21] H. Okabe, *J. Chem. Phys.* **1968**, *49*, 2726–33.
[22] H. Okabe, *Photochemistry of Small Molecules*, Wiley, New York, **1978**, pp. 232–4.
[23] J.R. McDonald, R.G. Miller, A.P. Baronavski, *Chem. Phys. Lett.* **1977**, *51*, 57–60.
[24] A.P. Baronavski, R.G. Miller, J.R. McDonalsd, *Chem. Phys.* **1978**, *30*, 119–31.
[25] F. Rohrer, F. Stuhl, *J. Chem. Phys.* **1988**, *88*, 4788–99.
[26] (a) W.S. Drozdovski, A.P. Baronavski, J.R. McDonald, *Chem. Phys. Lett.* **1979**, *64*, 421–5. (b) K-H. Gericke, R. Theinl, F. Comes, *Chem. Phys. Lett.* **1989**, *164*, 605–11. (c) K-H. Gericke, R. Theinl, F. Comes, *J. Chem. Phys.* **1990**, *92*, 6548–55. (d) J J. Chu, P. Marcus, P.J. Dagdigian, *J. Chem. Phys.* **1990**, *93*, 257–67. (e) K-H. Gericke, T. Haas, M. Lock, R. Theinl, F. Comes, *J. Phys. Chem.* **1991**, *95*, 6104–11.
[27] O. Kajimoto, T. Yamamoto, T. Fueno, *J. Phys. Chem.* **1979**, *83*, 429–35.
[28] J.C. Stephenson, M.P. Casassa, D.S. King, *J. Chem. Phys.* **1988**, *89*, 1378–87.
[29] M.H. Alexander, H-J. Werner, P.J. Dagdigian, *J. Chem. Phys.* **1988**, *89*, 1388–1400.
[30] (a) K-H. Gericke, T. Haas, M. Lock, F.J. Comes, *Chem. Phys. Lett.* **1991**, *186*, 427–30. (b) T. Haas, K-H. Gericke, C. Maul, F.J. Comes, *Chem. Phys. Lett.* **1993**, *202*, 108–14. (c) M. Lock, K-H. Gericke, F.J. Comes, *J. Chem. Phys.* **1996**, *213*, 385–96.
[31] (a) J.R. McDonald, R.G. Miller, A.P. Baronovski, *Chem. Phys.* **1978**, *30*, 133–45. (b) O. Kajimoto, T. Fueno, *Chem. Phys. Lett.* **1981**, *80*, 484–7.
[32] (a) J.A. Miller, C.T. Bowman, *Prog. Energy Combust. Sci.* **1989**, *15*, 287–38. (b) L.D. Smoot, S.C. Hill, H. Xu, *Prog. Energy Combust. Sci.* **1998**, *24*, 385–408.
[33] (a) M. Röhrig, H.G. Wagner, *Proc. Symposium (International) on Combustion*, **1994**, *25*, 975–81.

[34] J.C. Mackie, G.B. Bacskay, *J. Phys. Chem. A* **2005**, *109*, 11967–74.
[35] (a) M. Rohrig, H.J. Romming, H.G. Wagner, *Ber. Bunsenges Phys. Chem. Chem. Phys.* **1994**, *98*, 1332–4. (b) E. Henon, F. Bohr, *J. Mol. Struct. THEOCHEM* **2000**, *531*, 283–99.
[36] M. Rohrig, H.G. Wagner, *Ber. Bunsenges Phys. Chem. Chem. Phys.* **1994**, *98*, 858–63.
[37] W. Hack, in *N-Centered Radicals*, Z B. Alfassi, ed.; John Wiley & Sons, Inc., New York, **1998**.
[38] M. Rohrig, H.G. Wagner, *Ber. Bunsenges Phys. Chem. Chem. Phys.* **1994**, *98*, 864–8.
[39] C. Zetzsch, I. Hansen, *Ber. Bunsenges. Phys. Chem. Chem. Phys.* **1978**, *82*, 830–3.
[40] W. Hack, H. Kurzke, H.G. Wagner, *J. Chem. Soc., Faraday Trans. 2*, **1985**, *81*, 949–61.
[41] T. Fueno, K. Yokoyama, S. Takane, *Theor. Chim. Acta* **1992**, *82*, 299–308.
[42] E.D. Becker, J. Pimentel, M. Van Thiel, *J. Chem. Phys.* **1957**, *26*, 145–50.
[43] M. McCarty, G.W. Robinson, *J. Am. Chem. Soc.* **1959**, *81*, 4472–6.
[44] V.E. Bondybey, L.E. Brus, *J. Chem. Phys.* **1975**, *63*, 794–804.
[45] H. Esser, J. Langen, U. Schurath, Ber. *Bunsenges Phys. Chem. Chem. Phys.* **1983**, *87*, 636–43.
[46] A. Ramsthaler-Sommer, K.E. Eberhardt, U. Schurath, *J. Chem. Phys.* **1986**, *85*, 3760–9.
[47] C. Blindauer, N. van Riesenbeck, K. Seranski, *et al.*, *Chem. Phys.* **1991**, *150*, 93–108.
[48] S.L. Laursen, J.E. Grace, R.L. Dekock, S.A. Spronk, *J. Am. Chem. Soc.* **1998**, *120*, 12583–94.
[49] H.J. Himmel, M. Junker, H. Schnöckel, *J. Chem. Phys.* **2002**, *117*, 3321–6.
[50] P.H.H. Fischer, S.W. Charles, C.A. McDowell, *Can. J. Chem. Phys.* **1967**, *46*, 2162–5.
[51] F.D. Wayne, H.E. Radford, *Mol. Phys.* **1976**, *32*, 1407–22.
[52] H. Okabe, M. Lenzi, J. *Chem. Phys.* **1967**, *47*, 5241–6.
[53] J. Masanet, A. Gilles, C. Vermeil, *J. Photochem.* **1974/75**, *3*, 417–29.
[54] P.W. Fairchild, G.P. Smith, D.R. Crosly, J.B. Jeffries, *Chem. Phys. Lett.* **1984**, *107*, 181–6.
[55] P.C. Engelking, W.C. Lineberger, *J. Chem. Phys.* **1976**, *65*, 4323–4.
[56] E.P. Kyba, in *Azides and Nitrenes – Reactivity and Utility*; E.F.V. Scriven, ed.; Academic Press, New York, **1984**.
[57] F.D. Lewis, W.H. Saunders, in *Nitrenes*; W. Lwowski, ed.; John Wiley & Sons, Inc., New York, **1970**.
[58] R.M. Moriarty, R.C. Reardon, *Tetrahedron*, **1970**, *26*, 1379–92.
[59] F.C. Montgomery, W.H. Saunders, *J. Org. Chem.* **1976**, *41*, 2368–72.
[60] E.P. Kyba, R.A. Abramovitch, *J. Am. Chem. Soc.* **1980**, *102*, 735–40.
[61] R.E. Banks, D. Berry, M.J. McGlinchey, M.J. Moore, *J. Chem. Soc. C*, **1970**, 1017–23.
[62] A. Pancrazi, Q. Khuong-Huu, *Tetrahedron*, **1975**, *31*, 2049–56.
[63] D.E. Milligan, *J. Chem. Phys.* **1961**, *35*, 1491–7.
[64] M.E. Jaxcox, D.E. Milligan, *J. Mol. Spectrosc.* **1975**, *56*, 333–56.
[65] I.R. Dunkin, P.C.P. Thomson, *Tetrahedron Lett.* **1980**, *21*, 3813–16.
[66] J. Michl, J.G. Radziszewski, J.W. Downing, *et al.*, *Pure. Appl. Chem.* **1983**, *55*, 315–21.
[67] I.R. Dunkin, C.J. Shields, H. Quasi, B. Seiferling, *Tetrahedron Lett.* **1983**, *24*, 3887–90.
[68] R.S. Sheridan, G.A. Ganzer, *J. Am. Chem. Soc.*, **1983**, *105*, 6158–60.
[69] J.G. Radziszewski, J.W. Downing, M. Jawdosiuk, P. Kovacic, J. Michl, *J. Am. Chem. Soc.* **1985**, *107*, 594–603.
[70] J.G. Radziszewski, J.W. Downing, C. Wentrup, *et al.*, *J. Am. Chem. Soc.*, **1985**, *107*, 2799–2801.
[71] E. Wasserman, G. Smolinsky, W.A. Jager, *J. Am. Chem. Soc.* **1964**, *86*, 3166–7.
[72] (a) L. Barash, E. Wasserman, W.A. Jager, *J. Chem. Phys.* **1967**, *89*, 3931–5. (b) E. Wasserman, *Prog. Phys. Org. Chem.* **1971**, *8*, 319–36.
[73] P.J. Wagner, B.J. Scheve, *J. Am. Chem. Soc.* **1979**, *101*, 378–83.
[74] (a) S.M. Mandel, J.A.K. Bauer, A.D. Gudmundsdottir, *Org. Lett.* **2001**, *3*, 523–6. (b) S. Muthukrishnan, S.M. Mandel, J.C. Hackett, *et al.*, *J. Org. Chem.* **2007**, *72*, 2757–68.
[75] (a) P.N.D. Singh, S.M. Mandel, R.M. Robinson, *et al.*, *J. Org. Chem.* **2003**, *68*, 7951–60. (b) P.N.D. Singh, S.M. Mandel, S. Muthukrishnan, *et al.*, *J. Am. Chem. Soc.* **2007**, *129*, 16263–72.
[76] (a) F.D. Lewis, W.H. Saunders, *J. Am. Chem. Soc.* **1968**, *90*, 7031–3. (b) ibid. **1968**, *90*, 7033–8.

[77] R.F. Klima, A.D. Gudmundsdottir, *J. Photochem. Photobiol. A: Chemistry* **2004**, *162*, 239–47.
[78] H. Quast, P. Eckert, *Liebigs. Ann. Chem.*, **1974**, 1727–41.
[79] N.P. Gritsan, I. Likhotvorik, Z. Zhu, M.S. Platz, *J. Phys. Chem. A* **2001**, *105*, 3039–41.
[80] R.F. Ferrante, *J. Chem. Phys.* **1987**, *86*, 25–32.
[81] T. Franken, D. Perner, M.W. Bosnali, *Z. Naturforsch A* **1970**, *25*, 151–3.
[82] R.F. Ferrante, *J. Chem. Phys.* **1991**, *94*, 4678–9.
[83] H. Shang, C. Yu, L. Ying, X. Zhao, *J. Chem. Phys.* **1995**, *103*, 4418–26.
[84] L. Ying, Y. Xia, H. Shang, Y. Tang, *J. Chem. Phys.* **1996**, *105*, 5798–5805.
[85] P.G. Carrick, P.C. Engelking, *J. Chem. Phys.* **1984**, *81*, 1661–5.
[86] P.G. Carrick, C.R. Brazier, P.F. Bernath, P.C. Engelking, *J. Am. Chem. Soc.* **1987**, *109*, 5100–2.
[87] E.L. Chappell, P.C. Engelking, *J. Chem. Phys.* **1988**, *89*, 6007–16.
[88] C.R. Brazier, P.G. Carrick, P.F. Bernath, *J. Chem. Phys.* **1992**, *96*, 919–26.
[89] H. Shang, R. Gao, X. Zhao, Y. Tang, *Chem. Phys. Lett.* **1997**, *267*, 345–50.
[90] J.H. Glowina, J. Misewich, P.P. Sorokin, in *Supercontinuum Laser Sources*, R.R. Alfano, ed.; Springer Verlag, New York, **1989**.
[91] M.J. Travers, D.C. Cowles, E.P. Clifford, G.B. Ellison, P.C. Engelking, J. *Phys. Chem.* **1999**, *111*, 5349–60.
[92] (a) D.R. Yakony, H.F. Schaefer, S. Rothenberg, *J. Am. Chem. Soc.* **1974**, *96*, 5974–7. (b) J. Demuyanck, D.J. Fox, Y. Yamaguchi, H.F. Schaefer III, *J. Am. Chem. Soc.* **1980**, *102*, 6204–7. (c) J.A. Pople, K. Raghavachari, M.J. Frisch, J.S. Binckley, P. v R. Schleyer, *J. Am. Chem. Soc.* **1983**, *105*, 6389–98. (d) M.T. Nguyen, *Chem. Phys. Lett.* **1985**, *117*, 290–4. (e) Y. Xie, G.E. Scuseria, B.F. Yates, Y. Yamaguchi, H.F. Schaefer III, *J. Am. Chem. Soc.* **1989**, *111*, 5181–5.
[93] (a) C. Richards Jr., C. Meredith, S-J. Kim, G.E. Quelch, H.F. Schaefer III, *J. Chem. Phys.* **1994**, *100*, 481–9. (b) M.T. Nguyen, D. Sengupta, T-K. Ha, *J. Phys. Chem.* **1996**, *100*, 6499–6503. (c) J.F. Arenas, J.C. Otero, A. Sánchez-Gálvez, J. Soto, P. Viruela, *J. Phys. Chem.* **1998**, *102*, 1146–51.
[94] J.F. Arenas, J.I. Marcos, J.C. Otero, A. Sánchez-Gálvez, J. Soto. *J. Chem. Phys.* **1999**, *111*, 551–61.
[95] C.R. Kemnitz, G.B. Ellison, W.L. Karney, W.T. Borden, *J. Am. Chem. Soc.* **2000**, *122*, 1098–1101.
[96] A. Hassner, in *Azides and Nitrenes – Reactivity and Utility*; E.F.V. Scriven, ed.; Academic Press, New York, **1984**.
[97] (a) O. Meth-Cohn, N.J.R. Williams, A. MacKinnon, J.A.K. Howard, *Tetrahedron* **1998**, *54*, 9837–48. (b) J.R. Fotsing, K. Banert, *Synthesis* **2006**, 261–72.
[98] G. Smolinsky, *J. Org. Chem.* **1962**, *27*, 3557–9. (b) G. Smolinsky, C.A. Pryde, *J. Org. Chem.* **1968**, *33*, 2411–16.
[99] (a) A. Hassner, F.W. Fowler, *Tetrahedron Lett.* **1967**, *8*, 1545–8. (b) A. Hassner, F.W. Fowler, J. *Am. Chem. Soc.* **1968**, *90*, 2869–75.
[100] (a) K. Isomura, M. Okada, H. Taniguchi, *Tetrahedron Lett.* **1969**, *10*, 4073–7. (b) W. Bauer, K. Hafner, *Angew. Chem. Int. Ed. Engl.* **1969**, *8*, 772–3.
[101] (a) J.H. Boier, W.E. Krueger, G.J. Mikol, *J. Am. Chem. Soc.* **1967**, *87*, 5504–5. (b) J.H. Boier, W.E. Krueger, R. Modler, *J. Org. Chem.* **1969**, *34*, 1987–9.
[102] O.L. Chapman, J-P. Le Roux, *J. Am. Chem. Soc.*, **1978**, *100*, 282–5.
[103] F.W. Fowler, A. Hassner, L.A. Levy, *J. Am. Chem. Soc.* **1967**, *89*, 2077–82.
[104] (a) H. Bock, R. Dammel, S. Aygen, *J. Am. Chem. Soc.* **1983**, *105*, 7681–5. (b) L.L. Lohr Jr, M. Hanamura, K. Morokuma, *J. Am. Chem. Soc.* **1983**, *105*, 5541–7. (c) T. Yamabe, M. Kaminoyama, T. Minato, *et al.*, *Tetrahedron* **1984**, *40*, 2095–9.
[105] V. Parasuk, C.J. Cramer, *Chem. Phys. Lett.* **1996**, *260*, 7–14.
[106] W.L. Karney, W.T. Borden. *J. Am. Chem. Soc.* **1997**, *119*, 1378–87.
[107] (a) T. Curtius, *Ber.* **1890**, *23*, 3023–5. (b) T. Curtius, *Z. Angew. Chem.* **1914**, *27*, 111–14.
[108] W. Lwowski, in *Azides and Nitrenes – Reactivity and Utility*; E.F.V. Scriven, ed.; Academic Press, New York, **1984**.

[109] W. Lwowski, G.T. Tisue, *J. Am. Chem. Soc.* **1965**, *87*, 4022–3.
[110] G.T. Tisue, S. Linke, W. Lwowski, *J. Am. Chem. Soc.* **1967**, *89*, 6303–7.
[111] S. Linke, G.T. Tisue, W. Lwowski, *J. Am. Chem. Soc.* **1967**, *89*, 6308–10.
[112] J. Muller, *Kinetic und Mechanismus der Curtius Umlagerung*, PhD thesis, University München, **1962**.
[113] R.P. Tiger, *Polym. Sci., Ser. B (Engl. Transl.)* **2004**, *46*, 142–53.
[114] (a) M.V. Zabalov, R.P. Tiger, *Russ. Chem. Bull.* **2005**, *54*, 2270–80. (b) M.V. Zabalov, R.P. Tiger, *Russ. Chem. Bull.* **2007**, *56*, 7–13.
[115] J. Liu, S. Mandel, C.M. Hadad, M.S. Platz, *J. Org. Chem.* **2004**, *69*, 8583–93.
[116] R. Puttner, K. Hafner, *Tetrahedron Lett.* **1964**, *5*, 3119–25.
[117] E. Eibler, J. Sauer, *Tetrahedron Lett.* **1974**, *15*, 2569–72.
[118] V.P. Semenov, A.N. Studenikov, A.D. Bespalov, K.A. Ogloblin, *Zh. Organ. Khim. (Russ)* **1977**, *13*, 2202–7.
[119] Y. Hayashi, D. Swern, *J. Am. Chem. Soc.* **1973**, *95*, 5205–10.
[120] M. Inagaki, T. Shingaki, T. Nagai, *Chem. Lett.* **1981**, 1419–22.
[121] M. Inagaki, T. Shingaki, T. Nagai, *Chem. Lett.* **1982**, 9–12.
[122] T. Autrey, G.B. Schuster, *J. Am. Chem. Soc.* **1987**, *109*, 5814–20.
[123] M.E. Sigman, T. Autrey, G.B. Schuster, *J. Am. Chem. Soc.* **1988**, *110*, 4297–4305.
[124] I. Woelfle, B. Sauerwein, T. Autrey, G.B. Schuster, *Photochem. Photobiol.* **1988**, *47*, 497–501.
[125] T. Melvin, G.B. Schuster, *Photochem. Photobiol.* **1990**, *51*, 155–60.
[126] K.-U. Clauss, K. Buck, W. Abraham, *Tetrahedron*, **1995**, *51*, 7181–92.
[127] V. Desikan, Y. Liu, J.P. Toscano, W.S. Jenks, *J. Org. Chem.* **2007**, *72*, 6848–59.
[128] P.A.S. Smith, in *Organic Reactions*; R. Adams, ed.; John Wiley & Sons, Inc., New York, **1946**, *3*, 337–449.
[129] C.R. Hauser, S.W. Kantor, *J. Am. Chem. Soc.* **1950**, *72*, 4284–5.
[130] J. Hine, *Physical Organic Chemistry*; McGraw-Hill Book Co Inc., New York, **1962**.
[131] E.S. Gould, *Mechanism and Structure in Organic Chemistry*; Henry Holt & Co., New York, **1959**.
[132] J.D. Roberts, M.C. Caserio, *Basic Principles of Organic Chemistry*; W.A. Benjamin Inc., New York, **1964**.
[133] L.F. Fieser, M. Fieser, *Advanced Organic Chemistry*; Reinhold Publishing Corp., New York, **1961**.
[134] R. C. Woodworth, P. S. Skell, *J. Am. Chem. Soc.* **1959**, *81*, 3383-3386.
[135] N.P. Gritsan, E.A. Pritchina, *Mendeleev Commun.* **2001**, *11*, 94–6.
[136] E.A. Pritchina, N.P. Gritsan, A. Maltsev, *et al.*, *Phys. Chem. Chem. Phys.* **2003**, *5*, 1010–18.
[137] E.A. Pritchina, N.P. Gritsan, T. Bally, *Russ. Chem. Bull., Int. Ed. (Engl. Transl)* **2005**, *54*, 525–32.
[138] (a) A.I. Kitaigorodski, P.M. Zorki, V.K. Belski, *Structure of the Organic Compounds. Data of Structure Study. 1929–1970*; Science, Moscow, **1980**, p. 628. (b) ibid., p. 511.
[139] C. Wentrup, H. Bornemann, *Eur. J. Org. Chem.* **2005**, 4521–4.
[140] V. Desican, Y. Liu, J.P. Toscano, W. S. Jenks, *J. Org. Chem.* **2007**, *72*, 6848–59.
[141] S.M. Mandel, M.S. Platz, *Org. Lett.* **2005**, *7*, 5385–7.
[142] W. Lwowski, R. DeMauriac, *Tetrahedron Lett.* **1964**, *5*, 3285–8.
[143] W. Lwowski, F.P. Woerner, *J. Am. Chem. Soc.* **1965**, *87*, 5490–1.
[144] D.S. Breslow, T.J. Prosser, A.F. Marcantonio, C.A. Genge, *J. Am. Chem. Soc.* **1967**, *89*, 2384–90.
[145] J.S. McConaghy, W. Lwowski, *J. Am. Chem. Soc.* **1967**, *89*, 2357–64.
[146] J.S. McConaghy, W. Lwowski, *J. Am. Chem. Soc.* **1967**, *89*, 4450–6.
[147] (a) M. Jones, K.R. Rettig, *J. Am. Chem. Soc.* **1965**, *87*, 4013–15. (b) M. Jones, K.R. Rettig, *J. Am. Chem. Soc.* **1965**, *87*, 4015–16.
[148] C. Buron, M.S. Platz, *Organic Lett.*, 1003, *5*, 3383–5.
[149] R.E. Wilde, T.K.K. Srinivasan, W. Lwowski, *J. Am. Chem. Soc.* **1971**, *93*, 860–3.
[150] J.H. Teles, G. Maier, *Chem. Ber.* **1989**, *122*, 745–8.

[151] C. Wentrup, *Reactive Molecules*; Wiley-Interscience, New York, **1984**.
[152] M.S. Platz, V.M. Maloney, In *Kinetics and Spectroscopy of Carbenes and Biradicals*; ed. M.S. Platz, Plenum, New York, **1990**.
[153] M.F. Budyka, M.M. Kantor, M.V. Alfimov, *Russ. Chem. Rev. (Uspekhi Khimi)* **1992**, *61*, 48–74.
[154] M.P. Gritsan, D. Polshakov, M.-L. Tsao, M.S. Platz, *Photochem. Photobiol. Sci.* **2005**, *4*, 23–32.
[155] G. Burdzinski, T.L. Gustafson, J.C. Hackett, C.M. Hadad, M.S. Platz, *J. Am. Chem. Soc.* **2005**, *127*, 13764–5.
[156] G. Burdzinski, J.C. Hackett, J. Wang, *et al.*, *J. Am. Chem. Soc.* **2006**, *128*, 13402–11.
[157] R.D. McCulla, G. Burdzinski, M.S. Platz, *Org. Lett.* **2006**, *8*, 1637–40.
[158] G. Burdzinski, C.T. Middleton, T.L. Gustafson, M.S. Platz, *J. Am. Chem. Soc.* **2006**, *128*, 14804–5.
[159] J. Wang, J. Kubicki, M.S. Platz, *Org. Lett.* **2007**, *9*, 3973-–6.
[160] J. Wang, G. Burdzinski, Z. Zhu, *et al.*, *J. Am. Chem. Soc.* **2007**, *129*, 8380–8.
[161] J. Wang, G. Burdzinski, M.S. Platz, *Org. Lett.* **2007**, *9*, 5211–14.
[162] P.A.S. Smith, in *Azides and Nitrenes – Reactivity and Utility*; E.F.V. Scriven, ed.; Academic Press, New York, **1984**.
[163] (a) R. Huisgen, D. Vossius, M. Appl, *Chem. Ber.* **1958**, *91*, 1–12. (b) R. Huisgen, M. Appl, *Chem. Ber.* **1958**, *91*, 12–31.
[164] W.E. Doering, R.A. Odum, *Tetrahedron*, **1966**, *22*, 81–93.
[165] S.E. Carroll, B. Nay, E.F.V. Scriven, H. Suschitzky, R.R. Thomas, *Tetrahedron Lett.* **1977**, *18*, 3175–8.
[166] A.K. Schrock, G.B. Schuster, *J. Am. Chem. Soc.* **1984**, *106*, 5228–34.
[167] T.-Y. Liang, G.B. Schuster, *J. Am. Chem. Soc.* **1986**, *108*, 546–8.
[168] E. Leyva, M.S. Platz, G. Persy, J. Wirz, *J. Am. Chem. Soc.* **1986**, *108*, 3783–90.
[169] A. Reiser, F.W. Willets, G.C. Terry, V. Williams, R. Marley, *Trans Faraday Soc.* **1968**, *64*, 3265–75.
[170] A.K. Schrock, G.B. Schuster, *J. Am. Chem. Soc.* **1984**, *106*, 5234–40.
[171] T. Yamaoka, H. Kashiwagi, S. Nagakura, *Bull. Chem. Soc. Japan* **1972**, *45*, 361–5.
[172] E.A. Pritchina, N.P. Gritsan, *J. Photochem. Photobiol. A: Chemistry*, **1988**, *43*, 165–82.
[173] N.P. Gritsan, E.A. Pritchina, *J. Inf. Rec. Mat.* **1989**, *17*, 391–404.
[174] J.C. Brinen, B. Singh, *J. Am. Chem. Soc.*, **1971**, *93*, 6623–9.
[175] E.A. Pritchina, N.P. Gritsan, T. Bally, *Phys. Chem. Chem. Phys.* **2006**, *8*, 719–27.
[176] G. Smolinsky, E. Wasserman, Y.A. Yager, *J. Am. Chem. Soc.* **1962**, *84*, 3220–1.
[177] A. Reiser, G. Bowes, R. Horne, R. Trans. *Faraday Soc.* **1966**, *62*, 3162–9.
[178] T. Donnelly, I.R. Dunkin, D.S.D. Norwood, *et al.*, *J. Chem. Soc. Perkin. Trans.* 2 **1985**, 307–10.
[179] J.C. Hayes, R.S. Sheridan, *J. Am. Chem. Soc.* **1990**, *112*, 5879–81.
[180] I.R Dunkin, M.A. Lynch, F. McAlpine, D. Sweeney, *J. Photochem. Photobiol. A: Chemistry* **1997**, *102*, 207–12.
[181] E.A. Pritchina, N.P. Gritsan, T. Bally, unpublished results, **2003**.
[182] C.J. Shields, D.R. Chrisope, G.B. Schuster, *et al.*, *J. Am. Chem. Soc.* **1987**, *109*, 4723–6.
[183] Y.Z. Li, J.P. Kirby, M.W. George, M. Poliakoff, G.B. Schuster, *J. Am. Chem. Soc.* **1988**, *110*, 8092–8.
[184] (a) R.J. Sundberg, M. Brenner, S.R. Suter, B.P. Das, *Tetrahedron Lett.* **1970**, *11*, 2715–18; R.J. Sundberg, S.R. Suter, M. Brenner, *J. Am. Chem. Soc.* **1972**, *94*, 513–20. (b) B.A. DeGraff, D.W. Gillespie, R.J. Sundberg, *J. Am. Chem. Soc.* **1974**, *96*, 7491–6.
[185] (a) R. Warmuth, S. Makowiec, *J. Am. Chem. Soc.* **2005**, *127*, 1084–5. (b) R. Warmuth, S. Makowiec, *J. Am. Chem. Soc.* **2007**, *129*, 1233–41.
[186] (a) G. Porter, B. Ward, *Proc. Roy. Soc., London, A.* **1968**, *303*, 139–56. (b) K. Ozawa, T. Ishida, K. Fuke, K. Kaya, *Chem. Phys. Lett.* **1988**, *150*, 249–53.
[187] (a) D.W. Cullin, N. Soundarajan, M.S. Platz, T.A. Miller, *J. Phys. Chem.* **1990**, *94*, 3387–91. (b) D.W. Cullin, N. Soundarajan, M.S. Platz, T.A. Miller, *J. Phys. Chem.* **1990**, *94*, 8890–6.

[188] (a) W.D. Crow, C. Wentrup, *Tetrahedron Lett.* **1968**, *9*, 6149–52. (b) C. Wentrup, *J. Chem. Soc.D: Chem. Commun.* **1969**, 1386–7. (c) M. Kuzaj, H. Lüerssen, C. Wentrup, *Angew. Chem. Int. Ed. Engl.* **1986**, *25*, 480–2. (d) C. Wentrup, *Topics Current Chem.* **1976**, *62*, 173–87.
[189] S.J.I. Kim, T.P. Hamilton, H.F. Schaefer III, *J. Am. Chem. Soc.* **1992**, *114*, 5349–55.
[190] D. Hrovat, E.E. Waali, W.T. Borden, *J. Am. Chem. Soc.* **1992**, *114*, 8698–9.
[191] O. Castell, V.M. Carefa, C. Bo, R. Caballol, *J. Comput. Chem.* **1986**, *17*, 42–8.
[192] B.A. Smith, C.R. Cramer, *J. Am. Chem. Soc.* **1996**, *118*, 5490–3.
[193] M.J. Travers, D.C. Cowles, E.P. Clifford, G.B. Ellison, *J. Am. Chem. Soc.* **1992**, *114*, 8699–8701.
[194] R.N. McDonald, S.J. Davidson, *J. Am. Chem. Soc.* **1993**, *115*, 10857–62.
[195] N.P. Gritsan, T. Yuzawa, M.S. Platz, *J. Am. Chem. Soc.* **1997**, *119*, 5059–60.
[196] R. Born, C. Burda, P. Senn, J. Wirz, *J. Am. Chem. Soc.* **1997**, *119*, 5061–2.
[197] N.P. Gritsan, Z. Zhu, C.M. Hadad, M.S. Platz, *J. Am. Chem. Soc.* **1999**, *121*, 1202–7.
[198] M.-L. Tsao,. M.S. Platz, *J. Am. Chem. Soc.* **2003**, *125*, 12014–25.
[199] K. Anderson, *Theor. Chim. Acta* **1995**, *91*, 31–46.
[200] R.A. McClelland, M.J. Kahley, P.A. Davidse, G. Hadzialic, *J. Am. Chem. Soc.* **1996**, *118*, 4794–4803.
[201] J.C. Fishbein, R.A. McClelland, *Can. J. Chem.* **1996**, *74*, 1321–8.
[202] J. Wang, J. Kubicki, M.S. Platz, *Org. Lett.* **2007**, *9*, 3973–6.
[203] C.J. Cramer, F.J. Dulles, D.E. Falvey, *J. Am. Chem. Soc.* **1994**, *116*, 9787–8.
[204] A. Reiser, R. Marley, Trans. *Faraday Soc.* **1968**, *64*, 1806–15.
[205] M.W. Geiger, M.E. Elliot, V.D. Karacostas, *et al.*, *Photochem. Photobiol.* **1984**, *40*, 545–8.
[206] M.F. Budyka, M.M. Kantor, M.V. Alfimov, *Russ. Chem. Bull.* **1992**, *41*, 590–1.
[207] N.P. Gritsan, H.B. Zhai, T. Yuzawa, D. Karweik, J. Brooke, M.S. Platz, J. *Phys. Chem. A* **1997**, *101*, 2833–40.
[208] N.P. Gritsan, D. Tigelaar, M.S. Platz, *J. Phys. Chem. A* **1999**, *103*, 4465–9.
[209] N.P. Gritsan, A.D. Gudmundsdóttir, D. Tigelaar, M.S. Platz, M.S. *J. Phys. Chem. A* **1999**, *103*, 3458–61.
[210] M. Cerro-Lopez, N.P. Gritsan, Z. Zhu, M.S. Platz, *J. Phys. Chem. A* **2000**, *104*, 9681–6.
[211] D.A. Polshakov, Y.P. Tsentalovich, N.P. Gritsan, *Russ. Chem. Bull.* **2000**, *49*, 50–5.
[212] N.P. Gritsan, I. Likhotvorik, M.-L. Tsao, *et al.*, *J. Am. Chem. Soc.* **2001**, *123*, 1425–33.
[213] N.P. Gritsan, A.D. Gudmundsdóttir, D. Tigelaar, *et al.*, *J. Am. Chem. Soc.* **2001**, *123*, 1951–62.
[214] M.-L. Tsao, N.P. Gritsan, T.R. James, *et al.*, *J. Am. Chem. Soc.* **2003**, *125*, 9343–58.
[215] W.T.G. Johnson, M.B. Sullivan, C.J. Cramer, *Int. J. Quant. Chem.* **2001**, *85*, 492–508.
[216] T. Kobayashi, K. Suzuki, T. Yamaoka, *J. Phys. Chem.* **1985**, *89*, 776–9.
[217] T.-Y. Liang, G.B. Schuster, *J. Am. Chem. Soc.* **1987**, *109*, 7803–10.
[218] T. Donnelly, I.R. Dunkin, D. S.D. Norwood, *et al.*, *J. Chem. Soc. Perkin. Trans. 2*, **1985**, 307–10.
[219] (a) R.A. Odum, A.M. Aaronson, *J. Am. Chem. Soc.* **1969**, *91*, 5680–1. (b) R.A. Odum, G. Wolf, *J. Chem. Soc. Chem.Commun.* **1973**, 360–1.
[220] R.J. Sundberg, S.R. Suter, M. Brenner, *J. Am. Chem. Soc.* **1972**, *94*, 513–20.
[221] S. Murata, S. Abe, H. Tomioka, *J. Org. Chem.* **1997**, *62*, 3055–61.
[222] E. Leyva, R. Sagredo, R. *Tetrahedron* **1988**, *54*, 7367–74.
[223] K. Lamara, A.D. Redhouse, R.K. Smalley, J.R. Thompson, *Tetrahedron* **1994**, *50*, 5515–25.
[224] M.A. Berwick, *J. Am. Chem. Soc.* **1971**, *93*, 5780–6.
[225] W.L. Karney, W.T. Borden, *J. Am. Chem. Soc.* **1997**, *119*, 3347–50.
[226] I.R. Dunkin, T. Donnelly, T.S. Lockhart, *Tetrahedron Lett.*, **1985**, *26*, 359–62.
[227] (a) R.A. Abramovitch, S.R. Challand, E.F.V. Scriven, *J. Am. Chem. Soc.* **1972**, *94*, 1374–76. (b) R.A. Abramovitch, S.R. Challand, E.F.V. Scriven, *J. Org. Chem.* **1975**, *40*, 1541–7.
[228] (a) R.E. Banks, G.R. Sparkes, *J. Chem. Soc. Perkin Trans. I* **1972**, *1*, 2964–70. (b) R.E. Banks, A. Prakash, *Tetrahedron Lett.*, **1973**, *14*, 99–102. (c) R.E. Banks, A. Prakash, *J.*

Chem. Soc. Perkin Trans. I **1974**, *3*, 1365–71. (d) R.E. Banks, N.D. Venayak, *J. Chem. Soc. Chem. Commun.* **1980**, *9*, 900–1.

[229] R. Poe, J. Grayzar, M.J.T. Young, *et al.*, *J. Am. Chem. Soc.* **1991**, *113*, 3209–11.

[230] (a) P.J. Crocker, N. Imai, K. Rajagopalan, *et al.*, *Bioconjugate* **1990**, *1*, 419–24. (b) R.R. Drake, J.T. Slama, K.A. Wall, *et al.*, *Bioconjugate Chem.* **1992**, *3*, 69–73. (c) P.R. Kym, K.E. Carlson, J.A. Katzenellenbogen, *J. Med. Chem.* **1993**, *36*, 1111–19; (d) M.W. Reed, D. Fraga, D.E. Schwartz, J. Scholler, R.D. Hinrichsen, *Bioconjugate Chem.* **1995**, *6*, 101–8. (e) I. Kapfer, J.E. Hawkinson, J.E. Casida, M.P. Goeldner, *J. Med. Chem.* **1994**, *37*, 133–40. (f) I. Kapfer, P. Jacques, H. Toubal, M.P. Goeldner, *Bioconjugate Chem.* **1995**, *6*, 109–14.

[231] (a) R. Poe, K. Schnapp, M.J.T. Young, J. Grayzar, M.S. Platz, *J. Am. Chem. Soc.* **1992**, *114*, 5054–67. (b) K.A. Schnapp, R. Poe, E. Leyva, N. Soundararajan, M.S. Platz, *Bioconjugate Chem.* **1993**, *4*, 172–7. (c) K.A. Schnapp, M.S. Platz, *Bioconjugate Chem.* **1993**, *4*, 178–83. (d) A. Marcinek, M.S. Platz, *J. Phys. Chem.* **1993**, *97*, 12674–7. (e) A. Marcinek, M.S. Platz, S.Y. Chan, *et al.*, *J. Phys. Chem.* **1994**, *98*, 412–19.

[232] I.R. Dunkin, P.C.P. Thomson, *J. Chem. Soc. Chem. Commun.* **1982**, 1192–3.

[233] J. Morawietz, W. Sander, *J. Org. Chem.* **1996**, *61*, 4351–4.

[234] C. Carra, R. Nussbaum, T. Bally, *Chem. Phys. Chem.* **2006**, *7*, 1268–75.

[235] (a) R.A. McClelland, P.A. Davidse, G. Hadzialic, *J. Am. Chem. Soc.* **1995**, *117*, 4173–4. (b) R.A. McClelland, *Tetrahedron*, *52*, 6823–58. (c) P. Suchai, R.A. McClelland, *J. Chem. Soc. Perkin Trans. 2* **1996**, 1529–30.

[236] (a) R.A. McClelland, M.J. Kahkey, P.A. Davidse, *J. Phys. Org. Chem.* **1966**, *9*, 355–60. (b) D. Ren, R.A. McClelland, *Can. J. Chem.* **1998**, *76*, 78–84. (c) R.A. McClelland, A. Postigo, *Biophys. Chem.* **2006**, *119*, 213–18. (d) B. Cheng, R.A. McClelland, *Can. J. Chem.* **2001**, *79*, 1881–6.

[237] D.E. Falvey, in *Reactive Intermediate Chemistry*, R.A. Moss, M.S. Platz, M. Jons, eds.; Wiley-Interscience, Hoboken, **2004**.

[238] (a) J.A. Miller, *Cancer Res.* **1970**, *30*, 559–76. (b) E.C. Miller, J.A. Miller, *Cancer*, **1981**, *47*, 2327–45.

[239] M.S. Rizk, X. Shi, M.S. Platz, *Biochemistry* **2006**, *45*, 543–51.

[240] N.P. Gritsan, *Russ. Chem. Rev. (Uspkhi Khimii)* **2007**, *76*, 1139–60.

[241] R.L. Safiullin, S.L. Khursan, E.M. Chainikova, *Kinetics and Catalysis* **2004**, *45*, 640–8.

[242] (a) P.A.S. Smith, B.B. Brown, *J. Am. Chem. Soc.* **1951**, *73*, 2435–8. (b) P.A.S. Smith, B.B. Brown, *J. Am. Chem. Soc.* **1951**, *73*, 2438–41. (c) P.A.S. Smith, J.H. Hall, *J. Am. Chem. Soc.* **1962**, *84*, 1632–5.

[243] J. Swenton, T. Ikeler, B. Williams, *J. Am. Chem. Soc.* **1970**, *92*, 3103–9.

[244] (a) R.J. Sundberg, R.W. Heintzelman, *J. Org. Chem.* **1974**, *39*, 2546–52. (b) R.J. Sundberg, D.W. Gillespie, B.A. DeGraff, *J. Am. Chem. Soc.* **1975**, *97*, 6193–6.

[245] P.A. Lehman, R.S. Berry, *J. Am. Chem. Soc.* **1973**, *95*, 8614–20.

[246] (a) L.K. Dyall, J.E. Kemp, *J. Chem. Soc. B* **1968**, *9*, 976–9. (b) L.K. Dyall, *Aust. J. Chem.* **1975**, *28*, 2147–59. (c) S.P. Efimov, V.A. Smirnov, A.V. Pochinok, *High Ener. Chem.* **1983**, *17*, 347–9. (d) R. Purvis, R.K. Smalley, H. Suschitzky, M. Alkhader, *J. Chem. Soc., Perkin Trans. I* **1984**, 249–54.

[247] G. Rauhut, F. Eckert, *J. Phys. Chem. A* **1999**, *103*, 9086–92.

[248] S.E. Hilton, E.F.V. Scriven, H. Suschitzky, *J. Chem. Soc. Chem. Commun.* **1974**, 853–4.

[249] S.E. Carroll, B. Nay, E.F.V. Scriven, H. Suschitzky, *Tetrahedron Lett.* **1977**, *18*, 943–6.

[250] E. Leyva, M.S. Platz, *Tetrahedron Lett.* **1987**, *28*, 11–14.

[251] (a) E. Wasserman, *Prog. Phys. Org. Chem.* **1971**, *8*, 319–36. (b) J.A.R. Coope, J.B. Farmer, C.L. Gardner, C.A. McDowell, *J. Chem. Phys.* **1965**, *42*, 54–9. (c) M. Kazaj, H. Luerssen, C. Wentrup, C. *Angew. Chem. Int. Ed. Eng.* **1986**, *25*, 480–2. (d) R. Alvarado, J.-Ph. Grivet, C. Igier, J. Barcelo, J. Rigaudy, *J. Chem. Soc. Faraday Trans. 2*, **1977**, *73*, 844–57.

[252] (a) W. Lwowski, In *Reactive Intermediates*; eds. Jone, M; Moss R.A., John Wiley & Sons, Inc.: New York, **1981**; Vol. *1* and *2*. (b) C. Wentrup, *Adv. Heterocycl. Chem.* **1981**, *28*, 231–361.

[253] (a) J. Rigaudy, C. Igier, J. Barcelo, *Tetrahedron Lett.*, **1975**, *16*, 3845–8. (b) J. Rigaudy, C. Igier, J. Barcelo, *Tetrahedron Lett.*, **1979**, *20*, 1837–40.

[254] (a) S.E. Carroll, B. Nay, E.F.V. Scriven, H. Suschitzky, *Synthesis* **1975**, 710–11. (b) S.E. Carroll, B. Nay, E.F.V. Scriven, H. Suschitzky, D.R. Thomas, *Tetrahedron Lett.*, **1977**, *18*, 3175–8.
[255] I.R. Dunkin, P.C.P. Thomson, *J. Chem. Soc. Chem. Commun.* **1980**, 499–500.
[256] A. Maltsev, T. Bally, M.-L. Tsao, *et al.*, *J. Am. Chem. Soc.* **2004**, *126*, 237–49.
[257] A. Kuhn, M. Vosswinkel, C. Wentrup, *J. Org. Chem.* **2002**, *67*, 9023–30.
[258] M.-L. Tsao, M.S. Platz, *J. Phys. Chem. A* **2004**, *108*, 1169–76.
[259] J. Wang, J. Kubicki, G. Burdzinski, *et al.*, *J. Org. Chem.* **2007**, *72*, 7581–6.
[260] A. Reiser, G.C. Terry, F.W. Willets, *Nature (London)* **1966**, *211*, 410.
[261] M.-L. Tsao, M. S. Platz, *J. Phys. Chem. A* **2003**, *107*, 8879–84.
[262] T. Tsunoda, T. Yamoka, M. Takayama, *Nippon Kaguku Kaishi* **1975**, *12*, 2074–9.
[263] T. Tsunoda, T. Yamoka, Y. Osabe, Y. Hata, *Photogr. Sci. Eng.* **1976**, *20*, 188–94.
[264] M. Sumitani, S. Nagakura, K. Yoshihara, *Bull. Chem. Soc. Jpn.* **1976**, *49*, 2995–90.

12

Organoazides and Transition Metals

Werner R. Thiel

Fachbereich Chemie, Technische Universität Kaiserslautern, Erwin-Schrödinger-Str. Geb. 54, D-67663 Kaiserslautern, Germany

12.1 Introduction

In coordination chemistry, azido complexes wherein azide ions are coordinated to metal sites are known for long.[1] The azide ion was found in terminal as well as in bridging (1,1-μ, 1,1,1-μ, 1,3-μ) coordination geometries. It is well established in the literature that the stability of such compounds decreases with an increase of the covalence of the M-N_3 bond and with an increasing M/N_3 ratio. Binary systems of the type $M(N_3)_x$ usually decompose under formation of dinitrogen and elemental M.

As outlined by the corresponding mesomeric formulae shown in Scheme 12.1 organoazides are able to provide free electron pairs for the formation of metal complexes too.

For a deeper understanding of the donor properties of organoazides, quantum chemistry can help. Calculation of methyl azide and its adducts to a lithium cation by MP2/6-311G* results the structural parameters and charge distributions presented in Figure 12.1 and Table 12.1.

The energy difference (MP2 energies) between the two regio isomeric lithium complexes is small in the gas phase: coordination of Li^+ to the terminal nitrogen atom N3 – the isomer shown in Figure 12.1, right – is energetically favoured by just 1.13 kcal/mol. Taking solely the charge distribution between the nitrogen atoms of methyl azide into account, one could assume a preferred coordination to the carbon bound nitrogen atom N1 to take place. However, the terminal nitrogen atom N3, for which a pronounced participation of sp hybridization can be postulated, will provide a more directed electron pair for the coordination of the Lewis acid.

Organic Azides: Syntheses and Applications Edited by Stefan Bräse and Klaus Banert

Scheme 12.1

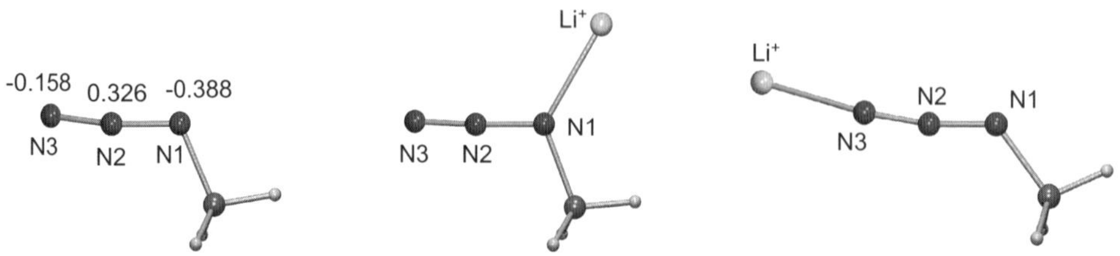

Figure 12.1 *Calculated structures and Mulliken charges of methyl azide (left) and calculated structures of the Li⁺ adducts to methyl azide; characteristic bond lengths [Å] and bond angles [deg]*

Table 12.1 *Calculated characteristic bond lengths [Å] and bond angles [deg] of methyl azide and of its Li⁺ adducts*

	Methyl azide	Li⁺ adduct at N1	Li⁺ adduct at N3
N1-N2	1.237	1.260	1.197
N2-N3	1.156	1.145	1.165
Li-N	–	1.959	1.903
N1-N2-N3	172.7	177.1	170.6
Li⁺-N-N		122.4	174.2
Li⁺-N1-C		125.9	
C-N1-N2		111.8	128.4

Coordination of Li^+ to the carbon bound nitrogen atom N1 (isomer shown in Figure 12.1, middle) stabilizes the negative charge at this position and therefore the mesomeric structure **B** shown in Scheme 12.1, which consequently results in a shortening of the terminal N2≡N3 bond and an elongation (weakening) of the bond between N1 and N2. Due to this, the loss of dinitrogen (N_2) is facilitated. The resulting nitrene species bound to a metal cation, will obviously be a highly reactive and highly oxidizing intermediate. On the other hand, σ donation from N3 as depicted in the structure on the right side of Figure 12.1 leads to a pronounced shortening of the N1=N2 bond, due to a stabilization of the negative charge at N3.

However, transition metal sites not only possess Lewis acidic properties, they are often redox active too. Metal compounds in low and middle oxidation states can deliver electrons to substrates like organoazides. By a formal two electron reduction, $[RN_3]^{2-}$ dianions are generated, which again can be described by at least two mesomeric forms (Scheme 12.2) and which should be stabilized due to the presence of electronegative nitrogen atoms.

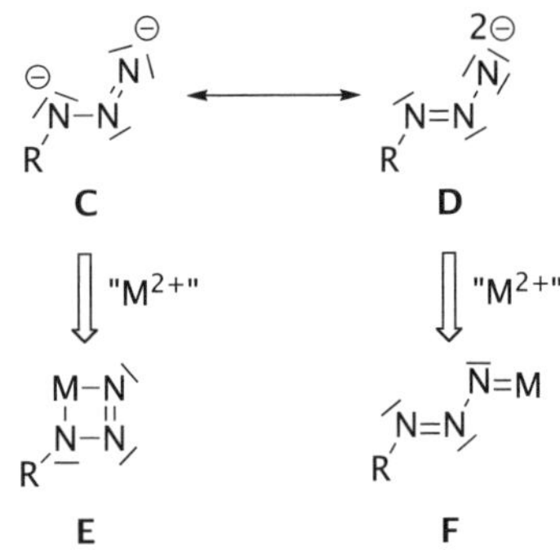

Scheme 12.2

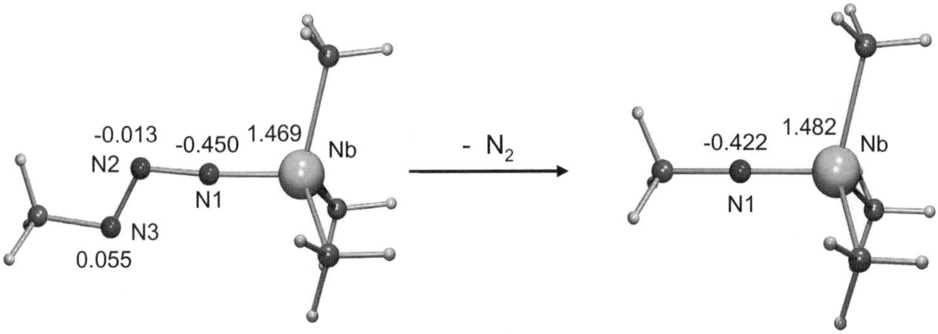

Figure 12.2 *Calculated structures and Mullikan charges of the methyl azide adduct to $NbMe_3$ (left) and the corresponding imino complex $MeN{=}NbMe_3$ (right)*

Obviously such species can act either as bidentate chelating ligands including the nitrogen atoms N1 and N3 as two electron donor sites (structures **C** and **E**) in a triaza-metalla cyclobutene type structure or as a monodentate ligand (structures **D** and **F**) wherein the terminal nitrogen atom N3 acts as a four electron donating site similar to the carbon atom of an alkylidene type ligand. The latter coordination mode thus should preferentially be realized with high valent transition metal centres. The triazametalla cyclobutene type structure **E** will have a destabilizing effect on the N1-N2 bond and can also be considered as a transition state in the description of the metal mediated decomposition of organo azides by dinitrogen elimination. In the case of high valent (transition) metal sites, this would lead to the formation of a stable imido complex, for metal sites in lower oxidation states, a more nitrene-like character of the resulting R-N fragment can be postulated. Quantum chemical calculations (B3LYP/LANL2DZ) on the model reaction $MeN_3{=}NbMe_3 \rightarrow MeN{=}NbMe_3 + N_2$ support these ideas (Figure 12.2, Table 12.2).

This reaction is strongly exothermic (ΔH_{calcd}: −47.7 kcal/mol) due to the formation of highly stable dinitrogen. Therefore, if the η^2-coordination of the organoazide (Scheme 12.2, structure **E**) is of relevance on the reaction pathway of the azide decomposition, the best way to prevent this is to avoid a free coordination site at the metal centre.

Table 12.2 Calculated characteristic bond lengths [Å] and bond angles [deg] of $MeN_3{=}NbMe_3$ and $MeN{=}NbMe_3$

	$MeN_3{=}NbMe_3$	$MeN{=}NbMe_3$
Nb-N1	1.796	1.771
N1-N2	1.355	
N2-N3	1.291	
Nb-N1-N2	175.4	
N1-N2-N3	115.5	
N2-N3-C	111.9	
Nb-N1-C		180.0

12.2 Metal Complexes Co-crystallized with an Organoazide

If the metal site of a coordination compound is in a sterically and/or electronically saturated situation, coordination of an organo azide will be hindered and thus metal centred reactivity of the N_3-fragment will not occur. If additionally the other ligands coordinated to the metal site are inert against an attack by the organoazide too, no reactivity will be observed at all. There must be a multitude of such intrinsically nonreactive combinations of organoazides and (transition) metal complexes which have not found their way into literature.

In 1965 Bailey *et al.* reported the structural elucidation of a co-crystallisate between bis(8-hydroxyquinolinato)copper(II) and picryl azide.[2] A similar system is known from free 8-hydroxyquinoline and picryl azide.[3] The driving force for the formation of these compounds is a charge transfer by π-interaction between the electron rich 8-hydroxyquinoline or the 8-hydroxyquinolinato ligand of the transition metal complex and the electron accepting picryl site resulting in deeply coloured products. Therefore, two equivalents of picryl azide per copper(II) complex were found in the solid state structure (Figure 12.3).

The authors did not report the thermal stability of the copper(II) complex. However, they mention, that it is 'noteworthy that picryl azide reacts with a variety of olefins, norbornene, pinene, cyclopentene, cyclooctene, and others, with exceptional ease'. The fact, that the copper(II) does not decompose this organoazide can clearly be attributed to the efficient shielding of the copper(II) centre by the two 8-hydroxyquinolinato ligands.

12.3 Cationic Metal Complexes with Organoazide Containing Anions

Another direct way to synthesize an inert system will be salt formation between a well-shielded metal cation and an organoazide containing anion. In a recent publication Cao *et al.* reported the synthesis of cationic cobalt(III) and nickel(II) ethylenediamine complexes with 4,4′-diazido-2,2′-stilbenedisulfonate ($dasb^{2-}$) as the counter ion.[4] The authors chose this special anion for the formation of hydrogen bound supramolecular structures, which are well established for organodisulfonates of the type $^{-}O_3S$-linker-SO_3^{-}. The NH_2 protons of the ethylenediamine ligands were acting as proton donors. Additionally the $dasb^{2-}$ ligand provides a rigid organic backbone and two azido substituents, which

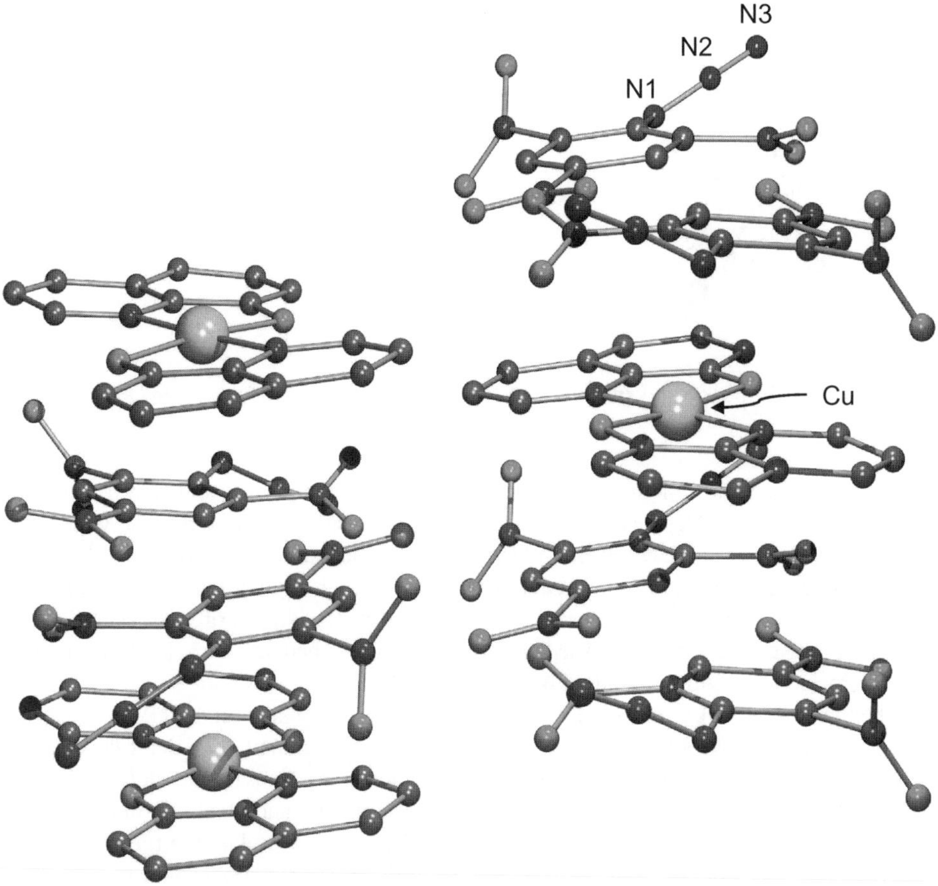

Figure 12.3 Solid state structure of bis(8-hydroxyquinolinato)copper(II)·(picrylazide)

can act as weak hydrogen bond acceptors too. For the nickel(II) compound, a sheet type of structure is built up by hydrogen bonds between the cation and the sulfonate groups of the anion and the sheets are packed to a 3D framework by weak hydrogen bonds involving the azido substituents. The high stability of these compounds was illustrated by TGA experiments: the cobalt(III) and nickel(II) frameworks are loosing only co-crystallized water molecules and ethylenediamine ligands up to temperatures of 177 °C and 212 °C, respectively.

12.4 Metal Complexes with Ligands Bearing a Non-coordinating Organoazide Unit

The next step to bring the organoazide 'closer' to the metal site would be to connect both fragments *via* a ligand backbone. The probably simplest examples for such type of

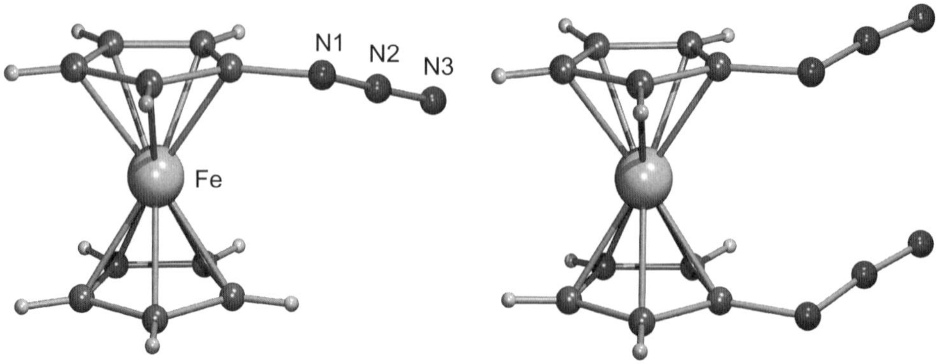

Figure 12.4 *Solid state structures of ferrocenyl azide and 1,1′-ferrocenylene azide*

compounds are azido functionalized ferrocenes. The mother compound ferrocenyl azide was reported first by Nesmeyanov *et al.* in 1962[5] by treating bromoferrocene with NaN_3 in aqueous DMF. This reaction provides excellent yields and opens up the chemistry of air sensitive ferrocenyl amine by reduction of the azido group with $LiAlH_4$. In a similar reaction, 1,1′-ferrocenylene azide can be obtained from 1,1′-dibromoferrocene and NaN_3. The solid state structure of ferrocenyl azide was published in the year 2006, that of 1,1′-ferrocenylene azide in 2000 (Figure 12.4).[6]

The thermal decomposition of ferrocenyl azide was investigated in detail. An activation energy of about 113 kcal/mol could experimentally be determined, the rate of decomposition is quite insensitive to the solvent, which indicates a relatively nonpolar transition state.[7] Dinitrogen, ferrocene, aminoferrocene, azoferrocene, and insoluble material were identified as the decomposition products, which allowed to postulate the formation of an intermediate ferrocenyl nitrene. The reaction products with cyclohexane, benzene and acetonitrile showed that this nitrene is nucleophilic. The Arrhenius parameters of this decomposition are comparable to those of other aromatic azides and do not offer any evidence for anchimeric assistance by the iron site. This finding is in contrast to the results of Azogu *et al.* who postulated, based on kinetic data, that 'lone pair e_g^2 electrons of the iron atom participate in the elimination of dinitrogen from ferrocenyl azide'.[8]

A chiral azide functionalized ferrocenophane (Figure 12.5) was generated by Erker *et al.* as a racemate by treating the corresponding lithiated ferrocenophane with tosyl azide and sodium pyrophosphate.[9] In principle such compounds can be synthesized in an enantiomerically pure form, which would allow to obtain valuable intermediates for chiral ligands.

The ferrocenyl fragment is a key example for a sterically and electronically saturated transition metal site. Therefore it is not surprising, that organoazides will not undergo reaction with this system under normal conditions. However, there is a series of other structurally well defined organoazides being in connection to a transition metal centre *via* a ligand backbone in the literature.

Modification of stable $18e^-$ complexes allows the generation of azide functionalized organometallic complexes. Among a series of other nucleophiles, trimethylsilyl azide is able to react with (η^4-6-hydroxy-7-methylnona-2,4,8-triene)tricarbonyliron(0) in the pres-

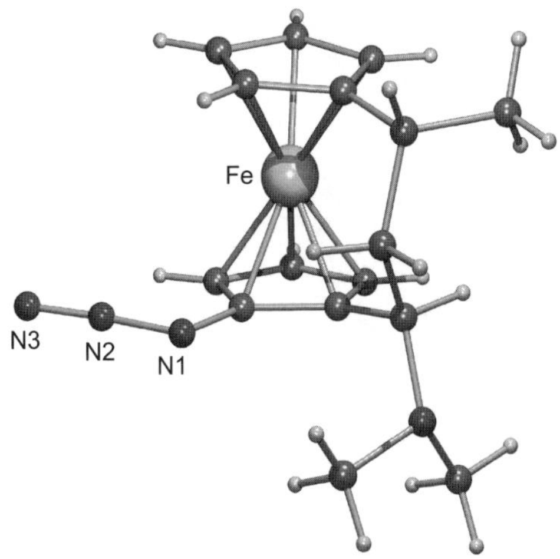

Figure 12.5 *Solid state structure of an azide functionalized chiral ferrocenophane*

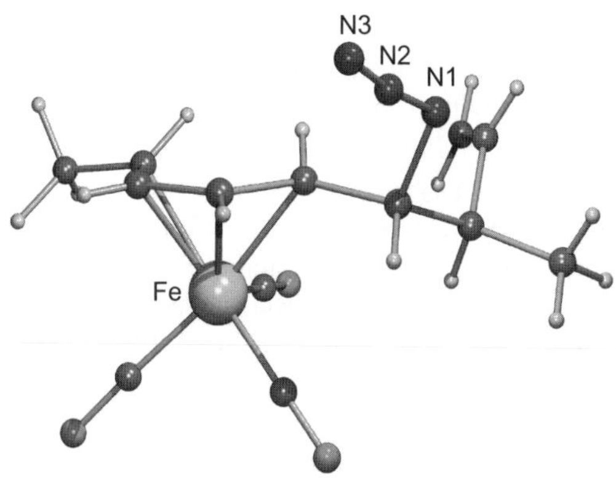

Figure 12.6 *Solid state structure of (η^4-6-azido-7-methylnona-2,4,8-triene)tricarbonyliron(0)*

ence of $BF_3{\cdot}Et_2O$ to form the corresponding organometallic azide (η^4-6-azido-7-methylnona-2,4,8-triene)tricarbonyliron(0) in excellent yields (Figure 12.6).[10]

A related reaction was published by Akita *et al.*: azide reacts with the cationic cyclopentadienyliron complex $[(\eta^5\text{-Cp})Fe(CO)_2(\eta^2\text{-Ph-C}\equiv\text{C-Ph})]^+$ by attacking the central C≡C triple bond to form the organoazide derivative dicarbonyl(η^5-cyclopentadienyl)(*trans*-1,2-diphenyl-2-azidoethenyl)iron(II) (Figure 12.7).[11] The *trans*-configuration of the C=C double bond seems to enhance the stability of the vinyl azide as shown by Fowler *et al.* for azido-1,2-diphenylethene.[12]

The intrinsic reactivity of the azido fragment in the presence of an inert metal site was used by Hahn *et al.* for the synthesis of benzannulated *N*-heterocyclic carbene ligands.[13] It is well established in the literature that 2-azido-phenyl isocyanide can be used as

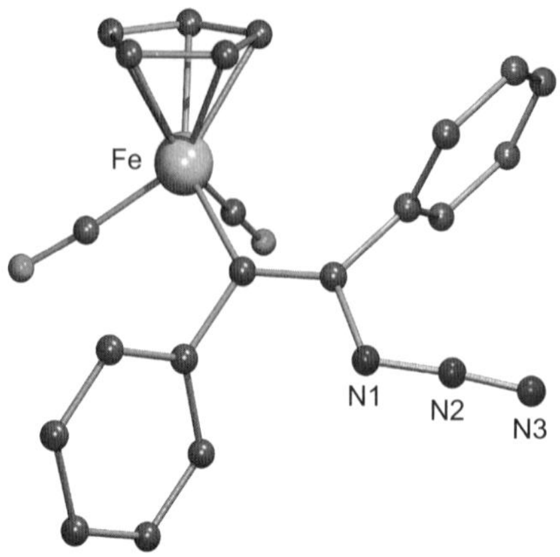

Figure 12.7 *Solid state structure of dicarbonyl(η^5-cyclopentadienyl)(*trans*-1,2-diphenyl-2-azidoethenyl)iron(II)*

precursor for the yet unknown 2-amino-phenyl isocyanide.[14] Isonitriles themselves are known to form stable complexes with low valent transition metal fragments. Reacting $[M(CO)_5(THF)]$ (M = Cr, W) with 2-azido-phenyl isocyanide yields by exchange of the thf ligand the corresponding stable isocyanide complexes (Figure 12.8). Those can be activated in a Staudinger reaction by addition of one equiv. of PPh_3 and loss of dinitrogen to give the stable iminophosphorane systems, which undergo the formation of the benzimidazolylidene complexes by treatment with catalytic amounts of HBr in aqueous methanol (Scheme 12.3). This methodology was further extended to a template synthesis of a cyclic tetracarbene platinum complex starting from $[Pt(PMe_3)_4]^{2+}$.[15] Subsequent replacement of the four PMe_3 ligands by 2-azido-phenyl isocyanide, followed by Staudinger reactions between the azide functions and liberated PMe_3 and acid catalyzed ring closure to platinum coordinated benzimidazolylidene ligands. The final cyclization is achieved with DMF in the presence of phosgene.

A square planar coordinated palladium(II) complex bearing a 4-azido-2,3,5,6-tetrafluorophenyl substituent was synthesized as a model system for a study on combined photo and radio labelling of cancer cells.[16] Here the azido substituent was introduced into a position far from the metal site to prevent any interaction. Nevertheless, the inert d^8-configured square planar coordinated metal centre again is typical for the class of organoazides discussed here. A further example of an inert metal site is gold(I) (d^{10}). Beck *et al.* used 3′-azido-3′-desoxythymidin, well known as the HIV drug AZT (Zidovudine, Retrovir) for coordination to gold(I) (Figure 12.9).[17]

The AZT gold(I) complex shows antiinflammatoric behaviour when tested with pig leukocytes. Additionally it is capable to inhibit HIV-I as pure AZT does it, which means that the substitution of the N3 proton of the thymine group by gold(I) has no significant influence on the virostatic activity of AZT. A zinc complex of the same ligand was published by Kimura *et al.*[18] Here the electronically inert zinc(II) cation (d^{10}) is perfectly shielded by a 1,4,7,10-tetraazacyclododecane ligand.

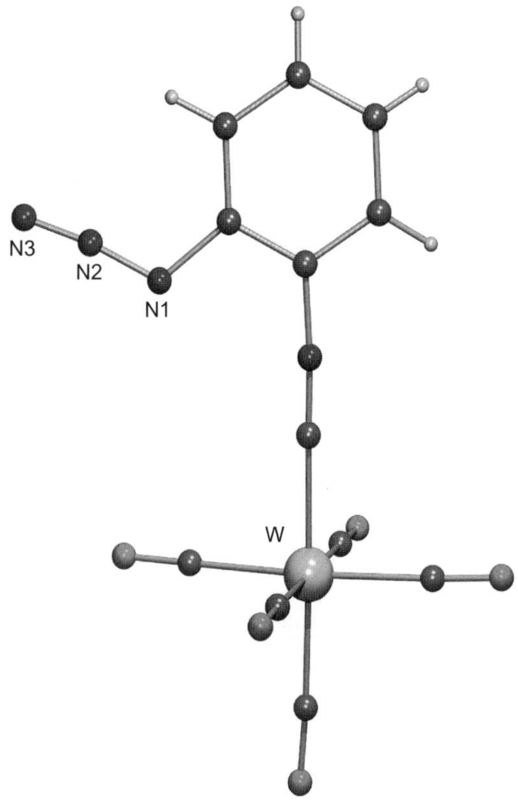

Figure 12.8 *Solid state structure of (2-azidophenylisocyanide)pentacarbonyltungsten(0)*

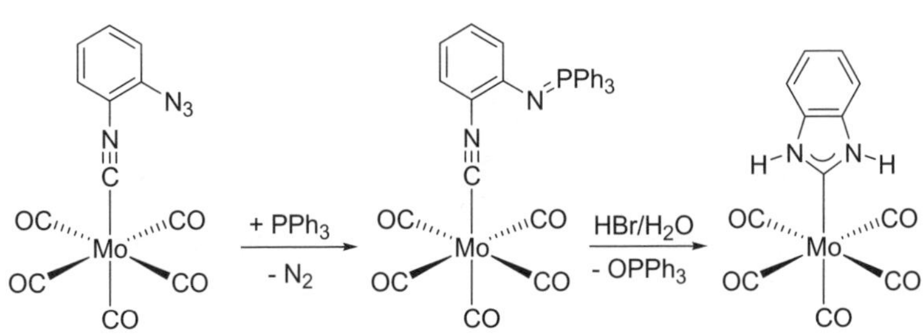

Scheme 12.3

This is also the case for the kinetically inert octahedrally coordinated chromium(III) centre (d^3) found in a salene chromium(III) complex with an organoazide in close proximity to the metal site (Figure 12.10).[19] Salene chromium(II) complexes are known to catalyze the enantioselective opening of prochiral epoxides with TMS-N_3.[20] The organoazide derivative was obtained by stoichiometric azide transfer from an azido chromium(III) salene complex to epoxycyclopentane.

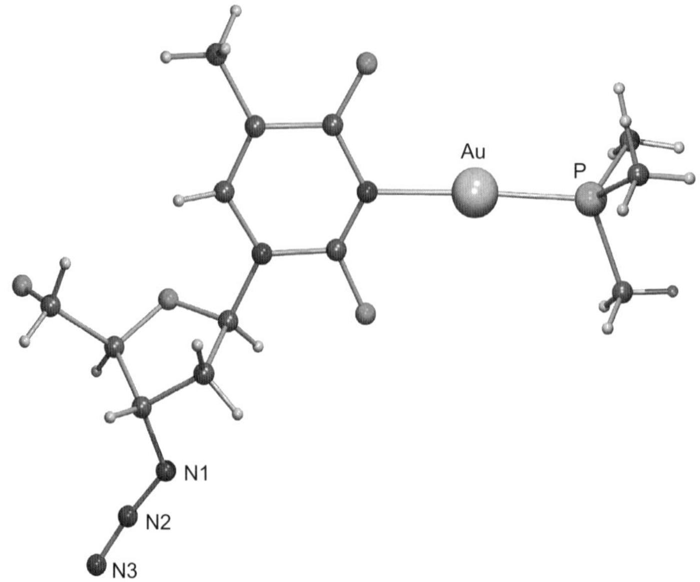

Figure 12.9 *Solid state structure of (3′-azido-3′-deoxythymidinyl)trimethylphosphinegold(I)*

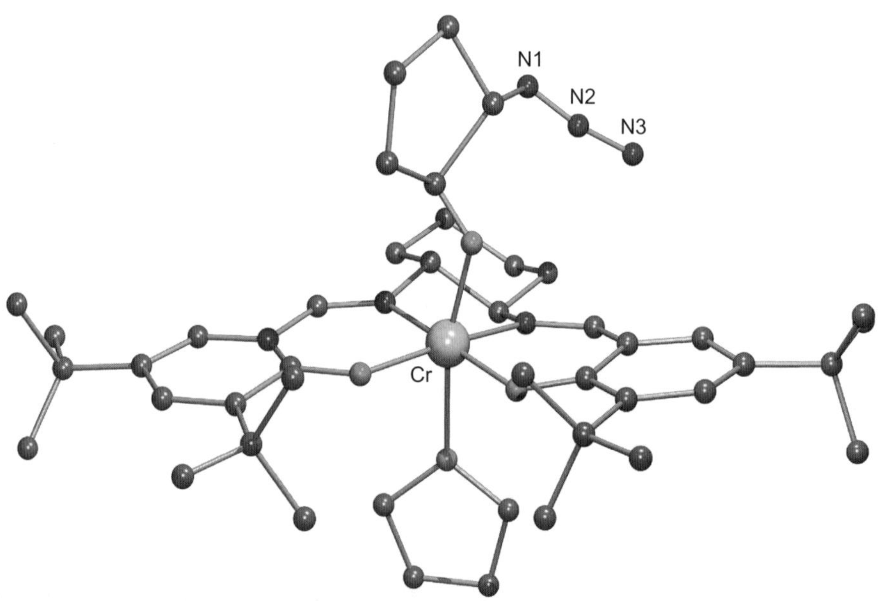

Figure 12.10 *Solid state structure of (2-azidocyclopentyloxy)(N,N′-cyclohexane-1,2-diyl-bis(3,5-di-tert-butylsalicylidenimine))(tetrahydrofuran)chromium(III)*

12.5 Metal Complexes with an Intact, Coordinating and Linear Organoazide Ligand

The discussion in Sections 12.2–12.4 should make it clear that preventing a metal mediated decomposition of an organoazide can be realized by hindering any interaction between the N_3 unit and the metal site. However, coordination of the organoazide must not obligatorily result in decomposition at all. There are several structurally characterized compounds in the literature, which allow to demonstrate the strategies to prevent this.

In 1998 Thiel *et al.* published the first structurally characterized linear organoazides coordinated to copper(II) and palladium(II).[21] Both elements are known among others to be efficient catalysts for the decomposition of organoazides.[22] The solid state structures of these compounds are shown in Figure 12.11. The strategy used here was to make the azido fragment a part of a chelating ligand which fixes the metal at the carbon bound nitrogen atom N1 of the azido unit. Additionally the positions *cis* to the organoazide motif are blocked by one or two further chloro ligands. This hinders the formation of the triazametallacyclobutene transition state for dinitrogen evolution and nitrene generation.

While the copper(II) complex is completely stable in the solid state and in solution, the palladium(II) compound decomposes slowly in dichloromethane solution to give the corresponding complex with an amine ($-NH_2$) function in the position of the azide ($-N_3$). The azide unit of the palladium complex is still capable of undergoing slow [2+3]-cycloaddition with EtOOC-C≡C-COOEt to yield the corresponding triazole. However, it is not yet clear whether this is due to decoordination of the azide ligand in solution or whether it takes place at the η^2-coordinated ligand.

Two even more impressive examples were published two years later by Dias, Marynick, *et al.*[23] Reacting $[HB(3,5\text{-}(CF_3)_2Pz)_3]Na(THF)$ (Pz = pyrazol-l-yl) with copper(I) trifluoromethanesulfonate in the presence of 1-azidoadamantane or reacting $[HB(3,5\text{-}(CF_3)_2Pz)_3]$

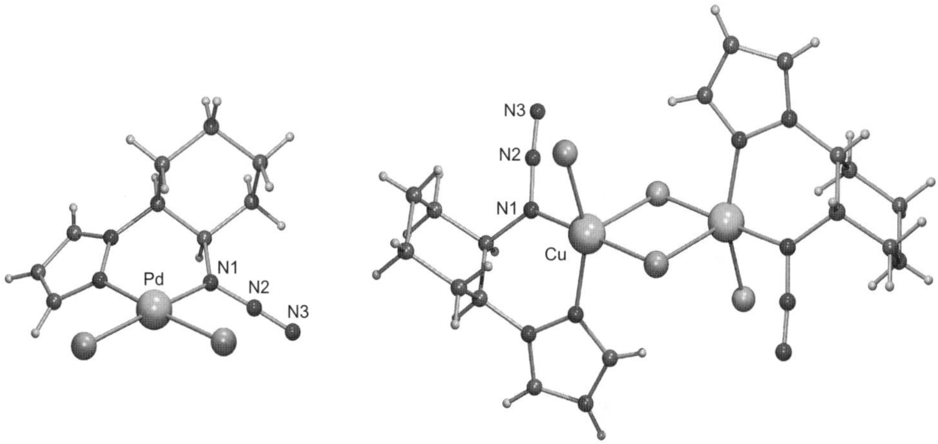

Figure 12.11 Solid state structures of (cis-2-azidopyrazol-1-ylcyclohexane) dichloridopalladium(II) and di(cis-2-azidopyrazol-1-ylcyclohexane)(chlorido)(μ-chlorido) copper(II)

Ag(THF) with 1-azidoadamantane gives the corresponding organoazide adducts to the M(I) centres in good to excellent yields. As in the palladium(II) and copper(II) complexes discussed above, the azide units are almost linear, indicating, that no electron transfer has taken place. Again, the metal sites are shielded efficiently, in these cases by the tridentate CF_3-functionalized trispyrazolylborato ligands. The most interesting aspect of this chemistry is that the organoazide undergoes two different coordination modi.

In the copper(I) compound, the organoazide coordinates with the terminal nitrogen atom N3. In silver(I), the carbon bound nitrogen atom N1 is used for coordination. The asymmetric stretching vibration of the azide group found for the silver complex is slightly lower (2120 cm^{-1}) than of the copper complex (2143 cm^{-1}). On the basis of theoretical calculations, the authors explained the different coordination modes of copper(I) and silver(I) by the presence or absence of metal to ligand back bonding: 'Theoretical calculations indicate that, in the absence of back bonding, the azide ligand prefers, if only slightly, binding through the more basic N1 site. (...) copper(I) exhibits enough π-donating ability to favour binding through the terminal nitrogen'. Steric effects were excluded. Here for the first time subtle electronic effects rise up, which will be dominant for the structure discussions in the following section.

12.6 Metal Complexes with an Intact, Coordinating but Bent Organoazide Ligand

According to Scheme 12.2 electron transfer from the metal centre to the organoazide will lead to a pronounced deviation from a linear arrangement of the N_3-unit. Such systems are better described as metal complexes of diazenylimido ligands.

In 1995 Bergman *et al.* described the first structurally characterized complex coordinating a bent organoazide obtained by reacting the tantalum(III) complex $Cp_2Ta(CH_3)(PMe_3)$ with phenyl azide (Figure 12.12).[24] Since the angle Ta-N-N (166.5 and 166.0°, two independent molecules in the unit cell) is large and the Ta=N bond distance (1.845 and 1.830 Å) is short, the Ta-N bond can be described as an intermediate between a double and a triple bond. The internal angle of the -N-N=N- unit (115.0 and 102°) is close to the expected value for a sp^2 hybridized nitrogen atom.

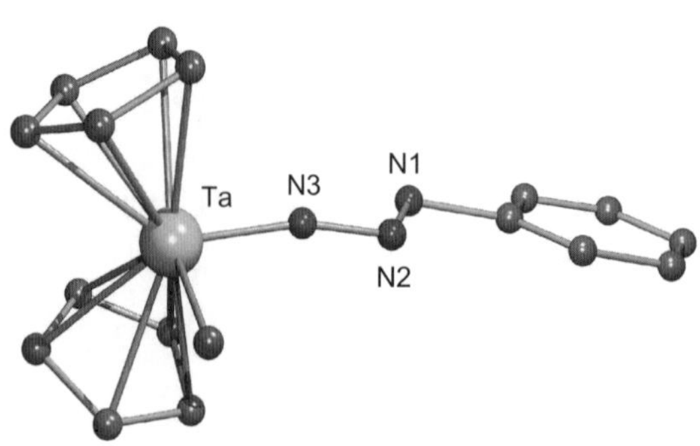

Figure 12.12 *Solid state structure of $Cp_2Ta(CH_3)(N_3C_6H_5)$*

Such complexes undergo loss of dinitrogen at elevated temperatures leading to the corresponding tantalum(V)imido complexes. The authors have investigated a series of such tantalum(V)diazenylimido complexes with differently substituted aryl groups at the N_3 unit. They have observed first order kinetics and no crossover of $^{13}CH_3$ and ^{15}N-aryl in double-labelling experiments. On the basis of their experiments Bergman *et al.* postulate a monomolecular pathway for the dinitrogen evolution including a triazatantalacyclobutene as an intermediate. A Hammett σ/ρ study suggests that not the loss of dinitrogen but the *anti/syn* isomerization of the Ta=N-N=N-R chain should be the rate determining step of this reaction.

Parallel to the work of Bergman *et al.*, Cummins *et al.* published a closely related system obtained from a vanadium(III) precursor and mesityl azide leading to a vanadium(V) diazenylimido complex, which loses dinitrogen already at room temperature.[25] Interestingly, the loss of dinitrogen from the vanadium(V)diazenylimido complex follows second-order kinetics and crossover was found in a double labelling experiment, which is in contrast to the observations of Bergman *et al.* with tantalum complexes.

Just a few years later, Floriani *et al.* described the reaction between a tungsten(IV) calixarene system and phenyl azide giving under oxidation to tungsten(VI) the corresponding diazenylimido complex, which again looses dinitrogen.[26]

In the three examples mentioned above, the two electrons, which are necessary to reduce the organoazide to a diazenylimido ligand, come from one single metal site. However, it is also possible that two metal sites each deliver one electron. This situation is of special interest for the description of azide decomposition at metallic surfaces. Again Bergman *et al.* published the reaction of a dinuclear zirconium-iridium system including a Zr-Ir single bond with phenyl azide.[27] Under cleavage of the metal-metal bond, a bridging diazenylimido ligand is formed, which again looses dinitrogen to give a bridging imido ligand at higher temperatures.

12.7 Organoazides Reacting with Other Metal Bound Ligands

Due to the reactive nature of organoazides a series of reactions concerning an attack at a metal coordinated ligand are known. Especially carbonyl complexes have been investigated for their reactivity with compounds of the type $R\text{-}N_3$. In most of these cases, a transfer of two electrons from the metal site to the organoazide will not lead to a stable diazenylimido complex but subsequent transformations will occur. One of the main routes of reactivity is the evolution of dinitrogen under primary formation of a nitrene species coordinated to the transition metal. For low valent transition metal sites such systems are instable in most cases, which may result in the typical nitrene type of reactivity. McElwee-White *et al.* have investigated the reaction of triphenylmethyl azide and triptycyl azide with $(CO)_5W(THF)$ and have solely found triphenyl imine and triptycyl amine as the major products.[28] Here no coordinating fragment of the organoazide could be detected. In contrast, reacting $Fe(CO)_5/MeN_3$, $CpCo(CO)_2/PhN_3$ or $Ni(COD)_2/C_6F_5N_3$ gives the η^4-tetraazadiene complexes (η^4-RN=N-N=NR)M(L)$_x$ (R = Me, C_6H_5, C_6F_5) resulting from an attack of the intermediately formed nitrene complexes to a further molecule of the organoazide.[29] If instead $Fe_2(CO)_9$ is treated with PhN_3 the azobenzene complex (η^2,μ^2-PhN=NPh)$Fe_2(CO)_6$ was obtained from the dimerization of two nitrene complexes.

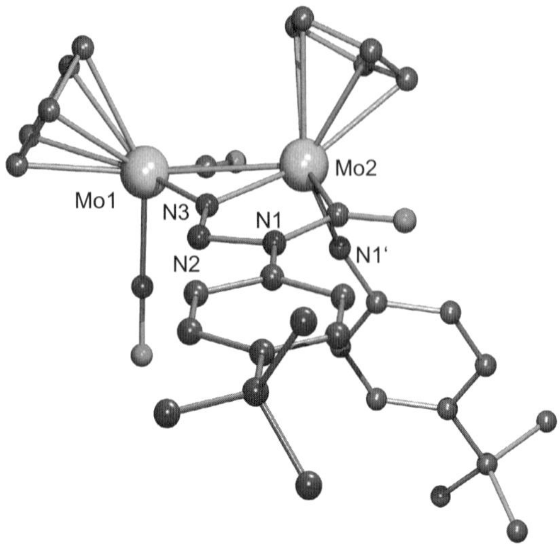

*Figure 12.13 Solid state structure of the adduct of two equivalents of 4-*tert*-butylphenyl azide to [CpMo(CO)$_2$]$_2$*

However, if the transition metal site has the chance to reach a stable oxidation state, carbonyl ligands can be included into the organoazide attack. This is the case for the reaction of $(PPh_3)_3Pd(CO)$ with p-$CH_3C_6H_4SO_2N_3$ (Ts-N_3) which gives the palladium(II) ureylene complex $(PPh_3)_3Pd(\eta^2$-TsNC(O)NTs)[30] and for the reaction of $(CO)_3Ru(PPh_3)_2$ with benzoyl azide or tosyl azide leading to the metal coordinated isocyanato derivatives C_6H_5CO-NCO and Ts-NCO or again to ruthenium(II) ureylene complexes.[31]

A series of quite interesting examples for the multiple reactivity of the dinuclear complexes $[CpMo(CO)_2]_2$ and $[Cp^*M(CO)_2]_2$ (M = Mo, W) against organoazides was published in the mid 1980s. Due to the d^5 configuration of the metal sites ($15\,e^-$), there are M≡M triple bonds in these compounds, which take part in the reaction.[32] The structure of the resulting reaction products could be elucidated by X-ray crystallography (Figure 12.13).

The reaction product can be understood by the deliverance of four electrons from the complex to two molecules of the organoazide. Hereby the metal sites are formally oxidized from M(I) to M(III) and thus the triple bond is transferred into a single bond. One of the organoazides looses dinitrogen by reduction and forms an imido R-N^{2-} ligand bound to one of the metal sites. The other RN_3^{2-} fragment, which can be considered to possess an electron distribution like structure C in Scheme 12.2, attacks the carbonyl ligand remaining at this metal site with the carbon bound nitrogen atom and bridges the two metal sites with the terminal nitrogen atom. This results in one short N3=N2 (126.0 pm) and one long N2-N1 (141.4 pm) distance and one longer Mo2=N3 (206.8 pm) and one shorter Mo1=N3 (196.4 pm) distance.

Reduction by hydride transfer ($H^- = H^+ + 2e^-$) occurs when the osmium cluster $[Os_3(\mu$-$H)_2(CO)_{10}]$ is reacted with a whole variety of organoazides: the hydride anion is transferred to the terminal nitrogen atom N3 and a monoanionic organoazide fragment is

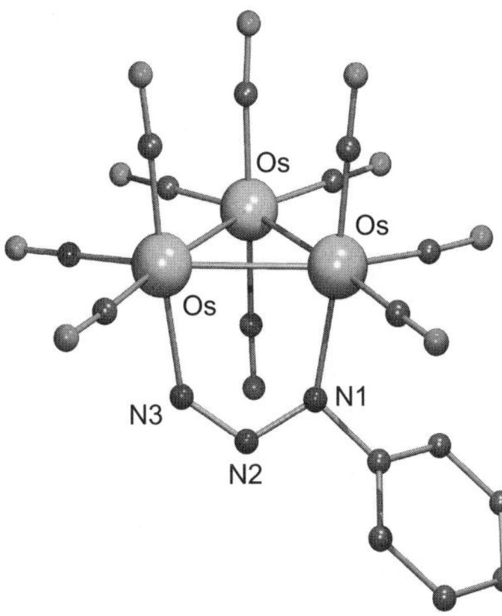

Figure 12.14 *Solid state structure of the adduct phenyl azide to [$Os_3(\mu\text{-}H)_2(CO)_{10}$], the bridging hydrido ligand and the proton at N3 were not localized*

obtained, which bridges two of the osmium sites by N1 and N3 (Figure 12.14).[33] At elevated temperatures, these clusters loose dinitrogen and the resulting nitrene fragment is stabilized by μ^3-coordination leading to [$Os_3(\mu\text{-}H)_2(CO)_9(\mu^3\text{-}NR)$].

One of the most prominent metal centred reactions of organoazides is the copper catalyzed 'click' triazole synthesis.[34] In a detailed computational study, Straub was able to elucidate the mechanism of this reaction: dicopper(I) μ-acetylide complexes were identified as the central intermediates.[35]

References

[1] K. Vrieze, G. van Koten, in Comprehensive Coordination Chemistry, vol. *2*, p. 225ff, Pergamon Press, **1987**.
[2] A.S. Bailey, C.K. Prout, *J. Chem. Soc.* **1965**, 4867–81.
[3] A.S. Bailey, R.J.P. Williams, J.D. Wright, *J. Chem. Soc.* **1965**, 2579.
[4] Y. Wang, R. Cao, W. Bi, X. Li, X. Li, Y. Wang, *Z. Anorg. Allg. Chem.* **2005**, *631*, 2309–11.
[5] A.N. Nesmeyanov, V.N. Drozd, V.A. Sazonova, *Dokl. Akad. Nauk SSSR* **1963**, *150*, 321–4.
[6] (a) P. Walla, V.B. Arion, U.H. Brinker, *J. Org. Chem.* **2006**, *71*, 3274–7. (b) A. Shafir, M.P. Power, G.D. Whitener, J. Arnold, *Organometallics* **2000**, *19*, 3978–82.
[7] (a) C. Steel, M. Rosenblum, A.S. Gehy, *Int. J. Chem. Kin.* **1994**, *26*, 631–41. (b) R.G. Sutherland, R.A. Abramovitch, C.I. Azogu, *J. Chem. Soc. D, Chem. Commun.* **1971**, 134–5.
[8] C.I. Azogu, M.N. Offor, *J. Organomet. Chem.* **1981**, *222*, 275–8.
[9] C. Nilewski, M. Neumann, L. Tebben, R. Fröhlich, G. Kehr, G. Erker, *Synthesis* **2006**, 2191–200.
[10] W.R. Roush, C.K. Wada, *Tetrahedron Lett.* **1994**, *35*, 7347–50.
[11] M. Akita, S. Kakuta, S. Sugimoto, M. Terada, M. Tanaka, Y. Moro-oka, *Organometallics* **2001**, *20*, 2736–50.
[12] F.W. Fowler, A. Hassner, L.A. Levy, *J. Am. Chem. Soc.* **1967**, *89*, 2077–82.

[13] F.E. Hahn, V. Langenhahn, N. Meier, T. Lügger, W.P. Fehlhammer, *Chem. Eur. J.* **2003**, *9*, 704–12.

[14] *Isonitrile Chemistry* (ed.: I. Ugi), Academic Press, New York, **1971**.

[15] F.E. Hahn, V. Langenhahn, T. Lügger, T. Pape, D. Le Van, *Angew. Chem. Int. Ed.* **2005**, *44*, 3759–63.

[16] R.S. Pandurangi, R.R. Kuntz, W.A. Volkert, C.L. Barnes, K.V. Katti, *J. Chem. Soc., Dalton Trans.* **1995**, 565–9.

[17] T. Pill, K. Polborn, A. Kleinschmidt, V. Erfle, W. Breu, H. Wagner, W. Beck, *Chem. Ber.* **1991**, *124*, 1541–8.

[18] M. Shionoya, E. Kimura, M. Shiro, *J. Am. Chem. Soc.* **1993**, *115*, 6730–7.

[19] K.B. Hansen, J.L. Leighton, E.N. Jacobsen, *J. Am. Chem. Soc.* **1996**, *118*, 10924–5.

[20] (a) L.E. Martinez, J.L. Leighton, D.H. Carsten, E.N. Jacobsen, *J. Am. Chem. Soc.* **1995**, *117*, 5897–8. (b) J.L. Leighton, E.N. Jacobsen, *J. Org. Chem.* **1996**, *61*, 389–90. (c) J.F. Larrow, S.E. Schaus, E.N. Jacobsen, *J. Am. Chem. Soc.* **1996**, *118*, 7420–1.

[21] M. Barz, E. Herdtweck, W.R. Thiel, *Angew. Chemie Int. Ed. Engl.* **1998**, *37*, 2262–5.

[22] (a) T. Sheradsky, in *The Chemistry of the Azido Group* (ed.: S. Patai), p. 332ff., Interscience Publishers, London, **1971**. (b) M. Mitani, M. Takayama, K. Koyama, *J. Org. Chem.* **1981**, *46*, 2226–7. (c) Y. Naruta, N. Nagai, Y. Arita, K Maruyama, *J. Org. Chem.* **1987**, *52*, 3956–7. (d) H. Kwart, A.A. Kahn, *J. Am. Chem. Soc.* **1950**, *89*, 1950–1.

[23] H.V.R. Dias, S.A. Polach, S.-K. Goh, E.F. Archibong, D.S. Marynick. *Inorg. Chem.* **2000**, *39*, 3894–3901.

[24] (a) G. Proulx, R.G. Bergman, *J. Am. Chem. Soc.* **1995**, *117*, 6382–3. (b) G. Proulx, R.G. Bergman, *Organometallics* **1996**, *15*, 684–92.

[25] M.G. Fickes, W.M. Davis, C.C. Cummins, *J. Am. Chem. Soc.* **1995**, *117*, 6384–5.

[26] G. Guillemot, E. Solari, C. Floriani, C. Rizzoli, *Organometallics* **2001**, *20*, 607–15.

[27] T.A. Hanna, A.M. Baranger, R.G. Bergman, *Angew. Chem., Int. Ed. Engl.* **1996**, *35*, 653–5.

[28] S.T. Massey, B. Mansour, L. McElwee-White, *J. Organometal. Chem.* **1995**, *485*, 123–6.

[29] (a) W.C. Trogler, *Acc. Chem. Res.* **1990**, *23*, 426–31. (b) M. Dekker, G.R. Knox, *Chem. Commun.* **1967**, 1243–4. (c) R.J. Doedens, *Chem. Commun.* **1968**, 1271–2. (d) S. Otsuka, A. Nakamura, *Inorg. Chem.* **1968**, *7*, 2542–4. (e) J. Ashley-Smith, M. Green, F.G.A. Stone, *J. Chem. Soc., Dalton Trans.* **1972**, 1805–9.

[30] W. Beck, W. Rieber, S. Cenini, F. Porta, G. La Monica, *J. Chem. Soc., Dalton Trans.* **1974**, 298–304.

[31] S. Cenini, M. Pizzotti, F. Porta, G. La Monica, *J. Organomet. Chem.* **1975**, *88*, 237–48.

[32] (a) J.J. D'Errico, L. Messerle, M.D. Curtis, *Inorg. Chem.* **1983**, *22*, 849–51. (b) W.A. Herrmann, G.W. Kriechbaum, R. Dammel, H. Bock, *J. Organometal. Chem.* **1983**, *254*, 219–41. (c) M.D. Curtis, J.J. D'Errico, W.M. Butler, *Organometallics* **1987**, *6*, 2151–7.

[33] (a) K. Burgess, B.F.G. Johnson, J. Lewis, P.R. Raithby, *J. Chem. Soc., Dalton Trans.* **1982**, 2085–92. (b) K. Burgess, B.F.G. Johnson, J. Lewis, P.R. Raithby, *J. Organomet. Chem.* **1982**, *224*, C40–C44.

[34] (a) V.V. Rostovtsev, L.G. Green, V.V. Fokin, K.B. Sharpless, *Angew. Chem. Int. Ed.* **2002**, *41*, 2596–9. (b) C.W. Tornøe, C. Christensen, M. Meldal, *J. Org. Chem.* **2002**, *67*, 3057–64.

[35] B.F. Straub, *Chem. Commun.* **2007**, 3868–70.

PART 3

Material Sciences

13

Azide-containing High Energy Materials

Thomas M. Klapötke and Burkhard Krumm

Department Chemie und Biochemie, Ludwig-Maximilians-Universität München, Butenandtstr. 5-13(D), D-81377 Munich, Germany

13.1 Introduction

The chemistry of azides has been explored for more than a century since the discovery of organic azides[1] by P. Griess and hydrazoic acid by T. Curtius.[2] The area of azides can formally be divided into organic and inorganic azides. Because of the rapidly growing number of publications on both categories, there have been various reviews on the preparation, properties, synthetic uses and applications for organic[3–12] and inorganic azides.[13–18] Most recently, the chemistry of HN_3 and its synthetic synthon trimethylsilyl azide Me_3SiN_3 has led also to novel, exciting polynitrogen compounds and more binary main group azides.[19,20] In this chapter newer developments on organic azides with a focus on energetic properties will be given. This survey can certainly not be exhaustive in scope, but rather will present some unusual molecules and/or highlights of this steadily developing field.

13.2 Organic Azides

The following sections will be subdivided into sections on alkyl, aryl and heteroaryl azides. There will be a focus especially on smaller molecules, or on those, which are larger but are synthesized for energetic purposes.

Organic Azides: Syntheses and Applications Edited by Stefan Bräse and Klaus Banert

13.2.1 Alkyl and Alkenyl Substituted Azides

13.2.1.1 Smallest Azidomethanes

The easiest organic azide and smallest member of azidomethanes, CH_3N_3, was first prepared by O. Dimroth in 1905 by simple methylation of sodium azide with dimethyl sulfate.[21] Methyl azide has been proven to be more explosive than originally reported (same accounts for ethyl azide).[22,23] The much more hazardous diazidomethane, $CH_2(N_3)_2$ and triazidomethane, $CH(N_3)_3$, are accessible by rather time-consuming slow reactions of dichloro/dibromomethane and tribromomethane with a polymeric ammonium azide reagent.[24] Several reports on the potential risk when working with azides in dichloromethane exist, and are attributed to the potential formation of diazidomethane (please see appropriate references cited in ref.[20]).

The synthesis of the ionic triazidocarbenium cation was accomplished as the hexachloroantimonate salt,[25] which was later also structurally characterized, being the first and up to date the only X-ray structurally characterized polyazidomethane compound.[26] Further triazidocarbenium salts with energetic counterions such as dinitramide and perchlorate, have been prepared and characterized in detail.[27]

The synthesis of the remaining elusive tetraazide of the azidomethane series, tetraazidomethane $C(N_3)_4$, was accomplished recently by treatment of trichloroacetonitrile with sodium azide.[28] Alternatively, CN_{12} can be prepared by reacting $[C(N_3)_3]SbCl_6$ with anhydrous LiN_3 or NaN_3. The fascinating tetraazidomethane, which is an extremely explosive liquid at ambient temperature (even superior to the lower azidomethanes), was shown to also undergo cycloaddition reactions to give triazoles and tetrazoles (Scheme 13.1). Traces of water hydrolyze tetraazidomethane to form the also explosive carbonyl diazide, a long-known sensitive carbonic acid derivative[29] (but now spectroscopically characterized) and hydrazoic acid.

Another very simple azidomethane is trinitroazidomethane, $N_3C(NO_2)_3$. This compound has been synthesized by the reaction of tetranitromethane with sodium azide in low yields, and is characterized and calculated (Figure 13.1) in detail. Furthermore, some indications for the existence of the rather explosive dinitrodiazidomethane, $(N_3)_2C(NO_2)_2$, have been reported.[30–32]

The trinitroazidomethane was one compound among some selected energetic molecules, which were used in a recent study to predict heats of formation of energetic materials from quantum mechanical calculations.[33]

13.2.1.2 Other Azidomethane Substituted Compounds

The very symmetrical hexakis(azidomethyl)benzene, $C_6(CH_2N_3)_6$, can be prepared easily by bromine/azide exchange, and the explosive properties have been investigated.[34]

The corresponding aliphatic per-azidomethyl derivative, pentaerythrityl tetraazide, $C(CH_2N_3)_4$, has been well-known for several years[35,36] and the crystal structure was recently reinvestigated.[37] The synthesis of the silicon analogue, tetrakis(azidomethyl)silane, $Si(CH_2N_3)_4$, was reported very recently and can be established in a similar fashion as the carbon analogue by treatment of the chloro derivative with excess sodium azide.[38]

The sila-azide derivative (Figure 13.2), a highly explosive liquid, is much more sensitive compared to $C(CH_2N_3)_4$, which also was proven by calculations.

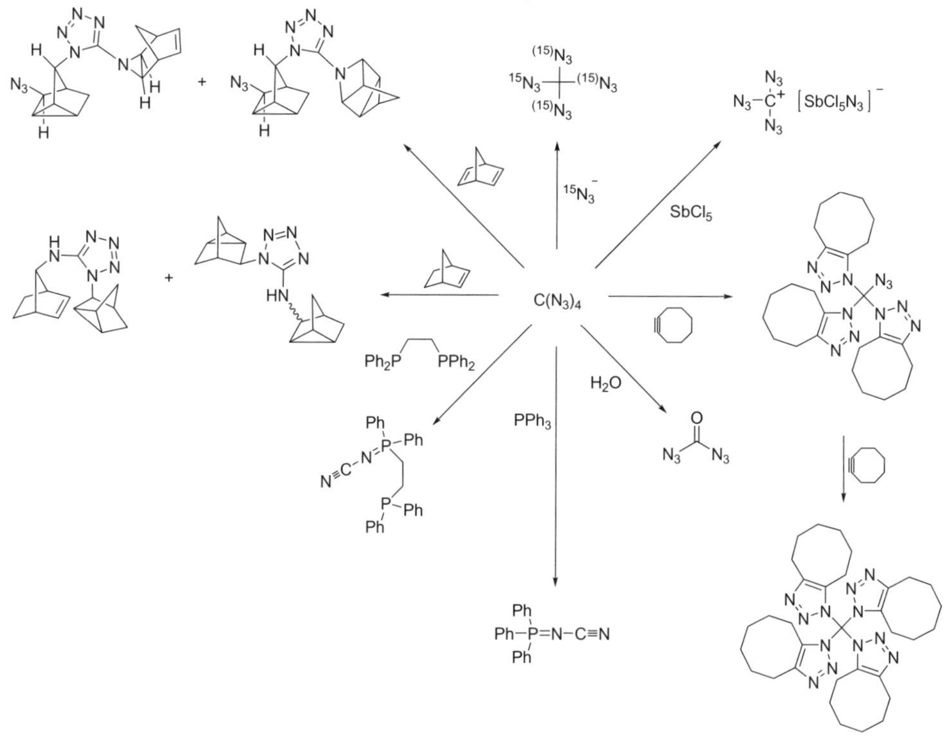

Scheme 13.1 *Reactions of tetraazidomethane*

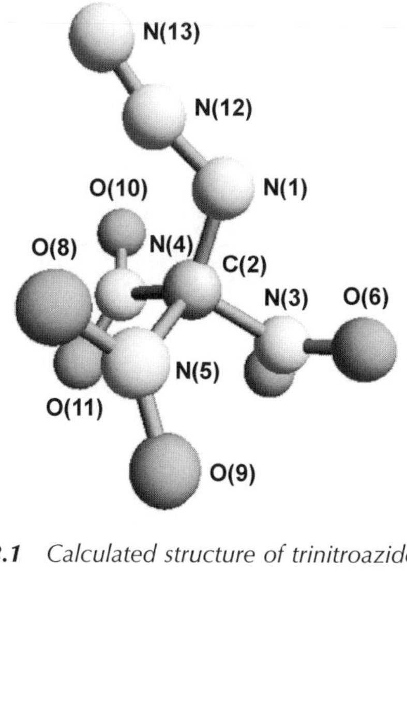

Figure 13.1 *Calculated structure of trinitroazidomethane*

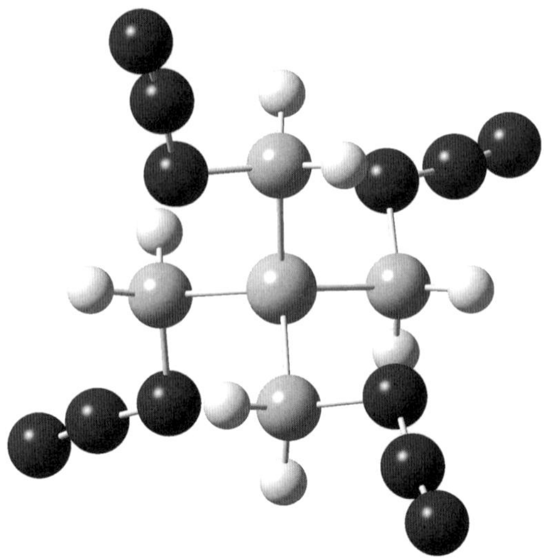

Figure 13.2 Calculated structure of tetrakis(azidomethyl)silane

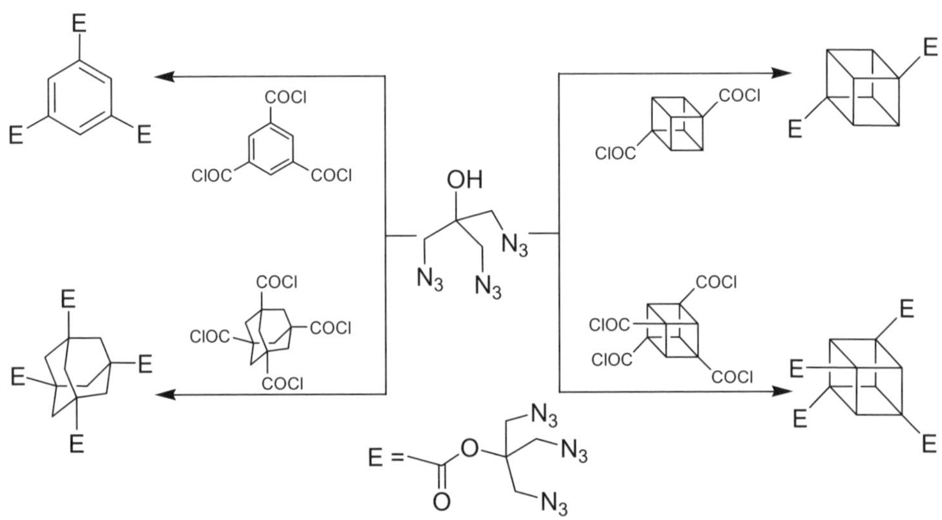

Scheme 13.2 Derivatives of tris(azidomethyl)methanol

Another series of novel compounds, deriving from the parent pentaerythrityl tetraazide containing the tris(azidomethyl)methyl moiety, were prepared for the purpose of cage based dendrimer synthesis.[39] An important precursor molecule is tris(azidomethyl)methanol (Scheme 13.2). With this precursor, interesting cubane (Figure 13.3), adamantane and benzene derivatives containing this functionality were synthesized.

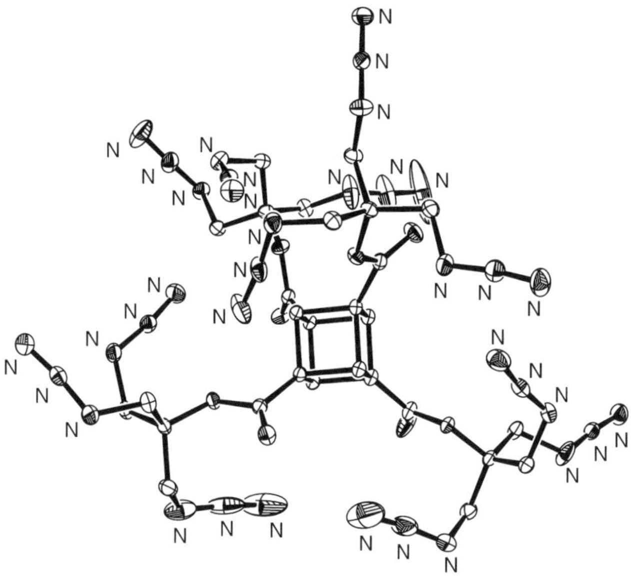

Figure 13.3 Molecular structure of tetrakis-1,3,5,7-tris(azidomethyl)methoxycarbonylcubane

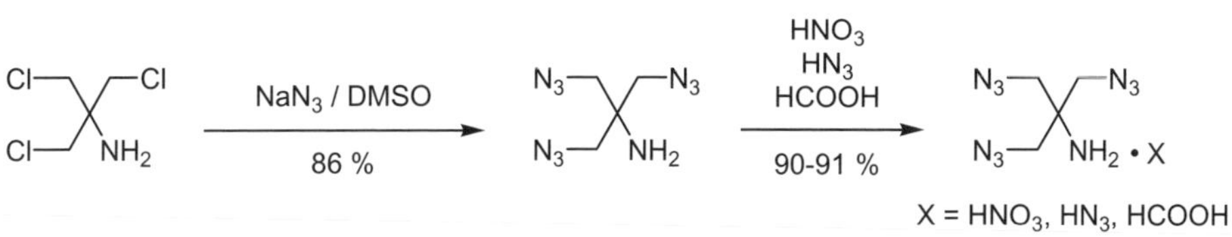

Scheme 13.3 Tris(azidomethyl)methylammonium salts

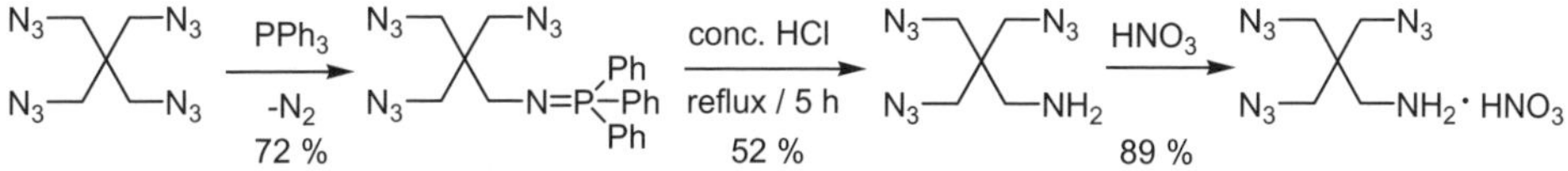

Scheme 13.4 2,2,2-Tris(azidomethyl)ethylammonium nitrate

The corresponding amine, $(N_3CH_2)_3CNH_2$, has been synthesized independently by two groups,[40,41] and was converted into the tris(azidomethyl)methylammonium salts (Scheme 13.3).

A Staudinger-type reaction with pentaerythrityl tetraazide leads to the amine $(N_3CH_2)_3CCH_2NH_2$, and further to the corresponding ammonium nitrate (Scheme 13.4).[41]

The corresponding alcohol $(N_3CH_2)_3CCH_2OH$ is prepared conveniently from 2,2,2-tris(chloromethyl)ethanol.[40–43] The alcohol serves as precursor for the new halogeno derivatives[41] (Scheme 13.5).

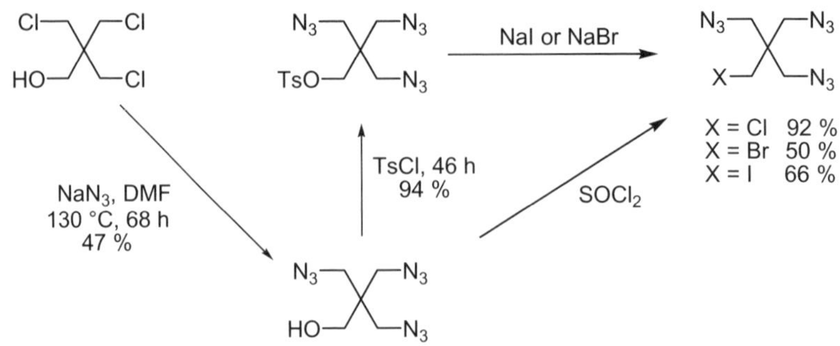

Scheme 13.5 *2,2,2-Tris(azidomethyl)ethanol and halogeno derivatives*

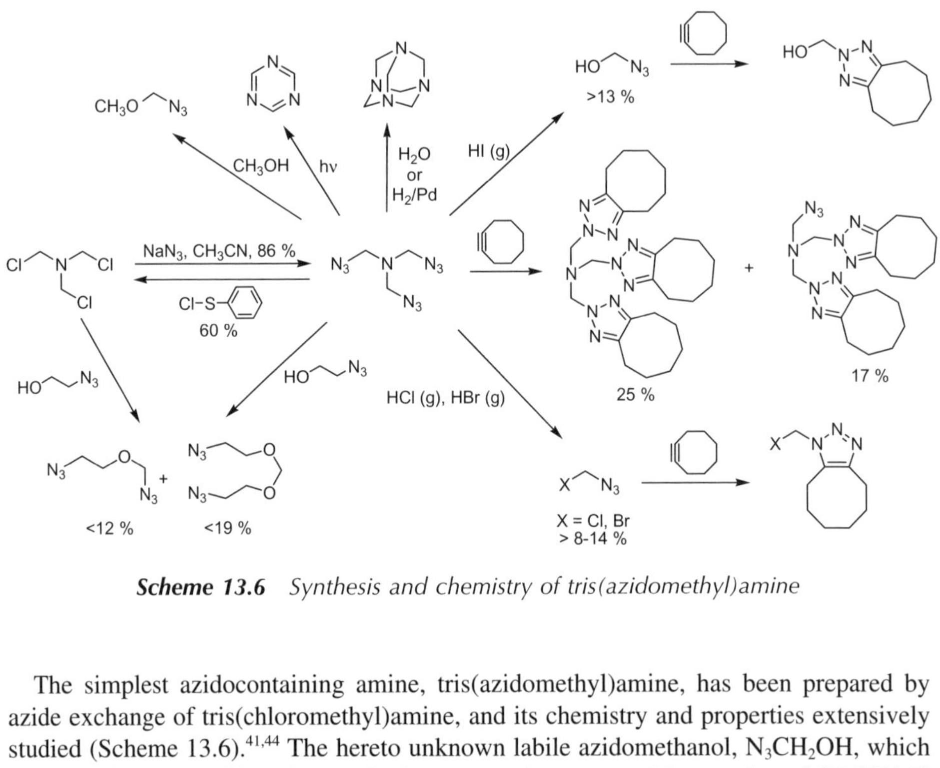

Scheme 13.6 *Synthesis and chemistry of tris(azidomethyl)amine*

The simplest azidocontaining amine, tris(azidomethyl)amine, has been prepared by azide exchange of tris(chloromethyl)amine, and its chemistry and properties extensively studied (Scheme 13.6).[41,44] The hereto unknown labile azidomethanol, N_3CH_2OH, which could not be prepared by other methods, can now be generated by reaction of $(N_3CH_2)_3N$ with hydrogen iodide, in an attempt to prepare the still unknown azidoiodomethane. The same accounts for azidochloromethane and azidobromomethane, N_3CH_2Cl/Br, which cannot be synthesized from the corresponding dihalogenomethanes. Both are formed by treatment of $(N_3CH_2)_3N$ with hydrogen chloride and hydrogen bromide, respectively.

The next larger homologue, tris(azidoethyl)amine, is known,[45] but has been re-investigated and prepared by a different route *via* the interesting liquid ammonium salt tetrakis(azidoethyl)ammonium azide (Scheme 13.7).[41] The Hofmann elimination proceeds under rather mild conditions without the use of a strong base.

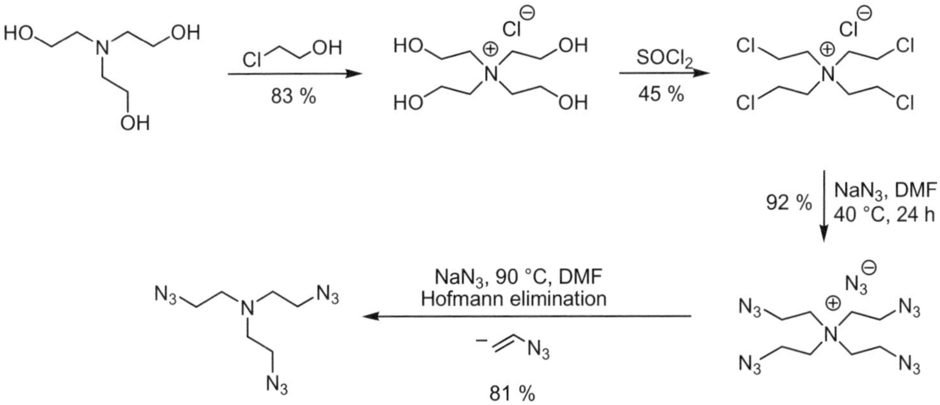

Scheme 13.7 *Synthesis of tetrakis(azidoethyl)ammonium azide and tris(azidoethyl)amine*

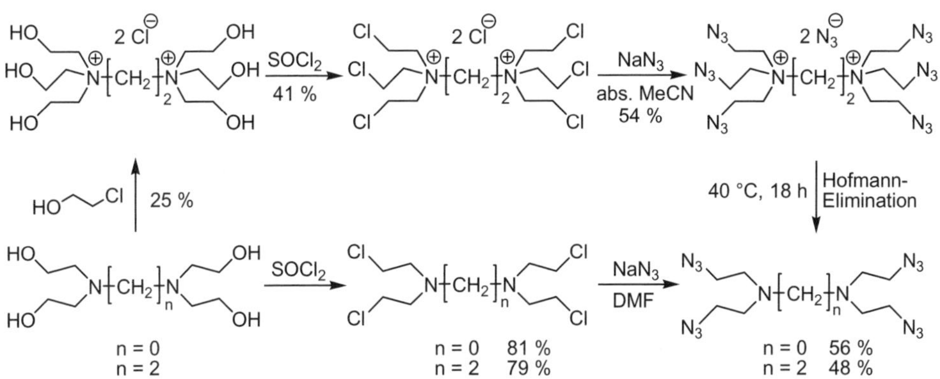

Scheme 13.8 *Synthesis of azidoethyl substituted diamine and hydrazine*

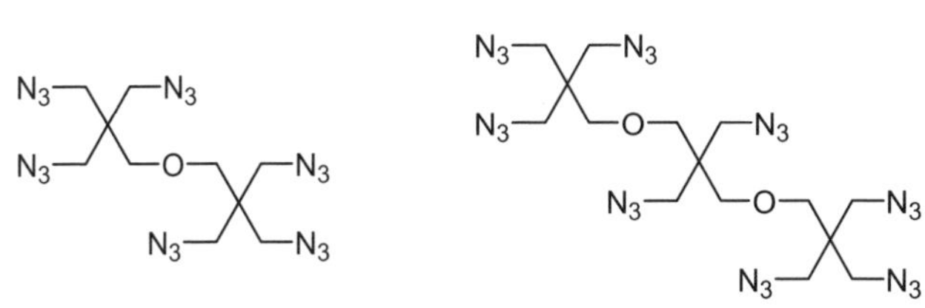

Figure 13.4 *Dipentaerythrityl hexaazide and tripentaerythrityl octaazide*

In addition, interesting polyazido diamine and hydrazine derivatives were prepared, either by Hofmann elimination of the corresponding diammonium and hydrazinium azide (again smooth conditions), or by reaction of sodium azide with the corresponding chloroethyl diamine or hydrazine (Scheme 13.8).[41]

Two larger derivatives of pentaerythrityl tetraazide, dipentaerythrityl hexaazide and tripentaerythrityl octaazide (Figure 13.4), both reported first in 1981,[46] have been re-examined and studied for their purpose in dendrimer synthesis with medicinal applications and energetic properties.[41,43,47]

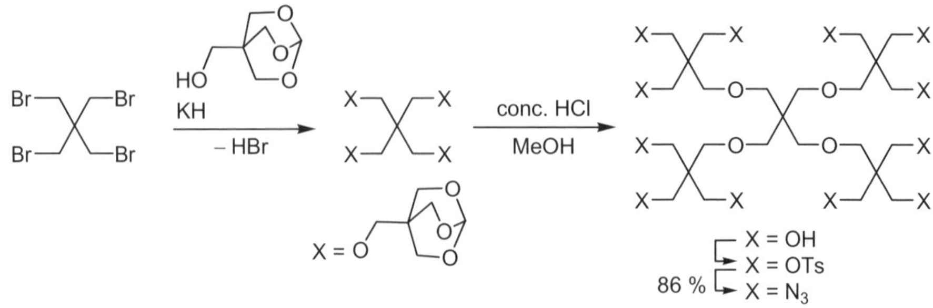

Scheme 13.9 *Synthesis of a dodecaazide*

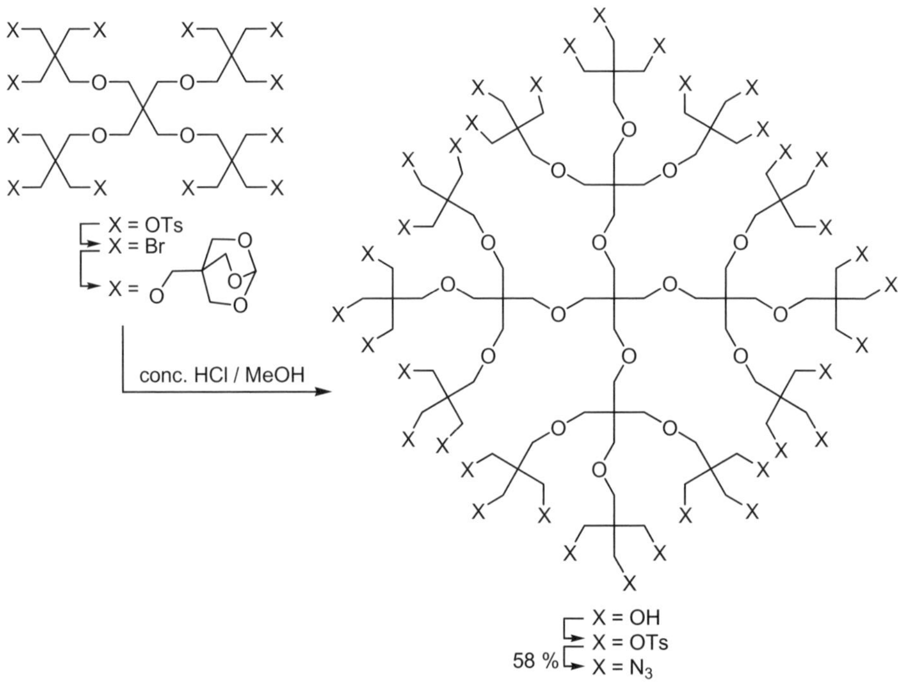

Scheme 13.10 *Synthesis of the hexatriacontaazide*

Even larger derivatives as the hexaazide and octaazide, a dodecaazide and a hexatriacontaazide, have been prepared by reaction of the tosylates as well, with azide (Schemes 13.9 and 13.10).[41] All four derivatives of pentaerythrityl tetraazide, are colorless liquids.

The first synthesis of 1,3-diazidoacetone, $(N_3CH_2)_2CO$, is described by two slightly different routes. This explosive liquid is also a precursor for other energetic materials, such as 2,2-dinitro-1,3-diazidopropane (Scheme 13.11).[39]

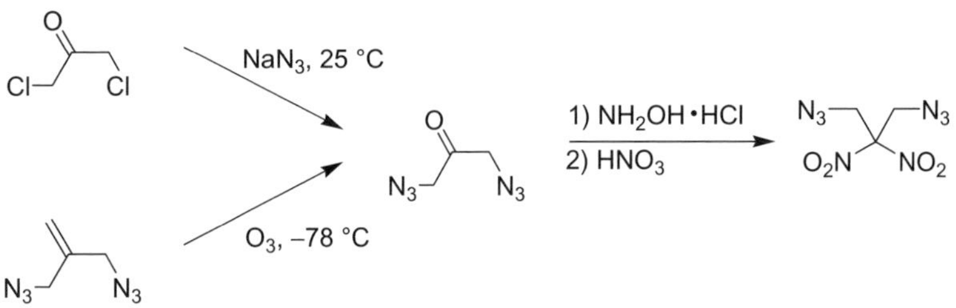

Scheme 13.11 *Preparation and nitration of 1,3-diazidoacetone*

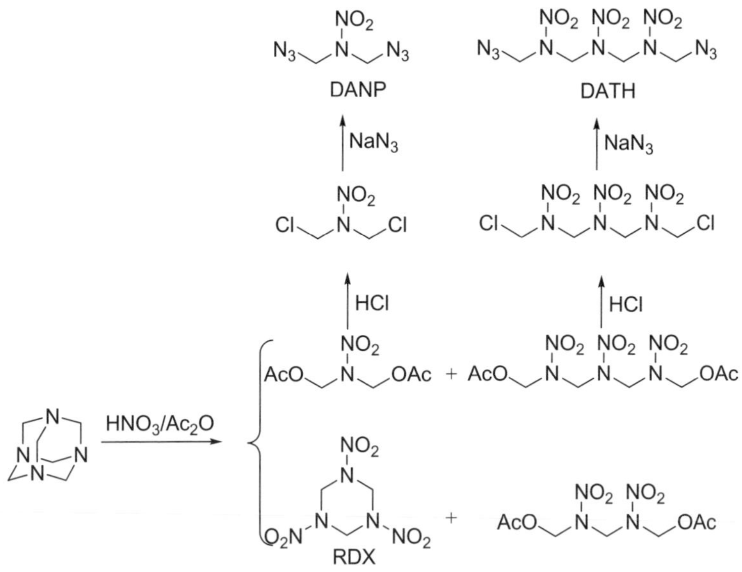

Scheme 13.12 *Synthesis of DANP and DATH from urotropine*

Another small organic azide similar to diazidoacetone, is azidoacetamide, $N_3CH_2CONH_2$, previously only sparsely mentioned, which can be easily prepared from chloroacetamide with sodium azide.[48,49]

A study of azidomethyl substituted nitramines and their sensitivity data was performed very recently.[50] The preparation of 1,3-diazido-2-nitro-2-azapropane (DANP) and 1,7-diazido-2,4,6-trinitro-2,4,6-triazaheptane (DATH) involves the easily available starting material urotropine (Scheme 13.12).

Table 13.1 shows the detonation parameters of DANP in comparison with well-known nitro explosives.

A polymer containing azidomethyl groups, glycidyl azide polymer (GAP), has been a subject of interest in the last years, and was originally developed in the US as an energetic binder for composite propellants. Because of the relatively high nitrogen content (42.4%), the large normal volume of detonation gases ($V_0 = 946\,\mathrm{L\,kg^{-1}}$) and the high thermal energy

Table 13.1 *Detonation parameters of DANP and commonly used nitro explosives*

		NG	RDX	PETN	TNT	DANP
Oxygen balance (%)	Ω	+3.5	−21.6	−10.1	−73.9	−37.2
Density (g/cm^3)	ρ	1.591[a]	1.82[a]	1.76[a]	1.654[a]	1.36
Heat of detonation (kJ/kg)	Q_v	−6671[a]	−6322[a]	−6322[a]	−4564[a]	−4863
Volume of detonation gases ($L/kg_{(explosive)}$)	V_0	716[a]	903[a]	780[a]	825[a]	805
Density (g/cm^3)	ρ	1.60[b]	1.80[b]	1.76[b]	1.64[b]	1.36
Detonation temperature (K)	T_{ex}	4260[b]	4354[b]	4349[b]	3744[b]	3954
Detonation pressure (kbar)	P	253[b]	345[b]	311[b]	202[b]	173
Detonation velocity (m/s)	D	7700[b]	8920[b]	8660[b]	7150[b]	7108

NG = nitroglycerine, RDX(Hexogen) = *s*-trimethylenetrinitramine; PETN = pentaerythrityl tetranitrate.
[a]cited in R. Meyer, J. Köhler, A. Homburg, Explosives, 5th Ed., Wiley-VCH, 2002
[b]cited in M. Suceska, Mat. Sci. For. **2004**, *465–466*, 325–330.

release (heat of explosion = 820 kcal kg^{-1}), in recent years GAP has also been used as an energetic binder in LOVA (low-vulnerability ammunition) gun propellants and in gas generating propellants.[51,52] In a recent combined experimental and theoretical study, the compatibility between GAP and various energetic fillers was investigated and decomposition pathways studied.[53,54]

13.2.1.3 Other Alkenyl and Cycloalkenyl Azido Substituted Compounds

Another small molecule is nitroguanyl azide, $(N_3)H_2NC{=}NNO_2$, a long known material,[55] but recently reinvestigated and structurally characterized.[56] It was found to have the nitrimino structure, and not as originally postulated, a nitramino structure, $(N_3)HN{=}CN(H)NO_2$. Energetic 5,5′-azotetrazolate salts, for example the azidoformamidinium salt, $[N_3C(NH_2)_2]_2[N_4CN{=}NCN_4]$, have been prepared, characterized, and their properties studied.[57]

The well-known explosive tetraazidoquinone[58] has also been reinvestigated and the crystal structure in a charge transfer adduct with tetrathiafulvalene determined (Figure 13.5).[59] The potential use of the adduct as a light detonant material under laser irradiation was examined.

13.2.2 Aryl Substituted Azides

13.2.2.1 Polyazido Benzenes

Since there are numerous examples of benzenes substituted with one azido group known, we will only take into consideration the much smaller number of compounds containing two or more azido groups, or molecules containing more than one azidoaryl group.

The use of a caged nitramine, CL-20, as a primary explosive, has stimulated interest to study further compounds containing this unusual cage system, referred to as hexaazaisowurtzitane, substituted with other potentially energetic groups. The formation of the cage system is accomplished by condensation of the corresponding benzyl amines with

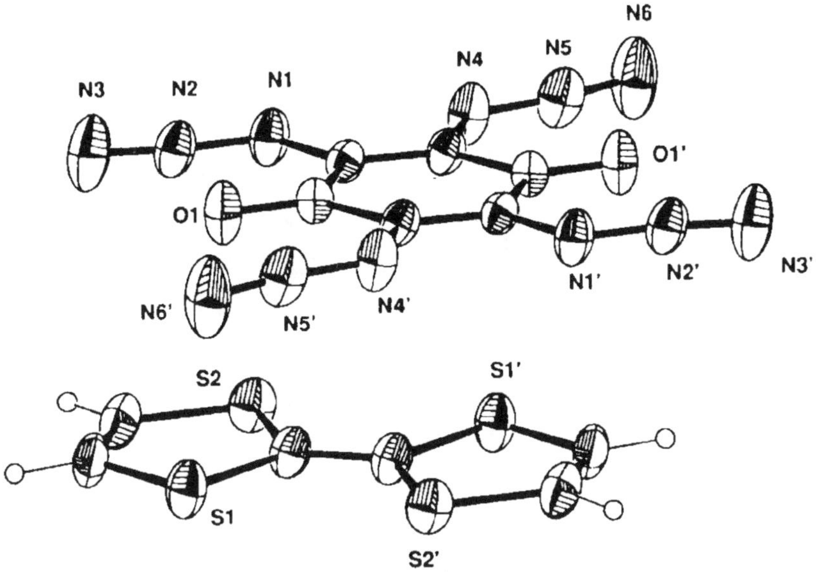

Figure 13.5 *Molecular structure of the tetraazidoquinone tetrathiafulvalene complex*

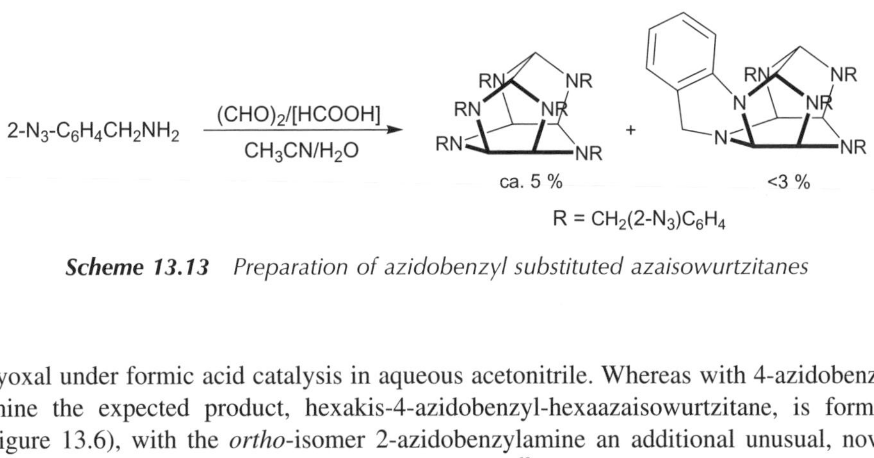

Scheme 13.13 *Preparation of azidobenzyl substituted azaisowurtzitanes*

glyoxal under formic acid catalysis in aqueous acetonitrile. Whereas with 4-azidobenzyl amine the expected product, hexakis-4-azidobenzyl-hexaazaisowurtzitane, is formed (Figure 13.6), with the *ortho*-isomer 2-azidobenzylamine an additional unusual, novel heterocycle is formed (Scheme 13.13, Figure 13.7).[60]

The preparation of a diazidosubstituted benzylamine, the 3,4-isomer, was achieved via a time-consuming multistep synthesis (Scheme 13.14). The precursor, 2-azido-4-toluidine, can be prepared by a four-step synthesis, with commercially available 2-nitro-4-toluidine as the starting material.[61]

The 3,4-diazidobenzylamine is a colorless, thermally labile liquid (explodes upon distillation).[62] Structural evidence exists for one of the precursors, 3,4-diazidobenzyl bromide, and is shown in Figure 13.8.

Some 1,3-diazidobenzene derivatives were prepared by double diazotation of the corresponding diaminobenzenes in low yields and studied for the purpose of dinitrene

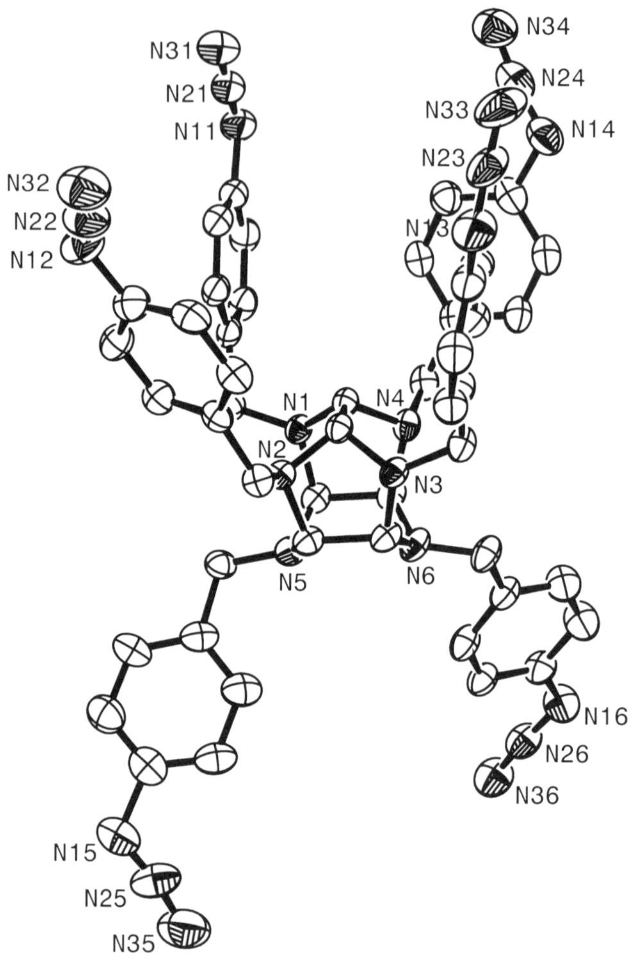

Figure 13.6 *Molecular structure of hexakis-4-azidobenzyl-hexaazaisowurtzitane*

generation.[63] The first crystal structure of a 1,3-diazidobenzene, 2,6-diazidotoluene, has been determined very recently and is shown in Figure 13.9.[64]

An interesting and unusual class of hydrogen free CNO-materials are the benzo-1,2,3,4-tetrazine 1,3-dioxides, such as the 6,8-diazido-5,7-dinitro derivative have been synthesized (Scheme 13.15).[65] This compound is very sensitive towards heating and should be handled very carefully.

The only structurally characterized benzene containing three azido groups is the sym-triazidotrinitrobenzene (molecular structure shown in a recent review[11]),[66] a surprisingly easy to handle, long-known material.[67,68]

The highest azide containing benzene derivative, containing four azide groups, which has been isolated in significant amounts, is 3,4,5,6-tetraazidophthalic anhydride. This reportedly sensitive material, was prepared for the controlled and careful thermolysis to give 3,4-dicyanomaleic anhydride (Scheme 13.16).[69]

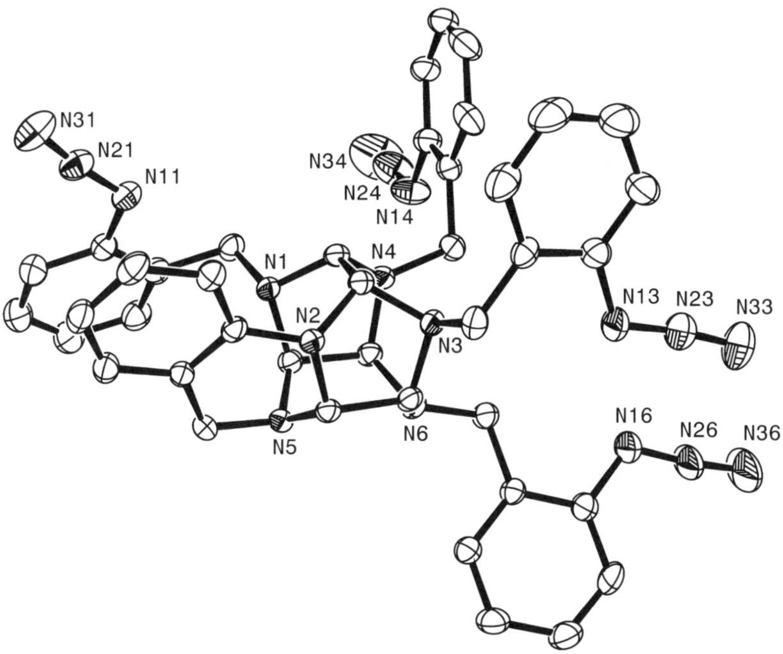

Figure 13.7 *Molecular structure of 2,10,12,14-tetrakis(2-azidobenzyl)-6,7-benzo-2,4,8,10,12,14-hexaazapentacyclo[7.5.1.0.0^{3,13}.0^{8,15}]pentadecane*

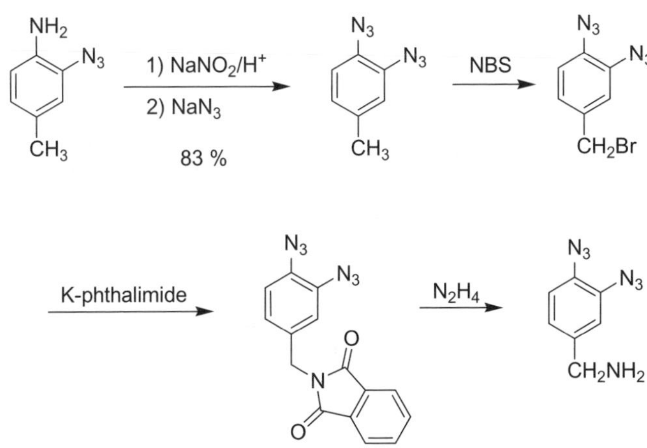

Scheme 13.14 *Preparation of 3,4-diazidobenzylamine*

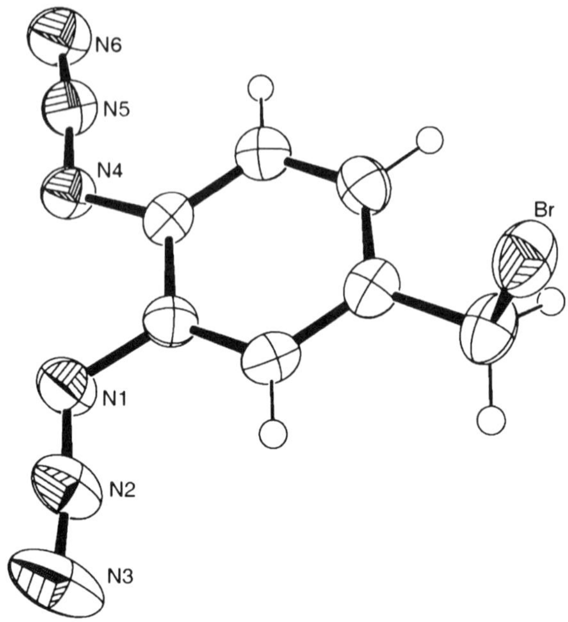

Figure 13.8 *Molecular structure of 3,4-diazidobenzyl bromide*

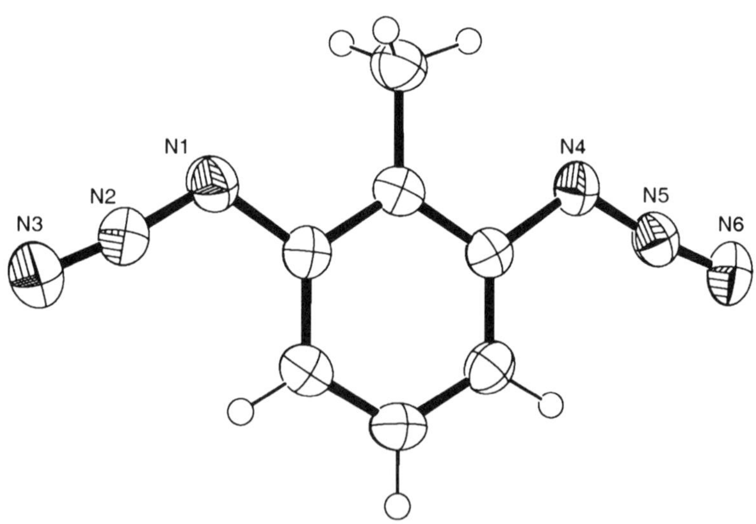

Figure 13.9 *Molecular structure of 2,6-diazidotoluene*

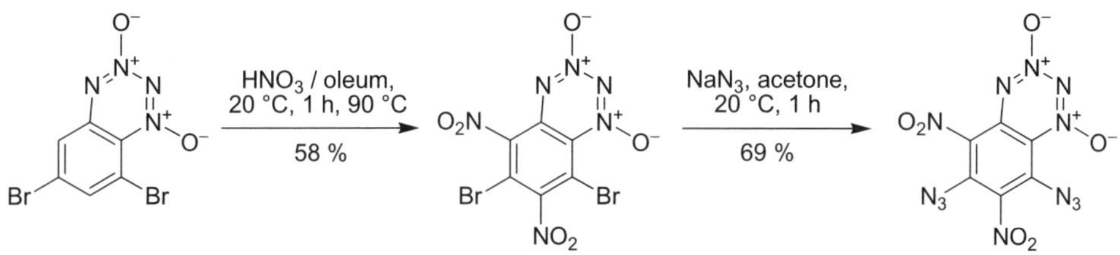

Scheme 13.15 *Synthesis of a diazido substituted benzo-1,2,3,4-tetrazine 1,3-dioxide*

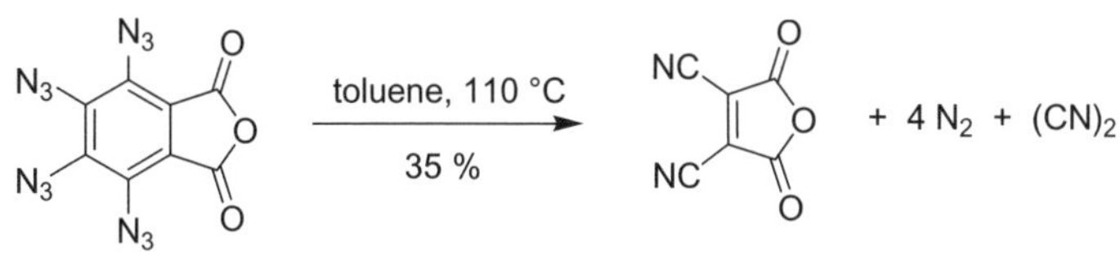

Scheme 13.16 *Thermolysis of 3,4,5,6-tetraazidophthalic anhydride*

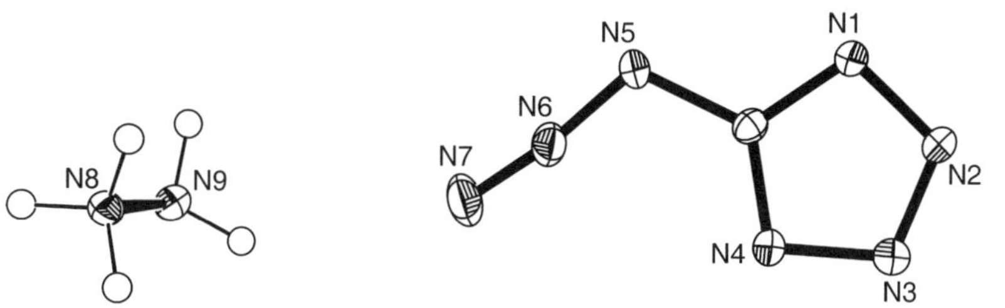

Figure 13.10 *Molecular structure of hydrazinium 5-azidotetrazolate*

The only report about a higher azide containing (penta- and hexasubstitution) benzene as a minor product is from a treatment of hexafluorobenzene with sodium azide and its stepwise substitution of fluorine by azide.[70] The major products from this reaction reportedly were those of di- and tetra-substitution, $C_6F_4(N_3)_2$ and $C_6F_2(N_3)_4$, but no further details are available.

13.2.3 Heterocycles Containing Azide Groups

13.2.3.1 5-Membered Rings – Imidazoles, Triazoles and Tetrazoles

3-Azido-1,2,4-triazoles and 3,5-diazido-1,2,4-triazole, as well as 2-azidoimidazole, have been known for some time. A recent account of these is given as well as a report of a study of some of their salts with energetic anions, such as 3-azido-1,2,4-triazolium nitrate, for the purpose of energetic ionic liquids.[71]

The determination of the crystal structure of 5-azido-1*H*-tetrazole, tetrazolyl azide CHN_7 (organic material with the highest nitrogen content of 88.3%),[72] has been reinvestigated very recently.[73] Several salts of CHN_7, azidotetrazolates, have been synthesized and their energetic properties studied. Figure 13.10 shows the crystal structure of the extremely sensitive hydrazinium salt (friction <5 N, impact <0.5 J), $[N_2H_5]CN_7$, which is at the present time the tetrazolate salt with the highest nitrogen content (88.1%).[74]

A rather surprising and unusual result was observed in an attempt to prepare 1-amino-5-azido-tetrazole from triaminoguanidinium chloride (Scheme 13.17).[75]

This 1-diazidomethylenamino substituted 5-azidotetrazole (structure proven by X-ray crystallography), can be obtained in good yields, and is extremely sensitive (friction <5 N, impact <0.5 J). Currently this C_2N_{14} is the tetrazole with the highest nitrogen content (89.1%) and approaches tetrazidomethane CN_{12} (93.3%[28]).

The hereto unknown and very sensitive 1,2-bis-(5-azido-1*H*-tetrazolyl)ethane has been prepared, spectroscopically and structurally characterized (Figure 13.11 and Scheme 13.18).[76] This substance is highly explosive, as the preliminary sensitivity data demonstrate (friction <5 N, impact >0.15 J). The synthesis is achieved in four steps all with good yields ranging from 40–80%, starting with triethyl orthoformate, ethylene diamine and sodium azide. The introduction of the azide group at the tetrazole moiety occurs at the last step; and therefore it can be regarded as a very safe synthesis to prepare a bis-tetrazolyl alkane.

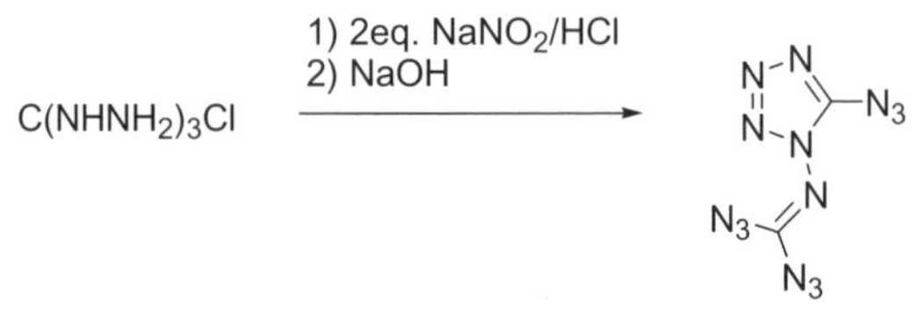

Scheme 13.17 *Synthesis of an azide enriched tetrazole*

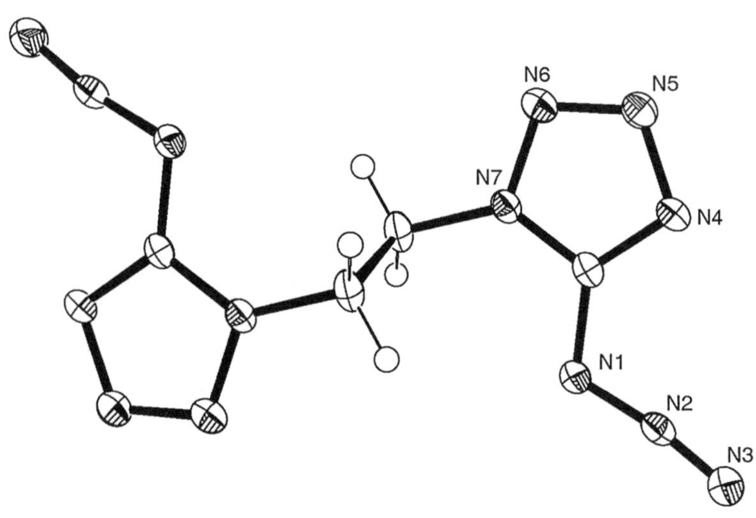

Figure 13.11 *Molecular structure of 1,2-bis-(5-azido-1*H*-tetrazolyl)ethane*

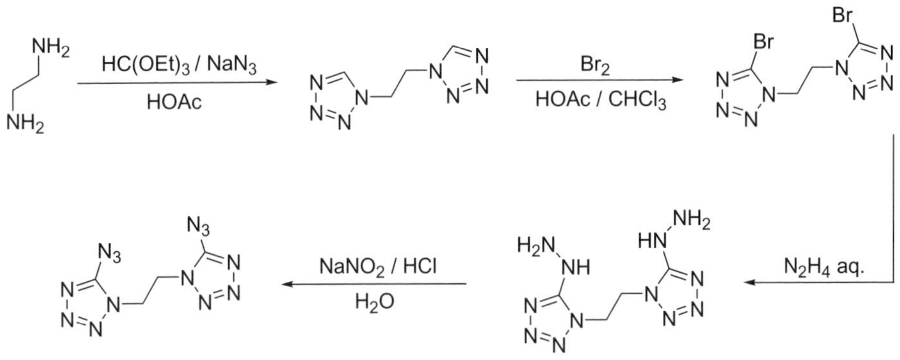

Scheme 13.18 *Synthesis of 1,2-bis-(5-azido-1H-tetrazolyl)ethane*

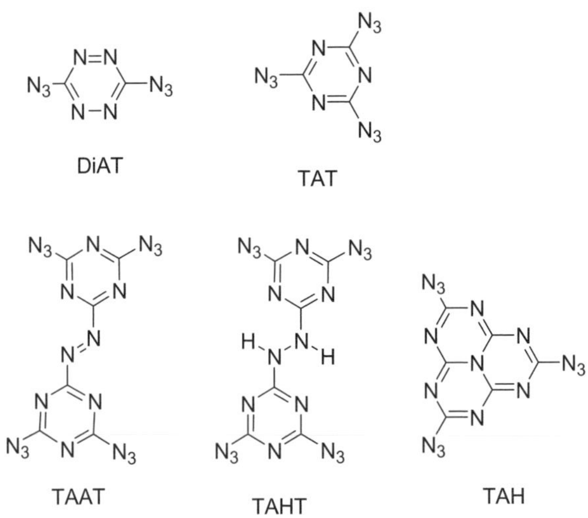

Scheme 13.19 *Various azido substituted polyazines*

13.2.3.2 6-Membered Rings – Triazines, Tetrazines, Heptazines, Pyrimidines

Recently there has been some progress on the synthesis of various azido substituted polyazines (Scheme 13.19).[77–79]

Those triazine derivatives are valuable precursors for the preparation of carbon nanospheres nitrogen-rich carbon nitride (CN_x) materials. One of the longest known compounds of that type is cyanuric azide (*s*-triazido-triazine, TAT), from which also some chemistry is reported.[80,81] Controlled decomposition studies and detonation studies of TAT under pressure were performed to give carbon nitrides and graphite particles, respectively.[82–84] The crystal structure of triazido-heptazine (TAH, Figure 13.12) has been determined recently.[85]

The azido-tetrazole ring-chain isomerism in polyazido-triazines, -tetrazines and heptazine has been discussed by extensive calculations.[86]

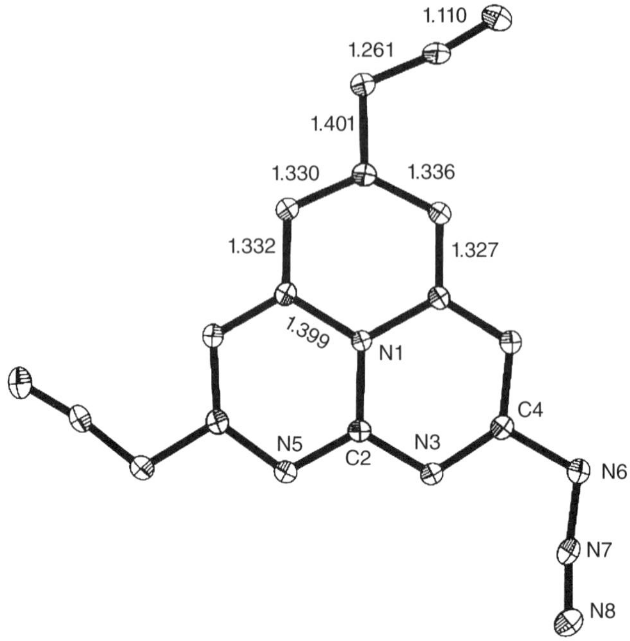

Figure 13.12 *Molecular structure of triazido-heptazine*

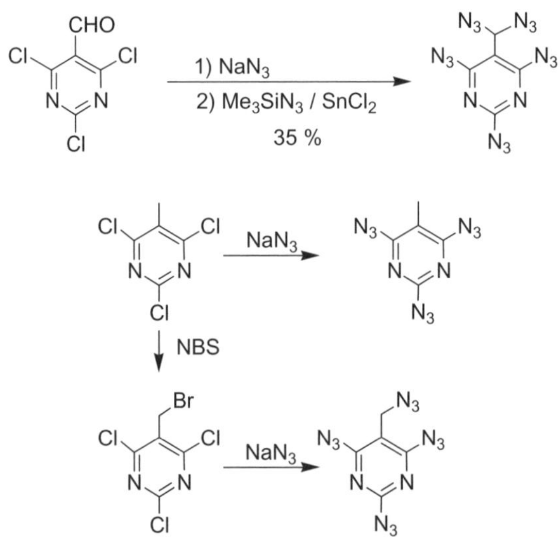

Scheme 13.20 *Azido-substituted pyrimidines*

The 1,3-system of diazine, the pyrimidines, has also been reported with azide substitution. 2,4,6-triazidopyrimidine (TAP) is known and was studied mainly for photochemical reactions.[87] Very recently, the first 5-methyl and 5-azidomethyl substituted derivatives of TAP have been prepared and also examined for their potential use on the route to carbon nanotubes (Scheme 13.20).[88]

For the diazido derivatives of 1,4-diazine, the pyrazine system, molecular surface electrostatic potential studies exist.[89] All three isomers of azidopyridine are well-known, but not further considered in this report; whereas polyazido substituted pyridines, such as diazidopyridines, remain unknown to present time.

Note added in proof: The sensitivity properties of tris(azidoethyl)amine have recently been determined experimentally,[90] because the azidomethyl derivative $(N_3CH_2)_3N$ was too dangerous. The values of $(N_3CH_2CH_2)_3N$ are comparable to those of RDX.

Acknowledgments

This work was supported by the University of Munich. The authors are grateful to F. Xaver Steemann, MSc and Mrs Carmen Nowak for redrawing several structures and schemes. We also would like to thank Prof. Dr Klaus Banert and Dr Matthias Scherr, for valuable suggestions and contributions.

References

[1] P. Griess, *Proc. R. Soc. London* **1864**, *13*, 375–84.
[2] T. Curtius, *Ber. Dtsch. Chem. Ges.* **1890**, *23*, 3023–33.
[3] G. L'abbé, *Chem. Rev.* **1969**, *69*, 345–63.
[4] *The Chemistry of the Azido Group* (ed.: S. Patai), Wiley, New York, **1971**.
[5] A. Hassner, *Acc. Chem. Res.* **1971**, *4*, 9–16.
[6] *The Chemistry of Halides, Pseudo-Halides and Azides* (eds.: S. Patai, Z. Rappoport), John Wiley & Sons Ltd, Chichester, **1983**.
[7] *Azides and Nitrenes: Reactivity and Utility* (ed.: E.F.V. Scriven), Academic Press, New York, **1984**.
[8] E.F.V. Scriven, K. Turnbull, *Chem. Rev.* **1988**, *88*, 297–368.
[9] *Chemistry of Halides, Pseudo-Halides and Azides, Part 1* (ed.: S. Patai), John Wiley & Sons Ltd, Chichester, **1995**.
[10] *Chemistry of Halides, Pseudo-Halides and Azides, Part 2* (Ed.: S. Patai), John Wiley & Sons Ltd, Chichester, **1995**.
[11] S. Bräse, C. Gil, K. Knepper, V. Zimmermann, *Angew. Chem.* **2005**, *117*, 5320–74; *Angew. Chem., Int. Ed.* **2005**, *44*, 5188–5240.
[12] S. Bräse, D. Keck, *Science of Synthesis* **2007**, *31b*, 1827–43.
[13] I.C. Tornieporth-Oetting, T.M. Klapötke, in *Combustion Efficiency and Air Quality* (eds.: I. Hargittai, T. Vidoczy), Plenum Press, New York, **1995**, pp. 51–62.
[14] I.C. Tornieporth-Oetting, T.M. Klapötke, *Angew. Chem.* **1995**, *107*, 559–68; *Angew. Chem., Int. Ed.* **1995**, *34*, 511–20.
[15] T.M. Klapötke, *Chem. Ber.* **1997**, *130*, 443–51.
[16] A. Kornath, *Angew. Chem.* **2001**, *113*, 3231–2; *Angew. Chem., Int. Ed.* **2001**, *40*, 3135–6.
[17] W. Fraenk, T.M. Klapötke, in *Inorganic Chemistry Highlights* (eds.: G. Meyer, D. Naumann, L. Wesemann), Wiley-VCH, Weinheim, **2002**, pp. 259–78.
[18] C. Knapp, J. Passmore, *Angew. Chem.* **2004**, *116*, 4938–41; *Angew. Chem., Int. Ed.* **2004**, *43*, 4834–6.
[19] K.O. Christe, *Propellants, Explos., Pyrotech.* **2007**, *32*, 194–204.
[20] T.M. Klapötke, B. Krumm, M. Scherr, R. Haiges, K.O. Christe, *Angew. Chem.* **2007**, *119*, 8840–5; *Angew. Chem., Int. Ed.* **2007**, *46*, 8686–90.
[21] O. Dimroth, W. Wislicenus, *Ber. Dtsch. Chem. Ges.* **1905**, *38*, 1573–6.
[22] C. Grundmann, H. Haldenwanger, *Angew. Chem.* **1950**, *62*, 410.
[23] *Nachr. Chem. Techn.* **1970**, *18*, 26–7.

[24] A. Hassner, M. Stern, H. E. Gottlieb, F. Frolow, *J. Org. Chem.* **1990**, *55*, 2304–6.
[25] U. Müller, K. Dehnicke, *Angew. Chem.* **1966**, *78*, 825; *Angew. Chem., Int. Ed. Engl.* **1966**, *5*, 841–2.
[26] U. Müller, H. Bärnighausen, *Acta Crystallogr.* **1970**, *B26*, 1671–9.
[27] M.A. Petrie, J.A. Sheehy, J.A. Boatz, *et al.*, *J. Am. Chem. Soc.* **1997**, *119*, 8802–8.
[28] K. Banert, Y.-H. Joo, T. Rüffer, B. Walfort, H. Lang, *Angew. Chem.* **2007**, *119*, 1187–90; *Angew. Chem., Int. Ed.* **2007**, *46*, 1168–71.
[29] W. Kesting, *Ber. Dtsch. Chem. Ges. B* **1924**, *57*, 1321–4.
[30] G.K. Khisamutdinov, V.I. Slovetsky, Y.M. Golub, S.A. Shevelev, A.A. Fainzil'berg, *Russ. Chem. Bull.* **1997**, *46*, 324–7.
[31] D.V. Levchenkov, A.B. Kharitonkin, V.A. Shlyapochnikov, *Russ. Chem. Bull.* **2001**, *50*, 385–9.
[32] V.A. Shlyapochnikov, D.V. Levchenkov, A.B. Kharitonkin, *Russ. Chem. Bull.* **2001**, *50*, 1173–80.
[33] E.F.C. Byrd, B.M. Rice, *J. Phys. Chem. A* **2006**, *110*, 1005–13.
[34] E.E. Gilbert, W.E. Voreck, *Propellants, Explos., Pyrotech.* **1989**, *14*, 19–23.
[35] E.B. Fleischer, A.E. Gebala, A. Levey, P.A. Tasker, *J. Org. Chem.* **1971**, *36*, 3042–4.
[36] W. Hayes, H.M.I. Osborn, S.D. Osborne, R.A. Rastall, B. Romagnoli, *Tetrahedron* **2003**, *59*, 7983–96.
[37] K.A. Lyssenko, Y.V. Nelubina, D.V. Safronov, *et al.*, *Mendeleev Commun.* **2005**, 232–4.
[38] T.M. Klapötke, B. Krumm, R. Ilg, D. Troegel, R. Tacke, *J. Am. Chem. Soc.* **2007**, *129*, 6908–15.
[39] P.R. Dave, R. Duddu, K. Yang, *et al.*, *Tetrahedron Lett.* **2004**, *45*, 2159–62.
[40] D.D. Diaz, S. Punna, P. Holzer, *et al.*, *J. Polym. Sci., A* **2004**, *42*, 4392–403.
[41] Y.-H. Joo, Ph.D. thesis, Technische Universität Chemnitz **2007**.
[42] E.R. Wilson, M.B. Frankel, *J. Org. Chem.* **1985**, *50*, 3211–12.
[43] M. Touaibia, T.C. Shiao, A. Papadopoulos, *et al.*, *Chem. Commun.* **2007**, 380–2.
[44] T.M. Klapötke, B. Krumm, M. Scherr, F.X. Steeman, K. Banert, Y-H. Joo, *Chem. Eur. J.* **2009**, *15*, 11341–5.
[45] E.F. Witucki, E.R. Wilson, J.E. Flanagan, M.B. Frankel, *J. Chem. Eng. Data* **1983**, *28*, 285–6.
[46] W.S. Anderson, H.J. Hyer, *CPIA Publ.* **1981**, *340*, 387–98.
[47] M. Touaibia, A. Wellens, T.C. Shiao, *et al.*, *ChemMedChem* **2007**, *2*, 1190–201.
[48] M. Kumasaki, K. Kinbara, Y. Wada, M. Arai, M. Tamura, *Acta Crystallogr.* **2001**, *E57*, o6–o8.
[49] J.M. Dyke, G. Levita, A. Morris, *et al.*, *J. Phys. Chem. A* **2004**, *108*, 5299–307.
[50] T.M. Klapötke, B. Krumm, F.X. Steemann, *Propellants, Explos., Pyrotech.* **2009**, *34*, 13–23.
[51] M.B. Frankel, L.R. Grant, J.E. Flanagan, *J. Propul. Power* **1992**, *8*, 560–3.
[52] A.N. Nazare, S.N. Asthana, H. Singh, *J. Energ. Mater.* **1992**, *10*, 43–63.
[53] M.A. Bohn, A. Hammerl, K. Harris, T.M. Klapötke, *Cent. Eur. J. Energ. Mater.* **2005**, *2*, 3–19.
[54] Ibid., pp. 29–44.
[55] E. Lieber, E. Sherman, R.A. Henry, J. Cohen, *J. Am. Chem. Soc.* **1951**, *73*, 2327–9.
[56] A.D. Vasiliev, A.M. Astachov, A.A. Nefedov, L.A. Kruglyakova, R.S. Stepanov, *Acta Crystallogr.* **2001**, *C57*, 625–6.
[57] A. Hammerl, M.A. Hiskey, G. Holl, *et al.*, *Chem. Mater.* **2005**, *17*, 3784–93.
[58] W.H. Gilligan, M.J. Kamlet, *Tetrahedron Lett.* **1978**, 1675–6.
[59] M. Fourmigue, K. Boubekeur, P. Batail, J. Renouard, G. Jacob, *New J. Chem.* **1998**, *22*, 845–50.
[60] T.M. Klapötke, B. Krumm, H. Piotrowski, K. Polborn, G. Holl, *Chem. Eur. J.* **2003**, *9*, 687–94.
[61] J.H. Hall, E. Patterson, *J. Am. Chem. Soc.* **1967**, *89*, 5856–61.
[62] T.M. Klapötke, B. Krumm, *unpublished results.*
[63] S.V. Chapyshev, H. Tomioka, *Bull. Chem. Soc. Jpn.* **2003**, *76*, 2075–89.
[64] T.M. Klapötke, B. Krumm, M. Scherr, G. Spiess, *Acta Crystallogr.* **2008**, *E64*, o348.

[65] A.Y. Tyurin, O.Y. Smirnov, A.M. Churakov, Y.A. Strelenko, V.A. Tartakovsky, *Russ. Chem. Bull.* **2006**, *55*, 351–6.
[66] D. Adam, K. Karaghiosoff, T.M. Klapötke, G. Holl, M. Kaiser, *Propellants, Explos., Pyrotech.* **2002**, *27*, 7–11.
[67] O. Turek, *Chim. Ind. (Paris)* **1931**, *26*, 781–94.
[68] R. Schmitt, *Z. Ges. Schiess-u. Sprengstoffwes.* **1943**, *38*, 148–9.
[69] K. Friedrich, R. Zimmer, *J. Prakt. Chem.* **1998**, *340*, 757–9.
[70] J.G. Morse, L.P. Kuhn, Ballistic Res. Lab., Aberdeen Proving Ground, MD, USA., **1970**, p. 20.
[71] H. Xue, Y. Gao, B. Twamley, J.M. Shreeve, *Chem. Mater.* **2005**, *17*, 191–8.
[72] A. Hammerl, T.M. Klapötke, H. Nöth, M. Warchhold, G. Holl, *Propellants, Explos., Pyrotech.* **2003**, *28*, 165–73.
[73] J. Stierstorfer, T.M. Klapötke, A. Hammerl, R.D. Chapman, *Z. Anorg. Allg. Chem.* **2008**, *634*, 1051–7.
[74] T.M. Klapötke, J. Stierstorfer, *J. Am. Chem. Soc.* **2009**, *131*, 1122–34.
[75] T.M. Klapötke, F. Martin, J. Stierstorfer, *unpublished results.*
[76] T.M. Klapötke, S. Sproll, *Eur. J. Org. Chem. Soc.* **2009**, 4284–9.
[77] M.-H.V. Huynh, M.A. Hiskey, E.L. Hartline, D.P. Montoya, R. Gilardi, *Angew. Chem.* **2004**, *116*, 5032–6; *Angew. Chem., Int. Ed.* **2004**, *43*, 4924–8.
[78] M.H.V. Huynh, M.A. Hiskey, J.G. Archuleta, E.L. Roemer, R. Gilardi, *Angew. Chem.* **2004**, *116*, 5776–9; *Angew. Chem., Int. Ed.* **2004**, *43*, 5658–61.
[79] M.H.V. Huynh, M.A. Hiskey, D.E. Chavez, D.L. Naud, R.D. Gilardi, *J. Am. Chem. Soc.* **2005**, *127*, 12537–43.
[80] E. Kessenich, T M. Klapötke, J. Knizek, H. Nöth, A. Schulz, *Eur. J. Inorg. Chem.* **1998**, 2013–16.
[81] E. Kessenich, K. Polborn, A. Schulz, *Inorg. Chem.* **2001**, *40*, 1102–9.
[82] E. Kroke, M. Schwarz, V. Buschmann, G. Miehe, H. Fuess, R. Riedel, *Adv. Mater.* **1999**, *11*, 158–61.
[83] E.G. Gillan, *Chem. Mater.* **2000**, *12*, 3906–12.
[84] T. Utschig, M. Schwarz, G. Miehe, E. Kroke, *Carbon* **2004**, *42*, 823–8.
[85] D.R. Miller, D.C. Swenson, E.G. Gillan, *J. Am. Chem. Soc.* **2004**, *126*, 5372–3.
[86] A. Hammerl, T.M. Klapötke, R. Rocha, *Eur. J. Inorg. Chem.* **2006**, 2210–28.
[87] V.Y. Pochinok, L.F. Avramenko, A.V. Pochinok, *et al.*, *Ukr. Khim. Zh.* **1979**, *45*, 1074–7.
[88] C. Ye, H. Gao, J.A. Boatz, *et al.*, *Angew. Chem.* **2006**, *118*, 7420–3; *Angew. Chem. Int. Ed.* **2006**, *45*, 7262–5.
[89] J.S. Murray, R. Gilardi, M.E. Grice, P. Lane, P. Politzer, *Struct. Chem.* **1996**, *7*, 273–80.
[90] F.X. Steemann, Ph.D. thesis, Ludwig-Maximilians-Universität, München, **2009**.

14

Azide Chemistry in Rotaxane and Catenane Synthesis

Stéphanie Durot,[1] Julien Frey,[2] Jean-Pierre Sauvage[1] and Christian Tock[3]

[1]Institut de Chimie, CNRS et Université de Strasbourg, UMR 7177 Laboratoire de Chimie-Organo-Minérale, 4 rue Blaise Pascal, BP 1032, F-67070 Strasbourg cedex, France; [2]CEA Saclay, iBiTec-S, Service de Bioénergétique, Biologie Structurale et Mécanismes. 91191 Gif-sur-Yvette, France; [3]BASF SE, 67056 Ludwigshafen, Germany

14.1 Introduction

The field of rotaxanes and catenanes is particularly active, mostly in relation to their potential applications as molecular machine prototypes[1–7] and new materials.[8–15] Through the years, the making of these mechanically interlocked compounds has evolved from a statistical approach to the use of efficient templating procedures.[16–18] The precursors, or 'pre-rotaxanes'[19] are built by taking advantage of precisely controlled non-covalent interactions such as hydrogen bonds, acceptor-donor and/or hydrophobic interactions between several organic fragments.[20–24] Alternatively, coordination chemistry has also demonstrated its power[25] with, in particular, the use of copper(I) as extremely efficient gathering and threading element,[32–37] able to induce the formation of pseudo-rotaxanes quantitatively. Generally speaking, such simple principles allow the preparation of rotaxane precursors in high yield and using simple experimental procedures. On the other hand, in most cases, when the final step is the introduction of one or more bulky groups, acting as stoppers, difficulties are encountered and this last step is often low-yielding and experimentally delicate.[24,36,38–42]

Organic Azides: Syntheses and Applications Edited by Stefan Bräse and Klaus Banert

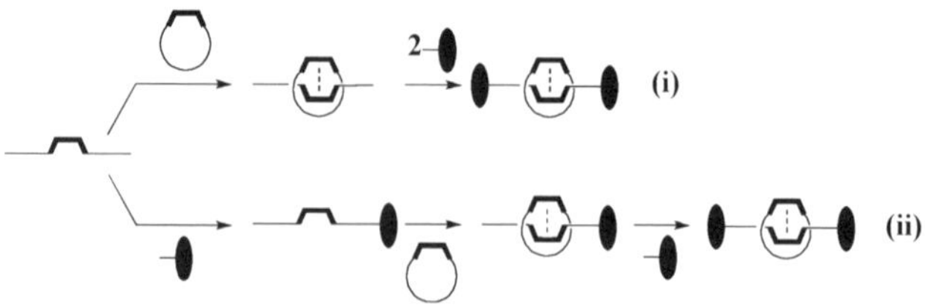

Scheme 14.1 *Schematic representation of possible strategies to prepare rotaxanes : (i) double-stoppering approach; (ii) stepwise approach. The dotted lines represent attractive interactions between given fragments belonging to the ring and to the thread respectively*

In general, two main strategies have been explored to synthesize rotaxanes (Scheme 14.1): the first one is the double-stoppering approach. After threading of a string-like fragment through a ring using non-covalent interactions, a double-stoppering reaction produces the desired rotaxane. Usually, this method leads to the formation of rotaxanes with modest or low yields, especially if the precursor is not very stable under the reaction conditions. To limit the unthreading reaction leading to dissociation of the precursor, a reasonably efficient strategy has been used which consists in a stepwise approach. It comprises three steps: (i) attachment of a first stopper to the string, followed by (ii) the threading reaction, and, finally, (iii) fixation of the second stopper to the thread. This procedure is well adapted when the pre-rotaxane is sensitive under the relatively harsh reaction conditions used to attach the stoppers, since a mono-stoppered precursor is less prompted to unthreading than a non-stoppered one. On the other hand, the double-stoppering approach can be very efficient in case of stable and robust pre-rotaxanes or mild reaction conditions which do not cause detrimental unthreading.

Until recently, azides were relatively rare precursors of interlocked molecules and most of the time they were meant to be linked to alkynes using 1,3-dipolar cycloaddition. In this review, we will restrict ourselves to their use in attachment of stoppers to pseudo-rotaxanes and present a selection of the most representative papers. They can be divided into purely organic molecular entanglements and into interlocked molecules prepared by transition metal templated approaches.

Huisgen's cycloaddition between azides and alkynes[43,44] is typically carried out at high temperature, at which unfortunately, labile molecules may not survive. Another reason for the scarcity of azides in rotaxane and catenane synthesis is the lack of regiospecificity of this reaction, which provides both 1,4- and 1,5-substituted 1,2,3-triazoles. Despite these drawbacks, a few interlocked molecules were prepared by Stoddart and co-workers by this *uncatalyzed* strategy using a symmetrical bulky alkyne (see Section 14.2.2). By contrast, the *catalyzed* versions of 1,3-dipolar cycloaddition are highly regiospecific and benefit from mild reaction conditions and excellent yields. The use of cucurbituril as catalyst will be presented in Section 14.2.1. The so-called 'click chemistry' reaction,[45,46] based on Cu(I)-catalyzed cycloaddition, discovered independently by Meldal[47,48] and Sharpless,[49] represents a promising possibility as stoppering reaction of the preliminary prepared pseudo-rotaxane. A limited number of such reactions has been recently reported,[32,33,50–58] a particularly impressive case being that of a copper-complexed rotaxane for which the copper(I) center is used both as template and catalyst.[32,33]

14.2 Purely Organic Rotaxanes and Catenanes

14.2.1 With Cucurbiturils (CB) and Cyclodextrins (CD) as Cyclic Components

Cucurbit[n]urils (CB[n]) (Scheme 14.2) and cyclodextrins (CD) are among the most popular macrocyclic molecules that can be used as hosts for various substrates in supramolecular chemistry. In particular they lead to the preparation of rotaxanes and catenanes. CD are cyclic oligosaccharides comprising six (α-CD), seven (β-CD) or eight (γ-CD) α-1,4-linked D-glucopyranose rings, creating a truncated cone-shape.[59,60] CB[n] (n = 6, 7 or 8) are a family of macrocyclic compounds obtained by an acid-catalyzed condensation of n-glycouril moieties and formaldehyde.[61,62] Thanks to a hydrophobic cavity and two hydrophilic carbonyl-bearing portals, CB[n] are able to form host-guest complexes, especially with protonated aminoalkanes through ion-dipole and hydrogen bonding interactions and hydrophobic effects.[63]

Mock's group discovered in 1989 CB[6]'s remarkable ability to catalyze 1,3-dipolar cycloaddition between encapsulated azide and alkyne in a regioselective manner.[64] During their kinetic investigations, the first triazole-assembled rotaxane **4**$^{2+}$ was synthesized (Scheme 14.3). This pioneer work could have opened the way for the synthesis of a number of more complex architectures, but surprisingly, Steinke and co-workers were almost the only ones to exploit this intriguing reaction.[65,66]

After having reproduced the synthesis of the [2]rotaxane **4**$^{2+}$, first prepared by Mock, in 76% yield, they designed and obtained [3]rotaxanes **6**$^{4+}$ and **8**$^{4+}$ (Scheme 14.4)[65,66] with a similar efficiency, as model compounds for polyrotaxanes (the number into brackets indicates the number of organic components). Encouraged by this successful trial, they

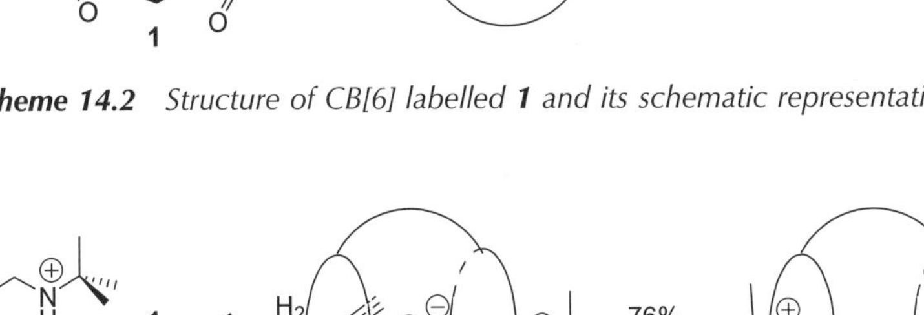

1

Scheme 14.2 *Structure of CB[6] labelled **1** and its schematic representations*

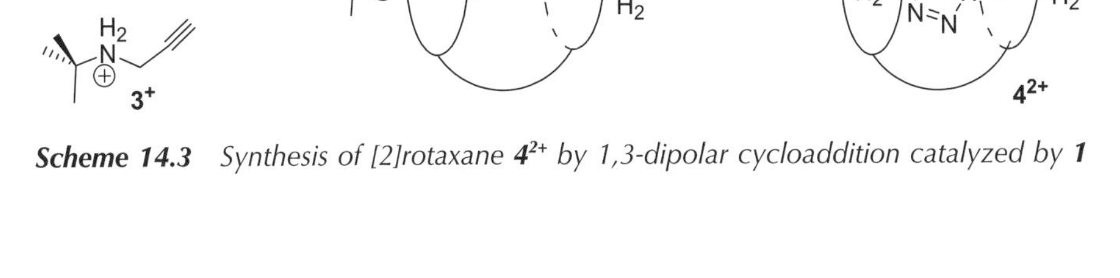

Scheme 14.3 *Synthesis of [2]rotaxane **4**$^{2+}$ by 1,3-dipolar cycloaddition catalyzed by **1***

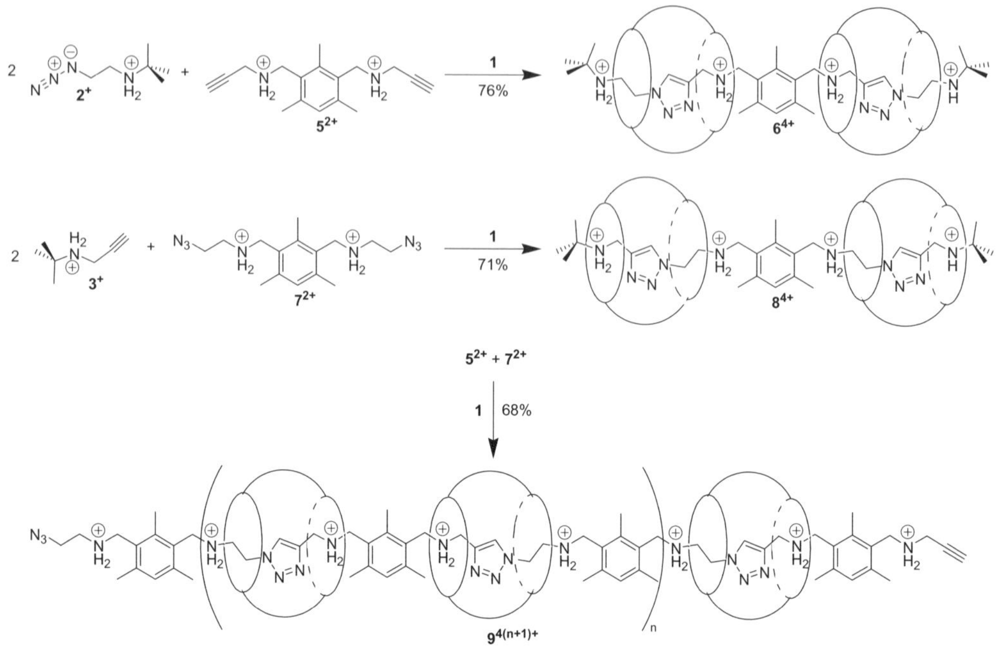

Scheme 14.4 *Synthesis of [3]rotaxanes* $\mathbf{6}^{4+}$ *and* $\mathbf{8}^{4+}$*, and of a perfect main-chain polyrotaxane*

performed the synthesis of a perfect linear main-chain polyrotaxane $\mathbf{9}^{4(n+1)+}$, including precisely two CB[6] rings threaded onto each polymer repeat unit. ^{1}H and ^{13}C NMR data as well as mass spectrometry measurements were in agreement with the proposed structures. During the process, polymerization and rotaxane formation occurred simultaneously, thanks to the presence of the catalytically-active self-threading macrocycle cucurbituril. Along with the original metal template concepts,[35,67] this efficient method partly inspired the catalytic 'active-metal' template strategy, later developed by Leigh[32,33] (see Section 14.3.2.). Using 1,3,5-functionalized 2,4,6-trimethylbenzene, they extended their architectural repertoire to (hyper)branched rotaxanes[66] and polyrotaxanes.[68] Replacement of the central 2,4,6-trimethylbenzene by a sufficiently long aliphatic spacer, namely dodecamethylene, yielded a bistable [3]rotaxane[69] and a polyrotaxane.[70] The pH-driven switching properties of these molecules were investigated by ^{1}H NMR spectroscopy. The movement of both CB[6] rings from the encapsulated triazole towards the aliphatic part of the thread could be triggered by base and the initial state could be recovered by addition of acid and heating when necessary.

A [5]rotaxane **10**, based on a meso-tetraphenyl porphyrin and synthesized in 85% yield, showed similar pH-driven switching properties, depicted in Scheme 14.5.[71] When all the nitrogen atoms of the amino groups are protonated, CB[6] prefers to encapsulate the protonated diaminotriazole site (state I) because of the strong ion-dipole interactions between ammonium ions and carbonyl functions at the portal of CB[6]. On the other hand, after complete deprotonation of the molecule, CB[6] resides mostly on the relatively more hydrophobic benzyl part, thanks to hydrophobic effect (state II).

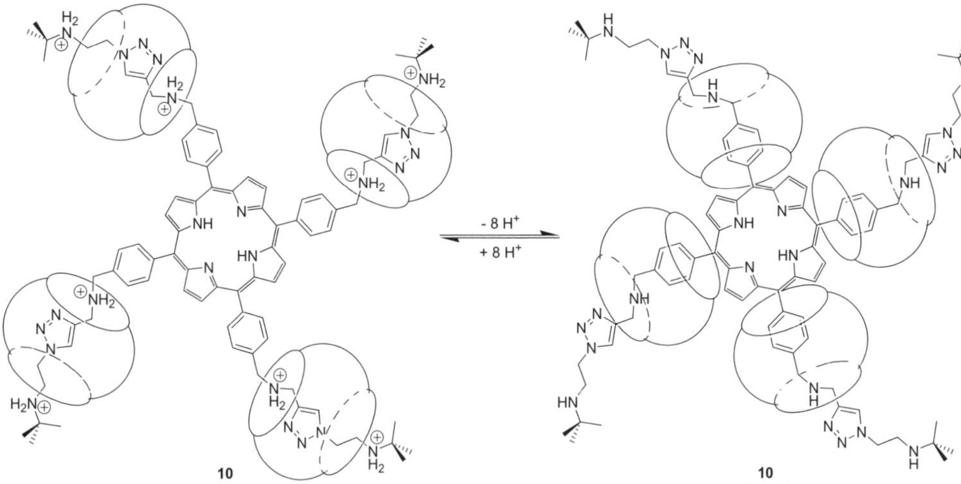

Scheme 14.5 *[5]rotaxane **10** and its pH-driven switching process*

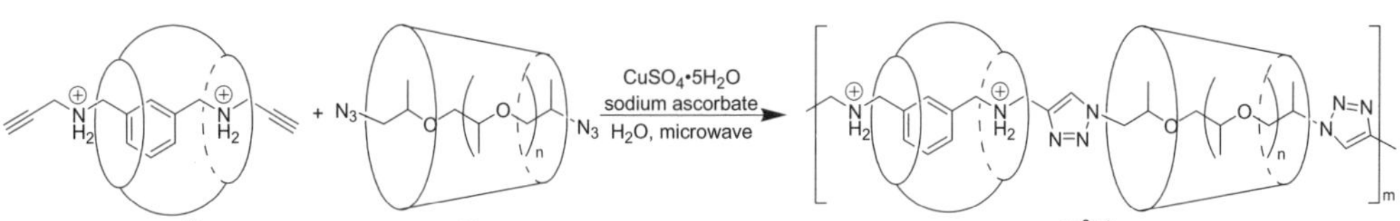

Scheme 14.6 *'Click' polymerization of two different [2]pseudorotaxane blocks **11**$^{2+}$ and **12***

Finally, very recently, Kim and Yui used the Cu(I)-catalyzed 1,3-cycloaddition of azides with terminal alkynes to synthesize a diblock polypseudorotaxane **13**$^{2m+}$ from two independent inclusion complexes, CB[7]/N,N′-[3-phenylenebis(methylene)]dipropargylamine (PMPA) **11**$^{2+}$ and 2,6-O-dimethyl-β-CD (DM-β-CD)/α,ω-bisazidopropylene glycol400 (PPG) [2]pseudorotaxanes **12** (Scheme 14.6).[72] A pH-responsive movement of the CB[7] units in the polypseudorotaxane was observed by ^{1}H NMR and 2D-ROESY NMR. At pH 2, the PMPA phenylene groups are still included in the cavity of CB[7] units, whereas at pH 11, CB[7] rings move towards the PPG units near the triazole group. By contrast, over a wide range of pH, the propylene glycol remains threaded by DM-β-CD, thus serving as a non-covalent blocking group with respect to CB[7].

14.2.2 Based on Hydrogen Bonding or on Organic Donor-Acceptor Complexes

Another wide family of purely organic interlocked systems has been developed by Stoddart's group, based on hydrogen bonding and on donor-acceptor interactions.[3] Suitably sized crown ethers can form pseudorotaxane complexes with appropriate secondary dialkylammonium ions which rely on N$^+$—H···O and C—H···O hydrogen bonds for stabilization. Similarly, a π-electron rich component, in interaction with a π-electron deficient moiety (traditionally cyclobis(paraquat-*p*-phenylene), CBPQT^{4+}, **27**$^{4+}$,

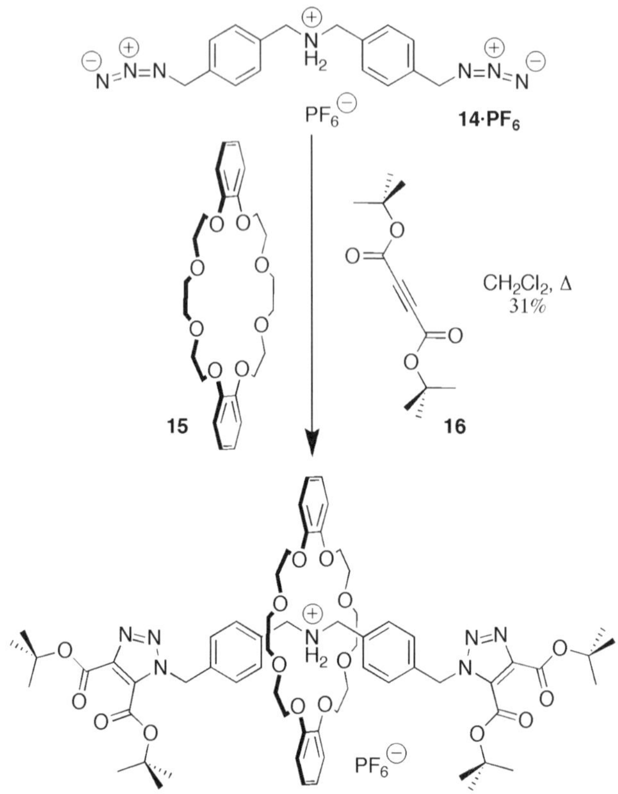

Scheme 14.7 *Synthesis of [2]rotaxane* $\mathbf{17}^{+}$ *by thermal 1,3-dipolar cycloaddition*

Scheme 14.10), can form a complex stabilized by a combination of electrostatic and dispersive forces, as well as hydrogen bonds.

Originally, Stoddart's group used azide-bearing thread-like fragments as a way to build the stoppers by the thermally allowed 1,3-dipolar cycloadditions between azides and bulky electron-deficient alkynes. By this 'threading-followed-by-stoppering' approach, the synthesis of the [2]rotaxane $\mathbf{17}^{+}$ (Scheme 14.7) was achieved with 31% yield, after reflux for seven days.[73] ^{1}H NMR spectroscopy, FAB mass spectrometry and X-ray crystal structure confirmed the interlocked nature of this molecule.

Willing to extend this methodology to the synthesis of a [2]rotaxane bearing two $R_2NH_2^+$ ion centres, they prepared the corresponding diazido precursor $\mathbf{18}^{2+}$ (Scheme 14.8). The expected 'two-station' [2]rotaxane, a potential degenerate molecular shuttle, could not be isolated under any circumstances from reaction between **15**, $\mathbf{18}^{2+}$ and di-*tert*-butyl acetylenedicarboxylate **16**.[74] The corresponding [3]rotaxane, $\mathbf{19}^{2+}$, was obtained instead, presumably because the solubility of $\mathbf{18}^{2+}$ in organic solvents was reliant upon the complexation of both their dialkylammonium centres by the crown ether ring. A protecting group approach had then been devised for the synthesis of the [2]rotaxane

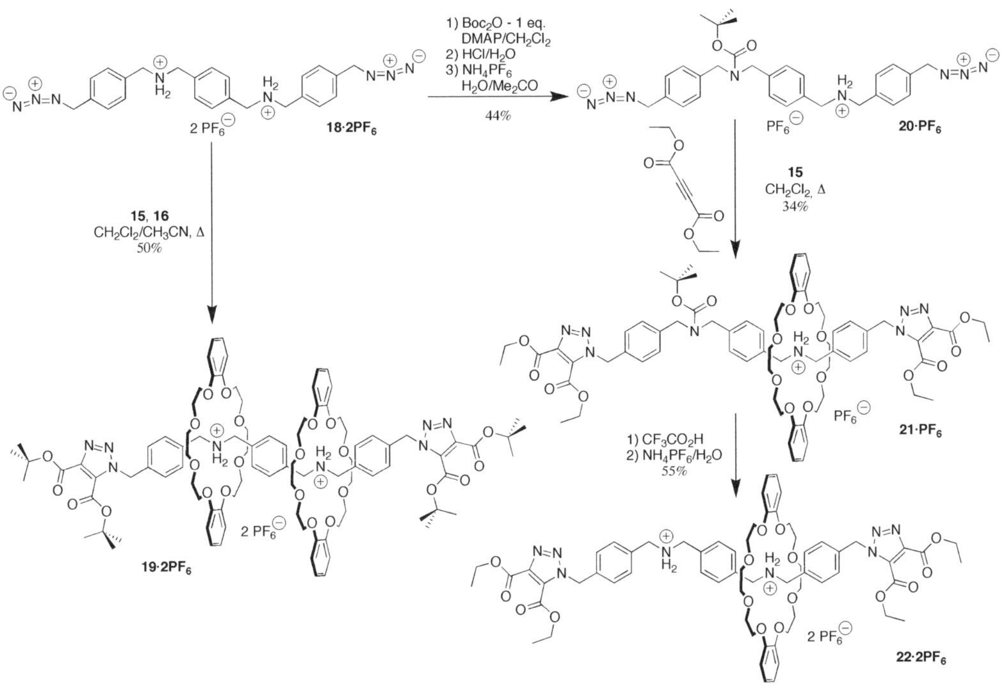

Scheme 14.8 *Synthesis of [3]rotaxane* $\mathbf{19}^{2+}$ *and [2]rotaxane* $\mathbf{22}^{2+}$ *by the protecting group approach*

$\mathbf{22}^{2+}$.[74] One dialkylammonium site remained protected by a *tert*-butoxycarbonyl (Boc) group while the other one was available for the template synthesis of the [2]pseudorotaxane. After formation of both stoppers by 1,3-dipolar cycloaddition between terminal azides and diethyl acetylenedicarboxylate, Boc could be removed, furnishing a two-station [2]rotaxane $\mathbf{22}^{2+}$. The shuttle properties of this molecule have been characterized using variable temperature ^{1}H NMR spectroscopy.

Lastly, the 'threading-followed-by-stoppering' methodology using thermal 1,3-dipolar cycloaddition was used to construct a mechanically interlocked bundle based on an intriguing triply-threaded topology.[75] The 1:1 triply threaded superbundle (Scheme 14.9) was assembled from the trifurcated cation $\mathbf{24}^{3+}$ and the tritopic crown ether **23**. The efficient conversion of the three azide functions at the ends of the arms of the threaded trication into bulky triazole stoppers afforded the interlocked structure $\mathbf{25}^{3+}$ in 40% yield. This interlocked bundle was a precursor for the design of a new molecular machine, namely a molecular elevator.[76]

In the meantime, Meldal[47,48] and Fokin-Sharpless[49] discovered independently that 1,3-dipolar cycloadditions of terminal alkynes to azides could be very efficient and regioselective when catalyzed by copper(I) salts. This 1,4-disubstituted triazole synthesis became very popular, as the ideal member of the family of 'click reactions'.[45,46] Its mild reaction conditions, its remarkable efficiency and its wide scope due to a high tolerance of other sensitive functional groups prompted a few research teams,[33,54] including ours (see Section

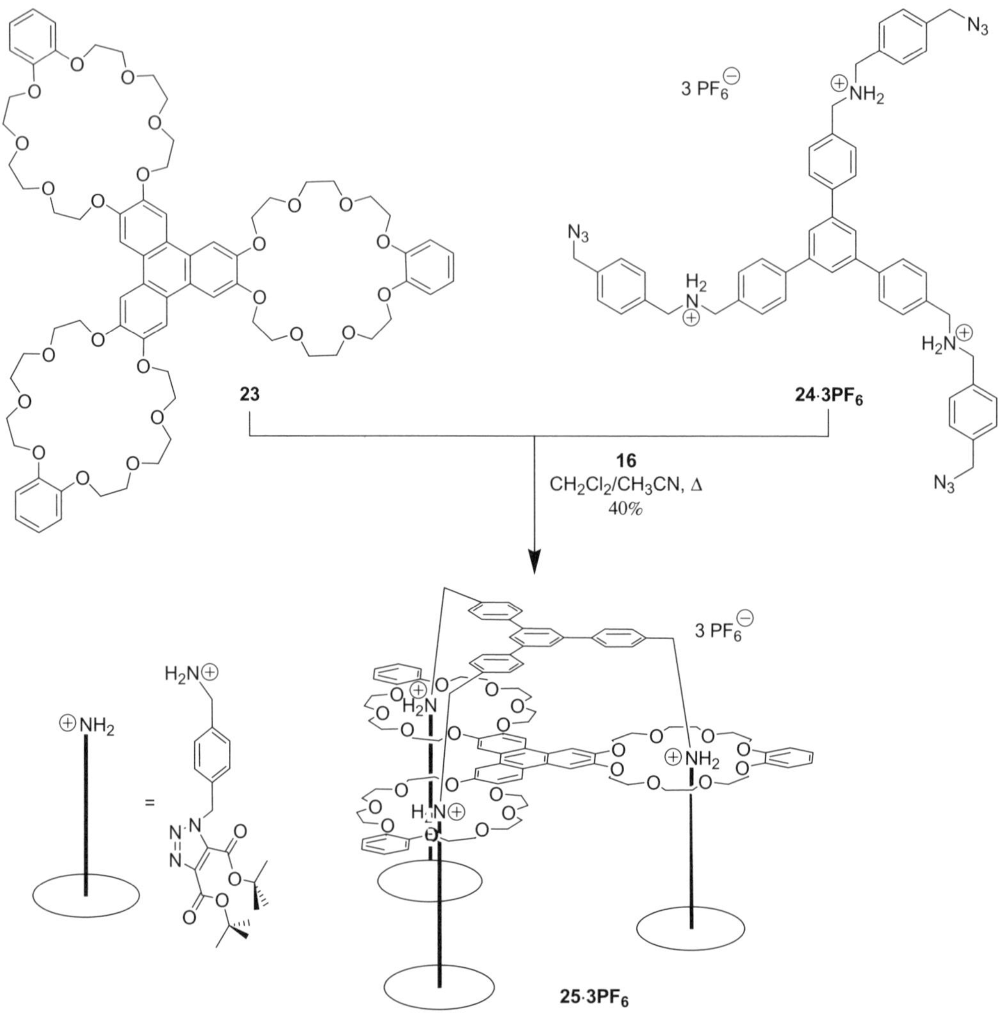

Scheme 14.9 *Template-directed synthesis of the mechanically interlocked bundle* $\mathbf{25}^{3+}$

14.3.1),[57] to use it for the synthesis of interlocked molecules such as rotaxanes and catenanes.

Stoddart *et al.* applied a 'threading-followed-by-stoppering' strategy for the synthesis of $CBPQT^{4+}$-based donor-acceptor systems, since this π-electron deficient ring is particularly sensitive to both nucleophiles and bases.[51] The inclusion complex $\mathbf{28}^{4+}$ (Scheme 14.10)[54] was formed by mixing $CBPQT^{4+}$ $\mathbf{27}^{4+}$ and dioxynaphthalene (DNP) derivative **26** carrying azide-terminated glycol chains in DMF at −10 °C. Addition of a slight excess of propargyl stopper **29**, along with $CuSO_4{\cdot}5H_2O$ and ascorbic acid gives the [2]rotaxane $\mathbf{30}^{4+}$ in 82% yield.

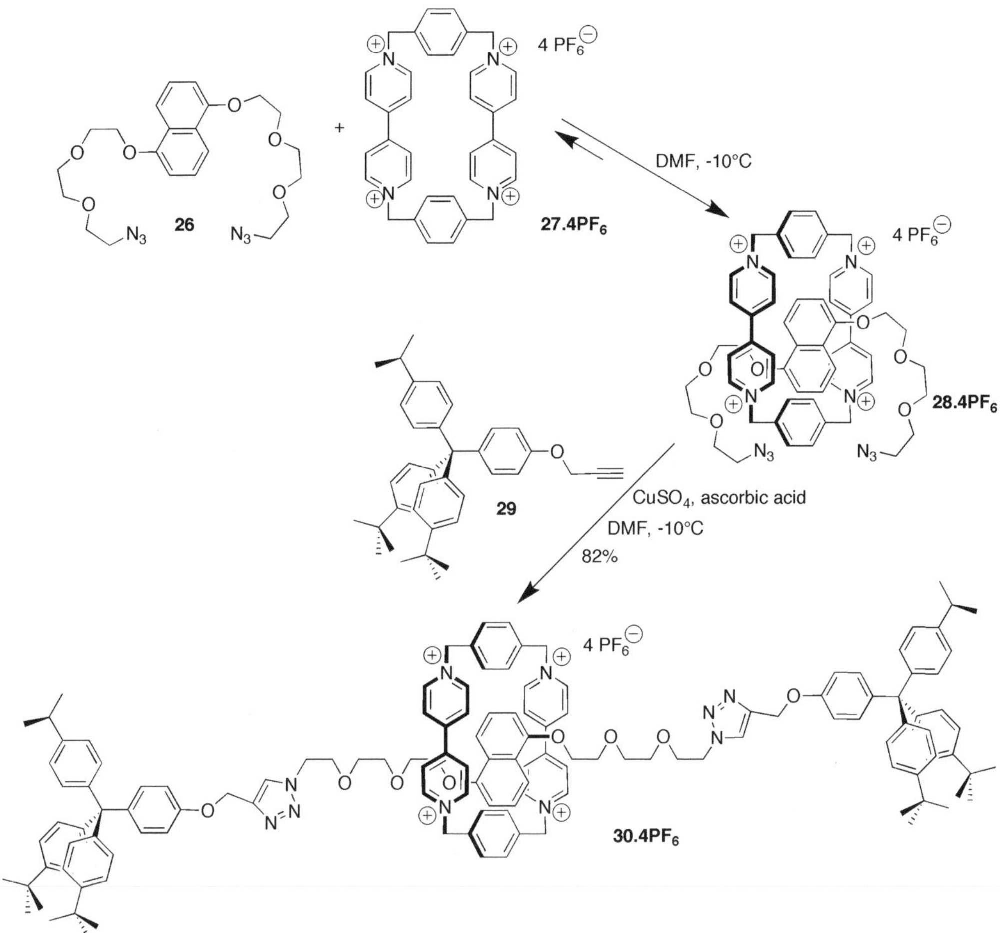

Scheme 14.10 *Synthesis of [2]rotaxane* ***30***$^{4+}$ *by Cu(I)-catalyzed 1,3-dipolar cycloaddition*

This modular approach could be extended to the synthesis of degenerate bistable [2] rotaxane **33**$^{4+}$, self-complex **35**$^{4+}$, branched [4]rotaxane **37**$^{12+}$ and [3]rotaxane **39**$^{8+}$ (Scheme 14.11) from a single precursor **31**.[53]

The first catenane synthesized *via* triazole formation, **42**$^{4+}$, was obtained in 41% yield after threading of a difunctional alkyne-azide moiety **40** through the preformed acceptor ring CBPQT^{4+} and subsequent cyclization (Scheme 14.12).[56] It had been well characterized by ^{1}H NMR and FAB mass spectroscopy, and the X-ray crystal structure was solved, as an additional proof of its interlocked nature.[55] Despite this moderate yield, the synthetic route was particularly attractive as it can be accomplished in one step.

Finally, bistable [2]rotaxanes **44**$^{4+}$ and **46**$^{4+}$ were also synthesized by the same method (Scheme 14.13). A thread-like fragment **43**, incorporating both tetrathiafulvalene (TTF) and DNP recognition units, and terminated by azide at both ends, was mixed with

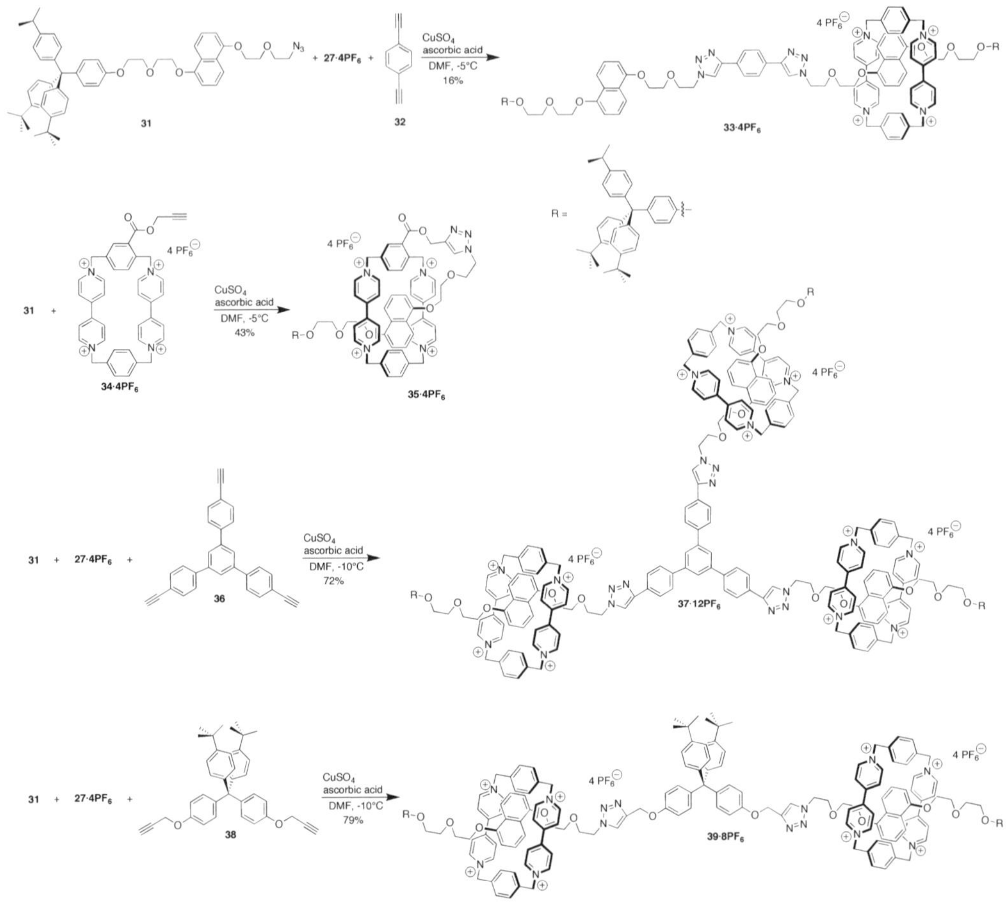

Scheme 14.11 *Preparation of degenerate bistable [2]rotaxane* $\mathbf{33^{4+}}$*, self-complex* $\mathbf{35^{4+}}$*, branched [4]rotaxane* $\mathbf{37^{12+}}$ *and [3]rotaxane* $\mathbf{39^{8+}}$ *by a modular approach*

CBPQT^{4+} $\mathbf{27^{4+}}$, leading to the formation of the inclusion complex. This pseudorotaxane was stirred for two days with two equivalents of the acetylenic stopper **29** and catalytic amounts of copper sulphate and ascorbic acid.[77] [2]rotaxane $\mathbf{44^{4+}}$ was obtained after chromatography in 60% yield.[77] NMR and mass spectra were in agreement with the interlocked structure. The shuttling of the CBPQT^{4+} ring induced by oxidation was demonstrated by using both cyclic and differential pulse voltammetry and UV-visible spectroelectrochemistry. These studies indicated that the 1,2,3-triazole rings did not act as recognition units since $\mathbf{44^{4+}}$ showed an electrochemical behaviour similar to those previously reported for TTF/DNP two-station [2]rotaxanes.[78,79] Since liquid crystals (LC) can be used as dynamic functional materials, the authors decided to try this approach for similar two-station [2]rotaxane with dendritic mesogenic stoppers like **45**.[52] Synthesis by more conventional reactions (DCC coupling, Mitsunobu reaction and triazine mediated coupling)

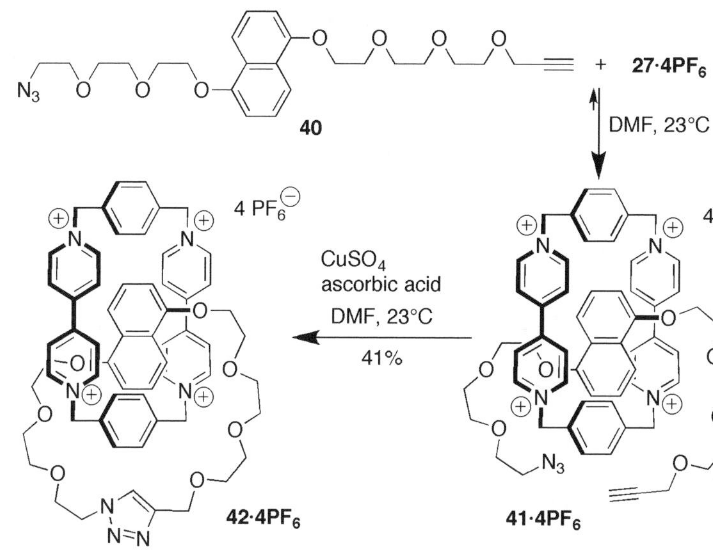

Scheme 14.12 *Template-directed synthesis of the first [2]catenane* $\mathbf{42}^{4+}$ *prepared using 'click' chemistry*

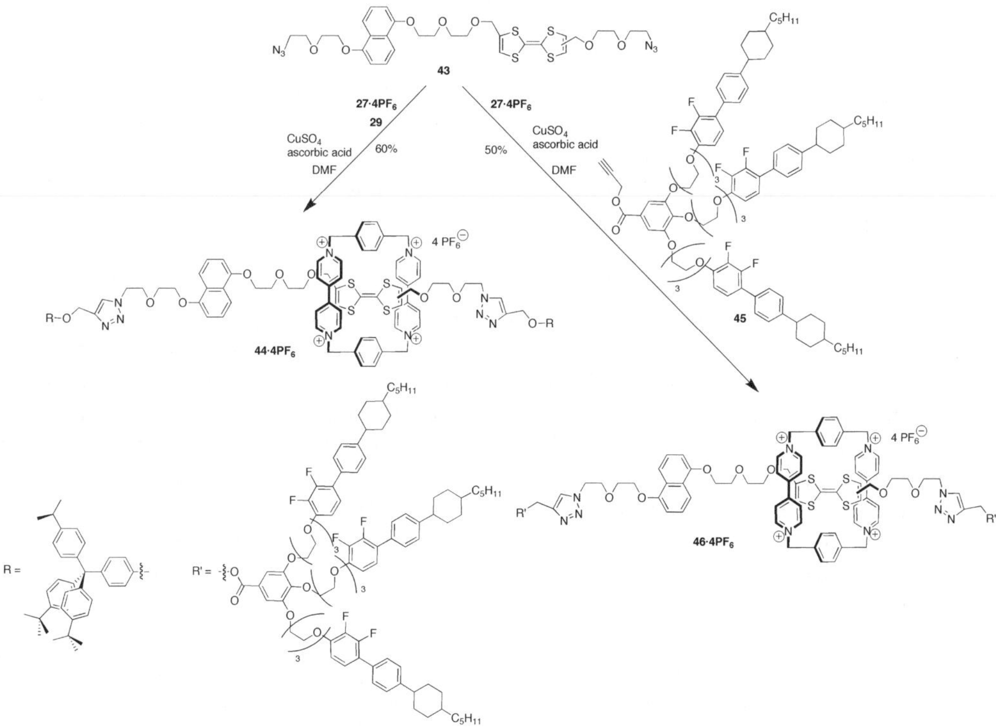

Scheme 14.13 *Synthesis of bistable [2]rotaxane* $\mathbf{44}^{4+}$ *and LC bistable [2]rotaxane* $\mathbf{46}^{4+}$

initially failed (this result is not necessarily general, since a LC [2]catenane could be prepared in 75% yield by double esterification mediated by EDC[80]) so that further synthetic transformations of precursors were made in view of using the Cu(I)-catalyzed 1,3-dipolar cycloaddition between azides and alkynes. This latter reaction yielded the LC bistable [2]rotaxane **46**$^{4+}$ in 50% yield. This successful preparation of the [2]rotaxane **46**$^{4+}$ using the click reaction demonstrates the power of this methodology since it can yield very efficiently molecules that could not be prepared by different reactions.

14.3 Transition Metal Templated Approaches

14.3.1 Cu(I) Assembled Rotaxanes

Our group has extensively used copper(I) as a template,[34,35] to gather the various pieces of the future rotaxane. If the string and the ring contain sterically hindering chelates of the 2,9-dianisyl-1,10-phenanthroline (dap) family, a very stable copper(I)-complex precursor is obtained which will most of the time survive, at least in part, the harsh stoppering reaction conditions of the Williamson procedure. The situation is very different when one of the two chelates coordinated to the copper(I) centre is non-encumbered. In fact when the steric hindrance around the metal is minimized, its lability is restored and it becomes unstable in basic media, leading to sensitive copper(I) species. However, such rotaxanes are particularly interesting as molecular machine prototypes since they are much faster-moving species than their highly shielded analogues.[81,82] 'Click chemistry' represents a promising possibility as stoppering reaction of the preliminarily prepared unstable pseudo-rotaxane. It had been explored for the Cu(I) templated synthesis of such a sterically non-hindering [2]rotaxane **47**$^{+}$[57] and analogous metallo-rotaxanes **48**$^{+}$ and **49**$^{+}$ (Scheme 14.14).

Although the synthesis of **47**$^{+}$ relied on a particularly unstable copper(I) complex, both stoppers **52** (Scheme 14.15) could be added at the same time, which allowed to circumvent the tedious two-step procedure, consisting in (i) threading the macrocycle on a one-stopper string-like fragment and (ii) attach the second stopper (Scheme 14.1). This strategy required the preparation of a macrocycle **50**, a thread-like fragment **51** bearing an azide function at each end and an acetylenic stopper **52** (Scheme 14.15). Following a typical procedure, the macrocycle **50** was dissolved in a degassed solution of dichloromethane and acetonitrile. Upon addition of 1 equivalent of copper(I) salt, the solution turned immediately to dark orange, while after addition of a stoichiometric amount of bipyridine **51**, it changed to deep red. This colour is characteristic of copper(I) complexes with two aromatic diimine ligands, and the precursor **53**$^{+}$ was obtained in quantitative yield. With a 5,5′-substituted 2,2′-bipyridine as a chelate, the pseudorotaxane **53**$^{+}$ was highly unstable; it could not be isolated and was immediately engaged in 'click' conditions.

Two equivalents of acetylenic stopper **52**, $[Cu(CH_3CN)_4](PF_6)$ as catalyst (0.5 eq.) and Na_2CO_3 as a base (0.4 eq.) were added. The progress of the reactions was followed by TLC analysis, which clearly showed the consumption of compound **53**$^{+}$. Nevertheless, supplementary additions of copper(I) (0.25 eq.), stopper **52** (1 eq.), and Na_2CO_3 and stirring overnight were necessary for the reaction to be complete.

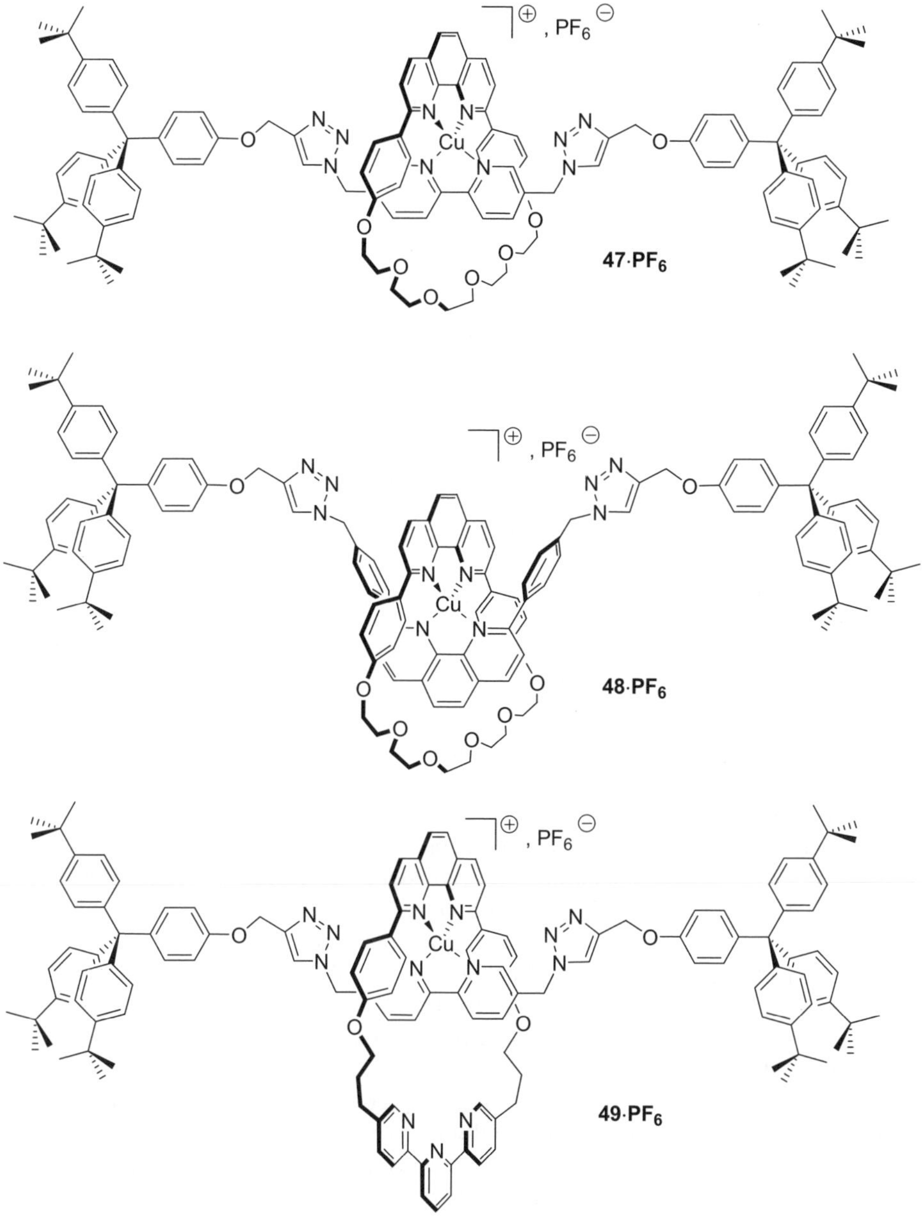

Scheme 14.14 *Chemical structures of copper(I)-complexed [2]rotaxanes* ***47***$^{+}$*,* ***48***$^{+}$ *and* ***49***$^{+}$ *prepared by Cu(I)-templated approaches and click chemistry*

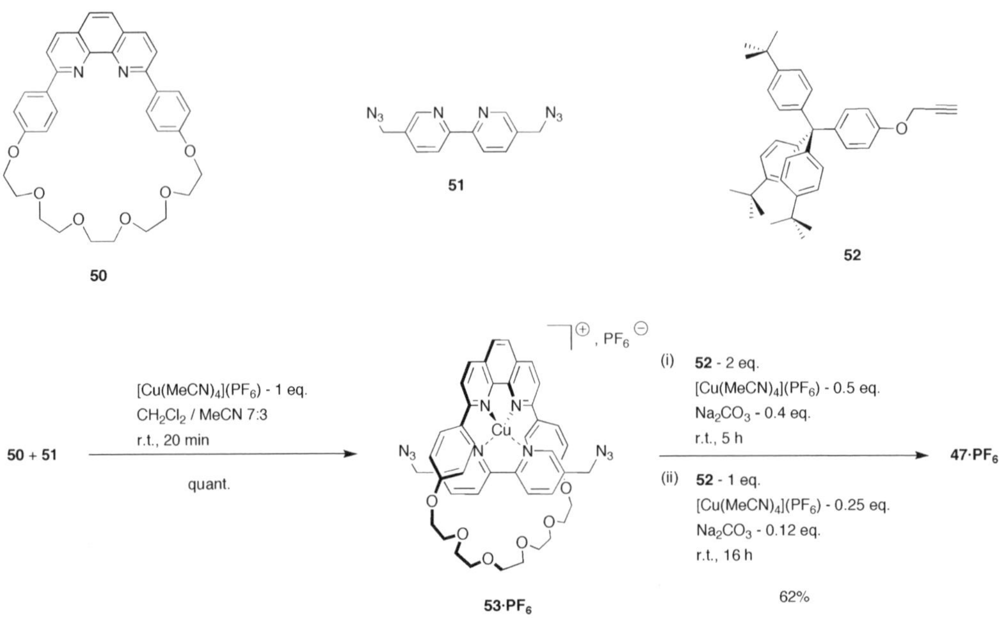

Scheme 14.15 *(top) Chemical structures of the building blocks. (bottom) Synthetic scheme leading to the [2]rotaxane* ***47***$^{+}$

Without any work-up, the reaction mixture was then subjected to column chromatography and produced the [2]rotaxane **47**$^{+}$ in high yield (62%). The compulsory use of a trace of hydrazine in degassed eluent has to be emphasized: the function of hydrazine is to inhibit formation of extremely labile copper(II) complexes, which would be detrimental and lead to lower yields (28% with no additional hydrazine). The isolated yield of rotaxane **47**$^{+}$ was 62%, which highlights the efficiency of the 'click' strategy even for a double copper(I)-catalyzed 1,3-dipolar cycloaddition. This method proved to be at least equally efficient in the case of the [2]metallo-rotaxane **48**$^{+}$ (Scheme 14.14), synthesized in 67% yield.[83] High resolution ES-MS measurements and ^{1}H NMR spectroscopy confirmed the postulated structures of metallo-rotaxanes **47**$^{+}$ and **48**$^{+}$.

Encouraged by these results, we extended this procedure to the formation of the bistable rotaxane **49**$^{+}$ (Scheme 14.16). In such pirouetting compounds, the ring **54** incorporates both a bidentate chelate, a 1,10-phenanthroline (phen), and a tridentate fragment, a 2,2′,6′,2″-terpyridine (terpy) and the axis contains a bipyridyl ligand. Macrocycle **54**[84] and copper(I) salt were mixed, and after addition of **51**, a red solution was obtained. It is known that the presence of the terpy moiety in the ring does not interfere with the threading step since copper(I) interacts preferably with bidentate chelate.[85] The intermediate **55**$^{+}$ was then subjected to 'click chemistry' conditions. In order to ease purification, copper was removed from the medium using excess KCN,[67] and pure rotaxane **56**, containing the bis-chelating ring **54**, was obtained in 12% yield. The chemical integrity of compound **56** was confirmed by one- and two-dimensional ^{1}H NMR spectroscopy as well as high-resolution ES-MS measurements. Remetalation with one equivalent of $Cu(CH_3CN)_4PF_6$ yielded quantitatively the metallo-rotaxane **49**$^{+}$.

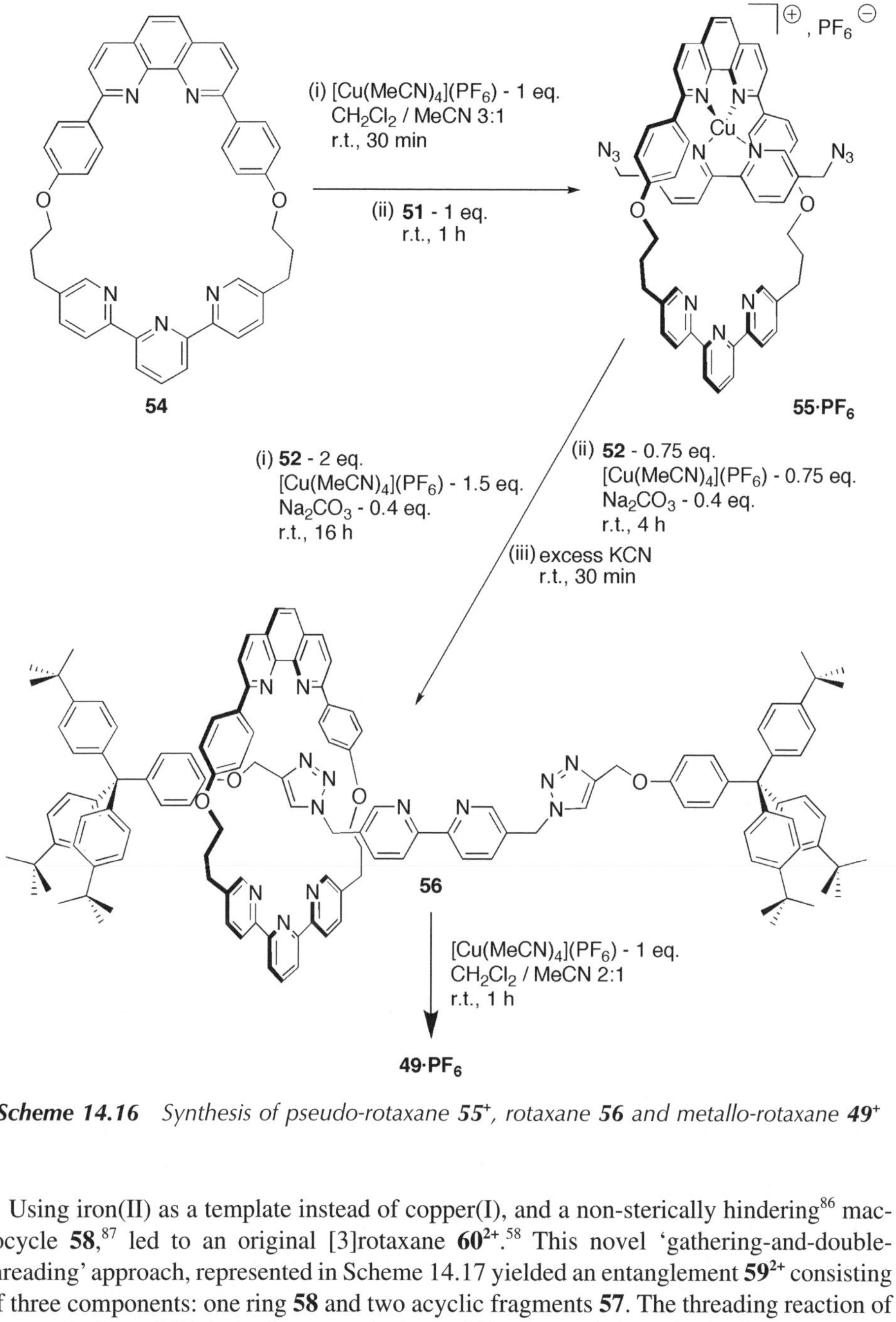

Scheme 14.16 *Synthesis of pseudo-rotaxane* ***55***$^{+}$*, rotaxane* ***56*** *and metallo-rotaxane* ***49***$^{+}$

Using iron(II) as a template instead of copper(I), and a non-sterically hindering[86] macrocycle **58**,[87] led to an original [3]rotaxane **60**$^{2+}$.[58] This novel 'gathering-and-double-threading' approach, represented in Scheme 14.17 yielded an entanglement **59**$^{2+}$ consisting of three components: one ring **58** and two acyclic fragments **57**. The threading reaction of two equivalent of **57** through one equivalent of **58** was carried out in the following way: an CH_3CN solution of $Fe(BF_4)_2{\cdot}6H_2O$ (1 eq.) was added to a CH_2Cl_2 solution of ring **58** (1.1 eq.) at ambient temperature. Then 2 equivalents of **57** dissolved in CH_2Cl_2 were added drop by drop to the mixture. The workup led to an 80% yield of the three-ligand iron(II)

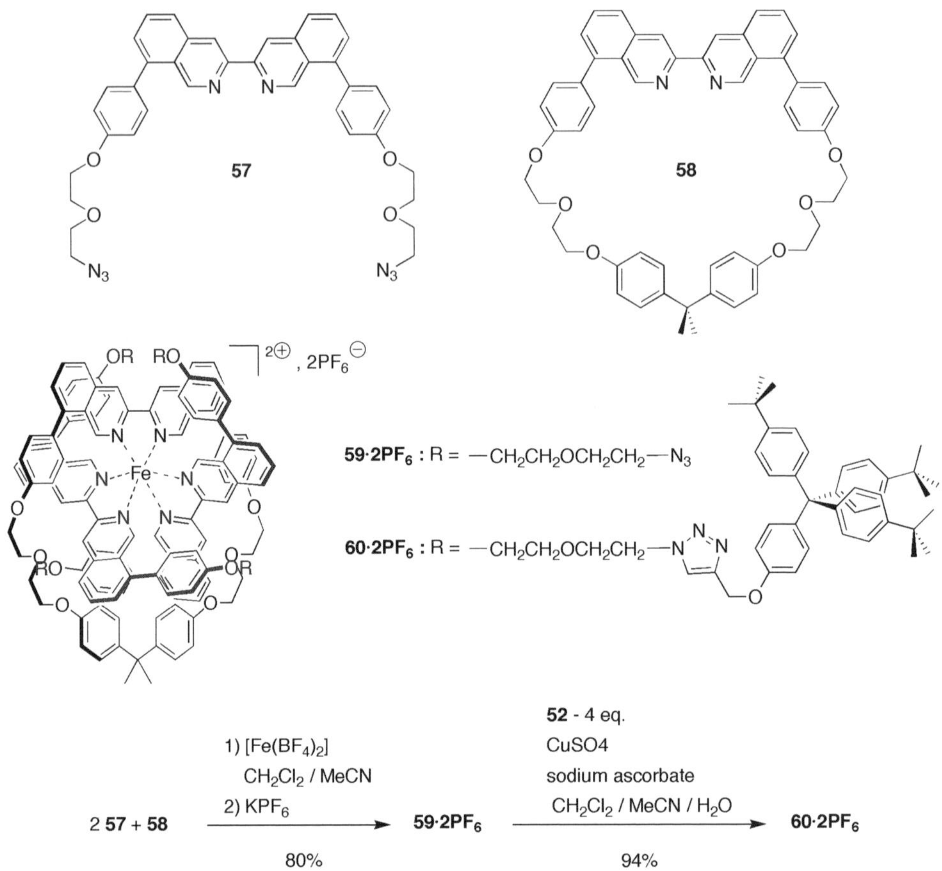

Scheme 14.17 *(top) Chemical structures of the building blocks and products. (bottom) Synthetic scheme leading to the [3]rotaxane **60^{2+}***

complex **59**$^{2+}$, [Fe(**58**)(**57**)$_2$]·2PF6, as a red solid. It was then subjected to 'click chemistry' conditions for end-functionalization of the two threads of complex **59**$^{2+}$. It was subsequently reacted with the propargyl derivative **52** in presence of copper(II) sulphate and sodium ascorbate, in a biphasic medium ($CH_2Cl_2/CH_2CN/H_2O$, 10:0.5:10). The tetratriazole derivative **60**$^{2+}$ was obtained in high yield (94%) after workup and chromatography, as a red solid. It has been fully characterized by the classical analytical techniques.

Once again, the use of azide precursors and their subsequent threading and transformation to 1,2,3-triazole rings provided an easy and high-yield access to new interlocked molecules, and more complicated systems[88,89] synthesized by this procedure have also recently been reported soon.

14.3.2 Cu(I) as Both a Template and a Catalyst

Leigh and co-workers have recently reported the first substoichiometric metal-template pathway to mechanically interlocked architectures.[32,33] Like in the Strasbourg-type chem-

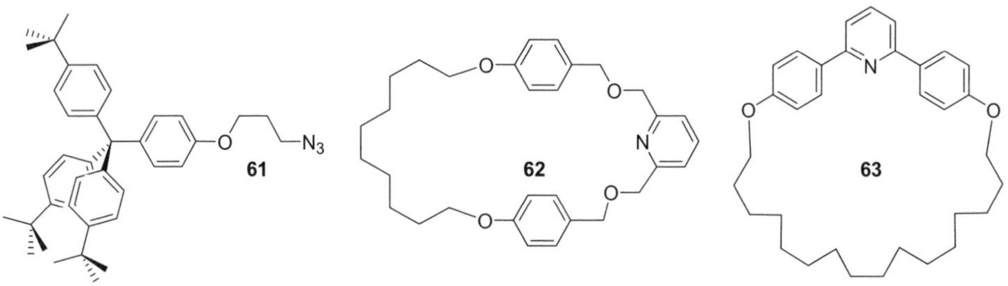

Scheme 14.18 *Rotaxane building blocks: Bulky azide **61** and macrocycles **62** and **63***

istry, the copper(I) ion functions as a template but in this case it also plays the role of a catalyst that turns over during the reaction, thus permitting substoichiometric amounts of the metal to be used.

The target molecule **68** was a [2]rotaxane containing one macrocycle and one thread (Scheme 14.19).[33] Two bulky groups, azide **61** and alkyne **52** (Scheme 14.18), were reacted overnight inside the cavity of a macrocycle **62** in presence of 1 equivalent of Cu(I). The 1,2,3-triazole, formed by cycloaddition, chemically connected the two entities, which then constituted the thread inside the ring. The role of the copper(I) ion was dual: first gather the two moieties inside the cavity of the macrocycle, and then catalyze the triazole formation. Demetalation with KCN yielded [2]rotaxane **68** in 57% yield, but also a non-negligible amount of non-interlocked thread **69** (41%), due to 1,3-dipolar addition having occurred also outside the macrocycle. The threaded character of **68** was proven by ^{1}H NMR. A typical shielding effect was observed for all non-stopper resonances of the thread, due to the ring current effect of the aromatic rings of the macrocycle, indicating also that the ring moves along the entire thread. By using five equivalents of azide **61** and alkyne **52** in the same conditions as described above, the yield for [2]rotaxane **68** could be increased up to 94% with respect to the amount of macrocycle used.

When only 20% of copper(I) salt was added to the reaction mixture, approximately the same amount of rotaxane (~20%) was formed, indicating that the copper(I) ions stayed coordinated to the freshly formed rotaxane and do not turn over. To enable this catalytic behaviour, addition of 3 equivalents of pyridine as competing ligand and higher temperatures (70 °C) were necessary. Under these conditions the [2]rotaxane **68** could be obtained in 82% yield by using only 4% copper(I) with respect to both azide **61** and alkyne **52**.

Leigh's group further investigated the effect of the macrocycle structure and the reaction mechanism.[32] To favour the cycloaddition inside the cavity of the ring over the outside reaction, the macrocycle needs to have an endotopic binding site. The rotaxane formation was most effective when using monodentate macrocycles. Bidentate and terdentate ligands tend to inactivate the metal centre, either by sequestering copper(I) or favouring its oxidation. Steric hindrance as well as electronic effects had dramatic consequences on rotaxane formation, probably by reducing the binding strength of the ring. Based on kinetic studies as well as general considerations on acetylides, the catalytic cycle shown in Scheme 14.19 was proposed.

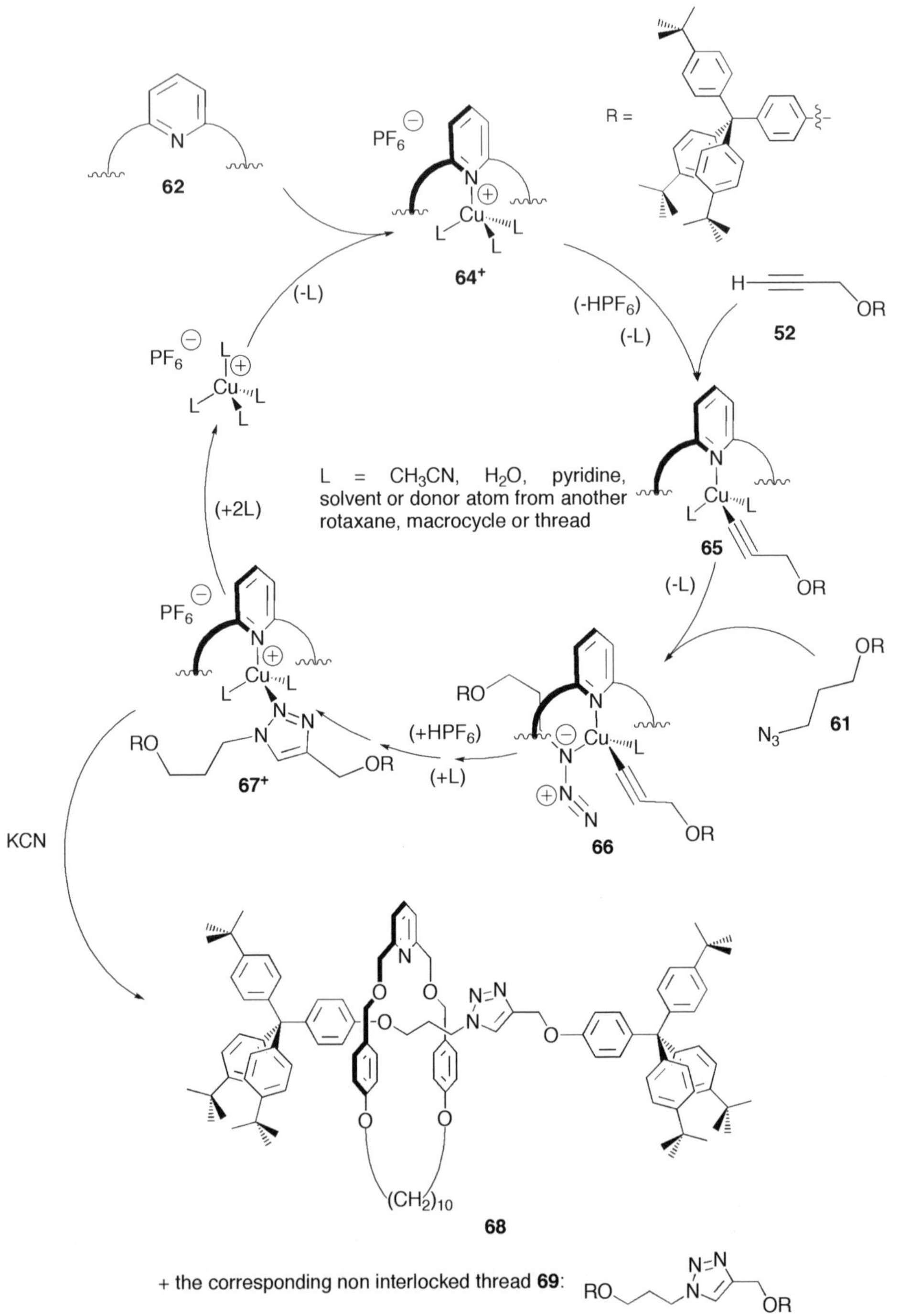

Scheme 14.19 *Proposed catalytic cycle for the copper(I) template synthesis of rotaxane* ***68***

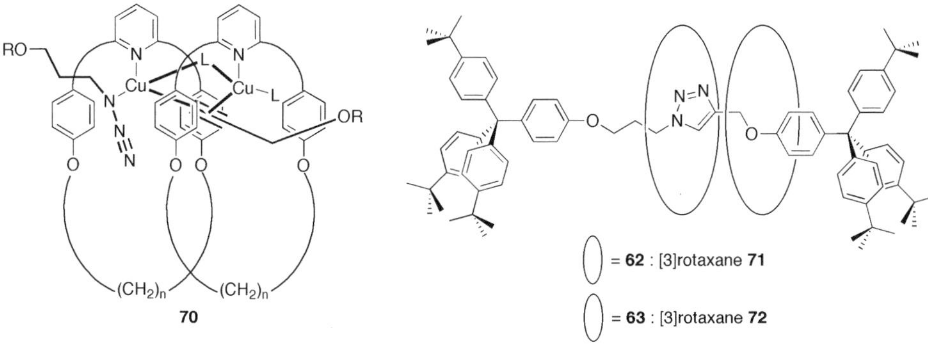

Scheme 14.20 *Reaction intermediate **70**, leading to the formation of [3]rotaxanes **71** or **72***

Different reaction intermediates can exist depending on the reaction conditions, but a bridged intermediate involving two copper centres being both bound to the alkyne (one σ-bonded and one π-bonded, Scheme 14.20) is probably involved in the dominant pathway in most Cu(I)-catalyzed formation of 1,2,3-triazoles. This assertion was further supported by the unexpected formation of [3]rotaxanes at high macrocycle/Cu(I) ratios. By using 10 equivalents of macrocycles **62** or **63**, with respect to copper(I), azide **61** and alkyne **52**, the formation of [3]rotaxanes **71** and **72**, in 5% and 33% yields respectively, could be observed (Scheme 14.20) (**72** is the homologue of **71** with ring **63** instead of **62**). This could be explained by the fact that during the catalytic cycle, the second copper(I) π-coordinated to the alkyne, would also be coordinated to a macrocycle, due to its abundance in the reaction media. The postulated intermediate **70** responsible for the double threading is shown in Scheme 14.20. The X-ray structure of [3]rotaxane **72** confirmed its constitution.

To see whether one macrocycle-copper(I) complex would act in procession to catalyze the formation of two triazole rings in a single molecule, systems requiring two cycloaddition reactions were further investigated. The catalytic reaction between diazide **73** and alkyne **52** in combination with macrocycle **62**, followed by demetalation by KCN gave rotaxane **74** in 81% yield (Scheme 14.21).[32]

Upon addition of 1 equivalent of copper(I) to the degenerate two-station bistriazole rotaxane **74**, ^{1}H-NMR analysis showed that both triazole rings were equivalent and coordinated to the metal centre, implicating fast shuttling of the macrocycle between the two triazole moieties. In contrast, when 1 equivalent of $Pd(CH_3CN)_2Cl_2$ was added, a much stronger complex was formed, so that no shuttling could be observed, as evidenced by two different triazole signals on the ^{1}H-NMR spectrum up to 343 K. The fast shuttling in presence of copper(I) could be effectively blocked by coordination to palladium.

By combining the catalytic and template effect of copper(I), Leigh *et al.* have developed a new powerful tool, which permits to use only catalytic amounts of the template agents for the synthesis of mechanically interlocked architectures. In theory, this concept can be applied to many other metal-mediated reactions, such as cross-couplings, condensations or other cycloaddition reactions.[32,90] This new development may be the seminal idea that could lead to a whole new range of catenane- and rotaxane-forming protocols.

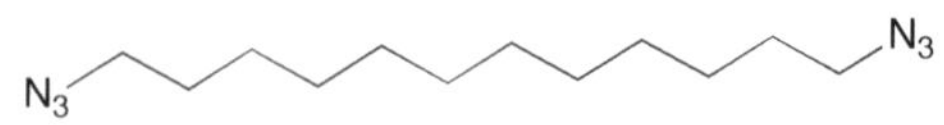

73

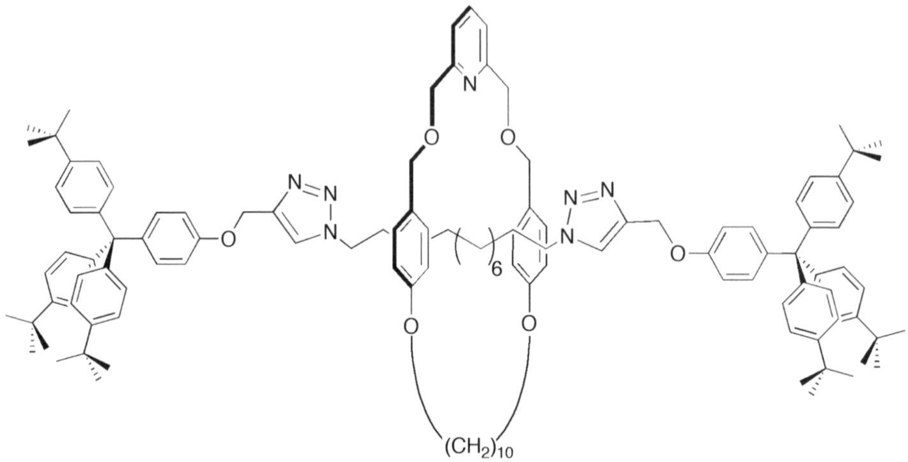

74

Scheme 14.21 *Diazide **73** and the corresponding rotaxane **74***

14.4 Conclusion

Generally speaking, one of the main difficulties of rotaxane synthesis is the potential instability of the pseudo-rotaxane precursors which, under various reaction conditions, could decompose and lead to the separate components, thus inhibiting to a large extent the formation of the desired compound. The mild reaction conditions of 'click chemistry' make it an extremely well adapted strategy for stoppering various threaded precursors and thus for preparing the corresponding rotaxanes.

In the first part of this review article, we have surveyed the click chemistry-based approaches to purely organic rotaxanes and discussed a few particularly efficient syntheses. Not too surprisingly, the copper(I)-catalyzed click reaction does not seem to be affected by the threaded nature of the precursors. In fact, since there is no interference between copper(I) and the functional groups borne by the organic fragments of the rotaxane precursor, rotaxane synthesis works as efficiently as classical click reactions with non-threaded precursors.

The situation could have been markedly different when the precursors contain coordinating moieties or transition metals, which could perturb the course of the reaction by interacting with the copper(I) catalyst or act as competitors, respectively. Interestingly, the presence of various transition metal-complexed fragments in the precursors is not particularly detrimental and does not seem to lower the preparative yields. Click chemistry has been successfully used with various transition metal complexes as precursors (Cu(I) or Fe(II)). The iron(II)-templated synthesis of a [3]rotaxane[58] is remarkably efficient since the yield of a 4-fold stoppering reaction is close to quantitative. An impressive example of click chemistry combined to transition metal-based rotaxanes, in which the metal has a dual function, was described recently.[33,90] In this particular case, copper(I) plays the 'classical' role of a gathering element, but at the same time, the same metal centre was used as a catalyst in the triazole formation reaction. Substoichiometric amounts of the metal could even be used.

Whereas the click chemistry-based synthesis of rotaxanes has been well illustrated in the course of the last few years, very different is the situation as far as the making of catenanes is concerned.[55,56] It is likely that this reaction will be more often used in the future for making rings and, in particular, catenanes. The field of catenanes, rotaxanes and molecular machines will certainly benefit in a significant way from the new synthetic methodology based on copper-catalyzed triazole synthesis.

References

[1] *Struct. Bonding: Molecular machines and motors., Vol. 99*, Sauvage, J.-P. ed., Springer, Berlin/Heidelberg, **2001**.

[2] V. Balzani, A. Credi, F. Raymo, J.F. Stoddart, *Angew. Chem. Int. Ed.* **2000**, *39*, 3348.

[3] V. Balzani, M. Gomez-Lopez, J.F. Stoddart, *Acc. Chem. Res.* **1998**, *31*, 405.

[4] V. Balzani, M. Venturi, A. Credi, *Molecular Devices and Machines – A Journey into the Nanoworld*, Wiley-VCH, Weinheim, **2003**.

[5] J.-C. Chambron, J.-P. Collin, V. Heitz, *et al.*, *Eur. J. Org. Chem.* **2004**, *8*, 1627.

[6] J.-P. Collin, C. Dietrich-Buchecker, P. Gavina, M.C. Jimenez-Molero, J.P. Sauvage, *Acc. Chem. Res.* **2001**, *34*, 477.

[7] J.-P. Sauvage, *Acc. Chem. Res.* **1998**, *31*, 611.

[8] D.J. Hoffart, S.J. Loeb, *Angew. Chem. Int. Ed.* **2005**, *44*, 901.

[9] M. Cavallini, F. Biscarini, S. Leon, *et al.*, *Science* **2003**, *299*, 531.

[10] J.-L. Weidmann, J.-M. Kern, J.-P. Sauvage, *et al.*, *Chem. Eur. J.* **1999**, *5*, 1841.

[11] T.J. Kidd, T.J.A. Loontjens, D.A. Leigh, J.K.Y. Wong, *Angew. Chem. Int. Ed.* **2003**, *42*, 3379.

[12] A. Belaissaoui, S. Shimada, A. Ohishi, N. Tamaoki, *Tetrahedron Lett.* **2003**, *44*, 2307.

[13] K.M. Huh, T. Ooya, S. Sasaki, N. Yui, *Macromolecules* **2001**, *34*, 2402.

[14] K.M. Huh, H. Tomita, T. Ooya, W.K. Lee, S. Sasaki, N. Yui, *Macromolecules* **2002**, *35*, 3775.

[15] H.D. Park, W.K. Lee, T. Ooya, K.D. Park, Y.H. Kim, N. Yui, *J. Biomed. Mater. Res.* **2002**, *60*, 186.

[16] For early work on catenanes and rotaxanes, see: G. Schill, *Catenanes, Rotaxanes and Knots*, Academic Press, New York, **1971**.

[17] C. Dietrich-Buchecker, J.-P. Sauvage, *Catenanes, Rotaxanes and Knots. A Journey Through the World of Molecular Topology*, Wiley-VCH, Weinheim, **1999**.

[18] For porphyrin-containing rotaxanes, see: L. Flamigni, V. Heitz, J.-P. Sauvage, in *Structure & Bonding : Non-Covalent Multi-Porphyrin Assemblies, Vol. 121* (Ed.: Springer), Springer, Berlin / Heidelberg, **2006**, p. 217.

[19] The notion of 'pre-rotaxane' was introduced by Schill to describe the ultimate precursor to a rotaxane in a long sequence of reactions; see ref. [16] pp. 151–4; a pseudorotaxane is a threaded species whose axis does not bear any stopper.

[20] D.B. Amabilino, J.F. Stoddart, *Chem. Rev.* **1995**, *95*, 2725.

[21] P.D. Beer, M.R. Sambrook, D. Curiel, *Chem. Commun.* **2006**, 2105.

[22] A. Harada, *Acc. Chem. Res.* **2001**, *34*, 456.

[23] E.R. Kay, D.A. Leigh, F. Zerbetto, *Angew. Chem. Int. Ed.* **2007**, *46*, 72.

[24] F. Vögtle, T. Dünnwald, T. Schmidt, *Acc. Chem. Res.* **1996**, *29*, 451.

[25] For rotaxanes or catenanes built with other transition metal templates than Cu(I), see: ref. [5] and [26–31]

[26] M. Fujita, *Acc. Chem. Res.* **1999**, *32*, 53.

[27] A.-M. L. Fuller, D.A. Leigh, P.J. Lusby, I.D.H. Oswald, S. Parsons, D.B. Walker, *Angew. Chem. Int. Ed.* **2004**, *43*, 3914.
[28] C. Hamann, J.-M. Kern, J.-P. Sauvage, *Inorg. Chem.* **2003**, *42*, 1877.
[29] L. Hogg, D.A. Leigh, P.J. Lusby, A. Morelli, S. Parsons, J.K.Y. Wong, *Angew. Chem. Int. Ed.* **2004**, *43*, 1218.
[30] D.A. Leigh, P.J. Lusby, S.J. Teat, A.J. Wilson, J.K.Y. Wong, *Angew. Chem. Int. Ed.* **2001**, *40*, 1538.
[31] D. Pomeranc, D. Jouvenot, J.-C. Chambron, J.-P. Collin, V. Heitz, J.-P. Sauvage, *Chem. Eur. J.* **2003**, *9*, 4247.
[32] V. Aucagne, J. Berna, J.D. Crowley, *et al.*, *J. Am. Chem. Soc.* **2007**, *129*, 11950.
[33] V. Aucagne, K.D. Hänni, D.A. Leigh, P.J. Lusby, B.D. Walker, *J. Am. Chem. Soc.* **2006**, *128*, 2186.
[34] C. Dietrich-Buchecker, J.-P. Sauvage, *Chem. Rev.* **1987**, *87*, 795.
[35] C. Dietrich-Buchecker, J.-P. Sauvage, J.-P. Kintzinger, *Tetrahedron Lett.* **1983**, *24*, 5095.
[36] C. Wu, P.R. Lecavalier, Y.X. Shen, H.W. Gibson, *Chem. Mater.* **1991**, *3*, 569.
[37] P. Bäuerle, M. Ammann, M. Wilde, *et al.*, *Angew. Chem. Int. Ed.* **2007**, *46*, 363.
[38] S. Anderson, H.L. Anderson, *Angew. Chem. Int. Ed. Engl.* **1996**, *35*, 1956.
[39] M.-J. Blanco, J.-C. Chambron, V. Heitz, J.-P. Sauvage, *Org. Lett.* **2000**, *2*, 3051.
[40] A. Harada, J. Li, M. Kamachi, *Chem. Commun.* **1997**, 1413.
[41] H. Ogino, *J. Am. Chem. Soc.* **1981**, *103*, 1303.
[42] G. Wenz, E. Von der Bey, L. Schmidt, *Angew. Chem., Int. Ed. Engl.* **1992**, *31*, 783.
[43] R. Huisgen, *1,3-Dipolar Cycloaddition Chemistry*, John Wiley & Sons, Inc., New York, **1984**.
[44] R. Huisgen, G. Szeimies, L. Moebius, *Chem. Ber.* **1967**, *100*, 2494.
[45] H.C. Kolb, M.G. Finn, K.B. Sharpless, *Angew. Chem. Int. Ed.* **2001**, *40*, 2004.
[46] H.C. Kolb, K.B. Sharpless, *Drug Discov. Today* **2003**, *8*, 1128.
[47] C.W. Tornøe, C. Christensen, M. Meldal, *J. Org. Chem.* **2002**, 3057.
[48] C.W. Tornøe, M. Meldal, *Peptidotriazoles: Copper(I)-catalyzed 1,3-dipolar Cycloadditions on Solid-phase*, American Peptide Society and Kluwer Academic Publishers, San Diego, **2001**.
[49] V.V. Rostovtsev, L.G. Green, V.V. Fokin, K.B. Sharpless, *Angew. Chem. Int. Ed.* **2002**, *41*, 2596.
[50] I. Aprahamian, W.R. Dichtel, T. Ikeda, J.R. Heath, J.F. Stoddart, *Org. Lett.* **2007**, *9*, 1287.
[51] I. Aprahamian, O.S. Miljanic, W.R. Dichtel, *et al.*, *Bull. Chem. Soc. Jpn.* **2007**, *80*, 1856.
[52] I. Aprahamian, T. Yasuda, T. Ikeda, *et al.*, *Angew. Chem. Int. Ed.* **2007**, *46*, 4675.
[53] A.B. Braunschweig, W.R. Dichtel, O.S. Miljanic, *et al.*, *Chem. Asian J.* **2007**, *2*, 634.
[54] W.R. Dichtel, O.S. Miljanic′, J.M. Spruell, J.R. Heath, J.F. Stoddart, *J. Am. Chem. Soc.* **2006**, *128*, 10388.
[55] O.S. Miljanic, W.R. Dichtel, S.I. Khan, S. Mortezaei, J.R. Heath, J.F. Stoddart, *J. Am. Chem. Soc.* **2007**, *129*, 8236.
[56] O.S. Miljanic, W.R. Dichtel, S. Mortezaei, J.F. Stoddart, *Org. Lett.* **2006**, *8*, 4835.
[57] P. Mobian, J.-P. Collin, J.-P. Sauvage, *Tetrahedron Lett.* **2006**, *47*, 4907.
[58] A.I. Prikhod'ko, F. Durola, J.-P. Sauvage, *J. Am. Chem. Soc.* **2008**, *130*, 448.
[59] D. Armspach, G. Gattuso, R. Koniger, J.F. Stoddart, in *Bioorganic Chemistry: Carbohydrates, Vol. 458–488*, Oxford University Press, Inc., New York, **1999**, p. 597.
[60] S.A. Nepogodiev, J.F. Stoddart, *Chem. Rev.* **1998**, *98*, 1959.
[61] J. Kim, I.-S. Jung, S.-Y. Kim, *et al.*, *J. Am. Chem. Soc.* **2000**, *122*, 540.
[62] K. Kim, *Chem. Soc. Rev.* **2002**, *31*, 96.
[63] W.L. Mock, N.Y. Shih, *J. Am. Chem. Soc.* **1988**, *110*, 4706.
[64] W L. Mock, T.A. Irra, J.P. Wepsiec, M. Adhya, *J. Org. Chem.* **1989**, *54*, 5302.
[65] D. Tuncel, J.H.G. Steinke, *Chem. Commun.* **1999**, 1509.
[66] D. Tuncel, J.H.G. Steinke, *Chem. Commun.* **2002**, 496.
[67] C. Dietrich-Buchecker, J.P. Sauvage, J.M. Kern, *J. Am. Chem. Soc.* **1984**, *106*, 3043.
[68] D. Tuncel, J.H.G. Steinke, *Macromolecules* **2004**, *37*, 288.
[69] D. Tuncel, O. Ozsar, T.H. Burak, S. Bekir, *Chem. Commun.* **2007**, 1369.
[70] D. Tuncel, H.B. Tiftik, B. Salih, *J. Mater. Chem.* **2006**, *16*, 3291.
[71] D. Tuncel, N. Cindir, U. Koldemir, *J. Inclusion Phenom. Macrocycl. Chem.* **2006**, *55*, 373.

[72] T. Ooya, D. Inoue, H.S. Choi, *et al.*, *Org. Lett.* **2006**, *8*, 3159.
[73] P.R. Ashton, P.T. Glink, J.F. Stoddart, P.A. Tasker, A.J.P. White, D.J. William, *Chem. Eur. J.* **1996**, *2*, 729.
[74] J. Cao, M.C.T. Fyfe, J.F. Stoddart, G.R.L. Cousins, P.T. Glink, *J. Org. Chem.* **2000**, *65*, 1937.
[75] J.D. Badjic, V. Balzani, A. Credi, J.N. Lowe, S. Silvi, J.P. Stoddart, *Chem. Eur. J.* **2004**, *10*, 1926.
[76] J.D. Badjic, V. Balzani, A. Credi, S. Silvi, J.F. Stoddart, *Science* **2004**, *303*, 1845.
[77] I. Aprahamian, W.R. Dichtel, T. Ikeda, J.R. Heath, J.F. Stoddart, *Org. Lett.* **2007**, *9*, 1287.
[78] Y. Liu, A.H. Flood, P.A. Bonvallet, *et al.*, *J. Am. Chem. Soc.* **2005**, *127*, 9745.
[79] H.-R. Tseng, S.A. Vignon, P.C. Celestre, *et al.*, *Chem. Eur. J.* **2004**, *10*, 155.
[80] E.D. Baranoff, J. Voignier, T. Yasuda, V. Heitz, J.-P. Sauvage, T. Kato, *Angew. Chem. Int. Ed.* **2007**, *46*, 4680.
[81] U. Létinois-Halbes, D. Hanss, J. Beierle, J.-P. Collin, J.-P. Sauvage, *Org. Lett.* **2005**, *7*, 5753.
[82] I. Poleschak, J.-M. Kern, J.-P. Sauvage, *Chem. Commun.* **2004**, 474.
[83] S. Durot, P. Mobian, J.-P. Collin, J.P. Sauvage, *Tetrahedron* **2008**, *64*, 2496.
[84] A. Livoreil, J.-P. Sauvage, N. Armaroli, V. Balzani, L. Flamigni, B. Ventura, *J. Am. Chem. Soc.* **1997**, *119*, 12114.
[85] C. Dietrich-Buchecker, J.-P. Sauvage, *Tetrahedron* **1990**, *46*, 503.
[86] F. Durola, J.-P. Sauvage, O.S. Wenger, *Chem. Commun.* **2006**, 171.
[87] F. Durola, J.-P. Sauvage, O.S. Wenger, *Helv. Chim. Acta* **2007**, *90*, 1439.
[88] J. Frey, PhD thesis, Université Louis Pasteur de Strasbourg (France), **2007**; C. Tock, PhD thesis, Université Louis Pasteur de Strasbourg (France), **2007**.
[89] J. Frey, C. Tock, J.-P. Collin, V. Heitz, J.-P. Sauvage, *J. Am. Chem. Soc.* **2008**, *130*, 4592.
[90] J. Berna, J.D. Crowley, S.M. Goldup, K.D. Hänni, A.-L. Lee, D.A. Leigh, *Angew. Chem. Int. Ed.* **2007**, *46*, 5709.

PART 4
Application in Bioorganic Chemistry

15

Aza-Wittig Reaction in Natural Product Syntheses

Francisco Palacios, Concepción Alonso, Domitila Aparicio, Gloria Rubiales and Jesús M. de los Santos

Departamento de Química Orgánica I, Facultad de Farmacia, Universidad del País Vasco. Apartado 450. 01080 Vitoria, Spain

15.1 Introduction

Staudinger and Meyers reported in 1919 the first example of an aza-Wittig reagent (Scheme 15.1).[1] These phosphorus reagents are named λ^5-phosphazenes, iminophosphoranes or phosphine imines although, in this account, we will use the general term, phosphazenes. Phosphazenes were first prepared at the beginning of the last century,[1] but it was not until Wittig's work, more than 30 years later, that the aza-Wittig reaction became accepted practice. In an analogous manner to phosphorus ylides in the Wittig reaction, phosphazenes can also react with carbonyl compounds to afford an excellent method for the construction of C=N double bonds (Scheme 15.1).[2]

Since then, the Wittig and aza-Wittig reactions have undergone tremendous development and have become a powerful tool in organic synthetic strategies directed towards the construction of acyclic and cyclic compounds, mainly because the reaction is conducted in neutral solvents in the absence of catalysts, generally at mild temperatures, and usually proceeds high yields.

Numerous research papers and several reviews[3] have appeared describing the general use of phosphazenes as reagents and intermediates in organic synthesis. This account describes the use of the aza-Wittig reaction for the preparation of natural products by means of intermolecular and intramolecular processes. Despite the aza-Wittig strategy

Organic Azides: Syntheses and Applications Edited by Stefan Bräse and Klaus Banert

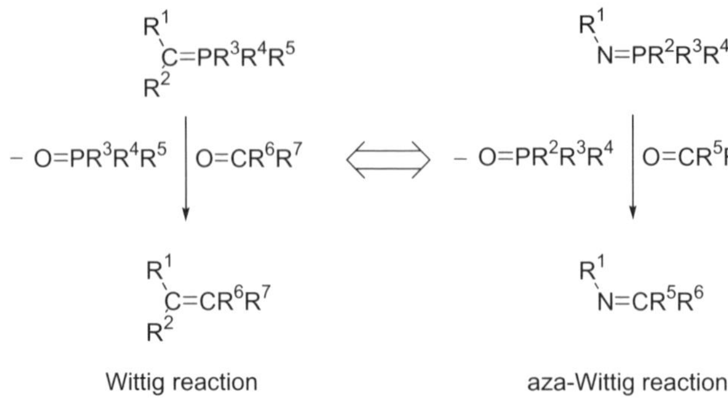

Scheme 15.1

has been developed for the synthesis of a wide type of heterocycles and their analogues, in this review we present those syntheses that exclusively lead to the preparation of natural products.

15.2 Intermolecular Aza-Wittig Reaction

15.2.1 Reaction with Carbonyl Compounds

The reaction of phosphazenes with carbonyl compounds is an excellent tool for the formation of carbon-nitrogen double bonds and this reaction seems to be one of the most efficient methods for the creation of the imine group under mild reaction conditions.[4]

15.2.1.1 Reaction of Phosphazenes with Aldehydes

The 5,11-methanomorphanthridine alkaloids are a small subclass of compounds of the *Amaryllidaceae* type first isolated by Wildman *et al.*[5] These natural products, produced by plants of various *Pancratium*, *Narcissus*, and *Brunsvigia* species, have a unique pentacyclic structural framework exemplified by the alkaloids (–)-pancracine, and (–)-coccinine. A very elegant synthesis of (–)-coccinine (**7**) and (–)-pancracine (**8**) has been described by Weinreb *et al.*[6] The corresponding imines were prepared by reaction between *N*-phenyl phosphazene **2** and functionalized aldehyde **1** (Scheme 15.2). Stereospecific cyclization was accomplished upon heating this imine/allenylsilane **3** in mesitylene at 162 °C by subsequent alkyne desilylation to afford amino acetylene **4**. Subsequent transformations led to the formation of the pentacyclic structural framework **6**, which can be reduced to (–)-pancracine **8** with sodium triacetoxyborohydride. In addition, enone **5** could be transformed into (–)-coccinine **7** after conversion of ketone group to the dimethyl ketal, followed by treatment with DIBALH (diisobutylaluminium hydride) for the reduction of the hydroxy ketal and cleavage of the TBS (tributylsilyl) group.

Conjugated phosphazenes have been widely used for the preparation of azadienes.[7] An important extension of the aza-Wittig/intramolecular electrocyclic ring closure (AW-IEC) methodology has been used in the construction of β-carboline alkaloids, which contain a

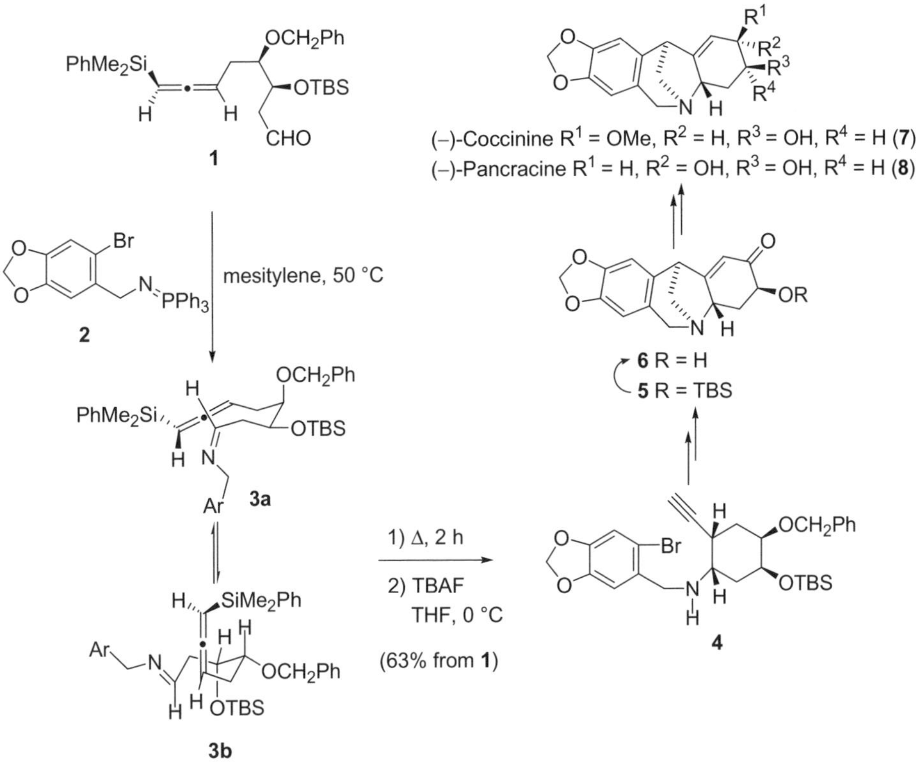

Scheme 15.2

phenyl acetyl or heteroaryl substituent at C-1. The reaction involves an initial formation of phosphazenes **10** by treatment of functionalized azides **9** with phosphine followed by aza-Wittig reaction with heterocyclic aldehyde to give the imine **11** (Scheme 15.3). Thermal treatment in the presence of Pd/C provides the corresponding 1-heteroarylsubstituted β-carbolines **12**. Finally, deprotection of the *N*-MOM (methoxymethyl) group, hydrolysis of the ester group and thermal decarboxylation afforded eudistomins A (**13a**) and M (**13b**),[8a] eudistomin U (**13c**),[8b] and nitramarine (**13d**).[8c] When this methodology is allowed using benzylglyoxal, the corresponding 1-phenylacetyl-β-carboline **12** was converted into eudistomins T (**13e**) and S (**13f**).[8d] The same approach has been extensively used for the synthesis of other alkaloids such as xestomanzamine A (**14**),[8d] fascaplysin (**15**),[8c] the unusual group of azafluoranthrene alkaloids rufescine (**16**),[8e] imeluteine (**17**),[8e] which are biosynthetically related to the tropoloisoquinolines imerubine and grandirubine and lavendamycin (**18**)[8a,8f] (Figure 15.1).

15.2.1.2 Reaction of Phosphazenes with Ketones

The preparation of a variety of vinylogous amides, precursors for a facile entry to the tetracyclic core of selaginoidine (**26**) (Scheme 15.4), was first attempted in different

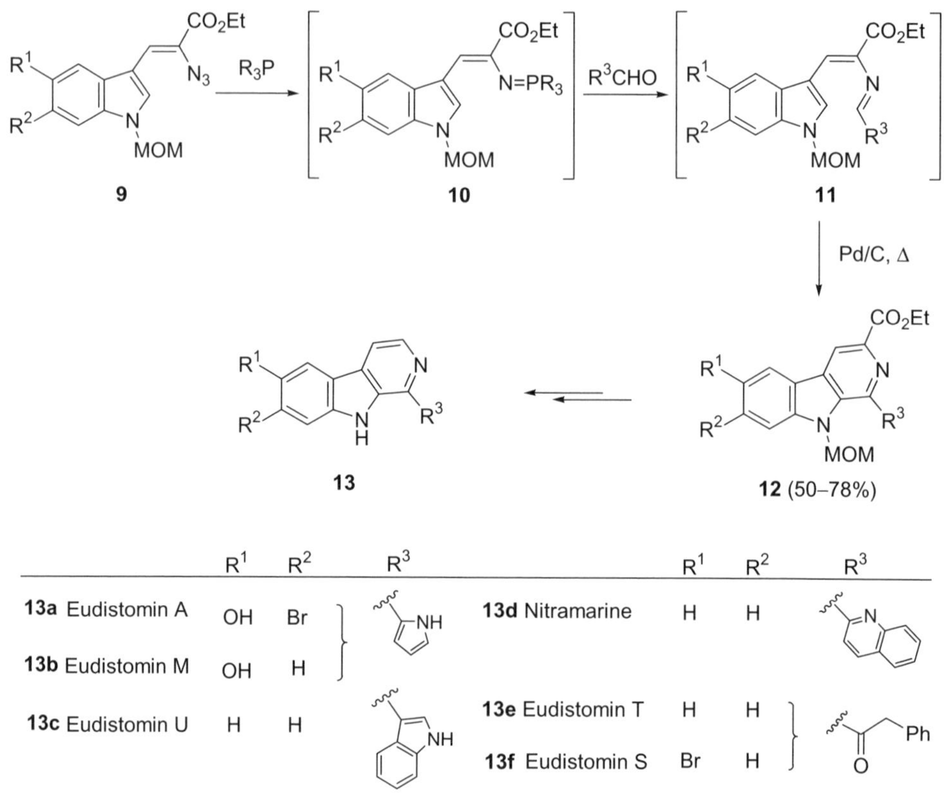

Scheme 15.3

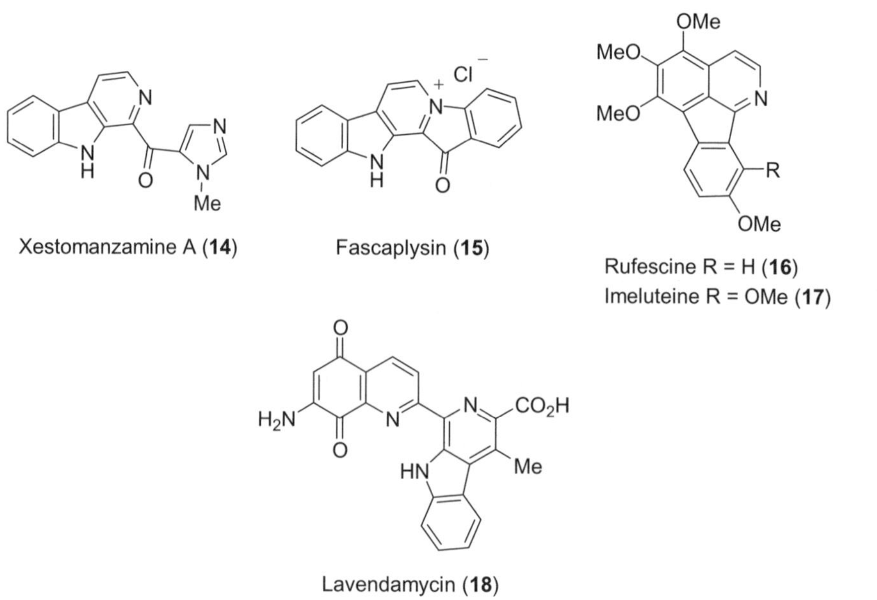

Figure 15.1

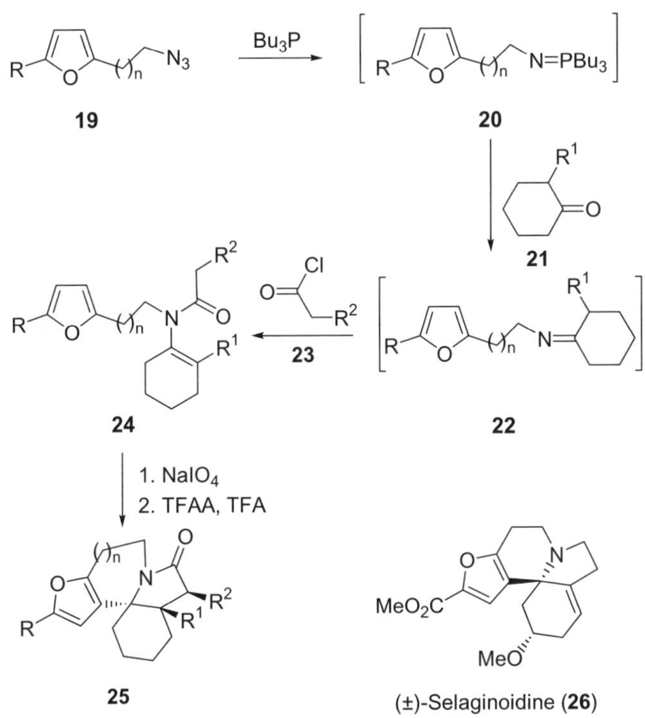

Scheme 15.4

ways.[9] However, only the aza-Wittig reaction gave the expected results, improved by the use of microwave technology. Reaction of phosphazenes **20**, derived from azides **19**, with cyclohexanone derivatives **21** gives the corresponding imines **22** (Scheme 15.4). Subsequent condensation with carboxylic acid derivatives **23** and rapid cyclization furnished the desired hexahydroindolinone systems **25**. By the moment, with a subsequent intramolecular electrophilic substitution of furanyl derivative, related homoerythrina alkaloids can be prepared.

Aza-Wittig reaction of the phosphazene **27** with methyl or benzyl glyoxalate, followed by *in situ* reduction of the intermediate aldimine **28a,b** with sodium cyanoborohydride, gives the cyclic enaminones **29a** or **29b**, which are the starting materials for the asymmetric synthesis of mycosporin I (**30**) and mycosporin-gly (**31**) (Scheme 15.5).[10]

15.2.1.3 Reaction of Phosphazenes with Carboxylic Acid Derivatives

The reaction of acyl halides with phosphazenes gives an mild method for the preparation of 2,5-disustituted oxazoles.[11] This method has been used in the preparation of naturally occurring oxazole alkaloids, such as pimprinine analogues **35a**,[12] *O*-methylhalfordinol (**35b**) and annuloline (**35c**).[13] The preparation of five pimprinine analogues through

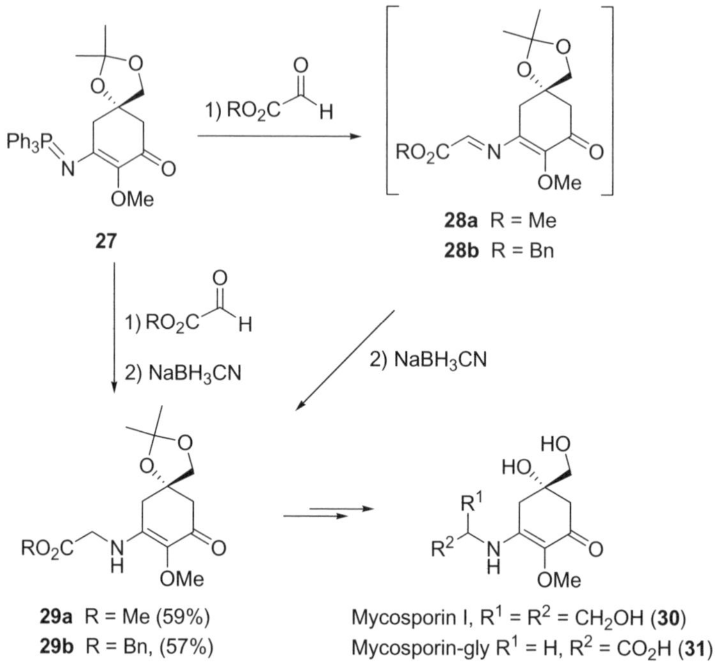

Scheme 15.5

aza-Wittig type reaction of phosphazenes with acyl chlorides has been reported. The one-pot reaction between an α-azidoketone **32**, trialkylphosphine and an acyl halide leads to oxazoles **35** (Scheme 15.6). The aza-Wittig reaction between the initially formed phosphazene **33** and the acyl chloride gives an imidoyl chloride **34** which cyclizes across the enol form of the carbonyl function to give the five-membered ring.

The first synthesis of the bis(indole) alkaloid rhopaladin D (**40**) isolated from the marine Okinawan tunicate *Rhopalaea sp.* was achieved.[14] Alkaloids rhopaladins A-D showed antibacterial activity against *Sarcina lutea* and *Corynebacterium xerosis* and inhibitory activity against cyclin dependent kinase 4 and *c-erb*β-2-kinase. The key step, construction of the central imidazolinone ring, is based on the aza-Wittig reaction of the phosphazene **37** derived from the α-azido-β-(3-indolyl)propenamide **36**, and indol-3-ylglyoxylyl chloride in the presence of a base such as polymer-supported BEMP (2-*tert*-butylimino-2-diethylamino-1,3-dimethylperhydro-1,3,2-diazaphosphorine) as a base (Scheme 15.7). Thus, the aza-Wittig reaction followed by intramolecular cyclization of the resulting imidoyl chloride **38** proceeded smoothly to afford the cyclized product **39** as a 6:4 mixture of *E*/*Z* isomers. The *N*-SEM ([β-(trimethylsilyl)ethox] methyl) deprotection of **39** with TBAF (tetrabutylammonium fluoride) in THF at reflux gave the rhopaladin D (**40**).

The same strategy has been reported for the synthesis of the marine alkaloid almazole C, isolated from the red seaweed *Heraldiophylum sp.*, which showed antibacterial activity against Gram-negative pathogens. The formation of the central 2,5-disubstituted oxazole

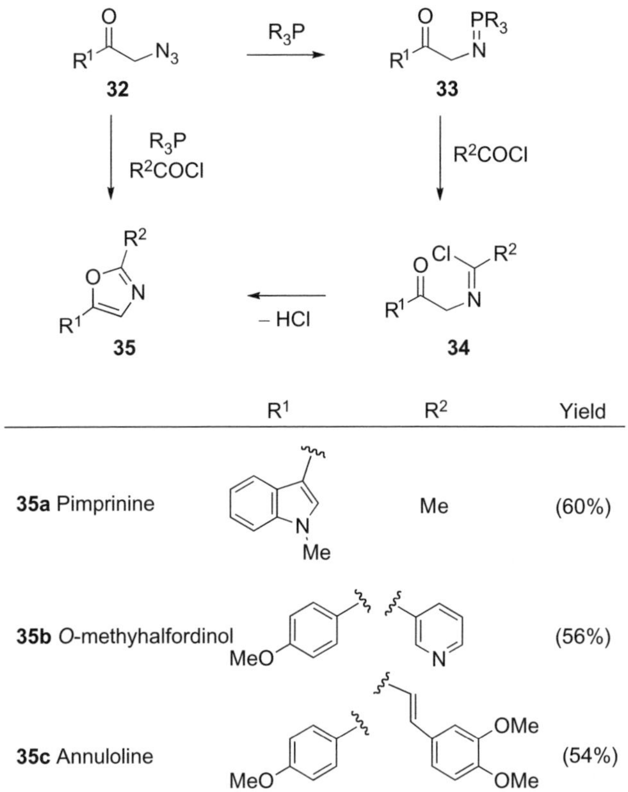

Scheme 15.6

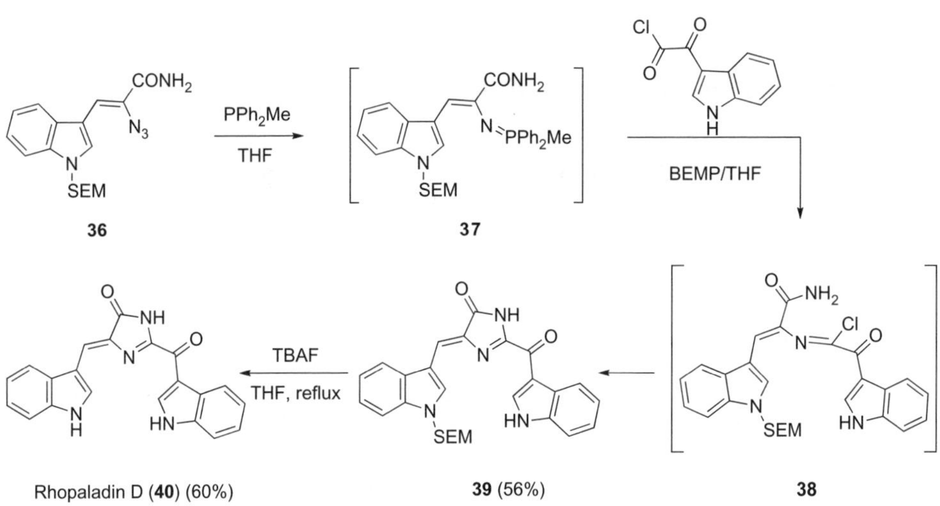

Scheme 15.7

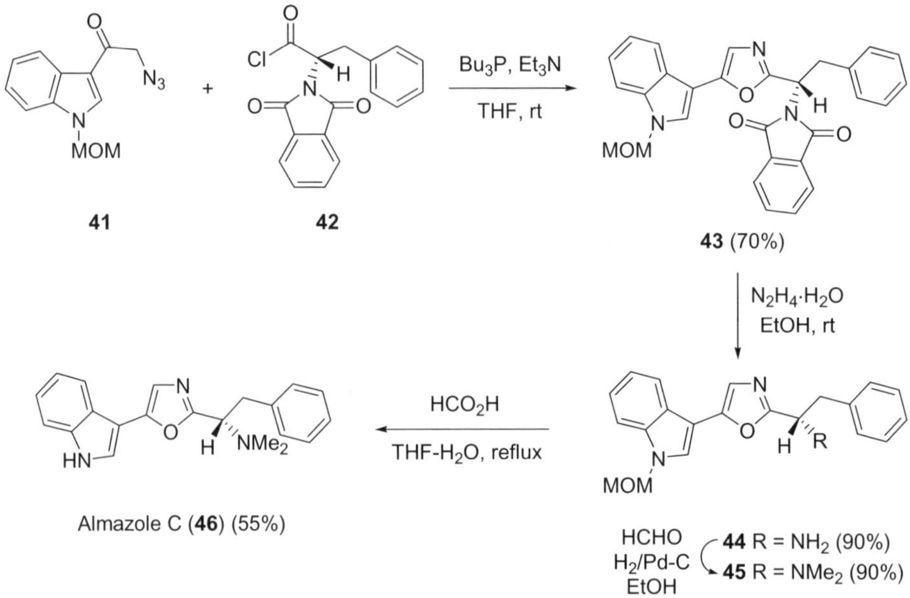

41

42

43 (70%)

Almazole C (46) (55%)

44 R = NH2 (90%)
45 R = NMe2 (90%)

Scheme 15.8

ring was achieved using the phosphazene methodology (Scheme 15.8).[15] The key intermediate **43** was prepared in 70% yield by reaction of α-azidoacetyl indole **41** with tributylphosphine in THF and subsequent addition of (*S*)-*N*-phthaloylphenylalanyl chloride **42**, followed by treatment with Et_3N. The *N*-phthaloyl group was removed with hydrazine to give amine **44**, which was dimethylated by reductive amination with hydrogen in the presence of formaldehyde and palladium as catalysis, to give compound **45** in 90% yield. Finally, compound **45** was converted into almazole C (**46**) by deprotection of the *N*-methoxymethyl substituent with formic acid in THF (Scheme 15.8).

15.2.2 Reaction with Heterocumulene Derivatives

$$R^1{-}N{=}PR^2R^3R^4 \xrightarrow[-\,R^2R^3R^4PO]{X{=}C{=}O} R^1{-}N{=}C{=}X$$

$$X = NR,\ CR_2$$

Scheme 15.9

The aza-Wittig reaction of phosphazenes with heterocumulenes[3a,16] such as isocyanates (X = NR) and ketenes ($X = CR_2$) has been extensively used for the formation of carbodiimides and ketenimines, respectively (Scheme 15.9).

15.2.2.1 Reaction of Phosphazenes with Isocyanates

From the range of general methods available for the construction of the carbodiimide functionality, the intermolecular aza-Wittig-type reaction of phosphazenes and isocya-

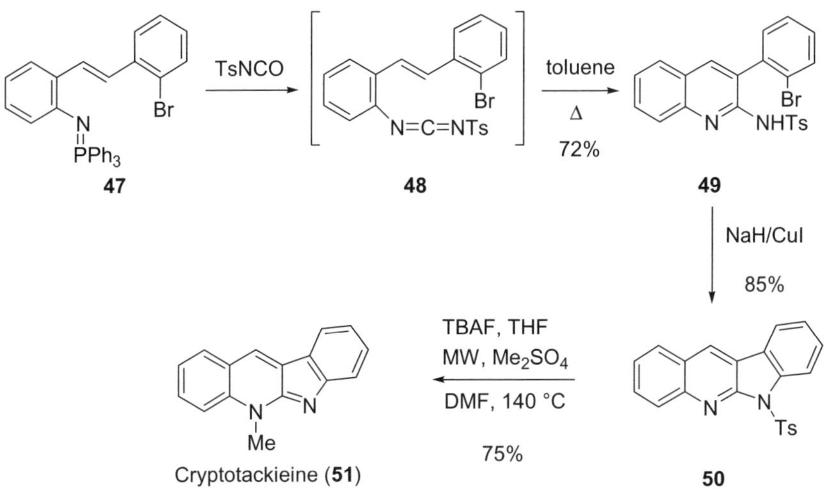

Scheme 15.10

nates seems to be an attractive method for preparation of these compounds, since it takes place under neutral conditions. In addition, this type of compounds can be subsequently used for the preparation of natural products by means of tandem or domino reactions. Aromatic phosphazenes have also been used in the synthesis of nitrogenated six-membered ring systems following the strategy of the aza-Wittig reaction and subsequent electrocyclic ring closure. The protocol has been used for the synthesis of indoloquinoline alkaloid cryptotackieine (**51**).[17] Thus, the phosphazenes **47**, containing an unsaturated side chain at the *ortho*-position (Scheme 15.10), participate in an aza-Wittig/electrocyclic ring closure, allowing the preparation of indolo[2,3-*b*]quinoline **50** through an electrocyclic process of **48** followed by the ring closure of **49**. Conversion of compound **50** into cryptotackieine **51** was achieved by deprotection of *N*-sulfonyl group with TBAF and subsequent microwave-promoted methylation with dimethyl sulfate followed by deprotonation.

N-Aromatic phosphazenes containing a triple bond reacted with heterocumulenes to give carbodiimides which can give heteropolycyclic compounds through an intramolecular [4+2] cycloaddition reaction. Thus, initial aza-Wittig reaction of phosphazene **52** (Scheme 15.11) with phenyl isocyanate and subsequent intramolecular cycloaddition of the formed carbodiimide **53** gave the quinindoline heterocycle **54**,[18] which was used in a straightforward formal total synthesis of cryptotackieine **51**[19] (neocryptolepine)[20] after methylation and subsequent deprotonation of the quinindoline **54**.

Usually, carbodiimides obtained by an aza-Wittig reaction of *N*-vinylic phosphazenes with isocyanates cannot be isolated.[3a,21] Therefore, the very reactive carbodiimides can be used as synthetic intermediates of polyheterocyclic natural products by domino processes involving aza-Wittig/intramolecular cyclization (AW-IC). In the synthesis of variolin B (**58**), the formation of the annulated 2-aminopyrimidine ring **57** is achieved from phosphazene **56** by a tandem aza-Wittig/carbodiimide-mediated intramolecular cyclization process[22] (Scheme 15.12).

Five-membered heterocycles were obtained when ammonia or amine nucleophiles reacted with carbodiimides, generated *in situ* from phosphazenes by a cascade process

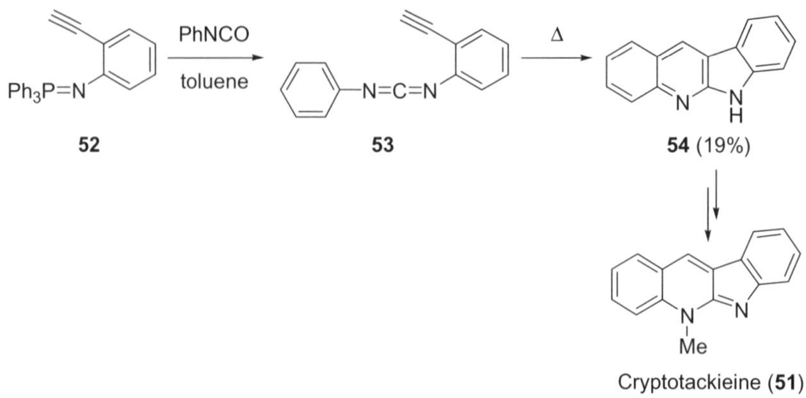

Scheme 15.11

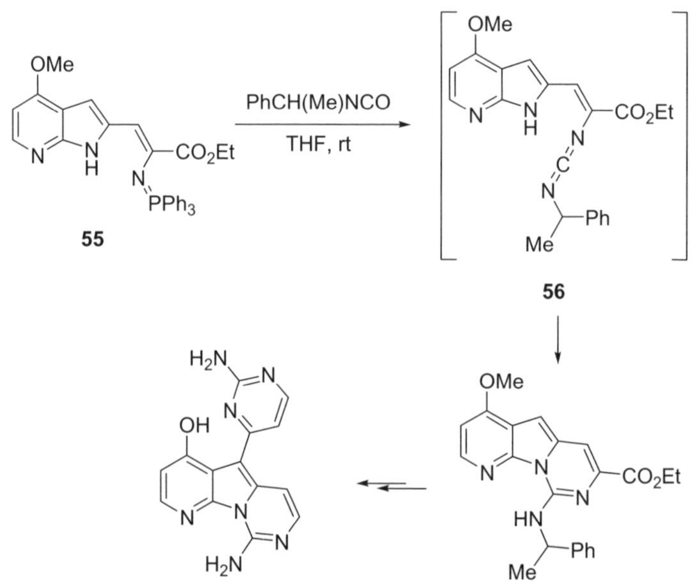

Scheme 15.12

involving aza-Wittig/intermolecular nucleophilic addition/intramolecular cyclization (AW-NA-IC). The strategy has also proved to be suitable for the preparation of leucettamine B (**62**).[23] Aza-Wittig reaction of phosphazene **59** with methyl isocyanate furnished the carbodiimide **60** in almost quantitative yield. Posterior treatment with ammonia yielded leucettamine B (**62**) through a guanidine-substituted intermediate **61**, which undergoes regioselective imidazole ring-formation across the ester and methyl amino functionality (Scheme 15.13).

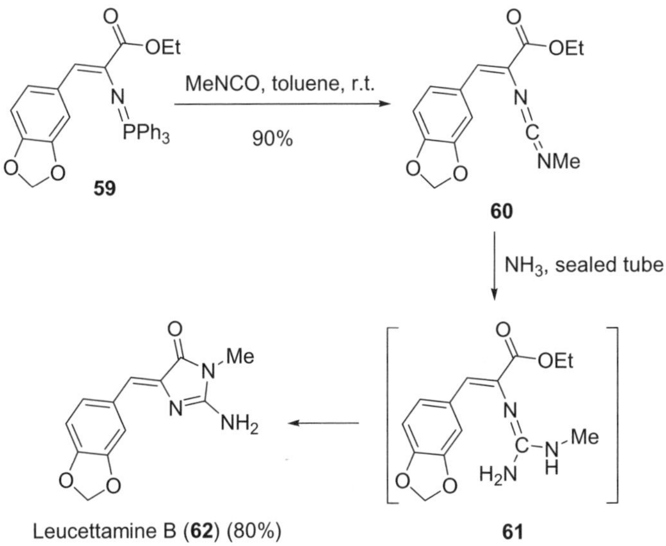

Scheme 15.13

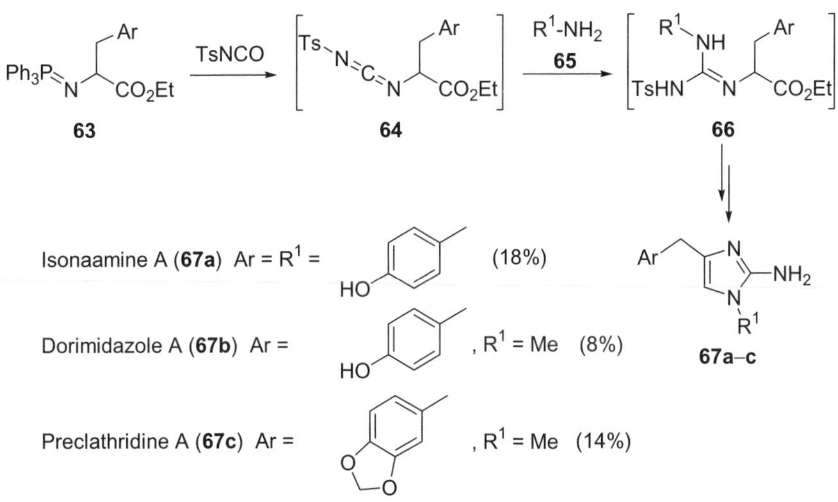

Scheme 15.14

This methodology has also been used for the preparation of several alkaloids (Scheme 15.14). Syntheses of the marine alkaloids, isonaamine A (**67a**), dorimidazole A (**67b**) and preclathridine A (**67c**),[24] have been described by aza-Wittig reactions of phosphazenes **63** with tosyl isocyanate, followed by addition of amine **65** to the carbodiimide intermediate **64** and subsequent intramolecular cyclization of compound **66**.

15.2.2.2 Reaction of Phosphazenes with Ketenes

The alkaloid (+)-cularine (**72**) was isolated by Manske in 1938 from plants belonging to the genera *Dicentra* and *Corydalis*.[25] Recently, Rodrigues *et al.* reported a diastereoselec-

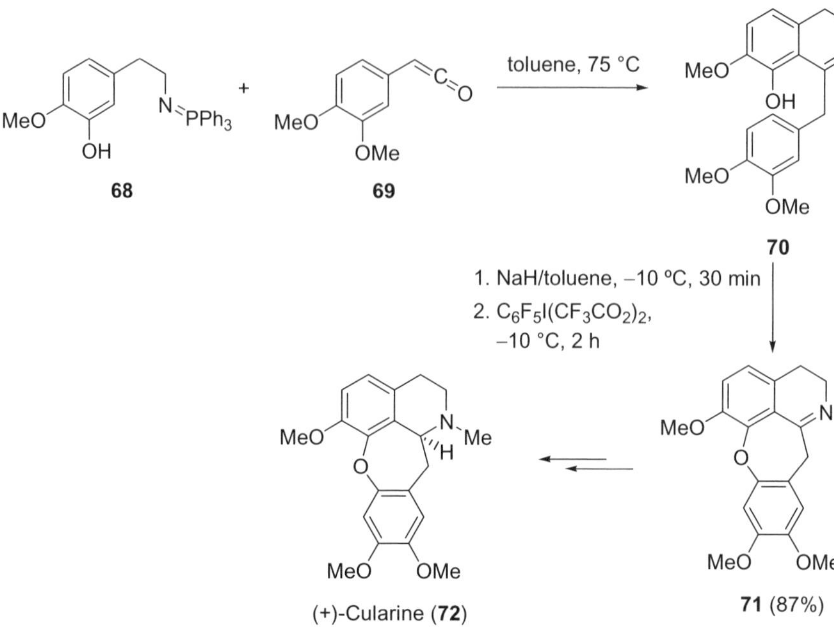

Scheme 15.15

tive synthesis of cularine alkaloids demonstrating the versatility of the aza-Wittig reaction for the construction of the isoquinoline core of the cularine alkaloids.[26] Thus, as shown in Scheme 15.15, the cascade aza-Wittig reaction/intramolecular cyclization (AW-IC) between phosphazene **68** and ketene **69** gave the 3,4-dihydroisoquinoline **70**. The corresponding aryloxenium or oxeniumoid was generated using a 1:1 ratio of the sodium salt of **70** and $C_6F_5I(CF_3CO_2)_2$, and cyclization product **71** (87%) was formed. Methylation of **71** followed by reduction with sodium borohydride at 0 °C for 4 h, yielded (+)-cularine (**72**) in 94% yield (Scheme 15.15). The synthesis of (+)-crassifoline (**73**), (+)-*O*-demethylcularine (**74**), (+)-sarcocapnine (**75**), and (+)-sarcocapnidine (**76**) has been reported by using the same approach (Figure 15.2).[26]

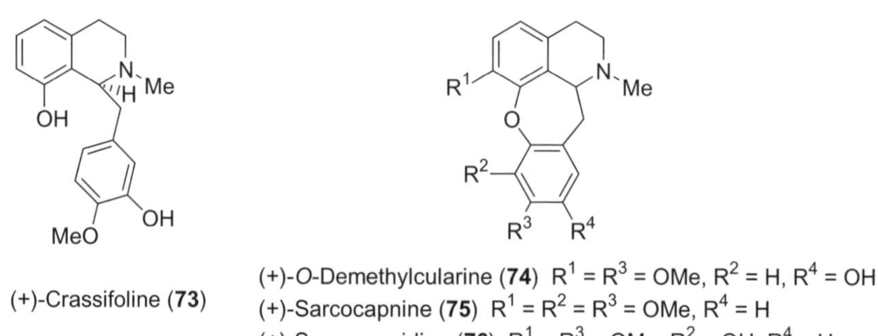

Figure 15.2

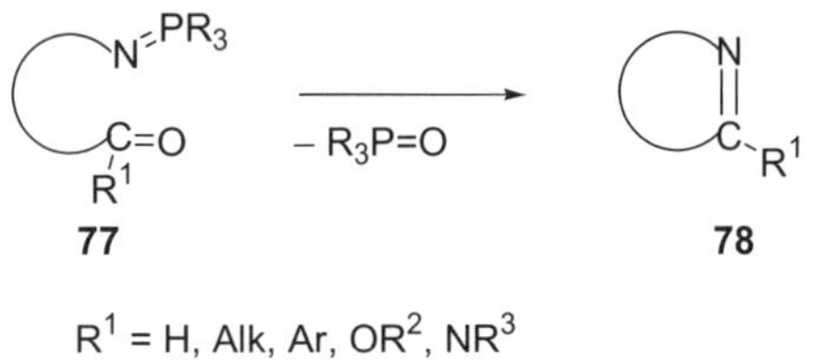

Scheme 15.16

15.3 Intramolecular Aza-Wittig Reaction

Special interest has been focused on those aza-Wittig reactions of compounds **77** (Scheme 15.16) where both the phosphazene, moiety and the carbon-oxygen double bond (C=O) (aldehydes, ketones, esters and amides) are found within one molecule.[3a,27] This strategy involving intramolecular aza-Wittig reactions allows a method for the preparation of five- to higher-membered heterocyclic compounds **78** under very mild reaction conditions, that are a structural feature in the skeleton of natural products.

15.3.1 Functionalized Phosphazenes Containing an Aldehyde Group

An intramolecular aza-Wittig reaction with an aldehyde function allowed the stereocontrolled total synthesis of the polycyclic stemona alkaloid which characteristically contains the 1-azabicyclo[5.3.0.]decane nucleous, such as (–)-stemospironine (**84**)[28] and (+)-croomine (**85**).[29] The required aldehyde **80** was prepared from the starting azide **79** by cleavage of benzyl ether and Dess-Martin oxidation of the obtained primary alcohol (Scheme 15.17). Subsequent addition of triphenylphosphine to give phosphazene **81** and *in situ* reduction of the formed imine bond of **82** in the intramolecular aza-Wittig reaction gave a seven-membered ring precursor **83** of the expected alkaloids **84** and **85**.

Phosphazenes containing an aldehyde group have been used by Yadav *et al.*[30] for the synthesis of the optically active (3*S*,4*S*)-hexahydroazepine core of balanol (**90**) and ophiocordin by ring expansion to the seven-membered azepine through an intramolecular aza-Wittig process as the key step. As shown in Scheme 15.18, treatment of functionalized azide **86** with triphenylphosphine in toluene at reflux temperature gave the crude imine **87** which was subjected to reduction with $NaBH_4$ in methanol followed by *in situ* protection with Boc_2O and TEA (triethylamine) to give the azepine segment, which was converted into the azide **88** *via* treatment with Tf_2O and 2,6-lutidine followed by NaN_3 displacement (Scheme 15.18). The final elaboration of the targeted hexahydroazepine moiety **89** of balanol is very straightforward. Hydrogenation of **88** in AcOEt liberated the amino group from its azido surrogate and simultaneously deprotected the benzyl ether function with PtO_2. Acylation of the resulting amino alcohol delivers the product **89**, which is directly amenable to the total synthesis of (3*S*,4*S*)-balanol.

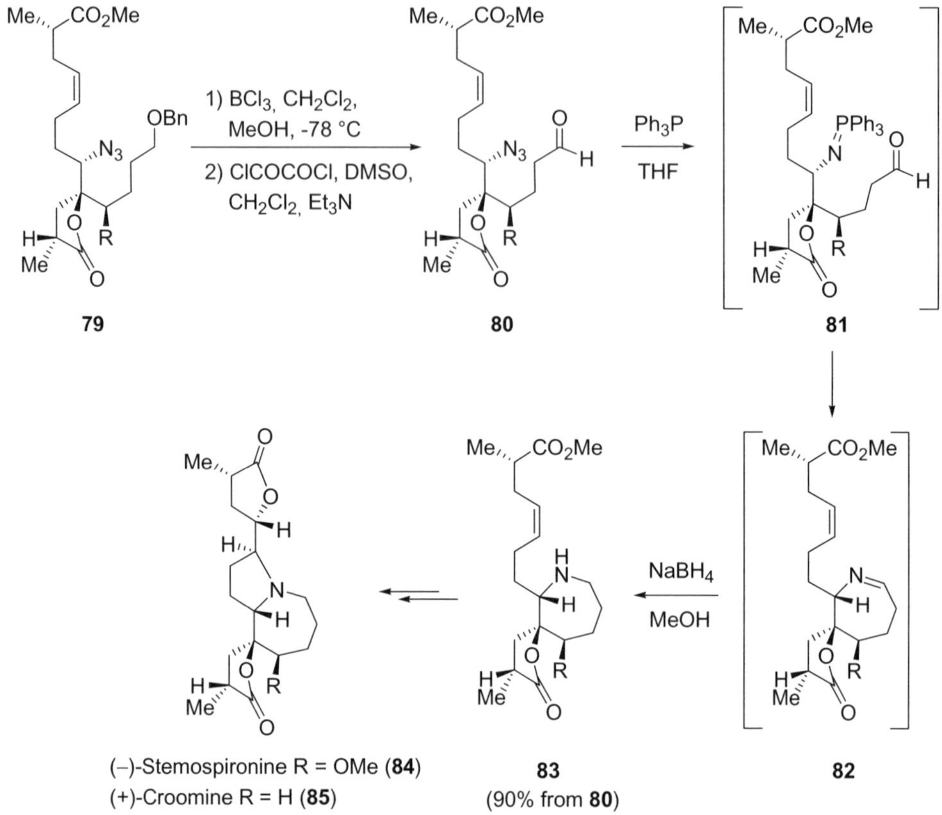

Scheme 15.17

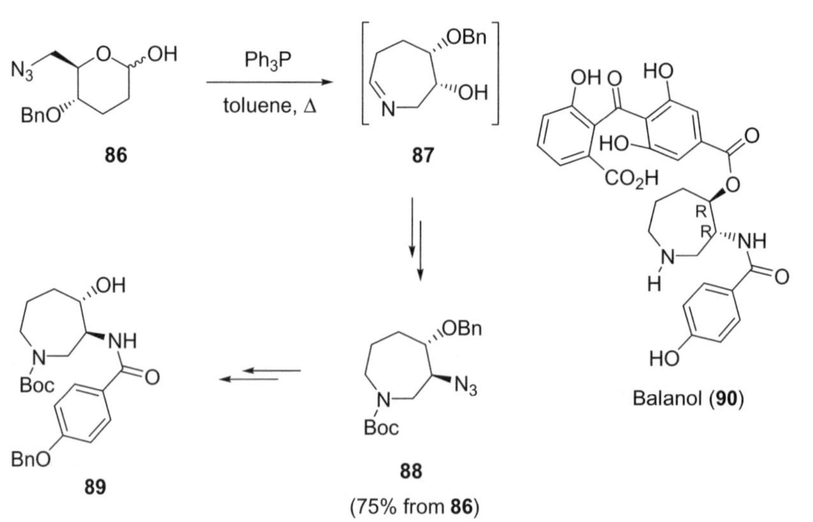

Scheme 15.18

The intramolecular aza-Wittig reaction has been successfully used for an elegant synthesis of the seven-membered nitrogen ring of (–)-stemonine (**97**) (Scheme 15.19). *Stemona* alkaloids represent a class of approximately 50 structurally novel, polycyclic metabolites isolated from monocotyledonous plants comprising the genera of *Stemona*, *Croomia*, and *Stichoneuron*. The total synthesis of (–)-stemonine (**97**) was reported for the first time by Williams *et al.*[31] nearly 75 years after its initial discovery. Cyclization to give perhydroazepine **92** *via* the Staudinger reaction of aldehydic azide **91** with ethyldiphenylphosphine generated the seven-membered imine by an intramolecular aza-Wittig process (Scheme 15.19). After imine formation, a reductive quench with $NaBH_4$ gave the amine **92** in 70% yield. Stereocontrolled formation of the pyrrolidino-butyrolactone C–D ring system occurred in a single step as a consequence of an iodine-induced cyclization. Thus, treatment of **92** with I_2 led to formation of **95** in reproducible yields of 42%. The synthesis of **97** was completed by simultaneous deprotection of TBS ethers followed by Dess–Martin oxidation, which resulted in isolation of the stable lactol **96**. Brief exposure to Jones reagent gave synthetic (–)-stemonine (**97**).

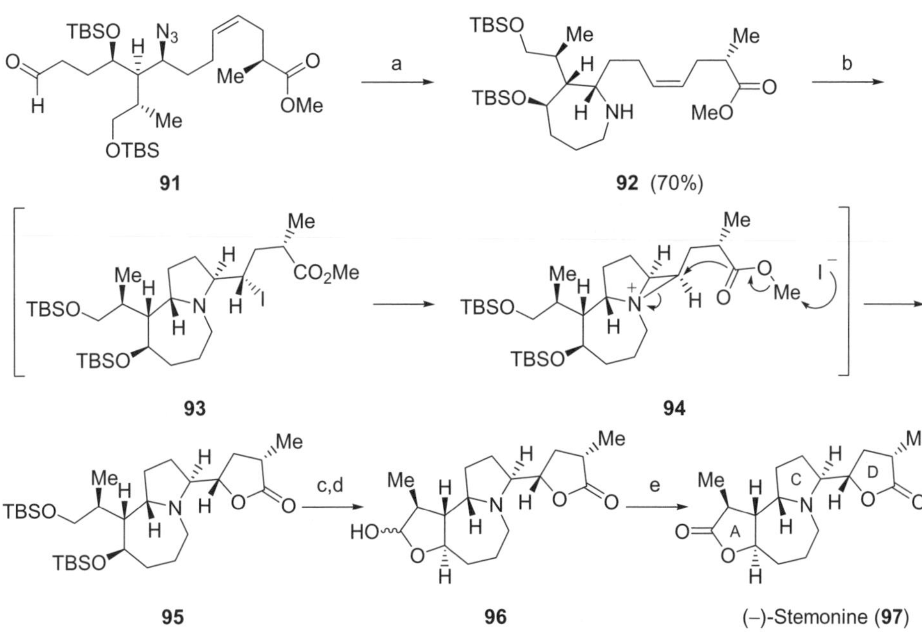

Reaction conditions: (a) $EtPPh_2$, benzene, rt, 18 h; the mixture was then concentrated in vacuo, and THF, $NaBH_4$, and MeOH were added, 70%. (b) I_2, CH_2Cl_2/Et_2O (2.5:1), rt, 48 h, 42%. (c) TBAF, THF, rt, 77%. (d) Dess-Martin periodinane, $NaHCO_3$, CH_2Cl_2, rt, 69%. (e) CrO_3, aq H_2SO_4, acetone, THF, rt, 68%.

Scheme 15.19

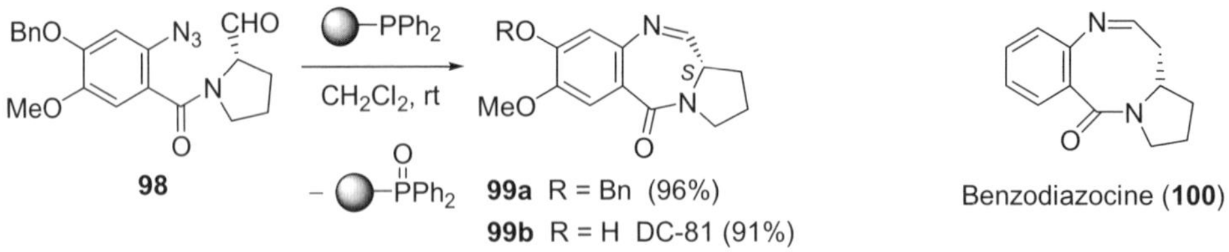

Scheme 15.20

Likewise, the antibiotic DC-81 (**99b**) can be synthesized by using the same strategy through an intramolecular reductive cyclization with polymer-supported triphenylphosphine.[32] Treatment of the azide **98** this polymer at room temperature afforded the intramolecular aza-Wittig compound **99a**, which, in turn, could be converted into the natural product DC-81 (**99b**) in a straightforward manner (Scheme 15.20). This strategy has also been used by Molina *et al.*[33] for the preparation of optically active (*R*)-enantiomer of antibiotic DC-81. A similar approach has been used for the synthesis of benzodiazocine (**100**), a novel eight-membered ring analogue to the anthramicin family of antibiotics, synthetized by O'Neil *et al.*[34]

15.3.2 Functionalized Phosphazenes containing a Ketone Group

Phosphazenes having a ketone substitution cyclize by an intramolecular aza-Wittig reaction to give ring systems of 5- or 6-membered heterocycles.[3a] In the total synthesis of (–)-dendrobine (**104**), the 5-membered nitrogen heterocycle can be formed by an intramolecular aza-Wittig reaction of the azido ketone **101** (Scheme 15.21).[35] Thus, treatment of **101** with triphenylphosphine gave polycyclic imine **103** *via* **102**. Reduction of the imine moiety with sodium cyanoborohydride from the less hindered α-face, followed by reductive methylation of the amine with paraformaldehyde and formic acid, afforded the enantiomerically pure (–)-dendrobine (**104**) (Scheme 15.21). In this synthesis, six stereogenic centers were induced, each in a stereoselective fashion, from a single chiral center of the starting material.

The formation of five-membered cyclic imines through a Staudinger/intramolecular aza-Wittig reaction can also be performed by solid-phase synthesis and has been applied for the first synthesis of lanopylin B_1 (**108**).[36] The total synthesis, which takes only four steps, starts with a phase-transfer alkylation of diethyl 2-oxopropylphosphonate **105** with a 2-iodoethyl azide, affording the azido phosphonate **106**, which undergoes a phase-transfer Horner-Emmons Wittig reaction with heptadecanal to provide the azido enone **107**. An intramolecular aza-Wittig reaction of the enone **107** with polymer-supported triphenylphosphine in toluene completed the first total synthesis of lanopylin B_1 (**108**) in 76% yield (Scheme 15.22).

A similar approach has been reported for the total synthesis of the marine cyanobacterial apratoxin A (**111**).[37] Apratoxin A, isolated from *Lyngbya spp.* cyanobacteria, is representative of a growing class of marine cyanobacterial cyclodepsipeptides wherein discrete polypeptide and polyketide domains are merged by ester and amide or amide-derived linkages. For the preparation of apratoxin A, the reaction of the azide **109** with

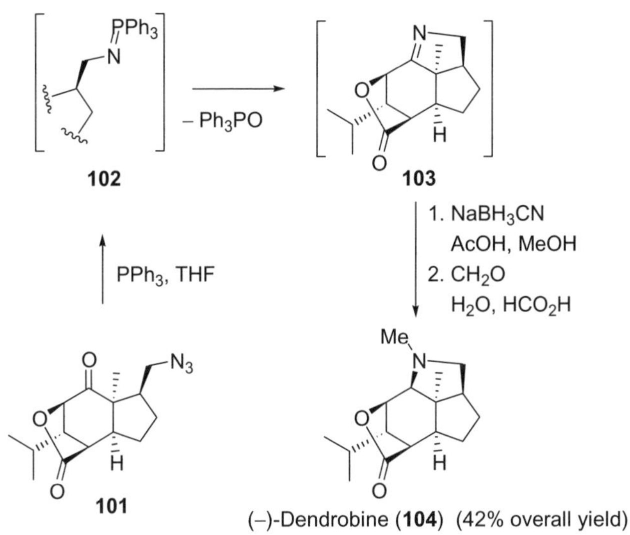

Scheme 15.21

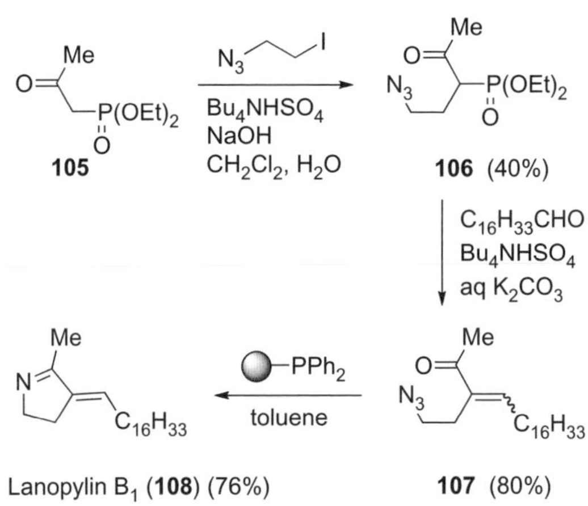

Scheme 15.22

Ph_3P in anhydrous THF effected thiazoline formation by the intramolecular Staudinger-aza-Wittig process to deliver C35 *O*-TBS ether **110**. Completion of the total synthesis of the cyclodepsipeptide apratoxin A (**111**) from thiazoline **110** would have required only two further operations: amide formation between proline amine and *iso*-leucine carboxylate residues and removal of the C35 *O*-TBS-protecting group (Scheme 15.23).

The formation of a tetrahydropyridine ring by means of the Staudinger reaction of azido-ketones and phosphines followed by intramolecular Aza-Witig reaction was applied

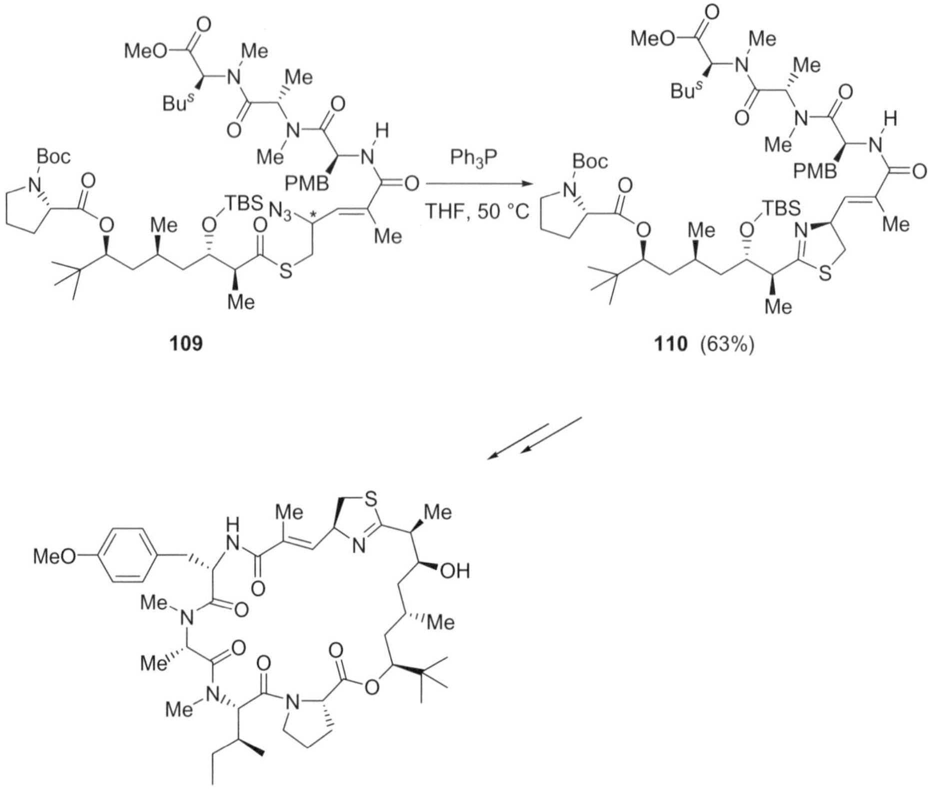

Scheme 15.23

for the synthesis of the alkaloid nigrifactine (**114a**) (Scheme 15.24), and represents the first example reported for the use of this synthetic strategy for the preparation of a natural product.[38] Polonicumtoxins A (**114b**), B (**114c**), and C (**114d**) are cyclic ketimine toxins isolated from the freshwater dinoflagellate *Peridinium polonicum*, which occasionally blooms in lakes and drinking water reservoirs. They exhibit extremely potent toxicity toward fish, making the dinoflagellate blooms a serious environmental problem. Yasumoto *et al.*[39] have reported the preparation of these compounds through a similar strategy involving an aza-Wittig procedure for the construction of the ketimine unit (Scheme 15.24). A microwave-assisted intramolecular aza-Wittig reaction was used by De Kimpe *et al.*[40] for the synthesis of the principal bread flavour component, 6-acetyl-1,2,3,4-tetrahydropyridine.

An optically active piperidine ring has been constructed by an intramolecular aza-Wittig reaction allowing a concise enantiospecific synthesis of nuphar piperidine alkaloids, among them (–)-anhydronupharamine (**118**), which has a sesquiterpenoid structure

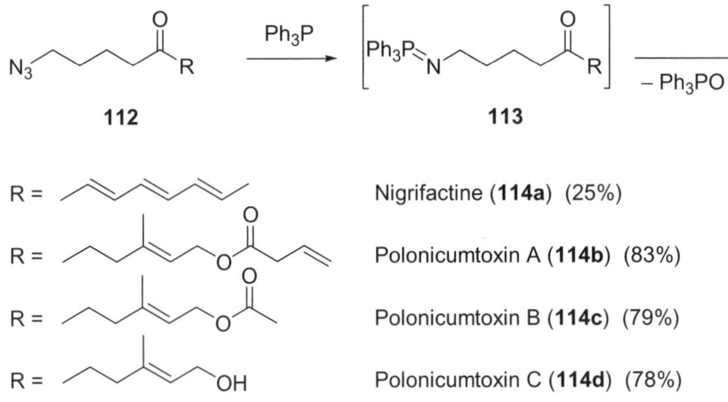

Scheme 15.24

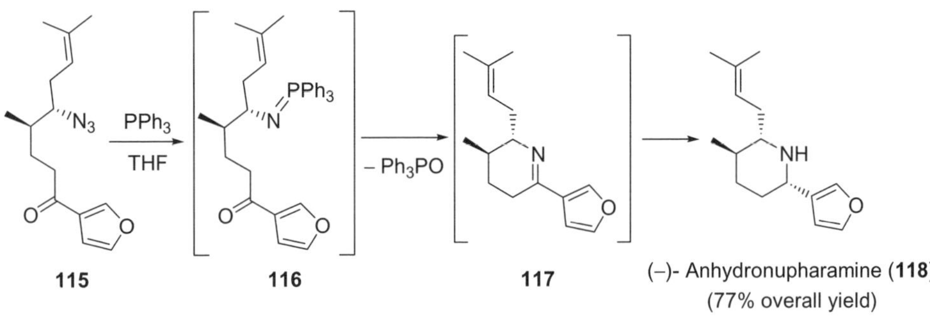

Scheme 15.25

with a furane and piperidine rings. Reaction of the phosphazene **116**, obtained by a Staudinger reaction of azide **115** with triphenylphosphine, in refluxing THF yielded the imine **117**. The non-isolated imine **117** was reduced with sodium borohydride in ethanol to give (–)-anhydronupharamine (**118**) stereoselectively (Scheme 15.25).[41]

A similar strategy has been applied for the preparation of tetrahydropyridine precursors in the synthesis of (–)-adenophorine (**123**), a rare example of a naturally occurring aza-sugar with hydrophobic susbtituents, 1-*epi*-adenophorine or deoxynojirimycin (DNJ) variants.[42] Ethyl ketimine **121** was synthesized through the Staudinger/aza-Wittig sequence from compound **119**, as shown in Scheme 15.26. The use of $LiAlH_4$ yielded tetrabenzyladenophorine **122**. Deprotection of **122** yielded (–)-adenophorine (**123**).

The azaspiracids **124** (Scheme 15.27) are the causative agents of a recently defined class of human poisoning resulting from consumption of tainted shellfish. The archetypal member of this novel class of marine toxins, azaspiracid-1 (AZA1, $R^1 = R^3 = H$, $R^2 = R^4 = Me$, Scheme 15.27), was reported by Yasumoto *et al.* as an isolated compound from the cultivated Irish mussel *Mytilus edulis*.[43] The azaspiracid natural products display

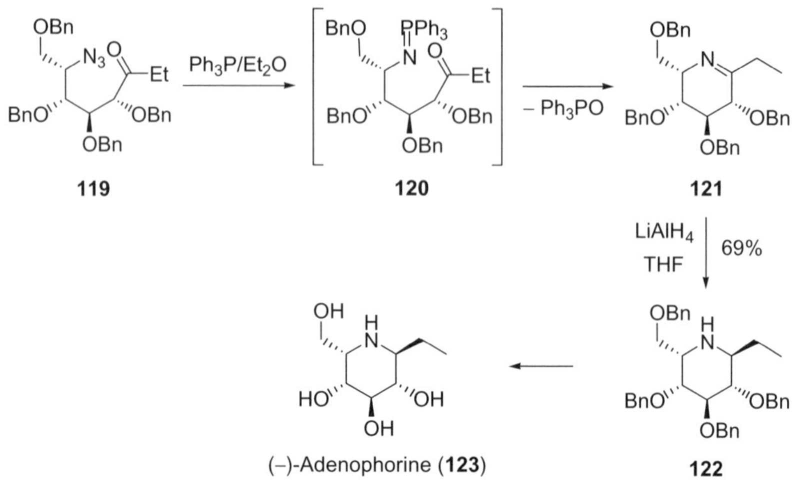

Scheme 15.26

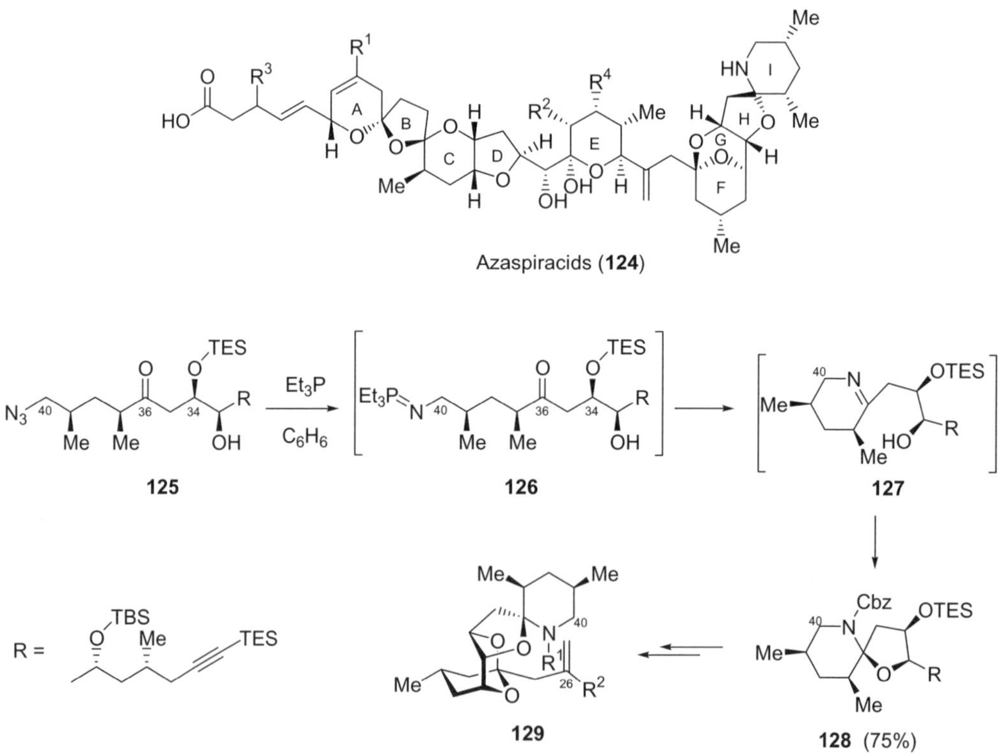

Scheme 15.27

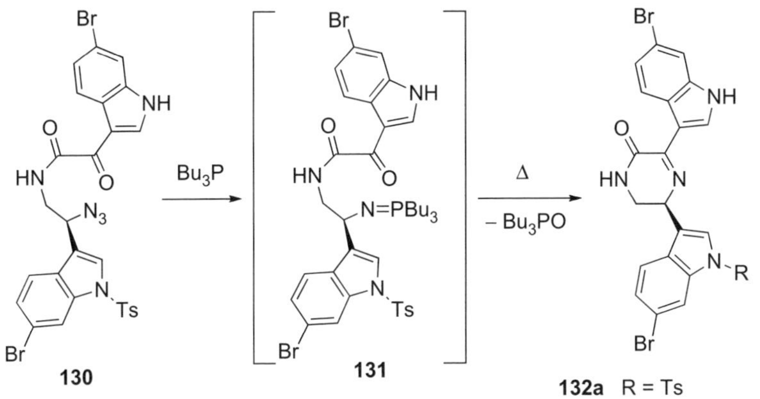

132a R = Ts

132b R = H, hamacanthin B (82%)

Scheme 15.28

a common spiroaminal-containing terminal domain that has inspired the development of its synthesis through a Staudinger reduction-aza-Wittig process.[44] The anticipated spiroaminal moiety in **128** was initially formed stereoselectively in 75% yield upon treatment of azide **125** with Et_3P in benzene (Scheme 15.27). The generated phosphazene **126** undergoes an intramolecular aza-Wittig reaction with the C36 ketone to form the six-membered cyclic imine **127**. Addition of the C33 hydroxyl group to the imine completes the cascade. Compound **128** represents the fully functionalized C27-C40 azaspiracid intermediate that is amenable to elaboration into the complete F-G-H-I ring domain.

This approach has been also used as the key step for the enantioselective synthesis of the marine indole alkaloid, hamacanthin B (**132b**)[45] (Scheme 15.28), and the antipode of hamacanthin A.[46] The central pyrazinone ring was achieved by the reaction of azide **130** with tributylphosphine in toluene at room temperature, to afford phosphazene intermediate **131**, followed by heating, to provide the expected cyclized product **132a**. Deprotection of **132a** led to the formation of hamacanthin B (**132b**) in 82% yield and keeping the configuration of the *C*–α of the starting azide.

15.3.3 Functionalized Phosphazenes Containing an Ester Group

It is well known that the carbonyl group of esters is less reactive than that of aldehydes and ketones in an aza-Wittig reaction. However, in recent years, some reports of the intramolecular aza-Wittig reaction of phosphazenes containing ester derivatives in the molecule for the preparation of heterocyclic systems[3a] have appeared. In the synthesis of siamine (**135**) the treatment of α-azidocinnamate **133** with triethylphosphine in benzene at room temperature gave the 1-ethoxyisoquinoline **134**.[47] The ester substituent of **134** was reduced in high yielding indirect sequence and finally treatment with boron tribromide resulted in simultaneous cleavage of ethyl and benzyl ethers to give siamine (**135**) (Scheme 15.29).

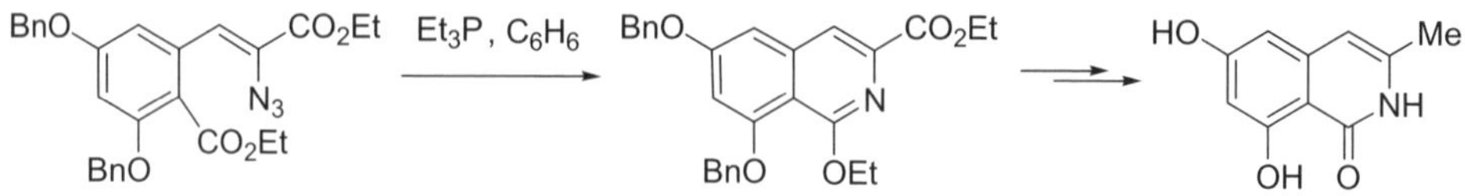

133 **134** (94%) Siamine (**135**)

Scheme 15.29

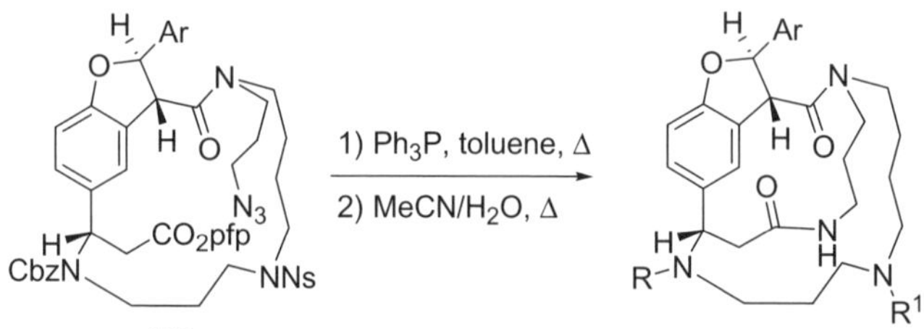

136

pfp: pentafluorophenyl
Ns: 2-nitrobenzenesulfonyl

137a R = Cbz, R^1 = Ns
Ar = *p*-BnO-C_6H_4
(73% from **136**)

137b R = R^1 = H,
Ar = *p*-OH-C_6H_4
(–)-Ephedradine A
(75% in two steps)

Scheme 15.30

The formation of complex 13-membered macrocycles through a Staudinger/intramolecular aza-Wittig reaction has been applied to the total synthesis of (–)-ephedradine A (orantine) (**137b**) (Scheme 15.30).[48,49] (–)-Ephedradine A is a complex macrocyclic spermine alkaloid, whose one of its two macrocycles can be constructed by an intramolecular aza-Wittig strategy. In this way the formation of the 13-membered iminoether **137a** was successfully obtained by treatment of the azide **136** (Ar = *p*-BnO-C_6H_4) with Ph_3P in refluxing toluene under high-dilution conditions (Scheme 15.30). Subsequent hydrolysis, removal of the Ns (2-nitrobenzenesulfonyl) group and simultaneous cleavage of the Cbz group and benzyl ether yielded (–)-ephedradine A (**137b**).

Methods for the preparation of seven-membered nitrogen-ring systems by the use of the intramolecular aza-Wittig reaction have increased in the last decade. This heterocycle is quite common in benzodiazepine derived alkaloids. This methodology has been applied for the first total synthesis of (–)-benzomalvin A (**142**) (Scheme 15.31).[50,51] Reaction of the starting azide **138** with tributylphosphine leads to the formation of the phosphazene intermediate **139**, which under the reaction conditions affords the benzodiazepine **141** in 58% yield *via* compound **140**. Benzodiazepine **141** suffered subsequent transformations to afford (–)-benzomalvin A (**142**).

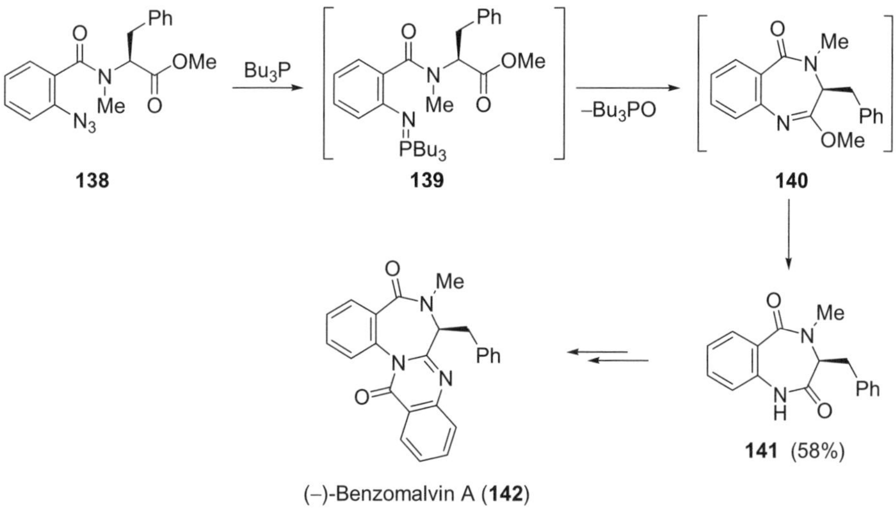

Scheme 15.31

15.3.4 Functionalized Phosphazenes Containing an Amide Group

Intramolecular aza-Wittig imination reactions involving less reactive amide carbonyl groups, which are known as the Eguchi protocol, have been reported, usually suffering from low yields. As previously reported (Section 15.3.3), intramolecular aza-Wittig reaction with ester substituents allowed the preparation of benzomalvin A. The last step in this total synthesis involves an intramolecular aza-Wittig reaction of a functionalized phosphazene containing an amide moiety. Thus, the azide **143** was treated with triphenylphosphine to generate the corresponding phosphazene, which reacted with the amide function to afford (–)-benzomalvin A (**142**) in 98% yield. Benzomalvin B (**144**) can also be obtained in two steps from (–)-benzomalvin A (Scheme 15.32).[51]

On the other hand a convenient combination of intramolecular aza-Wittig strategy and microwave technology for the preparation of the alkaloid, cryptotackieine, which has an indolo[2,3-*b*]quinoline core, has been described.[52,53] Thus, treatment of 3-(*o*-azidophenyl) quinolin-2-one **145** with trimethylphosphine in nitrobenzene under microwave irradiation between 150–180 °C, after five-membered ring construction, afforded cryptotackieine (**51**) in 40% yield *via* **146** (Scheme 15.33).

The intramolecular aza-Wittig reaction of functionalized phosphazenes containing an amide moiety has been used as the key-step for the preparation of the natural product deoxyvasicinone.[54] Azide **149** (R = H) was obtained from **148** and pyrrolidone **147** (R = H) in the presence of sodium hydride as a base at room temperature (Scheme 15.34). Then azide **149** (R = H) was treated with tributylphosphine and the natural product deoxyvasicinone (**150**) was successfully obtained in 99% yield even at room temperature for 2 h. More recently, fused [2,1-*b*]quinazolinones, namely vasicinone, deoxyvacisinone and related heterocycles have been prepared by solid-phase methods using the intramolecular aza-Wittig reaction of a phosphazene with an amide moiety.[55,56] Similarly, this

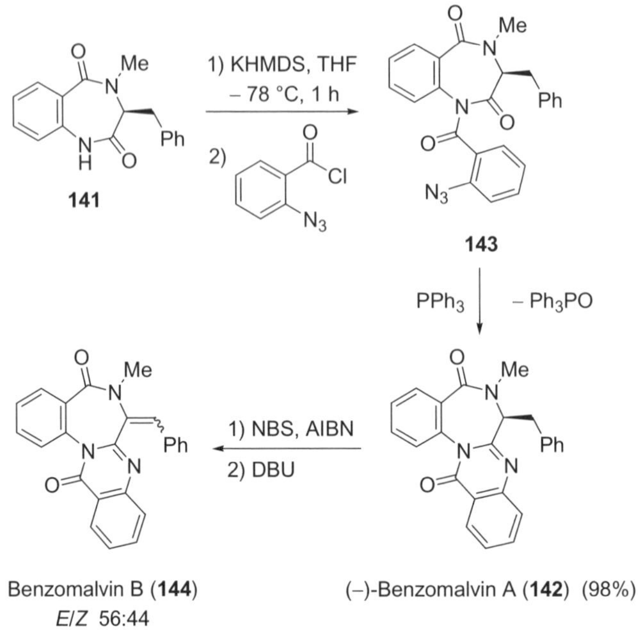

Scheme 15.32

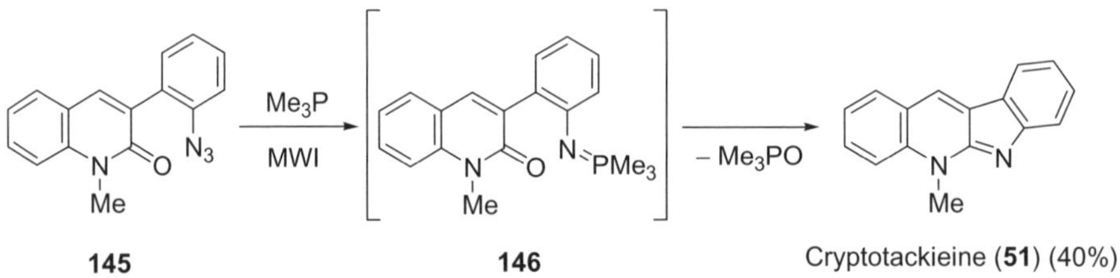

Scheme 15.33

strategy has also been used for the synthesis of optically active pyrrolo[2,1-*b*]-quinazoline alkaloid, (*S*)-(–)-vasicinone (**152**).[57] Fortunately, by this synthesis the authors have clarified that natural *L*-vasicinone has the (*S*)-configuration. After *O*-TBDMS (*O-tert*-butyldimethylsilyl) protection, *o*-azidobenzoylation followed by treatment of compound **149** (R = OTBDMS) with tributylphosphine afforded (*S*)-(–)-vasicinone (**152**) *via* the tandem Staudinger/intramolecular aza-Wittig reaction followed by TBDMS deprotection of **151** (Scheme 15.34).

Quinazoline alkaloids containing the indole skeleton such as tryptanthrin (**156**)[58] have been constructed *via* intramolecular aza-Wittig reaction of amide derivatives (Scheme 15.35). The fused quinazoline ring in tryptanthrin (**156**) could be synthesized efficiently in a one-pot procedure *via* the consecutive Staudinger/intramolecular aza-Wittig reaction of the corresponding azide **154** with tributylphosphine (Eguchi protocol).

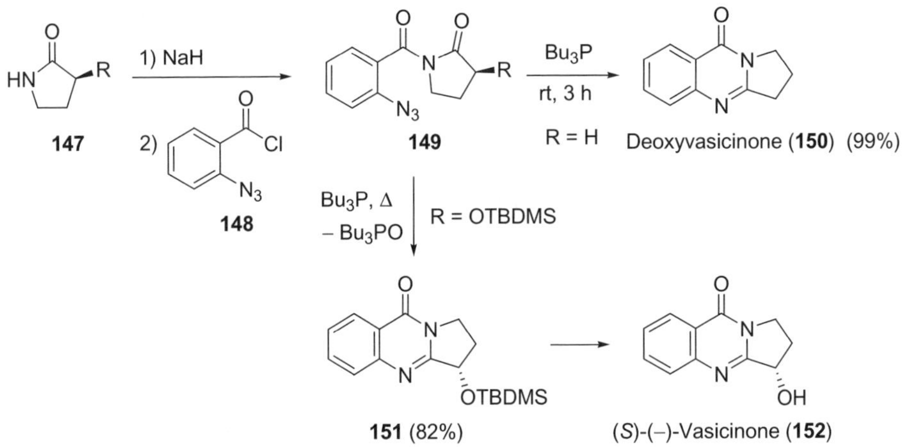

Scheme 15.34

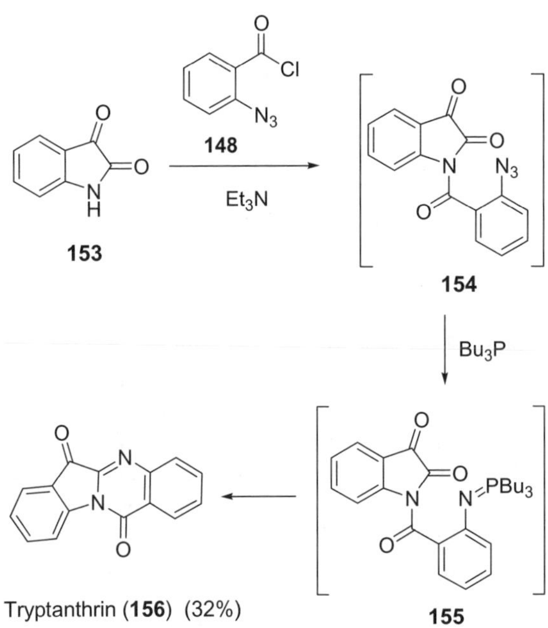

Scheme 15.35

Analogously, rutercarpine (**157**)[58] and alkaloids such as (–)-asperlicin C (**158**),[59] circumdatin F (**159**),[60] sclerotigenin (**160**),[60] and (–)-asperlicin (**161**)[59] (Figure 15.3) containing a quinazolino[3.2-*a*][1,4]benzodiazepinedione nucleus have been prepared through an intramolecular aza-Wittig procedure involving an amide moiety in very mild reaction conditions. The modified Eguchi protocol using polymer-supported phosphine-mediated intramolecular aza-Wittig reaction relied on an efficient formation of the fused

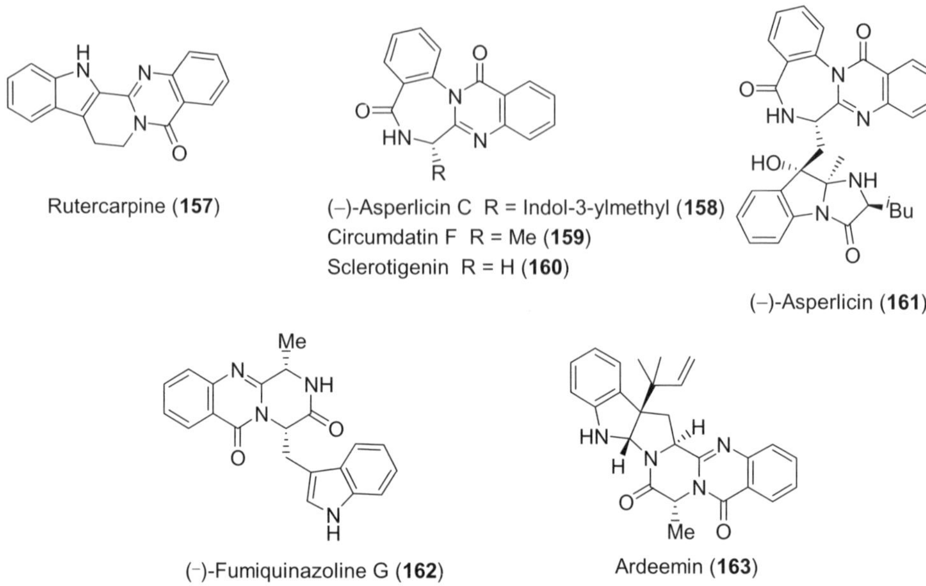

Figure 15.3

quinazoline ring system[61] yielding the simplest member of the benzodiazepine-quinazolinone family, sclerotigenin (**160**). In this manner, a multi-arrayed library generation strategy has been developed for the preparation of benzodiazepine-quinazolinone alkaloid structure of the circumdatin family of natural products. Also the key step in the synthesis of pyrazino[2,1-*b*]quinazoline core, found in the (−)-fumiquinazoline G (**162**)[60,62] and ardeemin (**163**)[63] (Figure 15.3), involves annulation of a quinazolin-4-one onto an amide by reaction with tributylphosphine following the Eguchi procedure.

This route has been adapted to the synthesis of both enantiomers of the alkaloid glyantrypine (**166**). In this case, the intramolecular aza-Wittig strategy allowed the preparation of pyrazino[2,1-*b*]quinazoline ring system present in this alkaloid and many others which exhibit very interesting biological properties.[64] The cyclization to pyrazino[2,1-*b*]quinazoline-3,6-dione derivatives was carried out through a Staudinger/aza-Wittig sequence by treatment of compound **164** with Bu_3P, which afforded compound **165** in 66% yield (Scheme 15.36). Subsequent deacetylation by addition of hydrazine hydrate gave compounds **166a,b** in good yields.

The antitumor antibiotic phloeodictine A1 (**171**) has been synthesized by Snider's group[65] (Scheme 15.37). The unstable azide derived from **167** was subjected to a polymer supported tandem Staudinger-aza-Wittig followed by a retro Diels-Alder reaction to afford intermediate **170**. Addition of 11-dodecenyl magnesium bromide followed by alkylation reaction and deprotection completes an efficient synthesis of phloeodictine A1 (**171**).

15.4 Conclusions

In summary, this review presents recent progress in the synthesis of some natural products based on the intermolecular and intramolecular aza-Wittig reaction of phosphazenes with

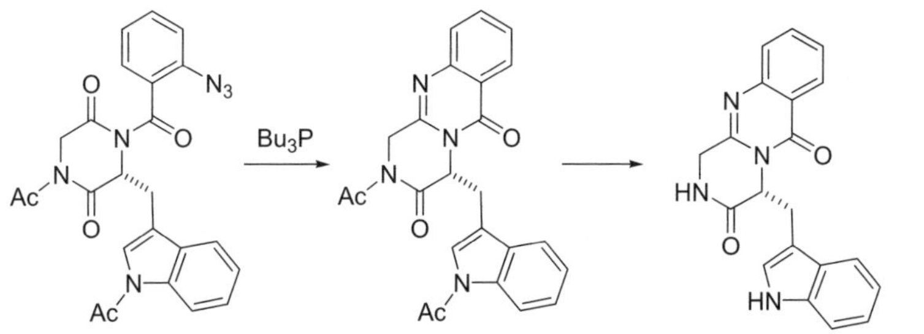

164 **165** (66%) (–)-Glyantrypine (**166a**) (+)-Glyantrypine (**166b**)

Scheme 15.36

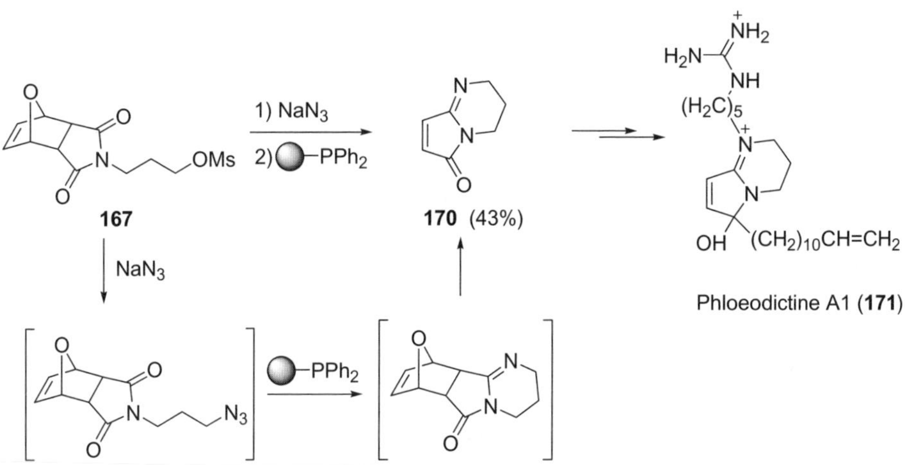

167 **170** (43%) Phloeodictine A1 (**171**) **168** **169**

Scheme 15.37

carbonyl compounds. These results indicate the importance and utility of these phosphazenes as versatile building blocks for the construction of C-N double bonds in very mild conditions, not only in the preparation of acyclic compounds, but also for heterocycle construction, ranging from simple monocyclic compounds to complex polycyclic and macrocyclic systems. In many cases, the synthesis is carried out stereoselectively and the resulting compounds are physiologically active or are potential intermediates in the synthesis of physiologically active compounds including analogues of natural products.

Acknowledgments

The present work has been supported by the *Dirección General de Investigación del Ministerio de Ciencia e Innovación* (MICINN, Madrid DGI, CTQ2006/09323) and *Departamento de Educación Universidades e Investigación (Gobierno Vasco) – Universidad del País Vasco* (GV, IT277-07; UPV, GIU06/51).

References

[1] H. Staudinger, J. Meyer, *Helv. Chim. Acta* **1919**, *2*, 635–46.

[2] H. Staudinger, E. Hauser, *Helv. Chim. Acta* **1921**, *4*, 861–6.

[3] (a) F. Palacios, C. Alonso, D. Aparicio, G. Rubiales, J.M. de los Santos, *Tetrahedron* **2007**, *63*, 523–75. (b) P.M. Fresneda, P. Molina, *Synlett* **2004**, 1–17. (c) S. Eguchi, T. Okano, T. Okawa, *Rec. Res. Dev. Org. Chem.* **1997**, 337–45. (d) H. Wamhoff, G. Richardt, S. Stølben, *Adv. Heterocyclic Chem.* **1995**, *64*, 159–249. (e) P. Molina, M.J. Vilaplana, *Synthesis* **1994**, 1197–1218. (f) Y.G. Gololobov, L.F. Kasukhin, *Tetrahedron* **1992**, *48*, 1353–1406. (g) J. Barluenga, F. Palacios, *Org. Prep. Proced. Int.* **1991**, *23*, 1–65.

[4] (a) F. Palacios, J. Vicario, A. Maliszewska, D. Aparicio, *J. Org. Chem.* **2007**, *72*, 2682–7. (b) F. Palacios, J. Vicario, D. Aparicio, *J. Org. Chem.* **2006**, *71*, 7690–6. (c) F. Palacios, E. Herrán, C. Alonso, *et al.*, *J. Org. Chem.* **2006**, *71*, 6020–30. (d) F.P. Cossío, C. Alonso, B. Lecea, *et al.*, *J. Org. Chem.* **2006**, *71*, 2839–47. (e) F. Palacios, C. Alonso, G. Rubiales, M. Villegas, *Tetrahedron* **2005**, *61*, 2779–94. (f) F. Palacios, E. Herrán, G. Rubiales, J.M. Ezpeleta, *J. Org. Chem.* **2002**, *67*, 2131–5.

[5] (a) W.C. Wildman, C.J. Kaufman, *J. Am. Chem. Soc.* **1955**, *77*, 1248–52. (b) Y. Inubushi, H.M. Fales, E.W. Warnhoff, W.C. Wildman, *J. Org. Chem.* **1960**, *25*, 2153–64. (c) W.C. Wildman, C.L. Brown, *J. Am. Chem. Soc.* **1968**, *90*, 6439–46.

[6] J. Jin, S.M. Weinreb, *J. Am. Chem. Soc.* **1997**, *119*, 2050–1.

[7] (a) F. Palacios, C. Alonso, D. Aparicio, G. Rubiales, J.M. de los Santos, *Curr. Org. Chem.* **2006**, *10*, 2371–92. (b) F. Palacios, E. Herrán, G. Rubiales, C. Alonso, *Tetrahedron* **2007**, *63*, 5669–76. (c) F. Palacios, E. Herrán, C. Alonso, G. Rubiales, *Tetrahedon* **2006**, *62*, 7661–6. (d) F. Palacios, E. Herrán, G. Rubiales, *J. Org. Chem.* **1999**, *64*, 6239–46.

[8] (a) P. Molina, P.M. Fresneda, M. Cánovas, *Tetrahedron Lett.* **1992**, *33*, 2891–4. (b) P. Molina, P.M. Fresneda, S. García-Zafra, *Tetrahedron Lett.* **1995**, *36*, 3581–2. (c) P. Molina, P.M. Fresneda, S. García-Zafra, P. Almendros, *Tetrahedron Lett.* **1994**, *35*, 8851–4. (d) P. Molina, P.M. Fresneda, S. García-Zafra, *Tetrahedron Lett.* **1996**, *37*, 9353–6. (e) P. Molina, S. García-Zafra, P.M. Fresneda, *Synlett* **1995**, 43–5. (f) P. Molina, F. Murcia, P.M. Fresneda, *Tetrahedron Lett.* **1994**, *35*, 1453–6.

[9] M.P. Cassidy, A.D. Özdemir, A. Padwa, *Org. Lett.* **2005**, *7*, 1339–42.

[10] J.D. White, J.H. Cammack, K. Sakuma, G.W. Rewcastle, R.K. Widener, *J. Org. Chem.* **1995**, *60*, 3600–11.

[11] (a) F. Palacios, A.M. Ochoa de Retana, E. Martínez de Marigorta, Marta Rodríguez, J. Pagalday, *Eur. J. Org. Chem.* **2003**, *10*, 913–91. (b) F. Palacios, M. Legido, I. Pérez de Heredia, G. Rubiales, *Heterocycles* **2000**, *52*, 1057–64.

[12] P. Molina, P.M. Fresneda, P. Almendros, *Synthesis* **1993**, 54–6.

[13] P. Molina, P M. Fresneda, P. Almendros, *Heterocycles* **1993**, *36*, 2255–8.

[14] P.M. Fresneda, P. Molina, M.A. Sanz, *Synlett* **2000**, 1190–2.

[15] P.M. Fresneda, M. Castañeda, M. Blug, P. Molina, *Synlett* **2007**, 324–6.

[16] (a) F. Palacios, A.M. Ochoa de Retana, J. Pagalday, *Tetrahedron* **2003**, *59*, 2617–23. (b) F. Palacios, M. Legido, I. Pérez de Heredia, G. Rubiales, *Heterocycles* **2001**, *55*, 1641–51.

[17] P. Molina, P.M. Fresneda, S. Delgado, *Synthesis* **1999**, 326–9.

[18] M. Alajarín, P. Molina, A. Vidal, *J. Nat. Prod.* **1997**, *60*, 747–8.

[19] M.H.M. Sharaf, P.L. Jr. Schiff, A.N. Tackie, C.H. Jr. Phoebe, G.E. Martin, *J. Heterocycl. Chem.* **1996**, *33*, 239–43.

[20] K. Cimanga, T. De Bruyne, L. Pieters, M. Claeys, A. Vlietinck, *Tetrahedron Lett.* **1996**, *37*, 1703–6.

[21] M.-W. Ding, S.-Z. Xu, J.-F. Zhao, *J. Org. Chem.* **2004**, *69*, 8366–71.

[22] P. Molina, P.M. Fresneda, S. Delgado, J.A. Bleda, *Tetrahedron Lett.* **2002**, *43*, 1005–7.

[23] P. Molina, P. Almendros, P.M. Fresneda, *Tetrahedron Lett.* **1994**, *35*, 2235–8.

[24] P. Molina, P.M. Fresneda, M.A. Sanz, *J. Org. Chem.* **1999**, *64*, 2540–4.

[25] R.H.F. Manske, *Can. J. Res.* **1940**, *18B*, 97–9.

[26] J.A.R. Rodrigues, R.A. Abraovitch, J.D.F. de Sousa, G.C. Leiva, *J. Org. Chem.* **2004**, *69*, 2920–8.

[27] F. Palacios, C. Alonso, P. Amezua, G. Rubiales, *J. Org. Chem.* **2002**, *67*, 1941–6.
[28] D.R. Williams, M.G. Fromhold, J.D. Earley, *Org. Lett.* **2001**, *3*, 2721–4.
[29] D.R. Williams, D.L. Brown, J.W. Benbow, *J. Am. Chem. Soc.* **1989**, *111*, 1923–5.
[30] J.S. Yadav, Ch. Srinivas, *Tetrahedron* **2003**, *59*, 10325–9.
[31] D.R. Williams, K. Shamin, J.P. Reddy, G.S. Amato, S.M. Shaw, *Org. Lett.* **2003**, *5*, 3361–4.
[32] A. Kamal, K.L. Reddy, V. Devaiah, D.N. Shankaraiah, *Synlett* **2004**, 2533–6.
[33] P. Molina, I. Díaz, A. Tárraga, *Tetrahedron* **1995**, *51*, 5617–30.
[34] I.A. O'Neil, C.L. Murray, A.J. Potter, *Tetrahedron Lett.* **1997**, *38*, 3609–10.
[35] C.-K. Sha, R.-T. Chiu, C.-F. Yang, *et al.*, *J. Am. Chem. Soc.* **1997**, *119*, 4130–5.
[36] B.B. Snider, J. Zhou, *J. Org. Chem.* **2005**, *70*, 1087–8.
[37] J. Chen, C.J. Forsyth, *Proc. Natl. Acad. Sci. U.S.A.* **2004**, *101*, 12067–72.
[38] M. Pailer, E. Haslinger, *Monatsh. Chem.* **1970**, *101*, 508–11.
[39] M. Yotsu-Yamashita, T. Yasumoto, V.H. Rawal, *Heterocycles* **1998**, *48*, 79–93.
[40] N. De Kimpe, C. Stevens, *Tetrahedron* **1995**, *51*, 2387–2402.
[41] T. Honda, F. Ishikawa, S.-I. Yamane, *J. Chem. Soc., Chem. Commun.* **1994**, 499–500.
[42] M.A.T. Maughan, I.G. Davies, T.D.W. Claridge, *et al.*, *Angew. Chem. Int. Ed.* **2003**, *42*, 3788–92.
[43] M. Satake, K. Ofuji, H. Naoki, *et al.*, *J. Am. Chem. Soc.* **1998**, *120*, 9967–8.
[44] S. Nguyen, J. Xu, C. J. Forsyth, *Tetrahedron* **2006**, *62*, 5338–46.
[45] B. Jiang, C.-G. Yang, J. Wang, *J. Org. Chem.* **2002**, *67*, 1396–8.
[46] B. Jiang, C.-G. Yang, J. Wang, *J. Org. Chem.* **2001**, *66*, 4865–9.
[47] M. Kennedy, C.J. Moody, C.W. Rees, J.J. Vaquero, *J. Chem. Soc., Perkin Trans. 1* **1987**, 1395–8.
[48] W. Kurosawa, T. Kan, T. Fukuyama, *J. Am. Chem. Soc.* **2003**, *125*, 8112–13.
[49] W. Kurosawa, H. Kobayashi, T. Kan, T. Fukuyama, *Tetrahedron* **2004**, *60*, 9615–28.
[50] T. Sugimori, T. Okawa, S. Eguchi, E. Yashima, Y. Okamoto, *Chem. Lett.* **1997**, 869–70.
[51] T. Sugimori, T. Okawa, S. Eguchi, A. Kakehi, E. Yashima, Y. Okamoto, *Tetrahedron* **1998**, *54*, 7997–8008.
[52] P.M. Fresneda, P. Molina, S. Delgado, *Tetrahedron Lett.* **1999**, *40*, 7275–8.
[53] P.M. Fresneda, P. Molina, S. Delgado, *Tetrahedron* **2001**, *57*, 6197–6202.
[54] H. Takeuchi, S. Hagiwara, S. Eguchi, *Tetrahedron* **1989**, *45*, 6375–86.
[55] C. Gil, S. Bräse, *Chem. Eur. J.* **2005**, *11*, 2680–8.
[56] A. Kamal, N. Shankaraiah, V. Devaiah, K.L. Reddy, *Tetrahedron Lett.* **2006**, *47*, 9025–8.
[57] S. Eguchi, T. Suzuki, T. Okawa, Y. *et al.*, *J. Org. Chem.* **1996**, *61*, 7316–19.
[58] S. Eguchi, H. Takeuchi, Y. Matsushita, *Heterocycles* **1992**, *33*, 153–6.
[59] F. He, B.M. Foxman, B.B. Snider, *J. Am. Chem. Soc.* **1998**, *120*, 6417–18.
[60] B.B. Snider, M.V. Busuyek, *Tetrahedron* **2001**, *57*, 3301–7.
[61] A. Grieder, A.W. Thomas, *Synthesis* **2003**, 1707–11.
[62] F. He, B.B. Sneider, *Synlett* **1997**, 483–4.
[63] S.P. Marsden, K.M. Depew, S.J. Danishefsky, *J. Am. Chem. Soc.* **1994**, *116*, 11143–4.
[64] Cledera, P., Avendaño, C., Menéndez, J. C., *J. Org. Chem.* **2000**, *65*, 1743–9.
[65] B.J. Neubert, B.B. Snider, *Org. Lett.* **2003**, *5*, 765–68.

16

Azides in Carbohydrate Chemistry

Henning S.G. Beckmann and Valentin Wittmann

Fachbereich Chemie, Universität Konstanz, Universitätsstr. 10, D-78457 Konstanz, Germany

16.1 Introduction

The first azide-containing sugar, a glycosyl azide, was reported in 1930 by Bertho.[1] Since that time various methods have been developed for the introduction of azides at different positions of sugars. A survey of available methods is given in Section 16.2. Until the late 1970s, azides contained in carbohydrate derivatives were simply used as accessible synthons for amines because of their easily performed reduction to amines. Due to their stability against a variety of reaction conditions, azides often can serve as masked amines during the course of carbohydrate synthesis. The development of the diazo transfer reaction facilitated the use of azides also as temporary protecting group for amines. This was extensively applied during the preparation of aminoglycoside derivatives (Section 16.3).

Over the last three decades, azides became an important tool especially for the synthesis of glycopeptides and -proteins. In 1978 Paulsen *et al.* developed the 'azide method' for the preparation of 1,2-*cis*-glycosides of glycosamine derivatives using 2-azido-2-deoxy-donors (Section 16.4). This reaction is widely used for the synthesis of *O*-linked glycosyl amino acid building blocks. In *N*-glycoproteins, the glycan chains are attached to the protein via a β-glycosyl amide. Staudinger-type reactions offer a convenient access to such structures and are applied since the 1990s for the synthesis of α- and β-glycosyl amides directly from glycosyl azides (Section 16.5).

An enormous impact on the field of glycobiology during the last decade had the development of two bioorthogonal reactions based on azides: the copper-catalyzed azide-alkyne [3+2] cycloaddition and the Staudinger ligation. Together with the possibility of *in vivo* incorporation of azide and alkyne tags into glycans and proteins, these reactions offer new options for selective labeling and manipulation of biomolecules even within

Organic Azides: Syntheses and Applications Edited by Stefan Bräse and Klaus Banert

living cells. Especially the azide-alkyne cycloaddition has been extensively applied for the chemical synthesis of neoglycoconjugates such as glycopeptide and glycoprotein mimics or multivalent glycoclusters (Section 16.6). Metabolic oligosaccharide engineering uses the biosynthetic pathways for the introduction of azide- (and alkyne-)tagged sugar moieties into the glycans of cells that can subsequently be labeled by a detectable probe. This approach is discussed in Section 16.7.

16.2 Synthesis of Azide-Containing Carbohydrates

A common way for the introduction of azides into carbohydrates is the nucleophilic replacement of leaving groups by the azide ion. These reactions can be divided into three groups: substitutions at the anomeric center leading to glycosyl azides, substitutions at primary, and substitutions at secondary carbon atoms.

A widely used method for the preparation of glycosyl azides[2–4] is the conversion of acetylated halogenoses, such as **1**, by treatment with sodium azide based on Bertho's initial work (Scheme 16.1A).[1] While homogeneous one-phase reactions in DMF often require elevated temperatures,[5] phase-transfer catalysis enables milder conditions.[6] One limitation of this methodology is the instability of glycosyl halides. Thus, sequential one-pot procedures have been developed that avoid the isolation of glycosyl halides.[7] An alternative, which circumvents the preparation of glycosyl halides completely, is the direct conversion of glycosyl acetates into the corresponding glycosyl azides using trimethylsilyl azide under Lewis acid catalysis (Scheme 16.1B).[8] Glycosyl azides with 1,2-*trans*-configuration are easily obtained by the described methods using acyl protecting groups due to their neighboring group participation. Glycosyl azides with 1,2-*cis*-configuration can be prepared from 1,2-*trans*-glycosyl halides in an S_N2-type reaction or from ether-protected glycosyl acetates by treatment with trimethylsilyl azide.[2–4]

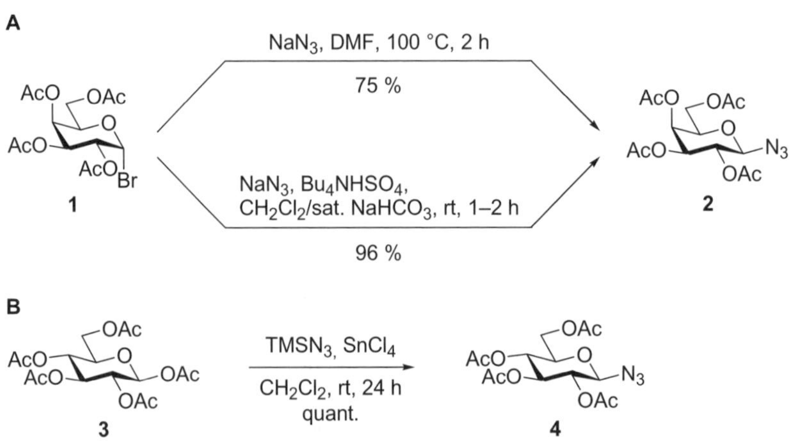

Scheme 16.1 *Preparation of glycosyl azides from (A) peracetylated glycosyl halides under classical homogeneous conditions[5] and under mild phase-transfer catalysis[6] and (B) from peracetylated sugars[9]*

The introduction of azides at the primary carbon of carbohydrates is conveniently carried out by an S_N2 reaction. The generation of a good leaving group, such as a sulfonate, is often possible in a selective way without need for protection of the secondary hydroxy groups as was shown for GlcNAc derivative **5** (Scheme 16.2).[10] Subsequent substitution with sodium azide usually proceeds at elevated temperatures with good yields.

In contrast, S_N2 reactions at secondary carbons of the sugar ring system are more complex. The success of such reactions is strongly dependent on the type of sugar (stereochemistry), the position at which the S_N2 reaction is carried out, anomeric configuration, and used protecting groups. Nevertheless, this approach is widely applied for the introduction of azido groups at the ring system. For instance, the mesylate of glucoside **7** was substituted yielding 4-azido galactoside **8** under inversion of configuration (Scheme 16.3).[11]

Epoxides are also useful precursors for the incorporation of azido groups by nucleophilic attack. According to the Fürst-Plattner rule,[12] ring opening of sugar epoxides by azide ions preferentially leads to the diaxial product. For instance, 2-azido compound **10** is obtained regioselectively by opening of Cerny epoxide **9** with sodium azide (Scheme 16.4).[13] **10** was further converted into the suitably protected glycosyl donor **11**, which was applied in the synthesis of a heparan sulfate synthon by 1,2-*cis*-glycosylation (cf. Section 16.4).

Azides can also be introduced by radical addition to glycals. The classical azidonitration, developed by Lemieux *et al.* in 1979, is a powerful method for the preparation of

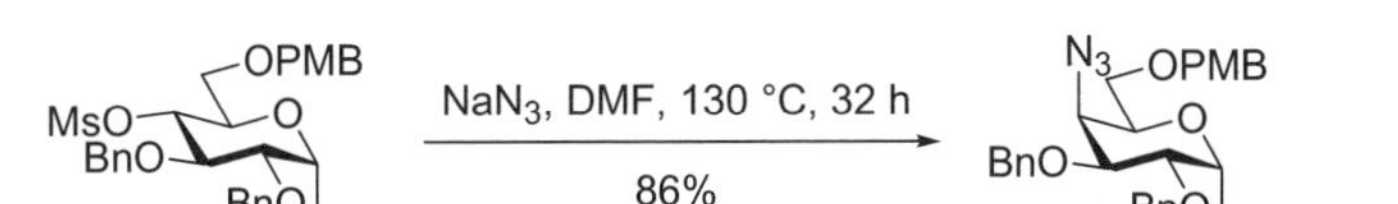

5 **6**

Scheme 16.2 *Regioselective introduction of an azido group at the primary carbon of **5** via nucleophilic replacement of a sulfonate intermediate*[10] *Py = pyridine*

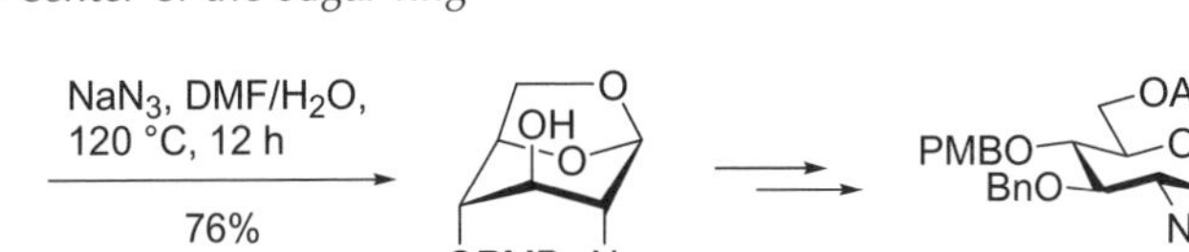

7 **8**

Scheme 16.3 *Replacement of a mesylate by an azido group under inversion of configuration at a secondary center of the sugar ring*[11]

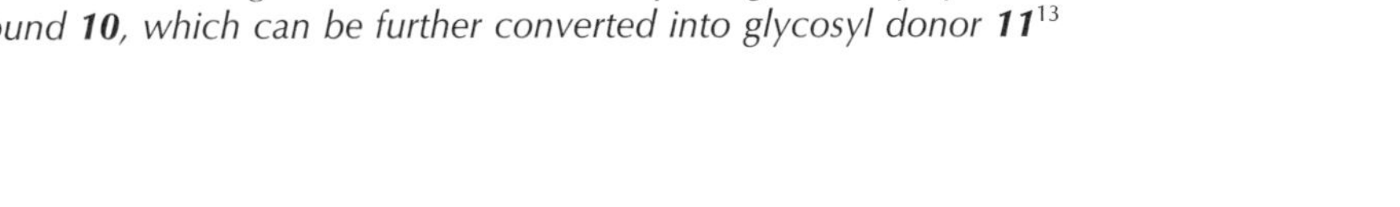

9 **10** **11**

Scheme 16.4 *Regio- and stereoselective opening of Cerny epoxide **9** leads to 2-azido compound **10**, which can be further converted into glycosyl donor **11***[13]

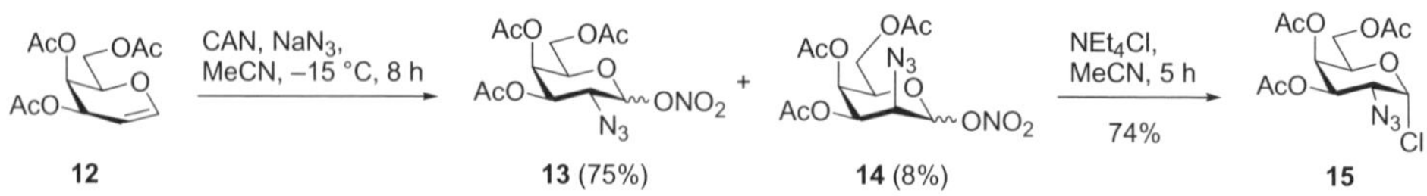

*Scheme 16.5 Azidonitration of galactal **12** leads to an epimeric mixture of the 2-azido-1-nitro-pyranoses **13** and **14** from which glycosyl donor **15** can be prepared directly.*[14] *CAN = cerium(IV) ammonium nitrate*

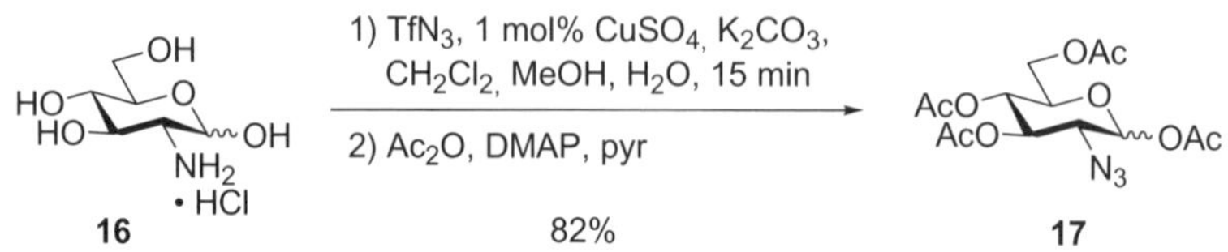

Scheme 16.6 Typical procedure for the Cu(II)-catalyzed diazo transfer.[26] *DMAP = 4-(dimethylamino)pyridine*

2-azido sugars that is still frequently used (Scheme 16.5).[14] It is especially useful for the synthesis of those 2-azido derivatives, whose corresponding glycosamines lack accessibility from natural sources as in the case of galactosamine. However, while the reaction is highly regioselective, in most cases epimeric mixtures of the 2-azido compounds are formed. The ratio of the epimers strongly depends on the employed glycal substrate.[15] The obtained 1-nitro-pyranoses can easily be converted into glycosyl donors, such as glycosyl halides,[14] trichloroacetimidates,[16] *n*-pentenyl glycosides,[17] or thioglycosides,[18–20] which are valuable building blocks for the preparation of 1,2-*cis* glycosides of *N*-acetyl-glycosamines (cf. Section 16.4). Similar methods for the synthesis of 2-azido sugars using radical addition to glycals are the azidochlorination[21] and the azidophenylselenation.[22,23]

Another possibility for the synthesis of organic azides is the diazo transfer using triflyl azide.[24] In contrast to the methods described above, not the entire azido group is incorporated into a molecule but an N_2 moiety is transferred onto an existing amine under retention of configuration. The first diazo transfer onto amino sugars was reported in 1991 by Vasella *et al.*[25] They treated different unprotected glycosamines with freshly prepared triflyl azide under basic conditions. After subsequent acetylation, the 2-azido sugars were isolated in good yields. This methodology was further improved by the addition of catalytic amounts of copper sulfate which leads to a much faster and more reliable reaction (Scheme 16.6).[26,27] Using the diazo transfer, it is possible to employ azides not only as amine synthons but also as temporary protecting groups for amines. This has been applied for example to the synthesis of aminoglycosides (Section 16.3), heparan sulfate fragments,[28] heparin fragments,[29,30] hyaluronan neoglycopolymers,[31] and *N*-acetyl-neuraminic acid derivatives.[32]

16.3 Azides as Protecting Groups during Aminoglycoside Synthesis

Protecting groups commonly employed for masking amino groups include alkyl carbamates such as benzyl-, *tert*-butyl-, and 9-fluorenylmethyl carbamates. If used for the protection of molecules containing multiple amino groups, however, carbamate protecting groups can seriously complicate the interpretation of NMR spectra. This is due to the occurrence of *E*/*Z* rotamers that are in slow interconversion leading to multiple sets of signals. The use of azides as protecting groups circumvents this problem. Azides are easily reduced to amines, for example by catalytic hydrogenation or by reaction with thiols or complex hydrides.[4,33,34] A widely applied method in carbohydrate chemistry is the Staudinger reduction using triaryl- or trialkylphosphines.[35] This mild procedure enables the selective reduction of azides in the presence of esters and benzyl ethers which are frequently used as OH-protecting groups. Furthermore, azides can be directly converted into carbamate-protected amines using a variant of the Staudinger reaction (cf. Section 16.5).[27,36,37]

Aminoglycosides are highly potent, broad-spectrum antibiotics, containing several amino groups presented on an oligosaccharide-like core.[38–40] Due to the appearance of bacterial strains resistant to these drugs and due to their relatively high toxicity, the synthesis of aminoglycoside derivatives with improved properties is of great interest.[41] Several syntheses of aminoglycoside derivatives using azides as amine protecting groups were reported,[42,43] for instance the preparation of analogs of neomycin B as shown in Scheme 16.7.[27,44,45] Starting from commercially available neomycin B (**18**), all six amino groups were converted into azides by diazo transfer. After chemical derivatization of the structure, amines were regenerated by Staudinger reduction.

In the course of these studies, it was observed that the regioselective reduction of a single azide of multiple azide-containing molecules is feasible if only one equivalent of phosphine is used.[27] Reduction of neamine derivative **23**, for example, gave mono-amine **24** in a yield of 46 % (Scheme 16.8). Strong evidence was presented that the selectivity is primarily determined by electronic factors with electron-deficient azides being reduced more rapidly and efficiently than electron-rich azides. In compound **23** this is the case for

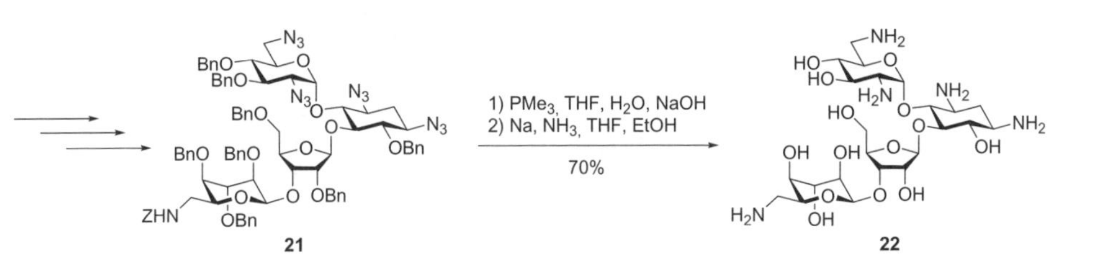

Scheme 16.7 *Synthesis of aminoglycoside derivative* ***22*** *using azides as protecting groups for amines. First, the amino groups of* ***18*** *were converted into azides by diazo transfer.*[27] *After chemical remodeling of the aminoglycoside (one amino group was replaced by a hydroxy group), the amines were regenerated by Staudinger reduction*[44]

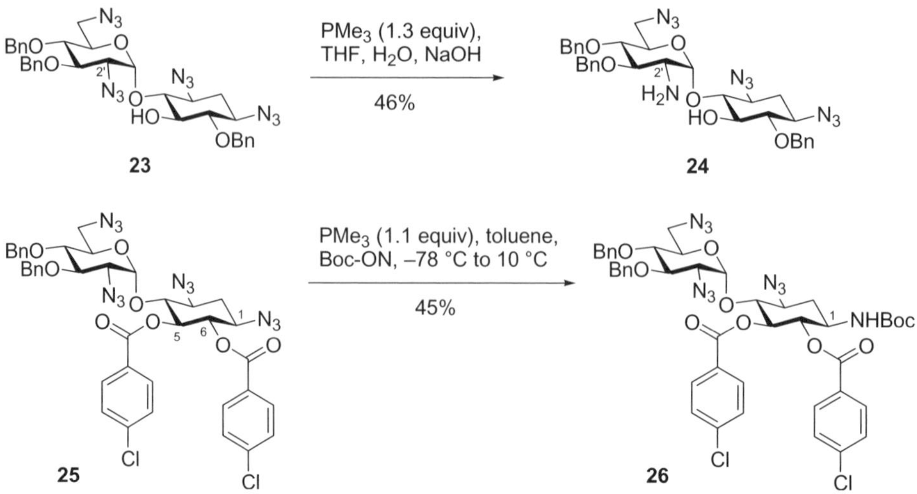

__Scheme 16.8__ Regioselective reduction of tetra-azides ***23***[27] *and* ***25***.[46,47] *Boc-ON = 2-(tert-butoxycarbonyloxyimino)-2-phenylacetonitrile*

the 2′-azide adjacent to the anomeric center. It was shown that the regioselectivity can be predicted on the basis of ^{15}N and, to some extent, ^{1}H NMR chemical shifts. Consequently, by introduction of electron withdrawing 4-chlorobenzoyl protecting groups in the 5- and 6-position, the selectivity can be tuned in favor for reduction of the 1-azide (**25** → **26**).[46,47]

16.4 Azides as Non-Participating Neighboring Groups in Glycosylations

Although 1,2-*cis* glycosides of 2-amino-2-deoxysugars are less frequently found in natural products compared to their 1,2-*trans* isomers, they are a common motive in important structures. In mucin-type *O*-glycoproteins, e.g. the glycan chains are attached to protein via an α-glycosidic linkage of *N*-acetyl-D-galactosamine to the β-hydroxy group of either serine or threonine, and α-glycosides of 2-acetamido-2-deoxy-D-glucose are found in the glycosaminoglycan heparan sulphate.[48–51] For the preparation of these 1,2-*cis* glycosides, the commonly employed *N*-acyl protecting groups are not suited because they lead to 1,2-*trans* products **29** via neighboring group participation (Scheme 16.9A).[15,52,53] In 1978 Paulsen *et al.* showed that 2-azido-2-deoxy-glycosyl halides are suitable donors in 1,2-*cis* glycosylations. This approach preferentially leads to α-glycosides **32** either directly from β-glycosyl halides **30** (Scheme 16.9B)[54,55] or by *in situ* anomerization[56] of α-glycosyl halides **33** (Scheme 16.9C).[57] Since then, the azide method has been widely used[15,52,58–62] and expanded by use of other glycosyl donors, such as trichloroacetimidates,[16] *n*-pentenyl glycosides,[17] and thioglycosides[18–20] just to name a few. The required 2-azido-2-deoxy sugars are usually prepared by azidonitration of glycals or by diazo transfer reaction of the corresponding glycosamines as described above. After glycosylation, the azide can

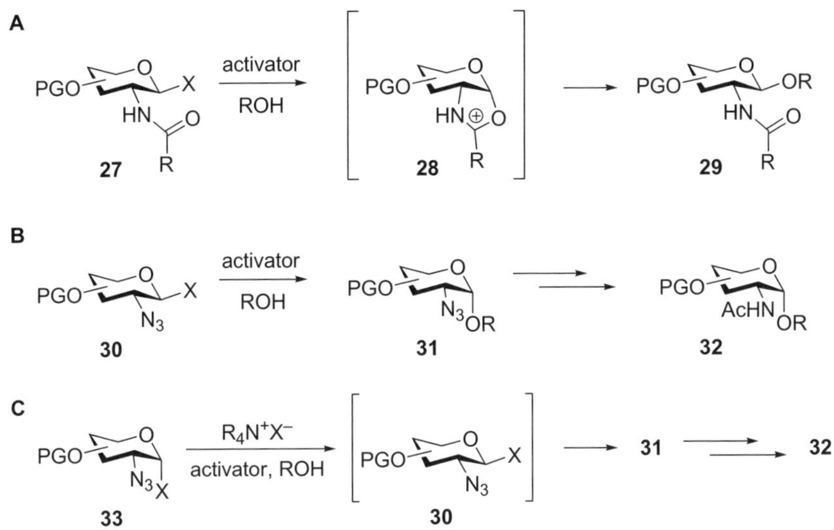

Scheme 16.9 *Preparation of O-glycosides of 2-amino-2-deoxysugars. (A) Use of N-acyl-protected donors* ***27*** *results in 1,2-trans glycosylation due to neighboring group participation. (B) 1,2-cis glycosylation products* ***32*** *from β-glycosyl halides* ***30***[54,55] *or (C) by in situ anomerization*[56] *of α-glycosyl halides* ***33***[57]

be transformed to the natural acetamido function either in two steps by reduction of the azide and subsequent acetylation or in one step by reductive acetylation using thioacetic acid.[63,64]

A successful approach for the synthesis of *O*-glycopeptides is the assembly of preformed, more or less complex glycosyl amino acid building blocks by solid phase peptide synthesis (SPPS).[60–62,65–67] Based on initial work of Ferrari[68] and Paulsen,[69] the azide method is extensively used for the preparation of such glycosyl amino acid building blocks. Especially the synthesis of complex glycosyl amino acids is challenging. Usually, glycosylation is performed with monosaccharides followed by attachment of further sugar residues because glycosylation reactions with oligosaccharide donors and serine or threonine acceptors often proceed with unpredictable stereochemistry. Nevertheless, oligosaccharides have been successfully used in many glycosylations as illustrated by the synthesis of glycosyl threonine building block **38** reported by Danishefsky and coworkers (Scheme 16.10).[64]

16.5 Glycosyl Azides as Precursors for Glycosyl Amides

Beside the *O*-linked glycoproteins, the more prevalent form of glycosylation of proteins is *N*-linked glycosylation.[48,70,71] *N*-Glycoproteins are characterized by a *β*-*N*-glycosidic linkage of the terminal *N*-acetylglucosamine of the pentasaccharide core structure to the amide nitrogen of asparagine. The conventional synthetic strategy for the preparation of such glycosyl amides starts from glycosyl amines which are reacted with activated and

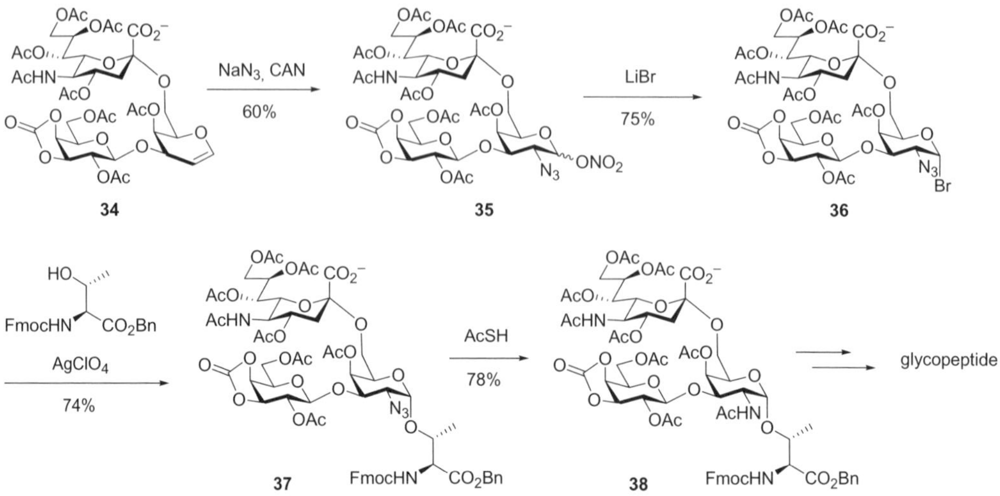

Scheme 16.10 *Synthesis of glycosyl threonine building block **38** using the azide method.*[64] *The 2-azido group is introduced by azidonitration of **34** followed by preparation of donor **36**. Glycosylation using threonine as acceptor leads to 1,2-cis glycoside **37**. After conversion of the azide group to an N-acetyl group by reductive acetylation, **38** was used as building block in glycopeptide synthesis*

suitably protected aspartic acid derivatives to form the amide linkage.[60–62,65–67] Glycosyl amines are commonly prepared either by reduction of glycosyl azides (cf. Section 16.3) or by amination of unprotected reducing sugars with saturated aqueous ammonium bicarbonate.[72] Recently, improved variants of the latter procedure employing microwave irradiation[73,74] and ammonium carbamate,[75,76] respectively, have been published. Drawbacks of this method are the instability of glycosyl amines and their propensity for dimerization and anomerization. Also, the preparation of α-glycosyl amides is a synthetic challenge.

While the classical Staudinger reaction[35] leads to iminophosphoranes which can be hydrolyzed to amines under aqueous conditions (Staudinger reduction, cf. Section 16.3), the addition of acyl donors under dry conditions results in amide formation.[77,78] This procedure was repeatedly applied for the synthesis of glycosyl amides, thus circumventing the preparation of glycosyl amines. Initially, three-component reactions employing glycosyl azide, activated carboxyl derivative and phosphine were reported (Scheme 16.11). The reaction starts from the β- (**39**) or α-glycoside **45** with the formation of an iminophosphorane (**40** and **42**, respectively), which is then trapped by an acylating agent in the second step. The resulting acylaminophosphonium salt (**43/46**) yields the corresponding glycosyl amide (**44/47**) upon hydrolysis. The intermediate iminophosphorane can undergo anomerization via open-chain form **41** preferring β-configuration. The degree of isomerization is dependent on the efficiency of iminophosphorane trapping by the acylating agent. Differently activated carboxylic acids, such as carboxylic halides,[79,80] anhydrides,[80,81] and carbodiimide-activated acids,[82,83] have been employed as acylating agents. While β-glycosyl amides **44** can be obtained easily from β-glycosyl azides **39**, the stereospecific

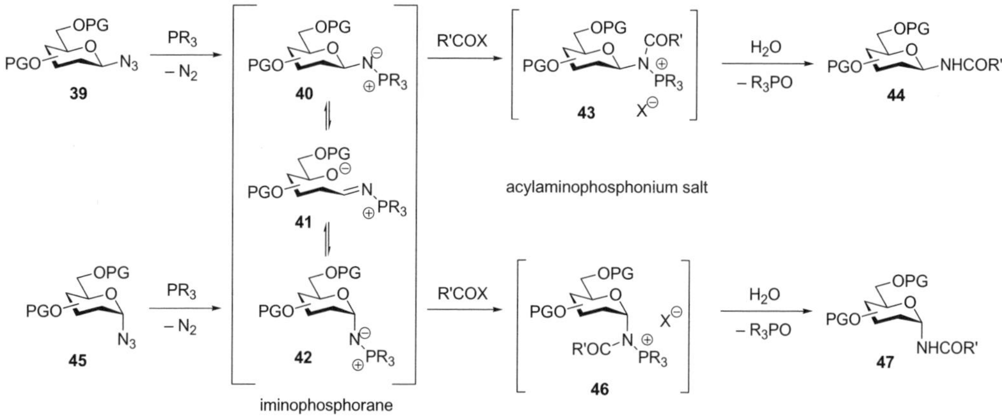

Scheme 16.11 *Mechanism of the Staudinger reaction with glycosyl azides*

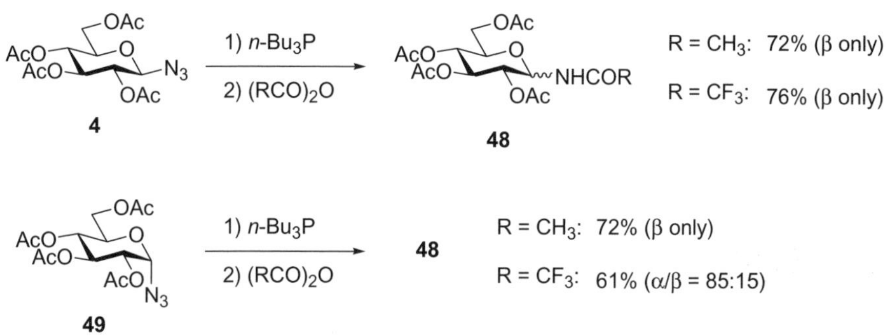

Scheme 16.12 *Three-component Staudinger-type reaction with β-glycosyl azide **4** stereoselectively leads to the β-glycosyl amides **48**. α-Glycosyl amides can only be obtained from α-glycosyl azide **49** with strong acylating agents to prevent complete anomerization of the intermediate iminophosphorane*[80]

synthesis of *α*-glycosyl amides **47** starting from *α*-glycosyl azides **45** is only possible with strong acylating agents which trap the intermediate iminophosphorane **42** before anomerization can take place.[80] Representative examples for the three-component Staudinger reaction are shown in Scheme 16.12. Rarely, the Staudinger reaction with reactive alkylphosphines and free carboxylic acids has been reported.[84,85] In this case, amide-bond formation is assumed to proceed in a concerted reaction without generation of an iminophosphorane intermediate.

Recently, the synthesis of glycosyl amides has also been achieved employing the traceless two-component Staudinger ligation[9,86,87] developed in the laboratories of Bertozzi[88] and Raines[89,90] (Scheme 16.13). Starting from glycosyl azides **50** and **55**, respectively, the initially formed iminophosphorane **52/57** reacts with an intramolecular (thio)ester group to form the acylaminophosphonium salt **53/58** from which the phosphine moiety is removed by hydrolysis with water. Using benzyl protected *α*-glycosyl azides such as **50**

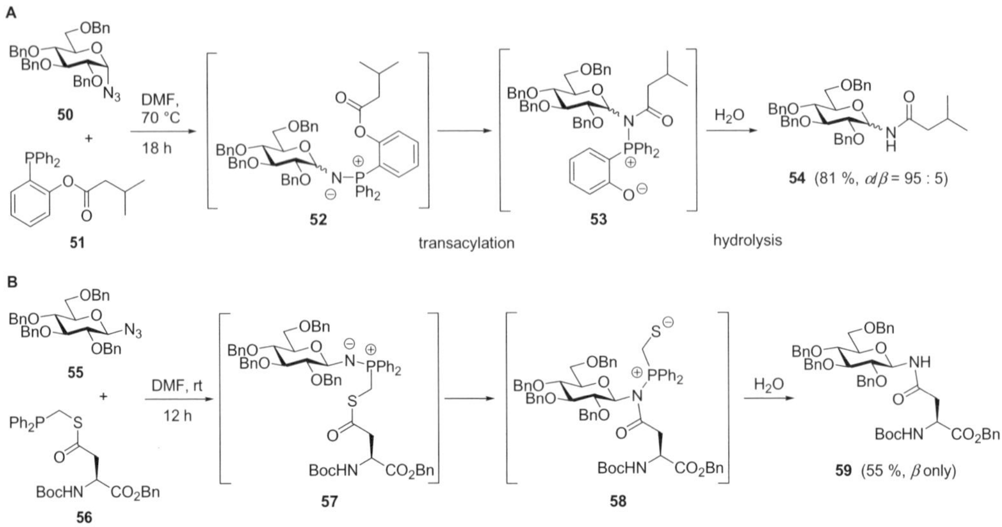

Scheme 16.13 *Two-component traceless Staudinger ligations using phosphine-derivatized ester* ***51*** *(A)*[9] *or thioester* ***56*** *(B)*[87]

and stable phosphine **51** or similar esters in polar aprotic solvents such as DMF, the reaction proceeded stereo conservatively to yield predominantly α-glycosyl amides (Scheme 16.13A).[9] The use of acetyl protected α-glycosyl azides, however, resulted only in β-glycosyl amides due to isomerization of the less reactive iminophosphorane.

All methods described above have been used for the preparation of the β-glycosyl amide linkage between *N*-acetylglucosamine and the side chain of asparagine in both three-component reactions using free[84,85] or activated[82,83] carboxylic acids and two-component reactions as shown in Scheme 16.13.[9,87] The obtained protected glycosyl amino acids can be used as building blocks in SPPS of *N*-linked glycopeptides.[91,92] It was also shown that deprotected sugars can be attached to amino acids and whole peptides using the three-component reaction.[92] Beside Staudinger-type reactions, another route towards the synthesis of glycosyl amides is the reaction of glycosyl azides with thiocarboxylic acids.[93]

16.6 Synthesis of Glycoconjugates via Azide-Alkyne [3+2] Cycloaddition

Although the azide-alkyne [3+2] cycloaddition[94] (cf. Chapter 9) is known in carbohydrate chemistry for more than 50 years,[95] its application for the preparation of glycoconjugates became particularly attractive with the development of the copper(I)-catalyzed variant by Meldal[96] and Sharpless.[97] The copper(I)-catalyzed azide-alkyne cycloaddition (CuAAC)[98,99] enables the regioselective formation of 1,4-disubstituted 1,2,3-triazoles under very mild conditions even in a biological context. However, the cellular toxicity of the copper catalyst precludes applications wherein cells must remain viable. Therefore, as an alternative

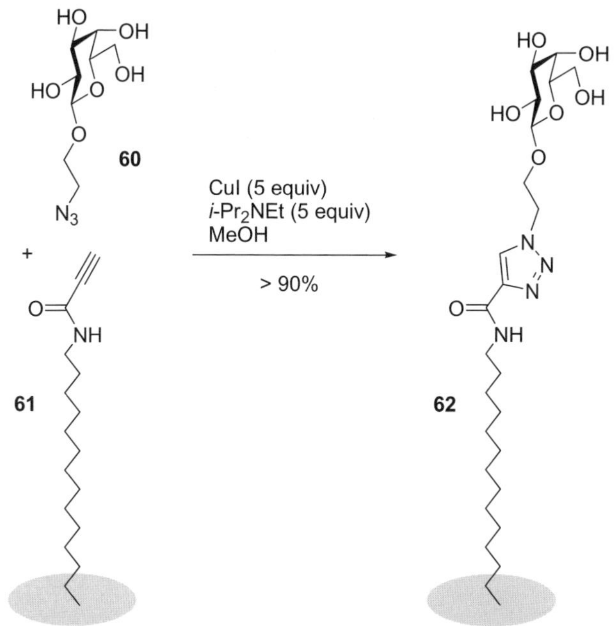

Scheme 16.14 *Coupling of azide-substituted galactoside **60** to alkyne-modified C_{14} hydrocarbon **61** noncovalently bound to the microtiter well surface*[108]

to CuAAC, strain-promoted azide-alkyne [3+2] cycloadditions have been developed that proceed at room temperature without the need for a catalyst.[100,101] These reactions are discussed in the next section dealing with metabolic oligosaccharide engineering. Another example of metal-free triazole formation by a tandem [3+2] cycloaddition-retro-Diels-Alder reaction has been developed by van Berkel *et al.* although no carbohydrate-related application was reported.[102]

CuAAC reactions have been extensively applied in carbohydrate chemistry including the synthesis of simple glycoside and oligosaccharide mimetics, glyco-macrocycles, glycoconjugates, glycoclusters, and for the attachment of carbohydrates to surfaces. The field has been thoroughly reviewed[98,103–107] and, therefore, we will focus on a few selected examples which are of special interest for glycobiology.

One of the first applications of CuAAC in carbohydrate chemistry was – beside the one in the seminal paper of Meldal[96] – the immobilization of azide-substituted sugars on microtiter plates (Scheme 16.14).[108] The surface-bound sugars such as **62** were screened with various lectins and could be elongated by glycosyltransferase-catalyzed fucosylation. The technique was later on improved by incorporation of a cleavable disulfide bond in the linker allowing mass spectrometric characterization of the carbohydrate array.[109]

Neoglycopeptides and -proteins[110] differ from naturally occurring structures by replacement of the natural carbohydrate-peptide linkage with a non-natural one. This not only allows studying the influence of distinct structural elements on biological activity, but has many practical applications as well. Use of chemoselective ligation reactions such as

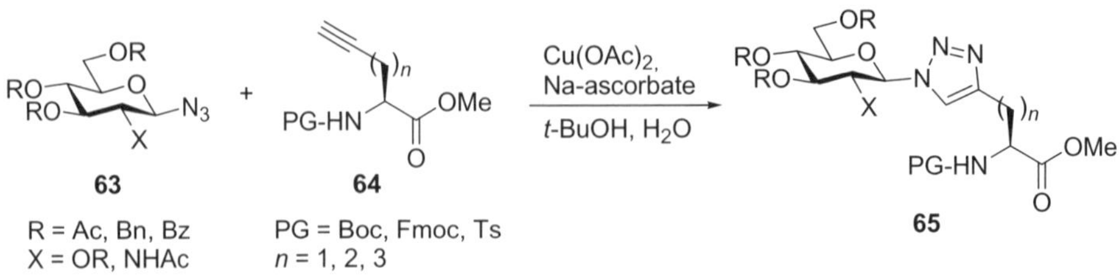

Scheme 16.15 *Application of CuAAC for the preparation of triazole-linked glycosyl amino acids* ***65***[111]

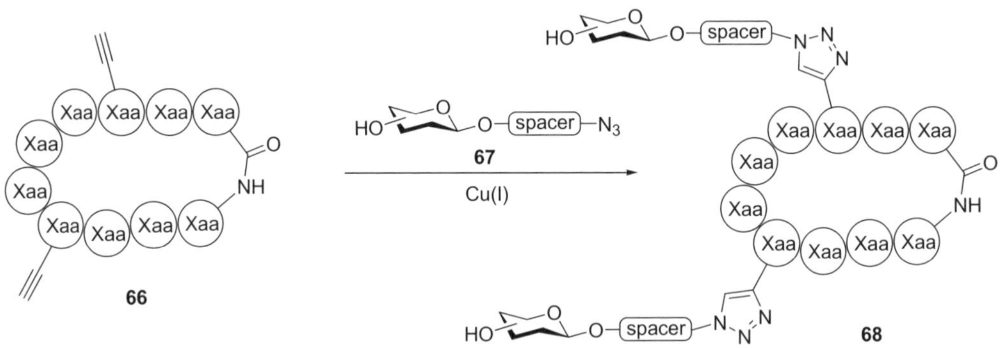

Scheme 16.16 *Preparation of tyrocidine derivatives* ***68*** *by CuAAC of propargylglycine-containing cyclic peptides* ***66*** *and azido-functionalized sugars* ***67***[113]

CuAAC makes glycoconjugates accessible to a broader community. Furthermore, the non-natural linkage often is more stable (both chemically and with respect to enzymatic degradation) which can lead to an increased half life of a glycoconjugate within a biological system. Scheme 16.15 depicts the synthesis of triazole-linked glycosyl amino acids **65** starting from glycosyl azides **63** and different alkyne-containing amino acids **64** which can be used as building blocks in peptide synthesis.[111,112]

Lin and Walsh applied CuAAC for the attachment of 21 different azido-functionalized monosaccharides **67** to 13 derivatives **66** of the cyclic decapeptide tyrocidine containing one to three propargylglycine residues at positions 3–8 (Scheme 16.16).[113] Head-to-tail cyclization of the peptides was accomplished using a thioesterase domain from tyrocidine synthetase. Antibacterial and hemolytic assays showed that the two best glycopeptide mimetics had a 6-fold better therapeutic index than the natural tyrocidine. CuAAC has further been used to attach carbohydrates to whole virus particles[114,115] and DNA.[116]

More challenging is the modification of bacterially expressed proteins by site-specific attachment of carbohydrates. Crucial step is the introduction of a chemical tag, which can be chemoselectively modified, into the protein. It has been shown that alkyne- and azido-modified amino acids, such as 2-amino-5-hexynoic acid (homopropargylglycine, Hpg),[117] 4-azidohomoalanine (Aha),[118,119] and with less efficiency also 3-azidoalanine, 5-azidonorvaline, and 6-azidonorleucine,[120] act as methionine surrogates that are activated by the methionyl-tRNA synthetase of *E. coli* and replace methionine in proteins expressed in methionine-depleted bacterial cultures. This, together with other methods for the incorporation of non-canonical amino acids into proteins,[121–123] offers the possibility to use azide-alkyne cycloaddition (and also Staudinger ligation[124–126]) not only for protein labeling within cells or on cell surfaces[119,120] but also for the preparation of neoglycoproteins.

Davis and coworkers expanded the diversity of chemical protein modification by a combination of this CuAAC-based labeling and disulfide bond formation via genetically engineered cysteine (Cys) residues.[127] Aha and Hpg, respectively, were introduced into engineered proteins by the auxotrophy-based residue-specific method. Subsequent CuAAC reactions with alkyne- and azide-substituted carbohydrates, respectively, resulted in homogeneous protein glycoconjugates. As second modification reaction, conjugation of Cys residues with substituted methanethiosulfonates was chosen. Applying these two orthogonal protein modification reactions, derivatives of the LacZ reporter enzyme carrying the tetrasaccharide sialyl Lewis X and a sulfotyrosine mimic were created that allowed detection of mammalian brain inflammation.

Recently, Merkel *et al.* reported efficient N-terminal glycoconjugation of proteins by the N-end rule.[128] Bulky amino acids at the second and third sequence position of the barnase inhibitor barstar efficiently prevent excision of N-terminal methionine analogue Aha introduced by the auxotrophy-based residue-specific method. The created azide tag at the protein's N-terminus was subsequently conjugated to propargyl glycosides of *N*-acetylglucosamine and *N,N′*-diacetylchitobiose, respectively, by CuAAC. The obtained glycoprotein mimetics show binding affinity to the lectin wheat germ agglutinin whereas the natural activity of barstar is conserved.

Lectins are carbohydrate-binding proteins other than immunoglobulins without enzymatic activity towards the recognized sugars.[129–131] Carbohydrate-lectin interactions are involved in numerous intra- and intercellular events during development, inflammation, immune response and cancer metastasis.[132–136] Multivalency appears to play an important role in lectin-mediated interactions,[137–140] and many lectins are found to recognize individual carbohydrate epitopes only with low affinity. Preparation of carbohydrate clusters, therefore, is a common strategy to obtain high-affinity lectin ligands.[141–144] Because of its robustness, CuAAC is excellently suited for the simultaneous attachment of several carbohydrate epitopes to a scaffold. Initially, Santoyo-Gonzáles and coworkers prepared different multivalent mannose clusters starting from propargyl mannosides and azide-containing scaffolds.[145] This strategy as well as the opposite approach based on azide-containing carbohydrates and alkyne-bearing scaffolds have been used intensively for the preparation of glycoclusters.[98,103–107] Glycosyl azides are easily produced, however, the direct attachment of a triazole to the sugar may interfere with the recognition of the carbohydrate by the protein and, therefore, linkers of varying length have been introduced between the sugar and the triazole moiety. It would be far beyond the scope of this chapter to mention all applications. Exemplarily, the asymmetrical, bifunctional dendrimer **69** containing 16 mannose units and two coumarin chromophores[146] and poly(methacrylate)-based glycopolymer **70**[147] are depicted in Scheme 16.17.

Although organic azides are stable against most reaction conditions, compounds containing multiple azide residues (like multivalent scaffolds) are potentially explosive. Therefore, several one-pot procedures to generate azides *in situ* followed by CuAAC have

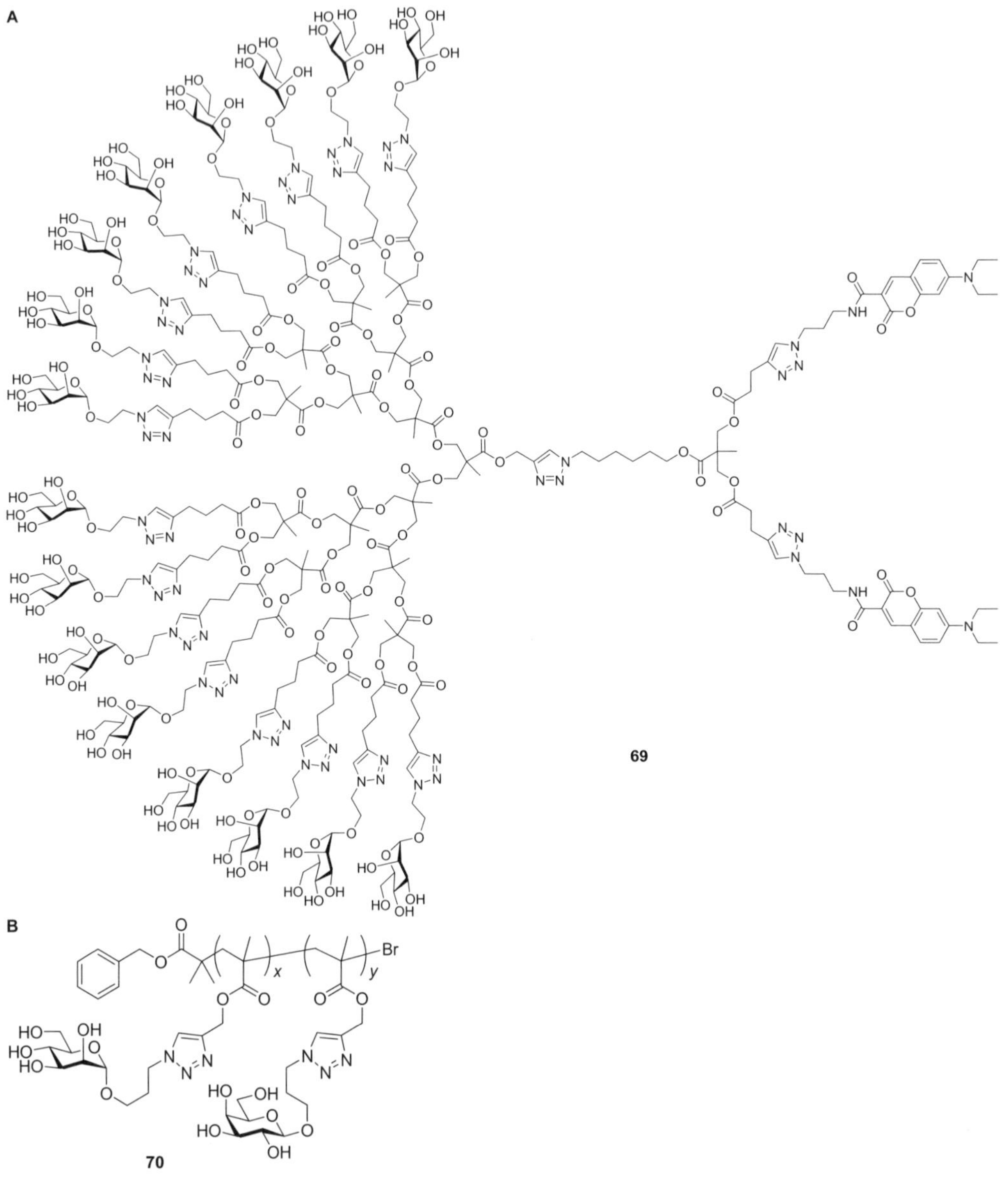

Scheme 16.17 *(A) Asymmetrical, bifunctional dendrimer **69** containing 16 mannose units and two coumarin chromophores*[146] *and (B) poly(methacrylate)-based glycopolymer **70***[147] *prepared by CuAAC and used for lectin binding studies with concanavalin A*

been reported.[148–154] While the azides in most of these procedures are introduced by a nucleophilic substitution of a leaving group in allyl, benzyl, glycosyl, or similar position,[148–152] aliphatic[154] and aromatic[153] amino groups may also serve as precursors. We reported, for example a one-pot procedure for diazo transfer and subsequent CuAAC which allows the preparation of multivalent structures starting from commercially avail-

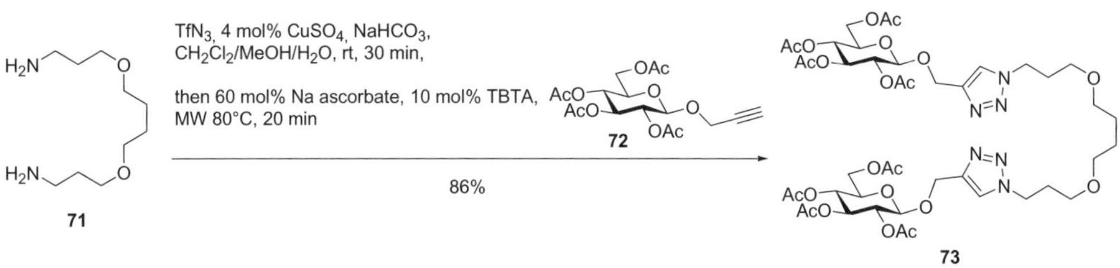

Scheme 16.18 *Sequential one-pot procedure for diazo transfer and CuAAC.[154] First, diamine **71** is transformed to the corresponding diazide by Cu(II)-catalyzed diazo transfer. After completion, Cu(I) required for subsequent CuAAC with **72** is generated by addition of reducing agent Na ascorbate. MW = microwave; TBTA = tris(benzyltriazolylmethyl)amine*

able amine scaffolds without need for isolation of the azide-containing intermediates.[154] As an example, divalent glycoconjugate **73** was synthesized from diamine **71** and propargyl glycoside **72** as shown in Scheme 16.18.

Azides can also undergo [3+2] cycloaddition reactions with nitriles giving access to 1,5-disubstituted tetrazoles. *Inter*molecular reactions, however, require electron deficient nitriles and very forcing conditions to occur with sufficiently high reaction rates.[155–157] High yields have been reported for the reaction of sulfonyl and acyl cyanides with unhindered aliphatic azides by neat, thermal fusion.[158,159] *Intra*molecular [3+2] cycloaddition reactions of organic azides to nitriles occur more readily.[160–164] Still, they require high reaction temperatures and yields are with few exceptions[165] not satisfactory. When precisely positioned on a rigid carbohydrate scaffold, however, azides can undergo cycloaddition reactions with nitriles under exceptionally mild conditions. Thus, 3-azido-1,2-*O*-cyanoethylidene-3-deoxy-allopyranose was shown to form a tetrazole embedded in a bridged tetracyclic ring system even at room temperature.[166]

16.7 Metabolic Oligosaccharide Engineering

Glycosylation of proteins is an important co- and posttranslational event that has been estimated to occur on more than 50% of eukaryotic proteins.[167] The glycan chains of cell-surface glycoproteins are involved in numerous recognition processes such as cell adhesion and attachment of bacteria or viruses. Inside cells, glycans direct protein trafficking and they modulate structure and activity of proteins.[132–136] Hence, in vivo monitoring of glycosylation processes is of utmost interest.[168] While fluorescent fusion proteins and other genetically encoded tags provide a means for labeling specific proteins in live cells, analogous techniques are not available for secondary gene products including glycans. Metabolic oligosaccharide engineering offers the possibility to introduce carbohydrates with unnatural structural elements into the glycans without genetic manipulation making use of the cell's biosynthetic machinery.[169] If suitable chemical reporter groups are introduced, subsequent addition of an exogenously delivered detectable probe allows for tagging of the glycans by a chemoselective ligation reaction. Examples for such reporter groups include ketones[170,171] and thiols.[172] However, the azido group is much more

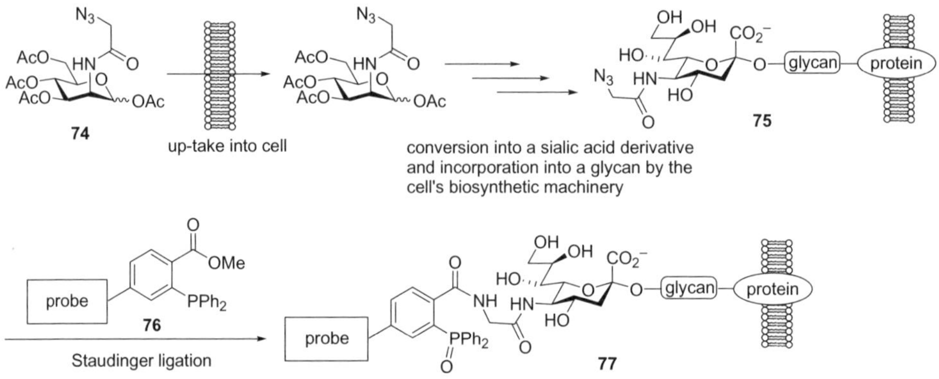

Scheme 16.19 *Metabolic oligosaccharide engineering: peracetylated ManNAz* ***74*** *is taken up by mammalian cells and converted into an azide-containing sialic acid derivative which is incorporated into sialic acid-bearing glycans* ***75****. In the next step, a detectable probe* ***76*** *can be attached to* ***75*** *via Staudinger ligation*[124,173]

suited for this approach because azides can take part in two important bioorthogonal ligation reactions, Staudinger ligation[124] (cf. Section 16.5) and azide-alkyne [3+2] cycloaddition (cf. Section 16.6 and Chapter 9).

Azide derivatization of monosaccharides represents a subtle structural change that is accepted by several metabolic pathways. Thus, azide derivatives of *N*-acetylmannosamine (i.e. *N*-(azidoacetyl)mannosamine, ManNAz), *N*-acetylgalactosamine (i.e. *N*-(azidoacetyl) galactosamine, GalNAz), *N*-acetylglucosamine (i.e. *N*-(azidoacetyl)glucosamine, GlcNAz), and L-fucose (i.e. 6-azido-L-fucose) have been explored.[168,169] Initially, ManNAz was employed to tag sialylated cell surface glycans of mammalian cells in vitro (Scheme 16.19).[124,173] Cells are grown in the presence of peracetylated ManNAz **74** which can be taken up by the cells more easily than ManNAz. After de-*O*-acetylation by cellular esterases, resulting ManNAz is metabolized similarly to its native counterpart *N*-acetylmannosamine and integrated into cellular glycans. Finally, the azide-labeled glycans are reacted with a detectable probe by Staudinger ligation. GalNAz can be metabolically introduced at the core position of mucin-type *O*-linked glycoproteins.[174] Thus, a selective labeling of mucin-type glycoproteins is possible. Both, the metabolic labeling of sialylated glycans with ManNAz[175] and labeling of mucin-type *O*-glycoproteins with GalNAz[176] can be carried out *in vivo*. Analogously, GlcNAz has been used for the labeling of *O*-GlcNAc glycosylated proteins.[177] Recently, cells were labeled simultaneously with an azide- and a ketone-containing sugar.[178] Using orthogonal ligation reactions, glycans bearing these sugar residues can be visualized in parallel on the same cells.

In the cases mentioned so far, fluorescence labeling has been achieved by a two-step procedure. First, a biotin label[124] or FLAG tag (octapeptide Asp-Tyr-Lys-Asp-Asp-Asp-Asp-Lys)[173,174,177] is covalently attached to the azide-containing glycan by Staudinger ligation at high concentration. In a second step, a fluorescently labeled receptor (avidin and anti-FLAG antibody, respectively) is added at lower concentration. To avoid the problem of high background fluorescence caused by the application of fluorescent dyes,

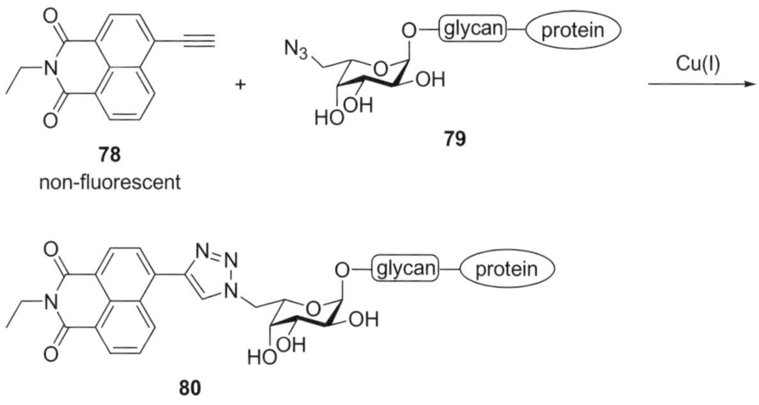

Scheme 16.20 *Generation of fluorescent triazole* ***80*** *by CuAAC of fluorogenic ethynyl-naphthalimide* ***78*** *and azide-labeled glycoproteins* ***79*** *applicable for intracellular localization of fucosylated glycoconjugates*[179]

Wong and coworkers developed a one-step labeling method based on CuAAC ligation using fluorogenic dyes (Scheme 16.20).[179] 6-Azido-L-fucose was applied for tagging of fucosylated proteins by metabolic oligosaccharide engineering. Reaction of alkyne-substituted naphthalimide **78** and azide-modified glycoprotein **79** results in formation of fluorescent triazole **80**. Since **78** is not fluorescent, it can be applied at high concentrations without producing a background signal. The method was used for cell surface glycoprotein analysis and intracellular localization of fucosylated glycoconjugates by using fluorescence microscopy.

Other examples for the application of CuAAC for labeling and visualization of glycoproteins in cells have been published by the research groups of Bertozzi[180] and Wong.[181] The main advantage of CuAAC over Staudinger ligation is its much faster reaction kinetics. However, the use of CuAAC for applications *in vivo* is limited due to the cellular toxicity of copper ions. This led to the development of copper-free variants of this cycloaddition. Based on observations made by Wittig who described the exothermic cycloaddition of cyclooctyne with phenyl azide,[182] Bertozzi and coworkers reported the copper-free, strain-promoted cycloaddition between azides and substituted cyclooctyne **81** for covalent modification of biomolecules in living systems (Scheme 16.21).[100] The reaction rates were lower than those of CuAAC but comparable to those of Staudinger ligation.[183] The validity of the approach was demonstrated by functionalization of modified Jurkat cells with a biotin derivative of **81**.[100] Reaction rates of the strain-promoted azide-alkyne cycloaddition could be dramatically improved by introduction of electron-withdrawing fluorine substituents in α position of the triple bond (Scheme 16.21, **82–84**) with the difluorinated cyclooctyne (DIFO) derivatives **83** and **84** possessing comparable kinetics to those of CuAAC.[183–185] Similar reaction rates were observed with dibenzocyclooctyne derivative **85**.[101] These reactions are not regioselective but proceed chemoselectively within minutes on live cells with no apparent toxicity.[101,184,185] Latest application of DIFO derivative **83** is the *in vivo* imaging of membrane-associated glycans in live

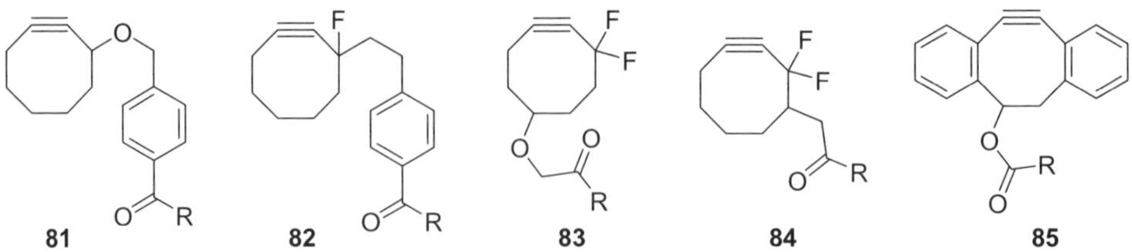

81 **82** **83** **84** **85**

Scheme 16.21 *Cyclooctyne derivatives for use in copper-free, strain-promoted azide-alkyne [3+2] cycloadditions designed by Bertozzi (**81**,*[100] ***82***,[183] ***83***,[184] ***84***[185]*) and Boons (**85**[101]). The second-generation difluorinated derivative **84** is easier to synthesize than **83***

developing zebrafish.[186] Using two derivatives **83** with different fluorophores attached, it was possible to perform a spatiotemporal analysis of glycan expression and trafficking.

References

[1] A. Bertho, *Ber. Dtsch. Chem. Ges.* **1930**, *63*, 836–43.
[2] Z. Györgydeák, L. Szilágyi, H. Paulsen, *J. Carbohydr. Chem.* **1993**, *12*, 139–63.
[3] M. Hayashi, H. Kawabata, in *Recent Devel. Carbohydrate Res., Vol. 1* (ed.: S.G. Pandalai), Transworld Research Network, Trivandrum, **2003**, pp. 195–208.
[4] Z. Györgydeák, J. Thiem, *Adv. Carbohydr. Chem. Biochem.* **2006**, *60*, 103–82.
[5] W. Pfleiderer, E. Bühler, *Chem. Ber.* **1966**, *99*, 3022–39.
[6] F.D. Tropper, F.O. Andersson, S. Braun, R. Roy, *Synthesis* **1992**, 618–20.
[7] R. Kumar, P. Tiwari, P.R. Maulik, A.K. Misra, *Eur. J. Org. Chem* **2006**, 74–9.
[8] H. Paulsen, Z. Györgydeák, M. Friedmann, *Chem. Ber.* **1974**, *107*, 1568–78.
[9] A. Bianchi, A. Bernardi, *J. Org. Chem.* **2006**, *71*, 4565–77.
[10] F. Sicherl, V. Wittmann, *Angew. Chem., Int. Ed.* **2005**, *44*, 2096–9.
[11] C. Vogel, P. Gries, *J. Carbohydr. Chem.* **1994**, *13*, 37–46.
[12] A. Fürst, P.A. Plattner, in *Abstracts of Papers, 12th International Congress of Pure and Applied Chemistry*, New York, **1951**, p. 409.
[13] S. Arndt, L.C. Hsieh-Wilson, *Org. Lett.* **2003**, *5*, 4179–82.
[14] R.U. Lemieux, R.M. Ratcliffe, *Can. J. Chem.* **1979**, *57*, 1244–51.
[15] A.F.G. Bongat, A.V. Demchenko, *Carbohydr. Res.* **2007**, *342*, 374–406.
[16] G. Grundler, R.R. Schmidt, *Liebigs Ann. Chem.* **1984**, 1826–47.
[17] S.A. Svarovsky, J.J. Barchi, Jr., *Carbohydr. Res.* **2003**, *338*, 1925–35.
[18] H. Paulsen, W. Rauwald, U. Weichert, *Liebigs Ann. Chem.* **1988**, 75–86.
[19] J. Hansson, P.J. Garegg, S. Oscarson, *J. Org. Chem.* **2001**, *66*, 6234–43.
[20] J.D.C. Codée, R.E.J.N. Litjens, R. den Heeten, *et al.*, *Org. Lett.* **2003**, *5*, 1519–22.
[21] N.V. Bovin, S.E. Zurabyan, A.Y. Khorlin, *Carbohydr. Res.* **1981**, *98*, 25–35.
[22] F. Santoyo-González, F.G. Calvo-Flores, P. García-Mendoza, *et al.*, *J. Org. Chem.* **1993**, *58*, 6122–5.
[23] S. Czernecki, E. Ayadi, D. Randriamandimby, *J. Org. Chem.* **1994**, *59*, 8256–60.
[24] C.J. Cavender, V.J. Shiner, *J. Org. Chem.* **1972**, *37*, 3567–9.
[25] A. Vasella, C. Witzig, J.-L. Chiara, M. Martin-Lomas, *Helv. Chim. Acta* **1991**, *74*, 2073–7.
[26] P.B. Alper, S.-C. Hung, C.-H. Wong, *Tetrahedron Lett.* **1996**, *37*, 6029–32.
[27] P.T. Nyffeler, C.-H. Liang, K.M. Koeller, C.-H. Wong, *J. Am. Chem. Soc.* **2002**, *124*, 10773–8.
[28] O. Gavard, Y. Hersant, J. Alais, *et al.*, *Eur. J. Org. Chem.* **2003**, 3603–20.

[29] Michael F. Haller, G.-J. Boons, *Eur. J. Org. Chem.* **2002**, 2033–8.
[30] H.A. Orgueira, A. Bartolozzi, P. Schell, *et al.*, *Chem. Eur. J.* **2003**, *9*, 140–69.
[31] S. Iyer, S. Rele, G. Grasa, S. Nolan, E.L. Chaikof, *Chem. Commun.* **2003**, 1518–19.
[32] N. Laurent, D. Lafont, P. Boullanger, J.M. Mallet, *Carbohydr. Res.* **2005**, *340*, 1885–92.
[33] S. Bräse, C. Gil, K. Knepper, V. Zimmermann, *Angew. Chem., Int. Ed.* **2005**, *44*, 5188–240.
[34] E.F.V. Scriven, K. Turnbull, *Chem. Rev.* **1988**, *88*, 297–368.
[35] H. Staudinger, J. Meyer, *Helv. Chim. Acta* **1919**, *2*, 635–46.
[36] X. Ariza, F. Urpi, J. Vilarrasa, *Tetrahedron Lett.* **1999**, *40*, 7515–17.
[37] X. Ariza, F. Urpi, C. Viladomat, J. Vilarrasa, *Tetrahedron Lett.* **1998**, *39*, 9101–2.
[38] J.G. Silva, I. Carvalho, *Curr. Med. Chem.* **2007**, *14*, 1101–19.
[39] D.S. Pilch, M. Kaul, C.M. Barbieri, *Top. Curr. Chem.* **2005**, *253*, 179–204.
[40] Q. Vicens, E. Westhof, *ChemBioChem* **2003**, *4*, 1018–23.
[41] M. Hainrichson, I. Nudelman, T. Baasov, *Org. Biomol. Chem.* **2008**, *6*, 227–39.
[42] Y. Ding, E.E. Swayze, S.A. Hofstadler, R.H. Griffey, *Tetrahedron Lett.* **2000**, *41*, 4049–52.
[43] M. Fridman, V. Belakhov, S. Yaron, T. Baasov, *Org. Lett.* **2003**, *5*, 3575–8.
[44] P.B. Alper, M. Hendrix, P. Sears, C.-H. Wong, *J. Am. Chem. Soc.* **1998**, *120*, 1965–78.
[45] W.A. Greenberg, E.S. Priestley, P.S. Sears, *et al.*, *J. Am. Chem. Soc.* **1999**, *121*, 6527–41.
[46] J. Li, H.-N. Chen, H. Chang, J. Wang, C.-W.T. Chang, *Org. Lett.* **2005**, *7*, 3061–4.
[47] J. Li, F.-I. Chiang, H.-N. Chen, C.-W.T. Chang, *J. Org. Chem.* **2007**, *72*, 4055–66.
[48] V. Wittmann, in *Glycoscience: Chemistry and Chemical Biology*, 2 ed. (eds.: B. Fraser-Reid, K. Tatsuta, J. Thiem), Springer-Verlag, Berlin, **2008**, pp. 1735–70.
[49] H.C. Hang, C.R. Bertozzi, *Bioorg. Med. Chem.* **2005**, *13*, 5021–34.
[50] P. Van den Steen, P.M. Rudd, R.A. Dwek, G. Opdenakker, *Crit. Rev. Biochem. Mol. Biol.* **1998**, *33*, 151–208.
[51] J.R. Bishop, M. Schuksz, J.D. Esko, *Nature* **2007**, *446*, 1030–7.
[52] J. Banoub, P. Boullanger, D. Lafont, *Chem. Rev.* **1992**, *92*, 1167–95.
[53] J. Debenham, R. Rodebaugh, B. Fraser-Reid, *Liebigs Ann./Recueil* **1997**, 791–802.
[54] H. Paulsen, W. Stenzel, *Chem. Ber.* **1978**, *111*, 2334–47.
[55] H. Paulsen, C. Kolar, W. Stenzel, *Chem. Ber.* **1978**, *111*, 2358–69.
[56] R.U. Lemieux, K.B. Hendriks, R.V. Stick, K. James, *J. Am. Chem. Soc.* **1975**, *97*, 4056–62.
[57] H. Paulsen, O. Lockhoff, *Tetrahedron Lett.* **1978**, 4027–30.
[58] H. Paulsen, *Angew. Chem., Int. Ed. Engl.* **1982**, *21*, 155–73.
[59] R.R. Schmidt, *Angew. Chem., Int. Ed. Engl.* **1986**, *25*, 212–35.
[60] H. Herzner, T. Reipen, M. Schultz, H. Kunz, *Chem. Rev.* **2000**, *100*, 4495–538.
[61] M.R. Pratt, C.R. Bertozzi, *Chem. Soc. Rev.* **2005**, *34*, 58–68.
[62] C. Haase, O. Seitz, *Top. Curr. Chem.* **2007**, *267*, 1–36.
[63] T. Rosen, I.M. Lico, D.T.W. Chu, *J. Org. Chem.* **1988**, *53*, 1580–2.
[64] J.B. Schwarz, S.D. Kuduk, X.-T. Chen, *et al.*, *J. Am. Chem. Soc.* **1999**, *121*, 2662–73.
[65] B.G. Davis, *Chem. Rev.* **2002**, *102*, 579–601.
[66] T. Buskas, S. Ingale, G.-J. Boons, *Glycobiology* **2006**, *16*, 113R–136R.
[67] *Glycopeptides and Glycoproteins: Synthesis, Structure, and Application (Topics in Current Chemistry, vol. 267)* (ed.: V. Wittmann), Springer-Verlag, Berlin, **2007**.
[68] B. Ferrari, A.A. Paviat, *Carbohydr. Res.* **1980**, *79*, C1–C7.
[69] H. Paulsen, J.-P. Holck, *Carbohydr. Res.* **1982**, *109*, 89–107.
[70] L. Lehle, S. Strahl, W. Tanner, *Angew. Chem., Int. Ed.* **2006**, *45*, 6802–18.
[71] P.M. Rudd, R.A. Dwek, *Crit. Rev. Biochem. Mol. Biol.* **1997**, *32*, 1–100.
[72] L.M. Likhosherstov, O.S. Novikova, V.A. Derevitskaya, N.K. Kochetkov, *Carbohydr. Res.* **1986**, *146*, c1–c5.
[73] M. Bejugam, S.L. Flitsch, *Org. Lett.* **2004**, *6*, 4001–4.
[74] M.A. Brun, M.D. Disney, P.H. Seeberger, *ChemBioChem* **2006**, *7*, 421–4.
[75] C.P.R. Hackenberger, M.K. O'Reilly, B. Imperiali, *J. Org. Chem.* **2005**, *70*, 3574–8.

[76] L.M. Likhosherstov, O.S. Novikova, V.N. Shibaev, *Dokl. Chem.* **2002**, *383*, 89–92.
[77] J. Garcia, F. Urpi, J. Vilarrasa, *Tetrahedron Lett.* **1984**, *25*, 4841–4.
[78] I. Bosch, A. González, F. Urpí, J. Vilarrasa, *J. Org. Chem.* **1996**, *61*, 5638–43.
[79] V. Maunier, P. Boullanger, D. Lafont, *J. Carbohydr. Chem.* **1997**, *16*, 231–5.
[80] L. Kovacs, E. Osz, V. Domokos, W. Holzer, Z. Gyorgydeak, *Tetrahedron* **2001**, *57*, 4609–21.
[81] J.J. García-López, F. Santoyo-González, A. Vargas-Berenguel, J.J. Giménez-Martínez, *Chem. Eur. J.* **1999**, *5*, 1775–84.
[82] J.P. Malkinson, R.A. Falconer, I. Toth, *J. Org. Chem.* **2000**, *65*, 5249–52.
[83] N. Rockendorf, T.K. Lindhorst, *J. Org. Chem.* **2004**, *69*, 4441–5.
[84] T. Inazu, K. Kobayashi, *Synlett* **1993**, 869–70.
[85] M. Mizuno, I. Muramoto, K. Kobayashi, H. Yaginuma, T. Inazu, *Synthesis* **1999**, 162–5.
[86] A. Bianchi, A. Bernardi, *Tetrahedron Lett.* **2004**, *45*, 2231–4.
[87] Y. He, R.J. Hinklin, J. Chang, L.L. Kiessling, *Org. Lett.* **2004**, *6*, 4479–82.
[88] E. Saxon, J.I. Armstrong, C.R. Bertozzi, *Org. Lett.* **2000**, *2*, 2141–3.
[89] B.L. Nilsson, L.L. Kiessling, R.T. Raines, *Org. Lett.* **2000**, *2*, 1939–41.
[90] B.L. Nilsson, L.L. Kiessling, R.T. Raines, *Org. Lett.* **2001**, *3*, 9–12.
[91] M. Mizuno, K. Haneda, R. Iguchi, *et al.*, *J. Am. Chem. Soc.* **1999**, *121*, 284–90.
[92] K.J. Doores, Y. Mimura, R.A. Dwek, *et al.*, *Chem. Commun.* **2006**, 1401–3.
[93] N. Shangguan, S. Katukojvala, R. Greenberg, L.J. Williams, *J. Am. Chem. Soc.* **2003**, *125*, 7754–5.
[94] R. Huisgen, *Angew. Chem., Int. Ed. Engl.* **1963**, *2*, 565–98.
[95] F. Micheel, G. Baum, *Chem. Ber.* **1957**, *90*, 1595–6.
[96] C.W. Tornøe, C. Christensen, M. Meldal, *J. Org. Chem.* **2002**, *67*, 3057–64.
[97] V.V. Rostovtsev, L.G. Green, V.V. Fokin, K.B. Sharpless, *Angew. Chem., Int. Ed.* **2002**, *41*, 2596–9.
[98] M. Meldal, C.W. Tornøe, *Chem. Rev.* **2008**, *108*, 2952–3015.
[99] R. Breinbauer, M. Köhn, *ChemBioChem* **2003**, *4*, 1147–9.
[100] N.J. Agard, J.A. Prescher, C.R. Bertozzi, *J. Am. Chem. Soc.* **2004**, *126*, 15046–7.
[101] X. Ning, J. Guo, Margreet A. Wolfert, G.-J. Boons, *Angew. Chem., Int. Ed.* **2008**, *47*, 2253–5.
[102] S.S. van Berkel, A.J. Dirks, M.F. Debets, *et al.*, *ChemBioChem* **2007**, *8*, 1504–8.
[103] A.D. Moorhouse, J.E. Moses, *ChemMedChem* **2008**, *3*, 715–23.
[104] I.S. Carrico, *Chem. Soc. Rev.* **2008**, *37*, 1423–31.
[105] R.J. Pieters, D.T.S. Rijkers, R.M.J. Liskamp, *QSAR & Comb. Sci.* **2007**, *26*, 1181–90.
[106] J.E. Moses, A.D. Moorhouse, *Chem. Soc. Rev.* **2007**, *36*, 1249–62.
[107] S. Dedola, S.A. Nepogodiev, R.A. Field, *Org. Biomol. Chem.* **2007**, *5*, 1006–17.
[108] F. Fazio, M.C. Bryan, O. Blixt, J.C. Paulson, C.-H. Wong, *J. Am. Chem. Soc.* **2002**, *124*, 14397–402.
[109] M.C. Bryan, F. Fazio, H.-K. Lee, *et al.*, *J. Am. Chem. Soc.* **2004**, *126*, 8640–1.
[110] D. Specker, V. Wittmann, *Top. Curr. Chem.* **2007**, *267*, 65–107.
[111] B.H.M. Kuijpers, S. Groothuys, A.R. Keereweer, *et al.*, *Org. Lett.* **2004**, *6*, 3123–6.
[112] S. Groothuys, B.H.M. Kuijpers, P.J.L.M. Quaedflieg, *et al.*, *Synthesis* **2006**, 3146–52.
[113] H. Lin, C.T. Walsh, *J. Am. Chem. Soc.* **2004**, *126*, 13998–14003.
[114] S. Sen Gupta, K.S. Raja, E. Kaltgrad, E. Strable, M.G. Finn, *Chem. Commun.* **2005**, 4315–17.
[115] S. Sen Gupta, J. Kuzelka, P. Singh, *et al.*, *Bioconjugate Chem.* **2005**, *16*, 1572–9.
[116] G.A. Burley, J. Gierlich, M.R. Mofid, *et al.*, *J. Am. Chem. Soc.* **2006**, *128*, 1398–9.
[117] J.C.M. van Hest, K.L. Kiick, D.A. Tirrell, *J. Am. Chem. Soc.* **2000**, *122*, 1282–8.
[118] K.L. Kiick, E. Saxon, D.A. Tirrell, C.R. Bertozzi, *Proc. Natl. Acad. Sci. U.S.A.* **2002**, *99*, 19–24.
[119] A.J. Link, D.A. Tirrell, *J. Am. Chem. Soc.* **2003**, *125*, 11164–5.
[120] A.J. Link, M.K.S. Vink, D.A. Tirrell, *J. Am. Chem. Soc.* **2004**, *126*, 10598–602.
[121] N. Budisa, *Engineering the Genetic Code*, Wiley-VCH, Weinheim, **2006**.
[122] L. Wang, P. Schultz, *Angew. Chem., Int. Ed.* **2005**, *44*, 34–66.

[123] A.J. Link, M.L. Mock, D.A. Tirrell, *Curr. Opin. Chem. Biol.* **2003**, *14*, 603–9.
[124] E. Saxon, C.R. Bertozzi, *Science* **2000**, *287*, 2007–10.
[125] F.L. Lin, H.M. Hoyt, H. Van Halbeek, R.G. Bergman, C.R. Bertozzi, *J. Am. Chem. Soc.* **2005**, *127*, 2686–95.
[126] M. Köhn, R. Breinbauer, *Angew. Chem., Int. Ed.* **2004**, *43*, 3106–16.
[127] S.I. van Kasteren, H.B. Kramer, H.H. Jensen, *et al.*, *Nature* **2007**, *446*, 1105–9.
[128] L. Merkel, H.S.G. Beckmann, V. Wittmann, N. Budisa, *ChemBioChem* **2008**, *9*, 1220–4.
[129] H. Lis, N. Sharon, *Chem. Rev.* **1998**, *98*, 637–74.
[130] D.C. Kilpatrick, *Handbook of Animal Lectins: Properties and Biomedical Applications*, John Wiley & Sons Ltd, Chichester, **2000**.
[131] H.-J. Gabius, H.-C. Siebert, S. André, J. Jiménez-Barbero, H. Rüdiger, *ChemBioChem* **2004**, *5*, 740–64.
[132] A. Varki, *Glycobiology* **1993**, *3*, 97–130.
[133] R.A. Dwek, *Chem. Rev.* **1996**, *96*, 683–720.
[134] D.H. Dube, C.R. Bertozzi, *Nat. Rev. Drug Discovery* **2005**, *4*, 477–88.
[135] H.-J. Gabius, *Crit. Rev. Immunol.* **2006**, *26*, 43–79.
[136] V. Wittmann, in *Glycoscience: Chemistry and Chemical Biology*, 2 ed. (eds.: B. Fraser-Reid, K. Tatsuta, J. Thiem), Springer-Verlag, Berlin, **2008**, pp. 1771–93.
[137] Y.C. Lee, R.T. Lee, *Acc. Chem. Res.* **1995**, *28*, 321–7.
[138] J.M. Rini, *Annu. Rev. Biophys. Biomol. Struct.* **1995**, *24*, 551–77.
[139] K. Drickamer, *Nat. Struct. Biol.* **1995**, *2*, 437–9.
[140] M. Mammen, S.-K. Choi, G.M. Whitesides, *Angew. Chem. Int. Ed.* **1998**, *37*, 2754–94.
[141] R.T. Lee, Y.C. Lee, *Glycoconjugate J.* **2000**, *17*, 543–51.
[142] J.J. Lundquist, E.J. Toone, *Chem. Rev.* **2002**, *102*, 555–78.
[143] V. Wittmann, in *Highlights in Bioorganic Chemistry: Methods and Applications* (Eds.: C. Schmuck, H. Wennemers), Wiley-VCH, Weinheim, **2004**, pp. 203–13.
[144] L.L. Kiessling, J.E. Gestwicki, L.E. Strong, *Angew. Chem., Int. Ed.* **2006**, *45*, 2348–68.
[145] F. Pérez-Balderas, M. Ortega-Muñoz, J. Morales-Sanfrutos, *et al.*, *Org. Lett.* **2003**, *5*, 1951–4.
[146] P. Wu, M. Malkoch, J.N. Hunt, *et al.*, *Chem. Commun.* **2005**, 5775–7.
[147] V. Ladmiral, G. Mantovani, G.J. Clarkson, *et al.*, *J. Am. Chem. Soc.* **2006**, *128*, 4823–30.
[148] P. Appukkuttan, W. Dehaen, V.V. Fokin, E. Van der Eycken, *Org. Lett.* **2004**, *6*, 4223–5.
[149] A.K. Feldman, B. Colasson, V.V. Fokin, *Org. Lett.* **2004**, *6*, 3897–9.
[150] S. Chittaboina, X. Fang, Q. Wang, *Tetrahedron Lett.* **2005**, *46*, 2331–6.
[151] S. Fukuzawa, E. Shimizu, S. Kikuchi, *Synlett* **2007**, 2436–8.
[152] K. Odlo, E.A. Høydahl, T.V. Hansen, *Tetrahedron Lett.* **2007**, *48*, 2097–9.
[153] K. Barral, A.D. Moorhouse, J.E. Moses, *Org. Lett.* **2007**, *9*, 1809–11.
[154] H.S.G. Beckmann, V. Wittmann, *Org. Lett.* **2007**, *9*, 1–4.
[155] W.R. Carpenter, *J. Org. Chem.* **1962**, *27*, 2085–8.
[156] H. Quast, L. Bieber, *Tetrahedron Lett.* **1976**, 1485–6.
[157] M.M. Krayushkin, A.M. Beskopyl'nyi, S.G. Zlotin, O.A. Luk'yanov, V.M. Zhulin, *Izv. Akad. Nauk SSSR, Ser. Khim.* **1980**, 2668.
[158] Z.P. Demko, K.B. Sharpless, *Angew. Chem., Int. Ed.* **2002**, *41*, 2110–13.
[159] Ibid., pp. 2113–16.
[160] P.A.S. Smith, J.M. Clegg, J.H. Hall, *J. Org. Chem.* **1958**, *23*, 524–9.
[161] R. Fusco, L. Garanti, G. Zecchi, *J. Org. Chem.* **1975**, *40*, 1906–9.
[162] L. Garanti, G. Zecchi, *J. Org. Chem.* **1980**, *45*, 4767–9.
[163] D. Korakas, A. Kimbaris, G. Varvounis, *Tetrahedron* **1996**, *52*, 10751–60.
[164] T.C. Porter, R.K. Smalley, M. Teguiche, B. Purwono, *Synthesis* **1997**, 773–7.
[165] B.G. Davis, T.W. Brandstetter, L. Hackett, *et al.*, *Tetrahedron* **1999**, *55*, 4489–500.
[166] M. Worch, V. Wittmann, *Carbohydr. Res.* **2008**, *343*, 2118–29.
[167] R. Apweiler, H. Hermjakob, N. Sharon, *Biochim. Biophys. Acta* **1999**, *1473*, 4–8.
[168] J.A. Prescher, C.R. Bertozzi, *Cell* **2006**, *126*, 851–4.
[169] J.A. Prescher, C.R. Bertozzi, *Nat. Chem. Biol.* **2005**, *1*, 13–21.
[170] L.K. Mahal, K.J. Yarema, C.R. Bertozzi, *Science* **1997**, *276*, 1125–8.

[171] H.-C. Tai, N. Khidekel, S.B. Ficarro, E.C. Peters, L.C. Hsieh-Wilson, *J. Am. Chem. Soc.* **2004**, *126*, 10500–1.
[172] S.-G. Sampathkumar, A.V. Li, M.B. Jones, Z. Sun, K.J. Yarema, *Nat. Chem. Biol.* **2006**, *2*, 149–52.
[173] E. Saxon, S.J. Luchansky, H.C. Hang, C. Yu, S.C. Lee, C.R. Bertozzi, *J. Am. Chem. Soc.* **2002**, *124*, 14893–902.
[174] H.C. Hang, C. Yu, D.L. Kato, C.R. Bertozzi, *Proc. Natl. Acad. Sci. U.S.A.* **2003**, *100*, 14846–51.
[175] J.A. Prescher, D.H. Dube, C.R. Bertozzi, *Nature* **2004**, *430*, 873–7.
[176] D.H. Dube, J.A. Prescher, C.N. Quang, C.R. Bertozzi, *Proc. Natl. Acad. Sci. U.S.A.* **2006**, *103*, 4819–24.
[177] D.J. Vocadlo, H.C. Hang, E.-J. Kim, J.A. Hanover, C.R. Bertozzi, *Proc. Natl. Acad. Sci. U.S.A.* **2003**, *100*, 9116–21.
[178] P.V. Chang, J.A. Prescher, M.J. Hangauer, C.R. Bertozzi, *J. Am. Chem. Soc.* **2007**, *129*, 8400–1.
[179] M. Sawa, T.-L. Hsu, T. Itoh, *et al.*, *Proc. Natl. Acad. Sci. U.S.A.* **2006**, *103*, 12371–6.
[180] D. Rabuka, S.C. Hubbard, S.T. Laughlin, S.P. Argade, C.R. Bertozzi, *J. Am. Chem. Soc.* **2006**, *128*, 12078–9.
[181] T.-L. Hsu, S.R. Hanson, K. Kishikawa, *et al.*, *Proc. Natl. Acad. Sci. U.S.A.* **2007**, *104*, 2614–19.
[182] G. Wittig, A. Krebs, *Chem. Ber.* **1961**, *94*, 3260–75.
[183] N.J. Agard, J.M. Baskin, J.A. Prescher, A. Lo, C.R. Bertozzi, *ACS Chem. Biol.* **2006**, *1*, 644–8.
[184] J.M. Baskin, J.A. Prescher, S.T. Laughlin, *et al.*, *Proc. Natl. Acad. Sci. U.S.A.* **2007**, *104*, 16793–7.
[185] J.A. Codelli, J.M. Baskin, N.J. Agard, C.R. Bertozzi, *J. Am. Chem. Soc.* **2008**, *130*, 11486–93.
[186] S.T. Laughlin, J.M. Baskin, S.L. Amacher, C.R. Bertozzi, *Science* **2008**, *320*, 664–7.

Index

Organic Azides: Syntheses and Applications Edited by Stefan Bräse and Klaus Banert